The Scattering of Light

AND OTHER ELECTROMAGNETIC RADIATION

This is Volume 16 of
PHYSICAL CHEMISTRY
A Series of Monographs
Editor: ERNEST M. LOEBL, *Polytechnic Institute of New York*

A complete list of titles in this series appears at the end of this volume.

THE SCATTERING OF LIGHT

AND OTHER ELECTROMAGNETIC RADIATION

MILTON KERKER

DEPARTMENT OF CHEMISTRY
CLARKSON COLLEGE OF TECHNOLOGY
POTSDAM, NEW YORK

AP 1969

ACADEMIC PRESS New York San Francisco London
A Subsidiary of Harcourt Brace Jovanovich, Publishers

ACADEMIC PRESS, INC.
111 Fifth Avenue, New York, New York 10003

United Kingdom Edition published by
ACADEMIC PRESS, INC. (LONDON) LTD.
24/28 Oval Road, London NW1

LIBRARY OF CONGRESS CATALOG CARD NUMBER: 69-26644

PRINTED IN THE UNITED STATES OF AMERICA

Preface

In writing this book, I have endeavored to summarize the theory of electromagnetic scattering, as well as to describe some of the practical applications, particularly to light scattering. The treatment is extensive, and yet it is hardly exhaustive, for the field is vast. The selection of topics, described in the introductory chapter, is biased very much in favor of my own interests.

A perusal of the list of references will show that although the theory is mainly more than half a century old, the applications have occurred almost entirely during the past two decades. This recent spate of research activity has opened up vast new possibilities for applications to the physics of particulate systems, the physical chemistry of solutions including those containing macromolecules, bio-colloids, and detergents, the morphology of solids, critical opalescence, low angle X-ray scattering, atmospheric and space optics, radar meteorology, and plasma physics including radiowave scattering by plasmas generated in the upper atmosphere by rapidly moving objects.

MILTON KERKER

Potsdam, New York
April, 1969

Acknowledgments

I want to express my gratitude to the late Victor K. LaMer, who introduced me to the study of light scattering, to David Atlas, who provided an opportunity to launch this project, and to Professor D. H. Everett and R. H. Ottewill in whose department at the University of Bristol it was completed during a tenure of the Unilever Professorship. D. H. Napper and W. A. Farone were helpful in pointing out a number of errors in the manuscript. The names of my collaborators and students are cited appropriately in the text and in the references. My debt to them is very great, as is my appreciation. The manuscript was expertly typed by Mrs. M. Frost, and the assistance of Mrs. Margaret Cayey in correcting proofs was invaluable. My work has been generously supported by the U.S. Atomic Energy Commission, the National Center for Air Pollution Control of the U.S. Public Health Service, and the Atmospheric Sciences Office of the U.S. Army Electronics Command.

Most important of all has been the support throughout this endeavor of my wife, Reva. This book is affectionately dedicated to her, to each of our parents, and to our children, Ruth Ann, Martin, Susan, and Joel.

Contents

Chapter 4. The Scattering Functions for Spheres

Chapter 5. Scattering by Stratified Spheres

Chapter 6. Scattering by Infinite Cylinders

Chapter 7. Analysis of Particle Size

Chapter 8. Rayleigh–Debye Scattering

Glossary of Principal Symbols

A	albedo
A	Helmholtz free energy
A_1	semi-axis of a polarization ellipse
A_{ij}	co-factor of the determinant $\lvert a_{ij}\rvert$
$\mathbf{B}$	magnetic induction
B	second virial coefficient
B_1	semi-axis of polarization ellipse
$C_{\text{sca}}, C_{\text{ext}}, C_{\text{abs}}$	cross sections for scattering, extinction, absorption
$C_\tau, C(\theta)$	Cabannes factors for turbidity and Rayleigh ratio
$\mathbf{D}$	dielectric displacement
d	diameter
D	divergence of a tube of rays
D	factor appearing in multi-component theory for poly-electrolytes
$\mathbf{E}$	electric field intensity
$F_a^2(h)$	intensity factor
G	gain
G	Gibbs free energy
G_{ij}	derivative of chemical potential
G_n	solution of radial equation for inhomogeneous sphere
$G(s)$	probability density function
$\mathbf{H}$	magnetic field intensity
H, H', H''	factors involving optical parameters
$H_n(kr)$	Hankel function
I	intensity
$J_n(kr)$	Bessel function
K	factor in expression for Rayleigh ratio
L	persistence length
$\mathbf{M}$	magnetic polarization
M	molecular weight
M_{w}	weight average molecular weight
N	number of particles or molecules per unit volume
N_{A}	Avogadro's number
$N_n(kr)$	Neumann function
O.D.	optical density
$\mathbf{P}$	electric polarization
P_{e}	depolarization factor of ellipsoid
P	radiation pressure
P	form factor
P_n	radial function in scattering by inhomogeneous cylinder
$P_n^m(\cos\theta)$	associated Legendre polynomial
$p(\theta)$	degree of polarization
Q_{pr}	efficiency for radiation pressure
$Q_{\text{sca}}, Q_{\text{ext}}, Q_{\text{abs}}$	efficiencies for scattering, extinction, absorption
R	molar gas constant
R_{g}	radius of gyration
R_θ	Rayleigh ratio
$\mathbf{S}$	Poynting vector
S	surface of interfacial area
S_1, S_2	scattering amplitudes for sphere
S_{sp}	specific surface
T	absolute temperature
T	transmission
T_1, T_2	scattering amplitudes for cylinder
T_n	radial function in scattering by inhomogeneous cylinder
V	volume
$V_{\text{h}}, H_{\text{v}}$	Rayleigh ratio with polarizer in the horizontal position and analyzer in the vertical position, and vice versa

V_v, H_h	Rayleigh ratio with polarizer and analyzer vertical and horizontal respectively
W_n	solution of radial equation for inhomogeneous sphere
Z_0	intrinsic impedance
Z_i	ionic charge
$Z_n(z)$	cylinder function
a	radius
a'	absorption coefficient
a_i	activity of ith component
a_{ij}	matrix element in multi-component theory
a_M	modal value of radius
a_n	scattering coefficient
b	radius of coated sphere
b_n	scattering coefficient
c	concentration in gm/ml.
c	velocity of light in free space
e	electronic charge
e_s	eccentricity of spheroid
f_i	activity coefficient (based on mole fraction)
$g(s)$	radial distribution function
i	$\sqrt{-1}$
i_1, i_2	angular intensity functions
g	concentration in grams per gram of water
g	wavelength exponent
g_{os}	rational osmotic coefficient
h	$(4\pi/\lambda)\sin\theta/2$
h	$k_0 \sin\phi$
j	$(m^2k_0^2 - h^2)^{1/2}$
k	Boltzmann constant
k	propagation constant
l	$k_0 \cos\phi$
l	path length
l	range of molecular forces
$\bar{l}$	characteristic length
l_c	coherence length
m_a	aggregation number for micelle, polyion
m	molality
m	refractive index
m_e	mass of the electron
$\mathbf{n}$	unit vector normal to surface
n	real part of refractive index
n_i	number of ions per unit volume
$\mathbf{p}$	dipole moment
p	micellar charge
p	pressure
$p(a)$	size distribution function
q	ratio of α/v or of a/b
r	radial distance
r_1, r_2	Fresnel reflection coefficients
s	intra-particle distance
s_0, s_1, s_2, s_3	Stokes' parameters
t	time
t_1, t_2	Fresnel transmission coefficients
u	any scalar component of E or H
u	ha
v	velocity of light in medium
x_i	mole fraction of ith component
$z(\theta)$	dissymmetry of angular scattering
α	degree of dissociation
α	dimensionless size parameter
α'	polarizability
α_M	modal value of size parameter α
α_n	phase angle
α_p	volume expansion coefficient
β	$m\alpha$
β_{ij}	interaction coefficient
β_n	phase angle
β_T	isothermal compressibility
γ	anisotropy factor
γ_i	activity coefficient
$\gamma(s)$, $\gamma_0(s)$	correlation function
δ	phase difference
ε	dielectric constant
$\zeta_n(kr)$	Ricatti–Bessel function
$\overline{\eta^2}$	mean square of the variation of the local dielectric constant
$\overline{\eta_A\eta_B}$	correlation distance
$\eta_n(kr)$	logarithmic derivative of Ricatti–Bessel function
θ	scattering angle
θ_i, θ_t	angles of incidence and refraction
κ	index of absorption
κ	reciprocal length in Debye–Hückel theory
κ_1, κ_2	factors of propagation constant
λ	wavelength
μ	magnetic permeability
μ	refractive index of a scattering medium
μ_i	chemical potential of ith component

ν	frequency
ν	size parameter of coated sphere
ν_{n}	number of ions into which an electrolyte dissociates
$\mathbf{\Pi}$, Π	Hertz vector, potential
π	osmotic pressure
$\pi_n(\cos\theta)$	angular function
ρ	density
ρ	dimensionless radial distance kr
ρ	phase shift parameter
ρ_{e}	electron density
$\rho_{\mathrm{u}}, \rho_{\mathrm{v}}, \rho_{\mathrm{h}}$	depolarization factors
$\rho(\theta)$	polarization ratio
σ	back scatter cross section
σ	specific conductance
σ	standard deviation or other measure of width of distribution
σ_0	breadth parameter of ZOLD
τ	turbidity
$\tau_n(\cos\theta)$	angular function
ϕ	tilt angle for cylindrical symmetry
ϕ_{os}	practical osmotic coefficient
ϕ, ϕ_i	volume fraction of scattering material
$\chi_n(kr)$	Ricatti–Bessel function
ψ	angle between scattering direction and incident electric vector
$\psi_{.}$	inclination of polarization ellipse
ψ_i	refractive index increment of ith component
$\psi_n(kr)$	Ricatti–Bessel function
ω	circular frequency
ω'	weight fraction of scattering material

The Scattering of Light

AND OTHER ELECTROMAGNETIC RADIATION

CHAPTER 1

Introduction

The optical properties of a medium are characterized by its refractive index, and as long as this is uniform, light will pass through the medium undeflected. Whenever there are discrete variations in the refractive index due to the presence of particles or because there are small scale density fluctuations, part of the radiation will be scattered in all directions.

The scattering of light is a ubiquitous natural phenomenon. We perceive the blue of the sky because of the scattering of the solar rays by the air molecules; were it not for this, the heavens would be black. Something close to the correct explanation was suggested more than four and a half centuries ago by da Vinci (ca. 1500) when he wrote:

> "I say that the blueness we see in the atmosphere is not intrinsic color, but is caused by warm vapor evaporated in minute and insensible atoms on which the solar rays fall, rendering them luminous against the infinite darkness of the fiery sphere which lies beyond and includes it.... If you produce a small quantity of smoke from dry wood and the rays of the sun fall on this smoke and if you place (behind it) a piece of black velvet on which the sun does not fall, you will see that the black stuff will appear of a beautiful blue color.... Water violently ejected in a fine spray and in a dark chamber where the sunbeams are admitted produces then blue rays.... Hence it follows, as I say, that the atmosphere assumes this azure hue by reason of the particles of moisture which catch the rays of the sun."

Other optical phenomena in the atmosphere such as the colors of the sunset, the rainbow, the glory, the corona, and the halo are due to scattering either by aerosols, by ice crystals, or by water droplets. The transparency of the atmosphere varies according to the extent that there is scattering of light by aerosol or fog. In interstellar space there are huge clouds of colloidal particles which scatter starlight directly to us, or by the same scattering process, alter the starlight which is transmitted through them. The zodiacal light seen in the western sky is due to scattering by interplanetary dust.

The turbidity of liquids and of solids, and in some cases their color, results from the scattering of the light in which they are viewed, either by their constituent molecules or by suspended particles. The brilliant colors of metal sols or of certain precious stones are derived from the preferential scattering and absorption of certain wavelengths by the suspended particles. The color of the sea is, in part, a light scattering phenomenon.

The scientific study of light scattering may be said to have commenced with the experiments on aerosols by Tyndall (1869), which were followed from 1871 onwards by Lord Rayleigh's great body of theoretical work. The problem is to relate the properties of the scatterer—its shape, its size, and its refractive index—to the angular distribution of the scattered light. The incident beam of known intensity and wavelength is usually taken to be parallel and linearly polarized. If the scatterer is absorptive, part of the light will be absorbed within it as heat, another part will be scattered, and the remainder will be transmitted unperturbed along the incident direction. A complete description of the scattered light entails a knowledge of the wavelength, amplitude, phase, and polarization of the radiation emanating in each direction from the scatterer. This also provides the information necessary to calculate the amount of absorption and the light pressure upon the particle.

Scattering is hardly restricted to the optical part of the spectrum, and the scattering laws apply with equal validity to all wavelengths. Interestingly, these depend upon the ratio of a characteristic dimension of the particle to the wavelength rather than explicitly upon the size. Thus, there is a built-in scaling factor. The scattering of radiowaves by artificial earth satellites, the scattering of microwaves by raindrops, and the scattering of light by aerosols are quite similar phenomena because in each case the wavelength is of the same magnitude as that of the scatterer.

The study of electromagnetic scattering is an interdisciplinary activity. The scattering of starlight by interstellar and interplanetary dust is of interest to astrophysicists. Meteorologists are concerned with the whole range of atmospheric optical phenomena mentioned earlier. In addition, the technique of observing the backscatter of microwaves by rain, snow, and hail has given rise to the science of radar meteorology. The radar technique is also utilized to observe meteors and artificial objects in the atmosphere as well as the plasmas created in the wake of these rapidly moving bodies. There are collateral laboratory studies using microwaves.

The transhorizon propagation of radiowaves along the surface of the earth is one of the classical problems of electromagnetic scattering, going back to the early days of radio at the turn of the century. In addition, there is considerable interest in the scattering and consequent attenuation of radiowaves due to density fluctuations in the atmosphere. The variety of scattering shapes encountered in both the microwave and the radiowave work has stimulated a considerable amount of theoretical activity by electrical engineers, mathematicians, and physicists.

Chemists, physicists, biochemists, and various engineers utilize light scattering to study a whole range of materials including gases, pure liquids, solutions of both ordinary molecules and particularly of macromolecules, colloidal suspensions, glasses, and polymers. In some cases X-ray scattering may be used. Also, there is considerable interest in the light scattering effects observed in the neighborhood of the critical point of pure substances or at the critical mixing point of partially miscible solutions.

There are two classes of problems—the direct problem and the inverse problem. The direct problem is to calculate theoretically or to observe experimentally the scattering by a known, well-defined system. The inverse problem is to characterize the system from a knowledge of the scattering, usually obtained by experiment or, in the case of natural phenomena, from observations. A most elusive and difficult example of the inverse problem is the goal of astrophysicists to describe the interstellar particles by analysis of the scattered and transmitted light. A much less ambitious but far from trivial program is to determine the size distribution of a suspension of colloidal spheres of known refractive index by light scattering experiments.

The treatment in this book reflects the bias of the author as a physical chemist. Most of the examples of practical applications and most of the experimental studies deal with light scattering from colloidal and macromolecular systems. However, the theoretical treatment in Chapters 2 to 6 is quite general and applies equally well to all parts of the electromagnetic spectrum. Furthermore, much of the discussion in this part of the book relates directly to topics in microwave and radiowave physics. Indeed, a perusal of the bibliography will show that a considerable body of electrical engineering, applied physics, and meteorology literature has been incorporated into the discussion, and we hope that at least this part of the book will be of some interest to workers in these and related fields.

Only single scattering and elastic scattering are treated. The latter condition means that there is no shift of frequency between the incident and the scattered radiation. This excludes quantum mechanical phenomena such as the Raman effect and fluorescence or Brillouin scattering which arises from the Doppler shifts associated with the motion of the scattering particles.

The restriction to single scattering implies that the scattering particle is unaffected by the presence of neighboring particles. Also, there is an absence of multiple scattering. After the encounter between the incident beam and the particle, the scattered radiation proceeds directly to the observer without any further scattering encounters. In the laboratory, the necessary conditions for single scattering can usually be attained by working with dilute systems and with small volumes. On the other hand, in the atmosphere and in space, multiple scattering can become the predominant effect.

Still a third gap is the absence of any consideration of laboratory procedures. The art of obtaining light scattering data is simple in principle but fraught with difficulties in practice and we have not attempted to treat laboratory technique in this volume.

1.1 Résumé

Chapter 2 is a review of electromagnetic waves and of optics. It introduces some of the physical concepts and quantities upon which the rest of the book will build. The treatment is brief and is hardly intended to substitute for a general introduction to the broader subject. Rather, it serves only to define and to interrelate the main physical quantities and the physical concepts of optics.

Chapters 3 to 6 deal with the theory of scattering by spheres and infinitely long cylinders. We are fortunate in having an exact theory in the sense of classical physics. The only restrictions are that the substance of which the particle is composed be isotropic and that any variations in the refractive index be radially symmetric.

In Chapter 3, after an historical introduction, there is an exposition of Rayleigh's theory of scattering by spheres which are small compared to the wavelength. This is followed by the theory of homogeneous spheres of arbitrary size. Although this is frequently called the Mie theory, it is quite clear that Mie's (1908) paper was preceded by the independent solutions of a number of workers, starting with the very elegant work of Lorenz (1890). Our treatment follows that of Debye (1909a). Chapter 3 continues with a discussion of the Bessel and Legendre functions needed to calculate the various scattering quantities and a very brief review of the point-matching method. This recently developed technique which solves for the boundary conditions at only a finite number of points upon the particle surface can readily be extended to nonspherical particles. Finally, there is an exposition of the theory of light pressure which is intimately connected with light scattering.

The dependence of the numerical values of the scattering functions upon the particle size, the refractive index, and the scattering angle is reviewed in Chapter 4. The first function dealt with is the scattering coefficient from which all of the other quantities are derived. The scattering cross section which is the total intensity scattered in all directions and the scattering efficiency which is the ratio of the scattering cross section to the geometrical cross section of the particle are considered next. When the particles are absorptive, there is a corresponding cross section for absorption and an efficiency for absorption. The cross section for extinction and the efficiency for extinction measure the total effect and each of these is the sum of the respective quantities for scattering and absorption. Finally, there is the intensity scattered at various angles with particular emphasis upon the backscatter. Two approximations are compared numerically with the exact theory. These are the theory of anomalous diffraction and ray optics. The latter interprets scattering as the result of reflection, refraction, and diffraction.

Stratified spheres, which are the subject of Chapter 5, may consist of a series of concentric spherical shells of different media or the radial profile of the refractive index may vary continuously in some arbitrary fashion. These configurations can be solved exactly. In some cases the solutions can be represented analytically in terms of standard functions; otherwise numerical, albeit exact, solutions are obtained. Chapter 6 contains a discussion of circular cylinders at both perpendicular and oblique incidence. Solutions are available, just as for spheres, for any arbitrary radially varying refractive index. In addition, there are solutions for two particular anisotropies—the gyroelectric and the gyromagnetic media.

The application of light scattering to the determination of the size distribution of colloidal particles is reviewed in Chapter 7. There is a digression to consider the preparation of some of the colloids of narrow size distribution which have served as model systems for much of the experimental work as well as a discussion of some of the distribution functions which have been used to characterize the particle sizes. One useful technique for obtaining the particle size distribution is based upon a comparison of the experimental light scattering data with results computed for a large number of size distributions. That distribution for which the calculated and experimental results agree is chosen to characterize the colloid. This technique is successful only when the distribution is not too broad and when the particle size is comparable to the wavelength.

If the refractive index of a particle is sufficiently close to that of the external medium and if the particle is not too large, each volume element behaves as a Rayleigh scatterer. Each of the scattered wavelets, in turn, mutually interfere. The scattering for such a model was calculated precisely by Lord Rayleigh. Although this is frequently termed Rayleigh–Gans scattering, we have proposed that it should more appropriately be called Rayleigh–Debye scattering. This is the subject of Chapter 8. Rayleigh–Debye scattering may be applied to particles of any shape and has been particularly successful in elucidating the configuration of macromolecules in solution. The angular distribution at the forward angles leads directly to a value of the radius of gyration without any prior knowledge of the particle shape. When the particle shape is known, the forward scattering can sometimes be reduced directly to give the particle size distribution without any prior knowledge of the form of the distribution. The Rayleigh–Debye theory is also particularly useful in interpreting the scattering by nonparticulate media. These may be solids for which the scattering arises from a more or less continuous variation of the refractive index. Such internal structure can be described by a correlation function which can, in turn, be deduced directly from the angular distribution of the forward scattered radiation. Small angle X-ray scattering as well as light scattering is commonly utilized for this approach.

Chapter 9 deals with scattering by liquids. These may be considered to scatter light by virtue of the microscopic fluctuations of the density from the macroscopic value. If the density fluctuation has associated with it a fluctuation of the polarizability or the dielectric constant or the refractive index (each of these quantities being manifestations of the same optical property), there will be light scattering. The fluctuation in each volume element is assumed to take place independently of those in the neighboring elements, and since these volumes are also chosen to be small compared to the wavelength, Rayleigh's law for small particles serves to describe the scattering. The problem, which was solved by Einstein (1910), is to calculate the magnitude of the density fluctuations by the methods of statistical thermodynamics. The theory was extended from single component and binary systems to multicomponent systems by Zernike (1915).

This fluctuation theory has been particularly successful in its application to the study of the molecular weight and the thermodynamic interactions in solutions of macromolecules, polyelectrolytes, proteins, and surfactants. In conjunction with the Rayleigh–Debye theory, it has also been possible to obtain information about the configuration of such species. In the region of the critical point, the density fluctuations in neighboring volume elements are no longer independent but the mutual interference can be accounted for with the aid of a correlation function quite the same as that introduced in the treatment of Rayleigh–Debye scattering from inhomogeneous media.

The final chapter is concerned with anisotropy. Lord Rayleigh derived the scattering by ellipsoidal particles which are small compared to the wavelength in a manner similar to his treatment of small spheres. Such anisotropic particles scatter more intensely than spheres of the same volume by a factor known as the Cabannes factor. This can actually be determined from a measurement of the depolarization of the scattered light at a single angle without any prior knowledge of the configuration of the anisotropic particles. This chapter also considers the effect upon the light scattering of partial orientation of anisotropic particles in electrical and magnetic fields and in viscous flow. Finally, there is a brief consideration of attempts to develop a general theory of scattering by ellipsoids.

1.2 Notation

The selection of a notation is always a problem. This is particularly so for this subject because of its interdisciplinary complexion. Every author must be torn between a desire to provide a completely consistent notation (i.e., one in which no symbol is utilized to designate more than one quantity) and a reluctance to abandon a widely used and recognized symbol.

We have adopted an intermediate stance, perhaps favoring what we have felt was the use of a recognized symbol over the maintenance of consistency. Thus, c has been used to denote both the velocity of light in free space and the concentration in grams per milliliter; k denotes both the propagation constant and the Boltzmann constant, etc. Such lapses will never occur in the same equation and hardly ever in the same section. It is hoped that the correct usage will always be apparent from the context.

The glossary lists only the principal symbols. There are numerous empirical constants or physical quantities which appear in only a single context in a small section. We have avoided cluttering the glossary with these, since it would have increased its length many fold and would have made it unwieldy. Also, there has been no special effort to establish consistency among these "one shot" symbols. Each such symbol is defined in its particular context.

In some few cases, more than one symbol has been used to denote the same physical quantity. This has been done, where the discussion in this book parallels a more extended and complicated discussion in the literature, in order to make it easier for the reader to transpose to the literature.

CHAPTER 2

Electromagnetic Waves

2.1 Maxwell's Equations

In 1865 Maxwell published his famous set of equations which formalized the results of the accumulated research on electromagnetism. Electromagnetic phenomena had been recognized as originating in a distribution of electric charge and current which gives rise to the electromagnetic field. This field in turn is the domain of four field vectors, **E**, the electric field intensity, **D**, the dielectric displacement, **H**, the magnetic field intensity, and **B**, the magnetic induction. At every point in space in whose neighborhood the physical properties are continuous, the field vectors are subject to Maxwell's equations:

$$\nabla \times \mathbf{E} = -\partial \mathbf{B}/\partial t \tag{2.1.1}$$

$$\nabla \times \mathbf{H} = \mathbf{J} + \partial \mathbf{D}/\partial t \tag{2.1.2}$$

$$\nabla \cdot \mathbf{D} = \rho \tag{2.1.3}$$

$$\nabla \cdot \mathbf{B} = 0 \tag{2.1.4}$$

Here t is time, **J** is the current density, and ρ is the charge density. Both **J** and ρ will be zero for cases treated in this volume.

The first two equations enunciate Faraday's law for the induction of an electric field (or current) by a varying magnetic flux and Ampère's law for the calculation of the magnetic field from a distribution of current. The last term in (2.1.2) is the famous "displacement current" conceived by Maxwell as contributing to the magnetic field in the same way as an ordinary current of moving charge. The third equation is a consequence of Coulomb's law of force between electric charges. The final equation indicates that there are no sources of the magnetic field corresponding to the electric charges which form the source of the electric field.

These field equations must be supplemented by the following material equations in order to allow a unique determination of the field vectors from a given distribution of current and charge.

$$\mathbf{J} = \sigma \mathbf{E} \tag{2.1.5}$$

$$\mathbf{D} = \varepsilon \mathbf{E} \tag{2.1.6}$$

$$\mathbf{B} = \mu \mathbf{H} \tag{2.1.7}$$

The factors σ, ε, and μ are the specific conductance, the electric inductive capacity, and the magnetic inductive capacity, respectively. When they are independent of direction, the medium is said to be isotropic; otherwise it is anisotropic and the coefficients of the linear transformations relating each of the above pairs of vectors are the components of a symmetric tensor. In a homogeneous medium, the properties are constant from point to point and in such a case ε and μ are referred to as the dielectric constant and the magnetic permeability. For free space, the values and dimensions of the inductive capacities, ε_0 and μ_0, will depend upon the system of units adopted. In order to avoid units, the dimensionless quantities

$$\kappa_e = \varepsilon/\varepsilon_0, \qquad \kappa_m = \mu/\mu_0 \tag{2.1.8}$$

are sometimes utilized, where κ_e is called the specific electric inductive capacity (relative dielectric constant) and κ_m is the specific magnetic inductive capacity (relative magnetic permeability). It is a direct consequence of the field equations that the quantity

$$c = (\varepsilon_0 \mu_0)^{-1/2} \tag{2.1.9}$$

shall have the dimensions of a velocity.

It is sometimes convenient to consider two auxiliary vectors, the electric and magnetic polarization defined by

$$\mathbf{P} = \mathbf{D} - \varepsilon_0 \mathbf{E} \tag{2.1.10}$$

$$\mathbf{M} = (1/\mu_0)\mathbf{B} - \mathbf{H} \tag{2.1.11}$$

Obviously [cf. (2.1.6) and (2.1.7)] these quantities vanish in free space. They represent the influence of matter upon the field. Indeed, the influence of material bodies on an electromagnetic field may be accounted for by equivalent distributions of charge and current densities completely defined by **P** and **M**. Such distributions can in turn be conceived as the superposition of electric and magnetic dipoles, respectively.

In isotropic media the material properties can be described by the electric and magnetic susceptibilities, χ_e and χ_m, defined by

$$\mathbf{P} = \chi_e \varepsilon_0 \mathbf{E} \tag{2.1.12}$$

and

$$\mathbf{M} = \chi_m \mathbf{H} \tag{2.1.13}$$

These, in turn, are related to the specific inductive capacities by

$$\chi_e = \kappa_e - 1 \tag{2.1.14}$$

and

$$\chi_m = \kappa_m - 1 \tag{2.1.15}$$

Maxwell's equations are valid for regions of space through which the physical properties of the medium, as characterized by σ, ε, and μ, are continuous. The surfaces bounding one body or medium from another may, on a macroscopic scale, comprise discontinuities across which the field vectors themselves may be discontinuous. The boundary conditions at such surfaces are:

(1) The normal component of **B** is continuous

$$(\mathbf{B}_2 - \mathbf{B}_1) \cdot \mathbf{n} = 0 \tag{2.1.16}$$

(2) There is a discontinuity in the normal component of **D** equivalent to K, the surface charge density

$$(\mathbf{D}_2 - \mathbf{D}_1) \cdot \mathbf{n} = K \tag{2.1.17}$$

(3) The tangential component of **E** is continuous

$$(\mathbf{E}_2 - \mathbf{E}_1) \times \mathbf{n} = 0 \tag{2.1.18}$$

(4) There is a discontinuity in the tangential component of **H** equal to **L**, the surface current density

$$(\mathbf{H}_2 - \mathbf{H}_1) \times \mathbf{n} = \mathbf{L} \tag{2.1.19}$$

The corresponding conditions for the normal components of **E** and **H** and the tangential components of **D** and **B** can also be written with the aid of (2.1.6) and (2.1.7).

It is well known that perfect conductors cannot sustain an electric field. Consequently the tangential component of **E** must vanish at the surface and it follows from this that the normal component of **H** must also vanish. However the normal component of **E** and the tangential components of **H** are unrestricted at the surface. Since these fields must vanish immediately within the perfect conductor, the discontinuities in these components are accounted for by equivalent surface charge and current distributions.

The boundary conditions enter into the solution of scattering problems. In general, the technique is one of writing down solutions of Maxwell's equations and then selecting that solution which satisfies the boundary conditions of the particular spatial configuration being studied. For all cases which we will consider, there will be no appreciable surface charges or surface currents. Accordingly, the normal components of **D** and **B** and the tangential components of **E** and **H** will each be continuous across the boundary.

2.2 Electromagnetic Waves

The most spectacular immediate result of Maxwell's electromagnetic theory was the prediction of electromagnetic waves propagated with a velocity which could be calculated from purely electrical measurements. That this velocity was the same as that of light, not only in free space but also in media, confirmed that light itself was a form of electromagnetic radiation. The discovery by Hertz (1889) of electromagnetic waves of longer wavelength and the subsequent filling in of the electromagnetic spectrum are now well known.

Actually, Maxwell's equations admit of an enormous variety of solutions and there is no general method for discovering useful ones which may describe natural phenomena. The clue for the existence of electromagnetic waves comes with the demonstration that **E** and **H** satisfy the well-known homogeneous equations for damped wave motion

$$\nabla^2 \mathbf{E} - \sigma\mu \frac{\partial \mathbf{E}}{\partial t} - \varepsilon\mu \frac{\partial^2 \mathbf{E}}{\partial t^2} = 0 \tag{2.2.1}$$

and

$$\nabla^2 \mathbf{H} - \sigma\mu \frac{\partial \mathbf{H}}{\partial t} - \varepsilon\mu \frac{\partial^2 \mathbf{H}}{\partial t^2} = 0 \tag{2.2.2}$$

In nondissipative media where $\sigma = 0$, the second term drops out and the homogeneous equation for an undamped wave results:

$$\nabla^2 \mathbf{E} - \varepsilon\mu \frac{\partial^2 \mathbf{E}}{\partial t^2} = 0 \tag{2.2.3}$$

and

$$\nabla^2 \mathbf{H} - \varepsilon\mu \frac{\partial^2 \mathbf{H}}{\partial t^2} = 0 \tag{2.2.4}$$

It is the existence of these equations, which describe waves propagating with a velocity $(\varepsilon\mu)^{-1/2}$, that led Maxwell to his bold assertion about the existence of electromagnetic radiation. For Maxwell, the fact that the velocity of light in various media was actually given by the corresponding values of $(\varepsilon\mu)^{-1/2}$, as determined from electrostatic and magnetostatic measurements, provided the substantial confirmation of his hypothesis that light itself was a form of electromagnetic radiation.

We now consider the simplest electromagnetic wave; a plane, monochromatic wave, harmonic in time and propagated in an unbounded, isotropic, homogeneous medium in the positive z-direction of a Cartesian coordinate system (Fig. 2.1). A light wave of such characteristics is an ideal which can only be approached in nature by passing natural light through a perfect polarizer, a perfect collimator, and a perfect monochromator. For such a wave, the vectors **E** and **H** are perpendicular to each other and to the direction of propagation. The scalar components of the vectors themselves are related by

$$\frac{E_x}{H_y} = \frac{-E_y}{H_x} = Z_0 = \left(\frac{\mu}{\varepsilon - i\sigma/\omega}\right)^{1/2} \tag{2.2.5}$$

so that it will normally be sufficient to describe only the behavior of **E**. The quantity Z_0 is called the intrinsic impedance of the medium for plane waves. The circular frequency ω will be defined below; i is $\sqrt{(-1)}$.

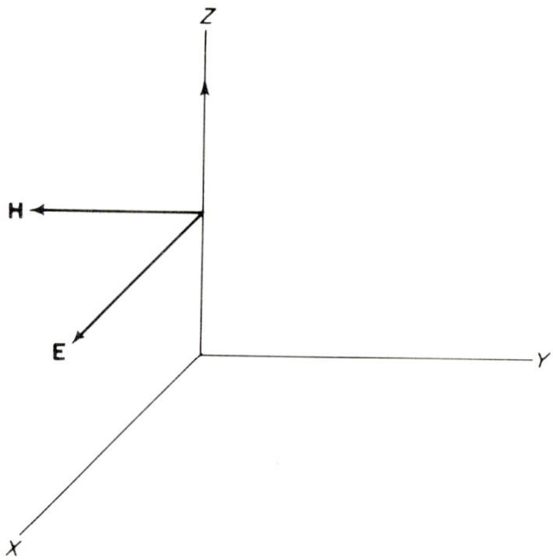

FIG. 2.1. Electric and magnetic vectors of a plane wave propagating in the positive z-direction.

The elastic medium which Maxwell conceived as the residence of the electromagnetic field was thought to undergo strain and vibration in the presence of the field and with this was associated the electrostatic and magnetostatic energies given by

$$U + T = \tfrac{1}{2} \int_V (\varepsilon|\mathbf{E}|^2 + \mu|\mathbf{H}|^2)\, dV \tag{2.2.6}$$

Though this mechanical view is no longer held, the electromagnetic energy is still regarded as localized in space. Its origin is the reversible isothermal work expended in bringing from infinity the various charges and currents which may be thought of as giving rise to the field. For a periodic field such as the electromagnetic wave there is a flux of energy crossing a unit area per unit time which is given by Poynting's vector

$$\mathbf{S} = \mathbf{E} \times \mathbf{H} \tag{2.2.7}$$

Since **E** and **H** in the wave zone (i.e., a region far removed from the source of the excitation) are at right angles to each other, and also to the direction of propagation, the Poynting vector is along this latter direction and measures the intensity of the wave. By taking the appropriate time average of Poynting's vector and utilizing (2.2.5), the intensity or the flow of energy crossing a unit area along the direction of propagation is

$$I = (2Z_0)^{-1}|\mathbf{E}|^2 \tag{2.2.8}$$

2.2.1 Refractive Index

The scalar quantity, u, which may represent one of the components of **E** or **H**, satisfies the nonhomogeneous scalar wave equation in dissipative media ($\sigma > 0$)

$$\nabla^2 u - \sigma\mu\frac{\partial u}{\partial t} - \varepsilon\mu\frac{\partial^2 u}{\partial t^2} = 0 \tag{2.2.9}$$

and the homogeneous scalar wave equation in dielectric media ($\sigma = 0$)

$$\nabla^2 u - \varepsilon\mu\frac{\partial^2 u}{\partial t^2} = 0 \tag{2.2.10}$$

Then the plane wave propagating along the positive z-axis may be described by

$$E_x = Ae^{i(\omega t - kz)} \tag{2.2.11}$$

where E_x is the component of **E** along the x-axis. E_x is harmonic in time and space, its maximum extent or amplitude being given by A. As written, (2.2.11) implies that E_x is a vector. However, it is understood that only the real part of the exponential is to be taken, i.e.

$$E_x = A\cos(\omega t - kz) \tag{2.2.12}$$

The quantity ω is the circular frequency,

$$\omega = 2\pi\nu \tag{2.2.13}$$

where ν is the vibrational frequency. The propagation constant, k, is related to the constitutive constants of the medium by

$$k^2 = \mu\varepsilon\omega^2 - i\mu\sigma\omega \tag{2.2.14}$$

and may be represented as

$$k = \alpha_1 - i\beta_1 \tag{2.2.15}$$

where

$$\alpha_1 = \omega\left\{\frac{\mu\varepsilon}{2}\left[\left(1 + \frac{\sigma^2}{\varepsilon^2\omega^2}\right)^{1/2} + 1\right]\right\}^{1/2} \tag{2.2.16}$$

and

$$\beta_1 = \omega\left\{\frac{\mu\varepsilon}{2}\left[\left(1 + \frac{\sigma^2}{\varepsilon^2\omega^2}\right)^{1/2} - 1\right]\right\}^{1/2} \tag{2.2.17}$$

For a conducting or dissipative medium ($\sigma > 0$), the propagation constant is complex. Its imaginary part, β_1, when multiplied by i in the exponential of (2.2.11) results in the factor

$$e^{-\beta_1 z} \tag{2.2.18}$$

This provides for the damping of the wave as it progresses in the z-direction and corresponds to the absorption of part of the radiant energy as Joule heat.

For an undamped wave in a nonconducting medium ($\sigma = 0$), k is real and

$$k = \alpha_1 = \omega(\mu\varepsilon)^{1/2} = \omega/v = 2\pi/\lambda \tag{2.2.19}$$

where v is the velocity and λ is the wavelength in the medium.

$$\lambda = \lambda_0/n \tag{2.2.20}$$

Here λ_0 is the wavelength in free space and n is the real refractive index

$$n = c/v = (\kappa_e\kappa_m)^{1/2} \tag{2.2.21}$$

In terms of the refractive index, the time independent factor in (2.2.11) now becomes $\exp[-ik_0nz]$ where

$$k_0 = 2\pi/\lambda_0 \tag{2.2.22}$$

The parameter k_0 is the propagation constant in free space.

For the general case when k is complex, a complex refractive index m can be defined by

$$E_x = A\exp[i(\omega t - k_0mz)] \tag{2.2.23}$$

where

$$m = k/k_0 = n(1 - i\kappa) \tag{2.2.24}$$

The imaginary part of the complex refractive index, $n\kappa$, is the damping factor while κ is called the index of absorption or the index of attenuation.[1]

Direct photometric measurement of absorption gives the attenuation of the intensity of the wave rather than attenuation of the electric field vector. The absorption coefficient is defined accordingly by the Beer–Lambert Law

$$I = I_0 \exp[-a'z] \tag{2.2.25}$$

where the intensity of the radiation decreases from I_0 to I over the path length z. Because the intensity is proportional to $E_x{}^2$ the absorption index κ and the absorption coefficient a' are related by

$$a' = \frac{4\pi n\kappa}{\lambda_0} = \frac{4\pi\kappa}{\lambda} \tag{2.2.26}$$

2.3 Polarization

Although the polarization of light had been discovered as early as 1690 by Huygens, it remained an isolated curiosity until the early nineteenth century. Later, Maxwell's electromagnetic theory provided a complete description of this phenomenon by associating the state of polarization with the localization of the field vectors within the plane perpendicular to the direction of propagation of the wave.

Once again we consider an electromagnetic wave propagated along the z-axis with the **E** and **H** vectors vibrating harmonically in the xy-plane. Because **E** and **H** are perpendicular to each other, it is only necessary to discuss the behavior of **E** which will define the direction of polarization. When there is no attenuation in the medium, the two Cartesian components are given by

$$E_x = A \exp[i(\tau + \delta_1)] = A \cos(\tau + \delta_1) \tag{2.3.1}$$

$$E_y = B \exp[i(\tau + \delta_2)] = B \cos(\tau + \delta_2) \tag{2.3.2}$$

Here again it is understood that only the real part of the exponential is to be taken. The phase factors consist of a variable part, $\tau = (\omega t - kz)$, and the phase angles δ_1 and δ_2 which are determined by the history of the wave. For natural light, these phase angles vary randomly with respect to

[1] Sometimes κ is called the extinction coefficient [cf. Born and Wolf (1959) p. 610].

each other but for polarized light, the phase difference, $\delta = \delta_2 - \delta_1$, is fixed. By elimination of τ from (2.3.1) and (2.3.2), the locus of **E** can be determined from the resulting equation

$$\left(\frac{E_x}{A}\right)^2 + \left(\frac{E_y}{B}\right)^2 - 2\frac{E_x E_y}{AB}\cos\delta = \sin^2\delta \tag{2.3.3}$$

which is the equation of an ellipse and is depicted in Fig. 2.2. This ellipse represents a projection on the xy-plane of the locus of the magnitude of the extreme values of the electric vector (the amplitude), and for this general case the radiation is said to be elliptically polarized.

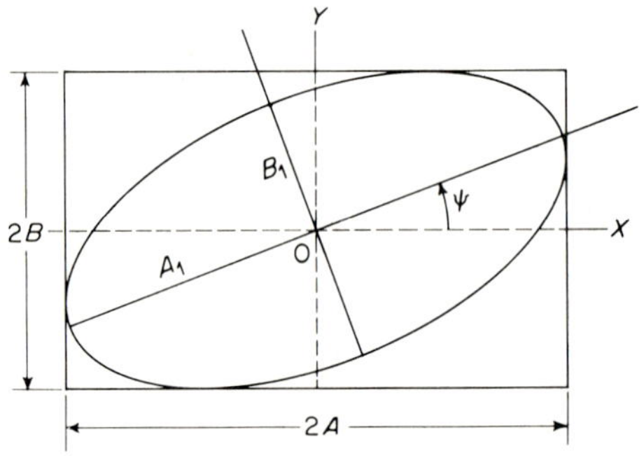

FIG. 2.2. Vibration ellipse for the electric vector.

The sides of the rectangle in which the ellipse is inscribed are parallel to the coordinate axes and have lengths of $2A$ and $2B$. The ellipse touches the sides at the points $(\pm A, \pm B\cos\delta)$ and $(\pm A\cos\delta, \pm B)$. The major axis of the ellipse is inclined at the angle ψ with respect to the x-axis $(0 \leqslant \psi < \pi)$. The semisides of the circumscribing rectangle define the auxiliary angle $\alpha\,(0 \leqslant \alpha \leqslant \pi/2)$

$$B/A = \tan\alpha \tag{2.3.4}$$

and the semiaxes of the ellipse define $\chi\,(-\pi/4 \leqslant \chi \leqslant \pi/4)$

$$\pm B_1/A_1 = \tan\chi \tag{2.3.5}$$

By the state of polarization is meant the eccentricity and orientation of the ellipse and these are defined either by the parameters α and δ or χ and ψ. Complete characterization of the radiation must also include the intensity so that a third parameter is needed, i.e., A, B, and δ or A_1, B_1, and ψ. The intensity is given by

$$I = \frac{1}{2Z_0}(A^2 + B^2) = \frac{1}{2Z_0}(A_1{}^2 + B_1{}^2) \tag{2.3.6}$$

The polarization is said to be right handed if the endpoint of the electric vector describing the ellipse rotates clockwise when the wave is "viewed" face-on by the observer. In this case $\sin\delta > 0$ and $0 < \chi \leqslant \pi/4$. Otherwise the polarization is left-handed.

There are two special cases of particular interest, namely when the polarization ellipse reduces to a straight line or to a circle. The first case gives linearly polarized light for which

$$\delta = j\pi \quad (j = 0, \pm 1, \pm 2, \ldots) \tag{2.3.7}$$

and (2.3.3) becomes

$$\frac{E_y}{E_x} = (-1)^m \frac{B}{A} \tag{2.3.8}$$

Here, the projection of the resultant vector oscillates in the xy-plane along the direction χ or $-\chi$ depending upon whether m is even or odd.

Similarly, when $A = B$ and

$$\delta = \tfrac{1}{2}m\pi \quad (m = \pm 1, \pm 3, \pm 5, \ldots) \tag{2.3.9}$$

the equation of a circle is obtained,

$$E_x^2 + E_y^2 = A^2 \tag{2.3.10}$$

Values of $m = 1, -3, 5, -7$, etc., correspond to right-handed polarized light.

2.3.1 Stokes Parameters

Although α, δ and χ, ψ are useful in visualizing the geometrical relations of the polarization ellipse, they often lead to unwieldy algebraic expressions when treating actual problems. There is still a third representation of polarized light, first introduced by Stokes (1852), which is frequently more useful. The Stokes parameters are

$$s_0 = A^2 + B^2 = A_1^2 + B_1^2 \tag{2.3.11}$$

$$s_1 = A^2 - B^2 = (A_1^2 + B_1^2)\cos 2\psi \cos 2\chi \tag{2.3.12}$$

$$s_2 = 2AB\cos\delta = (A_1^2 + B_1^2)\sin 2\psi \cos 2\chi \tag{2.3.13}$$

$$s_3 = 2AB\sin\delta = (A_1^2 + B_1^2)\sin 2\chi \tag{2.3.14}$$

Among these quantities there exists the relation

$$s_0^2 = s_1^2 + s_2^2 + s_3^2 \tag{2.3.15}$$

so that again the complete state of polarization is described by only three independent parameters. When $\delta = 0^0$ or 180^0, the radiation is linearly polarized and $s_0 = A^2 + B^2$, $s_1 = A^2 - B^2$, $s_2 = \pm 2AB$ and $s_3 = 0$. For circularly polarized light $\delta = 90°$ or $270°$ and $A = B$ so that $s_0 = \pm s_3 = 2A^2$ and $s_1 = s_2 = 0$.

The most important property of the Stokes parameters is their additivity for incoherent polarized beams. When such beams of light, each of which is elliptically polarized, are combined, the resultant beam is characterized by

$$s_0 = \sum_i s_{0i}, \qquad s_1 = \sum_i s_{1i}, \qquad s_2 = \sum_i s_{2i}, \qquad s_3 = \sum_i s_{3i} \tag{2.3.16}$$

It follows in this case that if the state of polarization is not identical for each of these beams, then (2.3.15) no longer holds, but

$$s_0^2 > s_1^2 + s_2^2 + s_3^2 \tag{2.3.17}$$

Such radiation is only partially polarized and, as was shown by Stokes, may be considered to consist of a mixture of completely polarized and of natural light. The vibrations corresponding to the natural light may be resolved into components which have the same amplitude in any direction in the plane perpendicular to the direction of propagation and whose phases rapidly vary in a random fashion with respect to each other. Accordingly, this radiation is not affected by the retardation of any rectangular component relative to the other. It follows from this that for natural light there is a finite value of s_0 and that the other Stokes' parameters of natural light are

$$s_1 = s_2 = s_3 = 0 \tag{2.3.18}$$

Partially polarized light obtained by the superposition of incoherent beams such as in (2.3.16) can be decomposed into a polarized and natural part

$$s_0 = s_0^p + s_0^n, \qquad s_1 = s_1^p, \qquad s_2 = s_2^p, \qquad s_3 = s_3^p \tag{2.3.19}$$

for which s_0^p obeys (2.3.15). The fraction of polarization F is defined as the ratio of the intensity of the polarized portion of the wave to the total intensity, from which it follows that

$$F = \frac{I_{\text{pol}}}{I_{\text{tot}}} = \frac{s_0^p}{s_0^p + s_0^n} = \frac{[(s_1^p)^2 + (s_2^p)^2 + (s_3^p)^2]^{1/2}}{s_0} \tag{2.3.20}$$

where F may vary from 0 to 1.

In recent years, the use of matrix calculus for the treatment of polarized radiation has become common. When an electromagnetic wave with arbitrary polarization passes through an optical device (which may be a scattering medium), both the incident and outgoing waves can be described by their respective Stokes' parameters $[s_0, s_1, s_2, s_3]$ and $[s_0', s_1', s_2', s_3']$.

Following a remark by Soleillet (1929), Perrin (1942) pointed out that the Stokes' parameters of the outgoing beam must be a homogeneous linear function of those of the incident beam.

$$\begin{aligned} s_0' &= a_{11}s_0 + a_{12}s_1 + a_{13}s_2 + a_{14}s_3 \\ s_1' &= a_{21}s_0 + a_{22}s_1 + a_{23}s_2 + a_{24}s_3 \\ s_2' &= a_{31}s_0 + a_{32}s_1 + a_{33}s_2 + a_{34}s_3 \\ s_3' &= a_{41}s_0 + a_{42}s_1 + a_{43}s_2 + a_{44}s_3 \end{aligned} \tag{2.3.21}$$

Mathematically, the 4×4 matrix $|a_{ij}|$ is an operator which transforms the 4-vector $[s_0, s_1, s_2, s_3]$ into the linearly related 4-vector $[s_0', s_1', s_2', s_3']$. Perrin (1942) and van de Hulst (1957) have shown for various scattering situations how symmetry considerations reduce the number of independent matrix elements a_{ij} to less than sixteen.

2.4 Geometrical Optics

By geometrical optics is meant the limiting theory based on the notion that the radiant energy is transported as light rays. In homogeneous media, these travel in straight lines independent of each other. In nonhomogeneous media, propagation of the rays is determined by Fermat's principle which asserts that the optical length along the path s

$$S = \int_{P_1}^{P_2} m\, ds \tag{2.4.1}$$

of an actual ray between any two points P_1 and P_2 is shorter than the optical length of any other curve which joins these points.

2.4.1 Reflection and Refraction

To the law of reflection, known since classical antiquity, the law of refraction was added in the seventeenth century by R. Descartes and W. Snell.

$$m_1 \sin\theta_i = m_2 \sin\theta_t \tag{2.4.2}$$

where θ_i and θ_t are the usual angles of incidence and refraction for rays traversing a plane boundary between homogeneous media characterized by refractive indices m_1 and m_2. The relative amplitudes of the transmitted and reflected beams which are polarized perpendicular and parallel, respectively, to the plane of incidence are

$$t_1 = \frac{2 \sin \theta_t \cos \theta_i}{\sin(\theta_i + \theta_t)} = \frac{2m_1 \cos \theta_i}{m_1 \cos \theta_i + m_2 \cos \theta_t} \tag{2.4.3}$$

$$t_2 = \frac{2 \sin \theta_t \cos \theta_i}{\sin(\theta_i + \theta_t) \cos(\theta_i - \theta_t)} = \frac{2m_1 \cos \theta_i}{m_2 \cos \theta_i + m_1 \cos \theta_t} \tag{2.4.4}$$

$$r_1 = -\frac{\sin(\theta_i - \theta_t)}{\sin(\theta_i + \theta_t)} = \frac{m_1 \cos \theta_i - m_2 \cos \theta_t}{m_1 \cos \theta_i + m_2 \cos \theta_t} \tag{2.4.5}$$

$$r_2 = \frac{\tan(\theta_i - \theta_t)}{\tan(\theta_i + \theta_t)} = \frac{m_2 \cos \theta_i - m_1 \cos \theta_t}{m_2 \cos \theta_i + m_1 \cos \theta_t} \tag{2.4.6}$$

The plane of incidence is determined by the normal to the surface and the incident direction. For normal incidence these reduce to

$$t_1 = t_2 = 2m_1/(m_1 + m_2) \tag{2.4.7}$$

$$r_1 = -r_2 = (m_1 - m_2)/(m_1 + m_2) \tag{2.4.8}$$

These equations were first deduced by Fresnel in 1820. The relative intensities may be obtained by squaring each of these relative amplitudes, taking into account the effect of the change of cross section of the refracted beam [cf. Vasicek (1960), p. 38]. Although first developed for dielectrics for which the refractive index is real, their more general form given here involves the complex refractive indices m_1 and m_2.

Since each of the two polarized components behaves differently, there will be a change in polarization upon reflection and refraction. When linearly polarized incident radiation proceeds from an optically rarer to an optically denser dielectric medium ($n_2 > n_1$), both reflected and refracted beams will remain linearly polarized but the direction of the electric vibration undergoes rotation as given by

$$\tan \alpha_r = -\frac{\cos(\theta_i - \theta_t)}{\cos(\theta_i + \theta_t)} \tan \alpha_i \tag{2.4.9}$$

$$\tan \alpha_t = \cos(\theta_i - \theta_t) \tan \alpha_i \tag{2.4.10}$$

The azimuth denoted by α is the angle between the direction of the electric vector and the reflection plane.

When the incident light is natural, it may be resolved into two equal components, one linearly polarized in the incident plane and the other polarized perpendicular to this plane. Since each of these components will be reflected to a different extent, the reflected light will now be partially polarized with a fraction of polarization

$$F_r = (r_2^2 - r_1^2)/(r_2^2 + r_1^2) \tag{2.4.11}$$

The corresponding quantity for the transmitted beam is

$$F_t = (t_2^2 - t_1^2)/(t_2^2 + t_1^2) \tag{2.4.12}$$

For the special case that $(\theta_i + \theta_t) = \pi/2$, the denominator of (2.4.6) becomes infinite and r_2 goes to zero. For incident light under these conditions, the reflected light is completely polarized with the electric vector perpendicular to the plane of incidence. This polarizing angle or Brewster's angle is given by

$$\tan \theta_B = n_2/n_1 \tag{2.4.13}$$

When the light is propagated from an optically denser dielectric medium into one which is optically less dense ($n_2 < n_1$), the Snell–Descartes law gives a complex angle of refraction whenever

$$\sin \theta_t = (n_1/n_2) \sin \theta_i > 1 \tag{2.4.14}$$

This determines the condition of total reflection. Actually, as we shall see later, not all of the energy is reflected but some is propagated along the boundary in the plane of incidence as a surface wave. For the totally reflected ray itself, the two components undergo phase shifts of different amounts so that linearly polarized light will become elliptically polarized upon total reflection.

For absorbing media such as metals, the mathematical formalism is the same as above. The refractive index is now complex. If an incident ray proceeds from a dielectric into an absorbing medium in which m_2 is complex, the angle θ_t in (2.4.2) also becomes complex. The refracted ray is now described by an inhomogeneous wave which is dissipated within the absorbing medium. However the complex quantity θ_t when substituted in Fresnel's formulas also affects the reflected ray. The relative amplitudes r_1 and r_2 become complex, indicating that characteristic phase changes occur on reflection; thus incident linearly polarized light generally becomes elliptically polarized. Analysis of this elliptically polarized light can, in turn, lead to evaluation of the optical constants n_2 and κ_2 comprising the complex refractive index (2.2.24). Thus, if the incident light is linearly polarized in the azimuth α_i, the azimuthal angle of the reflected ray [cf. (2.4.9), (2.4.10)] will be given by

$$\tan \alpha_r = \frac{r_1}{r_2} \tan \alpha_i = \left[-\frac{\cos(\theta_i - \theta_t)}{\cos(\theta_i + \theta_t)} \right] \tan \alpha_i = P \exp(-i\Delta) \tan \alpha_i \tag{2.4.15}$$

where P gives the ratio of the absolute values of each of the reflection coefficients and Δ is the phase difference of the reflected rays. Solution of the following equations lead to explicit values for the optical constants

$$m_2{}^2 = n_2{}^2(1 - \kappa_2{}^2) = \sin^2\theta_i\left\{1 + \frac{\tan^2\theta_i(\cos^2 2\psi - \sin^2 2\psi \sin^2\Delta)}{(1 + \sin 2\psi \cos\Delta)^2}\right\} \tag{2.4.16}$$

$$2n_2{}^2\kappa_2 = \frac{\sin^2\theta_i \tan^2\theta_i \sin 4\psi \sin\Delta}{(1 + \sin 2\psi\cos\Delta)^2} \tag{2.4.17}$$

where

$$\tan\psi = P \tag{2.4.18}$$

2.4.2 Reduction from Electromagnetic Theory

It is possible to show that the law of reflection, the Snell–Descartes law, and the Fresnel formulas may be deduced directly from electromagnetic theory, provided that the plane boundary is infinite in extent or at least large compared to the wavelength. Indeed, this suggests immediately that all of geometrical optics may be obtained by a reduction of electromagnetic theory under the condition that the wavelength is small compared to the dimensions of the geometrical constructions involved. The most widely accepted argument for such a connection between electromagnetic theory was given by Sommerfeld and Runge (1911), who derived the eiconal differential equation from the scalar wave equation under the assumption that the wavelength approaches zero. The eiconal differential equation

$$(\nabla S)^2 = m^2 \tag{2.4.19}$$

is the fundamental equation from which all the laws of geometrical optics can be derived. The eiconal function, S, is "the optical path" of a ray. The solutions of (2.4.19) are the wave fronts of geometrical optics normal to which the rays travel. However, although it is generally accepted that Maxwell's equations should reduce to geometrical optics for small wavelengths, the manner in which this reduction actually takes place is still under vigorous discussion (Kline and Kay, 1965).

2.5 Interference and Diffraction

The superposition of two light beams results in interference, provided that these beams are similarly polarized, have the same wavelength, and possess a fixed correlation between their amplitude and phases. Except for the recently invented laser, light produced by physical sources does not approach these criteria sufficiently for interference between separate beams to occur. Although there are devices for controlling the monochromaticity and polarization with sufficient precision, the irregular fluctuations in amplitude and phase are so rapid that the interference effects cannot be followed. However, when rays from the same source are divided and recombined the beams do interfere. This phenomenon was first studied by R. Hooke and R. Boyle in connection with the colors exhibited by thin films (Newton's rings).

2.5.1 Coherent and Incoherent Beams

Consider the resultant field for the addition of monochromatic, linearly polarized waves. Since the direction of polarization can be specified separately, the vector notation may be omitted and the total electric field described by

$$E = \sum_k A_k \cos(\omega t - \alpha_k) \tag{2.5.1}$$

or

$$E = \left(\sum_k A_k \cos \alpha_k\right) \cos \omega t + \left(\sum_k A_k \sin \alpha_k\right) \sin \omega t \tag{2.5.2}$$

where the α's arise from the various optical paths traversed by each wave.

The intensity of the radiation at the point under investigation is proportional to E^2. If all the phases are the same, say $\alpha_k = 0$, then the intensity is proportional to $(\sum_k A_k)^2$ which is greater than the sum of the intensities of each of the beams separately, viz., $\sum_k (A_k{}^2)$. Should the amplitude of each of the N beams be equal, the intensity would be N^2 times that of a single beam. This corresponds to perfectly constructive interference. On the other hand, it is possible to have all other combinations of α_k so that the intensity can have any value between this maximum and zero. The zero value corresponds to completely destructive interference.

There is a particular case that will be important in most of the scattering problems in which we will be interested. This is where a random variation in α_k is introduced because the scattering is from a random array of particles. In evaluating the intensity from E^2 as given by (2.5.2), this quantity must be averaged at any time over all possible values of α_k.

$$\begin{aligned} E^2 = \Bigg[& \sum_k A_k{}^2 \overline{\cos^2 \alpha_k} + \sum_k A_k{}^2 \overline{\sin^2 \alpha_k} \\ & + 2 \sum_{k \neq l} A_k A_l \overline{\cos \alpha_k \cos \alpha_l} + 2 \sum_{k \neq l} A_k A_l \overline{\sin \alpha_k \sin \alpha_l} \\ & + 2 \sum_k A_k{}^2 \overline{\cos \alpha_k \sin \alpha_k} + 2 \sum_{k \neq l} A_k A_l \overline{\cos \alpha_k \sin \alpha_l} \Bigg] \end{aligned} \tag{2.5.3}$$

The last four summations are each zero because the indicated averages are zero. Also, since

$$\overline{\cos^2 \alpha_k} = \overline{\sin^2 \alpha_k} = \tfrac{1}{2} \tag{2.5.4}$$

the final result is

$$E^2 = \left[\sum A_k{}^2\right] \tag{2.5.5}$$

This is the condition of incoherence for which the individual intensities add. For this case, if there are N equal amplitudes the intensity would be N times that of a single wave.

2.5.2 Diffraction

The term diffraction is applied to the bending of waves around the edge of an obstacle, the effect of which is usually restricted to a small region close to the forward direction. Although some diffraction problems can be treated rigorously by electromagnetic theory and others may be handled as an interference among geometrical optics rays, the main problems are treated in terms of Huygens' principle. Actually, there are no clear lines of demarcation and what we will term scattering by particles is often referred to as diffraction.

Huygens asserted that each element of a wave front may be regarded as the center of a secondary disturbance which in turn gives rise to a spherical wavelet. The position of the wave front at a later time is the envelope of all such wavelets. Fresnel applied this principle to the diffraction of light at an aperture, further postulating that the secondary wavelets mutually interfere.

A light wave propagates through an opening in a screen as if every element in the opening emitted a spherical wave whose phase and amplitude are given by the incident wave. The Huygens–Fresnel principle is an approximation valid for wavelengths sufficiently small compared to the dimensions of the opening since the boundary values at its surface are unknown. Its validity is intermediate between geometrical optics and the exact treatment based upon a solution of Maxwell's equations subject to the appropriate boundary conditions. A further limitation is the absence of any specification of the magnitude and polarization of the field as a function of time and position. Also it is confined to opaque screens and transparent openings.

Babinet's principle applies to complementary diffracting screens where the opening of one body is congruent with the screen of another, e.g., a disk and a circular hole in a screen. Two such complementary bodies produce diffraction patterns of equal intensity.

2.6 Surface Waves

When an oscillating dipole is located just above the surface of a sphere, large compared to the wavelength, part of the energy radiated by the dipole propagates along the spherical surface. A. Sommerfeld identified this ground wave or surface wave in connection with radio wave propagation over the surface of the earth in order to account for transhorizon propagation. We shall see that such waves play a central role in scattering by spheres and cylinders.

The theory of surface waves, which may be represented as solutions of Maxwell's equations, is potentially at least as rich as that of waves in space, although the development of this theory is as yet hardly as complete. The mathematical theory is related to that of elastic surface waves which have been of special interest in seismology (Jeffreys, 1959) since Rayleigh (1885) investigated them and pointed out their importance for the study of earthquakes. Indeed, it is often possible to apply the results of an investigation of elastic waves directly to electromagnetic waves. The reader is referred elsewhere for a formal introduction (Barlow and Brown, 1962); we will be content merely to sketch some of the physical ideas which later will provide insight into the scattering problem.

A surface wave propagates along an interface between two media. If the wave is not to radiate from the surface, the interface must be straight in the direction of propagation, although it can assume a variety of forms in the transverse direction. Thus a nonradiating surface wave may travel axially along the surface of a cylinder as well as along a plane. Where either or both of the media are dissipative, the wave will be attenuated by absorption as it propagates.

The excitation of a surface wave may be understood by considering a plane wave, polarized in the plane of incidence, transmitted through a dielectric medium and incident upon a plane surface beyond which is a second medium. The reflection coefficient at the surface is given by

$$r_2 = \frac{m^2 \cos \theta_i - (m^2 - \sin^2 \theta_i)^{1/2}}{m^2 \cos \theta_i + (m^2 - \sin^2 \theta_i)^{1/2}} \tag{2.6.1}$$

which is (2.4.6) expressed in terms of m, the refractive index of the second medium relative to that of the first medium, and of the angle of incidence, θ_i. We now consider the condition that $r_2 = 0$ which determines the Brewster angle. This condition is that

$$\tan \theta_B = m \tag{2.6.2}$$

There is no difficulty when both media are dielectrics so that (2.4.13) is obtained directly. When the second medium is absorptive and the angle of refraction is complex, we have already seen that the expressions for the reflection coefficients (2.4.5), (2.4.6) can still be treated analytically and that they can be explicitly expressed in terms of the optical constants (2.4.16), (2.4.17). However, in order that the condition of complete polarization be satisfied for the case that the second medium is absorptive, θ_B, the angle of incidence at the Brewster angle must also be complex. A complex angle determines an inhomogeneous wave whose planes of constant phase and amplitude do not coincide. The result is that at the Brewster angle a reflected field is generated which flows along the surface in the z-direction, i.e. a surface wave [Stratton (1941) p. 516].

A somewhat simpler visualization of the origin of a surface wave is provided by considering a wave traveling from a medium of higher to lower refractive index at the critical angle. At this angle, which is determined by the condition $\theta_t = 90°$, the refracted ray follows the surface. Figure 2.3 shows the incident ray to be AB and the tangentially refracted ray to be BCC'. The Fresnel coefficients predict perfect reflection of the incident ray and zero intensity for the refracted one. Accordingly, for this and still greater angles of incidence which lead to a complex angle of reflection, total reflection is said to occur.

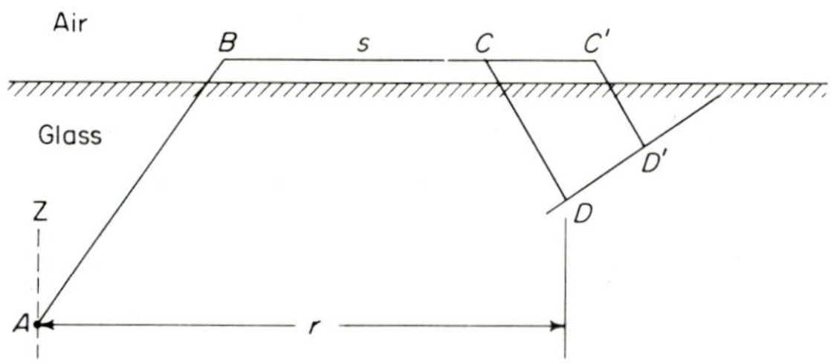

FIG. 2.3. Generation of a surface wave by a source A in a medium with a larger refractive index. AB is at the angle for total reflection.

Actually, Fresnel's formulas do not hold for the totally reflected ray and there is a ray of finite intensity transmitted along the surface (BCC') which travels at the faster speed corresponding to the rarer medium. From each point of this single refracted ray, there split off diffracted rays parallel to the "totally reflected ray" ($CD, C'D'$, etc.). When the optical paths of these rays are considered ($m_2 \cdot AB + m_1 \cdot BC + m_2 \cdot CD$, etc.), they determine a wave front of constant phase perpendicular to the reflected ray. The intensity of the energy transmitted along the rays ($CD, C'D'$, etc.) drops off rapidly as these move further from the reflected ray, B. However this effect can be observed and has been demonstrated experimentally by Maecker (1949) and follows the theoretical analysis of Ott (1942, 1949). The interested reader is also referred to later work by Friederichs and Keller (1955) and Felsen (1967).

CHAPTER 3

Scattering by a Sphere

3.1 Historical Introduction. The Color and the Polarization of Skylight

One impetus for developing a theory of light scattering was the attempt to explain the color and the polarization of skylight. The notion that the brightness of the daytime sky is due to "reflection" of sunlight by particles contained in the air is an old one, already formulated by Alhazen of Basra, the Arabic physicist who carried out much of his work in Cairo during the early eleventh century. Da Vinci (ca. 1500) described experiments in which fine water mists displayed an "azure hue" when illuminated against a black background, and during the nineteenth century other such artificial skies were prepared by Brücke (1853), in the form of an alcoholic suspension of mastic, by Govi (1860), with tobacco and alcohol smokes, and by Tyndall (1869), in a series of celebrated experiments with aerosols prepared by condensation of the products of gaseous reactions. Tyndall pointed out that

> "the blue color of the sky, and the polarization of skylight ... constitute, in the opinion of our most eminent authorities, the two great standing enigmas of meteorology. Indeed it was the interest manifested in them by Sir John Herschel, in a letter of singular speculative power, that caused me to enter upon the consideration of these questions so soon".

The experiments of each of these workers demonstrated that when the particles were small, they scattered blue light. Both Govi and Tyndall showed that the scattered light was polarized. Tyndall found that for the smallest particles, the light scattered at right angles to the incident beam was completely linearly polarized. These effects were independent of the nature of the scattering media and depended only upon the particles being sufficiently small. Accordingly, for such particles, the scattering takes place by quite different laws from those operating for the reflection of light from the surfaces of larger bodies.

The polarization of skylight at right angles to the sun, investigated by Arago in 1811, had posed the greatest stumbling block to an explanation of the origin of skylight. Newton (1706) had suggested that the blue of the sky was produced in the same manner as the colors of thin films and resulted from the interference of the rays reflected from the front and rear surfaces of suspended water droplets. However such specular reflection should have given rise to complete polarization at 74°, the Brewster angle for water [cf. (2.4.13)], rather than at the observed angle of reflection, 45°. In addition, Brücke noted that the blue of the sky and the color obtained by interference from thin films were vastly different, and somewhat earlier Clausius (1847a, b, 1848) had demonstrated that a cloud of droplets sufficiently large to exhibit interference effects would cause the stars to appear enormously magnified. Clausius attempted to save the hypothesis of Newton by proposing that the atmosphere was charged with water bubbles, but apart from the difficulty of seeing how such bubbles could be formed, these still could not account for the color and polarization of the skylight.

Tyndall's experiments showed quite decisively that scattering by small particles could account for skylight, and indeed he believed that this originated with "water particles ... in fine state of division ... in the higher regions of the atmosphere." He dismissed an earlier suggestion of Brewster that perfect polarization would occur at right angles to the incident beam if there were reflection in air upon air and asserted categorically that "the law of Brewster does not apply to matter in this condition; and it rests with the undulatory theory to explain why."

This mandate was carried out by Lord Rayleigh, who became interested in color perception early in his career (1871a). He noticed that his ability to match colors reproducibly on different days was influenced by whether the illumination arose from an unusually blue sky or from a cloudy one. He first fell into the not uncommon trap of "suspecting the light of the sky would be similar in composition to that of dilute solutions of copper, which acquire their light blue tint by a partial suppression of the extreme red."

However, Tyndall's experiments soon made clear to Lord Rayleigh that the light received from the sky was due to scattering, and he gave a simple physical explanation to explain why the light scattered at 90° to the incident beam is completely polarized. It can be visualized with the aid of Fig. 3.1. A beam of unpolarized light traveling along the z-axis impinges upon a spherical particle located at the origin of a Cartesian coordinate system. The unpolarized incident beam can be resolved into two linearly polarized components and each of these can be considered to act independently of the other. The particle is isotropic, homogeneous, and small compared to the wavelength. This latter condition assures that the instantaneous electromagnetic field is uniform over the extent of the particle which will become polarized in the same direction as the field. The effect is to create a dipole which oscillates synchronously and in the same direction as the vibrating electromagnetic field.

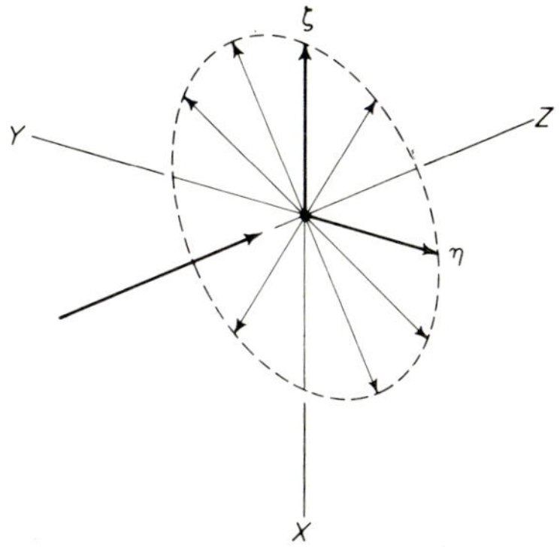

FIG. 3.1. Rayleigh scattering at 90°. Incident beam is along the z-axis and is polarized in the xy-plane. Scattered beam is along the η-direction in the xy-plane. Electric vector of scattered wave is parallel to that in incident wave and is resolved into ζ and η components.

The oscillating dipole will radiate electromagnetic energy and it is this secondary emission which constitutes the scattering. The scattered radiation will be polarized in the same sense as the dipole. For the moment, we will concern ourselves only with those rays traveling outwards in the xy-plane since this is perpendicular to the incident direction; it is this radiation which is to be demonstrated as being linearly polarized. Because the initial unpolarized radiation has no component in the z-direction, which is the direction of its propagation, neither does the oscillating particle nor the scattered radiation have a component in this direction. For any arbitrary scattering direction η in the xy-plane, the oscillation can be resolved into two components. These correspond to the two components into which the unpolarized incident radiation has been resolved. One of these, ζ, will be chosen perpendicular to the scattering direction and the other, η, will be taken parallel to this direction. The latter component of the oscillation will not contribute to the scattering because an electromagnetic wave does not have a longitudinal component. The scattered radiation consists of only the transverse component which is perpendicular to the scattering direction and by virtue of its being in the xy-plane also perpendicular to the incident direction. By this argument we see that this mechanism accounts for the complete polarization of the 90° scattered light.

Actually, this very argument, which Rayleigh used to demonstrate that the polarization of skylight was the result of a scattering process, had been used in reverse nearly twenty years earlier by Stokes (1852) to support his statement that the ether particles vibrated perpendicularly to the plane of polarization (along the direction of the electric vector).

The explanation for the color of the scattered light can be arrived at by an equally simple argument based upon dimensional analysis. The scattered intensity is proportional to the intensity of the incident radiation. Rayleigh assumed that the other variables which enter as unknown functions are the volume V of the spherical particle, the distance r to the point of observation, the wavelength λ, and the refractive indices of the particle and medium n_1 and n_2. Assuming that neither of the indices is complex, then

$$I = f(V, r, \lambda, n_1, n_2)I_0 \tag{3.1.1}$$

Obviously $f(V, r, \lambda, n_1, n_2)$ is dimensionless. Since the dipole radiates energy in all directions, the intensity must fall off as r^{-2}. The field of the dipole is proportional to the dipole moment, which for a given uniform exciting field is in turn proportional to the volume of the particle. The intensity is proportional to the square of the field so that it will vary as V^2. Whatever the functional dependence upon n_1 and n_2, there will be no effect upon the dimensional analysis since these quantities are themselves dimensionless. Therefore, the wavelength dependence will be given by

$$I = f'(n_1, n_2)(V^2/r^2\lambda^4)I_0 \tag{3.1.2}$$

It is this inverse fourth power dependence upon wavelength which causes the blue end of the visible spectrum to be more strongly scattered, giving the sky its characteristic color.

Rayleigh had actually compared the spectral composition of the blue of the sky with that of direct sunlight prior to developing his theory. By some remarkably simple, yet accurate experiments, he was able to show that the intensity of skylight did vary with the inverse fourth power of the wavelength rather than the inverse second power as required for interference by thin plates.

As for the composition of the scattering particles in the atmosphere, previous writers had taken for granted that these consisted of water or ice. Rayleigh took exception to this and suggested at first that they might be common salt. Shortly after publication, in 1871, of his first paper on scattering, Maxwell directed Rayleigh's attention to the possibility that the molecules of air itself might be the scattering particles and that if this were the case, information about the size of the molecules might be obtained.[1]

[1] Rayleigh (1899) reported: Under date August 28, 1873, he (Maxwell) wrote:

"I have left your papers on the light of the sky, etc. at Cambridge, and it would take me, even if I had them, some time to get them assimilated sufficiently to answer the following

Rayleigh (1899) later published calculations which indicated "that the light scattered from molecules would suffice to give us a blue sky, not so very greatly darker than that actually enjoyed." He understood the limitations of his results, not only due to the uncertainty of the meteorological data but also because of the complications due to the presence of colloidal particles, to multiple scattering, to dispersion, and to nonsphericity of the molecules. In specifying these difficulties, he laid down the guidelines for research in meteorological optics that today is carried out with the assistance of aircraft, balloons, and rockets (Gates, 1966).

3.2 The Rayleigh[2] Theory of Scattering by Small Dielectric Spheres

These simple arguments of Rayleigh were buttressed by a rigorous derivation of scattering by a small dielectric sphere based upon the elastic solid concept of ether vibrations and later (1881) upon the electromagnetic theory.

The isolated sphere is illuminated by a parallel beam of linearly polarized radiation. It becomes polarized in the electromagnetic field due to the displacement of the electrons with respect to the nuclei and also due to the partial orientation of any permanent dipoles that may be present. The latter effect will be insignificant for high frequency optical fields. The basic premise is, because the particle is small compared to the wavelength, that the instantaneous field which it experiences due to the electromagnetic wave is uniform over its extent. Only the effect of the electric vector is considered so that this reduces to the standard electrostatic problem of an isotropic, homogeneous, dielectric sphere in a uniform field. The solution is well known [Stratton (1941) p. 205]. Within the sphere, the field is uniform and parallel to the external field (see Fig. 3.2) with the electric field intensity given by

$$\mathbf{E}_{\text{int}} = [3\varepsilon_2/(\varepsilon_1 + 2\varepsilon_2)]\mathbf{E}_0 \tag{3.2.1}$$

question, which I think will involve less expense to the energy of the race if you stick the data into your formula and send me the result....

"Suppose that there are N spheres of density ρ and diameter S in unit volume of the medium. Find the refractive index of the compound medium and the coefficient of extinction of light passing through it.

"The object of the enquiry is, of course, to obtain data about the size of the molecules of air."

[2] Unfortunately the attachment of Lord Rayleigh's name to the theory of scattering by small isotropic spheres has tended to obscure his vast contribution to many other aspects of scattering. Twersky (1964) has described these in broad outline and Wait (1965a) has assembled a bibliography of pertinent papers.

where ε_1 and ε_2 are the electric inductive capacities of the sphere and the external medium, respectively. Outside the sphere the field is composed of two parts—the initial uniform field, $\mathbf{E}_0$, that would have existed in the absence of the particle, and superimposed upon this is an induced field identical with that which would be given by a simple dipole oriented parallel to the incident field with dipole moment

$$\mathbf{p} = 4\pi\varepsilon_2 a^3[(\varepsilon_1 - \varepsilon_2)/(\varepsilon_1 + 2\varepsilon_2)]\mathbf{E}_0 \tag{3.2.2}$$

where a is the radius. Part of the factor preceding $\mathbf{E}_0$ is called the polarizability, α', and is given by

$$\alpha' = a^3[(\varepsilon_1 - \varepsilon_2)/(\varepsilon_1 + 2\varepsilon_2)] \tag{3.2.3}$$

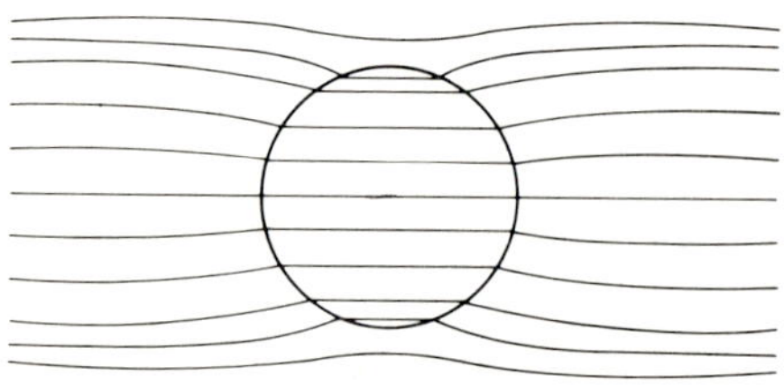

FIG. 3.2. Perturbation of a uniform electric field by a homogeneous sphere. The field within the sphere is uniform.

If the incident field oscillates harmonically, then to a close approximation the induced dipole will follow synchronously so that

$$\mathbf{p}e^{i\omega t} = 4\pi\varepsilon_2 a^3[(\varepsilon_1 - \varepsilon_2)/(\varepsilon_1 + 2\varepsilon_2)]\mathbf{E}_0 e^{i\omega t} \tag{3.2.4}$$

where the exponential factor describes the time dependence. Thus, the spherical particle acts as an oscillating electric dipole which now radiates secondary or scattered waves in all directions.

The geometry can be visualized with the aid of Fig. 3.3. The linearly polarized incident wave propagating along the positive z-axis has its electric vector parallel to the x-axis. The induced dipole is at the origin and is also oriented along the x-axis. The scattering direction is taken from the origin through the point defined by the polar coordinates r, θ, ϕ.

$$x = r \sin\theta \cos\phi; \quad y = r \sin\theta \sin\phi; \quad z = r\cos\theta \tag{3.2.5}$$

The angle measured from the scattering direction to the dipole is ψ. The angle of observation, θ, is measured from the forward to the scattered directions and defines the scattering plane or the plane of observation. For an incident wave of unit intensity, the intensity of the scattered wave at a distance r from the particle is given by [Stratton (1941) p. 436]

$$I = \frac{16\pi^4 a^6}{r^2 \lambda^4} \left(\frac{\varepsilon_1 - \varepsilon_2}{\varepsilon_1 + 2\varepsilon_2} \right)^2 \sin^2 \psi \tag{3.2.6}$$

or

$$I = \frac{16\pi^4 a^6}{r^2 \lambda^4} \left(\frac{n^2 - 1}{n^2 + 2} \right)^2 \sin^2 \psi \tag{3.2.7}$$

where n is the relative refractive index, i.e., the ratio, n_1/n_2, and λ is the wavelength in the medium.

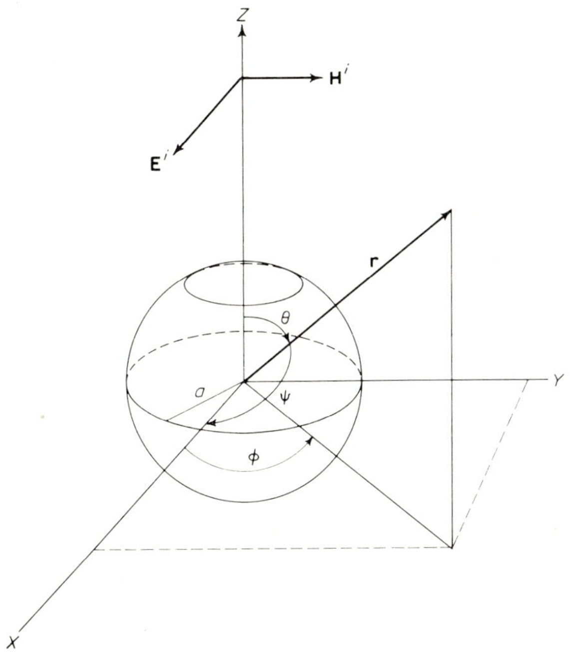

FIG. 3.3. Geometry for Rayleigh scattering. Incident wave travels along positive z-axis with electric vector polarized along x-axis. Particle with radius a has its center at the origin. Direction of scattered wave is defined by polar angles θ and ϕ.

If the scattering particle is absorptive, it will be characterized by a complex relative refractive index, m. When the imaginary part is small compared to the real part, the appropriate expression is

$$I = \frac{16\pi^4 a^6}{r^2 \lambda^4} \left| \frac{m^2 - 1}{m^2 + 2} \right|^2 \sin^2 \psi \tag{3.2.8}$$

Otherwise, as in the case for highly reflecting or highly absorbing particles, a somewhat different expression, to be considered later, must be used.

There are two special cases of interest. For Case 1 where the scattering is in the yz-plane and $\psi = 90°$, the incident beam is perpendicularly polarized with respect to the scattering plane. The scattered radiation, also perpendicularly polarized, has an intensity

$$I_1 = \frac{16\pi^4 a^6}{r^2 \lambda^4}\left(\frac{n^2 - 1}{n^2 + 2}\right)^2 \tag{3.2.9}$$

This is independent of the angle of observation since all directions in this plane are equivalent with respect to the dipole.

Case 2 is for scattering in the xz-plane. Here the polarization is parallel to the scattering plane and ψ, which may now take on all values, is related to θ by

$$\psi = (\pi/2) - \theta \tag{3.2.10}$$

so that

$$I_2 = \frac{16\pi^4 a^6}{r^2 \lambda^4}\left(\frac{n^2 - 1}{n^2 + 2}\right)^2 \cos^2 \theta \tag{3.2.11}$$

An alternative geometrical view is sometimes convenient, in which the scattering plane is fixed as the yz-plane and in which the direction of the incident electric vector and hence of the induced dipole is in the xy-plane. This is depicted in Fig. 3.4 where χ is the angle between the induced dipole and the y-axis. In this configuration, the yz-plane can be visualized as the

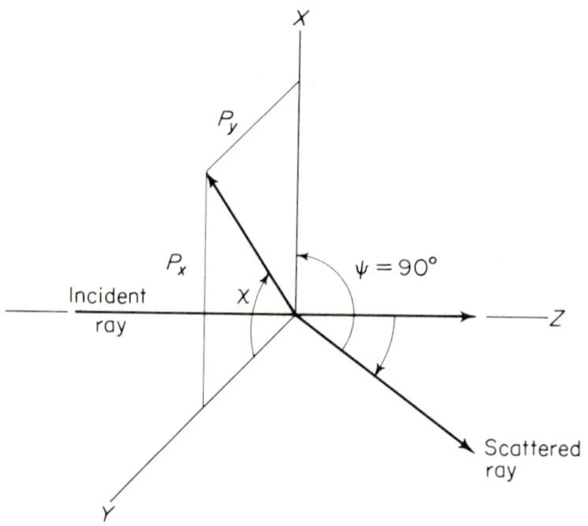

FIG. 3.4. Geometry for Rayleigh scattering; "bench-top" view with yz as the horizontal plane. Electric vector of incident wave in xy-plane. Components of induced dipole are p_x and p_y.

horizontal plane or as the laboratory bench top. The x-axis now depicts the vertical direction. The two components of the dipole, p_x and p_y, correspond to Cases 1 and 2 and may be referred to alternatively as the vertical and horizontal components as well as the perpendicular and parallel components. For incident light of unit intensity linearly polarized with azimuthal angle χ, the scattered light will consist of two linearly polarized components

$$I_V(\chi) = \frac{16\pi^4 a^6}{r^2\lambda^4}\left(\frac{n^2 - 1}{n^2 + 2}\right)^2 \sin^2 \chi \tag{3.2.12}$$

and

$$I_H(\chi) = \frac{16\pi^4 a^6}{r^2\lambda^4}\left(\frac{n^2 - 1}{n^2 + 2}\right)^2 \cos^2 \chi \cos^2 \theta \tag{3.2.13}$$

There will be no phase difference introduced between the two components by the scattering so that the azimuthal angle of the scattered radiation, χ', is obtained directly from the ratio

$$\left[\frac{I_V(\chi)}{I_H(\chi)}\right]^{1/2} = \tan \chi' = \frac{\tan \chi}{\cos \theta} \tag{3.2.14}$$

The azimuth of polarization of the scattered light will be rotated for all angles of observation except forward and back scattering. The right side of (3.2.14) is infinite for scattering at 90° which corresponds to complete polarization of the scattered light, with the electric vector vibrating perpendicular to the scattering plane, something which has already been established.

When the incident light, again of unit intensity, is elliptically polarized with a phase difference δ, the Stokes parameters for the scattered light will be

$$s_0 = I_V(\chi) + I_H(\chi) \tag{3.2.15}$$

$$s_1 = I_V(\chi) - I_H(\chi) \tag{3.2.16}$$

$$s_2 = 2(I_V(\chi)I_H(\chi))^{1/2} \cos \delta \tag{3.2.17}$$

$$s_3 = 2(I_V(\chi)I_H(\chi))^{1/2} \sin \delta \tag{3.2.18}$$

where $I_V(\chi)$ and $I_H(\chi)$ are given by (3.2.12) and (3.2.13).

An unpolarized incident wave can be resolved into two incoherent linearly polarized components which are parallel and perpendicular to the scattering plane. In this case

$$I_U = \frac{I_1 + I_2}{2} = \frac{8\pi^4 a^6}{r^2\lambda^4}\left(\frac{n^2 - 1}{n^2 + 2}\right)^2 (1 + \cos^2 \theta) \tag{3.2.19}$$

In his original paper, Lord Rayleigh [1871b, p. 113] failed to normalize for unpolarized incident light by dividing by two as has been done above and, as pointed out by Sinclair (1947) and by Herman (1961), this error has often been repeated.

The main features of Rayleigh scattering are now apparent, including the dependence of the scattering upon the inverse fourth power of the wavelength and the complete polarization at 90°. The polarization is often described by a radiation pattern or polar diagram of the scattered intensity as shown in Fig. 3.5. The radius vector to the curves designated I and II gives the intensity scattered in the direction θ for incident beams polarized vertically and horizontally, respectively, each of equal intensity. The curve labeled III is for unpolarized incident light. The degree of polarization for unpolarized incident light is

$$P = (I_1 - I_2)/(I_1 + I_2) = (1 - \cos^2\theta)/(1 + \cos^2\theta) \qquad (3.2.20)$$

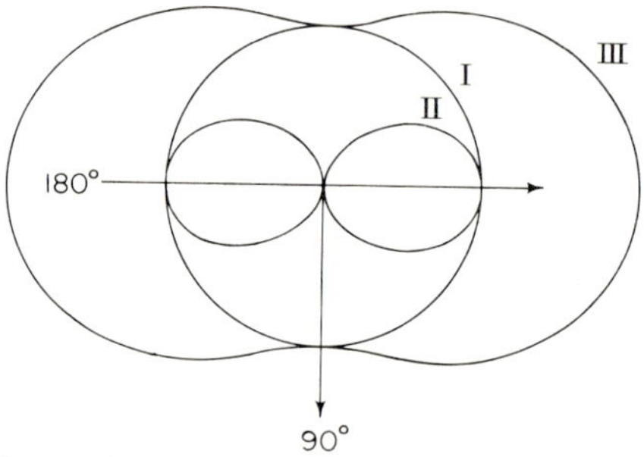

FIG. 3.5. Radiation diagram for Rayleigh scattering. The radius vector to each curve is proportional to the intensity scattered at the corresponding angle. Curve I is for Case I; curve II is for Case II; curve III is the superposition of the other two and is for unpolarized incident radiation or for linearly polarized incident radiation with $\chi = 45°$.

The polarization ratio defined as the ratio of the intensity of the horizontal to that of the vertical component of the scattered light is

$$\rho_u(\theta) = \cos^2\theta \qquad (3.2.21)$$

The Tyndall effect, which describes the increase in scattering power when a particular volume of material is more coarsely dispersed, also follows from the Rayleigh equation. In a random array of identical particles, the scattering is incoherent, so that the intensity scattered per unit volume of the medium is the sum of the effect from each individual particle. If (3.2.19) is written in terms of the particle volume V instead of the radius a, the intensity scattered per unit volume is

$$NI_u = \frac{9\pi^2 N V^2}{2r^2\lambda^4}\left(\frac{n^2 - 1}{n^2 + 2}\right)^2 (1 + \cos^2\theta) \qquad (3.2.22)$$

where N is the number of particles per unit volume. Accordingly, the intensity scattered by a given total volume of scattering material, NV, dispersed as small particles, is directly proportional to the volume of the individual particles; the larger these are, the more intense is the scattering. The upper limit of particle radius for which the Rayleigh equation is presumed to be "valid" is generally set at $a/\lambda \leqslant 0.05$. This will be discussed further after the general theory of scattering by spheres of arbitrary size has been considered.

The total energy scattered by a particle in all directions or the scattering cross section can be obtained by integration of (3.2.6) over the surface of a sphere

$$C_{\text{sca}} = \int_0^{\pi} \int_0^{2\pi} Ir^2 \sin \psi \, d\psi \, d\phi \tag{3.2.23}$$

where ϕ is designated in Fig. 3.3. This is called the scattering cross section of the particle since it has the dimension of an area. Upon integration with the aid of (3.2.7), the above expression becomes

$$C_{\text{sca}} = \frac{128\pi^5 a^6}{3\lambda^4} \left(\frac{n^2 - 1}{n^2 + 2} \right)^2 = \frac{24\pi^3 V^2}{\lambda^4} \left(\frac{n^2 - 1}{n^2 + 2} \right)^2 \tag{3.2.24}$$

The efficiency factor for scattering is obtained from the cross section by dividing by the actual geometrical cross section, which is πa^2 for a sphere. It represents the fraction of energy geometrically incident upon the particle which is scattered in all directions. For a Rayleigh scatterer this is

$$Q_{\text{sca}} = \frac{128\pi^4 a^4}{3\lambda^4} \left(\frac{n^2 - 1}{n^2 + 2} \right)^2 \tag{3.2.25}$$

It will be convenient to introduce the dimensionless size parameter

$$\alpha = 2\pi a/\lambda \tag{3.2.26}$$

which is the radius of the sphere in units of $2\pi/\lambda$. Then

$$Q_{\text{sca}} = \frac{8}{3} \alpha^4 \left(\frac{n^2 - 1}{n^2 + 2} \right)^2 \tag{3.2.27}$$

These equations for C_{sca} and Q_{sca} are valid for any combination of polarized and unpolarized incident radiation provided the total incident intensity is unity. The formulas for these quantities given by Rayleigh (1899) are in in error by the factor of 2.

The scattering cross section may be related to the transmission of a beam through a dispersion of Rayleigh scatterers of equal size. For N particles per unit volume, the attenuation due to scattering is

$$-dI/dx = NC_{\text{sca}}I \tag{3.2.28}$$

The transmission is

$$T = I_l/I_0 = \exp(-NC_{\text{sca}}l) = \exp(-\tau l) \tag{3.2.29}$$

where I_0 is the incident intensity and I_l is the intensity of the beam emerging at the distance l. The attenuation coefficient τ is called the turbidity. It represents the total energy scattered by a unit volume of the scattering medium for unit incident intensity. It is important in carrying out transmission measurements to ensure that the detecting system does not intercept a significant amount of the radiation scattered near the forward direction.

For Rayleigh scattering, the relation between the turbidity and the scattering at any particular angle is sufficiently simple so that only one of these quantities need be determined. For example, from (3.2.22), (3.2.24), and (3.2.29)

$$\tau = 16\pi r^2 N I_{\text{u}}/3(1 + \cos^2\theta) \tag{3.2.30}$$

where I_{u} is the scattered intensity per particle for unpolarized incident light of unit intensity. An auxiliary quantity defined by

$$R_\theta = r^2 N I_{\text{u}} = (3/16\pi)\tau(1 + \cos^2\theta) \tag{3.2.31}$$

is called the Rayleigh ratio. It is the energy scattered by a unit volume in the direction θ, per steradian, when the medium is illuminated with unit intensity of unpolarized light. The first term on the right of (3.2.31) gives the vertical component and the second term gives the horizontal component of the Rayleigh ratio.

The light which propagates through a dispersion of scatterers in the incident direction consists of two parts. On the one hand, there is the transmitted beam which is unperturbed by the presence of the scatterers. Its phase is determined by the optical path through the medium. The second part consists of the radiation which is scattered in the forward direction and very close to that direction. This has suffered a phase retardation due to the scattering process, in addition to that due to the propagation through the medium. In the forward direction, the retardation of the radiation scattered by each particle will be the same and will be independent of the positions of the various particles in the dispersion.

Rayleigh (1899) considered the resultant transmitted beam obtained by adding the fields of the transmitted and the forward scattered wave. This resultant wave has a phase that is different from the phase of a beam transmitted in the absence of the scatterers and this increment can be interpreted as a refractive index increment due to the presence of the scatterers. The result for Rayleigh scatterers is

$$\mu - n_2 = \frac{3NV}{2}\left(\frac{n^2 - 1}{n^2 + 2}\right) \tag{3.2.32}$$

where n is the refractive index of the scatterers relative to that of the medium, n_2 is the refractive index of the medium, and μ is the apparent refractive index of the dispersion. For a gas, n is a hypothetical refractive index of the gas molecules, the medium is a vacuum for which $n_2 = 1$, and μ is the refractive index of the gas. This leads in turn to the following expression for the turbidity:

$$\tau = (32\pi^3/3N\lambda^4)(\mu - 1)^2 \tag{3.2.33}$$

thus completing Maxwell's injunction (cf. footnote 1 in this chapter) to "find the refractive index of the compound medium and the coefficient of extinction of light passing through it."

From estimates of atmospheric turbidity, Rayleigh arrived at a value of $N = 7 \times 10^{18}$ particles per cubic centimeter which corresponds to a value of 1.6×10^{23} for Avogadro's number. Later, Abbot and Fowle (1914) and King (1913) calculated Avogadro's number to be 6.23×10^{23}, after making an allowance for the effect of dust on the attenuation of solar radiation by the atmosphere. The agreement of this with the accepted value lent considerable support to Rayleigh's theory of the blue color of the sky.

Actually, the above equation was anticipated several years earlier by Lorenz (1890, 1898b) through the use of the Lorenz–Lorentz formula

$$\left(\frac{\mu^2 - 1}{\mu^2 + 2}\right)V = \left(\frac{n^2 - 1}{n^2 + 2}\right)V_0 \tag{3.2.34}$$

where V_0 is the specific volume occupied by molecules in a volume of gas, V. This leads precisely to the same equation as (3.2.32). Using other data for the turbidity of air, Lorenz obtained 3.7×10^{23} for Avogadro's number.

3.3 General Theory of Scattering by a Sphere[3]

We will proceed directly to the exact solution of the scattering of a plane electromagnetic wave by an isotropic, homogeneous sphere of arbitrary size, and then later will consider some historical aspects of this problem.

[3] In addition to the original sources which are reviewed in the historical section following this section, various treatments of this theory may be found in Bateman (1914), Born (1933), Stratton (1941), Shifrin (1951a), van de Hulst (1957), Born and Wolf (1959), and Newton (1966).

Whenever a plane wave is incident upon an object possessing a discrete boundary, and with optical constants different from those of the medium, a scattered wave is generated. The field vectors which describe the electromagnetic properties of space may be resolved into three parts—the incident wave $\mathbf{E}_i$, $\mathbf{H}_i$, the wave inside the particle, $\mathbf{E}_r$, $\mathbf{H}_r$, and the scattered wave, $\mathbf{E}_s$, $\mathbf{H}_s$. In addition to obeying the field equations (2.1.1) to (2.1.4), these quantities also satisfy the vector wave equations (2.2.1) and (2.2.2). A particular solution of the vector wave equation is sought for which the field inside the object, $\mathbf{E}_r$, $\mathbf{H}_r$, and the external field, $\mathbf{E}_i + \mathbf{E}_s$, $\mathbf{H}_i + \mathbf{H}_s$, satisfy the four boundary conditions formulated by (2.1.16) to (2.1.19). Once this solution is obtained, not only is the scattered wave completely defined, but the electromagnetic conditions within the object are known as well.

Often the shape of the object bears some simple relationship to a particular coordinate system which enables the boundary conditions to be expressed in tractable form. The wave equation is then described using these coordinates. For a sphere, spherical coordinates r, θ, ϕ, as depicted in Fig. 3.3 provide such a "natural" coordinate system.

3.3.1 Hertz–Debye Potentials; TM and TE Modes

Rather than dealing directly with the vector wave equation [Hansen (1935), Stratton (1941) p. 392], it is possible to work with the scalar wave equation (2.2.9). One device is to introduce two auxiliary functions, the electric Hertz vector, $\boldsymbol{\pi}_1$, and the magnetic Hertz vector, $\boldsymbol{\pi}_2$, which may be defined by [Stratton (1941) p. 29]

$$\mathbf{B}_1 = \mu\varepsilon\nabla \times \partial\boldsymbol{\pi}_1/\partial t \tag{3.3.1}$$

$$\mathbf{E}_1 = \nabla\nabla\cdot\boldsymbol{\pi}_1 - \mu\varepsilon\, \partial^2\boldsymbol{\pi}_1/\partial t^2 \tag{3.3.2}$$

$$\mathbf{D}_2 = -\mu\varepsilon\nabla \times \partial\boldsymbol{\pi}_2/\partial t \tag{3.3.3}$$

$$\mathbf{H}_2 = \nabla\nabla\cdot\boldsymbol{\pi}_2 - \mu\varepsilon\, \partial^2\boldsymbol{\pi}_2/\partial t^2 \tag{3.3.4}$$

This definition is not unique, and by subjecting the above to transformations which leave the field vectors invariant, other Hertz vectors may also be obtained. However, these always satisfy the following forms of the vector wave equation:

$$\nabla^2\boldsymbol{\pi}_1 - \sigma\mu\frac{\partial\boldsymbol{\pi}_1}{\partial t} - \varepsilon_0\mu\frac{\partial^2\boldsymbol{\pi}_1}{\partial t^2} = -\frac{\mathbf{P}}{\varepsilon_0} \tag{3.3.5}$$

$$\nabla^2\boldsymbol{\pi}_2 - \sigma\mu_0\frac{\partial\boldsymbol{\pi}_2}{\partial t} - \varepsilon\mu_0\frac{\partial^2\boldsymbol{\pi}_2}{\partial t^2} = -\mathbf{M} \tag{3.3.6}$$

where **P** and **M** [(2.1.10) and (2.1.11)], the electric and magnetic polarizations, arise from respective distributions of electric and magnetic dipoles, and ε_0, μ_0 are the free space inductive capacities. Thus, there are two sets of field vectors obtained directly from each of these Hertz vectors, and the total field may then be obtained by addition of them. The component fields are of physical interest. The one which is derived from the electric Hertz vector, called the electric wave or the transverse magnetic wave (TM), is characterized by a zero value of the radial component of its magnetic field intensity, $H_{1r} = 0$. For the magnetic wave or transverse electric wave (TE), the radial component of the electric field intensity is zero, $E_{2r} = 0$. These component waves from which the total solution is constructed can each be conceived of as arising from a distribution in space of oscillating electric dipoles (TM) and of oscillating magnetic dipoles (TE), or alternatively from the superposition of oscillating electric and magnetic multipoles, located at the origin, possessing a distribution of multipole moments. This aspect of the solution will be discussed again later.

What is of interest now is that the Hertz vectors can, in turn, be derived in a simple way from corresponding scalar potential functions known either as Hertz potentials or as Debye potentials, depending precisely upon how the original vector quantity has been defined.

$$-\nabla \cdot \boldsymbol{\pi}_1 = \pi_1 \tag{3.3.7}$$

$$-\nabla \cdot \boldsymbol{\pi}_2 = \pi_2 \tag{3.3.8}$$

The Hertz–Debye potentials, π_1 and π_2, in turn, are solutions of the scalar wave equation. Accordingly, if the latter can be solved in the appropriate coordinate system so that the boundary conditions can be applied, the field vectors for the TM and TE waves can be obtained and the final result is then achieved directly by addition of these component waves.

In terms of the Debye potentials, the components of the field vectors in spherical coordinates are given by (Debye, 1909a; Born and Wolf, 1959)

$$E_r = E_{1r} + E_{2r} = \frac{\partial^2(r\pi_1)}{\partial r^2} + k^2 r\pi_1 + 0 \tag{3.3.9}$$

$$E_\theta = E_{1\theta} + E_{2\theta} = \frac{1}{r}\frac{\partial^2(r\pi_1)}{\partial r\,\partial\theta} + \kappa_2 \frac{1}{r\sin\theta}\frac{\partial(r\pi_2)}{\partial\phi} \tag{3.3.10}$$

$$E_\phi = E_{1\phi} + E_{2\phi} = \frac{1}{r\sin\theta}\frac{\partial^2(r\pi_1)}{\partial r\,\partial\phi} - \kappa_2 \frac{1}{r}\frac{\partial(r\pi_2)}{\partial\theta} \tag{3.3.11}$$

$$H_r = H_{1r} + H_{2r} = 0 + \frac{\partial^2(r\pi_2)}{\partial r^2} + k^2 r\pi_2 \tag{3.3.12}$$

$$H_\theta = H_{1\theta} + H_{2\theta} = -\kappa_1 \frac{1}{r \sin\theta} \frac{\partial(r\pi_1)}{\partial\phi} + \frac{1}{r} \frac{\partial^2(r\pi_2)}{\partial r\, \partial\theta} \tag{3.3.13}$$

$$H_\phi = H_{1\phi} + H_{2\phi} = \kappa_1 \frac{1}{r} \frac{\partial(r\pi_1)}{\partial\theta} + \frac{1}{r \sin\theta} \frac{\partial^2(r^2\pi_2)}{\partial r\, \partial\phi} \tag{3.3.14}$$

where the propagation constant (2.2.14)

$$k^2 = -\kappa_1\kappa_2 \tag{3.3.15}$$

and

$$\kappa_1 = i\omega\varepsilon + \sigma \tag{3.3.16}$$

and

$$\kappa_2 = i\omega \tag{3.3.17}$$

Since all media are considered to be nonmagnetic, μ has been dropped.

3.3.2 Solution of the Wave Equation

For sinusoidal time dependence, $e^{i\omega t}$, the nonhomogeneous scalar wave equation (2.2.9) reduces to the homogeneous form

$$\nabla^2 u' + k^2 u' = 0 \tag{3.3.18}$$

where

$$u = u' e^{i\omega t} \tag{3.3.19}$$

The Hertz–Debye potentials are solutions of this equation which can be solved by the method of separation of the variables. In spherical coordinates, the wave equation becomes

$$\frac{1}{r}\left(\frac{\partial^2 r\pi}{\partial r^2}\right) + \frac{1}{r^2 \sin\theta} \frac{\partial}{\partial\theta}\left(\sin\theta \frac{\partial\pi}{\partial\theta}\right) + \frac{1}{r^2 \sin^2\theta} \frac{\partial^2\pi}{\partial\phi} + k^2\pi = 0 \tag{3.3.20}$$

Here the exponential time dependence has been factored out of the potential function. The potential, π, is now considered to be a product of three functions; one of these is a function of r, another is a function of θ, and the third is a function of ϕ, i.e.

$$\pi = R(r)\Theta(\theta)\Phi(\phi) \tag{3.3.21}$$

Each of these functions satisfies well-known ordinary differential equations

$$\frac{d^2 rR(r)}{dr^2} + \left[k^2 - \frac{n(n+1)}{r^2}\right] rR(r) = 0 \tag{3.3.22}$$

$$\frac{1}{\sin\theta}\frac{d}{d\theta}\left(\sin\theta\frac{d\Theta(\theta)}{d\theta}\right) + \left[n(n+1) - \frac{m^2}{\sin^2\theta}\right]\Theta(\theta) = 0 \tag{3.3.23}$$

$$\frac{d^2\Phi(\phi)}{d\phi^2} + m^2\Phi(\phi) = 0 \tag{3.3.24}$$

where n is integral and m can assume the integral values $-n, \ldots, 0, \ldots, +n$.

The solutions of the radial equation (3.3.22) are the Ricatti–Bessel functions defined as

$$\psi_n(kr) = (\pi kr/2)^{1/2} J_{n+\frac{1}{2}}(kr) \tag{3.3.25}$$

$$\chi_n(kr) = -(\pi kr/2)^{1/2} N_{n+\frac{1}{2}}(kr) \tag{3.3.26}$$

where $J_{n+\frac{1}{2}}(kr)$ and $N_{n+\frac{1}{2}}(kr)$ are the half integral order Bessel and Neumann functions. We will note at this time that the linear combination

$$\zeta_n(kr) = \psi_n(kr) + i\chi_n(kr) = (\pi kr/2)^{1/2} H^{(2)}_{n+\frac{1}{2}}(kr) \tag{3.3.27}$$

where $H^{(2)}_{n+\frac{1}{2}}(kr)$, the half integral order Hankel function of the second kind has the property of vanishing when kr becomes infinite and will be useful for this reason.

The solutions to (3.3.23) are the associated Legendre polynomials subject to the above restrictions on m and n.

$$\Theta = P_n^{(m)}(\cos\theta) \tag{3.3.28}$$

and for (3.3.24) the solutions are $\sin(m\phi)$ and $\cos(m\phi)$.

The properties of each of these functions are discussed in some detail later in this chapter.

The general solution of the scalar wave equation in spherical coordinates may now be obtained by a linear superposition of all of the particular solutions, each multiplied by a constant coefficient, thus

$$\begin{aligned} r\pi &= r\sum_{n=0}^{\infty}\sum_{m=-n}^{n}\pi_n^{(m)} \\ &= \sum_{n=0}^{\infty}\sum_{m=-n}^{n}\{c_n\psi_n(kr) + d_n\chi_n(kr)\}\{P_n^{(m)}(\cos\theta)\}\{a_m\cos(m\phi) \\ &\quad + b_m\sin(m\phi)\} \end{aligned} \tag{3.3.29}$$

We now require the potential functions corresponding to the three parts into which the original field vectors have been resolved, viz. those for the incident wave $\pi_1{}^i$ and $\pi_2{}^i$, those for the waves inside the particle, $\pi_1{}^r$ and $\pi_2{}^r$, and those for the scattered waves $\pi_1{}^s$ and $\pi_2{}^s$. The isotropic homogeneous sphere is characterized by a propagation constant k_1 which may or may not be complex. The isotropic, homogeneous medium will be considered to be a dielectric so that its propagation constant, k_2, is real. The ratio of these quantities defines the relative refractive index.

$$m = k_1/k_2 = m_1 k_0/m_2 k_0 = m_1/m_2 \tag{3.3.30}$$

The particle of radius a is located at the origin of the spherical coordinate system so that its boundary corresponds to the constant coordinate surface, $r = a$. The geometry is the same as for Rayleigh scattering (Fig. 3.3). The plane polarized wave, propagating along the positive z-axis, has its electric vector of unit amplitude vibrating parallel to the x-axis.

$$|\mathbf{E}^i| = |\exp(-ik_2 z)| = 1 \tag{3.3.31}$$

When this is expanded in the form of Eq. (3.3.29), it becomes

$$r\pi_1{}^i = \frac{1}{k_2{}^2}\sum_{n=1}^{\infty} i^{n-1}\frac{2n+1}{n(n+1)}\psi_n(k_2 r)P_n^{(1)}(\cos\theta)\cos\phi \tag{3.3.32}$$

$$r\pi_2{}^i = \frac{i}{k_2\kappa_2^{(2)}}\sum_{n=1}^{\infty} i^{n-1}\frac{2n+1}{n(n+1)}\psi_n(k_2 r)P_n^{(1)}(\cos\theta)\sin\phi \tag{3.3.33}$$

where $P_n^{(1)}(\cos\theta)$ is the associated Legendre function of the first kind. The functions $\chi_n(k_2 r)$ have been dropped from this expression since they become infinite at the origin through which the incident wave must pass. Therefore only the Ricatti–Bessel function $\psi_n(k_2 r)$ is utilized. The above equations describe the unperturbed incident wave.

In order to match these potentials with those of the internal and scattered waves, the latter must be expressed in a series of similar form but with arbitrary coefficients. Again, only the function $\psi_n(k_1 r)$ may be used in the expression for the potential inside the particle since $\chi_n(k_1 r)$ becomes infinite at the origin. On the other hand, the scattered wave must vanish at infinity and the Hankel functions, $\zeta_n(k_2 r)$, will impart precisely this property. Accordingly, it will be used in the expression for the scattered wave so that

$$r\pi_1{}^s = -\frac{1}{k_2{}^2}\sum_{n=1}^{\infty} i^{n-1}\frac{2n+1}{n(n+1)}a_n\zeta_n(k_2 r)P_n^{(1)}(\cos\theta)\cos\phi \tag{3.3.34}$$

$$r\pi_2{}^s = \frac{-i}{k_2\kappa_2^{(2)}}\sum_{n=1}^{\infty} i^{n-1}\frac{2n+1}{n(n+1)}b_n\zeta_n(k_2 r)P_n^{(1)}(\cos\theta)\sin\phi \tag{3.3.35}$$

and

$$r\pi_1{}^r = \frac{1}{k_1{}^2}\sum_{n=1}^{\infty} i^{n-1}\frac{2n+1}{n(n+1)}c_n\psi_n(k_1 r)P_n^{(1)}(\cos\theta)\cos\phi \tag{3.3.36}$$

$$r\pi_2{}^r = \frac{i}{k_1\kappa_2^{(1)}}\sum_{n=1}^{\infty} i^{n-1}\frac{2n+1}{n(n+1)}d_n\psi_n(k_1 r)P_n^{(1)}(\cos\theta)\sin\phi \tag{3.3.37}$$

The boundary conditions are that the tangential components of **E** and **H** be continuous across the spherical surface $r = a$. From (3.3.9) through (3.3.14), it is apparent that for the Debye potential this is equivalent to

$$(\partial/\partial r)[r(\pi_1{}^i + \pi_1{}^s)] = (\partial/\partial r)[r\pi_1{}^r] \tag{3.3.38}$$

$$(\partial/\partial r)[r(\pi_2{}^i + \pi_2{}^s)] = (\partial/\partial r)[r\pi_2{}^r] \tag{3.3.39}$$

$$\kappa_1^{(2)} r(\pi_1{}^i + \pi_1{}^s) = \kappa_1^{(1)} r\pi_1{}^r \qquad \text{at } r = a \tag{3.3.40}$$

$$\kappa_2^{(2)} r(\pi_2{}^i + \pi_2{}^s) = \kappa_2^{(1)} r\pi_2{}^r \tag{3.3.41}$$

Since the terms in the series expansions of the potential functions are independent of each other, the above equalities expressing the boundary conditions must hold for each corresponding term in the series, leading to the following four linear equations in the coefficients, a_n, b_n, c_n, and d_n.

$$m[\psi_n'(k_2a) - a_n\zeta_n'(k_2a)] = c_n\psi_n'(k_1a) \tag{3.3.42}$$

$$[\psi_n'(k_2a) - b_n\zeta_n'(k_2a)] = d_n\psi_n'(k_1a) \tag{3.3.43}$$

$$\psi_n(k_2a) - a_n\zeta_n(k_2a) = c_n\psi_n(k_1a) \tag{3.3.44}$$

$$m[\psi_n(k_2a) - b_n\zeta_n(k_2a)] = d_n\psi_n(k_1a) \tag{3.3.45}$$

The relative complex refractive index of the sphere appears in these equations through its relation to the propagation constants by

$$k_1 = m_1k_0; \qquad \kappa_1^{(1)} = im_1{}^2k_0; \qquad \kappa_2^{(1)} = ik_0 \tag{3.3.46}$$

and

$$k_2 = m_2k_0; \qquad \kappa_1^{(2)} = im_2{}^2k_0; \qquad \kappa_2^{(2)} = ik_0 \tag{3.3.47}$$

where, as usual, $k_0 = 2\pi/\lambda_0$ is the propagation constant in free space.

These equations can now be solved for the four sets of coefficients a_n, b_n, c_n, and d_n. Only the first two are of interest here, and these are given by

$$a_n = \frac{\psi_n(\alpha)\psi_n'(\beta) - m\psi_n(\beta)\psi_n'(\alpha)}{\zeta_n(\alpha)\psi_n'(\beta) - m\psi_n(\beta)\zeta_n'(\alpha)} \tag{3.3.48}$$

$$b_n = \frac{m\psi_n(\alpha)\psi_n'(\beta) - \psi_n(\beta)\psi_n'(\alpha)}{m\zeta_n(\alpha)\psi_n'(\beta) - \psi_n(\beta)\zeta_n'(\alpha)} \tag{3.3.49}$$

where

$$\alpha = k_2a = 2\pi a/\lambda = 2\pi m_2 a/\lambda_0 \tag{3.3.50}$$

$$\beta = k_1a = 2\pi m_1 a/\lambda_0 = m\alpha \tag{3.3.51}$$

Here λ_0 is the wavelength *in vacuo*, λ is the wavelength in the medium, and $m = m_1/m_2$ is the refractive index of the particle relative to that of the medium. The addition of a prime to the Ricatti–Bessel functions denotes differentiation with respect to their arguments.

This completes the formal solution of the problem. The Debye potentials of the scattered wave, $\pi_1{}^s$ and $\pi_2{}^s$, as given by (3.3.34) and (3.3.35) are now completely determined. The scattering coefficients, a_n and b_n [(3.3.48) and (3.3.49)] are expressed in terms of the parameters m and α. The field vectors describing the scattered wave may now be obtained directly from (3.3.9) to (3.3.14).

3.3.3 The Far-Field Solution

We now consider the scattered field at distances sufficiently far from the particle so that $k_2 r \gg n$ where n is the order of the Ricatti–Bessel function. This is the far field or wave zone. For light scattering, all observations are, in practice, carried out in the far-field zone. However, it is often practical with microwaves and longer wavelength radiation to carry out near-field measurements.

The expressions describing the scattered field are somewhat simpler in the far-field zone. First, the Hankel functions in (3.3.34) and (3.3.35) reduce as follows (Mie, 1908):

$$\zeta_n(k_2 r) = i^{(n+1)} \exp(-ik_2 r) \tag{3.3.52}$$

and

$$\zeta_n'(k_2 r) = i^n \exp(-ik_2 r) \tag{3.3.53}$$

A further simplification in the far-field zone results from the scattered wave becoming a transverse wave as a result of the rapid decay of the longitudinal component. The transverse components of the field vectors (E_θ, E_ϕ, H_θ, H_ϕ) decay with λ/r in accordance with the inverse square dependence of a spherical wave upon the radial distance. The radial components E_r and H_r fall off as $(\lambda/r)^2$ so that they may be neglected in the far-field zone. The final result is

$$E_\phi = \frac{H_\theta}{(m_2)} = -\frac{i\exp(-ik_2 r)}{k_2 r}\sin\phi \sum_{n=1}^{\infty} \frac{2n+1}{n(n+1)} \times \left\{ a_n \frac{P_n^{(1)}(\cos\theta)}{\sin\theta} + b_n \frac{dP_n^{(1)}(\cos\theta)}{d\theta} \right\} (-1)^{n+1} \tag{3.3.54}$$

$$E_\theta = -\frac{H_\phi}{(m_2)} = \frac{i\exp(-ik_2 r)}{k_2 r}\cos\phi \sum_{n=1}^{\infty} \frac{2n+1}{n(n+1)} \times \left\{ a_n \frac{dP_n^{(1)}(\cos\theta)}{d\theta} + b_n \frac{P_n^{(1)}(\cos\theta)}{\sin\theta} \right\} (-1)^{n+1} \tag{3.3.55}$$

The components of **E** and **H** orthogonal to each other are in the correct relation for an electromagnetic wave (2.2.5) and their inverse dependence upon r indicates that in the radiation zone, the scattered wave is the usual type of spherical wave. Since the phase relation between the two complex quantities E_θ and E_ϕ is arbitrary, the scattered wave will, in general, be elliptically polarized. It will be convenient to designate the quantities in the brackets above as the amplitude functions

$$S_1 = \sum_{n=1}^{\infty} \frac{2n+1}{n(n+1)} \{a_n \pi_n(\cos\theta) + b_n \tau_n(\cos\theta)\} (-1)^{n+1} \quad (3.3.56)$$

$$S_2 = \sum_{n=1}^{\infty} \frac{2n+1}{n(n+1)} \{a_n \tau_n(\cos\theta) + b_n \pi_n(\cos\theta)\} (-1)^{n+1} \quad (3.3.57)$$

where the angular functions are

$$\pi_n(\cos\theta) = \frac{P_n^{(1)}(\cos\theta)}{\sin\theta} \quad (3.3.58)$$

$$\tau_n(\cos\theta) = \frac{d}{d\theta} P_n^{(1)}(\cos\theta) \quad (3.3.59)$$

Now, using Poynting's theorem, the energy flow in the scattered wave is given by

$$S = \tfrac{1}{2}\mathrm{Re}(E_\theta H_\phi{}^* - E_\phi H_\theta{}^*) \quad (3.3.60)$$

where the asterisk denotes the complex conjugate. The intensity of scattered radiation polarized in the θ and ϕ azimuths is

$$I_\phi = \frac{\lambda^2}{4\pi^2 r^2} |S_1|^2 \sin^2\phi = \frac{\lambda^2}{4\pi^2 r^2} i_1 \sin^2\phi \quad (3.3.61)$$

$$I_\theta = \frac{\lambda^2}{4\pi^2 r^2} |S_2|^2 \cos^2\phi = \frac{\lambda^2}{4\pi^2 r^2} i_2 \cos^2\phi \quad (3.3.62)$$

where i_1 and i_2 will be called the intensity functions. These components are perpendicular and parallel, respectively, to the scattering plane. This plane contains the incident direction and the direction of the scattered wave (θ, ϕ).

Each of these components of the scattered light can be thought of as arising from that component of the incident beam polarized in the same sense, i.e., I_ϕ arises from an incident beam of intensity $\sin^2\phi$ polarized perpendicularly to the scattering plane, and I_θ from a beam of intensity $\cos^2\phi$ polarized parallel to the scattering plane. There will be a phase difference between these components of the scattered beam given by

$$\tan\delta = \frac{\mathrm{Re}(S_1)\,\mathrm{Im}(S_2) - \mathrm{Re}(S_2)\,\mathrm{Im}(S_1)}{\mathrm{Re}(S_1)\,\mathrm{Re}(S_2) + \mathrm{Im}(S_1)\,\mathrm{Im}(S_2)} \tag{3.3.63}$$

where Re and Im designate the real and imaginary parts of the indicated complex amplitude functions.

Just as for Rayleigh scattering, we now consider two special cases. When yz is chosen as the scattering plane so that the direction of the electric vector of the incident radiation is perpendicular to this plane, $\phi = 90°$, and

$$I_\phi = I_1 = (\lambda^2/4\pi^2 r^2) i_1 \tag{3.3.64}$$

corresponding to Case 1 considered earlier for Rayleigh scattering. On the other hand, with xz as the scattering plane, the incident radiation has its electric vector parallel to the plane, $\phi = 0°$, and

$$I_\theta = I_2 = (\lambda^2/4\pi^2 r^2) i_2 \tag{3.3.65}$$

corresponding to Case 2.

In a similar way, the configuration corresponding to Fig. 3.4 can be used where yz is the horizontal plane and x is the vertical direction. Now the scattering is observed only in the yz-plane and the direction of the incident polarization in the xy-plane is defined by the angle χ between the electric vector of the incident radiation and the y-axis. In this case

$$I_1 = I_\mathrm{V}(\chi) = (\lambda^2/4\pi r^2) i_1 \sin^2\chi \tag{3.3.66}$$

$$I_2 = I_\mathrm{H}(\chi) = (\lambda^2/4\pi r^2) i_2 \cos^2\chi \tag{3.3.67}$$

where $I_\mathrm{V}(\chi)$ and $I_\mathrm{H}(\chi)$ designate the vertical and horizontal components of the intensity of the scattered radiation for unit incident radiation polarized along χ.

When the incident light is linearly polarized as above, the Stokes parameters of the elliptically polarized scattered light are

$$s_0 = (\lambda^2/4\pi^2 r^2)(i_1 \sin^2\chi + i_2 \cos^2\chi) \tag{3.3.68}$$

$$s_1 = (\lambda^2/4\pi^2 r^2)(i_1 \sin^2\chi - i_2 \cos^2\chi) \tag{3.3.69}$$

$$\begin{aligned} s_2 &= (\lambda^2/2\pi^2 r^2)(i_1 i_2)^{1/2} \sin\chi \cos\chi \cos\delta \\ &= (\lambda^2/4\pi^2 r^2)[\mathrm{Re}(S_1)\,\mathrm{Re}(S_2) + \mathrm{Im}(S_1)\,\mathrm{Im}(S_2)] \sin 2\chi \end{aligned} \tag{3.3.70}$$

$$\begin{aligned} s_3 &= (\lambda^2/2\pi^2 r^2)(i_1 i_2)^{1/2} \sin\chi \cos\chi \sin\delta \\ &= (\lambda^2/4\pi^2 r^2)[\mathrm{Re}(S_1)\,\mathrm{Im}(S_2) - \mathrm{Re}(S_2)\,\mathrm{Im}(S_1)] \sin 2\chi \end{aligned} \tag{3.3.71}$$

For unpolarized incident radiation of unit intensity, the scattered light is given by

$$I_{\mathrm{U}} = (\lambda^2/8\pi^2 r^2)(i_1 + i_2) \tag{3.3.72}$$

and the degree of polarization is

$$P = \left| \frac{i_1 - i_2}{i_1 + i_2} \right| \tag{3.3.73}$$

Cross Sections and Efficiencies. We now consider the cross section or the total energy abstracted from the beam. The problem is somewhat more complicated than in the case of Rayleigh scattering because, when the index of refraction for the particle is complex, energy is lost within the particle by absorption in addition to that lost by scattering. Accordingly, the total cross section for the particle comprises energy abstracted from the incident beam both by scattering and absorption. This is designated as the extinction cross section, C_{ext}, and is the sum of the scattering cross section and the absorption cross section

$$C_{\mathrm{ext}} = C_{\mathrm{sca}} + C_{\mathrm{abs}} \tag{3.3.74}$$

In order to evaluate these cross sections, we consider a concentric sphere outside the scattering particle whose radius is large compared to that of the particle. Here the field is the sum of the incident and scattered fields, and the energy flow is found in the usual way from the real part of the time average of Poynting's vector

$$\mathbf{S} = (\mathbf{E}_i + \mathbf{E}_s) \times \mathbf{H}_i + \mathbf{H}_s) \tag{3.3.75}$$

We have already seen that in the wave zone the radial components of the field vectors disappear and that the scattered radiation flows radially outward as a spherical wave. Accordingly, our concern is with the radial flow of energy. The integral of this over the large concentric spherical surface gives the total outward flow of energy. When the above vector product is expanded in terms of the scalar components of the field vectors and the integration performed, the result can be resolved in three parts, viz.

$$\mathrm{I} = \mathrm{Re} \int_0^{\pi} \int_0^{2\pi} \tfrac{1}{2}(E_{i\theta} H_{i\phi}^* - E_{i\phi} H_{i\theta}^*) r^2 \sin\theta \, d\theta \, d\phi \tag{3.3.76}$$

$$\mathrm{II} = \mathrm{Re} \int_0^{\pi} \int_0^{2\pi} \tfrac{1}{2}(E_{s\theta} H_{i\phi}^* + E_{i\theta} H_{s\phi}^* - E_{s\phi} H_{i\theta}^* - E_{i\phi} H_{s\theta}^*) r^2 \sin\theta \, d\theta \, d\phi \tag{3.3.77}$$

$$\mathrm{III} = \mathrm{Re} \int_0^{\pi} \int_0^{2\pi} \tfrac{1}{2}(E_{s\theta} H_{s\phi}^* - E_{s\phi} H_{s\theta}^*) r^2 \sin\theta \, d\theta \, d\phi \tag{3.3.78}$$

The first part measures the net outflow of energy in the unperturbed incident wave and gives zero as long as the medium is a dielectric ($\sigma_2 = 0$). The third part measures the energy of the scattered field and is C_{sca}. Obviously, if energy is to be conserved, the second part must be $-C_{ext}$. This is because the net energy flow must be $-C_{abs}$, since by virtue of the absorption within it, the particle acts as an energy sink of this magnitude. These integrals have been evaluated (Mie, 1908), leading to

$$C_{sca} = (\lambda^2/2\pi) \sum_{n=1}^{\infty} (2n + 1)\{|a_n|^2 + |b_n|^2\} \tag{3.3.79}$$

$$C_{ext} = (\lambda^2/2\pi) \sum_{n=1}^{\infty} (2n + 1)\{\mathrm{Re}(a_n + b_n)\} \tag{3.3.80}$$

The corresponding efficiency factors for scattering and extinction are obtained by division by the geometric cross section of the particle, πa^2,

$$Q_{sca} = (2/\alpha^2) \sum_{n=1}^{\infty} (2n + 1)\{|a_n|^2 + |b_n|^2\} \tag{3.3.81}$$

$$Q_{ext} = (2/\alpha^2) \sum_{n=1}^{\infty} (2n + 1)\{\mathrm{Re}(a_n + b_n)\} \tag{3.3.82}$$

The derivation of C_{ext} has been obtained differently by van de Hulst (1949) [also cf. Jones (1955)] in a way that provides a rather interesting physical insight into the process. The wave scattered in, and very close to, the forward direction has the function of interfering with the incident wave, thereby producing the shadow. It is this reduction in intensity of the incident wave which constitutes the extinction cross section of the particle, which in turn is related to the scattering amplitude in the forward direction ($\theta = 0°$) by

$$C_{ext} = (\lambda^2/\pi)\,\mathrm{Re}\{S(0)\} \tag{3.3.83}$$

This formula is independent of the form of the particle. Indeed it can be generalized for any type of scattering (Lax, 1950). The only requirement is that $S(0)$ correspond to the component of forward scattered radiation polarized in the same direction as the incident beam, since it is only this that will interfere. For a spherical particle

$$\mathrm{Re}\, S_1(0) = \mathrm{Re}\, S_2(0) = \tfrac{1}{2} \sum_{n=1}^{\infty} (2n + 1)\,\mathrm{Re}(a_n + b_n) \tag{3.3.84}$$

so that (3.3.83) reduces properly to (3.3.80).

3.3.4 Multipole Expansion

The formal scattering theory does not, at first sight, seem to offer much in the way of a physically intuitive picture of the mechanism of the scattering process. However, some qualitative insights can be obtained. The scattered wave arises from oscillations of the electrons in the particle excited by the incident wave. Here this has been described by an infinite series of electric and magnetic Debye potentials. These functions can be imagined to have their origin in density distributions of oscillating electric and magnetic dipoles in the particle. Thus the electric wave (TM) arises from the electric dipoles and the magnetic wave (TE) from the magnetic dipoles.

One can associate a partial electric and a partial magnetic wave with each term in the expansion of the electric and magnetic waves [(3.3.34) and (3.3.35)], viz. with each of the electric and magnetic Debye potentials. The amplitude and the phase of each of these partial waves are determined by the scattering coefficients a_n and b_n. Accordingly, each of the terms in the various expansions used to describe the scattered wave [Debye potentials, (3.3.34), (3.3.35); field vectors, (3.3.54), (3.3.55); amplitude functions, (3.3.56), (3.3.57); intensity functions, (3.3.61), (3.3.62)] corresponds to a particular partial wave. Mie has sketched out a plane projection of the electric field lines on the surface of a large sphere concentric with the particle for the first four electric and magnetic partial waves. These sketches have been reproduced by Born and Wolf (1959) and by Stratton (1941).

It is possible to proceed further and to specify each partial wave more particularly in terms of oscillatory electric and magnetic multipoles. The reader is referred elsewhere for a detailed exposition of the general theory of multipoles [Stratton (1941) p. 179]. The first four electric multipoles are visualized in Fig. 3.6. These may be generated one from the other, as follows. A dipole may be generated from a monopole of magnitude q (point charge) by displacement of a charge of opposite sign in a direction and over a distance indicated by the vector $\mathbf{l}_0$. The dipole moment is

$$\mathbf{p} = q\mathbf{l}_0 \tag{3.3.85}$$

We are interested in the field of this dipole at an arbitrarily large distance or alternatively the field at a finite distance in the limit when the two point charges are allowed to coalesce while the dipole moment remains constant. It is this singularity that constitutes the precise definition of the dipole.

The quadrupole is generated from the dipole by locating a second dipole of equal moment but opposite sign at the same position and then displacing this second dipole by the vector $\mathbf{l}_1$. The octupole may be generated from the quadrupole in a similar manner. Whereas the dipole moment is a vector (tensor of the first rank), the quadrupole and octupole must be characterized by moments which are tensors of the second and third ranks respectively. In each of these cases too, the multipole is defined as the singularity that occurs in the limit as the charges coalesce while the multipole moment remains constant.

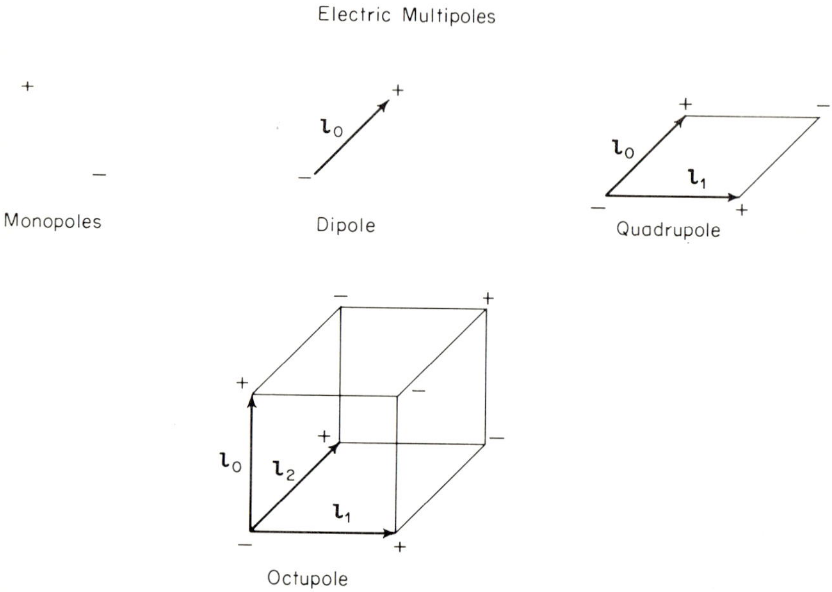

FIG. 3.6. First four electric multipoles. The multipole moments are $\mathbf{l}_0, \mathbf{l}_1, \mathbf{l}_2$.

It is not possible to visualize the arrays of charges in hyperspace corresponding to multipoles of higher order than the three dimensional octupole. However, each of the higher multipoles may be generated analytically from the preceding one in a general way—rather the multipole moments and the fields associated with them may be so generated—so that, by induction, expressions for multipoles of arbitrary order may be obtained. The magnetic multipoles can be generated by a similar procedure. Even though the magnetic monopoles (north and south poles) may not have physical reality, it is still possible to consider hypothetically a magnetic charge analogous to the electrostatic case. An alternative viewpoint leading to equivalent results is to treat magnetization in terms of a current distribution where the current loop provides the equivalent of the magnetic dipole.

Any distribution of electric charges and magnetic dipoles may be represented by a superposition of electric and magnetic multipoles located at some origin with arbitrary multipole moments. If, as in the case of scattering, the distribution of charges and currents is oscillating synchronously with the exciting wave, the scattered radiation arises from the corresponding oscillating multipoles. The important result in which we are interested is that there is a close relationship between the representation of electromagnetic fields by Debye potentials and their representation by multipole expansion (Wilcox, 1957). The oscillating electric multipoles give rise to the partial electric waves and the oscillating magnetic multipoles to the partial magnetic waves. Consider the Debye potentials corresponding to the partial waves, π_{1n}^s and π_{2n}^s, defined by

$$r\pi_1{}^s = \sum_{n=1}^{\infty} r\pi_{1n}^s \tag{3.3.86}$$

and

$$r\pi_2{}^s = \sum_{n=1}^{\infty} r\pi_{2n}^s \tag{3.3.87}$$

Then π_{11}^s describes the radiation by an oscillating electric dipole whose dipole moment is proportional to the scattering coefficient a_1. π_{21}^s is a Debye potential which can be attributed to an oscillating magnetic dipole. Here the magnetic dipole moment is given by b_1. Each of the higher terms in the expansion of the Debye potentials are similarly related to corresponding multipoles, and the scattering coefficients a_n and b_n are the multipole moments (Bowkamp and Casimir, 1954). This gives these coefficients a clear physical significance. The scattered radiation is a superposition of multipole radiations, each weighted by its appropriate multipole moment. The frequency of the oscillations is equal to the exciting frequency and the partial fields corresponding to each mode mutually interfere to produce the total effect. Each scattering coefficient determines the magnitude of the wavelet associated with each particular multipole.

3.3.5 Summary

This completes the theory of scattering of a monochromatic plane polarized wave by an isotropic, homogeneous sphere of arbitrary size and optical constants (except that $\mu = 1$). The practical formulas are (3.3.61), (3.3.62), (3.3.79), and (3.3.80), giving I_θ, I_ϕ, C_{sca} and C_{ext}.

The geometry is shown in Fig. 3.3. The incident wave propagates along the z-axis with its electric vector polarized along the x-axis. The scattered wave is observed in the direction θ, ϕ. This direction and that of the incident wave constitute the scattering plane. The scattering angle or angle of observation is θ.

The incident wave is decomposed into two components; one of these with intensity $\sin^2 \phi$ is polarized perpendicular to the scattering plane; the other with intensity $\cos^2 \phi$ is polarized in the scattering plane (parallel). Then the scattered light can be resolved into two components I_ϕ and I_θ which are polarized perpendicular and parallel to the scattering plane respectively. There is a phase difference δ between these (3.3.63), so that the scattered light is elliptically polarized.

When $\phi = 90°$, both the incident and scattered beams are polarized perpendicular to the scattering plane, then $I_\phi = I_1$ and $I_\theta = I_2 = 0$. Alternatively when $\phi = 0°$, both the incident and scattered beams are polarized parallel to the scattering plane with $I_\phi = I_1 = 0$ and $I_\theta = I_2$.

For two units of natural incident light, the scattered light polarized perpendicular to the scattering plane is I_1 and that polarized parallel to this plane is I_2.

C_{sca} represents the intensity scattered in all directions for an incident wave of unit intensity. It is independent of the state of polarization of the incident wave.

The efficiency factor Q_{sca} for a particular sphere (3.3.81) gives the ratio of the scattering cross section to the geometrical cross section. C_{ext} is the total intensity abstracted from the incident beam of unit intensity both by scattering and absorption and like C_{sca}, it is independent of the state of polarization of the incident beam. Q_{ext} is the corresponding efficiency factor.

3.4 Historical Postscript

The solution for scattering by a sphere is generally referred to as the "Mie theory," although the above exposition, except for the consideration of the scattering and extinction cross sections, has followed the treatment of Debye (1909a). Debye's work was based upon his 1908 thesis, which dealt with the closely related problem of the light pressure upon a spherical particle. He utilized a potential function (Debye potential) derived from a Hertz vector rather than working directly with the components of the field vectors as did Mie.

Mie's paper (1908) entitled "Considerations on the optics of turbid media, especially colloidal metal sols" is not only an exposition of the scattering formulas but also is concerned with a variety of both computational and experimental aspects of the problem. Mie addressed himself to the long-standing problem of accounting for the brilliant colors exhibited by colloidally dispersed metal particles which had been studied by Faraday (1857). Depending upon the mode of preparation, any one of a variety of colors may be displayed by a particular metal sol, even though the dispersion consists only of the elemental metal suspended in a colorless transparent medium such as glass (Maxwell-Garnett, 1904, 1906), water (Steubing, 1908), or gelatin (Kirchner and Zsigmondy, 1904).

Maxwell-Garnett (1904) attempted to explain the phenomenon by assuming that the particles were considerably smaller than the wavelength and that the effect was due to the alteration of the polarizability of the medium by the presence of the particles. His approach was to use the Lorentz (1879) relation between polarizability and dielectric constant. This latter, in turn, is related to the complex refractive index from which the absorption index can be obtained. On the other hand, the relation between polarizability and light scattering is given by (3.2.3) and (3.2.6).

Mie carried out his analysis after being introduced to the problem by the experimental dissertation on gold sols of a student at the Physical Institute of Greifswald, Steubing (1908). He proposed that these effects might be due to the particular manner in which the individual particles absorb light rather than to their effect upon the optical constants of the sol although, phenomenologically, the two points of view should be equivalent. He pointed out that particles of the same material, but of different size and form, might absorb and scatter light quite differently. The crucial factor was that he did not restrict the particles to sizes smaller than the wavelength. While recognizing that the gold sols which he proposed to explore were undoubtedly not composed of homogeneous, isotropic spheres, he utilized this simple model to explore the influence of particle size upon the absorption spectrum of the dispersion.

The only earlier work on scattering to which Mie makes reference is that by Thomson (1893) on perfectly reflecting spheres, that by Rayleigh on small dielectric spheres, and that by Lorenz (1880, 1898a) on small absorbing spheres. However, the theory had already been worked out by several workers prior to Mie and there is even a prehistory that goes back to the mid-nineteenth century.

Logan (1962, 1965) has traced the roots to a remarkable memoir entitled "Concerning reflection on a spherical surface" by Clebsch, which was submitted on October 30, 1861 and published in 1863, a year before the electromagnetic theory of light was proposed by Maxwell. In this paper, Clebsch obtained the general solution for the elastic wave equation in terms of the vector wave functions used by modern writers (Stratton, 1941). Indeed, Lamb (1906) in a later discussion of his solution of the vector wave equation asserts that "it was long overlooked that substantially the same analysis had been given by Clebsch in the paper ... to which reference has already been given on pp. 110, 512." Both Lorenz (1890, 1898b) and Debye (1909a) also cite Clebsch's work.

The elastic wave problem is much more complex than either the acoustic or the electromagnetic wave problems. The solution to the latter may be obtained from Clebsch's analysis by letting the velocity of propagation of the longitudinal waves tend to infinity. On the other hand, Rayleigh's solution for the scattering of sound waves by a perfectly rigid sphere may be obtained from Clebsch's solution by letting the velocity of propagation of the transverse waves tend to zero.

For the special case of transverse waves, Clebsch used a set of potential functions identical with the Debye potentials and he applied these to the problem of light scattering by a sphere of arbitrary size. Unable to specify the boundary conditions, Clebsch assumed the simplest situation of a perfectly rigid sphere. Under this assumption, a physically significant result was obtained only when the radius is small relative to the wavelength. Indeed it is this case that Rayleigh discussed (1871b), using the same elastic wave theory of light, and Rayleigh's law follows from results explicitly derived in Clebsch's memoir.

The work of Clebsch was consummated by the brilliant Danish physicist Lorenz (1890, 1898b) many years before the appearance of the papers of Mie and Debye. The contributions of Lorenz have been discussed recently by Logan (1962, 1965) and by Pihl (1963). His solution is based upon his own theory of electromagnetism rather than on that of Maxwell, and it is possibly this, as well as the fact that the original paper was published in Danish, that has caused his work to be overlooked. Although the physical interpretations of the two theories are different, the mathematical description of scattering and propagation is identical. However, even Lorenz does not appear to have perceived the equivalence of the two theories.

Lorenz actually used what are now termed the Debye potentials. His expressions for the scattering coefficients, for the amplitude functions, and for the scattering cross section [Lorenz (1898b) pp. 492–493], are identical with those given here (3.3.48, 49, 56, 57, 79). Furthermore, the Debye asymptotic estimates of the Hankel functions for large values of the arguments, which will be discussed below in connection with the computations, were also anticipated completely by Lorenz. He did not discuss the case of an absorbing sphere, but this, in the main, would only have involved permitting the refractive index to become complex.

Maxwell-Garnett (1904) was aware of Lorenz's 1890 paper since he points out that it gives the Rayleigh scattering formula. Later in his paper, Maxwell-Garnett utilized this expression for the polarizability of a Rayleigh scatterer in Lorentz's (1879) theory, relating the distribution of dipoles in a medium to the polarization and dielectric constant of the medium. In a curious mix-up, Mie attributes to Maxwell-Garnett not the use of Lorentz's 1879 paper in Wieddemann's Annalen but rather a related paper by Lorenz published in 1880 in the same journal. However, Mie appears not to have been aware of Lorenz's 1890 light scattering paper, which was the one actually cited by Maxwell-Garnett in connection with light scattering.

Despite the fact that the research papers of the leading workers on scattering are completely blank on Lorenz's contribution to scattering by a sphere, two of the early English language textbooks on electromagnetic waves are quite explicit on this point. MacDonald (1902) stated clearly that "the problem of the diffraction of waves by a transparent body has been solved for the case of a circular cylinder (Rayleigh, 1881) and for that of a sphere (Lorenz, 1890), the velocity of the radiation in the body differing by a finite amount from that in the surrounding medium. The problem has been solved for the general case, when the difference between the velocities of radiation is very small (Rayleigh, l.c.). The body of the diffraction of waves by a perfectly conducting body has been solved for the case of a circular cylinder (J. J. Thomson, 1893), a sphere (ibid.) and an indefinitely thin wedge in the form of a semiinfinite plane (Poincaré, 1892)."

Bateman (1914) was quite familiar with Lorenz's work and refers to it in his treatise (Chapter IV, p. 79). However, he prefers to follow Mie's treatment in his general exposition of the sphere problem. Like so many of the authors who followed him, he reproduced Mie's diagrams of the electric lines of force of the first four partial electric and magnetic waves. In addition, he discussed in detail the question of the colors of metal sols along the lines of Mie and the workers who immediately followed him.

In 1893, Thomson published, using the electromagnetic theory, the formulas for scattering by a perfectly conducting sphere with no restrictions on the size. Instead of using the Lorenz–Debye potentials, he used a solution of the vector wave equation which had been given in 1881 by Lamb. Because the particle is assumed to be a perfect conductor, there is no internal wave and the boundary conditions now require the disappearance of the tangential components of **E** and the normal component of **H**. After obtaining the final result in the form of the field components of the scattered wave, Thomson proceeded to approximate these in the wave zone for the special case of a small perfectly reflecting particle by retaining only the lowest terms in a series expansion of the functions appearing in the final expression. He then contrasted the polarization of the light scattered by such a sphere to that obtained in the dielectric case by Rayleigh. Thomson was unaware of Lorenz's work. He refers to Rayleigh's paper of 1881 but states that he does "not know of any papers which discuss the special problem of scattering by metal spheres." Thomson also mentions that "the incidence of a plane wave on a sphere" was the subject of a dissertation sent to Trinity College, Cambridge, by Professor Michell in 1890. The librarians of the University Library, Trinity College, and of the Cavendish Laboratory have been unable to trace this, but they note that a John Henry Michell became a Fellow of Trinity College in 1890. Possibly the paper referred to was a dissertation for a prize fellowship which would not have been preserved. Somewhat later

Schwarzschild (1901) analyzed the problem of the perfectly conducting sphere for the purpose of application to light pressure effects upon comets.

Several years later Love (1899) extended Thomson's solution to dielectric and partially conducting spheres. His treatment is completely general and although the title of his paper is "The Scattering of Electromagnetic Waves by a Dielectric Sphere," Love also considered the case when the material of the sphere is absorptive. Love's presentation is extremely concise. He clearly describes each of the steps to be effected but does not write down the explicit formulas giving the final results. Love refers the reader to a series of papers by Lamb for a detailed discussion of the requisite analysis.

Still another derivation of the scattering of electromagnetic waves by a sphere was given by Walker (1900a, b). Walker's work followed a suggestion by Thomson that it would be interesting to explore the general case and then to ascertain the manner in which this solution approached that for a perfect conductor and for a perfect insulator.

Both Thomson and Love mention Rayleigh's paper of 1881, and indeed Love explicitly presents his work as a generalization of Rayleigh's approximate treatment. In his 1881 paper, Rayleigh not only rederived the scattering formula for small dielectric spheres using the electromagnetic theory, but also considered scattering by particles no longer small compared to the wavelength. Tyndall (1869) had already described experiments in which the scattered light was no longer blue and completely polarized at 90° from the incident beam, and in his paper, Rayleigh (1871b) had also observed such deviations from small particle scattering in his experimental work with sulfur hydrosols. Rayleigh had published the exact solution for the scattering of sound waves by a sphere in 1872 but was unable to generalize his methods to the scattering of electromagnetic waves. In 1881, Rayleigh treated the large sphere problem under the assumption that the refractive index of the particle was close to that of the medium so that the disturbances inside and outside of the particle are essentially the same. The problem then is no longer a boundary value problem and the scattering becomes a simple interference effect.

What initially appears to be a latecomer on the scene is a paper by Bromwich (1919). He points out that Thomson's and Love's derivations, which were carried out in Cartesian coordinates, can be more simply effected by using spherical coordinates and then proceeds to do this. Bromwich also uses the Debye potential functions and his work is similar to the Lorenz–Debye derivation.

Bromwich stated that his solution was worked out in 1899, but that he delayed publication since his primary interest was in obtaining a simplified approximation valid for large spheres. This was only effected in 1910 when he utilized the asymptotic estimates of the Hankel functions for large arguments. Bromwich located these in a paper by MacDonald (1910), who in turn obtained them from Lorenz's 1890 paper. Bromwich also cites Lorenz's paper but only in connection with the asymptotic expansions, and it is not clear whether he actually read it himself. It does seem possible that if he did first read it in 1910, he may have revised his solution of the scattering problem along the lines of Lorenz. However, this is mere conjecture.

Bromwich makes no mention of either Mie's or Debye's work. Perhaps this is not so remarkable. Mie's paper was entitled "Optics of turbid media especially for metal colloidal sols" and might, if opened at certain pages, give the impression of being concerned exclusively with the experimental aspects of absorption by metal sols, a subject which was being intensively explored in both the German and British literature. Rayleigh, too, was unaware of Mie's solution when in 1910 he provided a fuller discussion of Love's (1899) results and carried out numerical computations. Debye's paper may also have been missed by Bromwich because of its title, "The light pressure upon a sphere of arbitrary material," although Bromwich does refer to Debye's (1909b) independent derivation of the asymptotic expansion of the Hankel functions. This work of Debye was not cited by MacDonald.

Although he presented his results in his 1910 lectures at Cambridge and read the paper in 1916, Bromwich continued to delay publication until he could check his approximation by calculations. When these were finally effected (Proudman *et al.*, 1918) and the pressure of war work was over, the publication finally appeared.

Nicholson (1910, 1912) is still another Englishman who referred to Lorenz's paper. He considered Thomson's general treatment of the perfectly conducting sphere for the case that the radius is large compared with the wavelength, and he was able to show that the theory is in accord, to a first approximation, with the results obtained by the ordinary methods of geometrical optics. In order to accomplish this, he utilized the asymptotic estimates of the Hankel functions as developed by Lorenz, and it is in this connection that he cited Lorenz's paper on the scattering by a dielectric sphere. Although Nicholson refers to Love's general treatment of the dielectric sphere, he makes no reference to the fact that Lorenz's paper is primarily concerned with the same problem. However, he also cites MacDonald's (1904) calculations of the Hankel functions, so that it is not clear whether he actually read Lorenz's paper or merely extracted the reference from MacDonald.

It is not the intention of this author to arbitrate the questions of priority raised here nor to identify the theory of scattering by a sphere with any one man's name. Indeed, coincident and consecutive discoveries are common occurrences in science. But certainly if this theory is to be associated with the name or names of individuals, at least that of Lorenz, in whose paper are to be found the practical formulas so commonly used today, should not be omitted.

TABLE 3.1 COMPARISON OF NOTATION

	(a) This work (b) van de Hulst (1957)	Mie (1908)
1. Radius of sphere	a	ρ
2. Propagation constant in the medium	(a) $k_2 = \dfrac{2\pi}{\lambda} = \dfrac{2\pi m_2}{\lambda_0}$	$\dfrac{2\pi}{\lambda'}$
	(b) $k = \dfrac{2\pi}{\lambda}$	—
3. Size parameter	(a) $\alpha = k_2 a$	α
	(b) $x = ka$	
4. Relative refractive index	$m = \dfrac{m_1}{m_2} = \dfrac{k_1}{k_2}$	m'
5. Size parameter multiplied by relative refractive index	(a) $\beta = m\alpha = k_1 a$	β
	(b) $y = mx$	
6. Scattering angle	θ	$180 - \theta$
7. Ricatti–Bessel functions where $J_{n+\frac{1}{2}}(z)$, $N_{n+\frac{1}{2}}(z)$, and $H_{n+\frac{1}{2}}(z)$ are the half integral order Bessel, Neumann and Hankel functions while $j_n(z)$, $n_n(z)$ and $h_n(z)$ are the corresponding spherical Bessel, Neumann and Hankel functions. The superscripts indicate Hankel functions of the first or second kind.	$\psi_n(z) = zj_n(z) = \left(\dfrac{\pi z}{2}\right)^{1/2} J_{n+\frac{1}{2}}(z)$	$I_n(z)$
	$\chi_n(z) = -zn_n(z) = -\left(\dfrac{\pi z}{2}\right)^{1/2} N_{n+\frac{1}{2}}(z)$	—
	$\zeta_n(z) = zh_n^{(2)}(z) = \left(\dfrac{\pi z}{2}\right)^{1/2} H_{n+\frac{1}{2}}^{(2)}(z) = \psi_n(z) + i\chi_n(z)$	$i^{n+1}K_n(-z)$
	$\zeta_n^{(1)}(z) = zh_n^{(1)}(z) = \left(\dfrac{\pi z}{2}\right)^{1/2} H_{n+\frac{1}{2}}^{(1)}(z)$	—
8. Scattering coefficients[a]	$a_n = \dfrac{\psi_n'(\beta)\psi_n(\alpha) - m\psi_n(\beta)\psi_n'(\alpha)}{\psi_n'(\beta)\zeta_n(\alpha) - m\psi_n(\beta)\zeta_n'(\alpha)}$	$\dfrac{a_n}{(2n+1)i^{2n+1}}$
	$b_n = \dfrac{m\psi_n'(\beta)\psi_n(\alpha) - \psi_n(\beta)\psi_n'(\alpha)}{m\psi_n'(\beta)\zeta_n(\alpha) - \psi_n(\beta)\zeta_n'(\alpha)}$	$\dfrac{-p_n}{(2n+1)i^{2n+1}}$

IN COMMON USE FOR SCATTERING BY SPHERES

Debye (1909a)	Born and Wolf (1959)	Stratton (1941)	(a) Lowan (1948) (b) Gumprecht and Sliepcevich (1951a, b) (c) Pangonis *et al.* (1957)
a	a	a	r
$k_a = 2\pi/\lambda$	$k^{(I)} = 2\pi/\lambda^{(I)}$	$k_2 = 2\pi/\lambda$	$2\pi/\lambda$
$k_a a$	q	ρ	α
k_i/k_a	$\hat{n} = k^{(II)}/k^{(I)}$	$N = k_1/k_2$	m
$k_i a$	$\hat{n}q = k^{(II)}a$	Np	β
$180 - \theta$	θ	—	$180 - \gamma$
$\psi_n(z)$	$\psi_n(z)$	$zj_n(z)$	$S_n(z)$
$\chi_n(z)$	$\chi_n(z)$	$zn_n(z)$	$C_n(z)$
$\zeta_n(z)$	—	$zh_n^{(2)}(z)$	$\phi_n(z)$
$\eta_n(z)$	$\zeta_n(z)$	—	—
$\alpha_n{}^1 = \dfrac{A_n{}^1}{-i^{n-1}\left(\dfrac{2n+1}{n[n+1]}\right)}$	${}^eB_n \Big/ \left[i^{n+1}\left(\dfrac{2n+1}{n[n+1]}\right)\right]$	$-b_n{}^r$	$C_n{}^1 = a_n/(-1)^{n+\frac{1}{2}}(2n+1)$
$\alpha_n{}^2 = \dfrac{A_n{}^2}{-i^{n-1}\left(\dfrac{2n+1}{n[n+1]}\right)}$	${}^mB_n \Big/ \left[i^{n+1}\left(\dfrac{2n+1}{n[n+1]}\right)\right]$	$-a_n{}^r$	$C_n{}^2 = p_n/(-1)^{n+\frac{3}{2}}(2n+1)$
			$A_n = a_n/n(n+1) \quad P_n = p_n/n(n+1)$

TABLE 3.1 (*continued*)

	(a) This work (b) van de Hulst (1957)	Mie (1908)
9. Legendre functions	$\pi_n(\cos\theta) = \frac{1}{\sin\theta} P_n^{(1)}(\cos\theta)$	$(-1)^{n+1}\pi_n(\cos\gamma)$
	$\tau_n(\cos\theta) = \frac{d}{d\theta} P_n^{(1)}(\cos\theta)$	$(-1)^n \cos\gamma\pi_n(\cos\gamma)$ $-(1-\cos^2\gamma)\pi_n'(\cos\gamma)$
10. Amplitude functions.[a] Where these are not explicitly denoted, we shall designate them as a quantity within a bracket.	$S_1(\theta) = \sum_{n=1}^{\infty} \frac{2n+1}{n(n+1)}[a_n\pi_n(\cos\theta)$ $+ b_n\tau_n(\cos\theta)]$	$i[\ ]_{\mathrm{I}}$
	$S_2(\theta) = \sum_{n=1}^{\infty} \frac{2n+1}{n(n+1)}[a_n\tau_n(\cos\theta)$ $+ b_n\pi_n(\cos\theta)]$	$-i[\ ]_{\mathrm{II}}$
11. Intensity functions	$i_1 = \lvert S_1(\theta)\rvert^2$	—
	$i_2 = \lvert S_2(\theta)\rvert^2$	—
12. Intensity for incident natural light of unit intensity	$I = \frac{\lambda_2^2}{8\pi^2 r^2}(i_1 + i_2)$	$\frac{J_{\mathrm{I}} + J_{\mathrm{II}}}{2}$
13. Extinction cross-section	$Q_{\mathrm{ext}} = (\lambda_2^2/2\pi) \sum_{n=1}^{\alpha} (2n+1)$ $\mathrm{Re}(a_n + b_n)$	$(\lambda'^2/2\pi)\,\mathrm{Im} \sum_{n=1}^{\alpha} (-1)^n(a_n - p_n)$
14. Efficiency for extinction	$Q_{\mathrm{ext}} = C_{\mathrm{ext}}/\pi a^2$	
15. Scattering cross-section	$C_{\mathrm{sca}} = (\lambda^2/2\pi) \sum_{n=1}^{\alpha} (2n+1)$ $\times \{\lvert a_n\rvert^2 + \lvert b_n\rvert^2\}$	$\frac{k^1}{N} = \frac{\lambda'^2}{2\pi} \sum \frac{\lvert a_n\rvert^2 + \lvert p_n\rvert^2}{2n+1}$
16. Efficiency for scattering	$Q_{\mathrm{sca}} = C_{\mathrm{sca}}/\pi a^2$	—

[a] The coefficients obtained by solution of the boundary equations given by Born and Wolf

Debye (1909a)	Born and Wolf (1959)	Stratton (1941)	(a) Lowan (1948) (b) Gumprecht and Sliepcevich (1951a, b) (c) Pangonis *et al.* (1957)
$(-1)^{n+1}\frac{1}{\sin\theta}P_n^{(1)}(\cos\theta)$	$\frac{1}{\sin\theta}P_n^{(1)}(\cos\theta)$	—	$(-1)^{n+1}\pi_n(\cos\gamma)$
$(-1)^n\frac{d}{d\theta}P_n^{(1)}\cos\theta$	$(-1)P_n^{(1)'}(\cos\theta)\sin\theta$		$(-1)^n\{\cos\gamma\,\pi_n(\cos\gamma)$ $-(1-\cos^2\gamma)\pi_n'\cos\gamma\}$ [Ref. (c)]
			$(-1)^{n+1}\frac{1}{\sin\gamma}P_n^{1}(\cos\gamma)$
			$(-1)^n\frac{d}{d\gamma}P_n^{1}(\cos\gamma)$
—	$-i[\ ]$	—	$+i\cdot i_1{}^*$
—	$i[\ ]$	—	$-i\cdot i_2{}^*$
—	—	—	$i_1=\lvert i_1{}^*\rvert^2$
—	—	—	$i_2=\lvert i_2{}^*\rvert^2$
—	$\frac{1}{2}(I_{\text{I}}+I_{\text{II}})$	—	$\frac{\lambda^2}{8\pi^2r^2}(i_1+i_2)$
—	Q	Q_t	—
—	—	—	K
—	$Q^{(s)}$	Q_s	—
—	—	—	$\bar{K}$

(1959) actually lead to the negative of the given coefficients.

3.5 Notation

A variety of notations commonly used in connection with scattering by spheres has been assembled in Table 3.1. We have adopted mainly the notation of van de Hulst (1957), with the exception of the size parameters α and β which van de Hulst designates as x and y and the propagation constant in the medium, k_2, designated by van de Hulst as k. Van de Hulst assumes that the medium is a vacuum ($m = 1$), so that it is unnecessary for him to introduce separately m_2 and m_1 for the medium and particle with the corresponding propagation constants k_2 and k_1 as well as k_0 for the vacuum. We prefer to retain these separately. As for the size parameters, α has been used by so many workers that it has become nearly standard. The selection of notation assembled in Table 3.1 is arbitrary, but we believe the authors selected provide a representative cross section.

3.6 Bessel Functions

A review of some properties of Bessel functions applicable to the theory of light scattering will be given here. The reader is referred to the standard treatises (Watson, 1944; Hobson, 1931) for a systematic treatment.

Bessel's equation is

$$\frac{d^2 Z_n(z)}{dz^2} + \frac{1}{z}\frac{dZ_n(z)}{dz} + \left(1 - \frac{n^2}{z^2}\right) Z_n(z) = 0 \tag{3.6.1}$$

As a second order differential equation, it has two independent solutions, the Bessel function $J_n(z)$ and the Neumann function $N_n(z)$ each of order n and of argument z. The term Bessel function is frequently used as a generic name for both solutions which are then called Bessel functions of the first and second kind, respectively. Otherwise the general solution may be called a cylinder function and denoted $Z_n(z)$ as above.

The Bessel function is defined by

$$J_n(z) = \sum_{m=0}^{\infty} \frac{(-1)^m}{m!\Gamma(m+n+1)}\left(\frac{z}{2}\right)^{n+2m} \tag{3.6.2}$$

or

$$J_n(z) = \frac{(z/2)^n}{\Gamma\frac{1}{2}\Gamma(n+\frac{1}{2})}\int_0^{\pi} \cos(z\cos w)\sin^{2n} w\, dw \tag{3.6.3}$$

Whenever n is not an integer, the Neumann function is constructed as

$$N_n(z) = [\sin(n\pi)]^{-1}[J_n(z)\cos(n\pi) - J_{-n}(z)] \tag{3.6.4}$$

There is no simple expansion of Neumann's function comparable to (3.6.2). The linear combination of Bessel and Neumann functions as follows leads to Bessel functions of the third kind or Hankel functions of the first and second kind.

$$H_n^{(1)}(z) = J_n(z) + iN_n(z) \tag{3.6.5}$$

$$H_n^{(2)}(z) = J_n(z) - iN_n(z) \tag{3.6.6}$$

The following recurrence relations which are general for all cylinder functions will be found useful in carrying out computations:

$$Z_{n-1}(z) + Z_{n+1}(z) = (2n/z)Z_n(z) \tag{3.6.7}$$

$$dZ_n(z)/dz = \tfrac{1}{2}Z_{n-1}(z) - \tfrac{1}{2}Z_{n+1}(z) \tag{3.6.8}$$

For the Ricatti–Bessel functions as defined earlier:

$$f_{n-1}(z) + f_{n+1}(z) = [(2n+1)/z]f_n(z) \tag{3.6.9}$$

$$(2n+1)\, df_n(z)/dz = nf_{n-1}(z) - (n+1)f_{n+1}(z) \tag{3.6.10}$$

3.6.1 Half Integral Orders

The differential equation

$$\frac{d^2Z_n(z)}{dz^2} + \frac{2}{z}\frac{dZ_n(z)}{dz} + \left[1 - \frac{n(n+1)}{z^2}\right]Z_n(z) = 0 \tag{3.6.11}$$

is the form of Bessel's equation whose solutions are half-integral order cylinder functions. As we have already seen, this is precisely the form of the radial part of the wave equation expressed in spherical coordinates so that the half-integral order cylinder functions are involved in the scattering equations for the sphere. In addition to the above recursion relations, the following useful relation specialized for these half-integral order functions is noted. From (3.6.4)

$$N_{n+\frac{1}{2}}(z) = (-1)^{n+1}J_{-n-\frac{1}{2}}(z) \tag{3.6.12}$$

The Ricatti–Bessel functions have already been defined in terms of the half-integral order cylinder functions. These can now be expressed as infinite series:

$$\psi_n(z) = 2^n z^{n+1} \sum_{m=0}^{\infty} \frac{(-1)^m(n+m)!}{m!(2n+2m+1)!} z^{2m} \tag{3.6.13}$$

and

$$\chi_n(z) = -\frac{1}{2^n z^n} \sum_{m=0}^{\infty} \frac{\Gamma(2n-2m+1)}{m!\Gamma(n-m+1)} z^{2m} \tag{3.6.14}$$

These functions may also be written as a terminating series (Gumprecht and Sliepcevich, 1951b)

$$\psi_n(z) = \sin(z - (n\pi/2)) \sum_{m=0}^{\leqslant n/2} [(-1)^m (n+2m)!/(2m)!(n-2m)!(2z)^{2m}]$$
$$+ \cos\left(z - \frac{n\pi}{2}\right) \sum_{m=0}^{\leqslant (n-1)/2} \frac{(-1)^m (n+2m+1)!}{(2m+1)!(n-2m-1)!(2z)^{2m+1}} \tag{3.6.15}$$

and

$$\chi_n(z) = (-1)^n \left[\cos\left(z + \frac{n\pi}{2}\right) \sum_{m=0}^{\leqslant n/2} \frac{(-1)^m (n+2m)!}{(2m)!(n-2m)!(2z)^{2m}} \right.$$
$$\left. - \sin\left(z + \frac{n\pi}{2}\right) \sum_{m=0}^{\leqslant (n-1)/2} \frac{(-1)^m (n+2m+1)!}{(2m+1)!(n-2m-1)!(2z)^{2m+1}} \right] \tag{3.6.16}$$

Finally they may be represented in the following derivative form

$$\psi_n(z) = z^{n+1}(-d/z\,dz)^n(\sin z/z) \tag{3.6.17}$$

and

$$\chi_n(z) = (-1)^n z^{n+1}(d/z\,dz)^n(\cos z/z) \tag{3.6.18}$$

Obviously a variety of techniques now present themselves for computing these functions. The terminating series [(3.6.15), (3.6.16)] are more appropriate for evaluating the functions $\psi_n(z)$ and $\chi_n(z)$ than the infinite series [(3.6.13), (3.6.14)], because of the slow convergence of the latter. A still simpler procedure is to evaluate the first two orders of each function from

$$\psi_0(z) = \sin z \qquad \text{and} \qquad \psi_1(z) = [(\sin z)/z] - \cos z \tag{3.6.19}$$

and

$$\chi_0(z) = \cos z \qquad \text{and} \qquad \chi_1(z) = [(\cos z)/z] + \sin z \tag{3.6.20}$$

The higher orders can then be obtained by use of the recursion formulas (3.6.9) and (3.6.10)

Repeated use of the recursion formula magnifies the rounding-off error so that extreme care must be used when large orders are involved. Various stratagems may be devised by the programmer in order to obviate this difficulty, e.g. (1) utilizing the recursion formula until the rounding-off error becomes significant, evaluating the function at this point by the series formula, and then continuing with the recursion formula until the situation repeats itself; (2) using the series evaluation of the function for the highest order needed and then carrying out the recursion downwards [Todd (1962) p. 92];

(3) utilizing double or even triple precision arithmetic in order to carry a sufficient number of significant figures at the beginning in order to obtain adequate precision in the highest order.

When the argument of the cylinder function is complex as it is for absorbing spheres, the computation becomes somewhat more tedious. Aden (1951) has proposed the use of the logarithmic derivatives of the Ricatti–Bessel functions for which he has derived simple recursion formulas. These functions $\eta_n^{(1)}(z)$ and $\eta_n^{(3)}(z)$ are defined by

$$\eta_n^{(1)}(z) = \psi_n'(z)/\psi_n(z) \tag{3.6.21}$$

and

$$\eta_n^{(3)}(z) = \zeta_n'(z)/\zeta_n(z) \tag{3.6.22}$$

or alternatively by

$$\eta_n^{(1)}(z) = [\psi_{n-1}(z)/\psi_n(z)] - (n/z) \tag{3.6.23}$$

and

$$\eta_n^{(3)}(z) = [\zeta_{n-1}(z)/\zeta_n(z)] - (n/z) \tag{3.6.24}$$

In terms of the logarithmic derivative functions, the amplitude functions for scattering by a sphere become

$$a_n = \frac{\psi_n(\alpha)}{\zeta_n(\alpha)}\left[\frac{\eta_n^{(1)}(\beta) - m\eta_n^{(1)}(\alpha)}{\eta_n^{(1)}(\beta) - m\eta_n^{(3)}(\alpha)}\right] \tag{3.6.25}$$

$$b_n = \frac{\psi_n(\alpha)}{\zeta_n(\alpha)}\left[\frac{\eta_n^{(1)}(\alpha) - m\eta_n^{(1)}(\beta)}{\eta_n^{(3)}(\alpha) - m\eta_n^{(1)}(\beta)}\right] \tag{3.6.26}$$

Obviously it is only $\eta_n^{(1)}(z)$, involving the Bessel functions but not the Neumann functions, for which the argument may become complex. The following recursion formula permits computation of any logarithmic derivative function,

$$\eta_n^{(1)}(z) = \frac{z^2 + nz\eta_{n-1}^{(1)}(z) - n^2}{nz - z^2\eta_{n-1}^{(1)}(z)} \tag{3.6.27}$$

With the value for the first order of $\eta_n^{(1)}(z)$ given by

$$\eta_0^{(1)}(z) = \cot(z) = \cot(a - bi) = \frac{\sin 2a + i \sinh 2b}{\cosh 2b - \cos 2a} \tag{3.6.28}$$

higher order terms may be found. It should be pointed out that the computational scheme described by Deirmendjian *et al.* (1961) for complex arguments is identical to that of Aden, the quantity designated by the former as A_n being identical with the logarithmic deviation function, $\eta_n^{(1)}(z)$.

Stephens and Gerhardt (1961a, b) have pointed out that for small arguments ($\alpha < 0.5$) Aden's method is susceptible to computational rounding-off error. However, for these cases the truncated series approximations for the scattering functions which will be discussed below may be used [(3.9.12), (3.9.15)].

Finally, we consider the asymptotic values that the Bessel functions assume for large arguments. Provided that the argument is larger than the order; $|z| \gg 1, |z| \gg |n|$

$$J_n(z) \simeq \left(\frac{2}{\pi z}\right)^{1/2} \cos\left(z - \frac{2n+1}{4}\pi\right) \tag{3.6.29}$$

$$N_n(z) \simeq \left(\frac{2}{\pi z}\right)^{1/2} \sin\left(z - \frac{2n+1}{4}\pi\right) \tag{3.6.30}$$

We have already seen how in the representation of the outgoing scattered wave in the radiation zone this gave

$$\zeta_n(k_2 r) = i^{n+1} \exp(-ik_2 r) \tag{3.3.52}$$

thereby leading to considerable simplification in the resulting scattering functions.

The following approximate formulas apply when both z and n are large. They are given for different ranges of z with respect to n for real n [Morse and Feshbach (1953) p. 631].

$$J_n(z) \simeq \frac{\exp[n(\tanh\alpha - \alpha)]}{(2\pi n \tanh\alpha)^{1/2}}; \quad z < n; \quad \tanh\alpha = [1 - (z/n)^2]^{1/2} \tag{3.6.31}$$

$$J_n(z) \simeq \frac{\sin(\pi/3)\,\Gamma(\frac{1}{3})}{3\pi(z/6)^{1/3}} + \frac{\sin(2\pi/3)\,\Gamma(\frac{2}{3})}{3\pi(z/6)^{2/3}}(z - n); \quad z \simeq n \tag{3.6.32}$$

$$J_n(z) \simeq \left(\frac{2}{\pi n \tan\beta}\right)^{1/2} \cos[n\tan\beta - n\beta - \tfrac{1}{4}\pi]; \qquad z > n; \quad \tan\beta = [(z/n)^2 - 1]^{1/2} \tag{3.6.33}$$

$$N_n(z) \simeq -\frac{2\exp[n(\alpha - \tanh\alpha)]}{(2\pi n \tanh\alpha)^{1/2}}; \quad z < n \tag{3.6.34}$$

$$N_n(z) \simeq -\frac{2\sin(\pi/3)\,\Gamma(\frac{1}{3})}{3\pi(z/6)^{1/3}} + \frac{2\sin(2\pi/3)\,\Gamma(\frac{2}{3})}{3\pi(z/6)^{2/3}}(z - n); \quad z \simeq n \tag{3.6.35}$$

$$N_n(z) \simeq \left(\frac{2}{\pi n \tan\beta}\right)^{1/2} \sin[n\tan\beta - n\beta - \tfrac{1}{4}\pi]; \quad z > n \tag{3.6.36}$$

The half integral order Bessel and Neumann functions have been extensively tabulated by Lowan (1947a, b, c) in the form of spherical Bessel functions defined as

$$j_{\pm n} = (\pi/2z)^{1/2} J_{\pm(n+\frac{1}{2})}(z) \tag{3.6.37}$$

where negative orders lead through (3.6.12) to the appropriate Neumann functions. These tables cover the arguments from $z = 0$ to 10 in steps of 0.01 for integral orders from $n = -22$ to 21 and the arguments from 10 to 25 in steps of 0.1 for integral orders from -31 to 30. Gumprecht and Sliepcevich (1951b) have tabulated the Ricatti–Bessel functions $\psi_n(z)$, $\chi_n(z)$ [(3.3.25), (3.3.26)] for the following arguments:

1, 2, 3, 4, 5, 6, 8

10(5)100

100(10)200, 250, 300, 350, 400

In addition to these results, values were also tabulated for the above arguments, each multiplied by 1.20, 1.33, 1.40, 1.50, 1.60. These numbers cover a useful range of the refractive index so that if the first set of numbers corresponds to possible values of parameter α, the second will give the necessary values of parameter $\beta = m\alpha$ which appears in (3.3.48), and (3.3.49). The calculations were carried out for values of the order n, sufficiently high for the convergence of the series expressions for the amplitude functions [(3.3.56), (3.3.57)]; viz. several orders, larger than the value of the arguments. For a listing of additional tabulations of these and other functions, the reader should consult the compendium of Fletcher *et al.* (1962).

3.6.2 Integral Orders

Finally, we shall give some consideration to the integral order cylinder functions. These are solutions of (3.6.1) for the case that n assumes integral values. It is this form of Bessel's equation which corresponds to the radial part of the wave equation expressed in cylindrical coordinates. These functions will appear in the solution for scattering by a cylinder which is to be considered later. When n is integral, J_{-n} is no longer a solution independent of J_n, so that the Neumann function cannot be constructed by Eq. (3.6.4). This arises because the gamma function $\Gamma(n + m + 1)$ in (3.6.2) is replaced by the factorial $(m + n)!$ so that

$$J_{-n}(z) = (-1)^n J_n(z) \quad \text{for } n \text{ integral} \tag{3.6.38}$$

and (3.6.4) becomes indeterminate. This may be evaluated in the usual way by differentiating numerator and denominator with respect to n and then passing to the limit $n \rightarrow$ integral values. The result (Lowan, 1947c) is given as follows:

$$N_n(z) = \frac{2}{\pi}\{J_n(z)(\gamma + \ln \tfrac{1}{2}z)\} - (1/\pi) \sum_{m=0}^{n-1} \frac{(n-m-1)!}{m!}(\tfrac{1}{2}z)^{2m-n}$$

$$+ (1/\pi) \sum_{m=0}^{\infty} \frac{(-1)^{m+1}(\tfrac{1}{2}z)^{n+2m}}{m!(m+n)!} \{\phi(k) + \phi(n+k)\} \qquad (3.6.39)$$

where $\gamma = 0.5772156649$ is Euler's constant and

$$\phi(n) = 1 + \tfrac{1}{2} + \tfrac{1}{3} + \cdots \frac{1}{n} \quad \text{and} \quad \phi(0) = 0 \qquad (3.6.40)$$

There are very extensive tabulations of integral order Bessel functions compiled by the Harvard University Computation Laboratory in twelve huge volumes (Aiken, 1947–51). For $J_0(z)$ through $J_{15}(z)$, these include the arguments $z = 0.001(0.001)25.000(0.01)\ 99.99$ and for $J_{16}(z)$ through $J_{135}(z)$ the arguments are $z = 0.01(0.01)99.99$. The corresponding tables of Neumann functions have been compiled by Chistova (1959) for $N_0(z)$ and $N_1(z)$. Other tables of $J_n(z)$ not as extensive as the Harvard tables have been prepared by the British Association for the Advancement of Science (1950, 1952) and Cambi (1948). Tables of $J_0(z)$, $J_1(z)$, $N_0(z)$, and $N_1(z)$ have been tabulated by Lowan (1947b, c) for complex arguments. The complex number is denoted by $z = \rho e^{i\phi}$, and these tables cover the range $\rho = 0.01(0.01)10.00$, and $\phi = 0°(5°)90°$.

If the functions are programmed for machine computation, tables such as those just described are of use primarily to provide values for checking the programs. Again the programmer must be warned about the pitfalls due to rounding-off errors. The problems in connection with repeated use of the recursion formulas have already been mentioned. In addition, with integral order functions, there is excessive rounding-off error for low orders and large arguments when the series definition is used ($z > 10$). A stratagem to obviate this difficulty is to compute for each argument two successive cylinder functions of sufficiently high order to give precise results and then calculate the lower orders by the usual recurrence relation (Farone *et al.*, 1963). If the two orders for the series calculation are selected sufficiently high in each case, this will provide Bessel functions for all the necessary terms in the expression for the amplitude functions.

As a matter of fact, Todd (1962) has pointed out that if the backward recurrence is started at a sufficiently high order, it is possible to initialize the calculation with any two small randomly selected values for the Bessel functions of the two highest orders. The values for the lower orders will then be in error by a scale factor, s, which can be determined by comparing the summation over the trial functions

$$J_0{}^{\mathrm{T}}(z) + 2 \sum_{n=1}^{\infty} J_n{}^{\mathrm{T}}(z) = s \tag{3.6.41}$$

since for integral order Bessel functions

$$J_0(z) + 2 \sum_{n=1}^{\infty} J_n(z) = 1 \tag{3.6.42}$$

Stegun and Abramowitz (1957) have outlined the method for application in a similar manner to other sets of cylinder functions.

3.7 Legendre Functions

Legendre's associated equation (MacRobert, 1945; Hobson, 1931) is

$$(1 - \eta^2)\frac{d^2 P_n{}^m(\eta)}{d\eta^2} - 2\eta\frac{dP_n{}^m(\eta)}{d\eta} + \left[n(n+1) - \frac{m^2}{1-\eta^2}\right]P_n^{(m)}(\eta) = 0 \tag{3.7.1}$$

When the parameters n and m are integers, the solution is the associated Legendre polynomial, $P_n^{(m)}(\eta)$ of order n and degree m. This can be represented as a terminating series in powers of η or more compactly as

$$P_n^{(m)}(\eta) = \frac{1}{2^n n!}(1 - \eta^2)^{m/2}\frac{d^{n+m}}{d\eta^{n+m}}(\eta^2 - 1)^n \tag{3.7.2}$$

For the special case that $m = 0$, (3.7.1) reduces to Legendre's equation and the corresponding solutions are the Legendre polynomials. The associated Legendre polynomials can be obtained from these with

$$P_n^{(m)}(\eta) = (1 - \eta^2)^{m/2}\, d^m P_n(\eta)/d\eta^m \tag{3.7.3}$$

A variety of recursion and differential relations are available since interchange both of order and degree must be considered. The reader is referred to Stratton [(1941) p. 401–402] for the principal relations.

Legendre's associated equation was the θ-dependent equation obtained when the wave equation expressed in spherical coordinates was solved by the method of separation of the variables (3.3.23). There $\eta = \cos\theta$ and $P_n^{(m)}(\eta) = \Theta(\theta)$. Because the direction of the incident wave was made coincident with a coordinate axis (z-axis), $m = 1$. Accordingly in the expansion of the Debye potentials, the associated Legendre polynomials are of degree 1. The final field equations (3.3.54) and (3.3.55) are expressed in terms of $P_n^{(1)}(\cos\theta)$ and $d/d\theta\, P_n^{(1)}(\cos\theta)$. The necessary relations for these will be given here. From (3.7.3) we find

$$P_n^{(1)}(\cos\theta) = -\frac{dP_n(\cos\theta)}{d\theta} = \sin\theta\,\pi_n(\cos\theta) \tag{3.7.4}$$

and by differentiation

$$d/d\theta[P_n^{(1)}(\cos\theta)] = \cos\theta\,\pi_n(\cos\theta) - \sin^2\theta\,\pi_n'(\cos\theta) = \tau_n(\cos\theta) \tag{3.7.5}$$

where $P_n(\cos\theta)$ is the Legendre polynomial (associated Legendre polynomial of degree zero). The amplitude functions [(3.3.56), (3.3.57)] have been expressed in terms of $\pi_n(\cos\theta)$ and $\tau_n(\cos\theta)$. The differential function $\pi_n(\cos\theta)$ may be defined by the terminating series

$$\pi_n(\cos\theta) = \frac{1\cdot3\cdot5\ldots(2n-1)}{n!}\left[n(\cos\theta)^{n-1} - \frac{n(n-1)(n-2)}{2(2n-1)}(\cos\theta)^{n-3} + \frac{n(n-1)(n-2)(n-3)(n-4)}{2\cdot4(2n-1)(2n-3)}(\cos\theta)^{n-5} + \cdots\right] \tag{3.7.6}$$

and π_n' $(\cos\theta)$ by the corresponding series obtained upon differentiation. Computations may be facilitated by using the following recursion relations:

$$\pi_n(\cos\theta) = \cos\theta[(2n-1)/(n-1)]\pi_{n-1}(\cos\theta) - [n/(n-1)]\pi_{n-2}(\cos\theta) \tag{3.7.7}$$

$$\pi_n'(\cos\theta) = (2n-1)\pi_{n-1}(\cos\theta) + \pi_{n-2}'(\cos\theta) \tag{3.7.8}$$

The first few values may be obtained directly from the above series and its derivative, e.g.,

$$\pi_0(\cos\theta) = 0; \qquad \pi_0'(\cos\theta) = 0 \tag{3.7.9}$$

$$\pi_1(\cos\theta) = 1; \qquad \pi_1'(\cos\theta) = 0 \tag{3.7.10}$$

$$\pi_2(\cos\theta) = 3\cos\theta; \qquad \pi_2'(\cos\theta) = 3 \tag{3.7.11}$$

The values of $\pi_n(\cos\theta)$ and $\tau_n(\cos\theta)$ [computed from (3.7.5)] are then used directly in (3.3.56) and (3.3.57).

Extensive tabulations of $\pi_n(\cos\theta)$ and $\tau_n(\cos\theta)$ have been compiled by Gumprecht and Sliepcevich (1951c) for $n = 0(1)420$ and for $\theta = 0°(1°)10°(10°)180°$. In addition, Gucker *et al.* (1964), following earlier work by Gucker and Cohn (1953), have published values for the first 43 orders at intervals of 1°. Clark and Churchill (1957) have compiled tables of $P_n(\cos\theta)$ for $n = 0(1)80$ and $\theta = 0°(10)180°$. Each of these workers actually use the supplement of θ

in the formulations for the amplitude [(3.3.56), and (3.3.57)] and intensity functions [(3.3.61), and (3.3.62)]. Accordingly, the following relations which connect $\pi_n(\cos\theta)$ and $\tau_n(\cos\theta)$ for an angle and its supplement will be found useful:

$$\pi_n \cos(180° - \theta) = (-1)^{n+1}\pi_n(\cos\theta) \tag{3.7.12}$$

$$\tau_n \cos(180° - \theta) = (-1)^{n}\tau_n(\cos\theta) \tag{3.7.13}$$

A bibliography of earlier tabulations has been given by Lowan (1945).

An alternative scheme for the computation of the amplitude functions [(3.3.56), (3.3.57)] which avoids the explicit calculation of $\pi_n(\cos\theta)$ and $\tau_n(\cos\theta)$ has been proposed by Gucker *et al.* (1964). The amplitude functions are expressed[4] as a power series in $\cos\theta$.

$$S_1(\theta) = -i \sum_{p=0}^{\infty} \alpha_p(\cos\theta)^p \tag{3.7.14}$$

$$S_2(\theta) = i \sum_{p=0}^{\infty} \beta_p(\cos\theta)^p \tag{3.7.15}$$

in which α_p and β_p involve a series of values of a_n and b_n as follows:

$$\alpha_p = \sum_{n=p,p+2,\ldots}^{\infty} R_n{}^p b_n + \sum_{n=p+1,p+3,\ldots}^{\infty} Q_n^{p+1} a_n \tag{3.7.16}$$

$$\beta_p = \sum_{n=p,p+2,\ldots}^{\infty} R_n{}^p a_n + \sum_{n=p+1,p+3,\ldots}^{\infty} Q_n^{p+1} b_n \tag{3.7.17}$$

The quantities $R_n{}^p$ and Q_n^{p+1} are defined by

$$R_n{}^p = \frac{(2n+1)(n^2+n-p)}{n(n+1)}\mu_n{}^p \tag{3.7.18}$$

$$Q_n^{p+1} = \frac{(2n+1)(p+1)}{n(n+1)}\mu_n^{p+1} \tag{3.7.19}$$

and $\mu_n{}^p$ is the coefficient in the terminating series expansion of the zeroth degree Legendre polynomial

$$P_n(\cos\theta) = \sum_{p=\delta,\delta+2}^{n} \mu_n{}^p(\cos\theta)^p \tag{3.7.20}$$

Here $\delta = 0$ for n even, 1 for n odd, and

$$\mu_n{}^p = (-1)^{(n-p)/2}\frac{1\cdot 3\cdot 5\ldots(n+p-1)}{2^{(n-p)/2}[(n-p)/2]!\,p!} \tag{3.7.21}$$

[4] The factors $-i$ and i in (3.7.14) and (3.7.15) are omitted in the original reference.

Two alternative formulations have also been proposed (Gucker *et al.* 1968) which may have computational advantages. These are

$$S_1(\theta) = -i \sum_{\nu=1,2}^{\infty} [A_\nu(\sin \nu\theta/\sin \theta) + \nu B_\nu \cos \nu\theta] \tag{3.7.22}$$

$$S_2(\theta) = i \sum_{\nu=1,2}^{\infty} [\nu A_\nu \cos \nu\theta + B_\nu(\sin \nu\theta/\sin \theta)] \tag{3.7.23}$$

and

$$S_1(\theta) = -i \sum_{\nu=0,1}^{\infty} \alpha_\nu \cos \nu\theta \tag{3.7.24}$$

$$S_2(\theta) = i \sum_{\nu=0,1}^{\infty} \beta_\nu \cos \nu\theta \tag{3.7.25}$$

The first set of coefficients are

$$A_\nu = \sum_{n=\nu,\nu+2}^{\infty} Y_n^\nu a_n \tag{3.7.26}$$

$$B_\nu = \sum_{n=\nu,\nu+2}^{\infty} Y_n^\nu b_n \tag{3.7.27}$$

where

$$Y_n^\nu = 2\nu \frac{2n+1}{n(n+1)} \left[\prod_{j=1,2}^{n} \frac{2j-1}{2j} \right] \left[\prod_{j=1,2}^{(n-\nu)/2} \frac{(2j-1)(n-j+1)}{j(2n-2j+1)} \right] \tag{3.7.28}$$

The second set of coefficients are

$$\alpha_\nu = \sum_{n=\nu,\nu+2}^{\infty} (Z_n^\nu b_n + X_{n+1}^\nu a_{n+1}) \tag{3.7.29}$$

$$\beta_\nu = \sum_{n=\nu,\nu+2}^{\infty} (Z_n^\nu a_n + X_{n+1}^\nu b_{n+1}) \tag{3.7.30}$$

where

$$Z_n^\nu = \nu Y_n^\nu \tag{3.7.31}$$

$$X_n^\nu = 2^\varepsilon \sum_{k=\nu+1,\nu+3}^{n} Y_n^k \tag{3.7.32}$$

Here, $\varepsilon = 0$ for $\nu = 0$; $\varepsilon = 1$ for $\nu \neq 0$.

Finally, it can be noted that the expressions for $S_1(\theta)$ and $S_2(\theta)$ given by (3.7.22) and (3.7.23) can be derived directly from the exact solution of Maxwell's equations in the same way that $\pi_n(\cos \theta)$ and $\tau_n(\cos \theta)$ were derived above. The essential difference is in the original assumption that the angular dependence on θ can be expressed in terms of $\cos \nu\theta$, rather than $P_n^{(1)}(\cos \theta)$ which leads to $\pi_n(\cos \theta)$ and $\tau_n(\cos \theta)$.

3.8 Tabulations of Scattering Functions for Spherical Particles

Before the advent of the electronic digital computer, the considerable labor of hand computations impeded utilization of the theory of scattering by spheres for the many problems to which it was applicable. These computations consist of two main parts; evaluation of the scattering coefficients, a_n and b_n [(3.3.48) and (3.3.49)] from the appropriate Ricatti–Bessel functions, and then formation of the amplitude functions $S_1(\theta)$ and $S_2(\theta)$ [(3.3.56) and (3.3.57)] by combination of a_n and b_n with the Legendre functions. The various cross sections and efficiencies [(3.3.79) to (3.3.82)] follow directly from a_n and b_n.

These computations can become very laborious, especially when the size parameter α becomes large. A number of terms somewhat larger than the magnitude of α must be evaluated before the series for the amplitude functions converges. The computation is further complicated when the particle has a complex refractive index, in which case the Ricatti–Bessel functions have complex arguments.

Early workers simply avoided the full computation. Mie (1908) resorted to reduction of the expressions for a_n and b_n to a series approximation rather than computing them directly, and he limited his investigation to only three partial waves, i.e., a_1, a_2, and b_1.

The first extensive publication of scattering functions was by Blumer (1925, 1926). Since then a huge but diffuse collection of scattering functions has been generated. These have appeared in three periods. During the first period, the calculations were carried out with the aid of desk calculators and nearly all of these results appeared in the published literature. With the advent of high speed digital computers, the production of computed results increased but the volume was still sufficiently manageable so that they were published in books and journals or were tabulated in widely circulated research reports. More recently, the speed at which the computations can be produced has increased so fantastically and the computers have become so readily accessible that it is no longer feasible to publish the huge output. Indeed, in many cases it is more convenient to retain the computed results in the memory of the computer or on punched cards or magnetic tape and then to utilize them as required. Despite the facility with which the individual research worker can generate his own computations, the existing compendium of published results still provides a useful reservoir for many purposes. In what follows, some guidance through this literature will be offered.

The computed results have been tabulated in a variety of forms which are usually dictated by the particular interests of the individual worker and the applications he has in mind. Tables of the scattering coefficients, a_n and b_n, provide the greatest potential source of information since the amplitude functions, $S_1(\theta)$ and $S_2(\theta)$, and the intensity functions, i_1 and i_2, at any particular angle may be obtained from these. However, the extraction of these latter quantities with the aid of the $\pi_n(\cos\theta)$ and $\tau_n(\cos\theta)$ functions is still a major computation.

The most directly useful forms for tabulation are $S_1(\theta)$, $S_2(\theta)$, i_1, and i_2. However since interpolation between angles is precarious, especially for large values of α, very extensive tabulations are often needed for optimum utility.

Chu and Churchill (1955) and Clark *et al.* (1957) have proposed tabulation of functions designated as the angular distribution coefficients f_n, rather than a_n and b_n.

$$f(\theta) = \frac{i_1 + i_2}{2\pi\alpha^2 Q_{\text{sca}}} = \frac{1}{4\pi} \sum_{n=0}^{\infty} f_n P_n(\cos\theta) \tag{3.8.1}$$

These coefficients may be calculated directly from a_n and b_n with the aid of some rather complicated formulas. There may be a slight advantage over the tabulation of a_n and b_n in that they lead to the angular distribution function $f(\theta)$, through the more familiar Legendre polynomial, $P_n(\cos\theta)$, rather than through $\pi_n(\cos\theta)$ and $\tau_n(\cos\theta)$. On the other hand, as presently formulated there is no separation of i_1 and i_2, and it is doubtful whether any real computational advantages are introduced by this approach. However, these functions provide the basis for the Hartel (1940) theory of multiple scattering and have been used in this connection (Smart *et al.*, 1965).

Penndorf (1963) has compiled a very useful bibliography of scattering functions. This represents an updating of earlier ones published by Kerker (1955) and van de Hulst (1957). A supplementary bibliography is given in Tables 3.2 and 3.3 for real and complex refractive indices, respectively. These tables do not include sources already cited by van de Hulst, which is quite complete up to 1957. Also, not all the newer sources cited by Penndorf are included. Unpublished tables which are not readily available, at least in widely circulated research reports, have been omitted as well as results which are presented in graphical form. A number of results not in Penndorf's report are also included here. In some cases, the reader will find that the source we have cited may refer back to extensive computations on deposit with the American Documentation Institute of the Library of Congress. These are readily available on either microfilm or photoduplicates.

The tables list the refractive indices and α values for which the computations have been carried out as well as the quantities tabulated. Where S

or i is designated, this indicates that S_1, S_2, and i_1, i_2 are each given. For $\theta = 0$ and 180°, the two polarized components are identical. In some cases, the tabulated quantities may not correspond precisely to those indicated, but the connection will be apparent if the reader will carefully examine the notation of each author. Thus Pangonis and Heller (1960) tabulate i_1/α_3 rather than i_1. Workers in radar will usually use the radar backscattering cross section, σ, which is defined by

$$\sigma = (\lambda^2/\pi)|S(180)|^2 \tag{3.8.2}$$

For some of the references in Table 3.3 the separate listing of each of the complex refractive indices for which computations have been published would be too voluminous. In these cases, the substance and wavelengths for which the calculations have been carried out are given and the reader is referred to the original sources for the values of the refractive index. These mostly apply to scattering by metals in the visible, by water in the infrared, and by water and ice in the microwave parts of the spectrum.

Despite the large number of computations which have been published, the domain of m and α is hardly complete. Obviously, publication of computations for all parameters of physical interest is impractical. Many workers who have access to high speed computers have generated so many numerical results that these can hardly be coped with in the usual tabular form, much less published in the normal way. Thus Donn and Powell (1963) have computed but have not published the nearly one million intensity functions for

$$m = 1.2(0.2)2.4$$

$$\theta = 0°(5°)180°$$

$$\alpha = 0.1(0.1)100$$

and also the integrated function

$$F(\alpha, \theta, m) = \int_{0.1}^{\alpha} I(\alpha, \theta, m)\, d\alpha \tag{3.8.3}$$

The end result of a computation is often not the scattering function for a particular value of α and m but some quantity which may involve integration over a distribution of radii or of wavelengths or even over both of these parameters. Then, it is no longer convenient to tabulate individual scattering functions. Instead, these may then be stored on punched cards or magnetic tapes in a computer facility. In the author's laboratory, one particular program has resulted in the accumulation of nearly half a million intensity functions integrated over a size distribution, each representing a particular combination of size distribution, refractive index, and angle of observation (Kerker *et al.*, 1964c).

TABLE 3.2

SUPPLEMENTARY BIBLIOGRAPHY OF SELECTED SCATTERING FUNCTIONS FOR DIELECTRIC SPHERES

No.	Ref.	Refractive index	Range of α-values	Tabulated quantities
1	Shifrin (1951b)	1.33	60	a_n, b_n, Q_{sca} i; $\theta = 0, 1, 2, 5$
2	Shifrin (1955a)	∞	1.5, 2, 4, 6.2, 7, 8, 9.3	$a_n, b_n, i; \theta = 90$
3	Clark *et al.* (1957)	1.33	1(1)6(2)10(5)30	$f(\theta)$
4	Chu *et al.* (1957)	0.90	1(1)6(2)10, 20, 25	$Q_{\text{sca}}, f(\theta)$
		0.93	1(1)5(5)30	
		1.05, 1.10, 1.30, 1.44, 1.50, 1.55, 1.60, 2.00	1(1)6	
		1.15, 1.25	1(1)6(2)10, 15	
		1.20, 1.33, 1.40	1(1)6(2)10(5)30	
5	Walter (1957)	1.33	6(2)12, 15, 18, 25, 30, 45, 60	i $\theta = 0, 90, 170(1)180$
			90, 120, 180, 250	$\theta = 0, 90, 168(2)180$
		∞	6, 10, 18, 30, 60	a_n, b_n, i $\theta = 0, 90, 168(2)180$
6	Mori and Kikuchi (1957)	1.33	5.5(0.5)10	$S; \theta = 0(5)180$
7	Mori and Kikuchi (1958)	1.33	10(0.5)18.5, 10.2, 10.8, 14.3, 14.7, 16.2, 16.3	$S; \theta = 0(2.5)180$
8	Ashley and Cobb (1958)	1.20	1, 2, 3, 5, 8, 10, 15, 20, 30, 35	$S; \theta = 0(10)180$
9	Boll *et al.* (1958)	0.6, 0.7, 0.75, 0.8, 0.9	1(1)10(2)20(5)100 (10)160(20)200	a_n, b_n, Q_{sca}
		0.93	1(1)5(5)80, 95(20) 135, 160, 200	
10	Walter (1959)	1.33	10(5)80	i $\theta = 0(10)170(1)180$
			85(5)100(10)200(50) 400	$\theta = 0, 90, 150, 160, 170(1)180$
11	Meehan and Beattie (1960)	1.75	0.1–4.0(22) 0.6–3.6(10)	Q_{sca} $i_1; \theta = 0, 40, 90, 140$ $i_2; \theta = 90$
12	Heller *et al.* (1959)	1.05(0.05)1.30	1(1)15	$i; \theta = 0, 180$
13	Pfleiderer (1959)	1.33	85(5)100(10)200(50) 400	$i; \theta = 10(10)140$
14	Chromey (1960)	0.5(0.25)3.0	0.2(0.2)2.0	$Q_{\text{sca}}; (a_n, b_n - \text{L.C.})$

TABLE 3.2 (*continued*)

No.	Ref.	Refractive index	Range of α-values	Tabulated quantities
15	Pangonis and Heller (1960)	1.05(0.05)1.30	0.2(0.2)7.0	$i; \theta = 0(5)180$
16	Kerker and Matijevic (1961b)	2.105	0.2(0.4)5.8(0.2)15 0.2(0.4)5.8	$Q_{sca}, i; \theta = 0(10)180$ $\theta = 45, 135$
17	Kerker *et al.* (1961b)	1.60(0.04)2.08	0.1(0.1)10.0	Q_{sca} (L.C.)
18	Kerker *et al.* (1962)	1.4821, 2.1050	0.1(0.1)23(1)53	Q_{sca}
19	Giese *et al.* (1962)	1.5	0.2(0.2)159	$Q_{sca}, i;$ $\theta = 0(1)10(10)180$
20	Remy-Battiau (1962)	1.25	0.1(0.1)10(1)20(3) 48, 50	$i; \theta = 0(2)10(10)170(2)180$
21	Napper and Ottewill (1963d)	1.7067	0.1(0.1)10.0	$Q_{sca}, i; \theta = 35(5)145$
22	Deirmendjian (1963)	1.29, 1.315, 1.525 1.44 1.54, 1.55, 1.56 2.2	0.5(0.5)15.0 0.5(0.5)7.0 0.5(0.5)10 0.5(0.5)10, 12(4)40	$S; \theta = 0, 180$
23	Wakashima and Takata (1963)	1.33	0.1(0.1)30.0	$i; \theta = 90$
24	Rheinstein (1963)	∞	$\alpha/2\pi = 0.01(0.01)19.0$	$S; \theta = 180$
25	Denman *et al.* (1963)	1.05(0.05)1.30 1.333	0.2(0.2)25.0	$i; \theta = 0(5)180$
26	Atlas *et al.* (1963)	1.60, 1.61	0.1(0.1)20.0(1)50.0	$i; \theta = 180$
27	Dettmar *et al.* (1963)	1.11, 1.23, 1.60, 1.64, 1.78, 1.85, 2.06	0.1(0.1)20.0	Q_{sca}
		1.44, 1.55, 1.72, 1.91, 2.00, 2.37, 2.54, 2.74	0.1(0.1)10.0	Q_{sca}

With a desk calculator, the investment behind each scattering function (e.g. i_1, i_2, or Q) may represent from several man hours to man months. However, with modern machines the computation proceeds so rapidly that it may not be feasible to include even the integrated scattering functions in the machine output. Rather these may be computed internally as needed and discarded after being utilized in the overall program just as any simple intermediate function such as the square or the logarithm of a number. This was the procedure utilized by Kerker *et al.*, (1966b) in a study of the color of the scattered light which required an integration over both particle size distribution and wavelength.

TABLE 3.3
SUPPLEMENTARY BIBLIOGRAPHY OF SELECTED SCATTERING FUNCTIONS FOR ABSORBING SPHERES

No.	Ref.	Refractive index	Compound	Spectral range	Range of α-values	Tabulated quantities
1	Chu (1952)	After Saxton (1946) and Saxton and Lane (1946)	Water	0.1 cm 0.3 cm 0.5 cm 0.75 cm 1.00 cm	0.05(0.05)0.5(0.1) 1.5(0.25)5	Q_{sca}, i; $\theta = 180$
2	Kennaugh and Sloan (1952)	$3.41 - 1.94i$ $4.21 - 2.51i$ $5.55 - 2.85i$ $7.20 - 2.65i$ $8.18 - 1.96i$ $8.90 - 0.69i$	Water	0.28 cm 0.45 cm 0.8 cm 1.6 cm 2.8 cm 10.0 cm	0.1(0.05)1(0.1)5 0.1(0.05)1(0.1)3 0.1(0.05)1(0.1)2 0.1(0.025)1(0.05)1.3 0.1(0.025)1 0.1(0.01)0.3(0.005)0.43 (0.01)0.6	i; $\theta = 180$
3	Shifrin (1954, 1955b, c, 1961)	After Shifrin (1951a)	Water	3–117 μ [19] 9, 10, 13, 15, 18 μ	6.265 μ in radius 12.53 μ in radius	Q_{ext}, Q_{sca} Q_{ext}, Q_{sca}
4	Chromey (1960)	$n = 0.5(0.25)3$ $\kappa = 0(0.1)1$			0.2(0.2)2	Q_{ext}, Q_{sca}
5	Havard (1960)	Varies with λ	Water	3.6–13.5 μ	1, 2, 4, 9, 12, 15 μ in radius	Q_{ext}, Q_{abs} i; $\theta = 0(10)180$
6	Stephens (1961a)	After Dorsey (1940) and McDonald (1960)	Water	4.0(0.5)10 μ 10(2)30 μ 30(5)90 μ	1.0(0.5)10 μ in diameter. Also large number of α-values over the range 0.01 to 8	Q_{ext}, Q_{sca}

7	Stephens (1961b)	After Saxton (1946) and Kerr (1951)	Water	0.43, 0.86, 1, 1.24, 2, 3.2, 4, 5, 6, 7, 8, 9, 10, 16.23 cm	Various α-values from 0.04 to 5.2	$i;\theta = 180$
			Ice	3.2 cm		
8	Stephens and Gerhardt (1961a, b)	After Dorsey (1940) and McDonald (1960)	Water	5, 6.2, 7.8, 10, 12, 13, 15, 18 μ	0.05(0.05)1.1(0.1)2(0.2)5.2	Q_{abs}
9	Herman and Battan (1961a)	$1.78 - 0.0024i$	Ice	Microwaves	0.1(0.1)3.0(0.2)30	$i;\theta = 180$
10	Herman, Browning and Battan (1961)	After Saxton (1946) and Kerr (1951)	Water	0.62, 0.86, 1.24, 1.87, 3.21, 4.67, 5.5, 10.0 cm	0.1(0.1)5.0	Q_{ext}, Q_{sca}, $i;\theta = 180$
11	Giese (1961)	$1.27 - 1.37i$	Fe	440 mμ	1(1)40	Q_{ext}, Q_{sca}, Q_{pr}
		$1.38 - 1.50i$	Fe	508 mμ	1(1)37	
		$1.70 - 1.84i$	Fe	668 mμ	1(1)30	
		$1.50 - 3.10i$	Ni	—	1(1)17	
12	Deirmendjian and Clasen (1962)	$1.315 - 0.0143i$	Water	5.3 μ	0.5(0.5)15.0	$S:\theta = 180$ Q_{sca} is given in Deirmendjian (1963)
		$1.315 - 0.1370i$	Water	6.05 μ	0.5(0.5)15.0	
		$1.315 - 0.4298i$	Water	15.00 μ	0.5(0.5)15.0	
		$1.525 - 0.0682i$	Water	3.07 μ	0.5(0.5)25.0	
		$1.353 - 0.0059i$	Water	3.90 μ	0.5(0.5)20.0	
13	Deirmendjian (1963)	$1.29 - 0.0427i$	Water	8.15 μ	0.5(0.5)15.0	Q_{sca}, S: $\theta = 0, 180$
		$1.29 - 0.0645i$			0.5(0.5)15.0	
		$1.29 - 0.4720i$			0.5(0.5)15.0	
		$1.212 - 0.060i$	Water	10.0 μ	0.5(0.5)10.0	
		$1.111 - 0.183i$	Water	11.5 μ	0.5(0.5)10.0	
		$1.44 - 0.400i$	Water	16.6 μ	0.5(0.5)7.0	
		$2.20 - 0.0220i$			0.5(0.5)10, 12(4)40	
		$2.20 - 0.220i$			0.5(0.5)10, 12(4)40	
		$1.55 - 0.0155i$			2(2)12(4)40	
		$1.55 - 0.155i$			2(2)12(4)40	

TABLE 3.3 (*continued*)

No.	Ref.	Refractive index	Compound	Spectral range	Range of α-values	Tabulated quantities
		$1.28 - 1.37i$	Fe	441 mμ	0.1(0.1)1(0.25)2(0.5)10	
		$1.51 - 1.63i$	Fe	589 mμ	0.1(0.1)1(0.25)2(0.5)10	
		$1.70 - 1.84i$	Fe	668 mμ	0.1(0.1)1(0.25)2(0.5)10	
		$1.78 - 0.0024i$	Ice	Microwaves	0.5(0.5)4.0(2.0)16.0	
		$2.56 - 0.895i$	Water	0.2 cm	0.5(0.5)4.0(2.0)16.0	
		$3.19 - 1.766i$	Water	0.5 cm	0.25(0.25)1(1)7	
		$5.84 - 3.005i$	Water	2.0 cm	0.1(0.1)1, 1.5, 2	
14	Adler and Johnson (1962)	$1.46 - 0.0190i$	Teflon	Microwaves	0.2(0.1)5	$i; \theta = 180$
		$1.63 - 0.0195i$	Lucite	Microwaves		
		$1.99 - 0.0248i$	Bakelite	Microwaves		
		$2.01 - 0.0392i$	Bakelite	Microwaves		
		$8.18 - 1.96i$	Water	Microwaves		
15	Plass (1964)		Al_2O_3	2 μ	0.1–9.9 μ in radius	Q_{ext}, Q_{sca}
				5 μ	0.1–9.9 μ in radius	
			MgO	2 μ	0.1–9.9 μ in radius	
				5 μ	0.1–9.9 μ in radius	
16	Doyle and Agarwal (1965)	After Ives and Briggs (1936, 1937)	Na	320(10)460 mμ	0.025, 0.100, 0.200, 0.300, 0.400, μ radius	Q_{ext}
			K	460(10)600 mμ		
			Rb	520(10)660 mμ		
			Cs	600(10)710 mμ		
17	Meehan (1968)	$2 - ki$; values of k are				
		0, 10^{-3}, 10^{-2}, 10^{-1} 0.5, 1, 1.5, 2, 3			0.05, 0.1, 0.2, 0.5, 1.0	Q_{ext}
		Same 0, 10^{-2}, 10^{-1}			0.1, 0.2, 0.3, 0.4, 0.5	$i; \theta = 0$
		1, 2, 3			0.1, 0.5, 1	$i; \theta = 0, 180$

3.9 Approximations

In this section, we are concerned with approximations based upon direct mathematical reduction of the general theory without the introduction of simplifying assumptions about the model. Thus we exclude at present such treatments as the Rayleigh–Debye theory in which the scattering is considered to be a simple interference effect of the volume elements of the scatterer or the consideration of scattering by large spheres in terms of ray optics as a combination of reflection, refraction, and diffraction.

The impetus for the search for approximate formulas came from the desire to seek a more rapid means of computation, at least in the limiting case of small particles. The main effort has been an attempt to express the scattering functions as power series in α rather than in terms of the Bessel and Legendre functions. Kleinman (1965) has characterized all such solutions for which convergence is sufficiently rapid so that they may be approximated by the first term as the Rayleigh region and has reviewed recent work.

The direct reduction of the complete solution has a venerable history starting with Lorenz's (1890, 1898b), Mie's (1908), and Debye's (1909a) extensive work along these lines. When the arguments, α and β, of the Ricatti–Bessel functions are small, the expansion of these functions as a power series in the argument converges very rapidly [(3.6.13), (3.6.14)]. The scattering coefficients can then be cast into the following form:

$$a_n = i\frac{n+1}{n(2n+1)}\frac{\alpha^{2n+1}}{1\cdot 3^2\ldots(2n-1)^2}u_n\frac{m^2 - v_n}{m^2 + [(n+1)/n]w_n} \tag{3.9.1}$$

$$b_n = -i\frac{n+1}{n(2n+1)}\frac{\alpha^{2n+1}}{1^2\cdot 3^2\ldots(2n-1)^2}u_n\frac{1 - v_n}{1 + [(n+1)/n]w_n} \tag{3.9.2}$$

where u_n, v_n, and w_n are series, in terms of α^2 and β^2, which converge rapidly for small values of α and β. Mie (1908) gives the explicit expressions. The next step is to clear the denominator and to expand the scattering coefficients themselves as power series in α. The first terms in the final expansion of the first three scattering coefficients are (Stratton, 1941)

$$a_1 = \frac{2}{3}i\left(\frac{m^2 - 1}{m^2 + 2}\right)\alpha^3 \tag{3.9.3}$$

$$a_2 = \frac{1}{15}i\left(\frac{m^2 - 1}{2m^2 + 3}\right)\alpha^5 \tag{3.9.4}$$

$$b_1 = -\frac{1}{45}i(m^2 - 1)\alpha^5 \tag{3.9.5}$$

The remaining coefficients are proportional to still higher powers of α. Obviously when both α and m are sufficiently small, the term containing a_1 becomes the leading term in the expressions for the amplitude functions [(3.3.56), (3.3.57)]. With

$$\pi_1(\cos\theta) = 1 \tag{3.9.6}$$

and

$$\tau_1(\cos\theta) = \cos\theta \tag{3.9.7}$$

the Rayleigh formula

$$I_u = \frac{8\pi^4 a^6}{r^2\lambda^4}\left|\frac{m^2 - 1}{m^2 + 2}\right|^2 (1 + \cos^2\theta) \tag{3.9.8}$$

is obtained. This differs from (3.2.19) in that m may be complex. However, the derivation brings out a limitation placed upon Rayleigh scattering that may not have been evident earlier. Since retention of only the first term in the expansion of the Ricatti–Bessel functions requires that the arguments be small, it is necessary that both α and $m\alpha$ be small. This restriction upon m applies to both the real and imaginary parts.

3.9.1 Range of Validity of the Rayleigh Equation

Although the range of validity of the Rayleigh equation has long been given by a rough rule which states that the radius should not exceed about one-twentieth of the wavelength, a detailed, quantitative comparison of this equation with the full theory has been discussed only recently by Jaycock and Parfitt (1962) and Heller (1965). The latter analysis is based primarily upon the refractive index range $m = 1.00$ to 1.30, but it also includes some calculations of Lowan (1948) and of Gumprecht and Sliepcevich (1951a) for higher refractive indices.

The results are presented both as $\Delta\alpha$ and as $\Delta\tau$. The former is the percent error committed in calculating the size parameter α from the specific turbidity when the Rayleigh equation is used, while $\Delta\tau$ is the error in calculating the specific turbidity from a given value of α. The specific turbidity is

$$\tau/\phi = (3/4a)Q_{\text{sca}} \tag{3.9.9}$$

where ϕ is the volume of scattering material per unit volume and τ is the turbidity. The corresponding results involving the Rayleigh ratio at 90° are not appreciably different from those based on turbidity and will not be discussed here.

In Fig. 3.7, $\Delta\tau$ is plotted as a function of α for refractive indices up to 1.30. The results for $m = 1.00$ were obtained by extrapolation. For these refractive indices, the turbidity obtained from the Rayleigh equation is always too high. We can now judge the validity of the conventional working rule in a quantitative way. For $m = 1.30$, the turbidity is in error by 2% when $a/\lambda = 0.06$ and by 5% when $a/\lambda = 0.1$. The Rayleigh equation is less accurate at smaller refractive indices so that at $m = 1.10$ the error is 2% for $a/\lambda = 0.04$ and 5% for $\alpha/\lambda = 0.06$.

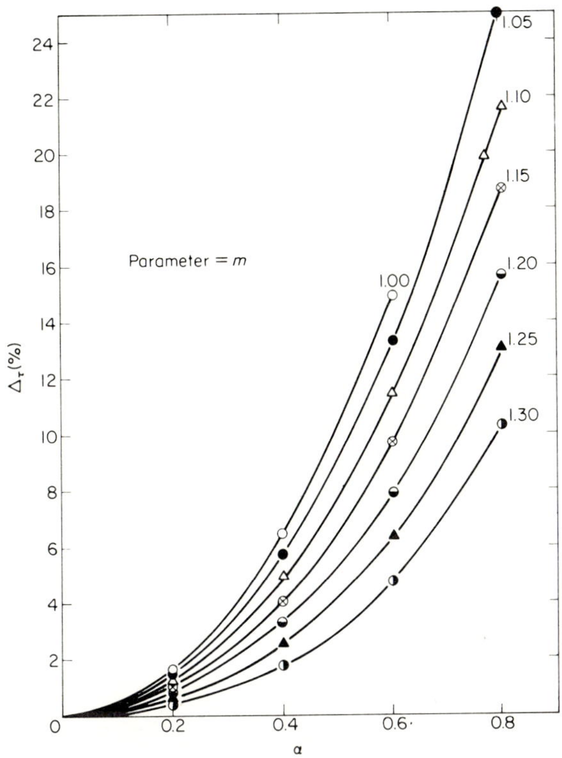

FIG. 3.7. Percentage deviation between specific turbidity calculated by the Rayleigh formula and the exact value plotted against α for $m = 1.00$ to 1.30 (Heller, 1965).

Heller has explored the decrease in $\Delta\tau$ with increasing refractive index even though he had only a limited number of computations at higher values of m available. The results are shown in Fig. 3.8 where $\Delta\tau$ is plotted against m for various values of α. The decrease of $\Delta\tau$ with m continues until there is a crossover ($\Delta\tau = 0$) to negative values. With further increase of m, $\Delta\tau$ goes through a minimum and then rises sharply through a second crossover. Between these two crossover points is a region where the size parameter is relatively high and yet the Rayleigh equation is quite accurate. Heller has

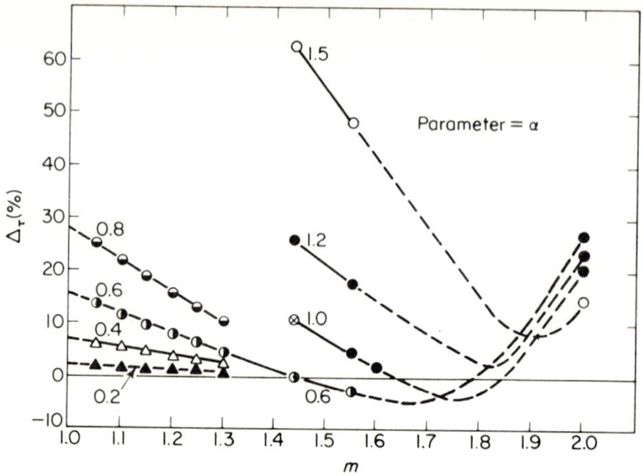

FIG. 3.8. Percentage deviation between specific turbidity calculated by the Rayleigh formula and the exact value plotted against m for $\alpha = 0.2$ to 1.5 (Heller, 1965).

termed this a "second range" of validity of the Rayleigh equation. This is clearly shown by the error contour chart in Fig. 3.9. The four lines running from the left of the chart at $m = 1.0$ delineate the regions of α and m for which $\Delta\tau$ is less than 1, 2, 5, and 10% respectively. The crossover ($\Delta\tau = 0$)

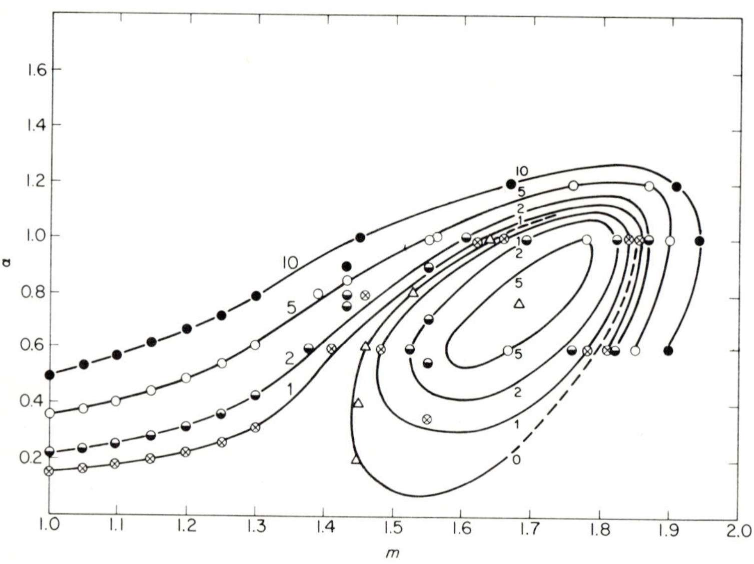

FIG. 3.9. Contours, on the $m\alpha$-plane, of percentage deviation (1, 2, 5, and 10%) between specific turbidity calculated by the Rayleigh formula and the exact value (Heller, 1965).

is depicted by the line labeled 0. Within this region the contour lines represent negative values of $\Delta\tau$ describing a shallow topographical "pit" whose bottom is somewhat below the 5% line. It is the lower slopes of this pit that constitute the second range of validity of the Rayleigh equation. Actually this is not due to any intrinsic validity of the theory but rather to a fortuitous mutual compensation of opposing deviations. Jaycock and Parfitt (1962) had also observed this crossover.

The same pattern emerges when considering the calculation of α from turbidimetric data, except that the percent errors ($\Delta\alpha$) are considerably smaller than those obtained for the specific turbidity for the corresponding values of m and α. This follows from the functional relation between α and τ. In Fig. 3.10, $\Delta\alpha$ is plotted against α over the range $m = 1.00$ to 1.30. For $m = 1.30$, the value of α is too low by 2% when $a/\lambda = 0.1$ and by 5% when $\alpha/\lambda = 0.14$. For $m = 1.10$, the corresponding values of a/λ are 0.07 and 0.11.

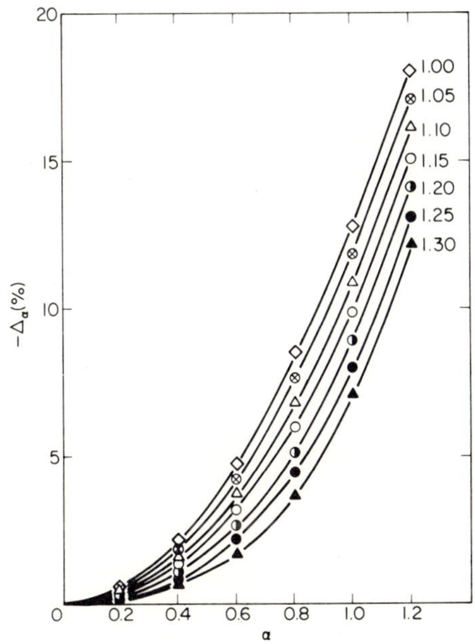

FIG. 3.10. Percentage deviation of α between value calculated from turbidimetric data using the Rayleigh formula and that calculated from the exact theory plotted against α for $m = 1.0$ to 1.30 (Heller, 1965).

Obviously the range of validity will depend upon the error which is tolerable for a particular investigation. The rough guide that the Rayleigh equation is valid when $a/\lambda < 0.05$ is certainly vindicated by the above detailed considerations. It should be pointed out that λ refers to the wavelength in the medium, so that for dispersions in liquid media the permissible radius is a correspondingly smaller fraction ($\lambda = \lambda_0/m_2$) of the vacuum wavelength.

3.9.2 Expansion as a Power Series in α

Higher approximations will now be considered in which additional terms in the expansion of the Ricatti–Bessel functions are retained as well as a greater number of terms in the expansion of the scattering coefficients a_n and b_n. Quite detailed analyses have been carried out by Penndorf (1960, 1962a, b, c) and by Goodrich *et al.* (1961). Walstra (1964b) has made some observations on the validity of these and some other approximations for estimating values of Q_{sca}.

Penndorf has expanded the first five scattering coefficients a_1, b_1, a_2, b_2, and a_3 for complex as well as real refractive indices, including also the case of $m = \infty$. The real part of the scattering coefficients has been carried out to powers of 10 in α; the imaginary part to powers of 7. Also he has noted a considerable number of errors by earlier workers. These results have then been collected into the formulas for the efficiency factors. The result for dielectric media is

$$Q_{\text{sca}} = \frac{8\alpha^4}{3}\left(\frac{n^2-1}{n^2+2}\right)^2\left[1 + \frac{6}{5}\alpha^2\left(\frac{n^2-2}{n^2+2}\right) + \alpha^4 \cdot \left\{\frac{3}{175}\left(\frac{n^6+41n^4-284n^2+284}{(n^2+2)^2}\right) + \frac{1}{900}\left(\frac{n^2+2}{2n^2+3}\right)^2[15+(2n^2+3)^2]\right\}\right] \tag{3.9.10}$$

Here Rayleigh's formula represents the leading term. For media characterized by a complex refractive index

$$m = n(1 - \kappa i) \tag{3.9.11}$$

Penndorf has obtained

$$Q_{\text{ext}} = \frac{24n^2\kappa}{Z_1}\alpha + \left[\frac{4n^2\kappa}{15} + \frac{20n^2\kappa}{3Z_2} + 4.8n^2\kappa \times \frac{7(n^2+n^2\kappa^2)^2 + 4(n^2-n^2\kappa^2-5)}{Z_1{}^2}\right]\alpha^3 + \frac{8}{3}\left\{\frac{[(n^2+n^2\kappa^2)^2 + (n^2-n^2\kappa^2-2)]^2 - 36n^4\kappa^2}{Z_1{}^2}\right\}\alpha^4 \tag{3.9.12}$$

where

$$Z_1 = (n^2 + n^2\kappa^2)^2 + 4(n^2 - n^2\kappa^2) + 4 \tag{3.9.13}$$

$$Z_2 = 4(n^2 + n^2\kappa^2)^2 + 12(n^2 - n^2\kappa^2) + 9 \tag{3.9.14}$$

The efficiency factor for scattering is given by

$$\begin{aligned} Q_{\text{sca}} &= (8/3Z_1^2)\{[(n^2 + n^2\kappa^2)^2 + n^2 - n^2\kappa^2 - 2]^2 - 36n^4\kappa^2\}\alpha^4 \\ &\quad \cdot \{1 + (6/5Z_1)[(n^2 + n^2\kappa^2)^2 - 4]\alpha^2 - (24n^2\kappa\alpha^3/3Z_1)\} \end{aligned} \tag{3.9.15}$$

The leading term of this equation is identical with

$$Q_{\text{sca}} = \frac{8}{3}\alpha^4\left|\frac{m^2 - 1}{m^2 + 2}\right|^2 \tag{3.9.16}$$

In (3.9.10) the term in α^4 is identical with the corresponding term above and must accordingly represent a contribution of scattering to the extinction. The terms in α and α^3 vanish when $\kappa \to 0$ so that they represent the absorption. The term in α which is the leading contribution to the absorption may also be written as

$$Q_{\text{abs}} = \text{Im}\left\{-4\alpha\left(\frac{m^2 - 1}{m^2 + 2}\right)\right\} \tag{3.9.17}$$

Each of these expressions corresponds to Rayleigh scattering by absorbing spheres since they may be derived by substituting a_1, the scattering coefficient for an oscillating dipole, directly into the general equations for the efficiencies as given by (3.3.81) and (3.3.82). Although formally Q_{ext} is obtained in the latter case rather than Q_{abs}, it is obvious that the scattering term is not included in this approximation, and that this result actually is the limiting value of Q_{abs}.

The full approximation given above for dielectric spheres can be used up to $\alpha = 1.4$ and $m = 2$. The error between the value obtained and that from the exact formula is less than 2% up to $m = 1.5$ although it may reach as high as 15% at $m = 2$. The formulas for absorbing spheres are useful up to $\alpha = 0.8$ in the range $n = 1.25$ to 1.75 and $n\kappa \leqslant 1$. Obviously, these equations provide only a limited advantage over the exact computations which, for such small values of α, require no more than two or three terms in the series expression.

Doyle and Agarwal (1965) have noted a special limitation on the above series expansions for metallic particles in the wavelength range where a sharp optical resonance occurs. For such cases, although direct use of the Mie expansions (3.9.1) and (3.9.2) gave correct results, the expansions in powers of α derived from these, such as (3.9.12), failed. It turns out that each of the series in the expressions for a_n and b_n given by (3.9.1) and (3.9.2) continue to converge as long as $\alpha^2 \ll 1$. However, the series in powers of α which is obtained after clearing the denominator may fail to converge unless $|\alpha^2/(m^2+2)|^2 < 1$. This condition may break down in the region of optical resonance unless α is very small.

3.9.3 Totally Reflecting Spheres

The condition that $m \to \infty$ corresponds to the very practical cases of scattering of infrared waves and microwaves by metallic spheres. While the physical problem is quite different whether it is n or κ which approaches infinity, the results are identical for both situations. The particle becomes a perfect reflector. There is no internal field. This reduction was also considered by Mie (1908) who pointed out the similarity of the resulting formulas with those obtained for this special case by Thomson (1893). The scattering coefficients become simply

$$a_n = \psi_n'(\alpha)/\zeta_n'(\alpha) \tag{3.9.18}$$

and

$$b_n = \psi_n(\alpha)/\zeta_n(\alpha) \tag{3.9.19}$$

Penndorf has expanded these for small arguments in the same manner as for a finite refractive index, obtaining

$$Q_{\text{sca}} = Q_{\text{ext}} = \tfrac{10}{3}\alpha^4(1 + 0.24\alpha^2 - 0.022614\alpha^4) \tag{3.9.20}$$

The error in using this formula is less than 1% if $\alpha < 0.5$. However, it increases rapidly and should not be used for $\alpha > 0.7$. For sufficiently small values of α

$$Q_{\text{sca}} = Q_{\text{ext}} = \tfrac{10}{3}\alpha^4 \tag{3.9.21}$$

This is valid to within 2% for $\alpha < 0.2$.

This limit is lower than that found for Rayleigh scattering by dielectric bodies. This is because the contribution of the magnetic dipole can no longer be neglected compared to the electric dipole terms, even for the limiting case of very small particles. For a small perfect reflector, the induced surface charges and currents correspond to oscillating electric and magnetic dipoles with moments a^3 and $-a^3/2$ respectively. The corresponding scattering coefficients are

$$a_1 = \tfrac{2}{3}i\alpha^3 \qquad \text{and} \qquad b_1 = -\tfrac{1}{3}i\alpha^3 \tag{3.9.22}$$

which lead in turn to the above result. The scattering patterns will be discussed in some detail later.

3.9.4 Forward Scattering from the Efficiency

Penndorf (1962c) has called attention to a useful approximate relation between the intensity functions in the forward direction and the efficiency for extinction. In the case of forward scattering

$$\pi_n = \tau_n = \tfrac{1}{2} \tag{3.9.23}$$

which we have already seen leads to (3.3.84). Since the real parts of the scattering coefficients, a_n and b_n, are always positive while the imaginary parts may be positive or negative, it follows that for sufficiently large values of α when there are numerous terms of comparable magnitude

$$\mathrm{Re}\, S_1(0°) \gg |\mathrm{Im}\, S_1(0°)| \tag{3.9.24}$$

and

$$i_1(0°) \cong [\mathrm{Re}\, S_1(0°)]^2 \tag{3.9.25}$$

Now if the cross section theorem given by (3.3.83) is utilized, it follows that in the forward direction

$$i_1(0°) \cong [(\alpha^2/4)Q_{\mathrm{ext}}] \tag{3.9.26}$$

At 0°, the other polarized component, i_2, is identical. Penndorf has checked this for $m = 1.33$ and finds that over the range $\alpha = 5$ to 30 the average error is 2.5%.

This approximation is particularly useful because there are much more extensive calculations of Q_{ext} than there are of $i_1(0°)$, thus permitting this to be calculated from existing tables. The intensity function at other angles very close to 0° can be found from the above, since the angular variation is approximately the same as that for Fraunhofer diffraction by a sphere and can be found from

$$i_1(\theta) \cong i_1(0°)[J_1(\alpha \sin\theta)/\alpha \sin\theta]^2 \tag{3.9.27}$$

3.10 Point Matching Method

Although the point matching method is an approximation to the general boundary value solution considered in Section 3.3, there are no limitations on the size parameter nor on the accuracy which can be attained. In this regard it is equivalent to the exact theory. However, unlike most approximations, it hardly provides a computational short cut, and its main advantage is the straightforward manner in which it can be extended to nonspherical shapes for which general solutions may not be available. It will be convenient to introduce the method here in connection with the discussion of the sphere. The known results for the sphere provide a test of the method.

This method was proposed in connection with scattering by infinite cylinders of arbitrary geometrical cross section by Mullin *et al.* (1965) and by Greenberg *et al.* (1967) in connection with spheres and spheroids. Watson (1964) had presented the same basic ideas somewhat earlier. Yee (1965a, b) and Harrington (1965) have considered the range of validity.

The point of departure is the expansion of the solution of the scalar wave equation in a standard form such as (3.3.32) to (3.3.37). This series in spherical harmonics is appropriate for completely bounded bodies of any shape including spheres, spheroids, cubes, etc., for which the excitation in the wave zone is an outgoing spherical wave. For infinite cylindrical structures, the appropriate expansion will be in cylindrical harmonics.

We know for spheres that these series converge according to the magnitude of α and that an accurate solution can be obtained when the series is truncated at a number of terms somewhat greater than this magnitude. We assume for other shapes that the series can also be truncated after a number of terms determined by the ratio of a characteristic dimension to the wavelength. When this number is N

$$r\pi_1{}^i = \frac{1}{k_2{}^2} \sum_{n=1}^{N} i^{n-1} \frac{2n+1}{n(n+1)} \psi_n(k_2 r) P_n^{(1)}(\cos\theta)\cos\phi \tag{3.10.1}$$

and there are the corresponding expressions for $\pi_2{}^i, \pi_1{}^s, \pi_2{}^s, \pi_1{}^r, \pi_2{}^r$. These equations must be valid in each of the regions of space for which they have been formulated (inside or outside of the particle) and in addition the boundary conditions must be satisfied at the boundary between the particle and the medium.

When the boundary conditions are applied to these truncated series, the set of four equations (3.3.38) through (3.3.41) are obtained, involving the $4N$ unknowns a_n, b_n, c_n, d_n ($n = 1, 2, 3, \ldots N$). In order to develop a set of simultaneous equations sufficient to solve for the unknowns, it is only necessary to supply the coordinates of N discrete points on the boundary. The basic assumption, if N has been chosen sufficiently large, is that when the solution is valid at N points on the boundary, it will be valid everywhere on the boundary. From the known geometry of the scatterer, the values of r, θ, and ϕ at N boundary points must be substituted into the set of equations expressing the boundary conditions. These points must be selected so that they lead to independent solutions. From the symmetry of a sphere, we know that the points on the circle defined by $r = a$ and $\theta =$ constant are not independent, so that the points selected must correspond to $r = a$, $\phi = 0$, and $\theta =$ variable.

Greenberg *et al.* (1967) have tested the method for spheres for $m = 1.3$ and for values of α up to 5 and have found that the values of the coefficients (a_n, b_n) converge to specific values very rapidly for $N \geqslant \alpha$. Also these values are independent of the choice of the points at which the boundary conditions are satisfied, and they agree with the correct values obtained from the exact theory, except where the coefficients themselves become so small as to be negligible. Mullin *et al.* (1965) have obtained similar agreement with the completely conducting infinite circular cylinder at perpendicular incidence.

3.11 Radiation Pressure

The experimental confirmation of the existence of light pressure was consummated in 1899 after a search of fully two centuries [Whittaker (1951) pp. 273–276]. During the eighteenth century, the adherents of the corpuscular theory of light believed that this hypothesis would be confirmed if it were demonstrated that light rays possess momentum. However, Maxwell eliminated this argument for the corpuscular theory by showing that electromagnetic waves also carry momentum. This is in the same direction as the energy flow and is given by

$$m = u/v \tag{3.11.1}$$

where u is the energy and v is the velocity of the radiation in the medium.

Experimentally, the electromagnetic momentum will manifest itself as a radiation force or pressure whenever the momentum of an incident field is changed by deflection or absorption, and it is this that links the light pressure on a sphere to the phenomenon of scattering. Debye (1909a) carried out the definitive study of the light pressure on spherical particles of arbitrary size and optical constants. The total energy per second removed from the incident beam of unit intensity by absorption and scattering is equal to the cross section for extinction, C_{ext}. However, of the momentum carried away by the scattered radiation, that part which is associated with the forward component is restored to the incident beam. This results in a time average force of

$$F = v^{-1}[C_{\text{ext}} - \overline{\cos\theta}\, C_{\text{sca}}] \tag{3.11.2}$$

where

$$\begin{aligned}\overline{\cos\theta} &= \left[(\pi/k^2)\int_{-1}^{1}(i_1 + i_2)\cos\theta\, d(\cos\theta)\right]\Big/\left[(\pi/k^2)\int_{-1}^{1}(i_1 + i_2)\, d(\cos\theta)\right] \\ &= \left[(\pi/k^2)\int_{-1}^{1}(i_1 + i_2)\cos\theta\, d(\cos\theta)\right]\Big/C_{\text{sca}} \end{aligned} \tag{3.11.3}$$

The quantity $\overline{\cos\theta}$ is called the asymmetry factor. It is the mean of $\cos\theta$ with the angular intensity as the weighting function. In addition, an efficiency for radiation pressure may be defined in analogy to the efficiencies for extinction and scattering by

$$Q_{\text{pr}} = Q_{\text{ext}} - \overline{\cos\theta} \cdot Q_{\text{sca}} \tag{3.11.4}$$

such that the pressure exerted on the particle of cross-sectional area πa^2 is

$$P = F/\pi a^2 = v^{-1} Q_{\text{pr}} \tag{3.11.5}$$

Debye has shown that the asymmetry function can be expressed in terms of the scattering coefficients by

$$\overline{\cos\theta} \cdot Q_{\text{sca}} = (4/\alpha^2) \sum_{n=1}^{\infty} \{[n(n+2)/(n+1)] \operatorname{Re}(a_n{}^* a_{n+1} + b_n{}^* b_{n+1}) + [(2n+1)/n(n+1)] \operatorname{Re} a_n{}^* b_n\} \tag{3.11.6}$$

where the asterisk indicates that the complex conjugate is to be taken.

Debye has also developed expressions for a number of special cases. For a small totally reflecting sphere, the Ricatti–Bessel functions can be expanded in a power series in α leading to

$$Q_{\text{pr}} = \frac{14}{3}\alpha^4\left[1 + \frac{1}{21}\alpha^2 - \frac{1409}{8820}\alpha^4 + \cdots\right] \tag{3.11.7}$$

A similar expansion for the extinction and scattering efficiencies is given by

$$Q_{\text{sca}} = Q_{\text{ext}} = \frac{10}{3}\alpha^4\left(1 + \frac{6}{25}\alpha^2 - \cdots\right) \tag{3.11.8}$$

so that

$$\overline{\cos\theta} \cdot Q_{\text{sca}} = -\frac{4}{3}\alpha^4\left(1 - \frac{13}{30}\alpha^2 + \cdots\right) \tag{3.11.9}$$

The negative value of the asymmetry factor reflects the fact that, for small perfectly conducting spheres, the scattering is predominantly in the backward direction.

In the case of a small dielectric sphere with refractive index, n, the following expansion is obtained:

$$Q_{\text{pr}} = \frac{8}{3}\left(\frac{n^2-1}{n^2+2}\right)\alpha^4\left[1 - \frac{\alpha^2}{15}\,\frac{n^6 - 29n^4 + 34n^2 + 120}{(n^2+2)(2n^2+3)}\right] \tag{3.11.10}$$

In the limiting case where α is sufficiently small so that the quantity in the bracket is unity, the efficiency for radiation pressure equals the extinction efficiency as given by the Rayleigh formula. Accordingly, $\overline{\cos\theta}$ is zero. This follows from the angular symmetry of Rayleigh scattering.

For very small absorbing spheres

$$Q_{pr} \cong Q_{abs} = -4\alpha \operatorname{Im}\left(\frac{m^2 - 1}{m^2 + 2}\right) \tag{3.11.11}$$

provided the scattering is sufficiently symmetrical so that the asymmetry factor is very close to zero and also that $Q_{sca} < Q_{abs}$. If only the former condition is valid which is the case of a small sphere composed of only a slightly absorbing material

$$Q_{pr} \cong Q_{ext} = -4\alpha \operatorname{Im}\left(\frac{m^2 - 1}{m^2 + 2}\right) + \frac{8}{3}\alpha^4 \left|\frac{m^2 - 1}{m^2 + 2}\right|^2 \tag{3.11.12}$$

Debye also derived the radiation pressure for the limiting case of large α from the exact formulas with the aid of his asymptotic expressions for the Ricatti–Bessel functions. For a large perfectly reflecting sphere, he showed that $\overline{\cos\theta} = \frac{1}{2}$ and $Q_{pr} = 1$. This follows directly if one considers that the scattered intensity, which is twice that contained in the beam incident upon the sphere, is equally divided between the specularly reflected and the diffracted radiation. The specularly reflected radiation is isotropically distributed with respect to angle and thus makes no contribution to $\overline{\cos\theta}$. The diffracted radiation, in the limit of large spheres, is entirely directed into the forward direction so that its contribution to $\overline{\cos\theta}$ is $\frac{1}{2}$.

The limiting case for large dielectric spheres is considerably more complicated. Debye introduces the angles τ_0 and τ_1 which are defined by

$$\cos\tau_0 = (n + \tfrac{1}{2})/\alpha \tag{3.11.13}$$

and

$$\cos\tau_0 = m\cos\tau_1 \tag{3.11.14}$$

These are inserted in the limiting expressions for a_n and b_n and after considerable reduction

$$Q_{pr} = 1 - \mathscr{I}_1(m) - \mathscr{I}_2(m) \tag{3.11.15}$$

where

$$\mathscr{I}_1(m) = \operatorname{Re}\int_0^{\pi/2} \exp(-2i\tau_0)\left[r_1^{\,2} + (1 - r_1^{\,2})\frac{\exp(2i\tau_1)}{1 - r_1^{\,2}\exp(2i\tau_1)}\right] \times \sin\tau_0\cos\tau_0\, d\tau_0 \tag{3.11.16}$$

$$\mathscr{I}_2(m) = \operatorname{Re}\int_0^{\pi/2} \exp(-2i\tau_0)\left[r_2^{\,2} + (1 - r_2^{\,2})\frac{\exp(2i\tau_1)}{1 - r_2^{\,2}\exp(2i\tau_1)}\right] \times \sin\tau_0\cos\tau_0\, d\tau_0 \tag{3.11.17}$$

The quantities r_1 and r_2 are the Fresnel reflection coefficients (2.4.5) and (2.4.6) for those rays falling on a surface with relative refractive index m at the angle of incidence $(\pi/2 - \tau_0)$.

As will be seen (Section 4.4.4), van de Hulst (1946) later invoked the notion of localization to provide a geometrical optics interpretation of these quantities. He pointed out that the contribution of each partial wave in the series expansion could be identified with a particular geometrical optics ray striking the surface of the sphere at the distance $(n + \frac{1}{2})a/\alpha$ from the ray through the origin. In such a case τ_0 and τ_1 are the complements of the angles of incidence and refraction, respectively, of this ray.

For an absorbing particle which is sufficiently large so that all of the energy associated with the refracted rays is absorbed within the particle, a geometrical optics approach can be used. (This is considered in detail in Section 4.2.3a.) In this case the deflected radiation consists of two parts—that which is diffracted and those rays undergoing specular reflection. The efficiency for scattering is

$$Q_{\text{sca}} = 1 + w \tag{3.11.18}$$

where by Babinet's principle the diffracted rays contribute unity and where the reflected rays contribute w. For large spheres all of the diffracted radiation is very close to the forward direction so that its contribution to the asymmetry factor is $1/(1 + w)$. Then it can be shown directly that

$$\overline{\cos\theta} = (1 + wg)/(1 + w) \tag{3.11.19}$$

and

$$Q_{\text{pr}} = 1 - wg \tag{3.11.20}$$

where the contribution of the specularly reflected radiation to the asymmetry factor is

$$\frac{gw}{1 + w} = \frac{\frac{1}{2}\int_0^\pi \{|r_1|^2 + |r_2|^2\} \cos\theta \, d(\cos^2\theta/2)}{1 + w} \tag{3.11.21}$$

CHAPTER 4

The Scattering Functions for Spheres

In this chapter we shall consider how the scattering functions for a homogeneous sphere vary with the various physical parameters such as the optical size α, the complex refractive index m, and the angle of observation θ. Despite the complexity of the equations, there are certain regularities in the computed results which offer interesting insights and are often useful. The quantities which will be considered are (1) the scattering coefficients a_n, b_n, (2) the efficiencies for extinction, absorption, and scattering, Q_{ext}, Q_{abs}, Q_{sca}, (3) the backscatter $i(180)$, (4) the lateral scattering i_1, i_2, and (5) the radiation pressure. Even though a huge number of functions have been tabulated, these are still insufficient for a completely comprehensive analysis. Indeed, such an analysis would be a monumental task. Accordingly, these considerations will be restricted to a selection from existing tabulations and will be especially concerned with those results which have been of practical interest.

4.1 Scattering Coefficients

4.1.1 Real Refractive Index

When the refractive index is real, corresponding to no absorption, the scattering coefficients may be formally simplified if we recall that

$$\zeta_n(z) = \psi_n(z) + i\chi_n(z) \tag{3.3.27}$$

Then, when the real and imaginary parts of the denominators of a_n and b_n are separated, these may be written (van de Hulst, 1946)

$$a_n = \tan\alpha_n/(\tan\alpha_n - i) = \tfrac{1}{2}[1 - \exp(-2i\alpha_n)] \tag{4.1.1}$$

$$b_n = \tan\beta_n/(\tan\beta_n - i) = \tfrac{1}{2}[1 - \exp(-2i\beta_n)] \tag{4.1.2}$$

where

$$\tan \alpha_n = -\frac{\psi_n'(\beta)\psi_n(\alpha) - m\psi_n(\beta)\psi_n'(\alpha)}{\psi_n'(\beta)\chi_n(\alpha) - m\psi_n(\beta)\chi_n'(\alpha)} \tag{4.1.3}$$

$$\tan \beta_n = -\frac{m\psi_n'(\beta)\psi_n(\alpha) - \psi_n(\beta)\psi_n'(\alpha)}{m\psi_n'(\beta)\chi_n(\alpha) - \psi_n(\beta)\chi_n'(\alpha)} \tag{4.1.4}$$

The loci of a_n and b_n can be represented as a circle in the complex domain with center at (0, 0.5) and radius 0.5. In Figs. 4.1 and 4.2 the circles correspond to a_1 and b_1 for $m = 1.29$ (Deirmendjian *et al.*, 1961). The coefficients trace the circles, clockwise with increasing α, the values of which are designated along the circles.

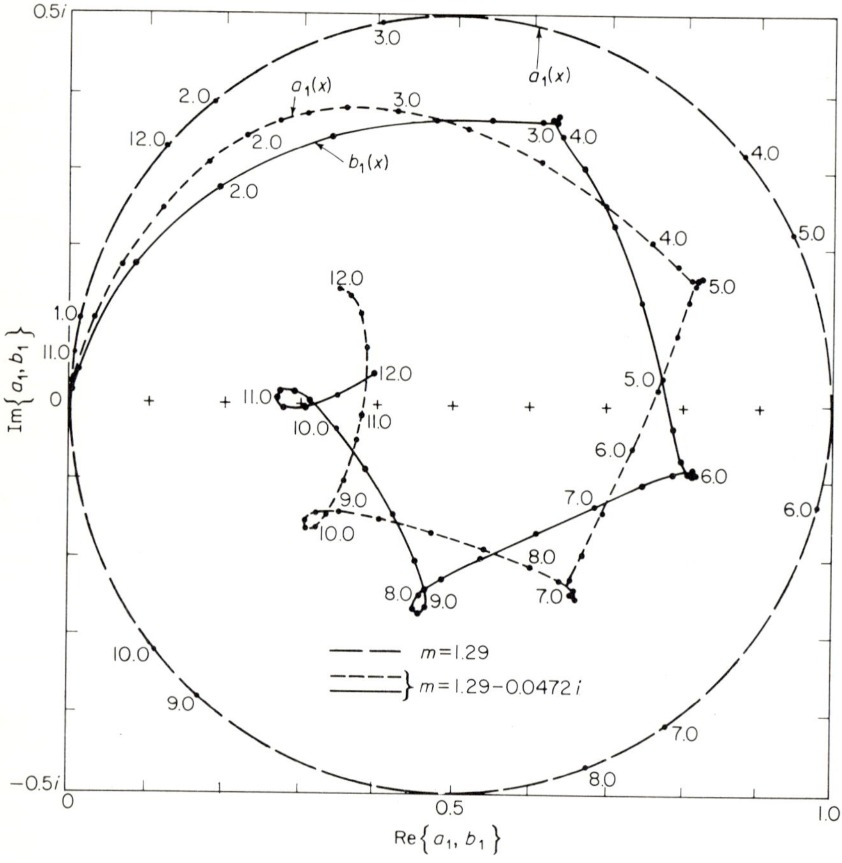

FIG. 4.1. Loci of the scattering coefficients a_1 and b_1 plotted in the complex domain. The circle is for a_1 and real refractive index $m = 1.29$. The numbers along the curves are the corresponding values of α. The noncircular curves are for a_1 and b_1 of absorbing substances with complex refractive index $m = 1.29 - 0.0472i$ (Deirmendjian *et al.*, 1961).

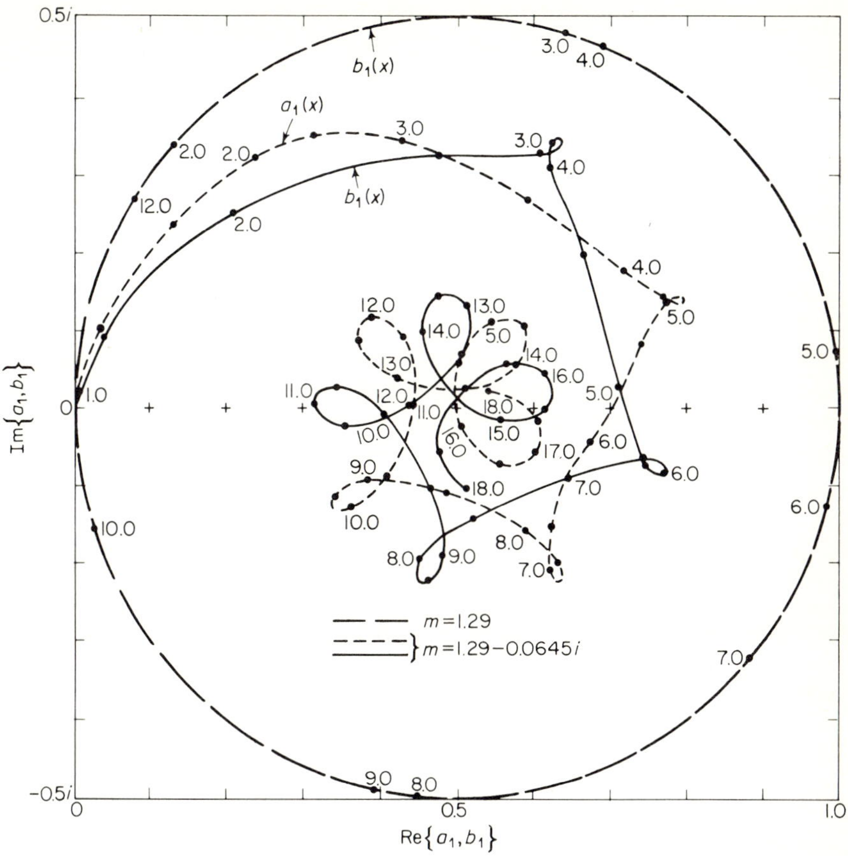

FIG. 4.2. Loci of the scattering coefficients a_1 and b_1 plotted in the complex domain. The circle is for b_1 and real refractive index $m = 1.29$. The numbers along the curves are the corresponding values of α. The noncircular curves are for a_1 and b_1 of absorbing substances with complex refractive index $m = 1.29 - 0.0645i$ (Deirmendjian *et al.*, 1961).

This property of the scattering coefficients can be used to check for possible computational errors as well as to assist in interpolation between available values. By plotting a_n, b_n in the complex domain, a value which is in error may not fit onto the circle in the correct sequence.

An alternative check can be obtained as follows. From (4.1.1)

$$\mathrm{Re}(a_n) = \tfrac{1}{2}(1 - \cos 2\alpha_n) \tag{4.1.5}$$

and

$$|a_n|^2 = \sin^2 \alpha_n \tag{4.1.6}$$

These trigonometric quantities are identical so that

$$\mathrm{Re}(a_n) = |a_n|^2 \tag{4.1.7}$$

for real m where $\mathrm{Re}(a_n)$ designates the real part of a_n. A corresponding expression holds for b_n. This could have been deduced directly from (3.3.81) and (3.3.82) since for dielectric spheres there is no absorption and

$$Q_{\mathrm{ext}} = Q_{\mathrm{sca}} \tag{4.1.8}$$

An alternative method of examining the behavior of the scattering coefficients has been used by Mevel (1958) and explored in greater detail by Metz and Dettmar (1963). The regularities in the variation of a_n and b_n with α are brought out in Figs. 4.3 and 4.4 where $|a_n|^2$ or $\sin^2 \alpha_n$ for $m = 2$ is plotted. In Fig. 4.3 the first six electric coefficients are shown up to $\alpha = 7.6$ and in Fig. 4.4 the first three of these are extended out to $\alpha = 23$. Each scattering coefficient undergoes a characteristic regular oscillation between

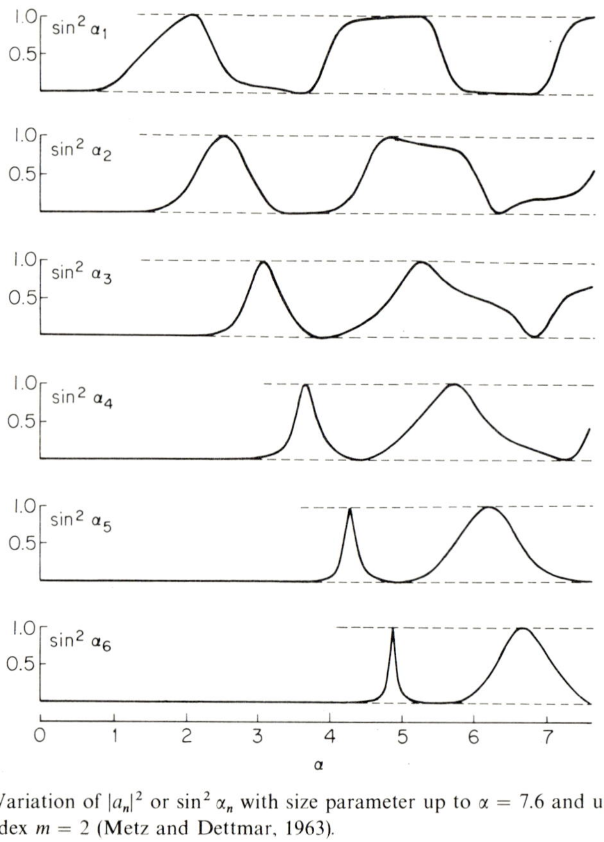

FIG. 4.3. Variation of $|a_n|^2$ or $\sin^2 \alpha_n$ with size parameter up to $\alpha = 7.6$ and up to $n = 6$ for refractive index $m = 2$ (Metz and Dettmar, 1963).

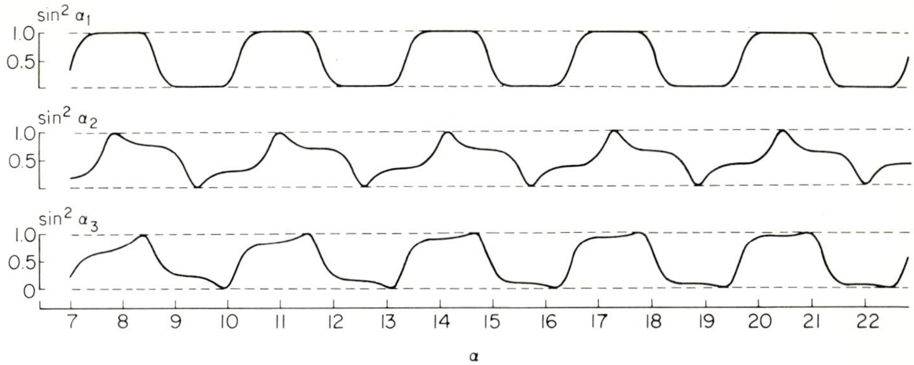

FIG. 4.4. Variation of $|a_n|^2$ or $\sin^2 \alpha_n$ with size parameter from $\alpha = 7$ to 23 and up to $n = 3$ for refractive index $m = 2$ (Metz and Dettmar, 1963).

the values 0 and 1. The first peak in these curves occurs at a value of α which increases almost linearly with the order n and inversely with refractive index m. This is brought out in Fig. 4.5 where the value of $m\alpha$, at which the first maximum of $\sin \alpha_n$ and $\sin \beta_n$ occurs, is plotted against n. The first peak becomes increasingly narrow with increasing order. The remarkable regularity with which successive peaks occur and their characteristic shapes are shown in Fig. 4.4. This regularity reflects the asymptotic approach of the Riccati–Bessel functions to simple sines and cosines when the arguments are large. The magnetic coefficients behave in analogous manner.

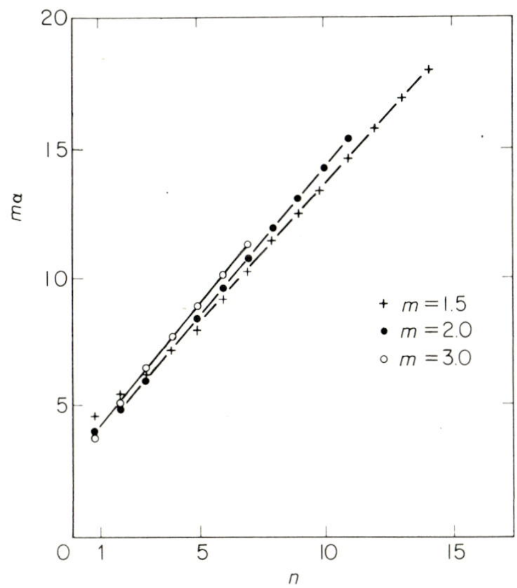

FIG. 4.5. Value of α at which the first resonance peak of $|a_n|^2$ or $\sin^2 \alpha_n$ occurs multiplied by the refractive index plotted against the index n of the partial wave (Metz and Dettmar, 1963).

4.1.2 Complex Refractive Index

When the refractive index of the sphere is complex, the above simple relations are no longer valid. In Figs. 4.1 and 4.2, a_1 and b_1 are also plotted in the complex plane for $m = 1.29 - 0.0472i$ and $m = 1.29 - 0.0645i$. With absorption, the curves spiral inward from the circle which describes the dielectric case, executing a series of small counterclockwise loops which increase in size and tend to converge toward an area near the center of the circle. The loops are amplified and the convergence of the spiral accelerated upon increasing the absorption from 0.0472 to 0.0645. The higher order coefficients behave in a similar manner.

With still higher absorption, the curves become smooth and at sufficiently high α the locus approaches a counterclockwise circle with a radius equal to one-half of the absolute value of the Fresnel reflection coefficient for perpendicular incidence

$$\frac{1}{2}|r| = \frac{1}{2}\left|\frac{m-1}{m+1}\right| \tag{4.1.9}$$

This is illustrated in Fig. 4.6 where the loci of a_1 and b_1 are traced for a highly absorbing material, $m = 3.41 - 1.94i$. In this case, the radius of the limiting circle is 0.32. For $\alpha = 5$, the radius vector of the locus of a_1 has reached the value 0.31 and that for b_1, 0.33. It is interesting that a_1 starts out in a clockwise sense but then executes a loop at about $\alpha = 1.4$ after which it proceeds in a counterclockwise sense just as b_1. We will see shortly that there is a similar pattern for $m = \infty$.

For small dielectric spheres, $a_1 > b_1$. This is apparent from the positions on Figs. 4.1 and 4.2 corresponding to $\alpha = 1$ and agrees with the Rayleigh theory which assumes, in the limit of small spheres, that the scattering can be entirely described by electric dipole radiation. We have already seen that a_1 represents the contribution of the electric dipole to the scattering; b_1 represents that of the magnetic dipole.

As the absorption increases, b_1 becomes comparable in magnitude to a_1 as can be seen from Fig. 4.6. These two coefficients are also of comparable magnitude for perfectly reflecting spheres ($m = \infty$) and as a consequence of this, the scattering pattern of small spheres is considerably changed.

4.1.3 Totally Reflecting Spheres, $m = \infty$

There are a number of features worth noting when the sphere becomes totally reflecting ($m = \infty$). The locus of a_n and b_n in the complex plane is a circle with center at (0, 0.5) and radius 0.5 just as for the nonabsorbing dielectrics. This is shown in Fig. 4.7. The corresponding values of α are designated outside the circle for a_1 and inside the circle for b_1.

As the value of α increases from zero, a_1 traces the circle from the origin in the clockwise sense until $a_1 = 0.5 + 0.5i$ at about $\alpha = 1.4$. At this point, the movement of a_1 with increasing α reverses and subsequently continues in the counterclockwise sense. The angular rotation of a_1 around the circular locus then proceeds at a constant rate of about one revolution per 3.2 units of α.

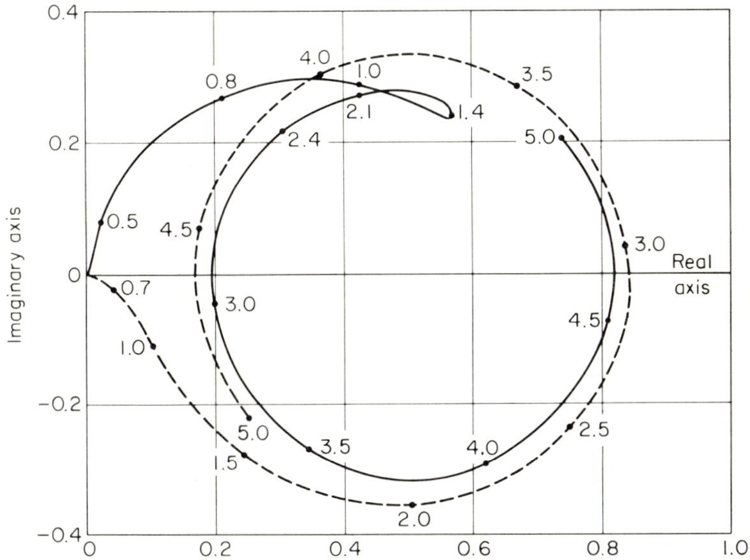

FIG. 4.6. Loci of the scattering coefficients a_1 (smooth line) and b_1 (dashed line) plotted in the complex domain for a highly absorbing substance with complex refractive index $m = 3.41 - 1.94i$. The numbers along the curve are the corresponding values of α.

Higher orders of a_n follow a similar pattern except that the initial velocity of a_n with increasing α proceeds more slowly as the order becomes larger in accord with what we already know about the convergence of the series expansion for the scattering functions. The turn-around point occurs at a value of α approximately equal to n, the order of the scattering coefficient. Furthermore, for sufficiently high orders, this occurs at the constant value $a_n = 0.254 + 0.435i$. Thereafter a_n cycles in the counterclockwise sense at the rate of one cycle per 3.2 units of α. The locus of b_n moves in the counterclockwise sense at the same rate as α increases from zero.

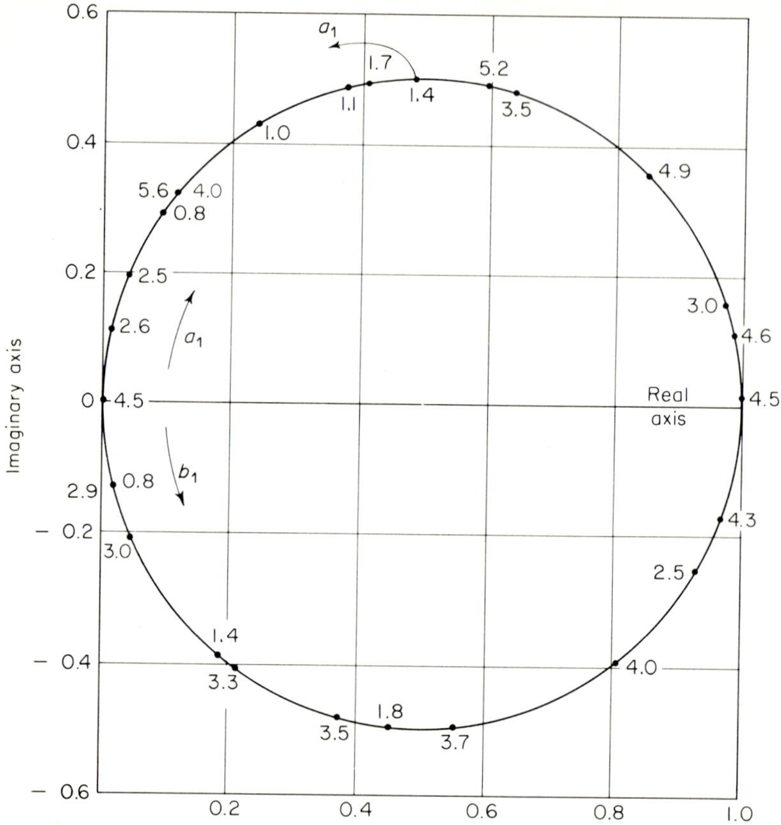

FIG. 4.7. Locus of scattering coefficients a_1 and b_1 plotted in the complex domain for a totally reflecting sphere, $m = \infty$. The numbers outside the circle are the corresponding values of α for a_1 and those on the inside of the circle are for b_1.

4.2 Efficiency Factors

4.2.1 Real Refractive Index

The scattering efficiency of dielectric spheres exhibits the characteristic patterns shown in Fig. 4.8. The parameter ρ against which Q_{sca} is plotted is

$$\rho = 2\alpha(m - 1) \tag{4.2.1}$$

This is the difference between the phase shift which the central ray experiences upon traversing the particle diameter and that obtained in the absence of the particle. Depending upon refractive index, Q_{sca} increases to a maximum

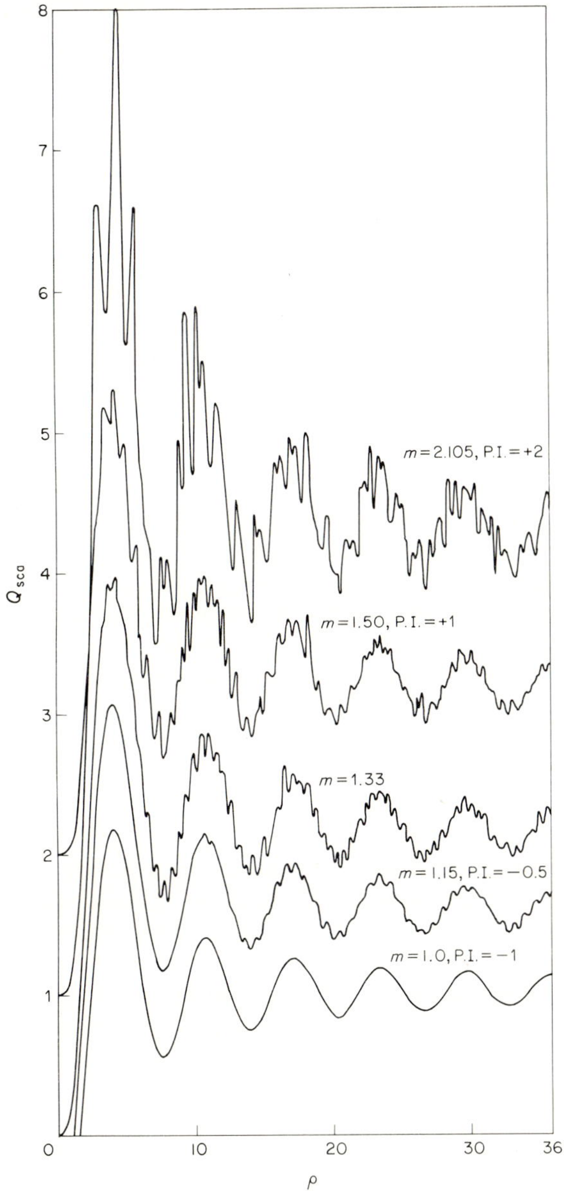

FIG. 4.8. Efficiency for scattering Q_{sca} plotted against parameter $\rho = 2\alpha(m - 1)$ for $m = 1$ (anomalous diffraction) and $m = 1.15$, 1.33, 1.50, and 2.105. P.I. designates the plotting increment.

value at $\rho \simeq 4.41$ of from about 3 to 6, and then undergoes a damped oscillation about the limiting value $Q_{\text{sca}} = 2$.

There are a number of features which are immediately apparent. Superimposed upon the main oscillation is a ripple structure which becomes increasingly irregular at higher refractive indices. The ρ position of the extremes of the main oscillation is approximately independent of refractive index. Also, the oscillation is not symmetrical about $Q_{\text{sca}} = 2$, dipping below 2 to a lesser extent than the maxima rise about this value.

a. Efficiencies Greater than Unity. The fact that Q_{sca} assumes values greater than unity may at first appear paradoxical. The efficiency represents the ratio of the energy scattered by the particle to the total energy physically intercepted by it; i.e., contained in the incident beam having the same cross-sectional area as the particle. Thus, for $m = 2.105$, this ratio is about 6 at the first maximum in the Q_{sca} vs. ρ curve. This means that the particle may perturb the electromagnetic field well beyond its physical confines in order that the field vary continuously from the condition of the undisturbed wave far from the particle to the boundary condition. Then it becomes apparent that, in addition to those rays directly impinging upon the particle, rays may be scattered from the entire region of the disturbance.

As the particle becomes larger, the value of the scattering efficiency undergoes a damped oscillation about 2 and approaches 2 in the limit of large particles. Yet the normal experience with macroscopic bodies is that in the geometrical shadow the light intensity is reduced by no more than the intensity geometrically intercepted, corresponding to a scattering efficiency of unity.

The explanation is the same as for the case of diffraction by an opaque circular disk whose scattering efficiency is also 2. In this case, all of the incident radiation in the beam intercepted by the disk is blocked, and this contributes unity to the extinction. By Babinet's principle, an equal amount of energy is diffracted by the rim of the disk so that the total energy lost to the incident beam is twice the cross-sectional area of the disk. In the case of scattering by a large sphere, an amount of energy equal to its cross-sectional area is lost to the incident beam as a result of the reflection and refraction of those rays physically impinging on the sphere. An equal amount is lost by diffraction around the edge of the sphere.

Practically all the light diffracted by the disk is contained within the angular divergence of the fifth black ring which is given by $\theta = 2.62\lambda/a$ where a is the radius of the disk. Obviously, for disks which are large compared to λ, this angle may be so small that it becomes difficult to distinguish this diffracted light from the incident beam, giving rise to the ordinary experience of the geometrical shadow. In order to separate the light scattered in a narrow forward cone from that directly transmitted, observations would have to be made at great distances from the particle. The same situation occurs in the scattering of light by a large sphere. That part of the scattered light which can be considered to arise from diffraction is constrained to an increasingly narrow forward cone as the particle becomes larger.

Brillouin (1949) has analyzed this problem by integrating the light scattered by a perfectly reflecting sphere into two equal parts: (1) that scattered into a small, solid angle about the forward direction, and (2) that scattered in all other directions. For a sphere with $\alpha = 160$, he found that the forward cone into which half the energy was scattered corresponded very closely to the angular divergence of the diffracted light of a disk of equal ratio of radius to wavelength.

This situation is depicted schematically in Fig. 4.9. If the scattering efficiency is 2, the energy removed from the incident wave would be equal to that contained in an incident beam of circular cross section having a radius $\sqrt{2}a$ where a is the particle radius. The angle 2θ describes the forward cone into which just half the radiation is scattered. A detector at D_1 would see all the radiation in this cone and would indicate an energy loss of πa^2 or $Q_{sca} = 1$. This would correspond to the geometrical shadow for an opaque object. However, if the detector were removed to D_2 where it is sufficiently distant from the scatterer so that it does not intercept a significant amount of forward scattered radiation, the energy loss would be $2\pi a^2$ leading to the true efficiency $Q_{sca} = 2$.

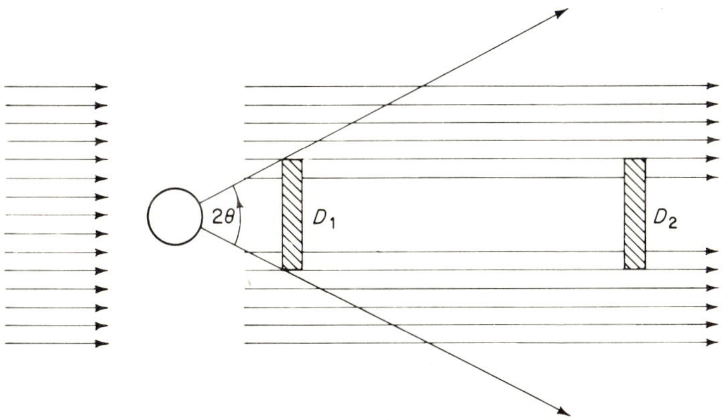

FIG. 4.9. Schematic representation of measurement of scattering efficiency when $Q_{sca} = 2$. Detector D_1 intercepts cone of scattered light (2θ) containing half of scattered energy. This gives measured $Q_{sca} = 1$. Detector D_2 is sufficiently removed so that virtually none of the scattered light is intercepted. This gives measured $Q_{sca} = 2$.

Sinclair (1947) studied this effect experimentally. He measured the transmission of lycopodium spores of radius 15.0 μ. The detector was varied from 6 in to 18 ft from the sample, over which distance Q_{sca} varied from 1 to 2. This poses an experimental problem for the measurement of scattering efficiency by transmission (Gumprecht and Sliepcevich, 1953a, b). It is important either to design the optics so that light scattered in the forward direction is not detected or to account for this light, if it is detected.

b. Anomalous Diffraction. The curve for $m \to 1$ in Fig. 4.8 is based upon the equation derived by van de Hulst (1946, 1957),

$$Q_{sca} = 2 - (4/\rho)\sin\rho + (4/\rho^2)(1 - \cos\rho) \tag{4.2.2}$$

for the limiting case that $(m - 1) \to 0$ and $\alpha \gg 1$. This formula has also been derived somewhat differently by Saxon [cf. Deirmendjian (1957)]. Van de Hulst calls this anomalous diffraction. It differs from Rayleigh–Gans scattering, which will be considered later, in that although $(m - 1) \to 0$, α remains sufficiently large so that ρ is also very large. The geometrical optical rays (Fig. 4.10) incident upon such a sphere pass through without

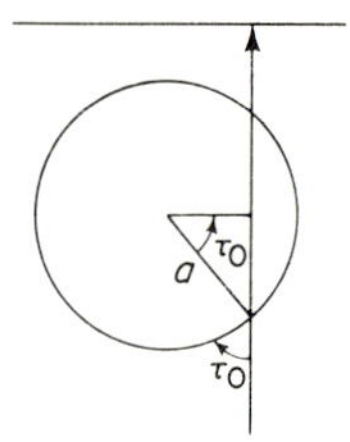

FIG. 4.10. Geometry illustrating the phase shift for an undeviated ray passing through a sphere (anomalous diffraction).

reflection or refraction, but they do undergo a significant phase shift because of the long path length through the large sphere. This leads to an alteration in the phase of the original wave at the plane V beyond the sphere. If the amplitude at V is equal to 1 at all points outside the geometrical shadow circle, it is $\exp(-i\rho \sin\tau_0)$ at a point that is $a \cos\tau_0$ from the center of this circle. This new field is considered to result from the interference between the original field and the forward scattered field $\exp(-i\rho\sin\tau_0) - 1$. An expression for the amplitude of the forward scattered field is obtained by integrating this over the area of the geometrical shadow and then the fundamental extinction formula (3.3.83) can be used to obtain (4.2.2). The derivation applies to complex as well as real values of m provided that the imaginary part of the complex refractive index is also small. However, the above formulation is for real values of the refractive index.

c. Empirical Approximation for Q_{sca}. The values of Q_{sca} at the extrema and the values of ρ at which these extrema occur are given in Table 4.1 for the anomalous diffraction case. Q_{sca} approaches 2 and intervals of ρ between extrema approach 2π as ρ increases. It is interesting that these values of Q_{sca} exhibit the same main features which appear in the general case. Although this curve of Q_{sca} vs. ρ is smooth and does not exhibit the ripple structure which becomes increasingly pronounced with increasing refractive index, the spacing of the extrema is quite similar.

Penndorf (1958) has plotted a large number of values of Q_{sca} vs. ρ for refractive indices varying over the interval $1.33 \leqslant m \leqslant 1.50$. The mean curve through these points and the two envelopes of the band which they describe, one drawn through the lowest values of Q_{sca} at each ρ and the other drawn through the highest values, are given in Fig. 4.11. The ρ-positions of the extrema of these curves correspond with fair accuracy to those for the anomalous diffraction case. Accordingly, ρ provides a generalized parameter for the location of the extrema.

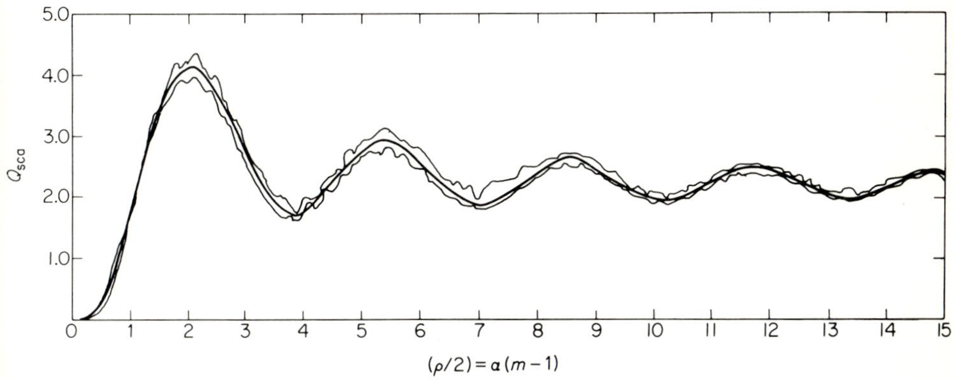

FIG. 4.11. Envelope and average values of Q_{sca} for 1500 computed points in the range $1.33 \leqslant m \leqslant 1.50$ plotted against the parameter $(\rho/2) = \alpha(m - 1)$ (Penndorf, 1958).

TABLE 4.1
EXTREME VALUES OF Q_{sca} FOR ANOMALOUS DIFFRACTION

Maxima	ρ-value	Q_{sca}	Minima	ρ-value	Q_{sca}
1	4.08	3.173	1	7.62	1.542
2	10.79	2.404	2	14.0	1.734
3	17.16	2.246	3	20.32	1.814
4	23.47	2.178	4	26.63	1.855
5	29.78	2.138	5	32.93	1.882
6	36.07	2.114	6	39.22	1.901

A similar study has also been made for lower values of the refractive index, i.e., $m = 1.05(0.05)1.30$ up to values of $\alpha = 25.6$ (Heller and Pangonis, 1957; Heller and McCarty, 1958). Over the more limited range of ρ values, the variation of the scattering efficiency followed the same pattern.

Two other generalized parameters

$$L = \frac{m^2 - 1}{m^2 + 2}\alpha \tag{4.2.3}$$

and

$$k = \tfrac{1}{4}(m^2 - 1)(\alpha - 1) \tag{4.2.4}$$

have been investigated by Mal'tsev (1959, 1960), but they do not appear to offer any advantage over the phase shift parameter, ρ.

Penndorf (1958) has developed empirical expressions both for the values of ρ at which the extrema occur and the values of Q_{sca} at these extrema. The slight dependence of the location of the extrema upon m is given by

$$\rho_y(m) = \rho_y(1) + 0.3(m - 1) \tag{4.2.5}$$

where $\rho_y(m)$ is the value of ρ at which the yth maximum or minimum occurs and $\rho_y(1)$ is the value of ρ at the corresponding extremum obtained from Table 4.1.

For $m \to 1$, the spacing between extrema is close to 2π so that

$$\rho_y(1) \cong 2\pi(y \mp \tfrac{1}{4}) \tag{4.2.6}$$

where the minus sign refers to the maxima and the plus sign to the minima. Substitution of this into (4.2.2) leads to

$$Q_y(1) = 2 \pm \frac{4}{\rho_y(1)} + \frac{4}{[\rho_y(1)]^2} \tag{4.2.7}$$

for the value of the scattering efficiency at the yth extremum.

For $m > 1$, Penndorf has proposed for the maxima

$$Q_y'(m) = 2 + \frac{4}{\rho_y(1)} + \frac{4}{[\rho_y(1)]^2} + \frac{29M}{\rho_y(1)} - \frac{51M}{[\rho_y(1)]^2} \tag{4.2.8}$$

and for the minima

$$Q_y''(m) = 2 - \frac{4}{\rho_y(1)} + \frac{4}{[\rho_y(1)]^2} + \frac{8.01M}{\rho_y(1)} - \frac{27.3M}{[\rho_y(1)]^2} \tag{4.2.9}$$

where

$$M = (m^2 - 1)/(m^2 + 2) \tag{4.2.10}$$

The value of $\rho_y(1)$ appearing in the denominator of the above corresponds to the maxima and minima, respectively. Obviously the above expressions are of the form

$$Q_y(m) = a + bM \tag{4.2.11}$$

The coefficients in these linear equations in M are given for the first 6 maxima and minima in Table 4.2. If a smooth curve is drawn between the values of the extrema, Q_{sca} may be approximated with an accuracy of 3% or better.

TABLE 4.2
NUMERICAL CONSTANTS IN THE FORMULA $Q_y(m) = a + bM$

Order	At the maxima		At the minima	
	a	b	a	b
1	3.173	4.02	1.542	0.579
2	2.404	2.25	1.734	0.433
3	2.247	1.52	1.813	0.328
4	2.178	1.14	1.855	0.262
5	2.139	0.92	1.822	0.218
6	2.114	0.77	1.901	0.188

A three dimensional view of the smoothed values of Q_{sca} is shown in Figs. 4.12 and 4.13. The three coordinates are α, m, and Q_{sca}. The upper limits are $\alpha = 30$ and $m = 1.60$. In Fig. 4.12, the thin grid lines designate constant values of α in steps of $\alpha = 2$. The broken lines in the upper left corner stand for $\alpha = 1$ and $\alpha = 3$. In Fig. 4.13 the grid lines designate values of Q_{sca} in steps of $Q_{sca} = 0.4$. In some regions, there are broken lines indicating intermediate steps of $Q_{sca} = 0.2$. These two figures serve to give a first impression of Q_{sca} over the $m\alpha$-domain.

d. Ripple Structure. The main impediment to estimating Q_{sca} by a smoothing technique such as the one which has just been described is the ripple structure. This is particularly pronounced at larger values of m. The effect can be seen clearly in Fig. 4.14 (Kerker *et al.*, 1961b) where the smoothed curve obtained by the above method is plotted along with the exact curve for $m = 2.00$. Although the general trends of these curves are the same, the ripples are so large compared to the magnitude of the major oscillation about the value 2, that they cannot be neglected without considerable loss of accuracy.

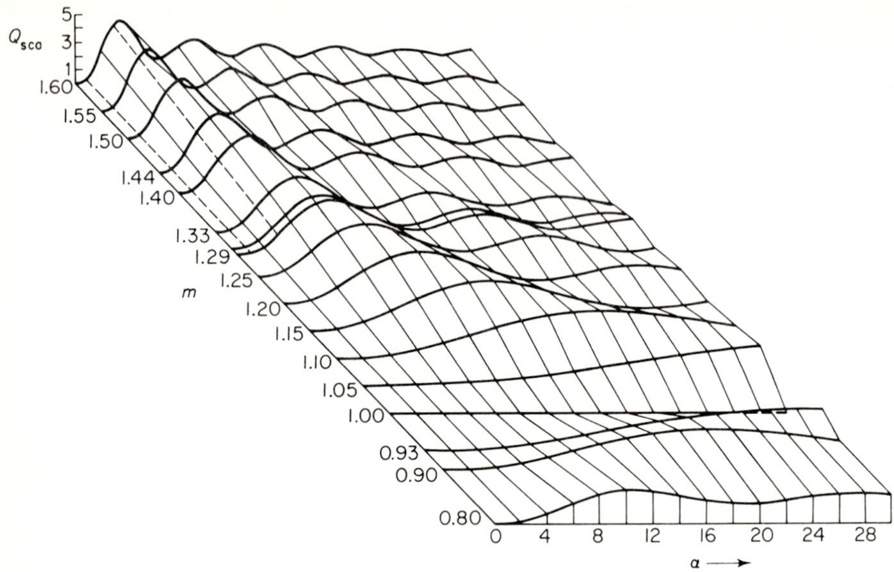

FIG 4.12. Three-dimensional view of smoothed values of Q_{sca} as a function of α and m. The grid lines designate constant values of α.

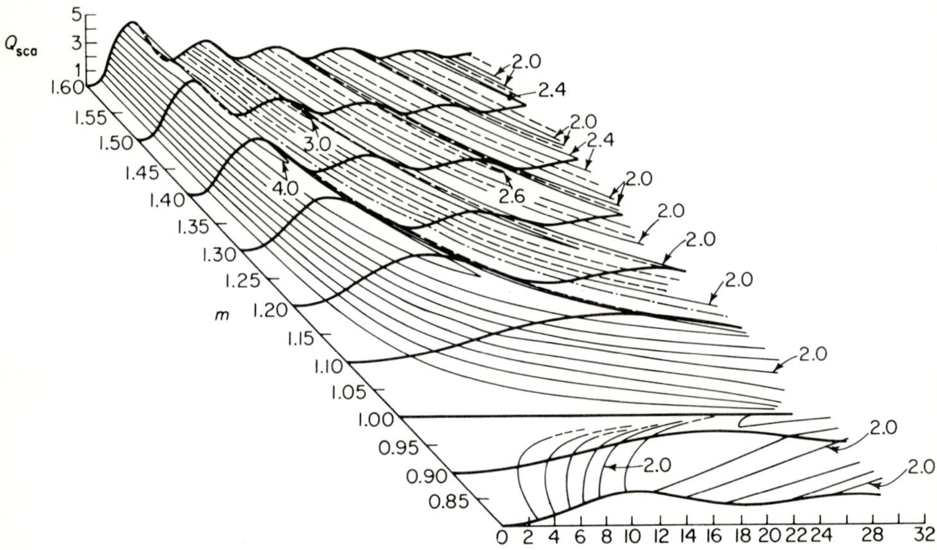

FIG. 4.13. Same as Fig. 4.12 in which the grid lines designate constant values of Q_{sca}.

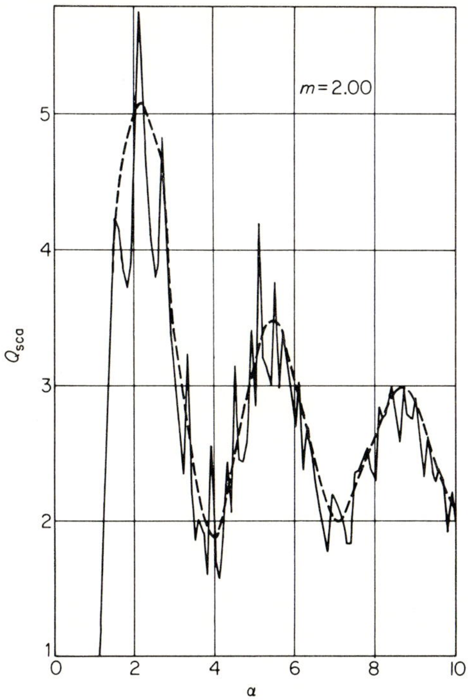

FIG. 4.14. Comparison of Q_{sca} for $m = 2.00$ (full curve) with smoothed curve obtained by Penndorf's method (dashed curve) (Kerker *et al.*, 1961b).

A more detailed picture of the ripples is shown in Fig. 4.15 where the differences between the smoothed values of Q_{sca} and the exact values are plotted vs. ρ for several low refractive indices (Walstra, 1964a). For these refractive indices, the ripple seems to be a fairly regular sine wave, except for some regions where double peaks occur. Irvine (1965) has presented results in even greater detail.

Smart and Vand (1964) have proposed the following empirical formula for estimating Q_{sca} which takes the ripple structure into consideration:

$$Q_{sca} = 2 + (Q_{sca}(1) - 2)f + R \tag{4.2.12}$$

where

$$f = 4m^2/(m + 1)^2 \tag{4.2.13}$$

and

$$R = [2(f - 1)/z][1 - \exp(-z)][1 - ay^3 \exp(-y) - bu^2 \exp(-u^2)] \tag{4.2.14}$$

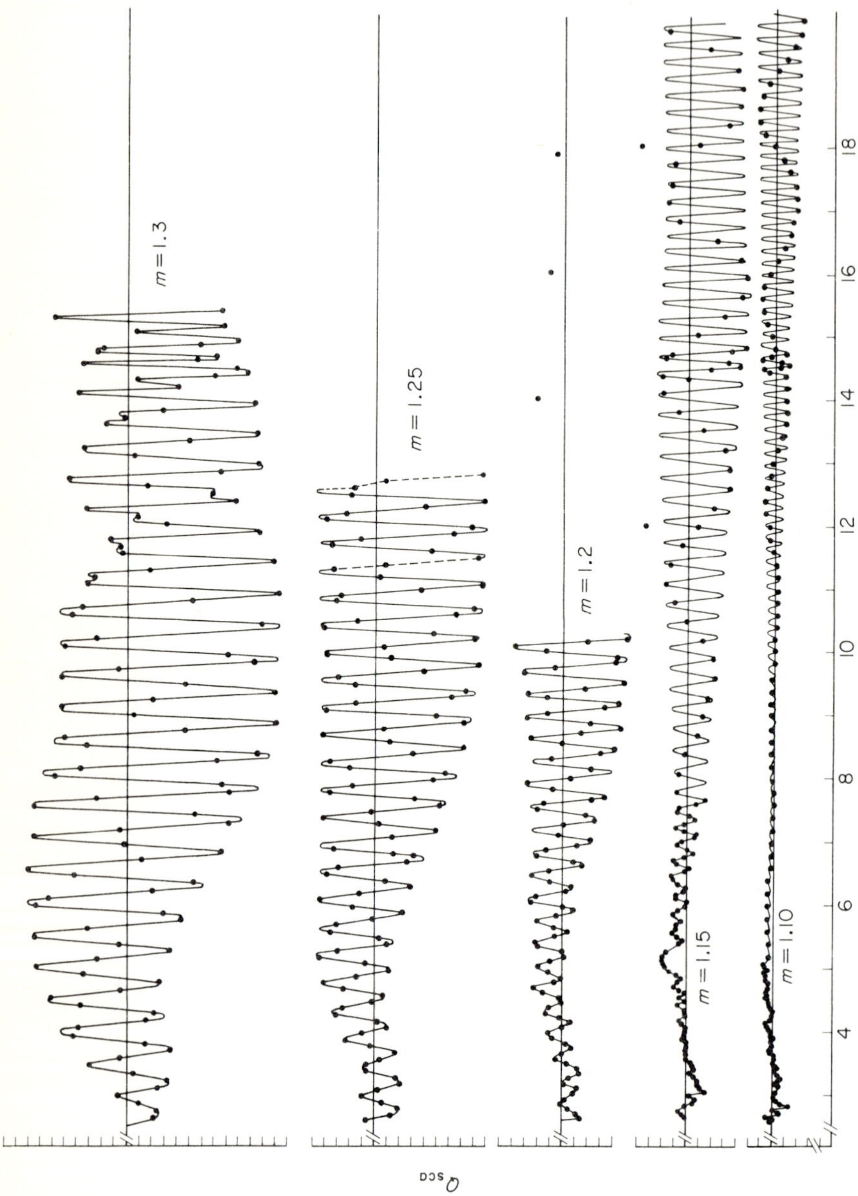

FIG. 4.15. Ripple structure of Q_{sca} plotted as the difference between the exact values and those obtained from the anomalous diffraction formula against ρ for $m = 1.10$ to 1.30 (Walstra, 1964).

Here

$$z = 0.115f\rho, \quad y = 2.38\rho/f, \quad u = [7.82(f-1)/\rho]/2(m-1)f^4,$$
$$a = 0.30[1.75(f-1) + (f-1)^{1/2}],$$

and $b = 0.230f$. This expression leads to values of Q_{sca} that agree with the exact values to within at least 2% over a range of refractive indices from 1.00 to 2.06.

Mevel (1958) and Metz and Dettmar (1963) have related the ripple structure, in a detailed way, to the properties of the individual scattering coefficients, which in turn represent the amplitudes of the electric and magnetic multipoles excited within the particle. The efficiency for scattering for dielectric spheres may be written

$$Q_{\text{sca}} = (2/\alpha^2) \sum_{n=1}^{\infty} (2n+1)(\sin^2 \alpha_n + \sin^2 \beta_n) = \sum_{n=1}^{\infty} q_n^{\ e} + q_n^{\ m} \qquad (4.2.15)$$

where $q_n^{\ e}$ and $q_n^{\ m}$ as defined above represent the contributions by the respective electric and magnetic multipoles. The variation of $q_n^{\ e}$ and $q_n^{\ m}$ is shown in Figs. 4.16 and 4.17 for a refractive index of 2. A number of general features appear. For smaller values of n, the contributions to Q_{sca} overlap. However, with increasing n, these become sharper and distinct. This tendency towards sharpness and distinctness is greater for the magnetic contributions than it is for the electric ones. Also this effect becomes more pronounced at large values of the refractive index and less pronounced at lower values. This is illustrated for a refractive index of 3 in Fig. 4.18. Here there are also contributions from the second peaks in the curves of $\sin^2 \beta_n$ vs. α which are designated β'_1, β'_2, etc.

The effect upon the ripple structure can be seen in Fig. 4.19 where each peak can be identified with the contribution of a particular mode. For a refractive index of 2, the effect of the first three electric modes cannot be resolved because of the overlap in the $q_n^{\ e}$ curves. When $\alpha > 6$, there are contributions from some of the second peaks of the $q_n^{\ m}$ curves. When the refractive index has a value of 3 the structure of both the $q_n^{\ e}$ and $q_m^{\ e}$ curves is so sharp and the peaks so discrete that all orders can be resolved on the curve for the scattering efficiency. Indeed with such a high refractive index, the so-called major oscillations are almost completely obscured by the ripple structure.

With still increasing refractive index, it would be expected that there would be so many discrete contributions that the overlap would eventually lead to a smooth curve. This is indeed the case. Ultimately, for $m = \infty$, the variation of Q_{sca} with α becomes quite smooth. This occurs whether it is the real part or the imaginary part of the refractive index which becomes infinite.

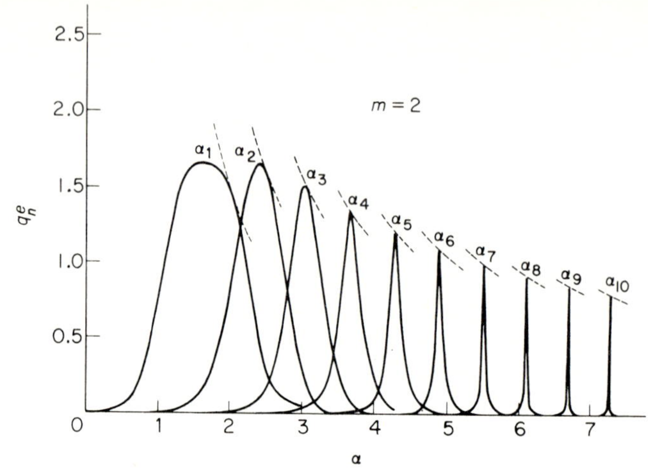

FIG. 4.16. Contribution of the electric multipoles to the scattering efficiency q_n^e for $m = 2.0$ plotted against α (Metz and Dettmar, 1963).

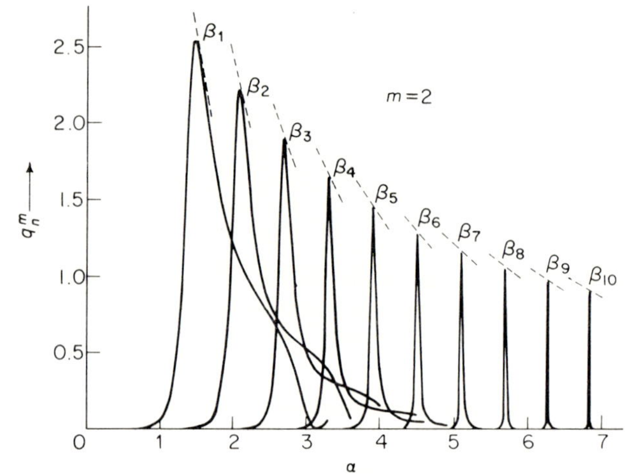

FIG. 4.17. Contribution of the magnetic multipoles to the scattering efficiency q_n^m for $m = 2.0$ plotted against α (Metz and Dettmar, 1963).

4.2.2 Totally Reflecting Spheres, $m = \infty$

Newly computed values of Q_{sca} for $m = \infty$ up to $\alpha = 40.0$ are given in Table 4.3 and are plotted in Fig. 4.20. As α increases from zero, Q_{sca} reaches

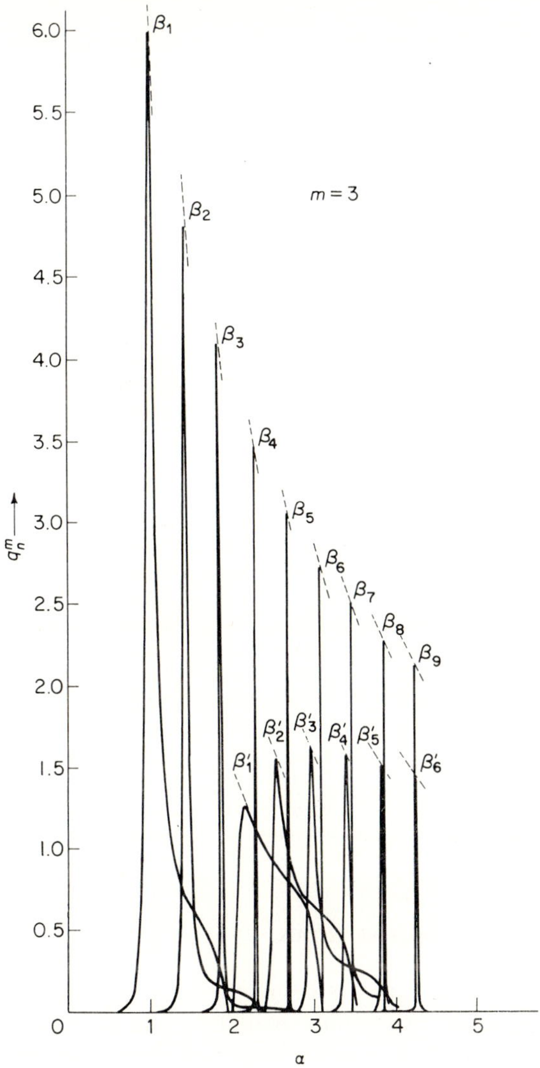

FIG. 4.18. Contribution of the magnetic multipoles to the scattering efficiency q_n^m for $m = 3.0$ plotted against α (Metz and Dettmar, 1963).

a maximum of 2.29 at $\alpha = 1.2$ and then undergoes a highly damped oscillation which approaches the limiting value of 2. Van de Hulst (1957) has proposed the following empirical formula for representing these results:

$$Q_{\text{sca}} = Q_{\text{ext}} = 2 + 0.50\alpha^{-8/9} \tag{4.2.16}$$

TABLE 4.3
EXTINCTION EFFICIENCY FOR TOTALLY REFLECTING SPHERES

α	Q_{ext}	α	Q_{ext}	α	Q_{ext}	α	Q_{ext}
0.2	0.0053	2.8	2.1636	5.8	2.1011	15.0	2.0429
0.3	0.0275	2.9	2.1691	6.0	2.0990	16.0	2.0404
0.4	0.0882	3.0	2.1725	6.2	2.0956	17.0	2.0382
0.5	0.2171	3.1	2.1716	6.4	2.0920	18.0	2.0363
0.6	0.4472	3.2	2.1664	6.6	2.0896	19.0	2.0345
0.7	0.7965	3.3	2.1587	6.8	2.0882	20.0	2.0329
0.8	1.2834	3.4	2.1505	7.0	2.0863	21.0	2.0315
0.9	1.6867	3.5	2.1440	7.2	2.0837	22.0	2.0302
1.0	2.0358	3.6	2.1402	7.4	2.0812	23.0	2.0290
1.1	2.2313	3.7	2.1390	7.6	2.0794	24.0	2.0279
1.2	2.2900	3.8	2.1395	7.8	2.0781	25.0	2.0269
1.3	2.2647	3.9	2.1401	8.0	2.0765	26.0	2.0259
1.4	2.2072	4.0	2.1396	8.2	2.0745	27.0	2.0250
1.5	2.1541	4.1	2.1373	8.4	2.0726	28.0	2.0242
1.6	2.1254	4.2	2.1336	8.6	2.0712	29.0	2.0235
1.7	2.1264	4.3	2.1291	8.8	2.0701	30.0	2.0228
1.8	2.1503	4.4	2.1248	9.0	2.0687	31.0	2.0221
1.9	2.1830	4.5	2.1214	9.2	2.0671	32.0	2.0215
2.0	2.2098	4.6	2.1193	9.4	2.0656	33.0	2.0209
2.1	2.2220	4.7	2.1183	9.6	2.0645	34.0	2.0204
2.2	2.2183	4.8	2.1178	9.8	2.0635	35.0	2.0198
2.3	2.2037	4.9	2.1173	10.0	2.0624	36.0	2.0193
2.4	2.1853	5.0	2.1161	11.0	2.0571	37.0	2.0189
2.5	2.1697	5.2	2.1114	12.0	2.0527	38.0	2.0184
2.6	2.1609	5.4	2.1060	13.0	2.0490	39.0	2.0180
2.7	2.1596	5.6	2.1026	14.0	2.0457	40.0	2.0176

4.2.3 COMPLEX REFRACTIVE INDEX

When the refractive index is complex, there are two distinct processes that contribute to the extinction—scattering and the absorption of energy within the particle as heat. The extinction efficiency may be separated as

$$Q_{ext} = Q_{sca} + Q_{abs} \tag{4.2.17}$$

where Q_{sca} and Q_{ext} are defined by (3.3.81), (3.3.82) and Q_{abs} is determined by difference. The effect of absorption which is related to the imaginary term in the refractive index is shown in Figs. 4.21, 4.22, and 4.23 where the efficiency factors are plotted against α. In the example chosen the real part of the refractive index is the same, or nearly the same ($n \sim 1.29$), while the imaginary part increases from zero to 1.37. The latter case corresponds to metallic iron in blue light ($\lambda = 0.441\ \mu$) and the other refractive indices[1] correspond to water in the infrared part of the spectrum ($1 - 100\mu$).

A number of features are apparent. The ripple decreases with the onset of appreciable absorption and the curve of Q_{ext} vs. α becomes smooth. There is only one principal maximum which shifts to smaller α values with increasing absorption and becomes progressively flatter, particularly at $m = 1.29 - 0.472i$. With still higher absorption ($m = 1.28 - 1.37i$), this maximum sharpens up somewhat. Beyond the first maximum, Q_{ext} drops more or less smoothly to the limiting value of 2.

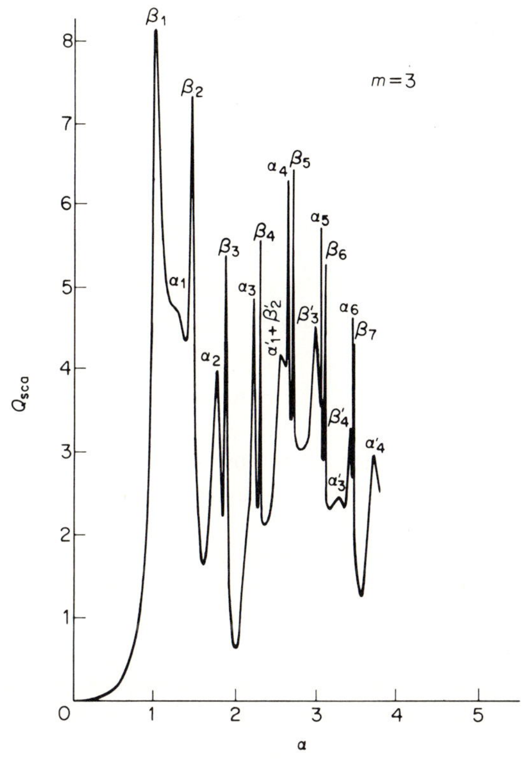

FIG. 4.19. Scattering efficiency for $m = 3$ plotted against α showing the contributions of the various resonance peaks α_1, β_1, etc., to the ripple structure (Metz and Dettmar, 1963).

[1] Herman (1962) has plotted similar curves corresponding to the refractive indices of water for the infrared wavelengths 4 to 10 μ in stages of 1 μ and 12 to 24 μ in stages of 2 μ. More recently Kattawar and Plass (1967) have plotted curves over a wide range of complex refractive indices.

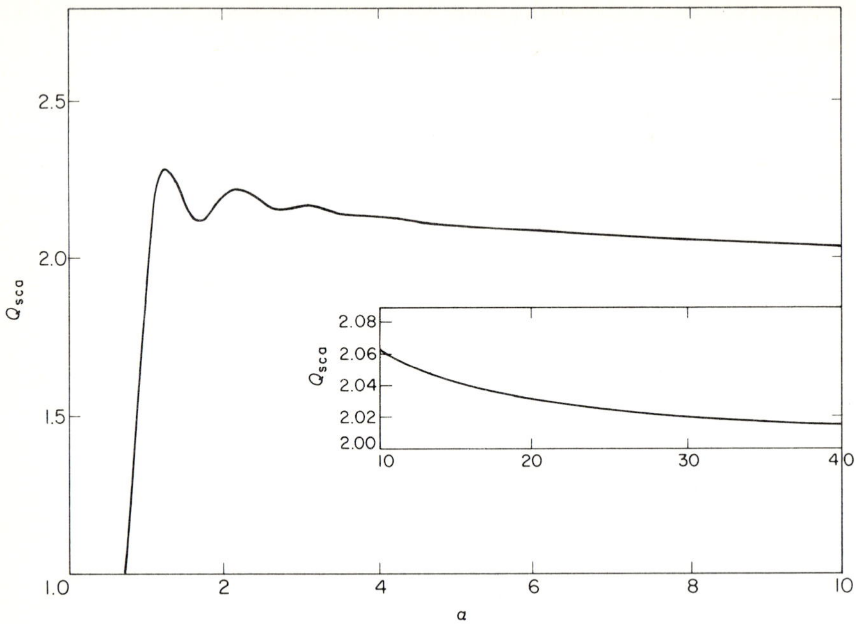

FIG. 4.20. Scattering efficiency for totally reflecting sphere ($m = \infty$) plotted against α.

a. Geometrical Optics Limit. The scattering efficiency and the absorption efficiency each approach limiting values as α increases. Some physical insight can be obtained if we consider the evaluation of these limits from the point of view of geometrical optics. The total extinction is due to the redistribution of the incident rays because of diffraction about the edge of the sphere, specular reflection at the surface, and refraction into the sphere. The refracted rays, which undergo successive internal reflections within the sphere and refractions out of the sphere, are attenuated while passing through the absorptive material. If the sphere is sufficiently large and if the material is sufficiently absorptive, most of the radiation refracted into the sphere will be absorbed and thus will represent the absorption efficiency.

The extinction efficiency for a large sphere is $Q_{ext} = 2$. Just as for a large opaque disk, the energy directly intercepted by the sphere contributes half of this extinction in the form of reflected and refracted rays while the other half is due to diffraction around the edge. Accordingly,

$$Q_{sca} = 1 + w \tag{4.2.18}$$

where the first term is due to the diffracted energy and w is due to the specularly reflected radiation. The absorption efficiency arises from the

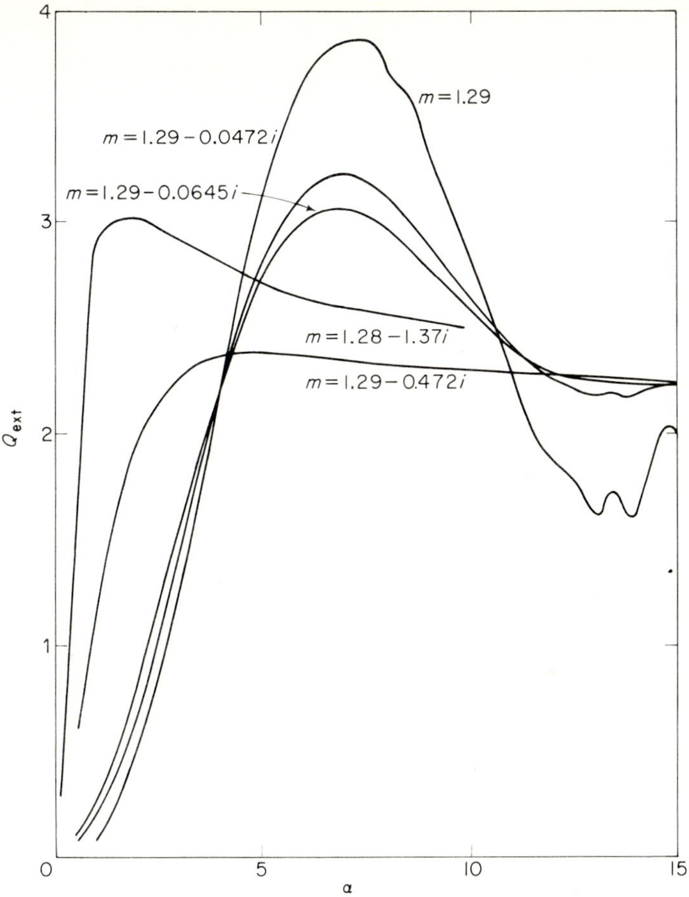

FIG. 4.21. Extinction efficiency for various absorption indices plotted against α.

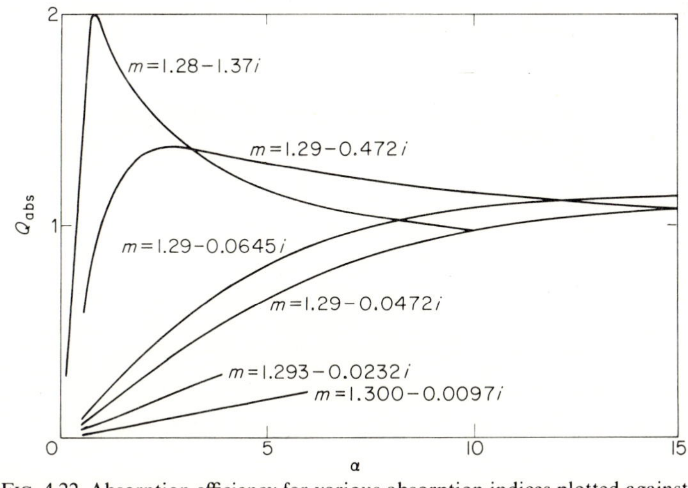

FIG. 4.22. Absorption efficiency for various absorption indices plotted against α.

refracted radiation, which in this case is completely absorbed, so that

$$Q_{abs} = 1 - w \tag{4.2.19}$$

The contribution of the specularly reflected rays to the scattering amplitude is (van de Hulst, 1957)

$$S_1(\theta) = \tfrac{1}{2}i\alpha r_1 \exp(2i\alpha \sin \theta/2) \tag{4.2.20}$$

$$S_2(\theta) = \tfrac{1}{2}i\alpha r_2 \exp(2i\alpha \sin \theta/2) \tag{4.2.21}$$

where r_1 and r_2 are the Fresnel reflection coefficients. The scattered angle, θ, is related to the angle of incidence, θ_i, by

$$\theta = 180° - 2\theta_i \tag{4.2.22}$$

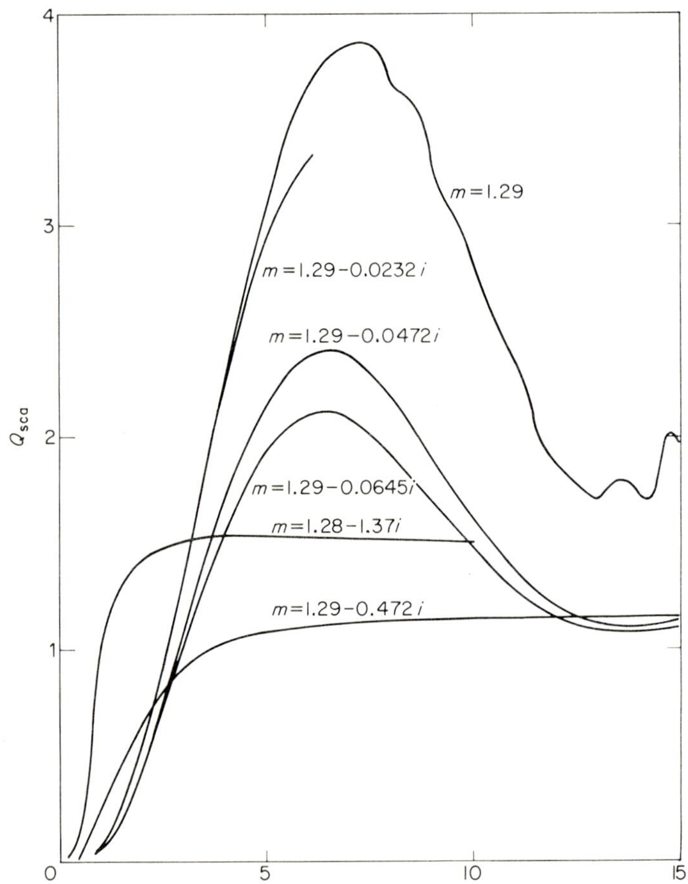

FIG. 4.23. Scattering efficiency for various absorption indices plotted against α.

Integration of the intensity functions over the surface of a sphere leads to

$$Q_{\text{sca}} = 1 + w = 1 + \int_0^{\pi} \tfrac{1}{2}(|r_1|^2 + |r_2|^2)\, d[\cos^2(\theta/2)] \qquad (4.2.23)$$

In Table 4.4 the values of the above integral, w, are listed for a broad range of complex refractive indices. The scattering efficiencies for these "large, absorbing" spheres are obtained by adding unity to these tabulated quantities. Calculations similar to these have been reported by Prishivalko (1963) and by Irvine (1965).

TABLE 4.4
CONTRIBUTION OF SPECULAR REFLECTION TO Q_{sca}

	w						
$n\kappa$ \ $n =$	1.2	1.4	1.6	1.8	2.0	3.0	4.0
0.025	0.0447	0.0770	0.1064	0.1341	0.1607	0.2762	0.3667
0.050	0.0458	0.0775	0.1067	0.1344	0.1608	0.2763	0.3667
0.075	0.0477	0.0784	0.1072	0.1348	0.1612	0.2764	0.3668
0.100	0.0501	0.0795	0.1080	0.1353	0.1616	0.2766	0.3669
0.3	0.0838	0.0992	0.1213	0.1453	0.1694	0.2800	0.3687
0.5	0.1277	0.1313	0.1453	0.1639	0.1844	0.2865	0.3723
0.7	0.1754	0.1704	0.1767	0.1894	0.2053	0.2961	0.3776
1.0	0.2506	0.2357	0.2323	0.2363	0.2451	0.3155	0.3886
2.0	0.4874	0.4556	0.4328	0.4172	0.4074	0.4091	0.4455
3.0	0.6553	0.6229	0.5961	0.5742	0.5567	0.5158	0.5188
4.0	0.7606	0.7331	0.7088	0.6876	0.6692	0.6114	0.5928

b. The Albedo. The albedo for single scattering is the fraction of energy lost from the incident beam due only to scattering. In terms of the efficiency factors this is

$$A = \frac{Q_{\text{sca}}}{Q_{\text{ext}}} = \frac{1}{(1 + Q_{\text{abs}}/Q_{\text{sca}})} \qquad (4.2.24)$$

For dielectrics and perfect reflectors ($m = \infty$), the albedo is unity. In Fig. 4.24, the albedo is plotted vs. α for those complex refractive indices whose efficiency factors are plotted in Figs. 4.21 to 4.23.

This clearly shows the effect of absorptivity. The albedo decreases with increasing absorption coefficient over most of the α range covered. However, at the highest value, $n\kappa = 1.37$, the trend is reversed and the albedo actually increases with increasing values of κ. This reversal is not surprising when one considers that as the imaginary part of the refractive index becomes larger, the specular reflectivity becomes greater. Thus relatively less energy gets into the particle where it will be absorbed. In the limit of infinite absorptivity, the particle becomes a perfect reflector and the albedo becomes unity.

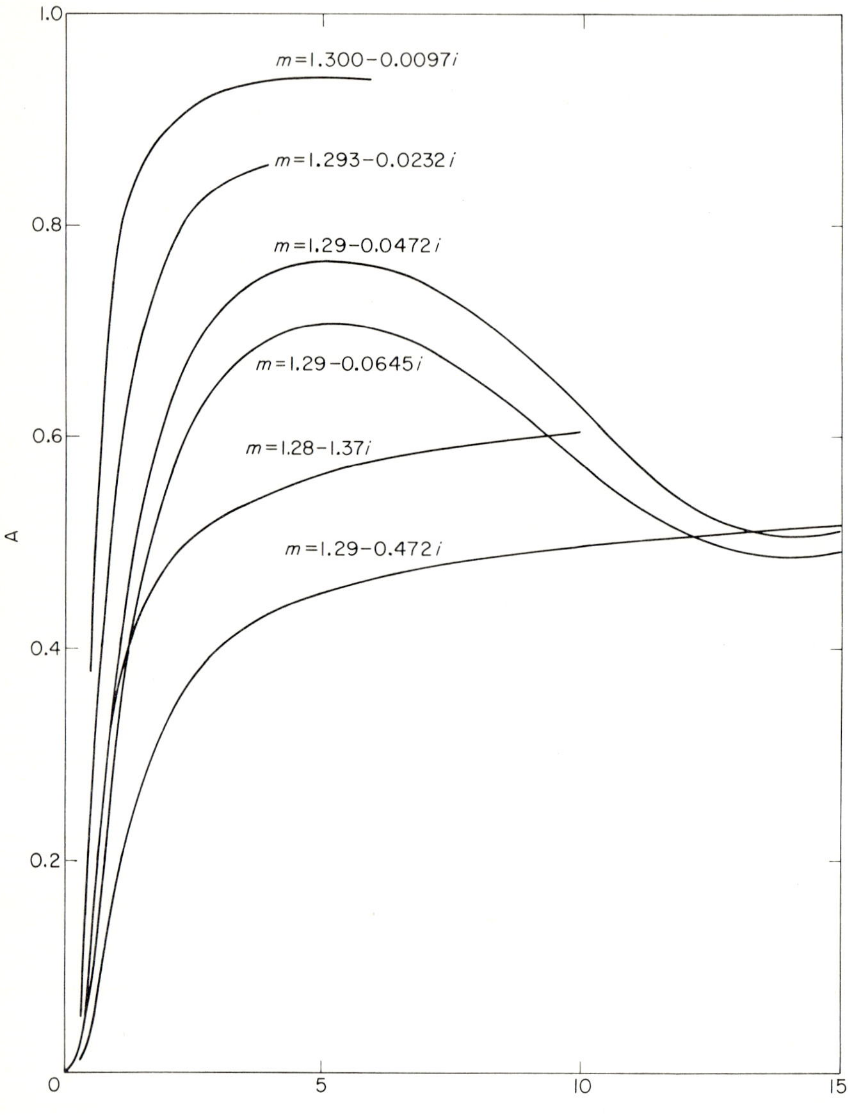

FIG. 4.24. Albedo for various absorption indices plotted against α.

The limiting value of the albedo for large α is

$$A = (1 + w)/2 \tag{4.2.25}$$

which can be calculated directly from Table 4.4. Some results for this case, based upon geometrical optics, are presented in Fig. 4.25 where A is plotted vs. $n\kappa$ for various values of n.

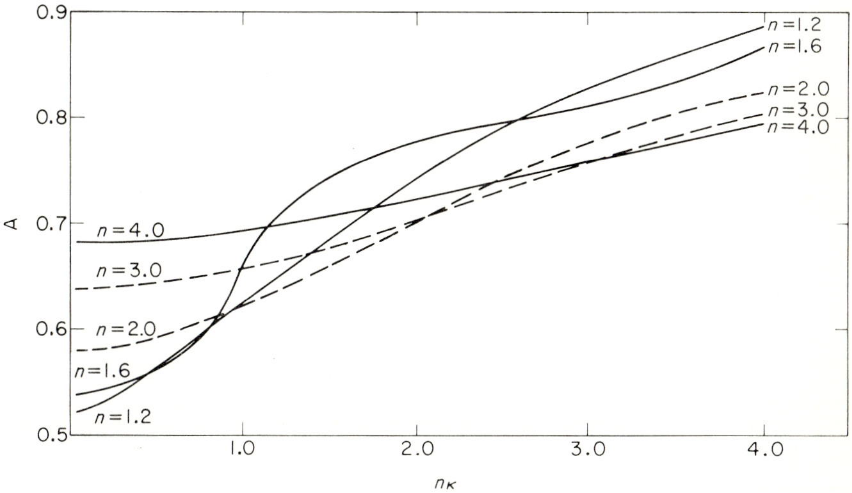

FIG. 4.25. Albedo, calculated according to geometrical optics limit (Table 4.4), for various complex refractive indices plotted against $n\kappa$.

For very small particles, the albedo is close to zero. The precise value in the small particle limit can be obtained from the series approximations for the efficiency factors [(3.9.16) and (3.9.17)] and is given by

$$\frac{1}{A} = 1 - \frac{3}{2\alpha^3} \frac{\operatorname{Im}[(m^2 - 1)/(m^2 + 2)]}{|(m^2 - 1)/(m^2 + 2)|^2} \tag{4.2.26}$$

c. Anomalous Diffraction. The theory of anomalous diffraction developed for large spheres (4.2.2) may also be applied when there is some absorption (van de Hulst, 1957). In addition to the restriction that the real part of m be close to unity, the imaginary part must also be small in order that the incident rays are not deviated to any appreciable extent. Then, the attenuation as well as the phase shift of the undeviated rays may be obtained directly. The extinction efficiency is

$$Q_{\text{ext}} = 2 - 4\exp(-\rho\tan\beta)(\cos\beta/\rho)\sin(\rho - \beta)$$
$$-4\exp(-\rho\tan\beta)[(\cos\beta)^2/\rho]\cos(\rho - 2\beta) + 4(\cos\beta/\rho)^2\cos 2\beta \tag{4.2.27}$$

where now only the real part of the refractive index (n) is contained in the definition of ρ and

$$\tan\beta = n\kappa/(n-1) \tag{4.2.28}$$

The absorption efficiency can be obtained by calculating directly the attenuation of the undeviated rays passing through the sphere and this leads to

$$Q_{\text{abs}} = 1 + \frac{\exp(-2\rho\tan\beta)}{\rho\tan\beta} + \frac{\exp(-2\rho\tan\beta) - 1}{2\rho^2\tan^2\beta} \tag{4.2.29}$$

The difference between these expressions gives Q_{sca}. Although these equations reproduce the general shape of the exact curves of the efficiencies vs. α such as those plotted in Figs. 4.21 to 4.23, there are systematic deviations, especially for Q_{abs} (Deirmendjian *et al.*, 1961). Problems associated with calculating Q_{ext} and Q_{abs} by these expressions from dispersion theory at wavelengths within and near an absorption band have been discussed by Latimer and Bryant (1965). Moore *et al.* (1968) have studied the dependence of the errors in Q_{sca}, Q_{abs}, and Q_{ext} upon both the real part (n up to 1.1) and the imaginary part ($n\kappa = 0.001, 0.01, 0.1$) of the refractive index over a range of α from 0.1 to 100.

Deirmendjian (1960) has developed empirical correction factors for these equations which are valid to within at least $\pm 4\%$ over each of four specified ranges of ρ for

$$1 < n \leqslant 1.50 \tag{4.2.30}$$

$$0 < n\kappa \leqslant 0.25 \tag{4.2.31}$$

The corrected value of the extinction efficiency is

$$Q'_{\text{ext}} = (1 + D_i)Q_{\text{ext}} \tag{4.2.32}$$

where Q_{ext} is calculated from (4.2.27) and the factors, D_i, are

$$D_1 = \frac{n-1}{2n}\cdot\frac{5(n-1)}{4.08}[f(\beta)+1] - \frac{5(n-1)-\rho}{5(n-1)f(\beta)};$$

$$\rho \leqslant 5(n-1) < 4.08/(1 + 3\tan\beta) \tag{4.2.33}$$

$$D_2 = \frac{n-1}{2n}[f(\beta)+1]\frac{\rho}{4.08}; \qquad 5(n-1) \leqslant \rho \leqslant \frac{4.08}{1+3\tan\beta} \tag{4.2.34}$$

$$D_3 = \frac{n-1}{2n}\,\frac{f(\beta)+1}{1+3\tan\beta}; \qquad \frac{4.08}{1+3\tan\beta} \leqslant \rho \leqslant \frac{4.08}{1+\tan\beta} \tag{4.2.35}$$

$$D_4 = \frac{n-1}{2n}\left(\frac{f(\beta)+1}{f(\beta)}\right)\frac{4.08}{\rho}; \qquad \rho > \frac{4.08}{1+\tan\beta} \tag{4.2.36}$$

Here

$$f(\beta) = (1 + \tan\beta)(1 + 3\tan\beta)$$

$$= 1 + 4\frac{n\kappa}{n-1} + 3\left(\frac{n\kappa}{n-1}\right)^2 \qquad (4.2.37)$$

These factors may also be used for the dielectric case, $\beta = 0$, as an approximation to Q_{sca} alternate to that developed by Penndorf (1958) and described above.

Napper (1967) has applied the anomalous diffraction theory to scattering by oriented cubes. When the incident light is normal to a cube face

$$Q_{sca} = 2 - 2\cos\rho_c \qquad (4.2.38)$$

where l is the edge of the cube and

$$\rho_c = (2\pi/\lambda)l(m-1) \qquad (4.2.39)$$

For edge incidence, the acute angle between the incident beam and each face of the cube is $\pi/4$ and

$$Q_{sca} = 2 - [\sqrt{2}\sin(\sqrt{2}\,\rho_c)/\rho_c] \qquad (4.2.40)$$

For corner incidence, the incident beam impinges upon a corner and the direction of propagation lies along the body diagonal of the cube. Then

$$Q_{sca} = 2 - (4/3\rho_c^2)[1 - \cos(\sqrt{3}\,\rho_c)] \qquad (4.2.41)$$

An array of randomly oriented cubes can be approximated by weighting the above expressions 3:6:4, respectively. The result is quite close to the values calculated with the anomalous diffraction theory for a sphere of radius $a = 0.64l$.

4.3 Backscatter

The special case of backscattering will be investigated before considering the angular variation of the intensity of the scattered radiation. The backscatter is observed whenever the radar technique is used. Although this has been mainly restricted to the microwave region of the spectrum, the development of lasers has made backscatter work in the optical region increasingly feasible.

The operation of a microwave radar is illustrated in Fig. 4.26. The transmitter produces the radiant energy; the antenna radiates this energy and also intercepts the backscattered energy; the receiver detects, amplifies, and transforms the received signals into video form; the indicator provides a visual display of the returned signals. The transmitted energy is in the form of pulses of short duration, usually 1 μsec repeated at intervals of 1 msec. The automatic switch closes off the receiver during the short interval when the transmitter is operating. Thus, during every interval of 1 msec, the radar is transmitting for 1 μsec and is receiving for 999 μsec. The distance of the target is determined by the time during the 999 μsec interval at which the backscattered signal is received. The intensity and polarization of this signal offers a clue to the size, shape, and optical properties of the target.

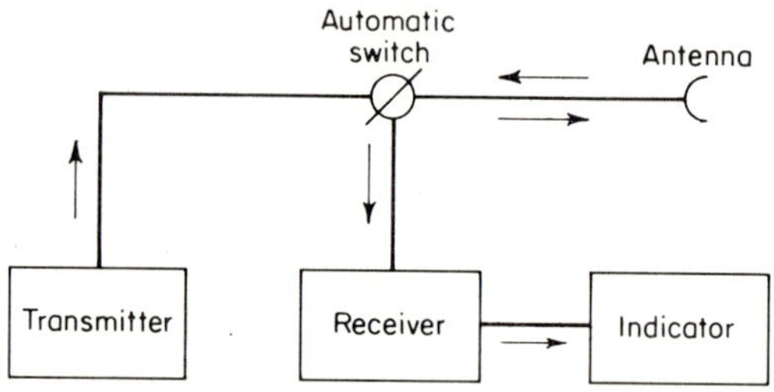

FIG. 4.26. Block diagram of radar set (Battan, 1959).

Radar targets may typically consist of cloud droplets, rain, snow, hail, airplanes, satellites, and even the moon and the planets. The application of radar to the study of atmospheric phenomena is termed "radar meteorology." The interested reader is referred elsewhere for reviews of this very active subject (Marshall *et al.*, 1955; Battan, 1959; Atlas, 1964; URSI, 1964). Here, we will consider only the fundamentals of backscatter from spheres.

The backscatter from spheres is the same for each of the plane polarized components of the incident radiation so that it will not be necessary to designate the polarization of the backscattered intensity, e.g., $I(180°)$. This is the intensity scattered back towards the incident direction when the sphere is illuminated with radiation of unit intensity. The backscatter cross section is

$$\sigma = 4\pi r^2 I(180°) = (\lambda^2/\pi)i(180°) \tag{4.3.1}$$

and the backscatter efficiency or backscatter gain is

$$\begin{aligned} G &= (\sigma/\pi a^2) = (4/\alpha^2)i(180°) \\ &= (4/\alpha^2)\left|\sum_{n=1}^{\infty}(n+\tfrac{1}{2})(-1)^n(b_n - a_n)\right|^2 \end{aligned} \tag{4.3.2}$$

The above expression follows because

$$-\pi_n(180°) = \tau_n(180°) = (-1)^n \tfrac{1}{2} n(n+1) \tag{4.3.3}$$

For a Rayleigh scatterer (dielectric or moderately absorbing small spheres), the limiting value of a_1 given by (3.9.3) leads to

$$G = 4\alpha^4|(m^2 - 1)/(m^2 + 2)|^2 \tag{4.3.4}$$

If the small sphere is a perfect reflector, then the limiting expressions for both a_1 and b_1 when $m = \infty$ must be used, (3.9.22) giving

$$G = 9\alpha^4 \tag{4.3.5}$$

The backscatter efficiency, G, is also called the backscatter gain or the normalized backscatter cross section. It is the ratio of the backscattered intensity to that which would be found if the sphere scattered the entire incident energy isotropically. Obviously the gain of an isotropic scatterer is unity. The backscatter gain of a perfectly focusing device such as a Luneberg lens reflector or of a perfectly reflecting flat plate is

$$G_f = 4\pi(\text{Area})/\lambda_2 = \alpha^2 \quad \text{(for circular cross section)} \tag{4.3.6}$$

4.3.1 Totally Reflecting Spheres, $m = \infty$

The backscattering from totally reflecting spheres will be considered first. For this case, the scattering is quite regular and lends itself to a simple interpretation in terms of ray optics. Rheinstein's very detailed calculations (1963) are plotted in Figs. 4.27 and 4.28. The abscissa is a/λ, corresponding to intervals of α from 0 to 9.4 and from 94 to 119, respectively. The interval between $\alpha = 9.4$ and 94 shows a similar damped oscillation. At the first maximum which is at $\alpha = 1.024$, the backscatter gain G is 3.65. The next two maxima are at $\alpha = 2.331$ and 3.557 and the succeeding ones occur at

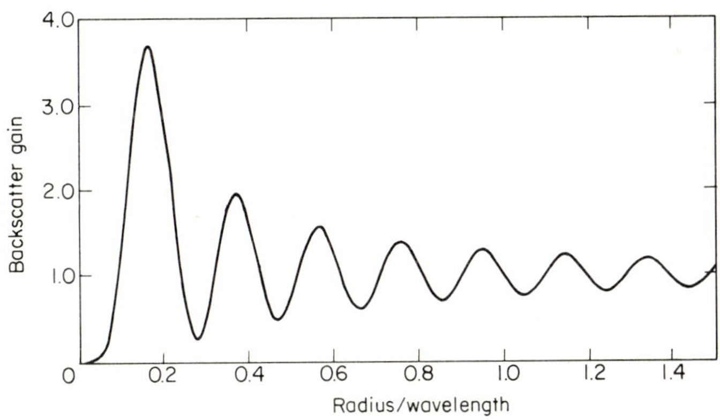

FIG. 4.27. Backscatter gain for a totally reflecting sphere ($m = \infty$) up to $a/\lambda = 1.4$ (Rheinstein, 1963).

regularly spaced intervals of about 1.21(4) in α. The backscatter efficiency rapidly damps to unity. After the first few extrema, the minima and maxima are quite symmetrical about this value, each minimum falling as much below unity as the next maximum rises above it.

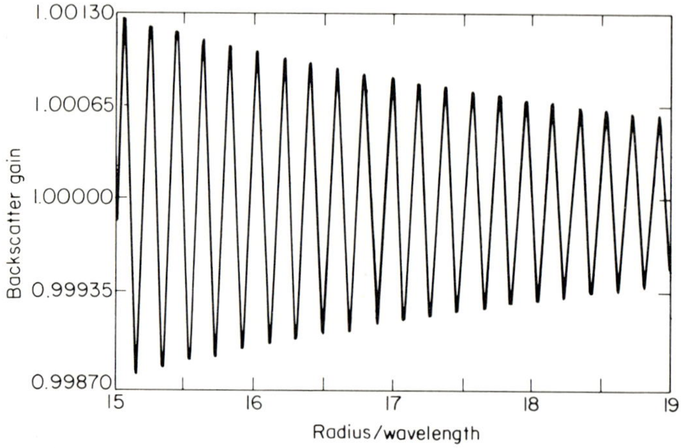

FIG. 4.28. Backscatter gain for a totally reflecting sphere ($m = \infty$) over range $a/\lambda = 15$ to 19 (Rheinstein, 1963).

For values of α greater than 2 these results can be accounted for in terms of geometrical optics. The limiting value of unity for a large totally reflecting sphere follows from a consideration of only the specularly reflected ray. For $m = \infty$ the Fresnel reflection coefficient r_1 is -1 so that Eq. (4.2.20) for the amplitude function of a reflected ray reduces to

$$S_1(\theta) = -\tfrac{1}{2}i\alpha \exp(2i\alpha \sin \theta/2) \tag{4.3.7}$$

The square of the modulus of this, which is $\alpha^2/4$, is proportional to the intensity due to specular reflection. There is no dependence upon θ since a large perfectly reflecting sphere reflects radiation isotropically. There is still the diffracted radiation to be considered but since this is absent in the backscatter, we obtain for the backscatter gain of a large perfectly reflecting sphere

$$G = (4/\alpha^2)|S_1(180°)|^2 = 1 \tag{4.3.8}$$

The oscillatory curve for finite values of α can be attributed to an interference effect between the "head on" ray, which is specularly reflected off the front surface and some other ray whose amplitude and phase vary with α. Let the gain of this latter ray be given by

$$G_s = |L_s e^{-i\delta}|^2 \tag{4.3.9}$$

The total gain is obtained by adding the complex amplitude function of this ray to that of the directly reflected ray (unity) and squaring the modulus. Accordingly δ represents the difference in phase between the two rays.

$$G = |1 + L_s e^{-i\delta}|^2 = 1 + G_s + 2L_s \cos \delta \tag{4.3.10}$$

Franz and Depperman (1952) were the first to attribute the interfering ray to the surface wave or creeping wave which travels along the interface between the sphere and the medium. This wave can be visualized (Fig. 4.29) as arising from the rays which impinge on the sphere at grazing incidence and then travel in the interfacial region. The surface wave is damped as it progresses around the surface of the sphere by continually spraying energy tangentially away from the surface. It is the rays sprayed off in the backward direction that interfere with the specularly reflected ray to give the total gain or backscatter efficiency. Actually the two polarized components must be considered separately. Exact expressions for the surface wave may be developed (Senior and Goodrich, 1964) from the usual series expansion and these are applicable to scattering in any direction as well as to the back-scatter. However, in what follows a somewhat empirical approach will be used.

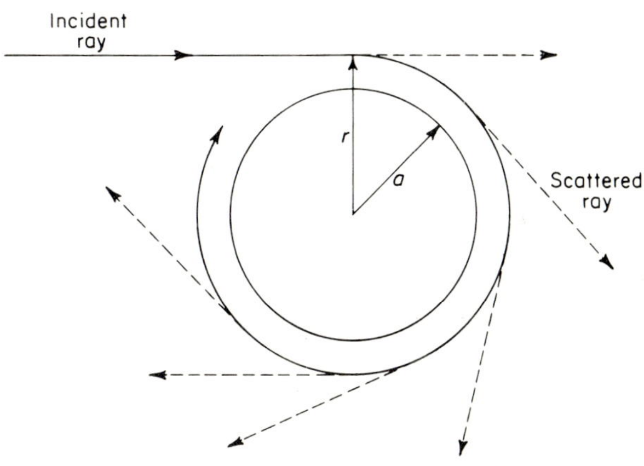

FIG. 4.29. Representation of surface wave which arises from ray incident on sphere of radius a at grazing incidence. Scattered rays are sprayed off tangentially.

The amplitude of the surface wave ejected in the backward direction after traveling an angular distance π from the point of injection may be represented by

$$L_s = B_1 \alpha^{1/3} \exp[-B_2 \alpha^{1/3} \pi] \tag{4.3.11}$$

The phase is obtained directly if it is assumed that the surface wave travels as if away from the surface by a distance proportional to $\alpha^{1/3}$. Alternatively this can be attributed to the wave traveling along the surface with its velocity relative to that in the medium reduced by the factor $1/(1 + A\alpha^{-2/3})$. For the wave sprayed backwards this gives

$$\delta_s = (\pi + 2)\alpha + \pi/2 - A\alpha^{1/3} \tag{4.3.12}$$

where the first term corresponds to the phase shift of a ray along the surface, the second term is due to the phase shift of the specular ray, and the last term corrects for the effective distance off the surface along which the surface ray is presumed to travel. Higher-order effects due to more than one trip around the sphere need not be considered since the damping is such that these are negligible.

Probert-Jones (1963) has fitted the parameters in the above expression to the actual results of the exact theory, obtaining

$$B_1 = 8.21; \qquad B_2 = 2.56/\pi; \qquad A = 0.42 \tag{4.3.13}$$

The backscatter gain from a totally reflecting sphere becomes

$$G = |1 + 8.21\alpha^{1/3} \exp\{-2.56\alpha^{1/3} - i(5.14\alpha - 0.42\alpha^{1/3} + 1.57)\}|^2 \tag{4.3.14}$$

This expression is compared with the exact computations in Fig. 4.30. There is quite good agreement except that the empirical expression has a somewhat longer period over the range plotted.

4.3.2 Real Refractive Index

Dielectric spheres show considerably more complicated variations of the backscatter gain with α than do those consisting of a totally reflecting material. Some results are shown in Figs. 4.31 and 4.32. This contrasts with the totally reflecting sphere whose backscatter efficiency has damped down to within 10% of the limiting value of $G = 1$ when $\alpha > 12$ and to within 1% when $\alpha > 36$. The backscatter is very sensitive to refractive index and at the higher values of α, even the curves for $m = 1.60$, 1.61, and $1.61 - 0.0025i$ are quite different (Atlas *et al.*, 1963).

Initially, the backscatter increases with α, and then oscillates sharply. The mean of this oscillation is higher with increasing refractive index. For $m = 1.60$, it is considerably greater than that for a perfectly reflecting sphere. Obviously, as m becomes increasingly larger, this trend with refractive index will be reversed since the backscatter efficiency must decrease towards the perfectly reflecting case. The maximum appears to occur at about $m = 1.8$.

We have already seen that, as α increases, the total extinction efficiency for all media approaches the geometrical optics limit of $Q_{ext} = 2$. This means that the scattering integrated over all directions must be comparable both for dielectric and for perfectly reflecting spheres. Accordingly, the high values of the backscatter gain for dielectric spheres must be associated with preferential scattering in the backward direction relative to the other directions.

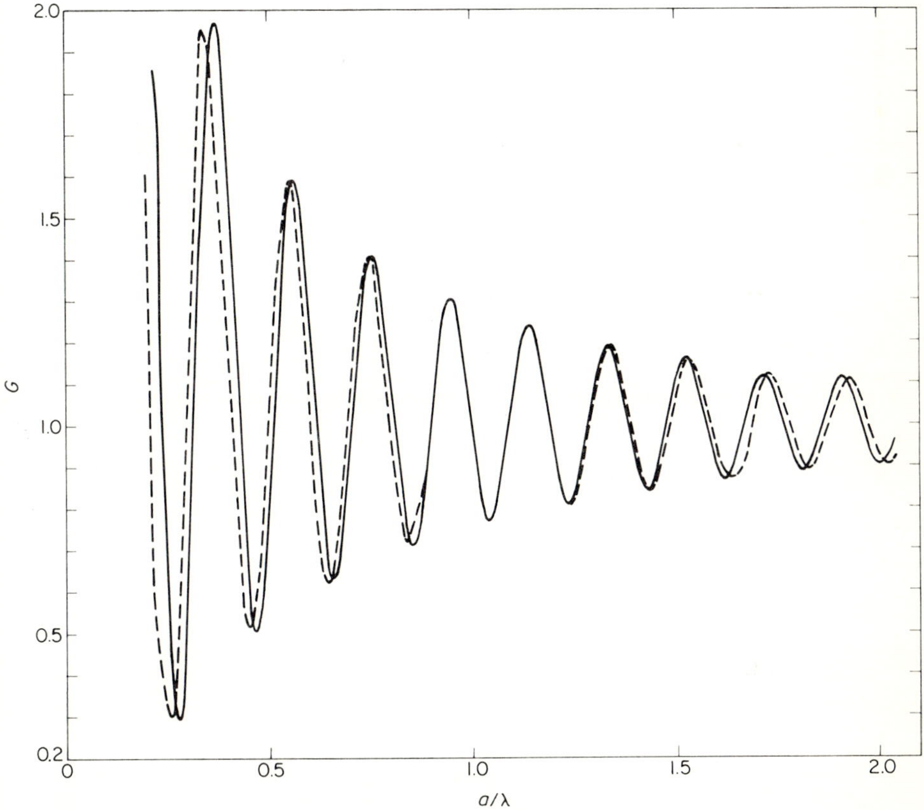

FIG. 4.30. Backscatter gain for totally reflecting sphere (full line) compared with values calculated (dashed line) according to Eq. (4.3.14) based upon interference between axial ray and surface wave. Exact data from Rheinstein (1963), full line; empirical data from Probert-Jones (1963), dashed line.

It is as if the dielectric sphere behaves as a backscatter focusing device. We will investigate how such focusing can occur in terms of ray optics but before doing so will first consider the behavior of absorbing spheres.

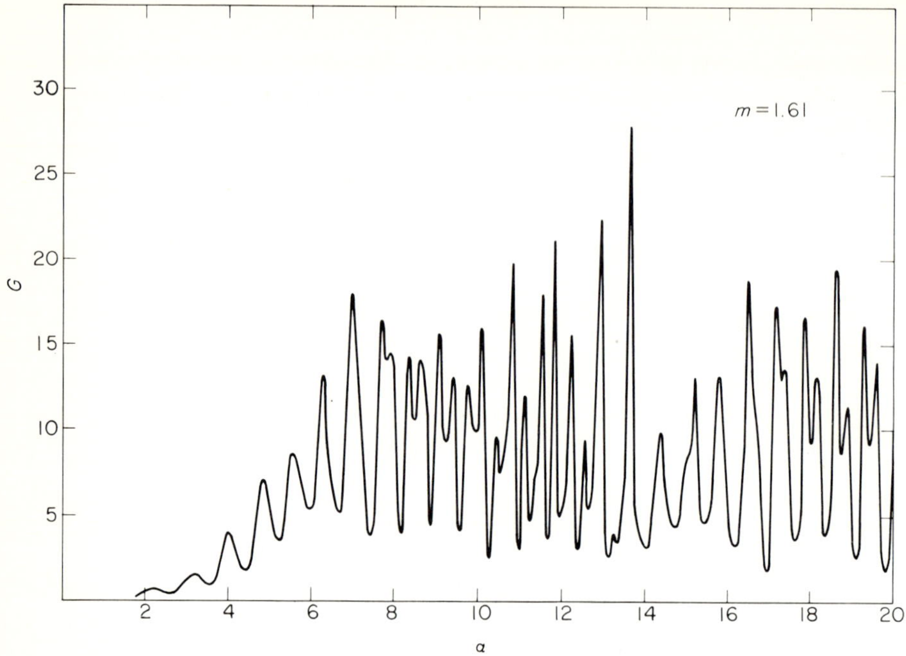

FIG. 4.31. Backscatter gain for dielectric sphere, $m = 1.61$.

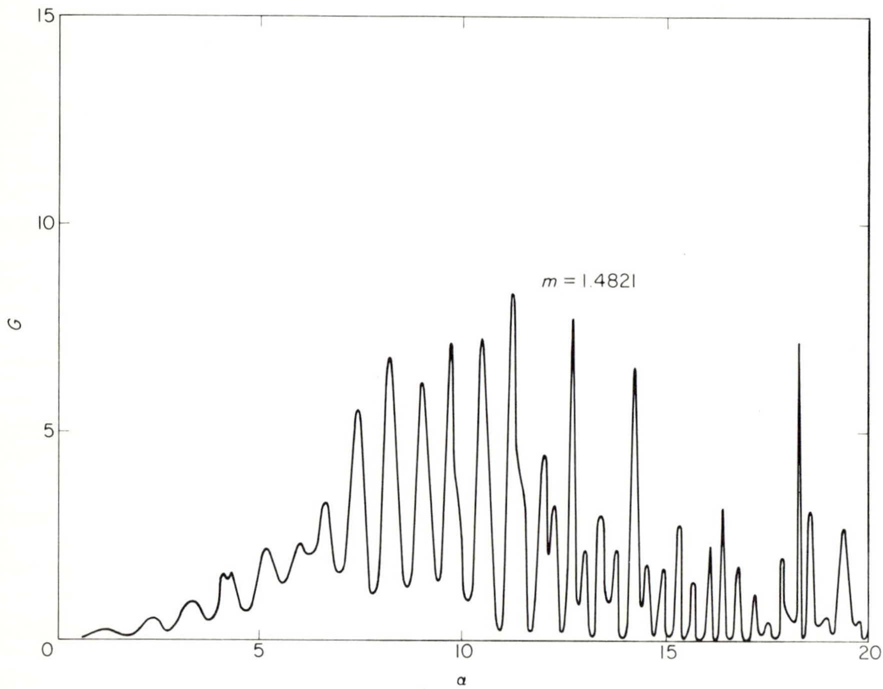

FIG. 4.32. Backscatter gain for dielectric sphere, $m = 1.4821$.

4.3.3 Complex Refractive Index

The backscatter efficiency of a slightly absorbing material eventually decreases with increasing α. Atlas *et al.* (1963) have compared the results for $m = 1.61$ and $m = 1.61 - 0.0025i$. These differ somewhat for values of α greater than about 3, although the backscatter for the slightly absorbing sphere does not become appreciably lower until α is greater than about 30. Actually, this should approach a geometrical optics limit equal to the reflectivity of the material at perpendicular incidence (McDonald, 1962a) which from (2.4.8) is

$$|r_2|^2 = |(m - 1)/(m + 1)|^2 \tag{4.3.15}$$

The limit arises because, for a sufficiently large sphere, all rays passing into the sphere will become completely attenuated by absorption. Accordingly, all of the scattering is due to reflection at the external surface and this is determined by the Fresnel reflection coefficients. Only the ray incident along the sphere diameter will be returned directly backwards, leading to (4.3.15).

For $m = 1.61 - 0.0025i$, which is the refractive index of plexiglas in the microwave region, $|r_2|^2 = 0.0545$. However, even for $\alpha = 90$ and 100, the backscatter efficiency is 1.139 and 0.326, considerably higher than the limiting value.

There has been considerable interest in just how $G(180°)$ approaches the limit for such slightly absorbing media, especially since, in addition to plexiglas, the refractive index of ice in the microwave region is in this range of absorptivity ($m = 1.78 - 0.0024i$). Herman and Battan (1961a, b) have carried out some calculations up to $\alpha = 500$. Still more extensive calculations have been made by Querfeld (1963). He has obtained $G(180°)$ for the following range of α:

0.01(0.01)480.00; 9.000(0.001)12.000; 500.00(0.01)510.00; 520.00(0.01)570.00; 580.00(0.01)600.00; 620.00(0.01)630.00; 640.00(0.01)694.00

The results up to $\alpha = 50$ are comparable to those for $m = 1.61 - 0.0025i$ (Atlas *et al.*, 1963). In the neighborhood of $\alpha = 50$, the mean level of the curve is at about $G(180°) = 34$ with peaks running as high as 46 and minima down to 26. At higher α, the mean values slowly decrease and the curve becomes increasingly regular. In fact beyond $\alpha = 260$, it becomes almost sinusoidal as can be seen from Fig. 4.33 where the intervals $\alpha = 75$ to 100, 175 to 200, and 275 to 300 are shown. However convergence to the plane reflectivity value of 0.079 is slow and even near $\alpha = 690$ the minima and maxima are at 0.04 and 0.125. An interesting feature of this curve is the remarkable constancy of its periodicity which is precisely 0.877α beyond values of $\alpha = 100$.

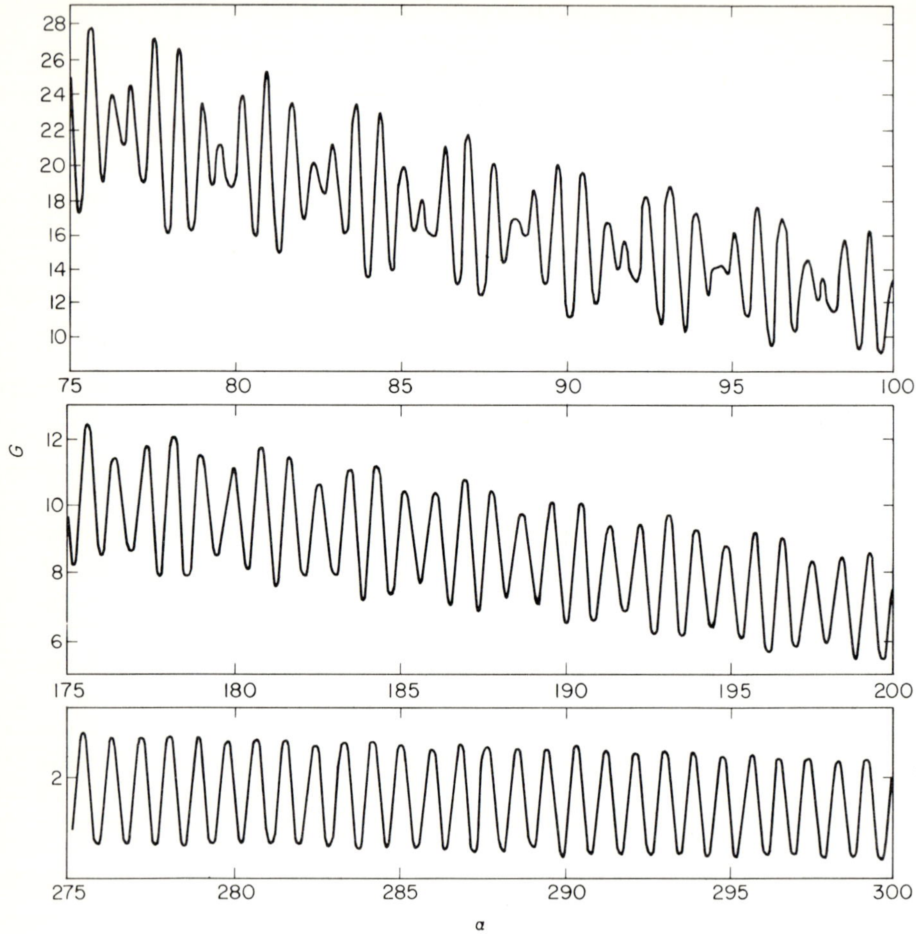

FIG. 4.33. Backscatter gain for slightly absorbing sphere, $m = 1.78 - 0.0024i$ for 3 intervals of α. Upper curve $\alpha = 75$ to 100, middle curve $\alpha = 175$ to 200, lower curve $\alpha = 275$ to 300. Scale of ordinate for the upper curve is half that of the middle and lower curves (Querfeld, 1963).

The effect of increasing the absorption while keeping the real part of the refractive index constant is shown in Fig. 4.34 (Deirmendjian, 1963). As noted earlier, the backscatter gain for the dielectric sphere ($m = 1.29$) oscillates in a complicated fashion. For $m = 1.29 - 0.0645i$ the oscillations are still quite strong, but now they range about the geometrical optics limit given by (4.3.15) which in this case is 0.0168. With still higher absorption, the oscillations are strongly damped so that for $m = 1.29 - 0.4720i$ the backscatter efficiency remains within 5% of the limiting value beyond

$\alpha = 13.0$ and for $m = 1.28 - 1.37i$ the limiting value is attained within this tolerance for $\alpha > 3.0$. The geometrical optics limit is approached at a lower value of α as the absorption increases.

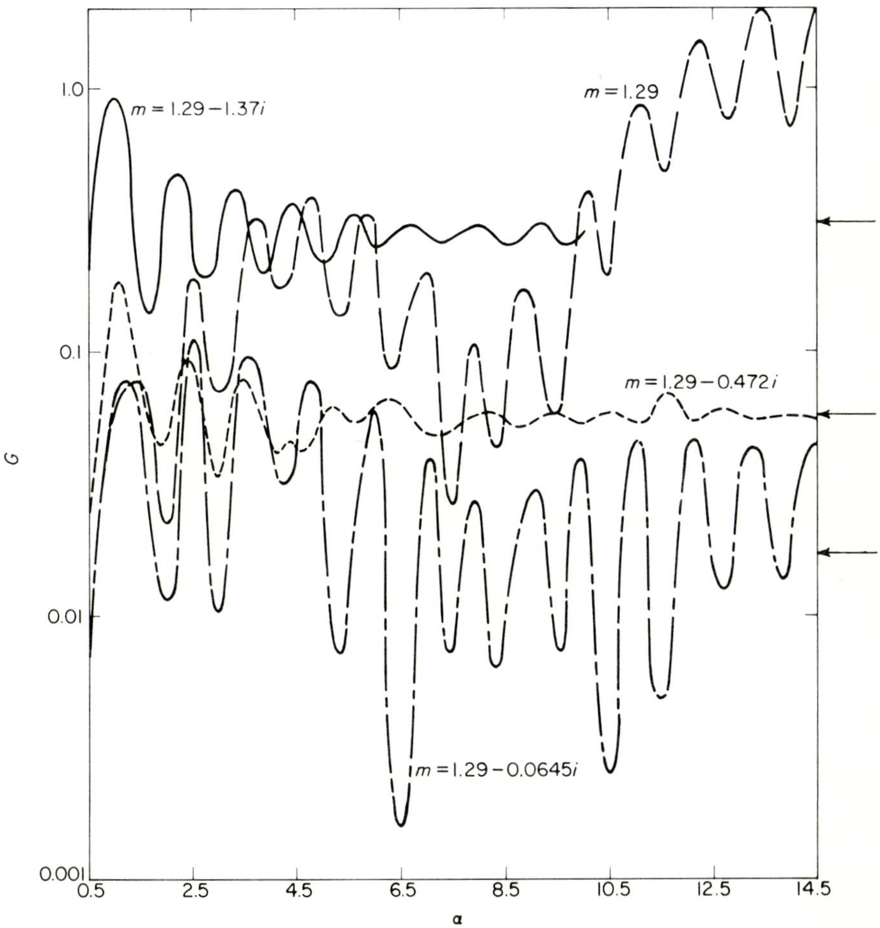

FIG. 4.34. Effect of absorption index on backscatter gain for spheres. The arrows indicate value of the gain obtained from the geometrical optics limit expression (4.3.15) for each of the absorbing cases.

Figure 4.35 (Herman and Battan, 1961a) shows the effect of increasing the real part of the refractive index. Here the imaginary part of the refractive index is 0.048 in all cases, while the real part takes on the values $n = 2.00$, 3.00, 4.00, 5.00, and 8.99. With increasing n, the backscatter efficiency assumes a more regular oscillation about the geometrical optics limit at

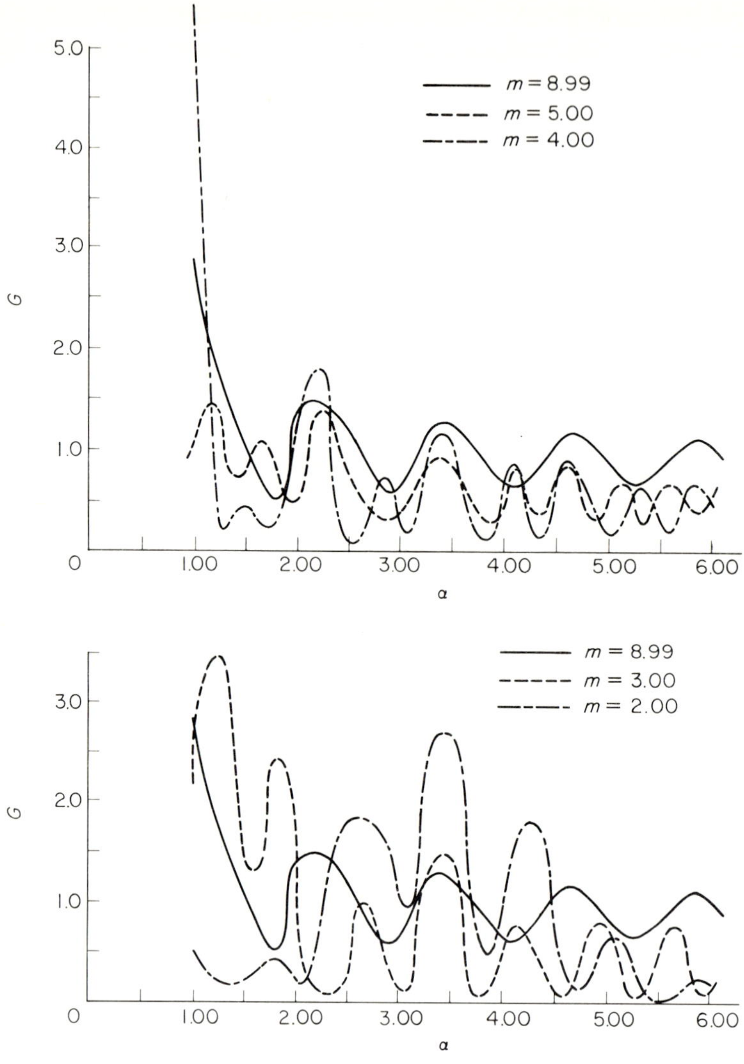

FIG. 4.35. Backscatter gain for spheres for which the imaginary part of the refractive index is 0.048 and for which the real part is 2.00, 3.00, 4.00, 5.00, and 8.99 (Herman and Battan, 1961a).

correspondingly lower values of α. In terms of ray optics, this is due to the fact that at higher n, relatively more energy is reflected at the external surface and less finds its way into the sphere by refraction. For this reason, the sphere does not have to grow to as large a size before the internal rays are sufficiently attenuated so that the scattering can be attributed solely to external reflection. The curve for $m = 8.99 - 0.048i$ is already quite similar to the corresponding one for totally reflecting spheres; indeed the period of the oscillation and the positions of the extrema are almost identical.

4.3.4 Ray Optics

We have already seen how the backscatter gain of a totally reflecting sphere can be attributed to interference between a specularly reflected ray and a ray which travels around the sphere as a surface wave. The possibility of accounting for effects encountered in dielectric and partially absorbing spheres in terms of similar constructions will now be considered. It will be necessary to include those rays which are refracted into the sphere from which they may either emerge by a second refraction or undergo internal reflection. This approach can be applied to scattering at any angle of observation and to a variety of shapes (Kouyoumjian *et al.*, 1963; Kawano and Peters, 1963) but will be specialized here to the case of backscattering by spheres. A principal aim of an analysis along these lines is to provide physical insight into the scattering process in terms of geometrical optics even in the range of small α, e.g. $\alpha \cong 5$, where geometrical optics might not be suspected of being useful. In addition, there is always the goal of seeking out simpler equations which will provide for easier computations than the full series formulation.

The energy incident upon the sphere is split into parts which are reflected, refracted, and diffracted. In addition, those rays which impinge at grazing incidence may travel along the interface between the two media as surface waves. The diffracted radiation need not be discussed here, since it will make no contribution to the backscatter.

The various rays are depicted in Fig. 4.36 where they are denoted as the incident ray, refracted ray, internally reflected ray, and emergent ray. The incident ray divides into a reflected and refracted ray; the latter in turn divides into an internally reflected and an emergent ray and for the internally reflected ray the process continues. The directions of each of these rays are determined by the law of reflection and Snell's law. For each encounter at the boundary, the two components polarized parallel and perpendicular to the surface must be considered separately.

The amplitude function which describes a ray is altered upon reflection or transmission at a surface by a factor given by the appropriate Fresnel coefficient [(2.4.3) to (2.4.6)]. In general, this quantity is complex so that both amplitude and phase will change. The effect of these Fresnel coefficients upon the emergent ray is given by a Fresnel factor, ε, defined by

$$\varepsilon = t(i)r(a)\ldots r(p)t(e) \tag{4.3.16}$$

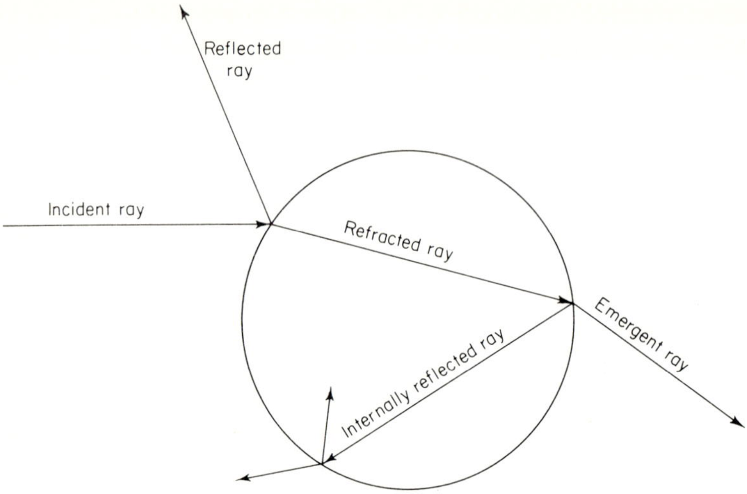

FIG. 4.36. Sphere geometry showing geometrical optics rays.

where $t(i)$ and $t(e)$ are the transmission coefficients of the incident and emergent rays and $r(a)$, etc., are the reflection coefficients of the internal rays. For a sphere, the transmission coefficients of the incident and emergent rays are equal to each other and are related to the reflection coefficient by

$$t = (1 - r^2)^{1/2} \tag{4.3.17}$$

so that

$$\varepsilon = (1 - r^2)(-r)^{p-1}, \qquad \text{for} \qquad p = 1, 2, 3 \tag{4.3.18}$$

$$\varepsilon = r, \qquad \text{for} \qquad p = 0 \tag{4.3.19}$$

where $(p - 1)$ is the number of internal reflections. For the externally reflected rays, $p = 0$, and for the rays refracted directly through the sphere, $p = 1$. The reflection coefficient which appears in this equation corresponds to that which is incident upon the sphere. For the internal ray, the reflection coefficient is the negative of the value for the incident ray. This follows, as an examination of (2.4.5) and (2.4.6) will verify, because the values of the angles of incidence and refraction for these two rays are interchanged.

In addition, whenever a tube of rays is incident upon a curved surface, the intensity associated with these will be altered by the spreading of the rays upon reflection or refraction. For a sphere, the effect upon the intensity of the emergent rays is

$$D = \frac{\sin \theta_i \cos \theta_i}{\sin \theta[2 - 2p(\tan \theta_i/\tan \theta_t)]} \tag{4.3.20}$$

where D is called the divergence. The angles θ, θ_i, θ_t are the scattering angle, angle of incidence, and angle of refraction.

The attenuation and phase shift which the rays experience upon propagation through the sphere are given by the usual exponential factor

$$\exp[-i(2pk_1 a \cos\theta_t)] = \exp[-i(2pm_1\alpha\cos\theta_t)] \tag{4.3.21}$$

where $2pa\cos\theta_t$ is the path length through the sphere; k_1 and m_1 are the propagation constant and refractive index of the sphere, respectively. An additional phase shift of $\pi/2$ must be added whenever the ray crosses a caustic (Keller and Kay, 1954).

The object is to determine the amplitude and phase of the field associated with each of the backscatter rays and to obtain the geometrical optics backscatter by adding these vectorially. The phase of each component must, of course, be related to a common equiphase point. The total backscattering efficiency can accordingly be expressed as

$$G = |\sum L_i \exp(-i\delta_i)|^2 \tag{4.3.22}$$

where each term gives the complex amplitude function from which a backscatter efficiency for each ray component could be calculated.

We will now consider each ray which contributes to the backscatter. The most obvious ones are the axial ray reflected from the front pole of the sphere and the axial ray which penetrates the sphere, is reflected from the rear pole, and finally emerges along the backward direction as shown in Fig. 4.37. Contributions from higher order axial reflections are sufficiently small so that they may be neglected. The front pole is chosen as the reference point for the phase so that the complex amplitude function for the front surface axial ray is simply

$$L_{fa}\exp(-i\delta_{fa}) = r_1 = -r_2 = (m-1)/(m+1) \tag{4.3.23}$$

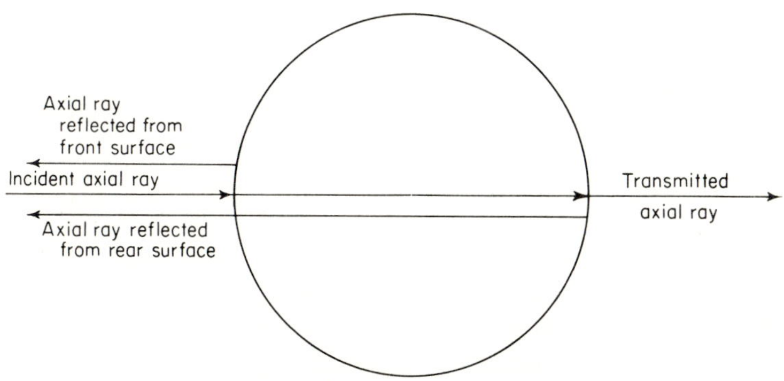

FIG. 4.37. Sphere geometry showing axial rays.

where r_1 and r_2 are the Fresnel reflection coefficients for normal incidence. The only phase shift here is that which may be introduced by the Fresnel reflection coefficient.

The amplitude function of the rear surface axial ray is obtained by considering the three reflection-refraction encounters at the interfaces as well as the propagation back and forth along the axis of the sphere. The Fresnel factor, ε, enters as the product of the transmission coefficient at the front surface on the way in, the reflection coefficient at the rear surface, and the transmission coefficient of the ray emerging from the front surface. The amplitude after each of these encounters must be corrected for the divergence of the ray bundle as the rays pass through or are reflected at each curved surface. For absorbing materials, there will be attenuation in the usual way as the ray propagates. In addition to any phase shifts accompanying the boundary encounters, there will be phase retardation with respect to the front surface axial ray by the optical distance traveled along the axis. Also an additional phase shift of π radians must be subtracted in order to account for the two caustics which are crossed along the axis. The final complex amplitude may be written as

$$L_{ra}\exp(-i\delta_{ra}) = [4m(1-m)/(m+1)^3][n/(2-n)] \times \exp[-4\alpha n\kappa - i(4m\alpha - \pi)] \qquad (4.3.24)$$

The first factor represents the Fresnel coefficients, the second the divergence, and finally the exponential includes both the damping due to absorption and the phase retardation due to the optical path length and the crossing of the caustics. For absorbing materials for which m is complex, the Fresnel factor is complex, and may thus make a contribution to the phase δ_{ra}. For $m = 2$, the divergence becomes infinite and must be calculated by a method to be described below in connection with the glory ray. Indeed at this point, the glory and axial rays are identical.

Thomas (1962) has found for refractive indices smaller than unity and over a limited range of α that the two axial rays can account for the backscatter efficiency quite well. In this case

$$G = |L_{fa}\exp(-i\delta_{fa}) + L_{ra}\exp(-i\delta_{ra})|^2 \qquad (4.3.25)$$

Thomas' results for $m^2 = 0.25$, 0.50, and 0.75 for α up to 12.5 are compared with the exact calculations in Fig. 4.38. The agreement is very good except at the lowest values of α. It is not obvious whether the results would continue to agree at still higher values. What is rather surprising is that such a simple model based upon the geometrical optics of two rays should describe the backscatter of such small spheres. The refractive index of 0.75 may correspond to gas bubbles in water or to certain plasmas.

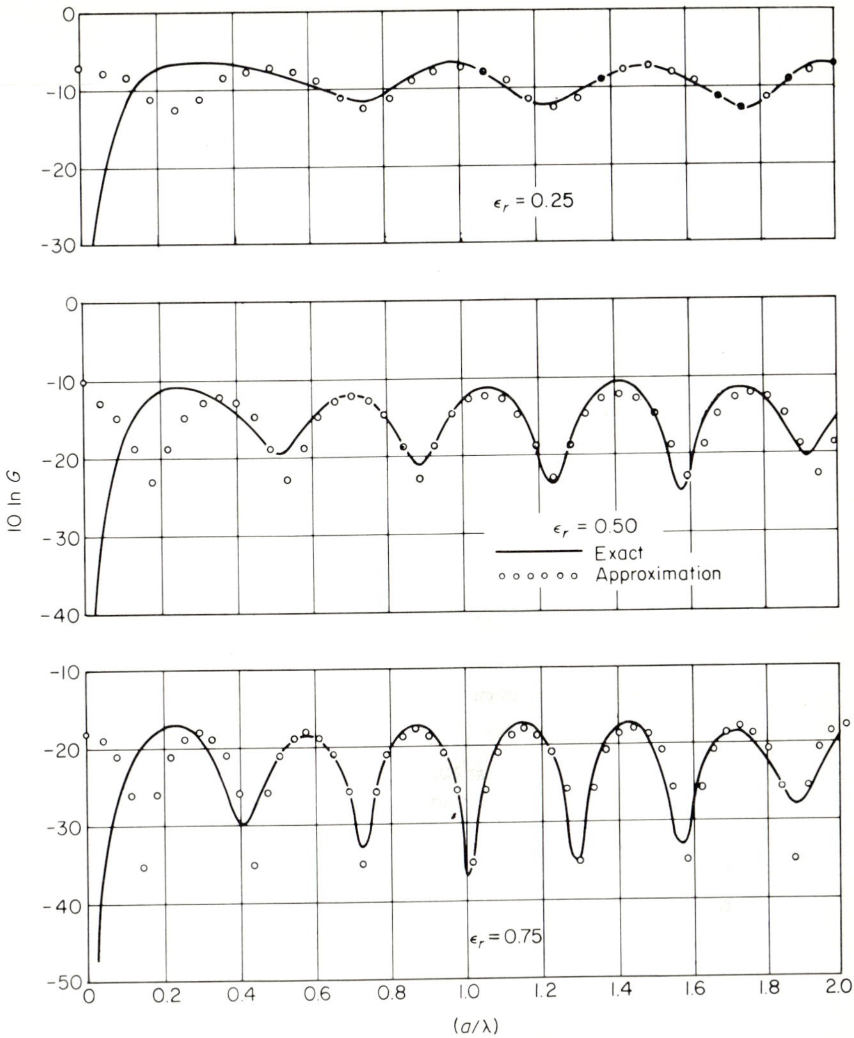

FIG. 4.38. Comparison of exact backscatter gain with approximation based upon interference between two axial rays for $\varepsilon_r = m^2 = 0.25$, 0.50, and 0.75 (Thomas, 1962).

Thomas found that for refractive indices greater than unity additional rays are necessary. These may be provided by the glory rays. The glory rays are emergent rays in the backward direction other than the axial rays. Those which emerge after one, two, or three internal reflections are shown in Fig. 4.39. For the case with one internal reflection

$$\theta_i = 2\theta_t \tag{4.3.26}$$

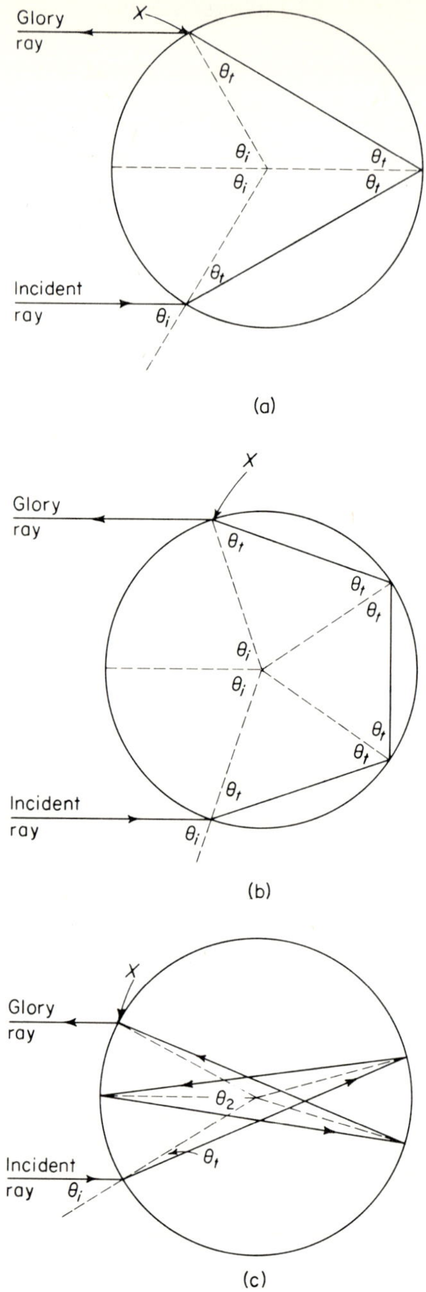

FIG. 4.39. Glory rays for one, two, and three reflections are depicted in (a), (b), and (c), respectively. The angles of incidence and refraction are θ_i and θ_t, respectively. The glory circle emerges at x.

or utilizing Snell's law

$$\sin 2\theta_t = m \sin \theta_t \tag{4.3.27}$$

It follows then that since θ_i falls between 0 and $\pi/2$, and θ_t between 0 and $\pi/4$, such a ray can only exist for $2 \geqslant m \geqslant \sqrt{2}$. For the glory ray with 2 internal reflections

$$\theta_i = 3\theta_t - \pi/2 \tag{4.3.28}$$

so that

$$-\cos 3\theta_t = m \sin \theta_t \tag{4.3.29}$$

In this case, $0 \leqslant m \leqslant 2/\sqrt{3} = 1.154$ as θ_i varies between 0 and $\pi/2$. For $m = 1$, there is a hypothetical glory ray with two internal reflections incident at $\theta_i = 45°$.

In general

$$\theta_i = p\theta_t + (2 - p)\pi/2, \qquad p = 2, 3, \ldots \tag{4.3.30}$$

and

$$\sin[p\theta_t + (2 - p)\pi/2] = m \sin \theta_t \tag{4.3.31}$$

where again $(p - 1)$ is the number of internal reflections. As p increases, the glory rays are restricted to smaller and smaller values of m given by

$$0 \leqslant m \leqslant \csc\{[(p - 1)/p]\pi/2\} \tag{4.3.32}$$

The two regimes of m corresponding to values greater than and less than unity are associated with angles of incidence given as follows:

$$1 \leqslant m \leqslant \csc\{[(p - 1)/p]\pi/2\}; \qquad [(p - 2)/(p - 1)]\pi/2 \leqslant \theta_i \leqslant \pi/2 \tag{4.3.33}$$

and

$$0 \leqslant m \leqslant 1, \qquad 0 \leqslant \theta_i \leqslant [(p - 2)/(p - 1)]\pi/2 \tag{4.3.34}$$

When there is more than one trip around the sphere by the internally reflected ray

$$\theta_i = p\theta_t + (2t + 2 - p)\pi/2 \tag{4.3.35}$$

where the rotation of the ray after emergence in this backward direction has been $(2t + 1) \cdot \pi$. The case for three internal reflections ($p = 4$) and a rotation of 3π ($t = 1$) is shown in Fig. 4.39c. Here

$$\theta_i = 4\theta_t \tag{4.3.36}$$

and the refractive index is constrained to

$$4 \geqslant m \geqslant \csc 22\tfrac{1}{2}^{\circ} = 2.613 \tag{4.3.37}$$

as θ_i varies from 0 to $\pi/2$.

The evaluation of the complex amplitude function of the glory is more complicated than for the axial rays. It can be noted that the locus of the glory rays upon emergence from the sphere is the glory circle. A ray incident anywhere on the glory circle will undergo internal reflections upon entering the sphere and will finally emerge at the conjugate point on the glory circle. Just as for the axial rays, the emerging glory rays will be determined by the following distinct processes: (1) entrance through the front, internal reflections, and emergence through the front, (2) attenuation due to absorption, (3) phase shifts along this optical path, and (4) divergence of the ray bundle. Here the appropriate Fresnel coefficients must be applied to each component of the ray parallel and perpendicular to the spherical surface at each refraction and reflection. At the glory angle for $p = 2$, these are given by

$$r_1 = (m^2 - 3)/(m^2 - 1) \tag{4.3.38}$$

and

$$r_2 = (-1)/(m^2 - 1) \tag{4.3.39}$$

In effect it is possible to use a Fresnel factor averaged over the circle which it turns out (Atlas and Glover, 1963) is given simply by

$$\varepsilon = (\varepsilon_1 + \varepsilon_2)/2 \tag{4.3.40}$$

where the subscripts 1 and 2 denote that the Fresnel coefficients r_1 and r_2 are to be used [(2.4.5), (2.4.6)] at the appropriate angle of incidence.

The real complexity arises because it is impossible to apply the geometrical optics formula (4.3.20) to obtain the divergence of a closed curve of rays such as the glory circle, which has emerged from the surface of a body in which the rays have followed a three-dimensional path. Atlas and Glover (1963) have used a stratagem involving diffraction theory which results in the following for the amplitude function

$$L_{\text{gl}} \exp(-i\delta_{\text{gl}}) = \left[\frac{3(4 - m^2)(m^2 - 2)}{2(m^2 - 1)^3}\right]\left[\left(\frac{\pi}{2}\right) m^4(m^2 - 2)\alpha\right]^{1/2} \times [\exp(-2m\alpha n\kappa)][\exp(-i[m^2 + 4]\alpha - \pi)] \tag{4.3.41}$$

These factors each represent the contributions respectively of (1) the Fresnel reflection and transmission coefficients, (2) the divergence, (3) the absorption, and (4) the phase retardation along the optical path. When the sphere is a dielectric, $\kappa = 0$ and $m = n$, so that

$$L_{\mathrm{gl}} \exp(-i\delta_{\mathrm{gl}}) = \left[\frac{3}{2}\left(\frac{\pi}{2}\right)^{1/2} \frac{(4 - n^2)(n^2 - 2)^{3/2}}{(n^2 - 1)^3} n^2 \alpha^{1/2}\right]$$

$$\times \exp(-i[n^2 + 4]\alpha - \pi) \qquad (4.3.42)$$

Finally, the backscatter efficiency, on the assumption that this can be accounted for in terms of the two axial rays and the glory, is

$$G = |L_{\mathrm{fa}} \exp(-i\delta_{\mathrm{fa}}) + L_{\mathrm{ra}} \exp(-i\delta_{\mathrm{ra}}) + L_{\mathrm{gl}} \exp(-i\delta_{\mathrm{gl}})|^2 \qquad (4.3.43)$$

where the first term is due to the front axial ray, the second term is due to the rear axial ray, and the last term is due to the glory rays.

Thomas and Atlas and Glover have compared this analysis with the exact theory. Thomas' (1962) results for $m = (1.25)^{1/2}$ are shown in Fig. 4.40.

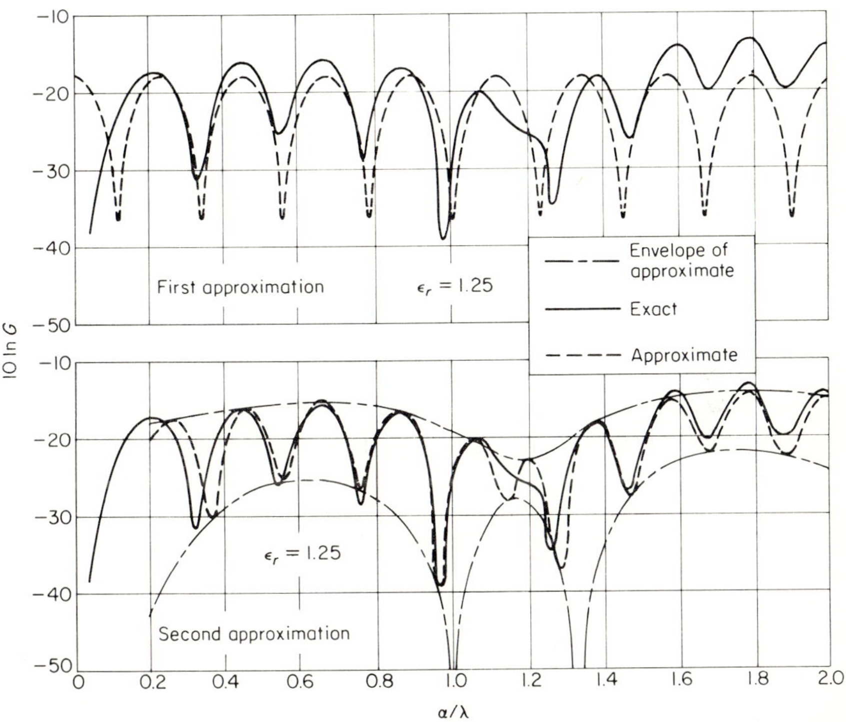

FIG. 4.40. Comparison of exact backscatter gain with two approximations at $\varepsilon_r = m^2 = 1.25$. The upper curve is for the approximation based upon the two axial rays. The lower curve shows the improvement in the approximation when the twice internally reflected glory ray is included (Thomas, 1962).

Once again, there is remarkably good agreement over a range of α-values which one might not expect to be amenable to such a geometrical optics approach. In this case, a glory ray undergoing two internal reflections was used.

Atlas and Glover (1963) carried out their analysis for $m = 1.61 - 0.0025i$, corresponding to the refractive indices of plexiglas in the microwave region of the spectrum, and then extended this work to $m = 1.78 - 0.0024i$ and $m = 1.732 - 0.0025i$ which are for ice and ice-simulating stycast (Glover and Atlas, 1963). It will be recalled that for these refractive indices the exact theory gives the remarkably high backscatter efficiencies which have already been discussed. For plexiglas and ice, Atlas and Glover derived

$$G = |-0.234 + 0.914\, e^{-0.01\alpha}\, e^{i6.44\alpha} + 0.796\alpha^{1/2}\, e^{-0.00805\alpha}\, e^{i6.60\alpha}|^2 \tag{4.3.44}$$

and

$$G = |-0.281 + 2.09\, e^{-0.0096\alpha}\, e^{i7.12\alpha} + 0.612\alpha^{1/2}\, e^{-0.00854\alpha}\, e^{i7.17\alpha}|^2 \tag{4.3.45}$$

Although these do not reproduce the detailed oscillations of the backscatter efficiency over the range $\alpha = 5$ to 20, they give values very close to those obtained from the exact theory over this range. However, from $\alpha = 20$ to 40 these results diverged significantly, the approximate theory overestimating for plexiglas and underestimating for ice.

Querfeld (1963) carried calculations for ice out to very high values of α, so that it is possible to compare (4.3.45) with the exact theory in a more detailed way. We have calculated G using the above ray optics formula for the following ranges of α: 75.0(0.1)100.0, 175.0(0.1)200.0, 275.0(0.1)300.0, 675.0(0.1)700.0, 975.0(0.1)1000.0. Although the periodicity of the approximate and the exact expressions agree precisely, and the extrema occur at the same values of α, the values of the backscatter efficiency obtained by (4.3.45) continue to be low, even out to $\alpha = 300$. The values of G at the extrema for various α are given in Table 4.5.

When α is as large as 691, the mean values obtained in each case are quite close to the geometrical optics limit of 0.079. The amplitude of the oscillation about this mean is not quite as great for the ray optics formulation. For $\alpha \sim 1000$, the latter oscillates precisely about the limiting value, exhibiting an amplitude of about 0.02 in G.

Obviously, in the range of greatest interest there are additional contributions to the backscatter besides the axial and glory rays. Before examining what these might be, let us first consider a remarkable series of experiments with a metal capped sphere designed to demonstrate the interaction between the front and rear axial rays with the glory ray (Atlas and Glover, 1963).

The backscatter target was a plexiglas sphere with diameter 7.20 cm which could be fitted with any one of a set of nine spherical aluminum caps covering a portion of the sphere starting with a 20° cap and ranging up to a 180° cap in 20° steps. The backscatter of 3.222 cm microwaves was measured with the caps in various positions on the sphere. We will consider here only the effect with the caps positioned on the front and rear surfaces of the sphere.

TABLE 4.5
COMPARISON OF VALUES OF G FOR ICE ($\lambda = 3$ cm) AT THE EXTREMA OBTAINED BY (4.3.45) AND BY THE EXACT THEORY

α	Type of extremum	G (4.3.45)	G (exact)
175.7	Max	3.22	12.42
176.2	Min	1.50	8.48
176.6	Max	3.16	11.42
177.0	Min	1.48	8.65
177.5	Max	3.12	11.85
177.9	Min	1.42	7.85
298.4	Max	1.06	2.75
298.8	Min	0.21	1.20
299.3	Max	1.04	2.75
299.7	Min	0.21	1.18
691.0	Max	0.104	0.125
691.4	Min	0.057	0.040
999.4	Max	0.081	—
999.8	Min	0.077	—

When it is on the rear surface, the reflection coefficient for the rear axial and glory rays becomes unity ($m = \infty$). When the cap is on the front surface, the rear axial ray is blocked and the reflection coefficient for the front axial ray becomes unity. The glory ray for plexiglas enters at $\theta_i = 73.8°$ so that it should be blocked only for the caps larger than 140°.

These considerations permit computation of the backscatter efficiency for the capped spheres with the aid of the ray optics approximation that has been discussed above. In Fig. 4.41, the amplitudes corresponding to the three primary contributions to the backscatter with and without a rear metal cap are shown. The contribution of the front axial ray is small over the range of α shown. However both the rear axial and glory rays have an exponential decrease with α due to absorption, so that eventually all of the backscatter will be due to the front axial ray. With the addition of the rear cap, there is a marked increase in the amplitudes of both the rear axial and glory rays. The expressions for the backscatter efficiency of both the rear and front capped spheres may be obtained in the original paper.

A comparison of the theoretical calculation with experiments is given in Table 4.6. The results are tabulated as the ratio of the backscatter efficiency of a plexiglas sphere to that of a standard metallic target in units of decibels.

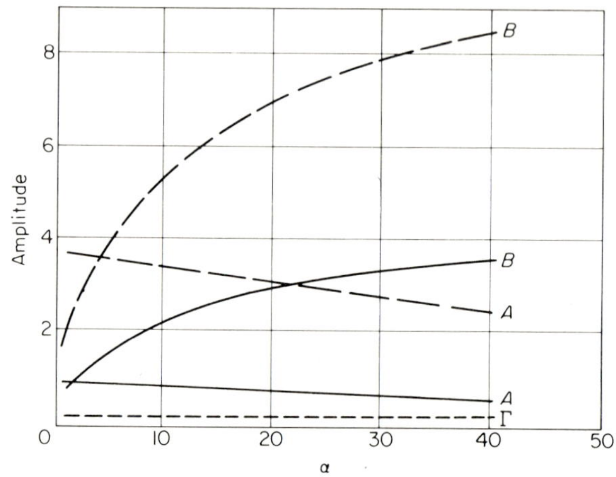

FIG. 4.41. Amplitudes of the three primary contributions to the backscatter from a plexiglas sphere with a rear metal cap (dashed) and uncapped (solid). The axial ray reflected from the rear surface is depicted by curves A and the glory ray by curves B. The curve corresponding to the front axial ray (short dashes) is the same with and without the rear cap (Atlas and Glover, 1963).

TABLE 4.6
THEORETICAL AND EXPERIMENTAL BACKSCATTER EFFICIENCIES[a]

Case	10 log(G/G_s)	
	Theory	Experiment
I Plexiglas sphere	7.3	9.5
	8.8	8.5
II Plexiglas sphere, rear cap	16.1	18.0
III Plexiglas sphere, front cap	8.1	11.0
IV Plexiglas sphere, front and rear caps	13.8	13.0

[a] Results are for a 7.2 cm plexiglas sphere with and without metal caps. The values are given as the ratio to an equal size metal sphere in units of decibels.

While the values calculated for this sphere ($\alpha = 7.02$) by the approximate theory are generally below the experimental ones (which agree with the exact theory for the uncapped sphere), the relative changes introduced by the caps are remarkably close to those expected. The theory predicts that addition of the rear cap to the plexiglas sphere should enhance the backscatter efficiency by 8.8 dB as compared to the measured enhancement of 8.5 dB. The effects for the other cases are similar.

Cases III and IV are for the front surface caps, so small that the glory ray has not yet been blocked. As the front caps grow larger than 80°, the backscatter intensity drops off until with the 140° cap it is only 2 dB above the all metal reference sphere. This and other experiments with the caps in lateral positions indicate that the energy associated with the glory rays enters the sphere over the angles of incidence between about 40° and 75°, rather than precisely at 73.8°. Glover and Atlas (1963) have carried out further experiments with stycast, obtaining similar results.

There is no glory ray predicted for $2/\sqrt{3} \leqslant m \leqslant \sqrt{2}$ and yet the backscatter curve cannot be approximated merely by the two axial rays. Thomas (1962) has proposed that there is a contribution here from a stationary ray. A stationary ray is one which upon emergence has undergone a minimum deviation compared to its neighboring rays in the incident beam. In Fig. 4.42, a stationary ray is traced along with the rays on either side. The effect of this is a concentration of rays in the emergent direction with a great increase in the intensity of the radiation scattered over a very small range of angle. The rainbow is due to such a stationary ray. For water ($m = 1.33$), the angle of minimum deviation occurs at $\theta = 138°$ for the ray undergoing a single internal reflection. This rainbow angle is slightly different for the various refractive indices corresponding to the wavelengths of the visible spectrum, thus enhancing the beauty of this phenomenon in nature.

Curiously, when a stationary ray occurs sufficiently close to the backward direction, it will make an appreciable contribution to the backscatter efficiency because of its intensity and of the divergence of the ray bundle. The divergence of the stationary ray, like that of the glory circle, cannot be treated quantitatively by geometrical optics, and again recourse may be had to diffraction theory. The theory was first developed by Airy in connection with the rainbow (Pernter and Exner, 1910). Thomas (1962) has derived the contribution to the backscatter efficiency and has then compared the results with those from the exact theory for $m = (1.50)^{1/2}$. Once again, reasonably good agreement was obtained for radii of the order of a wavelength except at some of the minimum points.

Ray optics would be expected to provide an increasingly accurate description as α becomes larger, yet the approximation discussed up to now still fails for $\alpha > 20$. Although the axial and glory rays provide the major components of the backscatter efficiency, it becomes necessary to find additional contributory rays which will interfere with these to give the detailed behavior. These are provided by rays emerging in the backward

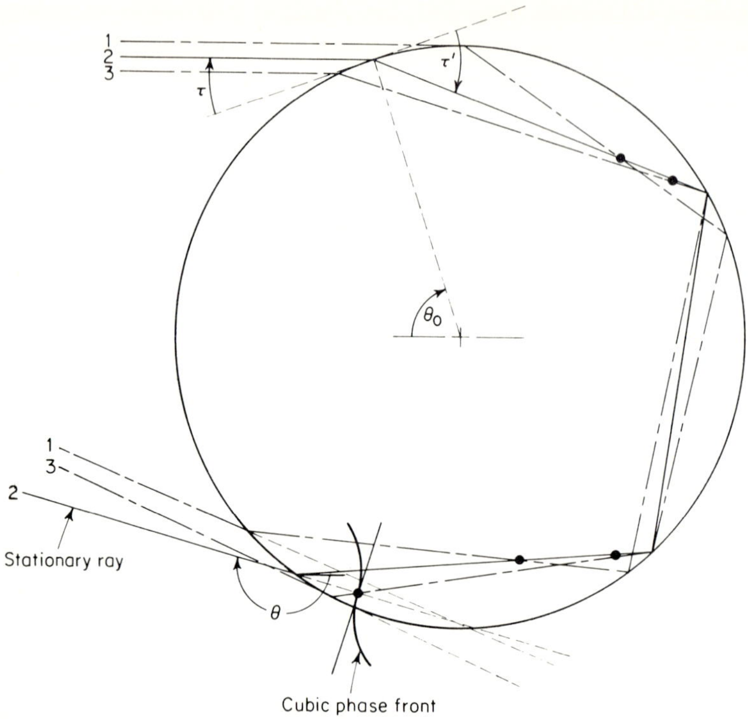

FIG. 4.42. Geometry of the stationary ray. Here τ and τ' are the complements of the angle of incidence ($\theta_0 = \theta_i$) and the angle of refraction. The emergent angle is θ. The stationary ray occurs when θ is a minimum; i.e. $(d\theta/d\tau) = 0$ (Thomas, 1962).

direction which have traveled at least part way around the sphere as a surface wave. The surface wave on a dielectric or partially absorbing sphere is considerably more complicated than on a conducting sphere. The intensity of the back sprayed ray in the sense of Fig. 4.29 is too small to figure significantly in the backscatter. However, in addition to those rays which spray off the surface, the surface wave loses energy by virtue of rays which are refracted into the sphere from the surface. These rays may then be reflected internally until they emerge in the backward direction. Such backscattered rays are denoted by $S(s, p)$ where the total angle which the emergent ray makes with the incident ray is $s \cdot \pi$ and again $(p - 1)$ is the number of internal reflections or p is the number of cuts through the sphere. An $S(3, 5)$ ray is sketched in Fig. 4.43.

In the absence of a complete theory for such combinations of surface waves and geometrical optics rays, Probert-Jones (1963) has been able to identify the most important of these rays in a particular case by a semi-empirical technique. He carried out a harmonic analysis of the difference between the total backscatter efficiency obtained from the exact theory and that obtained from ray optics for the combination of the glory and back surface axial rays. This was done for α between 4 and 30, for $m = 1.78 - 0.0024i$, corresponding to the refractive index of ice in the microwave region. The variation of the resultant with α could be accounted for by ten sinusoidal oscillations. This implies that there are interfering rays whose phase differences with the glory-front surface axial rays vary linearly with sphere size, which in turn indicates fixed geometrical paths in the sphere.

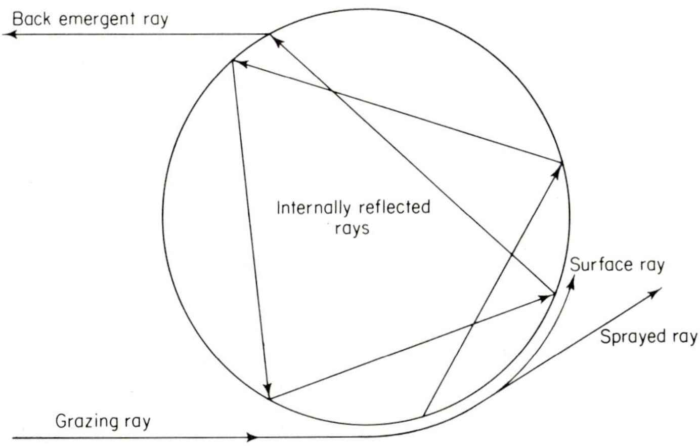

FIG. 4.43. Geometry of an $S(3, 5)$ ray which arises from the grazing incident ray, undergoes four internal reflections, and emerges in the back direction after having rotated 3π rad.

Each oscillation, described by its period and amplitude, is to be identified physically with a ray, which interferes with the glory-front surface axial rays. There is among these one ray which has already been treated completely by ray optics. It is the rear surface axial ray, which can be designated as $S(1, 1)$, and indeed the harmonic analysis does provide one particular oscillation whose period and amplitude are precisely equal to the values obtained by the ray optics treatment. This would seem to verify the efficacy of the analysis in identifying physically real rays. The period expected of a particular surface wave taking various cuts through the sphere and then emerging in the backward direction can be obtained approximately from the phase shift along the ray path. By comparison with the periods obtained from the harmonic analysis, a physical ray can be associated with each of the ten oscillations.

Peters (1965) has used a rather simple approach to analyze the backscatter from a plasma sphere with a purely imaginary dielectric constant. The backscatter is represented as due to the interference of two components—an axial ray and a surface ray. The axial ray results from a single specular reflection at the outer surface with an amplitude determined by the Fresnel reflection coefficient. The surface wave is similar to that utilized by Probert-Jones for a perfectly reflecting sphere, (4.3.11). However, in addition to attenuation by radiation, there is additional attenuation by absorption due to the absorptivity of the plasma medium. For this, Peters used the attenuation factor corresponding to a wave traveling along a planar interface between an absorbing and a nonabsorbing medium. Excellent agreement was obtained with the exact solution.

Bryant and Cox (1966) have provided further evidence for the contribution of surface waves to the backscatter of dielectric spheres in the form of growth curves. Two such curves are shown in Figs. 4.44 and 4.45 for $\alpha = 200.400$ and 500.235 respectively and for a refractive index $m = 1.333$.

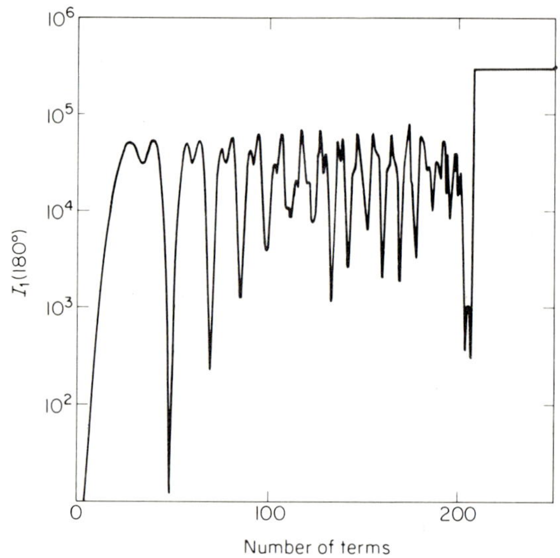

FIG. 4.44. "Growth curve" of backscatter intensity for $m = 1.333$ and $\alpha = 200.400$ (Bryant and Cox, 1966).

Each of these is a plot of the calculated intensity at 180° as a function of the number of terms included in the series expansion. It will be recalled that the series converges for a number of terms somewhat in excess of the value of α.

In both cases, the intensity rapidly rises to a maximum value which remains quite constant until the last few significant terms, although the oscillations are pronounced. The magnitude of the intensity in this region is comparable to that expected from the interference of the axial rays. For $m = 1.333$, there are no glory rays with less than four internal reflections which can contribute to the backscatter. There is an abrupt increase in the backscatter just before the series has converged to its final value. As will be pointed out in the next section, each term in the series of order n can be associated with geometrical optics rays incident upon the sphere at a distance $(n + \frac{1}{2})\lambda/2\pi$. Where n is of the order of α, this corresponds to rays nearly tangent to the surface. It is precisely these rays which give rise to the surface waves.

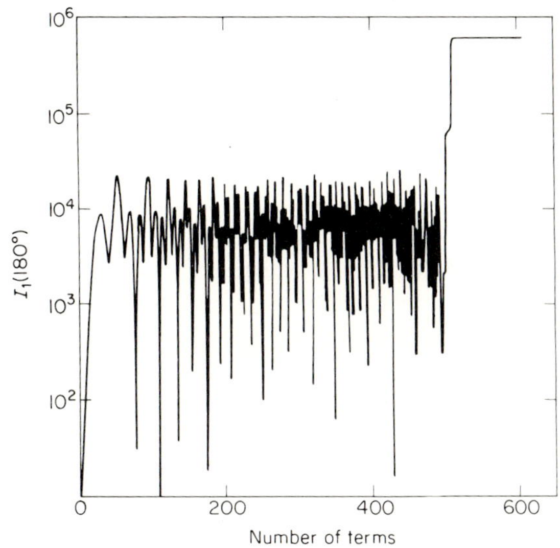

FIG. 4.45. "Growth curve" of backscatter intensity for $m = 1.333$ and $\alpha = 500.235$ (Bryant and Cox, 1966).

Experimental confirmation of the contributions of the axial rays and of the surface rays to the backscatter of water drops ranging from 0.5 to 1.5 mm diameter has been obtained by Fahlen and Bryant (1968) using a laser beam. With the narrow beam focused upon the center of the drop, the backscattered beam arises from the interference between the back and front surface axial rays. For an evaporating droplet this varied sinusoidally with time as expected from simple geometrical considerations. In a similar fashion, a backscatter ray is observed when the incident beam strikes the drop at its edge. The variation of the intensity of this backscattered ray with size was also followed as the water drop evaporated. This showed a complicated variation of periodic humps upon which are superimposed clusters of spikes. These possessed the same periodicity and were similar in shape to the intensity oscillations predicted by the theory, indicating that for such large spheres ($\alpha \sim 3000$) the surface wave makes the major contribution to the oscillatory behavior of the backscatter. It is possible to explain the circumferentially backscattered light by assuming that the surface ray takes various short cuts through the sphere as discussed.

4.4 Angular Intensity Functions

The intensity scattered in other directions will now be considered. Just as for the backscatter, it will be convenient to define an angular efficiency function or angular gain such that

$$G_1 = 4i_1/\alpha^2; \qquad G_2 = 4i_2/\alpha^2 \tag{4.4.1}$$

and

$$G_u = (G_1 + G_2)/2 = 2(i_1 + i_2)/\alpha^2 \tag{4.4.2}$$

for unit intensity of perpendicularly polarized, parallel polarized, and unpolarized incident radiation, respectively.

4.4.1 Totally Reflecting Spheres, $m = \infty$

When $m = \infty$, the total scattering or extinction efficiency and the backscatter efficiency (gain) vary more smoothly and in a more regular manner than for dielectrics. There is no ripple structure and each of these functions approaches the geometrical optics limit of 2 and 1, respectively, at relatively small values of α. It will be instructive, then, to commence our descriptive survey of the angular gain with this case.

The small sphere limit for total reflectors differs from the Rayleigh law for dielectrics. This is because the magnetic dipole now makes a contribution to the scattering comparable to that of the electric dipole; i.e., both a_1 and b_1, which are the electric and the magnetic dipole moments, must be considered. The limiting values of the series expansions for these scattering coefficients have already been given (3.9.22) and these lead directly to the angular gain for small totally reflecting spheres [(3.3.56), (3.3.57), (3.9.6), and (3.9.7)].

$$G_1 = 4\alpha^4(1 - \tfrac{1}{2}\cos\theta)^2 \tag{4.4.3}$$

$$G_2 = 4\alpha^4(\cos\theta - \tfrac{1}{2})^2 \tag{4.4.4}$$

The radiation patterns for G_1, G_2, and G_u are shown in Fig. 4.46. Here the radius vector from the center of the diagram to the curve is proportional to the intensity at the indicated scattering angle.

The most striking feature of this pattern is the predominance of scattering in the backward direction. The backscatter is nine times more intense than the forward scatter.

$$G(0°) = \alpha^4; \qquad G(180°) = 9\alpha^4 \tag{4.4.5}$$

At 0° there is a shallow minimum for each of the polarized components.

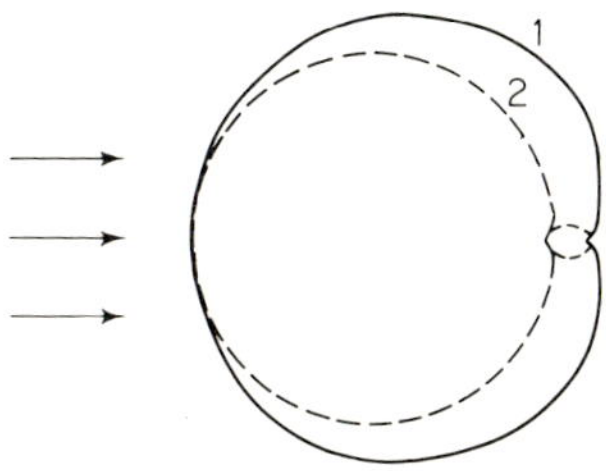

FIG. 4.46. Radiation pattern for small totally reflecting spheres. The angular patterns of the perpendicular and parallel components are given by the radial distances to curves 1 and 2, respectively. For unpolarized incident radiation, the radial distances for each curve are to be summed.

This is quite unlike Rayleigh scattering (Fig. 3.5) which is symmetrical about 90° or scattering by larger spheres, whether totally reflecting or dielectric, for which the forward scattering is predominant. At 60°, $G_2 = 0$, so that for unpolarized incident radiation the scattering at this angle is completely perpendicularly polarized. The gain for unpolarized incident radiation goes through a minimum at 36° 52′ where

$$G_u = 0.9\alpha^4 \tag{4.4.6}$$

We have seen earlier that with increasing size, the backscatter gain for perfectly reflecting spheres increases sharply to a maximum value of 3.65 at $\alpha = 1.02$ and then undergoes a rapidly damped oscillation about the geometrical optics limit of unity (Figs. 4.27 and 4.28). The variation of the angular gain with the size parameter is shown at other scattering angles ($\theta = 120°$, 90°, 60°, 10°) in Figs. 4.47 to 4.50. Just as for the backscatter, the angular gain exhibits a damped sinusoidal oscillation with increasing α which approaches the limiting value of unity. The vertical component, G_1, is more highly damped than G_2. The amplitude of the oscillation for each component decreases as the scattering angle decreases from the backward direction. However, this trend reverses near the forward direction so that for $\theta = 10°$ the gain at the first maximum ($G_1 = 42$ at $\alpha = 10$) is more than an order of magnitude greater than the maximum gain for the backscatter ($G = 3.64$ at $\alpha = 1.005$). The periodicity of the oscillation is the same for each polarized component and becomes larger as the scattering angle becomes less.

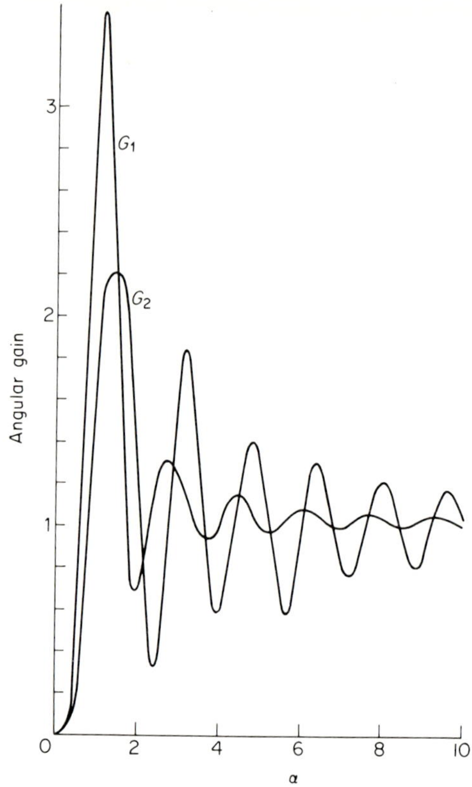

FIG. 4.47. Angular gain at $\theta = 120°$ for totally reflecting spheres.

Just as for the backscatter, the scattering at other angles can be interpreted as due to the interference between a specularly reflected ray and rays sprayed off the surface wave. In addition, at those angles near the forward direction there will be a contribution from the diffracted radiation. It is this which accounts for the high gain at 10° shown in Fig. 4.50. The phase difference between the reflected and sprayed rays is

$$\delta > 2a[\sin(\theta/2) + \theta/2] \tag{4.4.7}$$

The inequality is due to the fact that the surface wave behaves as if it were located somewhat above the surface. With increasing α, the energy carried by the grazing ray must be distributed over an increasingly large periphery so that eventually only the contribution of the specular reflection is significant. Since this is isotropically distributed, the angular gain of larger spheres is unity except in the forward directions.

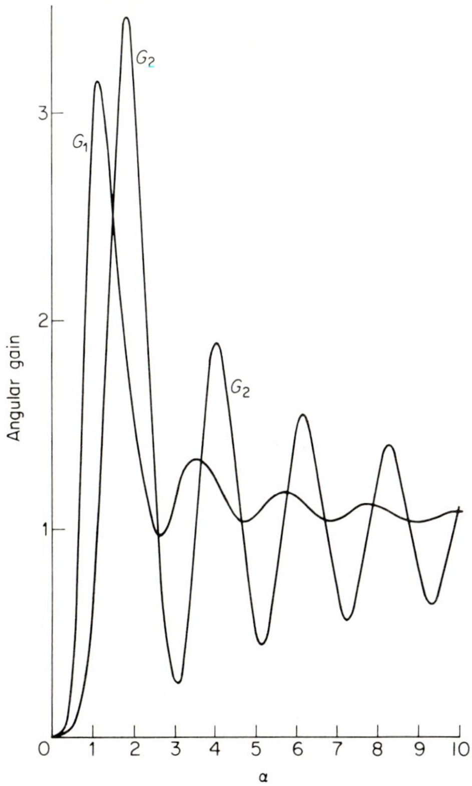

FIG. 4.48. Angular gain at $\theta = 90°$ for totally reflecting spheres.

In the forward directions, the diffracted energy makes a major contribution to the angular gain. This is

$$G_1{}^{\mathrm{F}} = G_2{}^{\mathrm{F}} = 4\alpha^2[J_1{}^2(\alpha \sin\theta)/(\alpha^2 \sin^2\theta)] = \alpha^2 F^2(u) \tag{4.4.8}$$

The function $F(u)$ has a maximum value of unity when the argument is zero. Its value decreases to zero at $u = 3.832$ and then undergoes a rapidly damped oscillation. For small u, the following approximation is valid

$$F(u) = 1 - (u^2/8) + \cdots \tag{4.4.9}$$

With increasing size, the main diffraction lobe, within which most of the diffracted energy is found, is constrained to a narrow cone around the forward direction given by

$$\sin\theta < 3.832/\alpha \tag{4.4.10}$$

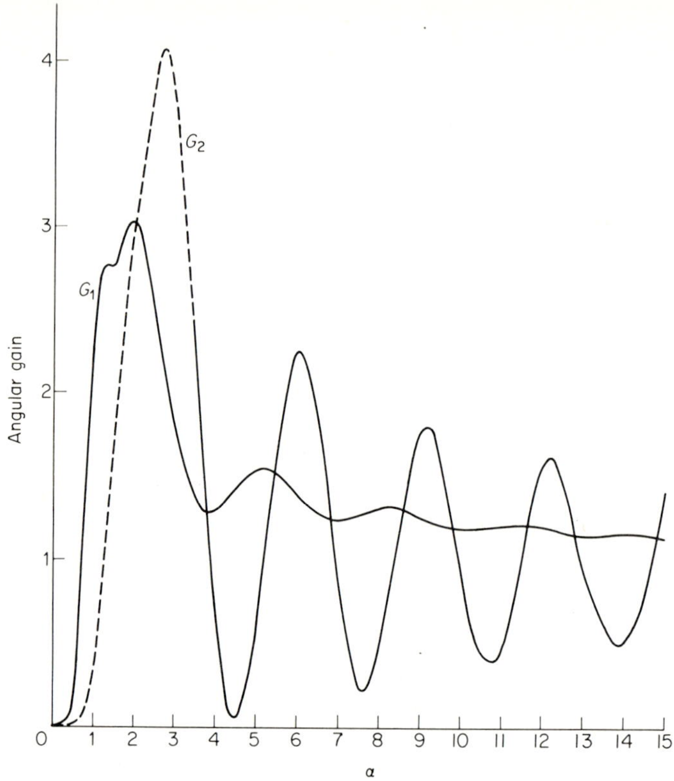

FIG. 4.49. Angular gain at $\theta = 60°$ for totally reflecting spheres.

The high angular gain obtained in the forward directions for large α results from this extreme anisotropy of that part of the scattering which arises from diffraction. The contribution of the diffraction to the angular gain at 10° is shown by the square points in Fig. 4.50. In this direction, the angular gain is substantially accounted for by diffraction.

The angular variation of the scattered intensity is depicted in Figs. 4.51 to 4.54 for $\alpha = 0.5$, 1, 2.8, 5, and 20. The qualitative features of the small sphere limit are still apparent for $\alpha = 0.5$. Thus, the scattered radiation is nearly completely polarized at about $\theta = 65°$. The gain at 0° and 180° is 0.099 and 0.53, respectively, rather than 0.0625 and 0.5625 as predicted by (4.4.5).

As α increases, the condition of complete polarization disappears and the forward scatter predominates over the backscatter. The angular gain is characterized by lateral lobes which increase in numbers with increasing α. We have already seen that the forward lobe may in the main be attributed to diffraction. The parallel component (G_2) is more highly oscillatory than the

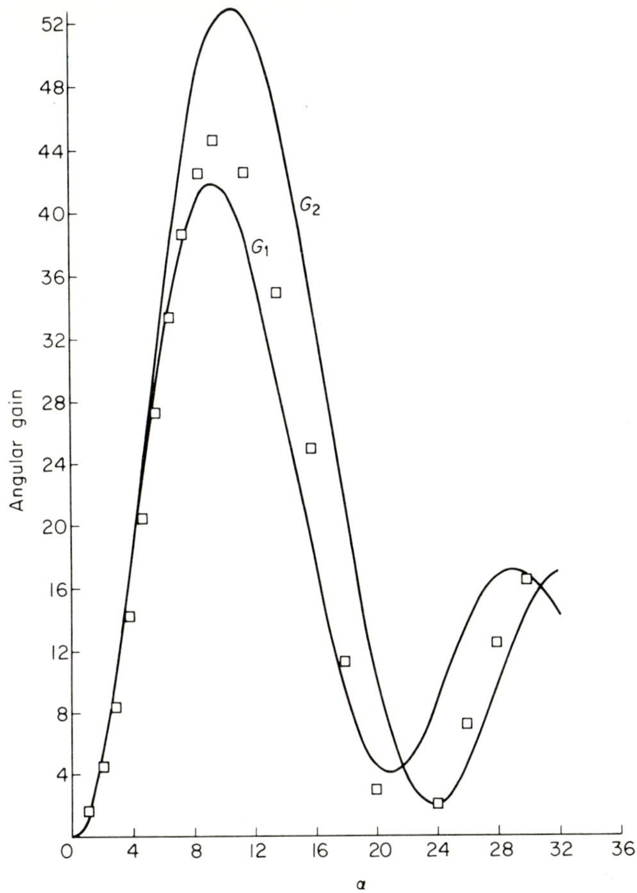

FIG. 4.50. Angular gain at $\theta = 10°$ for totally reflecting spheres. Points designated by squares are calculated according to diffraction equation (4.4.8).

perpendicular component. This oscillation is the result of interference between the specularly reflected rays and the surface wave. It damps down towards the limiting value of unity in the backward direction which can be attributed to the specularly reflected ray. In the intermediate region, all three effects must be considered: reflection, the surface wave, and diffraction.

4.4.2 Real Refractive Index

The scope of the published tabulations of the angular scattering functions has been outlined earlier. As extensive as these are, they still hardly cover the range of practical needs. Interpolation of intermediate values between those

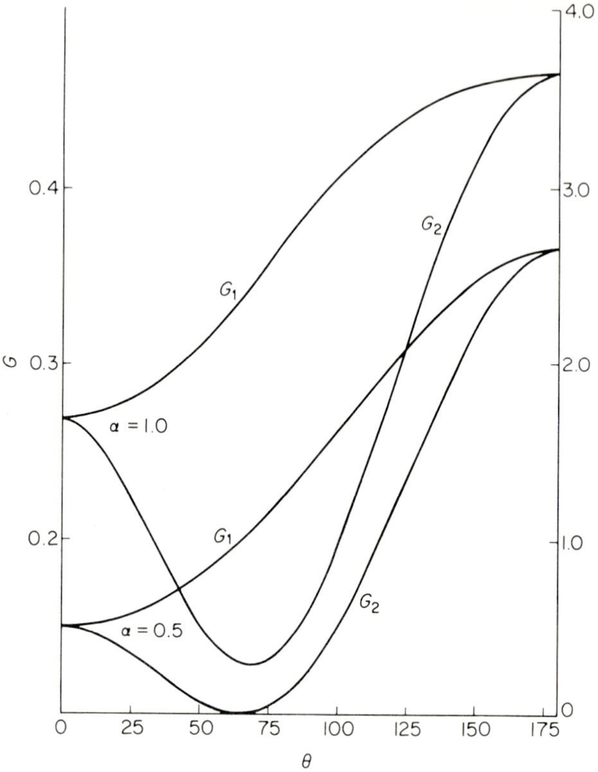

FIG. 4.51. Angular gain, G_1 and G_2, for totally reflecting spheres with $\alpha = 0.5$ and 1.0 plotted against angle of observation. Left ordinate scale is for $\alpha = 0.5$ and right ordinate scale is for $\alpha = 1.0$.

already tabulated is frequently precarious, so that the exact theory must be used to provide detailed results for particular applications. For dielectric and partially absorbing spheres, the variation of the angular scattering with the size parameter, with scattering angle, and with refractive index becomes quite complicated. Yet there are certain simple patterns that emerge for some features of the scattering and it will be instructive to examine these.

a. Altitude Chart. A complete description of the angular intensity functions for a particular refractive index can be provided by a contour diagram or "altitude chart" introduced by van de Hulst (1946) and modified somewhat by Penndorf (1963). The basic scheme of Penndorf's chart is given in Fig. 4.55. The coordinate system consists of the phase shift parameter, $\rho = 2\alpha(m - 1)$, and the diffraction parameter, $u = \alpha \sin \theta$. For a given refractive index,

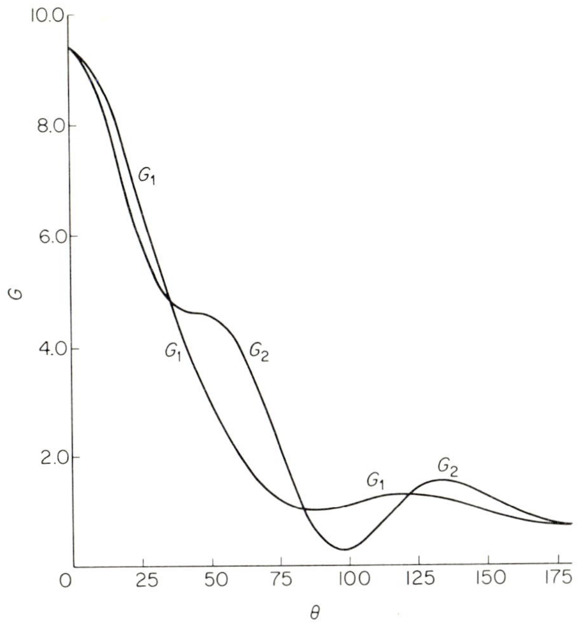

FIG. 4.52. Angular gain, G_1 and G_2, for totally reflecting sphere with $\alpha = 2.8$ plotted against angle of observation.

here chosen as $m = 1.33$, the scattering angles are straight lines in the $u\rho$-plane. For $\theta > 90°$, these lines are placed symmetrically about the 90° line. For example, the 90° line intercepts with $\rho = 10$ at $u = 15$, and the 180° line intercepts with it at $u = 30$. Each particular value of i_1 or i_2 is tabulated at the intersection of the appropriate $\alpha - \theta$ value and then contour lines for equal values are drawn. Fig. 4.56 shows the results for i_1 and $m = 1.33$ (Penndorf, 1963). The contour lines have been drawn for the scale 1, 2, 4, 8, 10^x with $x = -2$ to $+4$. The letters H and L indicate the highs and lows. The heavy dashed lines indicate the valleys between the ridges. The present chart is limited to values of α up to 15 ($\rho = 10$). It extends over all angles for $\alpha \leqslant 10$ (the upper right corner of the chart for $\rho = 6.67$ and $u = 20$ corresponds to $\alpha = 10$ and $\theta = 180°$). For $\alpha = 15$, the chart terminates at 138°10′, corresponding to $\rho = 10$ and $u = 20$ (lower right corner of the chart). Of course more detail can be obtained by blowing up any part of the chart and including more contour lines.

The chart shows that the ρu-terrain is a complex pattern of ridges and peaks, of valleys and trenches. In the forward directions ($\theta < 90$), the valleys run vertically, meandering around the classical diffraction minima at $u = 3.83$, $u = 7.02$, and $u = 10.17$. However, new valleys open up between,

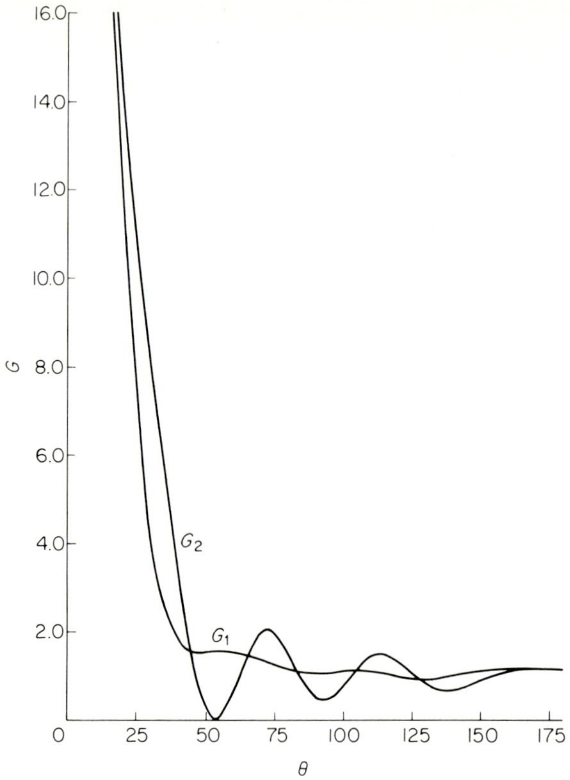

FIG. 4.53. Angular gain, G_1 and G_2, for totally reflecting sphere with $\alpha = 5.0$ plotted against angle of observation.

the main ones and, at larger values of ρ, these eventually interfere, leading to a pattern in which even the main features cannot be accounted for by simple diffraction theory. Although it cannot be seen from the topography shown here, the trenches along the valleys and the peaks situated on the ridges occur systematically at intervals of $\Delta\rho = \pi$, at least up to $\rho = 20$.

The scattering in the backward directions shows a different pattern of contours. The deep long valleys are considerably straighter and while not quite parallel to the lines of constant angle, they diverge only slightly from these. The trenches occur very regularly here. Along a particular valley, the center to center distance between trenches corresponds to an interval in α of about 1. Obviously these would be missed entirely if the computations were not carried out at sufficiently small intervals.

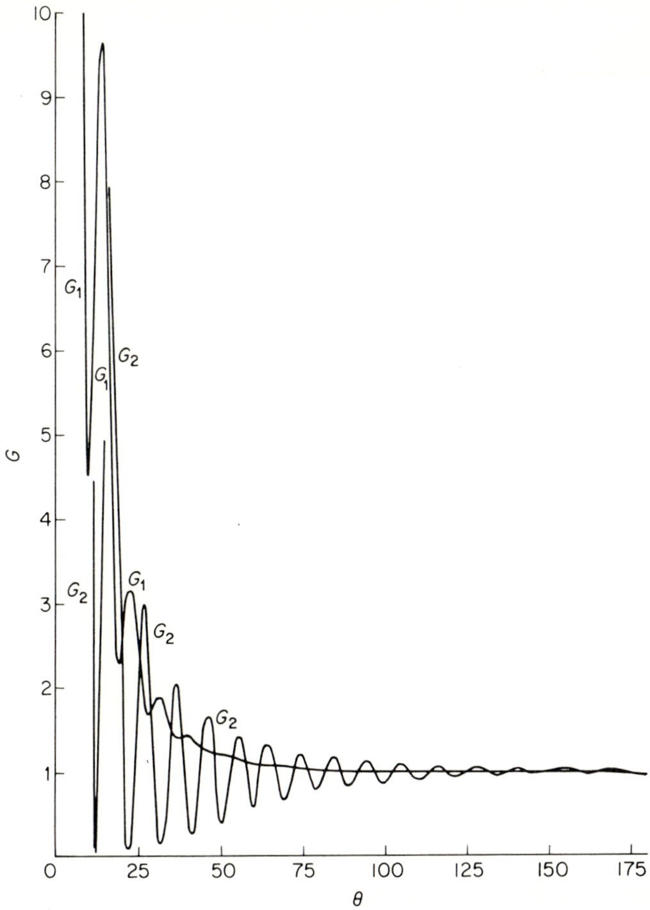

FIG. 4.54. Angular gain, G_1 and G_2, for totally reflecting sphere with $\alpha = 20$ plotted against angle of observation. The first peak for G_2 occurring at about $\theta = 16°$ is not drawn in order to avoid confusion with the first peak for G_1 at $\theta = 14°$. G_1 and G_2 nearly coincide below $\theta = 5°$.

Penndorf has found that, in addition to the general perspective which it affords, the altitude chart may be quite useful in the interpolation of scattering functions especially in the neighborhood of peaks and the trenches. Gucker *et al.* (1964) have represented the intensity surface by a contour diagram in which the coordinates are simply the size parameter α and the scattering angle θ. Such a diagram is easier to construct than the ρu-diagram but is somewhat less useful in correlating results for materials of different refractive indices.

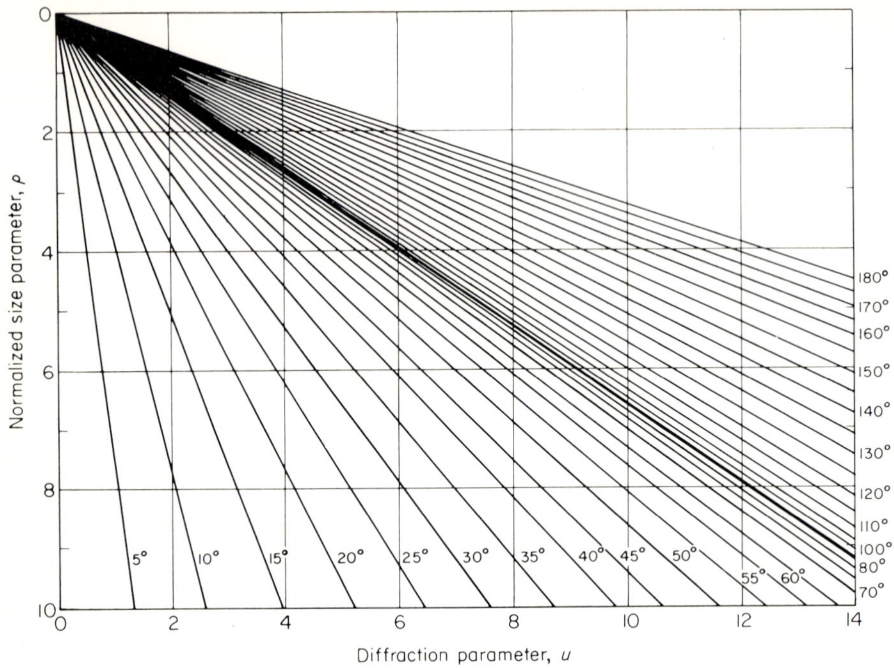

FIG. 4.55. Scheme of the altitude chart. The diffraction parameter $u = \alpha \sin \theta$ and normalized size parameter $\rho = 2\alpha(m - 1)$ (Penndorf, 1963).

b. Variation with Size. We now consider the variation of the angular intensity with size at a constant scattering angle. In Fig. 4.57, the angular gain at 90° of each polarized component is plotted up to $\alpha = 25$ for $m = 1.20$ (Pangonis *et al.*, 1961). These oscillating curves are quite regular and smooth and they have a rather constant periodicity. The envelopes drawn through the maxima and minima are also smooth. These exhibit an undulation with a large periodicity. The amplitude of the curve for the perpendicular component is considerably greater than for the parallel component. The pattern for refractive indices less than $m = 1.20$ is quite similar. The period of the oscillation increases somewhat as the refractive index decreases. The angular gain decreases with decreasing refractive index.

There is a similar pattern at smaller angles except that the parallel and perpendicular components become more similar and both the period of the oscillation and the median value of the gain increase. This increase in gain can be attributed to the contribution of the diffracted radiation as well as the greater reflectivity of those rays impinging upon the sphere at larger angles of incidence. The latter effect increases the contribution of the specularly reflected ray at the small scattering angles. These effects are

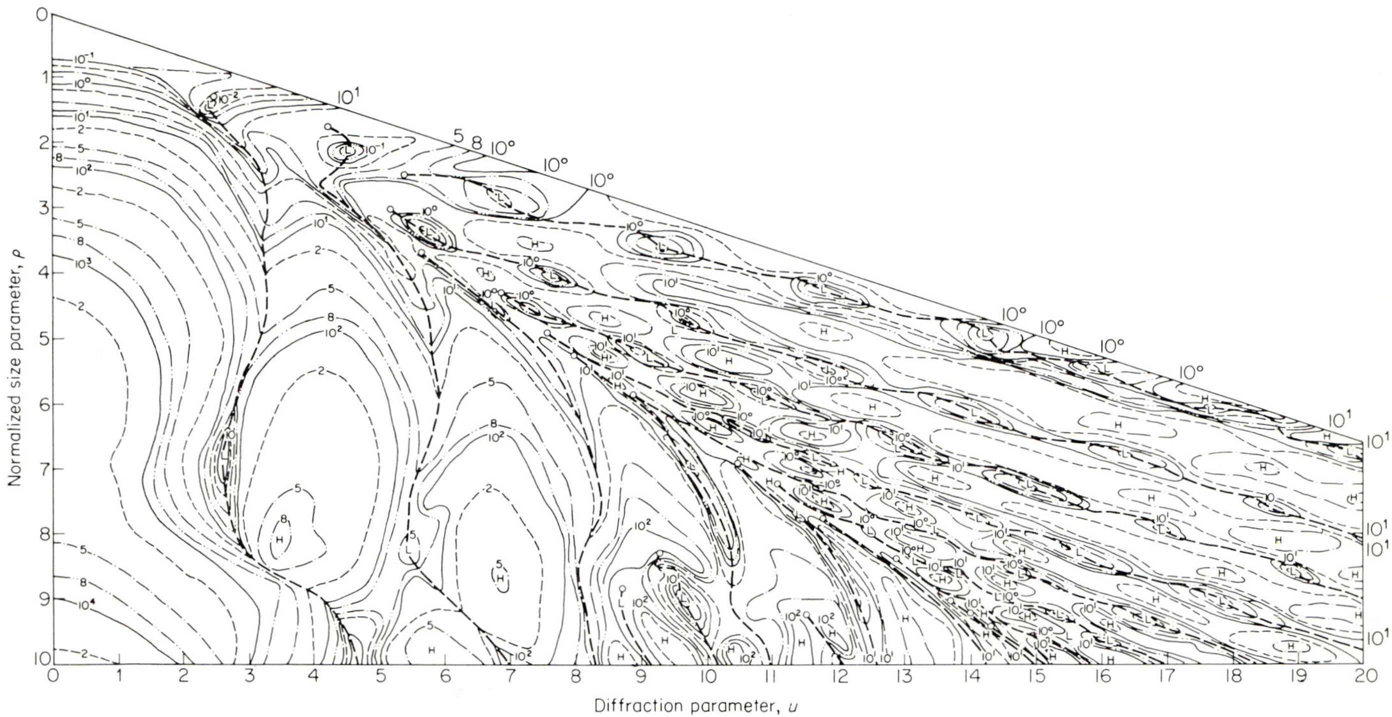

FIG. 4.56. Altitude chart for $m = 1.33$ and i_1. Contours for i_1 are drawn in a quasilogarithmic scale, namely 1×10^{-1}, 2×10^{-1}, 8×10^{-1}, 1×10^{0}, 2×10^{0}, and so on. Heavy dashed lines represent position of dark rings. Letters H (high) and L (low) represent peaks and trenches. For clarity, some contours, particularly around $\theta = 90°$, have been left out (Penndorf, 1963).

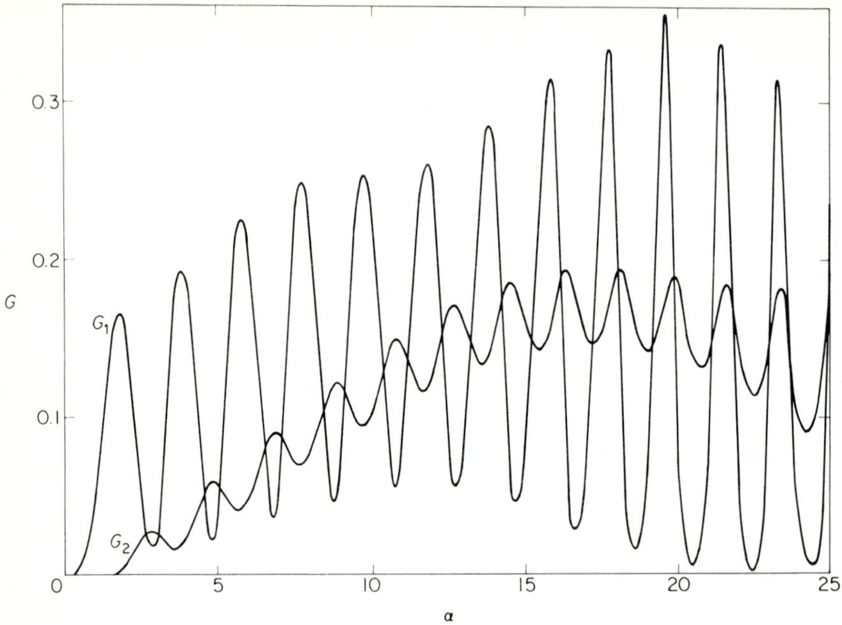

FIG. 4.57. Angular gain at 90° for $m = 1.20$.

illustrated for $m = 1.20$ and $\theta = 45°$ in Fig. 4.58 (Heller and Nakagaki, 1959). On the other hand, when the scattering angle is increased, the above trends are reversed and in addition the curves become irregular. This is illustrated in Fig. 4.59 where the angular gain for $m = 1.20$ is plotted against α at the scattering angle $\theta = 135°$.

At higher values of the refractive index, the angular gain varies with α in a less regular manner. This is shown in Fig. 4.60 where G_1 at 90° is plotted up to $\alpha = 5$ for $m = 1.482$, 1.68, 1.88, and 2.08. In Fig. 4.61, G_2 for $m = 1.60$ is plotted out to $\alpha = 20$. Farone *et al.*, (1963) proposed describing these curves by polynomial expansions such as

$$i_1 = \sum_{j=0}^{p} a_j \alpha^j \tag{4.4.11}$$

$$a_j = \sum_{k=0}^{q} b_{jk} m^k \tag{4.4.12}$$

so that a small number of tabulated coefficients, a_j and b_{jk}, would determine i_1 over a wide range of the parameters α and m. They utilized computations at 90° for $\alpha = 0.1(0.1)10.0$ and $m = 1.60(0.04)2.08$ to attempt to fit the light scattering functions for both spheres and cylinders with such polynomial

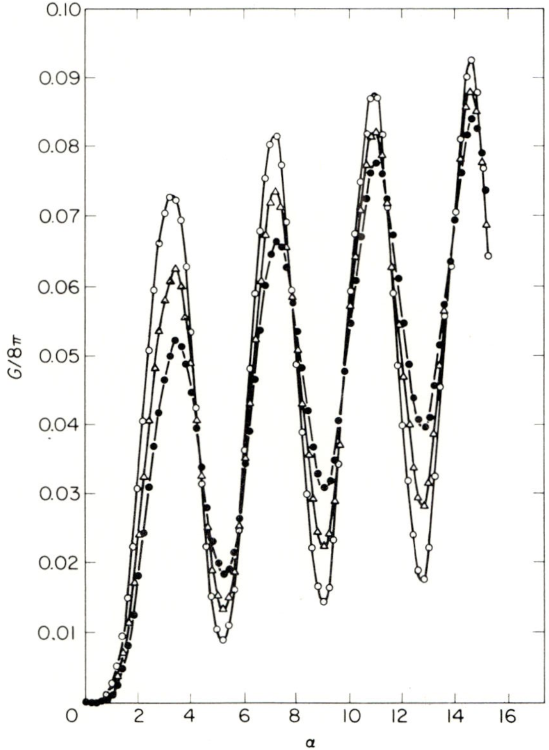

FIG. 4.58. Angular gain at 45° for $m = 1.20$. The points are used as follows: ○ for G_1; ● for G_2; and △ for G_u (Heller and Nakagaki, 1959).

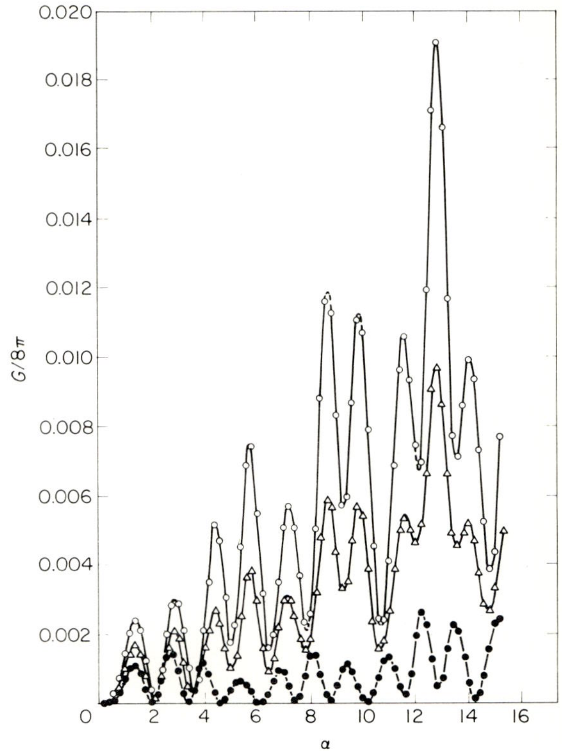

FIG. 4.59. Angular gain at 135° for $m = 1.20$. The points are used as follows: ○ for G_1; ● for G_2; and △ for G_u (Heller and Nakagaki, 1959).

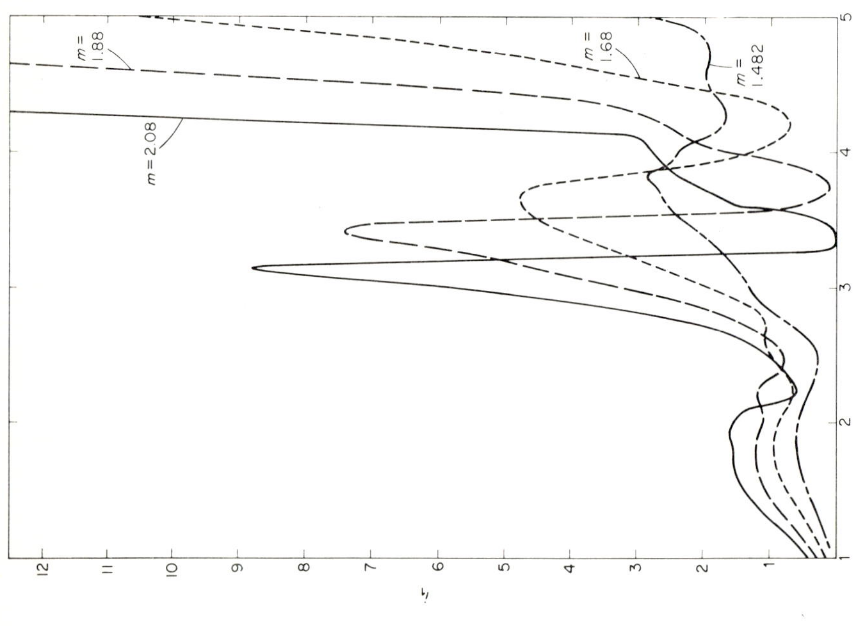

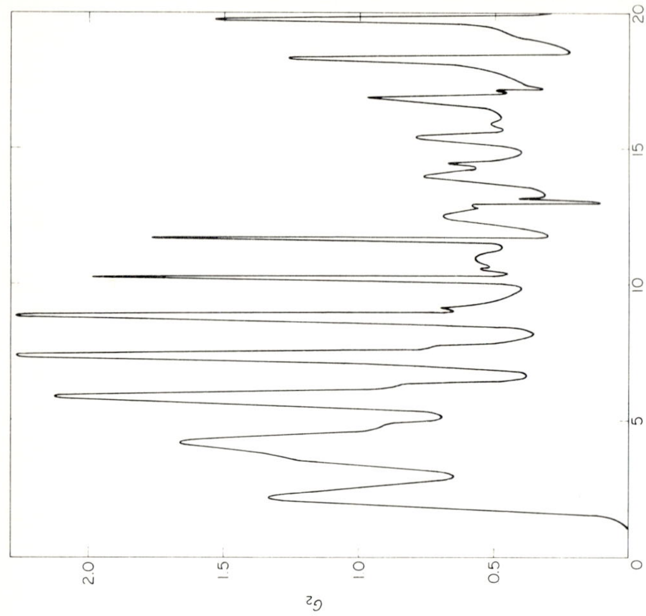

FIG. 4.61. Angular gain G_2 at 90° for $m = 1.60$ plotted against α up to $\alpha = 20$.

FIG. 4.60. Angular intensity function i_1 at 90° for $m = 1.4821, 1.68, 1.88$, and 2.08 plotted against α up to $\alpha = 5$.

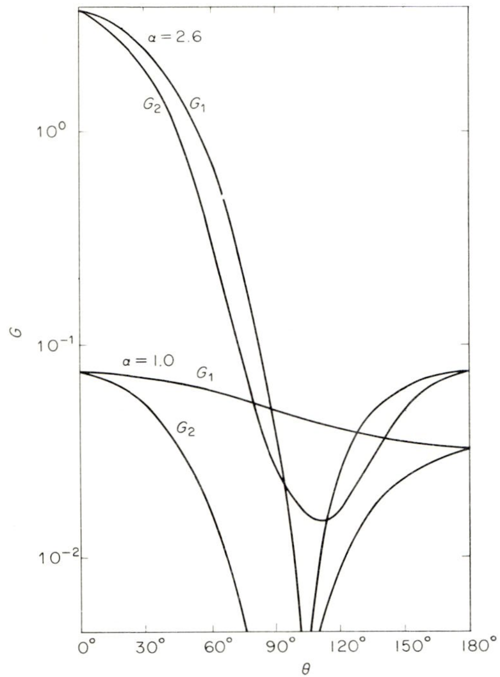

FIG. 4.62. Angular gain G_1 and G_2 for $\alpha = 1.0$ and 2.6 and for $m = 1.20$ plotted against θ.

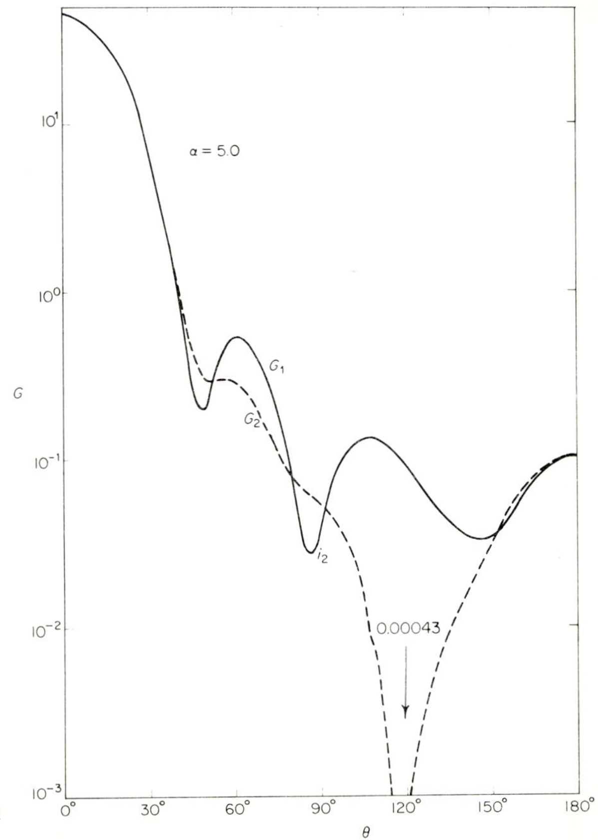

FIG. 4.63. Angular gain G_1 and G_2 for $\alpha = 5.0$ and for $m = 1.20$ plotted against θ.

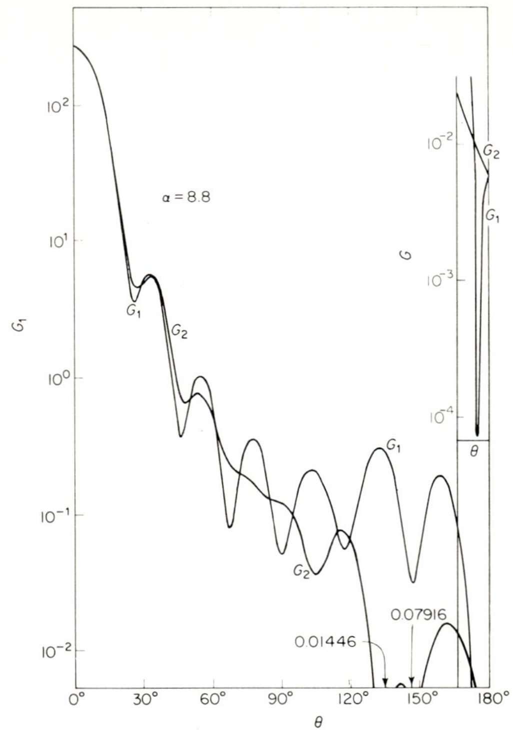

FIG. 4.64. Angular gain G_1 and G_2 for $\alpha = 8.8$ and for $m = 1.20$ plotted against θ.

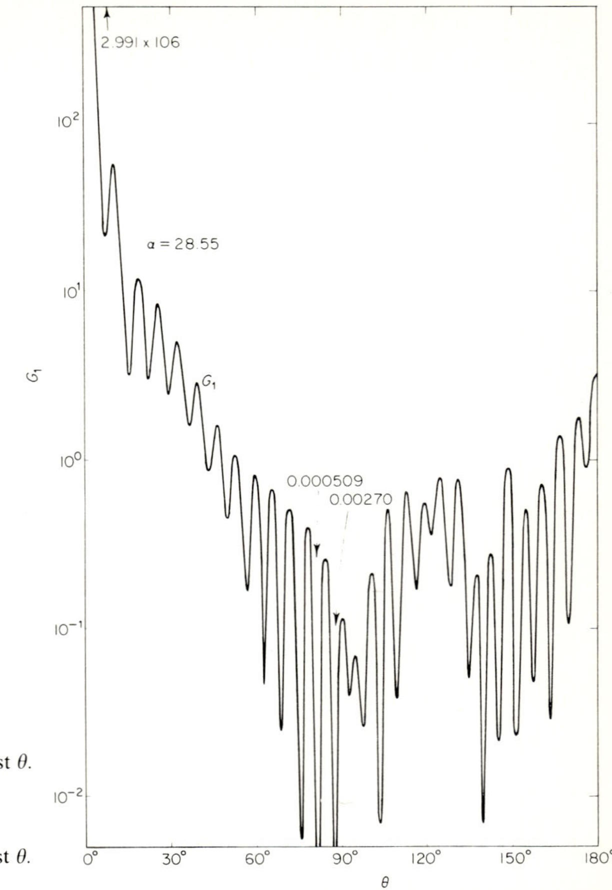

FIG. 4.65. Angular gain G_1 for $\alpha = 28.55$ and for $m = 1.20$ plotted against θ.

expansions, aiming for agreement with the exact computations of at least 2%. A variety of curve fitting techniques were used but even over the limited range of $\alpha = 1.6(0.1)4.0$, the scattering functions were too ill-conditioned to be fit to a high degree of accuracy by an array of coefficients as defined above.

c. Angular Variation. The angular variation of G_1 and G_2 is shown in Figs. 4.62 to 4.65 for $m = 1.20$ and for $\alpha = 1.0, 2.6, 5.0, 8.8$, and 28.55, and in Figs. 4.66 and 4.67, for $m = 1.33$ and $\alpha = 60$. A number of characteristic features are immediately apparent including the preponderance of forward scattering (the Mie effect) and the oscillation with scattering angle.

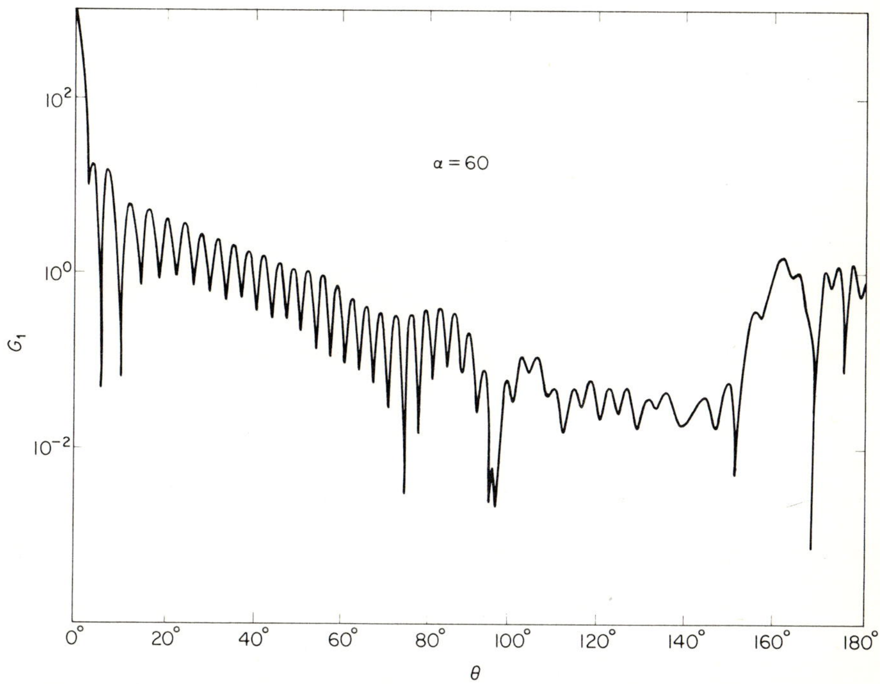

FIG. 4.66. Angular gain G_1 for $\alpha = 60$ and for $m = 1.33$ plotted against θ (Giese *et al.*, 1962).

We will first consider the progression of the extrema (for G_1) as α increases. At $\alpha = 1$, it is apparent that the scattering in the forward direction is somewhat more intense. The actual value of G_1 ranges from 0.031 at $\theta = 180°$ to 0.075 at $\theta = 0°$ compared to the constant value of 0.0655 which would be given for this size by the Rayleigh equation. With increasing α, a series of minima and maxima appear. The curve for $\alpha = 2.6$ shows an exceedingly deep minimum. The onset and development of this can be followed with the

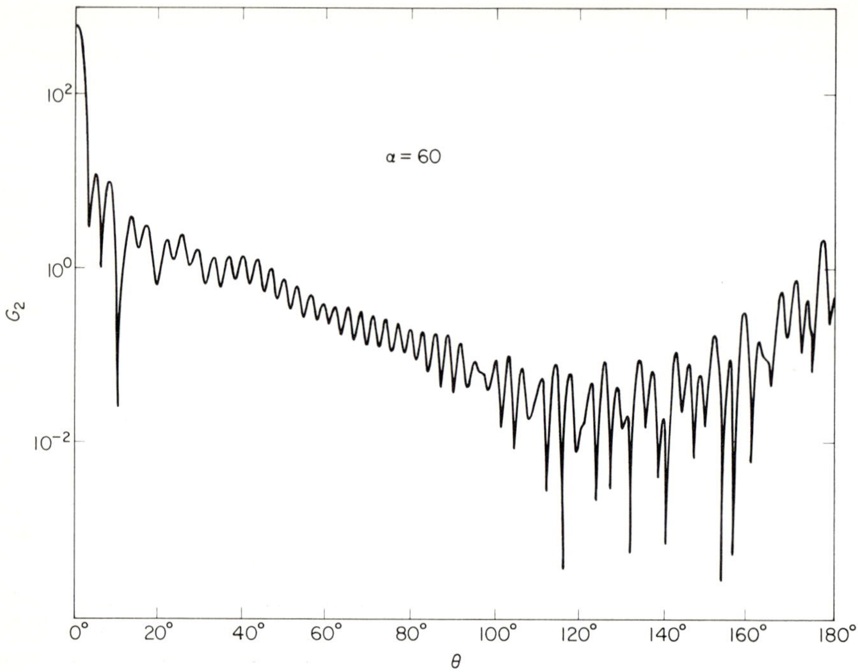

FIG. 4.67. Angular gain G_2 for $\alpha = 60$ and for $m = 1.33$ plotted against θ (Giese *et al.*, 1962).

aid of the information in Table 4.7. The second row of the table gives the scattering angle at which the first minimum appears for each of the values of α given in the top row. The third row gives the "depth" of the minimum.

TABLE 4.7
ANGULAR LOCATION OF EXTREMA IN G_2 UP TO $\alpha = 5$ FOR $m = 1.20$

α	2.0	2.2	2.4	2.6	2.8	3.0	3.2	4.0	5.0
$\theta_{min}(1)$	161°	129°	114°	103°	94°	83°	78°	63°	50°
Depth	1.1	8.2	23.8	22.8	9.4	5.9	1.6	—	—
$\theta_{max}(1)$	—	—	—	—	153°	118°	108°	81°	61°
$\theta_{min}(2)$	—	—	—	—	—	—	168°	116°	86°
$\theta_{max}(2)$	—	—	—	—	—	—	—	—	108°
$\theta_{min}(3)$	—	—	—	—	—	—	—	—	147°

This is defined as the ratio of the backscatter gain ($\theta = 180°$) to the gain at the minimum. The shallow minimum which first develops at about $\alpha = 2.0$ at 161° migrates toward the forward direction with increasing α.

At first it deepens very sharply. A maximum develops in the backward direction at $\alpha = 2.8$. With still increasing α, these extrema migrate toward the forward directions and additional maxima and minima arise in the backward directions. Thus for $\alpha = 5.0$, there are three minima and two maxima, as shown in Fig. 4.63 and listed in Table 4.7. The patterns for $\alpha = 8.8$ and 28.55 show a large number of more or less uniformly spaced extrema with the number of minima approximately equal to α so that the periodicity is $180/\alpha$. The envelope curves for $\alpha = 28.5$ drawn through the maxima and minima in the forward directions are rather smooth. However in the backward directions the amplitude of the oscillating curve is less regular. Some of the minima in the neighborhood of 90° are particularly deep so that the gain may undergo a change of as much as three orders of magnitude over an angular interval of only 3°. The oscillations are even less regular for $\alpha = 60$, even in the forward direction. This is due in part to the higher refractive index ($m = 1.33$) as well as to the larger value of α.

The parallel component G_2 exhibits a single minimum at $\alpha = 1$ which differs somewhat from Rayleigh scattering since it does not quite go to zero ($G_2 \cong 2 \times 10^{-5}$) and it occurs at an angle somewhat greater than 90°. For $\alpha = 2.6$, the minimum has shifted to 110° and is quite shallow. With still increasing α, this principal minimum continues to shift towards larger angles. Characteristic shoulders and extrema develop in the forward directions as can be seen for $\alpha = 5.0$ and 8.8. At sufficiently large α, the curve for G_2 exhibits a pattern of maxima and minima which is qualitatively similar to that for G_1 although the detailed structure is different. This is illustrated in Fig. 4.67.

d. Angular Positions of the Extrema. The scattering patterns corresponding to each particular value of m and α are so different that they provide a practically unique identification for a particular sphere. This offers the experimentalist a powerful tool for the determination of particle size by light scattering. Actually, there is more information in the scattering diagram than is usually needed, and it is often sufficient merely to locate the angular positions at which the maxima and minima occur in order to characterize a particle size. These positions frequently turn out to be remarkably regular. In Fig. 4.68 the angular locations of the extrema of G_1 obtained from Penndorf's (1961) calculations for $m = 1.33$, are plotted up to $\alpha = 15$. These originate near the backward direction and then migrate forward with increasing α. As each new extremum arises, it follows a parallel course of migration. Any particular value of α is characterized by the number and angular location of its extrema.

The behavior of G_2 is not as clear cut. The extrema are not as well defined, sometimes merging into shoulders, disappearing as α increases, and then reappearing at still higher values. In this case, new extrema arise in the lateral directions (50° to 130°) rather than near the backward direction. Some of these then migrate forwards and others migrate backwards as α increases.

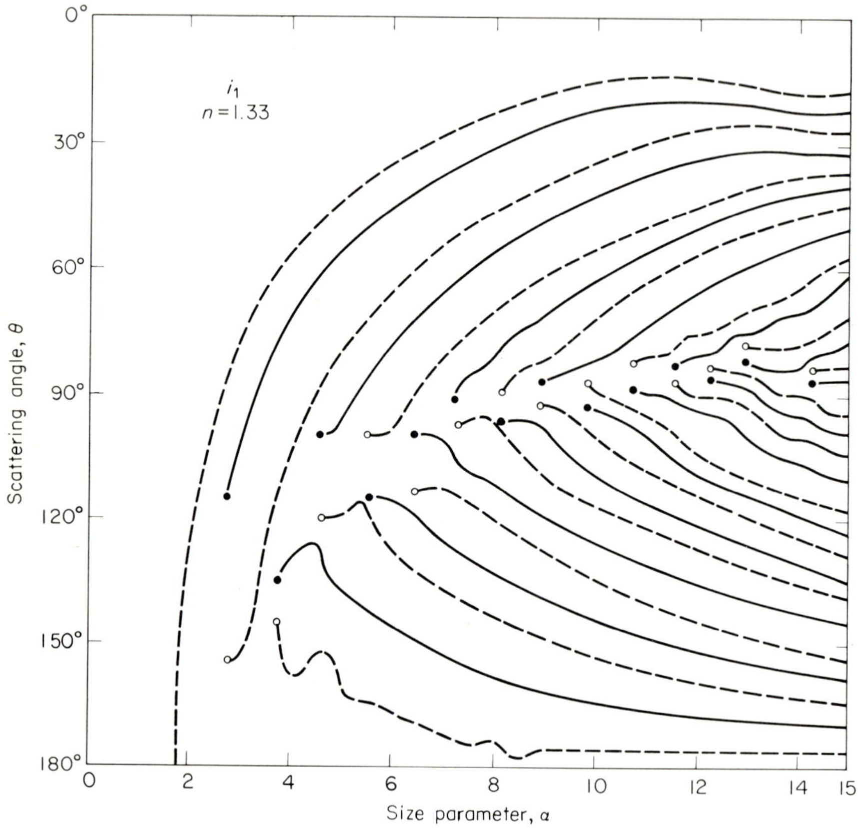

FIG. 4.68. Contours of the location of maxima (solid lines) and minima (dashed lines) of the angular gain G_1 for $m = 1.33$ on the size parameter ($\alpha = 0$ to 15)—angle of observation ($\theta = 0°$ to 180°) plane (Penndorf, 1961).

A typical plot for still higher m and α is shown in Fig. 4.69. This is for the parallel component G_2 and a refractive index of $m = 1.486$. It covers the range of α from 18.0 to 23.9 and θ from 0° to 180°. The calculations were obtained at intervals of 1° so that the extrema could be located very precisely (Rowell, 1962). Typically, the extrema at small scattering angles migrate towards the forward direction with increasing α. In the neighborhood of 10°, each maximum-minimum pair merges and disappears.

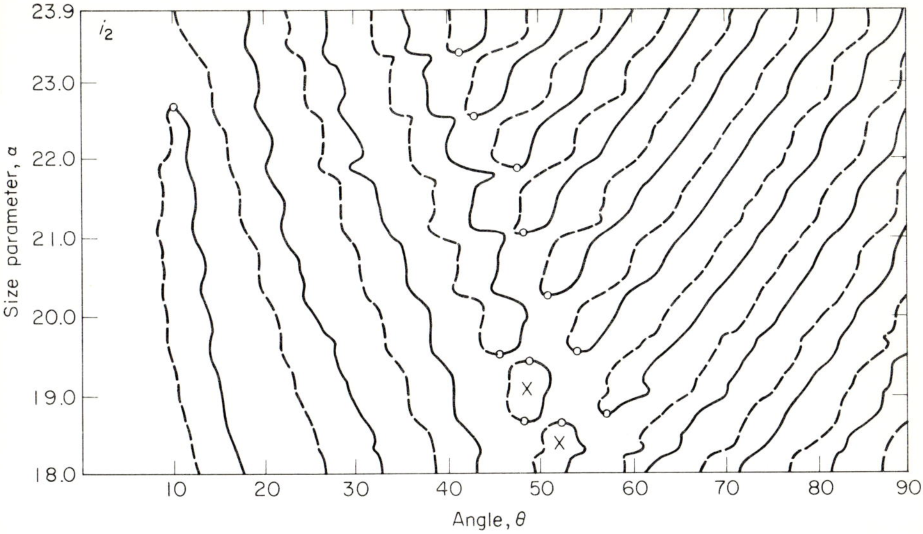

FIG. 4.69. Contours of the location of maxima (solid lines) and minima (dashed lines) of the angular gain G_2 for $m = 1.486$ on the size parameter ($\alpha = 18.0$ to 23.9)—angle of observation ($\theta = 0°$ to 90°) plane (Rowell, 1962).

There is a region of demarcation beyond which one of two things occurs. In the case depicted in Fig. 4.69, beyond 40° to 60° the migration of the extrema is towards the backward directions with increased α rather than forwards as at the smaller angles. In other cases, there is no clearcut pattern beyond the demarcation region. Minima and maxima arise and disappear over a very limited region so that the $\alpha\theta$-domain becomes filled by closed circles similar to those designated by X in the neighborhood of $\theta = 50°$ and $\alpha = 18.0$ to 19.0.

The extrema at low scattering angles vary quite regularly, provided m is not too large. The angular positions of the first several maxima and minima are given quite precisely (Maron and Elder, 1963a, b; Maron *et al.*, 1963b) by

$$\alpha \sin(\theta/2) = k_i(m) \qquad \text{or} \qquad K_i(m) \tag{4.4.13}$$

where the parameters $k_i(m)$ for the minima and $K_i(m)$ for the maxima depend only upon m. The value of θ is the scattering angle at which the ith maximum or minimum, counting from the forward direction, occurs. In the limit of $m \to 1$, these correspond to the minima and maxima in the Rayleigh–Debye formula (Chapter 8). Values of $k_i(m)$ and $K_i(m)$ based upon calculations of the angular intensity functions at 1° intervals by Kerker *et al.*, (1964a) will be discussed in more detail in Section 7.4.1.

4.4.3 Complex Refractive Index

The effect of optical absorption on the angular intensity has been studied by Olaf and Robock (1961), Deirmendjian and Clasen (1962), and Plass (1966). Plass has presented results for $\alpha = 1, 5, 8$, and $n = 1.33, 1.5, 2.0$ over a range of absorption indices. Some typical curves for $\alpha = 8$ and $n = 1.33$ are shown in Fig. 4.70.

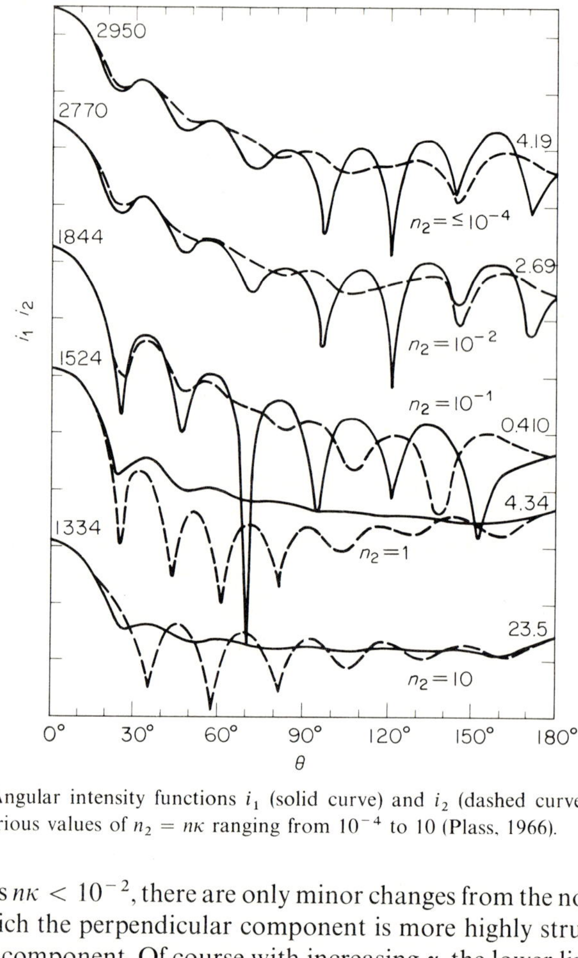

FIG. 4.70. Angular intensity functions i_1 (solid curve) and i_2 (dashed curve) for $n = 1.33$, $\alpha = 8$, and various values of $n_2 = n\kappa$ ranging from 10^{-4} to 10 (Plass, 1966).

As long as $n\kappa < 10^{-2}$, there are only minor changes from the nonabsorbing case for which the perpendicular component is more highly structured than the parallel component. Of course with increasing α, the lower limit at which the absorption begins to affect the scattering curves would be expected to decrease. At $n\kappa = 0.1$ there is a substantial effect. In this case the minima in the i_1 curve become more pronounced and sharper, although in other cases these may tend to become broader or disappear. With still increasing absorption, the oscillations in i_1 become damped out. On the other hand, i_2 seems always to develop deeper and more pronounced minima.

The values of the intensity at 0° and 180° are marked on the curves. The ratio of backward to forward scattering always decreases with increasing absorption to a minimum value in the range from $n\kappa = 10^{-2}$ to 1 and then increases.

Still another aspect of the scattering by highly absorbing spheres is the flatness of the scattering curve in the backward directions, particularly for i_1.

4.4.4 Ray Optics

We have already seen how the backscatter efficiency can be interpreted in terms of an interference among rays which are reflected from the surface of the sphere and those which are refracted into the sphere and then emerge after a second refraction. The latter rays may have undergone one or several internal reflections. For spheres which are not too large, there is an additional contribution from rays which have been sprayed off from the surface wave. Obviously this approach utilizing ray optics can also be used at any other angle. When considering the forward scatter, there will also be a contribution from the diffracted rays passing near the particle. Indeed by Babinet's principle, in the limit of large particles which have an efficiency for extinction of 2, half of the scattering arises from this diffraction. In this limit, the combination of diffraction theory and geometrical optics should approach the results of the exact theory.

The angular distribution of the diffraction pattern is described by (4.4.8). For spheres significantly larger than the wavelength, the diffracted radiation is concentrated very close to the forward direction. There are two interesting aspects to this part of the scattering. The polarization is the same as that of the incident beam and there is no dependence upon the refractive index of the sphere. As pointed out earlier, the scattering coefficients can be separated into two terms [(4.1.1) and (4.1.2)], one of which depends upon m and the other of which is independent of m. The latter will be associated with the diffraction. For a reason to be discussed shortly only scattering coefficients for $n + \frac{1}{2} < \alpha$ are considered. Then the contribution to scattering by diffraction is obtained by setting $a_n = b_n = \frac{1}{2}$ and by replacing the angular functions in (3.3.56) and (3.3.57) by the asymptotic formulas for large n and small θ

$$\pi_n(\cos\theta) = \tfrac{1}{2}n(n+1)\{J_0[(n+\tfrac{1}{2})\theta] + J_2[(n+\tfrac{1}{2})\theta]\} \tag{4.4.14}$$

$$\tau_n(\cos\theta) = \tfrac{1}{2}n(n+1)\{J_0[(n+\tfrac{1}{2})\theta] - J_2[(n+\tfrac{1}{2})\theta]\} \tag{4.4.15}$$

which leads to

$$G_1{}^{\mathrm{F}} = G_2{}^{\mathrm{F}} = 4\alpha^2[J_1{}^2(\alpha\theta)/(\alpha\theta)^2] \tag{4.4.16}$$

This differs from (4.4.8) by the argument $\alpha\theta$ compared to $\alpha \sin \theta$. It arises from the neglect of higher order terms in the expansion of $\sin \theta$.

Van de Hulst (1946) has proposed a localization principle which identifies each of the scattering coefficients, a_n and b_n, with those geometrical optics rays incident upon the sphere at a distance $(n + \frac{1}{2})\lambda/2\pi$ from the origin. This is illustrated in Fig. 4.71. For $(n + \frac{1}{2}) = \alpha$, this corresponds to a distance equal to the radius of the sphere. The series expansions for the scattering amplitudes [(3.3.56), (3.3.57)] converge very rapidly when n exceeds α by only a few integers, so that higher terms can be neglected. Only those scattering coefficients whose order is of this magnitude and less contribute to the scattering. By the localization principle, these coefficients are identified with those rays directly incident or passing very close to the sphere. This principle is suggested by the following remarkable finding. In the asymptotic expression for each of the scattering coefficients, there occurs the Fresnel reflection coefficient corresponding to the angle of incidence of that localized ray which is associated with that particular scattering coefficient.

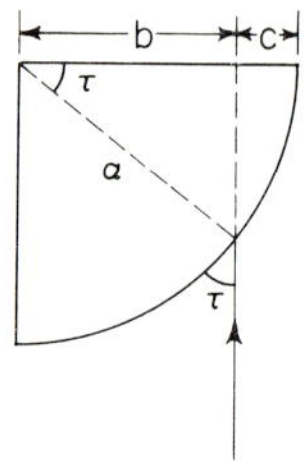

FIG. 4.71. Geometrical optics ray incident at angle τ associated with nth partial wave according to the localization principle. The distances $b = (n + \frac{1}{2})\lambda/2\pi$ and $c = a - (n + \frac{1}{2})\lambda/2\pi$.

Van de Hulst has carried through the complete reduction. The reflection-refraction contribution to the scattering is associated with that part of the scattering coefficient dependent upon the refractive index

$$a_n{}^1 = -\tfrac{1}{2}\exp(-2i\alpha_n); \qquad b_n{}^1 = -\tfrac{1}{2}\exp(-2i\beta_n) \tag{4.4.17}$$

The final result of the reduction from the exact scattering theory is equivalent to that obtained directly by tracing of the localized reflected-refracted rays.

Van de Hulst estimates that the approximation should be reasonably accurate for dielectrics in the range $\alpha = 10$ to 20 except in the neighborhood of the rainbow angle and the glory where it will fail no matter how large α. A definitive comparison of the asymptotic and the exact theory has not yet been carried out, although van de Hulst (1957) has outlined a program for this computation and has carried out some sample calculations. However, the labor involved in using the asymptotic formulas is by no means trivial. Any practical work is likely to be carried out by computer and since the exact theory lends itself just as readily to machine computation it is doubtful whether the asymptotic formulas offer any computational advantage.

Hodkinson and Greenleaves (1963) have proposed a simplified ray optics treatment which is useful in approximating the scattering at small angles. The intensity is calculated as though it arose from three effects: (a) diffraction of light passing near the particle, (b) specular reflection of light from the surface, and (c) refraction of light through the particle neglecting contributions from rays undergoing internal reflections. The treatment is applied to a collection of particles having a range of sizes so that phase effects are averaged out and also so that the average efficiency is 2. Half of the contribution to the efficiency arises from diffraction and the other half from reflection-refraction. The contribution to the angular intensity function from each of these sources is given by

$$i_1 + i_2 = i_a + i_b + i_c \tag{4.4.18}$$

where

$$i_a = i_{1a} + i_{2a} = 2\alpha^2\left[\frac{J_1(\alpha\sin\theta)}{\sin\theta}\right]^2 \tag{4.4.19}$$

$$\begin{aligned} i_b = i_{1b} + i_{2b} = {} & \frac{\alpha^2}{8}\left\{\frac{\sin(\theta/2) - [m^2 - 1 + \sin^2(\theta/2)]^{1/2}}{\sin(\theta/2) + [m^2 - 1 + \sin^2(\theta/2)]^{1/2}}\right\}^2 \\ & + \frac{\alpha^2}{8}\left\{\frac{m^2\sin(\theta/2) - [m^2 - 1 + \sin^2(\theta/2)]^{1/2}}{m^2\sin(\theta/2) + [m^2 - 1 + \sin^2(\theta/2)]^{1/2}}\right\}^2 \end{aligned} \tag{4.4.20}$$

$$\begin{aligned} i_c = i_{1c} + i_{2c} = {} & 2\alpha^2\left(\frac{m}{m^2 - 1}\right)^4 \frac{[m\cos(\theta/2) - 1]^3[m - \cos(\theta/2)]^3}{\cos(\theta/2)[m^2 + 1 - 2m\cos(\theta/2)]^2} \\ & \times (1 + \sec^4(\theta/2) \end{aligned} \tag{4.4.21}$$

The first term in each of the above expressions corresponds to the perpendicularly polarized component and the second to the parallel one. In Fig. 4.72 the result of a calculation for a distribution of spheres spanning the range $\alpha = 10$ to 15 and with $m = 2.105$ is compared with an exact computation over the angular range $\theta = 0$ to 60°. The agreement was generally at least this good. For absorbing spheres (Hodkinson, 1963a) it was necessary to consider only the contributions of diffraction and reflection, since the energy carried by the refracted rays was lost within the sphere.

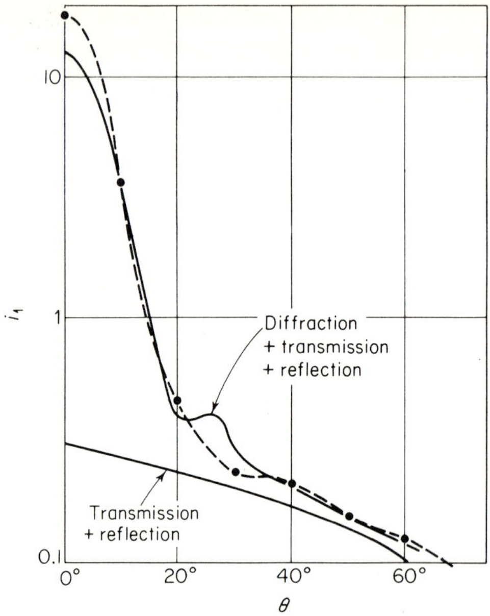

FIG. 4.72. Angular distribution function calculated as the sum of contributions from diffraction, reflection, and refraction [(4.4.21) to (4.4.24), plotted as full line] compared with the exact values (dashed line) for $m = 2.105$. A size distribution over the range $\alpha = 10$ to 15 is assumed for which the frequency of particles is proportional to α^{-2} (Hodkinson and Greenleaves, 1963).

4.5 Radiation Pressure

Irvine (1963, 1965) has made a detailed numerical study of the asymmetry factor and the efficiency for radiation pressure [see Shifrin and Zelmanovich (1964), Shifrin (1964) for additional results]. For the limiting case of large dielectric spheres, for which $Q_{sca} = Q_{ext} = 2$, the asymmetry factor is given in Table 4.8. The values drop smoothly from 1.0 to 0.5 as m increases from unity to infinity.

When the refractive index is unity, there are no reflected rays and all of the radiation incident upon the sphere passes through it undeviated. The condition that the particle is large ensures that the diffracted energy is also in the forward direction. The result of this is that the asymmetry factor is unity (3.11.3) and as a consequence the light pressure (3.11.2) is zero.

On the other hand for $m = \infty$, there are no refracted rays and the specularly reflected rays are isotropically distributed. Accordingly, they make no contribution to the asymmetry factor. There is only the contribution by the forward diffracted radiation resulting in $\overline{\cos\theta} = 0.5$.

TABLE 4.8

ASYMMETRY FACTOR FOR LARGE DIELECTRIC SPHERES

m	$\overline{\cos\theta}$	m	$\overline{\cos\theta}$
1.00	1.00000	2.00	0.71212
1.10	0.97308	2.10	0.69448
1.20	0.93407	2.30	0.66353
1.30	0.89612	2.50	0.63745
1.40	0.86134	2.75	0.61040
1.50	0.82990	3.0	0.58832
1.60	0.80153	3.5	0.55525
1.70	0.77587	4.0	0.53254
1.80	0.75259	5.0	0.50543
1.90	0.73142	∞	0.5

Some results for large absorbing spheres are given in Table 4.9. This corresponds to the case where all of the refracted rays are absorbed within the sphere. The absorption is expressed as $\tan\beta = n\kappa/(n-1)$ where n is the real part of the refractive index and κ is the absorption index. The values of Q_{pr} and $\overline{\cos\theta}$ are virtually independent of the absorption index as long as $\tan\beta < 0.1$. This condition corresponds to $n\kappa$ being small compared to n.

TABLE 4.9

ASYMMETRY FACTOR AND RADIATION PRESSURE FOR LARGE ABSORBING SPHERES

n	$\tan\beta$	Q_{pr}	$\overline{\cos\theta}$	n	$\tan\beta$	Q_{pr}	$\overline{\cos\theta}$
1.20	10^{-4}	0.97071	0.98563	1.50	10^{-4}	0.96066	0.95196
	10^{-2}	0.97071	0.98562		10^{-2}	0.96066	0.95196
	10^{-1}	0.97059	0.98551		10^{-1}	0.96057	0.95158
	1	0.96228	0.97461		1	0.95575	0.91802
	10	0.97706	0.68775		10	1.00800	0.55289
1.33	10^{-4}	0.96415	0.97177	1.75	10^{-4}	0.96059	0.92209
	10^{-2}	0.96415	0.97176		10^{-2}	0.96058	0.92208
	10^{-1}	0.96403	0.97153		10^{-1}	0.96054	0.92149
	1	0.95661	0.95086		1	0.96015	0.87134
	10	0.99904	0.59884		10	1.00970	0.52796

According to the Fresnel equations, the angular distribution of the reflected and the refracted rays will then be determined almost entirely by the value of n. Since the particle is so large that all of the refracted energy is lost by absorption and since the diffraction is independent of the optical constants, it follows that the asymmetry factor and the radiation pressure will depend primarily upon n and not upon κ.

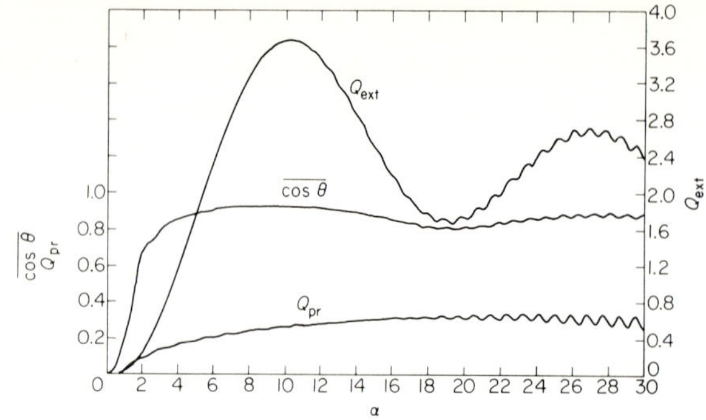

FIG. 4.73. Asymmetry factor $\overline{\cos\theta}$ and efficiencies for radiation pressure Q_{pr} and for extinction Q_{ext} for $m = 1.20$ (Irvine, 1965).

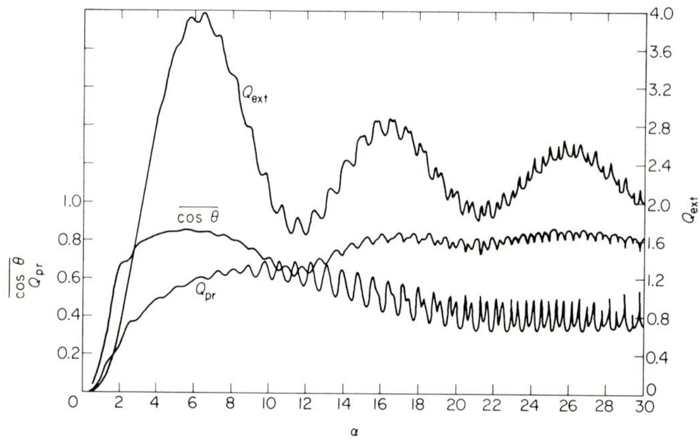

FIG. 4.74. Asymmetry factor $\overline{\cos\theta}$ and efficiencies for radiation pressure Q_{pr} and for extinction Q_{ext} for $m = 1.33$ (Irvine, 1965).

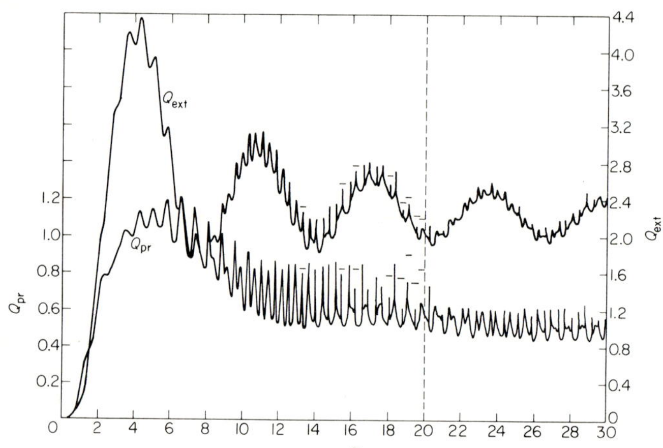

FIG. 4.75. Efficiencies for radiation pressure Q_{pr} and for extinction Q_{ext} for $m = 1.50$ (Irvine, 1965).

In Figs. 4.73 to 4.75, $\overline{\cos\theta}$ and Q_{pr} are plotted, along with Q_{ext}, up to $\alpha = 30$ for $m = 1.20$, 1.33, and 1.50. A very fine net of points has been used in order to delineate the ripple structure. This is more pronounced at the higher refractive indices. The ripples in $\overline{\cos\theta}$ are just out of phase with those in Q_{ext}, a peak in the latter corresponding to a trough in $\overline{\cos\theta}$. This means that the additional energy corresponding to such a peak is preferentially scattered to the backward directions compared to the main component which tends to be concentrated in the forward directions. However, as Orchard (1965) has pointed out, this effect depends strongly upon refractive index so that for $m = 1.87$ the major oscillations of $\overline{\cos\theta}$ are comparable to and in phase with Q_{ext}.

The effect of increasing the absorption index upon Q_{pr} is shown in Fig. 4.76 for $n = 1.20$, 1.35, 1.50 and for $\tan\beta = 0.01$, 0.1, 1.0, and 10. The sharp peaks in the ripple structure already are noticeably attenuated for $\tan\beta = 0.01$.

The asymmetry factor for $m = \infty$ is shown in Fig. 4.77. The negative values for small size are a result of the greater backward scattering for small, perfectly reflecting spheres, a phenomenon which has already been discussed.

4.6 Plasmas

A plasma is a neutral or nearly neutral system composed of a large number of free electrons and positive ions. Although Sir William Crookes, in studying the special properties of matter in electrical discharge tubes had advanced the idea that such gases should be considered a fourth state of matter, it was not until the work of Tonks and Langmuir (1929) that plasma physics has become a recognizable field of study.

Here, we are interested in the fact that in the presence of an applied alternating electrical field a plasma will exhibit conductivity and polarization so that it can be assigned a complex dielectric constant and a corresponding refractive index. Because the electrons in a plasma exhibit oscillations of finite amplitude, the plasma will acquire a net oscillating electric moment per unit volume which is out of phase with the applied electrical field. This results in a contribution of the electrons to the dielectric constant which is given by the vacuum contribution decreased by the bulk plasma polarization, explicitly

$$m^2 = \varepsilon = 1 - i\frac{(n_e e^2/\varepsilon_0 M_e)(1/\omega^2)}{(\omega_g/\omega) + i} \tag{4.6.1}$$

where n_e = electron density, e = electronic charge, M_e = electronic mass, ω_g = electron collision frequency, ω = circular frequency of the radiation,

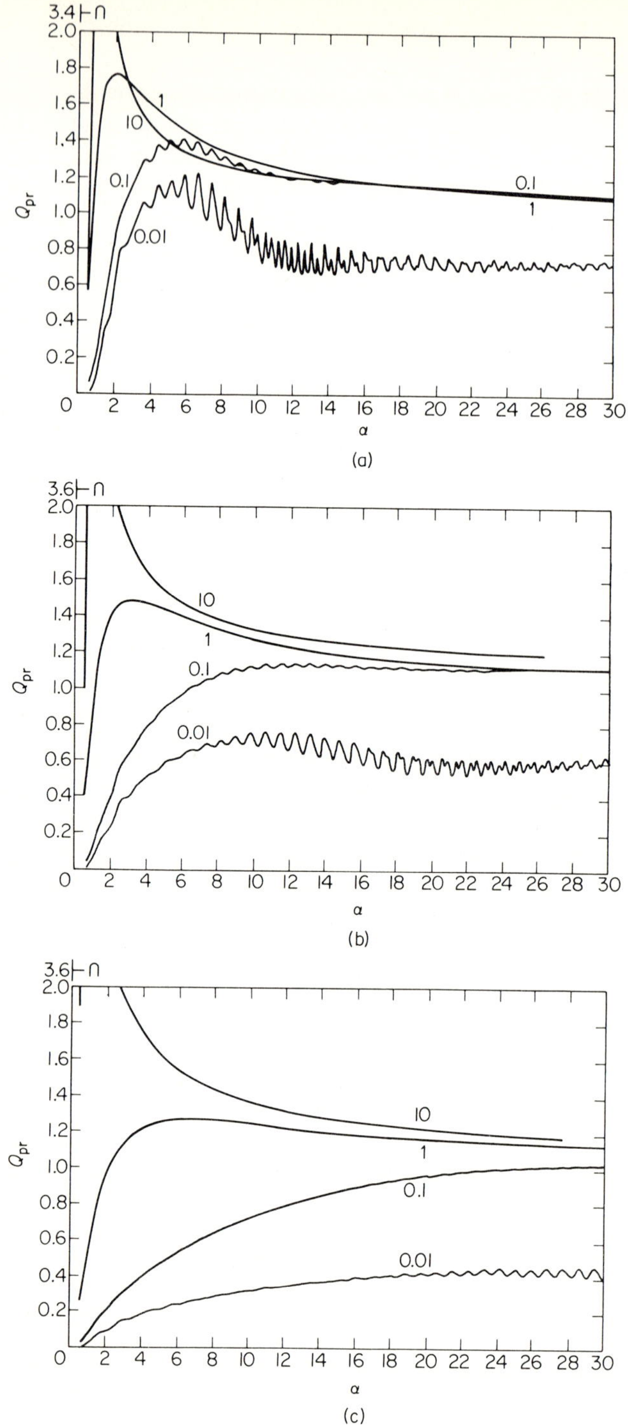

FIG. 4.76. Efficiency for radiation pressure Q_{pr} of absorbing spheres. Curves (a) are for $n = 1.50$, curves (b) are for $n = 1.33$, and curves (c) are for $n = 1.20$. The numbers labeling each curve designate values of $\tan \beta = n\kappa/(n - 1)$ (Irvine, 1965).

and ε_0 = permittivity of free space. It is convenient to define the parameters

$$\eta = \frac{n_e e^2}{\varepsilon_0 M_e \omega^2} = \left(\frac{\omega_p}{\omega}\right)^2 \tag{4.6.2}$$

and

$$\gamma = \omega_g/\omega \tag{4.6.3}$$

where ω_p is termed the plasma frequency. If collisions are few, as in the ionosphere, so that $\omega_g = 0$, it is possible to describe the plasma phenomenologically as a dielectric with a relative real dielectric constant of less than unity.

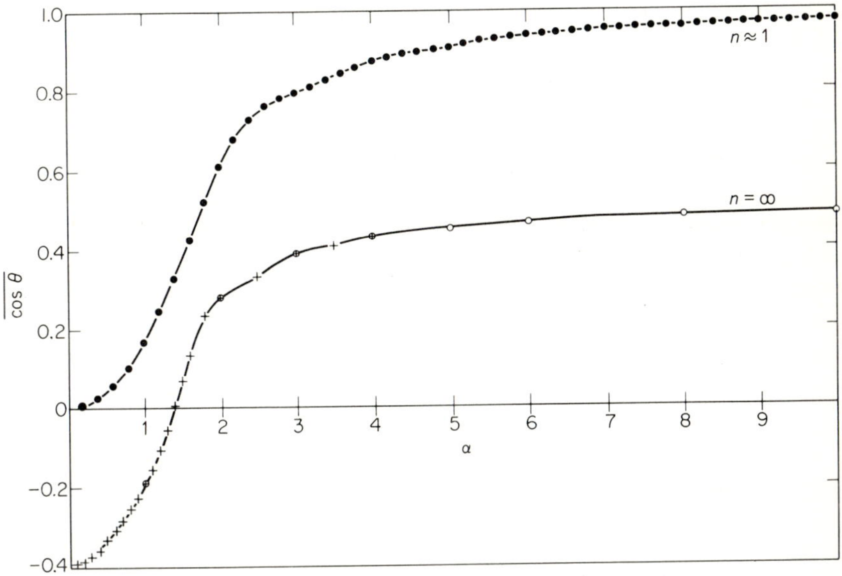

FIG. 4.77. Asymmetry factor $\overline{\cos\theta}$ for totally reflecting spheres (lower curve, $n = \infty$) and for Rayleigh–Debye spheres ($n = 1$) (Irvine, 1965).

It should be apparent that the scattering of electromagnetic radiation by plasma aggregates in the form of spheres and other shapes that are considered in this book may be treated formally just as any other substance, provided the appropriate refractive index is utilized. Accordingly, except for the two examples given below, illustrations of scattering by plasma media will be treated in the appropriate sections dealing with considerations of particle size, particle shape, refractive properties, etc.

Pedersen and Malmstrom (1964) have shown that small spheres with refractive indices comparable to those encountered in plasmas show strong absorption resonances. They observed that these are mainly associated with the first electric partial wave, viz. with a_1. By expanding a_1 as a power series in α, it was possible to show that for each value of α there is a unique complex dielectric constant at which absorption resonance occurs and that this is determined by

$$\eta = 3 + (12/5)\alpha^2 \tag{4.6.4}$$

and

$$\gamma = (2/3)\alpha^3 \tag{4.6.5}$$

Wyatt (1965a, b, 1966) has considered scattering by large spheres ($\alpha \cong 40$) comprised of overdense plasma media. Such media are characterized by a negligible collision frequency, i.e., $\gamma \ll 1$, and by a sufficiently large plasma frequency, i.e., $\eta > 1$, so that the effective dielectric constant is negative. A plane wave incident normally upon a semiinfinite slab of such an overdense medium is perfectly reflected and such objects are said to be equivalent to perfect conductors. However, overdense spheres do not, as a matter of fact, generally scatter as spheres for which $m = \infty$. Wyatt has shown that the backscatter can be interpreted as due to the interference between a specularly reflected axial ray and a surface wave. The surface wave attenuation is considerably less for the overdense plasma than for the corresponding infinite conductor. This results in the surface waves making more than one trip around the sphere before damping away.

CHAPTER 5

Scattering by Stratified Spheres

The problem discussed in this chapter is the scattering of electromagnetic radiation by a sphere whose optical properties are stratified, i.e., the complex refractive index may vary radially from the center to the outer surface. The exact solution of the wave equation will be considered here. Scattering by so-called "soft" obstacles for which the Rayleigh–Debye approximation is adequate will be separately treated later.

There are two principal cases. On the one hand, the sphere may consist of two or more concentric layers within each of which the refractive index is constant. When there are only two layers the object will be termed a coated sphere and will be said to consist of an inner spherical core and a coating which comprises a concentric spherical shell. For the second case the refractive index within any one layer, or indeed throughout the sphere, may vary continuously provided there is radial symmetry. Accordingly, this solution is the most general and will include the layered sphere and in turn the homogeneous sphere as special cases.

5.1 Coated Sphere

The theory of the coated sphere was first derived by Aden and Kerker (1951) and shortly thereafter by Güttler (1952). The notation and formalism here will follow that used earlier for the single sphere and is similar to that of Güttler. Aden and Kerker followed Stratton's (1941) method except that they utilized the logarithmic derivative functions (3.6.21) and (3.6.22).

5.1.1 General Theory

The spherical particle which scatters the electromagnetic radiation is depicted in Fig. 5.1. There are three propagation constants k_1, k_2, and k_3, corresponding to the material of the core, the shell, and the medium. The radius of the core is a, that of the total sphere is b.

The problem is formally very similar to that for the homogeneous sphere. As before, the field is represented by an expansion of the two Debye potentials corresponding to the TM and TE waves. For the incident field these will be

$$r\pi_1{}^i = \frac{1}{k_3{}^2} \sum_{n=1}^{\infty} i^{n-1} \frac{2n+1}{n(n+1)} \psi_n(k_3 r) P_n^{(1)}(\cos\theta)\cos\phi \tag{5.1.1}$$

$$r\pi_2{}^i = \frac{1}{k_3{}^2} \sum_{n=1}^{\infty} i^{n-1} \frac{2n+1}{n(n+1)} \psi_n(k_3 r) P_n^{(1)}(\cos\theta)\sin\phi \tag{5.1.2}$$

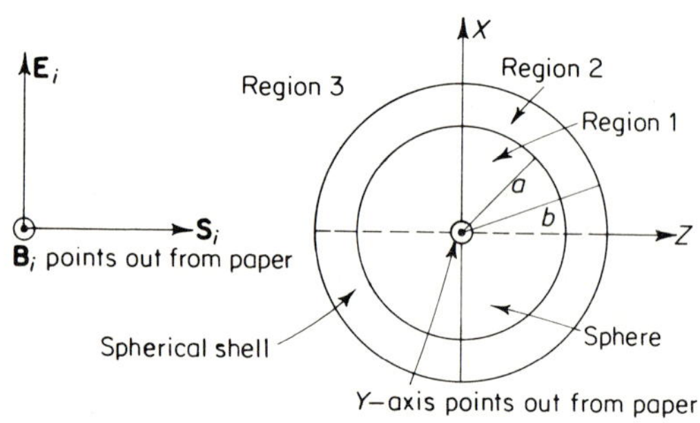

FIG 5.1. Geometry for scattering of sphere coated with spherical shell.

The Debye potentials for the induced secondary field must now be constructed in three parts, one applying to each of the three regions, viz. the scattered wave in the external medium,

$$r\pi_1{}^s = -\frac{1}{k_3{}^2} \sum_{n=1}^{\infty} i^{n-1} \frac{2n+1}{n(n+1)} a_n \zeta_n(k_3 r) P_n^{(1)}(\cos\theta)\cos\phi \tag{5.1.3}$$

$$r\pi_2{}^s = -\frac{1}{k_3{}^2} \sum_{n=1}^{\infty} i^{n-1} \frac{2n+1}{n(n+1)} b_n \zeta_n(k_3 r) P_n^{(1)}(\cos\theta)\sin\phi \tag{5.1.4}$$

the wave in the coating,

$$r\pi_1{}^u = -\frac{1}{k_2{}^2} \sum_{n=1}^{\infty} i^{n-1} \frac{2n+1}{n(n+1)} [c_n \psi_n(k_2 r) + d_n \chi_n(k_2 r)] P_n^{(1)}(\cos\theta)\cos\phi \tag{5.1.5}$$

$$r\pi_2{}^u = -\frac{1}{k_2{}^2} \sum_{n=1}^{\infty} i^{n-1} \frac{2n+1}{n(n+1)} [e_n \psi_n(k_2 r) + f_n \chi_n(k_2 r)] P_n^{(1)}(\cos\theta)\sin\phi \tag{5.1.6}$$

and the wave in the core,

$$r\pi_1{}^v = -\frac{1}{k_1{}^2}\sum_{n=1}^{\infty} i^{n-1}\frac{2n+1}{n(n+1)} g_n\psi_n(k_1 r)P_n^{(1)}(\cos\theta)\cos\phi \qquad (5.1.7)$$

$$r\pi_2{}^v = -\frac{1}{k_1{}^2}\sum_{n=1}^{\infty} i^{n-1}\frac{2n+1}{n(n+1)} h_n\psi_n(k_1 r)P_n^{(1)}(\cos\theta)\sin\phi \qquad (5.1.8)$$

As in the single sphere case, only ψ_n may be used for the incident wave and the wave in the core because ζ_n becomes infinite at the origin. On the other hand, only ζ_n drops off properly at $r = \infty$ and so it is used for the scattered wave. In the intermediate region of the shell, both ψ_n and χ_n are well behaved so that a linear combination of them is used.

The fields must be matched at each of the two boundaries so that at $r = a$:

$$\partial/\partial r\,(r\pi_1{}^u) = \partial/\partial r\,(r\pi_1{}^v) \qquad (5.1.9)$$

$$\partial/\partial r\,(r\pi_2{}^u) = \partial/\partial r\,(r\pi_2{}^v) \qquad (5.1.10)$$

$$\kappa_1^{(2)}(r\pi_1{}^u) = \kappa_1^{(1)}(r\pi_1{}^v) \qquad (5.1.11)$$

$$\kappa_2^{(2)}(r\pi_2{}^u) = \kappa_2^{(1)}(r\pi_2{}^v) \qquad (5.1.12)$$

and at $r = b$:

$$\partial/\partial r\,(r\pi_1{}^u) = \partial/\partial r\,(r\pi_1{}^s + r\pi_1{}^i) \qquad (5.1.13)$$

$$\partial/\partial r\,(r\pi_2{}^u) = \partial/\partial r\,(r\pi_2{}^s + r\pi_2{}^i) \qquad (5.1.14)$$

$$\kappa_1^{(2)}(r\pi_1{}^u) = \kappa_1^{(3)}(r\pi_1{}^s + r\pi_1{}^i) \qquad (5.1.15)$$

$$\kappa_2^{(2)}(r\pi_2{}^u) = \kappa_2^{(3)}(r\pi_2{}^s + r\pi_2{}^i) \qquad (5.1.16)$$

These lead to the following eight sets of linear equations:

$$m_1 c_n\psi_n'(m_2\alpha) + m_1 d_n\chi_n'(m_2\alpha) + m_2 g_n\psi_n'(m_1\alpha) = 0 \qquad (5.1.17)$$

$$c_n\psi_n(m_2\alpha) + d_n\chi_n(m_2\alpha) + g_n\psi_n(m_1\alpha) = 0 \qquad (5.1.18)$$

$$c_n\psi_n'(m_2 v) + d_n\chi_n'(m_2 v) + m_2 a_n\zeta_n'(v) = m_2\psi_n'(v) \qquad (5.1.19)$$

$$c_n\psi_n(m_2 v) + d_n\chi_n(m_2 v) + a_n\zeta_n(v) = \psi_n(v) \qquad (5.1.20)$$

$$m_1 e_n\psi_n'(m_2\alpha) + m_1 f_n\chi_n'(m_2\alpha) + m_2 h_n\psi_n'(m_1\alpha) = 0 \qquad (5.1.21)$$

$$m_1{}^2 e_n\psi_n(m_2\alpha) + m_1{}^2 f_n\chi_n(m_2\alpha) + m_2{}^2 h_n\psi_n(m_1\alpha) = 0 \qquad (5.1.22)$$

$$e_n\psi_n'(m_2 v) + f_n\psi_n'(m_2 v) + m_2 b_n\zeta_n'(v) = m_2\psi_n'(v) \qquad (5.1.23)$$

$$e_n\psi_n(m_2 v) + f_n\chi_n(m_2 v) + m_2{}^2 b_n\zeta_n(v) = m_2{}^2\psi_n(v) \qquad (5.1.24)$$

Assuming that the external medium is nonabsorbing, we have set

$$k_3 a = 2\pi a/\lambda = \alpha; \qquad k_3 b = 2\pi b/\lambda = v \tag{5.1.25}$$

and

$$m_1 = k_1/k_3; \qquad m_2 = k_2/k_3 \tag{5.1.26}$$

so that m_1 and m_2 are the refractive indices of the core and coating relative to that of the external medium. The above equations are now solved for a_n and b_n giving Eqs. (5.1.27) and (5.1.28).

In order to facilitate cross reference, the notation used here is compared in Table 5.1 with that of Aden and Kerker (1951) and of Güttler (1952). The former authors have used the spherical Hankel functions rather than the spherical Neumann functions as one of the two general Bessel functions needed to describe the potential functions in the coating. The results are equivalent although use of the Hankel function is somewhat redundant and involves slightly more computing.

$$
a_n = \frac{\begin{vmatrix} \psi_n'(m_2\alpha) & \chi_n'(m_2\alpha) & \psi_n'(m_1\alpha) & 0 \\ m_2\psi_n(m_2\alpha) & m_2\chi_n(m_2\alpha) & m_1\psi_n(m_1\alpha) & 0 \\ \psi_n'(m_2v) & \chi_n'(m_2v) & 0 & \psi_n'(v) \\ m_2\psi_n(m_2v) & m_2\chi_n(m_2v) & 0 & \psi_n(v) \end{vmatrix}}{\begin{vmatrix} \psi_n'(m_2\alpha) & \chi_n'(m_2\alpha) & \psi_n'(m_1\alpha) & 0 \\ m_2\psi_n(m_2\alpha) & m_2\chi_n(m_2\alpha) & m_1\psi_n(m_1\alpha) & 0 \\ \psi_n'(m_2v) & \chi_n'(m_2v) & 0 & \zeta_n'(v) \\ m_2\psi_n(m_2v) & m_2\chi_n(m_2v) & 0 & \zeta_n(v) \end{vmatrix}} \tag{5.1.27}
$$

$$
b_n = \frac{\begin{vmatrix} m_2\psi_n'(m_2\alpha) & m_2\chi_n'(m_2\alpha) & m_1\psi_n'(m_1\alpha) & 0 \\ \psi_n(m_2\alpha) & \chi_n(m_2\alpha) & \psi_n(m_1\alpha) & 0 \\ m_2\psi_n'(m_2v) & m_2\chi_n'(m_2v) & 0 & \psi_n'(v) \\ \psi_n(m_2v) & \chi_n(m_2v) & 0 & \psi_n(v) \end{vmatrix}}{\begin{vmatrix} m_2\psi_n'(m_2\alpha) & m_2\chi_n'(m_2\alpha) & m_1\psi_n'(m_1\alpha) & 0 \\ \psi_n(m_2\alpha) & \chi_n(m_2\alpha) & \psi_n(m_1\alpha) & 0 \\ m_2\psi_n'(m_2v) & m_2\chi_n'(m_2v) & 0 & \zeta_n'(v) \\ \psi_n(m_2v) & \chi_n(m_2v) & 0 & \zeta_n(v) \end{vmatrix}} \tag{5.1.28}
$$

TABLE 5.1
NOTATION FOR COATED SPHERES

Parameter	This book	Aden and Kerker (1951)	Güttler (1952)	Scharfman (1954)
Scattering coefficients	$a_n; b_n$	$-b_n^s; -a_n^s$	a_l, b_l	$-b_n^s, -a_n^s$
Propagation constants	k_1, k_2, k_3	K_1, K_2, K_3	k_1, k_2, k_3	k_3, k_2, k_1
Complex refractive index		Y_1, Y_2, Y_3		
Complex refractive index relative to the medium	$m_1 = k_1/k_3, m_2 = k_2/k_3$	N_1, N_2	m_1, m_2	—, $(\varepsilon_2/\varepsilon_1)^{1/2}$
Radii	a, b	a, b	R_0, R	$a, (a + \delta)$
Size parameters	$\alpha = k_3 a, v = k_3 b$	α, v	—, x	—, A
	$m_2\alpha, m_2 v$	$N_2\alpha, N_2 v$	$k_2 R_0, k_2 R$	C, B
Bessel functions	$\psi_n(x), \chi_n(x), \zeta_n(x)$	$xz_n^{(1)}(x)$, —, $xz_n^{(3)}(x)$	$\psi_l(x), \chi_l(x), \zeta_l(x)$	$xj_n(x)$, —, $xh_n^{(1)}(x)$
Bessel function derivatives	$\psi_n'(x), \chi_n'(x), \zeta_n'(x)$	$x\eta_n^{(1)}(x)$, —, $x\eta_n^{(3)}(x)$	$\psi_l'(x), \chi_l'(x), \zeta_l'(x)$	$[xj_n(x)]'$, —, $[xh_n^{(1)}(x)]'$

The following operators will be useful in solving the determinants:

$$[\chi\psi]'_{n,v} = \chi'_n(m_2 v)\psi_n(v) - m_2\chi_n(m_2 v)\psi'_n(v) \tag{5.1.29}$$

$$[\chi\psi]''_{n,v} = m_2\chi'_n(m_2 v)\psi_n(v) - \chi_n(m_2 v)\psi'_n(v) \tag{5.1.30}$$

$$[\chi\psi]'_{n,\alpha} = m_1\chi'_n(m_2\alpha)\psi_n(m_1\alpha) - m_2\chi_n(m_2\alpha)\psi'_n(m_1\alpha) \tag{5.1.31}$$

$$[\chi\psi]''_{n,\alpha} = m_2\chi'_n(m_2\alpha)\psi_n(m_1\alpha) - m_1\chi_n(m_2\alpha)\psi'_n(m_1\alpha) \tag{5.1.32}$$

Other symmetrical operators can be defined similarly by substituting for the χ and ψ in the square bracket and correspondingly in the expressions on the right hand side of the equations. This gives

$$a_n = \frac{[\psi\psi]'_{n,\alpha}\cdot[\chi\psi]'_{n,v} - [\chi\psi]'_{n,\alpha}\cdot[\psi\psi]'_{n,v}}{[\psi\psi]'_{n,\alpha}\cdot[\chi\zeta]'_{n,v} - [\chi\psi]'_{n,\alpha}\cdot[\psi\zeta]'_{n,v}} \tag{5.1.33}$$

$$b_n = \frac{[\psi\psi]''_{n,\alpha}\cdot[\chi\psi]''_{n,v} - [\chi\psi]''_{n,\alpha}\cdot[\psi\psi]''_{n,v}}{[\psi\psi]''_{n,\alpha}\cdot[\chi\zeta]''_{n,v} - [\chi\psi]''_{n,\alpha}\cdot[\psi\zeta]''_{n,v}} \tag{5.1.34}$$

From this point, all computations proceed identically as in the spherical case leading to the various parameters of physical interest such as the angular intensity functions, i_1 and i_2, and the efficiencies for extinction, absorption, and scattering, Q_{ext}, Q_{abs}, and Q_{sca}.

a. Reduction to Homogeneous Sphere. The above solution reduces to that given earlier for the single sphere whenever the parameters approach the appropriate limits. For example, when the optical properties of the core and the coating become identical

$$[\psi\psi]'_{n,\alpha} = [\psi\psi]''_{n,\alpha} = 0; \qquad m_1 = m_2 = m \tag{5.1.35}$$

$$a_n = [\psi\psi]'_{n,v}/[\psi\zeta]'_{n,v} \tag{5.1.36}$$

$$b_n = [\psi\psi]''_{n,v}/[\psi\zeta]''_{n,v} \tag{5.1.37}$$

These are equivalent to the single sphere formulas [(3.3.48), (3.3.49)] when now v is replaced by α and m_2 by m. If the optical properties of the coating approach those of the surrounding medium,

$$[\psi\psi]'_{n,v} = [\psi\psi]''_{n,v} = 0; \qquad m_2 = 1 \tag{5.1.38}$$

which, after some reduction of the denominator, can be shown to give

$$a_n = \frac{[\psi\psi]'_{n,\alpha}\cdot[\chi\psi]'_{n,v}}{[\psi\zeta]'_{n,\alpha}\cdot[\chi\psi]'_{n,v}} = \frac{[\psi\psi]'_{n,\alpha}}{[\psi\zeta]'_{n,\alpha}} \tag{5.1.39}$$

$$b_n = \frac{[\psi\psi]''_{n,\alpha}\cdot[\chi\psi]''_{n,v}}{[\psi\zeta]''_{n,\alpha}\cdot[\chi\psi]''_{n,v}} = \frac{[\psi\psi]''_{n,\alpha}}{[\psi\zeta]''_{n,\alpha}} \tag{5.1.40}$$

The cases $a \to b$ and $b \to a$ also reduce properly, if somewhat more laboriously. These cases correspond to the full sphere having the optical properties of the core material and the coating material, respectively. Here the reduction is carried out by substituting a for b in the one case and b for a in the other. The latter situation can also be approached by permitting a to go to zero. Then it follows that because

$$\psi_n(0) = 0 \tag{5.1.41}$$

$$[\psi\psi]'_{n,\alpha} = [\psi\psi]''_{n,\alpha} = 0; \qquad \alpha = 0 \tag{5.1.42}$$

leading to

$$a_n = \frac{[\chi\psi]'_{n,\alpha} \cdot [\psi\psi]'_{n,v}}{[\chi\psi]'_{n,\alpha} \cdot [\psi\zeta]'_{n,v}} = \frac{[\psi\psi]'_{n,v}}{[\psi\zeta]'_{n,v}} \tag{5.1.43}$$

$$b_n = \frac{[\chi\psi]''_{n,\alpha} \cdot [\psi\psi]''_{n,v}}{[\chi\psi]''_{n,\alpha} \cdot [\psi\zeta]''_{n,v}} = \frac{[\psi\psi]''_{n,v}}{[\psi\zeta]''_{n,v}} \tag{5.1.44}$$

Although this reduces analytically to the desired result, numerical calculation from the full formulas will necessitate computation of the χ-function for zero argument. This will lead to infinity so that a machine calculation for this cannot be effected unless appropriate precautions are taken in the computing program.

b. Dielectric Coated Reflecting Sphere. A rather important special case is that of the dielectric coated perfectly reflecting sphere, i.e., $m_1 = \infty$. This can be obtained directly from (5.1.33), and (5.1.34) in the usual way, leading to the following somewhat simplified expressions:

$$a_n = \frac{\psi'_n(m_2\alpha)[\chi\psi]'_{n,v} - \chi'_n(m_2\alpha)[\psi\psi]'_{n,v}}{\psi'_n(m_2\alpha)[\chi\zeta]'_{n,v} - \chi'_n(m_2\alpha)[\psi\zeta]'_{n,v}} \tag{5.1.45}$$

$$b_n = \frac{\psi_n(m_2\alpha)[\chi\psi]''_{n,v} - \chi_n(m_2\alpha)[\psi\psi]''_{n,v}}{\psi_n(m_2\alpha)[\chi\zeta]''_{n,v} - \chi_n(m_2\alpha)[\psi\zeta]''_{n,v}} \tag{5.1.46}$$

Scharfman (1954) has written equivalent expressions using the notation given in Table 5.1. Weston and Hemenger (1962) have developed approximate formulas for a large perfectly conducting sphere coated with a thin layer of material with large complex refractive index.

c. Bubble. Still another case of considerable physical interest is that of a bubble or shell. Now the core has the same properties as the medium; viz. $m_1 = 1$. Unfortunately this provides for no analytical simplification and the full formulation must be used. Of course, for a perfectly reflecting shell, $m_2 = \infty$, the equations reduce exactly to those given earlier for a perfectly conducting sphere [(3.9.18), (3.9.19)].

5.1.2 Small Sphere Limit

Güttler (1952) has developed a series approximation for the scattering coefficients of the coated sphere by expanding the Bessel functions in powers of their arguments in much the same fashion as described earlier for the single sphere. Again the assumption is made that both the refractive indices and the particle dimensions are sufficiently small so that these series can be terminated after only one or two terms. The results are expressed in the following form:

$$a_1 = -ip'_1 v^3 + i\bar{p}'_1 v^5 + \cdots \tag{5.1.47}$$

$$b_1 = i\bar{p}_1'' v^5 \tag{5.1.48}$$

$$a_2 = -ip'_2 v^5 \tag{5.1.49}$$

Explicit expressions for the parameters designated as p will be found in Güttler's paper.

In the extreme limiting case, comparable to Rayleigh scattering, it will be sufficient to consider only the first term a_1. Then

$$i_1 + i_2 = |\tfrac{3}{2}a_1|^2(1 + \cos^2\theta) = |-(3i/2)p'_1|^2 v^6(1 + \cos^2\theta) \tag{5.1.50}$$

where

$$p'_1 = -\frac{2}{3}\left[\frac{(m_2{}^2 - 1)(m_1{}^2 + 2m_2{}^2) + q^3(2m_2{}^2 + 1)(m_1{}^2 - m_2{}^2)}{(m_2{}^2 + 2)(m_1{}^2 + 2m_2{}^2) + q^3(2m_2{}^2 - 2)(m_1{}^2 - m_2{}^2)}\right] \tag{5.1.51}$$

Here q is the ratio of α/v. This reduces to the appropriate Rayleigh expression when $\alpha = 0$ or v or when $m_2 = 1$ or m_1. For $m_1 = 1$, corresponding to a hollow spherical shell,

$$p'_1 = -\frac{2}{3}\left[\frac{(1 - q^3)(m_2{}^2 - 1)(2m_2{}^2 + 1)}{(1 - q^3)(m_2{}^2 + 2)(2m_2{}^2 + 1) + 9q^3 m_2{}^2}\right] \tag{5.1.52}$$

If the shell is sufficiently thin ($q \to 1$) so that only the second term in the denominator need be considered and also if the refractive index is very close to unity, then

$$p'_1 = -(2/9)(1 - q^3)(m_2{}^2 - 1); \qquad q \to 1, \qquad m_2 \to 1 \tag{5.1.53}$$

and

$$i_1 + i_2 = (4/9)|m_2 - 1|^2[(1 - q^3)v^3]^2(1 + \cos^2\theta) \tag{5.1.54}$$

Since $(1 - q^3)$ represents the fraction of the total volume consisting of the material of the shell, the scattering by such a spherical shell reduces to that for an equivalent "soft" Rayleigh scatterer of the same volume of the shell itself. The size parameter of this sphere is

$$v' = (1 - q^3)^{1/3} v \tag{5.1.55}$$

so that for $m \to 1$

$$\begin{aligned} i_1 + i_2 &= |(m_2{}^2 - 1)/(m_2{}^2 + 2)|^2 (v')^6 (1 + \cos^2\theta) \\ &= (4/9)|m_2 - 1|^2 (v')^6 (1 + \cos^2\theta) \end{aligned} \tag{5.1.56}$$

Another limiting case would be the one in which q is sufficiently close to unity and m_2 is not too high, so that the first term in the denominator of (5.1.52) can be dropped. Yet, m_2 is sufficiently large so that

$$(m_2{}^2 - 1)(2m_2{}^2 + 1) \to 2m_2{}^4 \tag{5.1.57}$$

A water bubble in the microwave region would correspond to this. Then

$$i_1 + i_2 = (4/81)m_2{}^2(1 - q^3)^2 v^6 (1 + \cos^2\theta) = (4/81)m_2{}^2(v')^6(1 + \cos^2\theta) \tag{5.1.58}$$

so that the scattering is equivalent to that by a Rayleigh scatterer with size parameter v'' and refractive index m such that

$$[(m^2 - 1)/(m^2 + 2)]^2 (v'')^6 = (m_2{}^2/9)(v')^6 \tag{5.1.59}$$

5.2 Numerical Results for Coated Spheres; Q_{sca}

With the possibility of varying the size parameter and the refractive index of both the core and the coating, the variety of numerical results that might be explored is tremendous. Actually, there have been only a few numerical studies.

5.2.1 Real Refractive Index; $m_1 = 2.1050$, $m_2 = 1.4821$

In connection with experiments on aerosols consisting of silver chloride spheres coated with linolenic acid, Kerker *et al.* (1962) have computed the scattering efficiency and the intensity functions at 10° intervals for cores with $m_1 = 2.1050$ and with coatings of $m_2 = 1.4821$. The range of the total size parameter was $v = 0.1(0.1)23.0(2.0)53.0$ and the ratio of the radius of the core to the total particle radius was $q = \alpha/v = 0, 0.2, 0.4, 0.6, 0.8, 0.9, 0.95, 0.98, 0.99$, and 1.00. Only the scattering efficiency has been analyzed.

a. Variation with α/v. We first consider the transition of a sphere of fixed total size from where it consists only of the core material ($q = 0$) to where it consists only of the coating material ($q = 1$). For $v \leqslant 1.4$ the value of Q_{sca} varies smoothly with q as shown in Fig. 5.2.

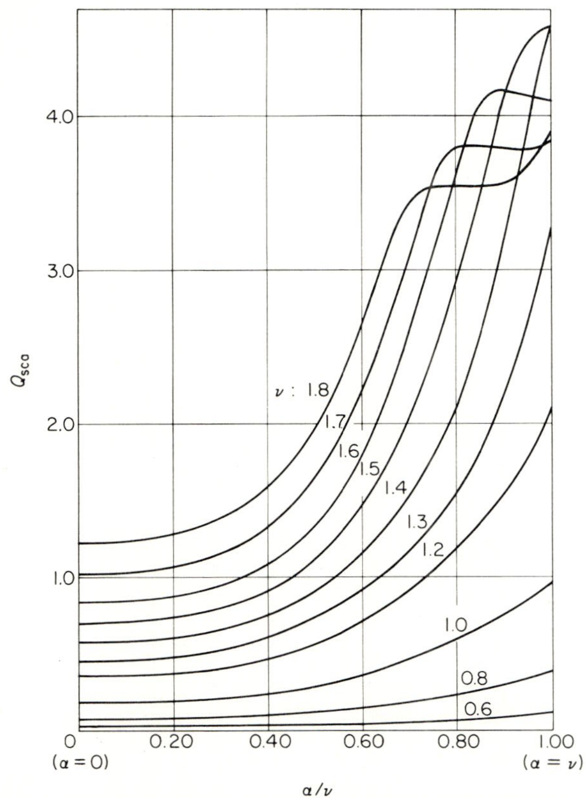

FIG. 5.2. Variation of scattering efficiency of coated sphere with α/ν for various values of ν. For the inner sphere, $\alpha = 2\pi a/\lambda$, $m_1 = 2.1050$; for the total sphere $\nu = 2\pi b/\lambda$, $m_2 = 1.4821$ (Kerker *et al.*, 1962).

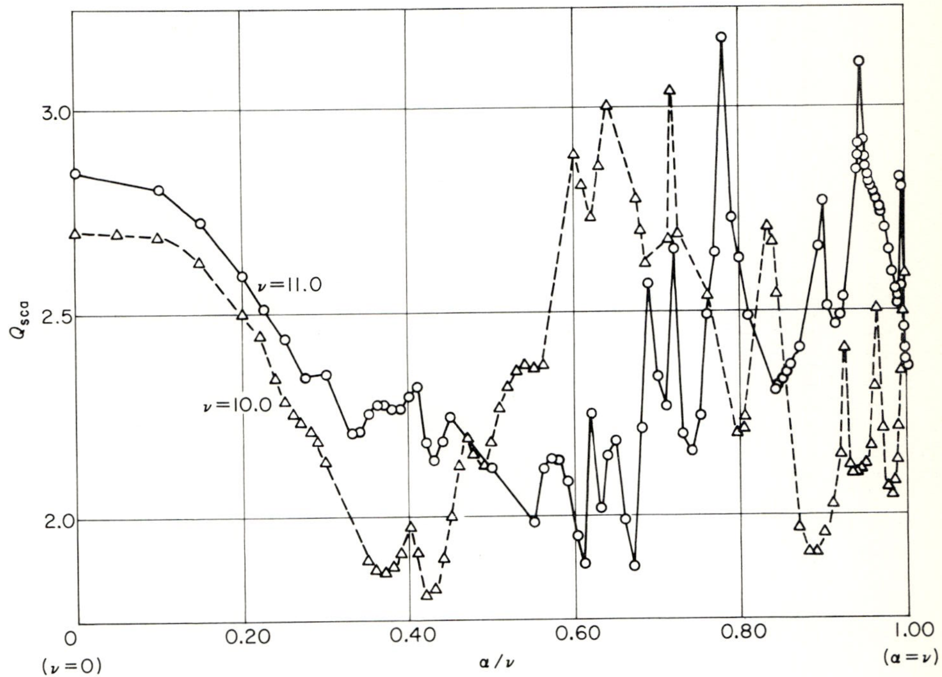

FIG. 5.3. Variation of scattering efficiency of coated sphere with α/ν for $\nu = 10.0$ and 11.0. For the inner sphere, $\alpha = 2\pi a/\lambda$, $m_1 = 2.1050$; for the total sphere $\nu = 2\pi b/\lambda$, $m_2 = 1.4821$ (Kerker *et al.*, 1962).

However, at larger sizes the curves become very irregular as exemplified by Fig. 5.3 for $v = 10.0$ and 11.0. Here, closely spaced intervals of q have been obtained wherever necessary to delineate the fine structure. The effect of thin coatings is particularly striking. For $v = 10.0$, the presence of a coating with thickness 2% ($q = 0.98$) of the total sphere radius causes a drop of Q_{sca} from 2.59 to 2.05. On the other hand, for $v = 11.0$ the effect is in the opposite direction. Q_{ext} for the homogeneous sphere increases from 2.36 to 2.82 for a coating thickness only 0.6% of the total radius ($q = 0.994$).

Although the published results of Kerker *et al.* (1962) indicated that a very tiny core ($\alpha/v = 0.005$) might have a marked effect upon scattering, this was later shown to be spurious, arising from a computational artifact (Espenscheid *et al.*, 1965). The Bessel functions had been evaluated by forward recursion from the known functions of order zero and unity with the aid of (3.6.7). This technique is suitable as long as the order of the Bessel function is not appreciably larger than the argument. The largest order corresponds to the number of terms needed for convergence of the series [(3.3.56), (3.3.57)] and this in turn is determined by the magnitude of v. Therefore those Bessel functions having v in their argument will be of an order which is never considerably greater than this argument. However, the concentric sphere will also have Bessel functions containing the argument α, corresponding to the size parameter of the core. When $\alpha \ll v$, the largest orders will be considerably greater than the argument and for this condition forward recursion leads to a very large round-off error. This round-off error may be avoided by following a procedure of backward recurrence. For each argument, the Bessel function is calculated for the two highest orders needed using the series definition (3.6.2). Then all lower orders are calculated using (3.6.7). The final result for $v = 10.0$ is that the effect of the core on the scattering efficiency becomes apparent only for $\alpha/v > 0.05$.

All particles which are somewhat larger than the wavelength have a value of approximately 2 for the scattering efficiency. It follows then that the variation of this quantity with coating thickness is not going to be huge. On the other hand, since this represents the scattering gain integrated over all angles, there may be particular directions in which the coating thickness may have a tremendous effect upon the scattering. Consider, for example, the results in Table 5.2 for i_1 and $\theta = 150°$. The abrupt changes for $q = 1.00$ and $q = 0.99$ are remarkable. The value of i_1 for the homogeneous sphere ($q = 1.00$) undergoes a drop and then an increase of more than an order of magnitude as v varies from 12.1 to 12.3. For $v = 12.2$ a 1% coating will increase the signal 12-fold whereas for $v = 12.3$ there will be a 68-fold decrease. Although this example illustrates an extreme situation, it does bring out the complexity that can be anticipated for coated spheres.

TABLE 5.2
INTENSITY FUNCTION i_1 AT $\theta = 150°$, $m_1 = 2.1050$, $m_2 = 1.1821$

q	$\nu = 12.0$	$\nu = 12.1$	$\nu = 12.2$	$\nu = 12.3$	$\nu = 12.4$
0.95	8.01	41.59	55.36	22.80	13.09
0.98	32.62	25.09	16.80	4.50	29.38
0.99	26.68	21.43	7.63	0.13	29.73
1.00	24.67	11.99	0.63	8.84	30.73

b. Some Empirical Expressions. It is possible to represent Q_{sca} by simple empirical expressions provided ν is not too large. One method is based upon Güttler's limiting expression for small particles (5.1.51) which leads to

$$Q_{sca} = 6p_1'^2\nu^4$$
$$= \frac{8}{3}\nu^4\left|\frac{(m_2^2-1)(m_1^2+2m_2^2)+q^3(2m_2^2+1)(m_1^2-m_2^2)}{(m_2^2+2)(m_1^2+2m_2^2)+q^3(2m_2^2-2)(m_1^2-m_2^2)}\right|^2 \quad (5.2.1)$$

For the refractive indices used here ($m_1 = 2.105$, $m_2 = 1.4821$), this is accurate to at least 4% provided $\nu \leqslant 0.3$. For $\nu = 0.5$, the error becomes as large as 100%. An expression valid over a wide range of ν is obtained by noting that up to $\nu = 1.4$ the ratio of the exact value of Q_{sca} to the value given by (5.2.1) is approximately the same for both homogeneous and coated spheres. This leads to

$$Q_{sca} \cong [p_1'/(p_1')_{q=1}]^2(Q_{sca})_{q=1} \quad (5.2.2)$$

where the bracketed quantities are to be evaluated for the homogeneous sphere with refractive index m_1. Table 5.3 lists the ratio of the exact value of the scattering efficiency to that calculated by means of (5.2.2). The approximation becomes poorer for small values of q but this can be adjusted by using the inverse relation where the homogeneous sphere has refractive index m_2.

$$Q_{sca} \cong [p_1'/(p_1')_{q=0}]^2(Q_{sca})_{q=0} \quad (5.2.3)$$

There is another approximation which, while not as accurate for thin coatings and small values of ν as the above, extends the range of values of ν. In this case the scattering efficiency of the coated sphere is approximated by

$$Q_{sca}(m) \cong Q_{sca}(m') \quad (5.2.4)$$

where $Q_{sca}(m')$ is the scattering efficiency for a homogeneous sphere with the same radius (b) and the volume weighted refractive index

$$m' = m_1q^3 + m_2(1-q^3) \quad (5.2.5)$$

TABLE 5.3
RATIO OF THE EXACT VALUE OF Q_{sca} FOR COATED SPHERES TO THAT CALCULATED BY EQ. (5.2.2)

q	$v = 0.3$	$v = 0.5$	$v = 1.0$	$v = 1.4$
0.0	0.965	—	—	—
0.2	0.966	0.908	—	—
0.4	0.971	0.924	—	—
0.6	0.984	0.957	0.907	—
0.8	0.996	0.989	1.000	—
0.9	0.999	0.998	1.010	0.887
0.95	1.000	1.000	1.006	0.971
0.98	1.000	1.000	1.002	0.998
0.99	1.000	1.000	1.002	1.001
1.00	1.000	1.000	1.000	1.000

A comparison with the exact values is given in Table 5.4. Since $Q_{sca}(m')$ was obtained by interpolation between values in existing tables, it is likely that the approximation is even better than indicated in Table 5.4.

TABLE 5.4
RATIO OF THE EXACT VALUE OF Q_{sca} FOR COATED SPHERES TO THAT CALCULATED BY EQ. (5.2.4)

q	$m' \backslash v$	0.1	0.5	1.0	1.5	2.0	2.3	2.5
0.0	1.482	1.000	1.000	1.000	1.000	1.000	1.000	1.000
0.2	1.487	1.002	1.000	1.102	1.017	1.010	1.015	1.015
0.4	1.522	—	—	1.051	1.094	1.073	—	1.038
0.6	1.617	0.972	1.000	1.121	1.194	1.176	1.017	0.929
0.8	1.801	0.968	0.998	1.123	1.228	0.986	0.999	0.859
0.9	1.936	0.978	0.998	1.075	1.137	0.987	0.910	0.843
0.95	2.016	0.986	0.999	1.040	1.047	1.018	0.922	0.899
0.98	2.068	0.994	0.999	1.016	1.083	1.013	0.964	0.954
0.99	2.087	0.997	1.000	1.008	1.004	1.006	0.982	0.977
1.00	2.105	1.000	1.000	1.000	1.000	1.000	1.000	1.000

The Ryde (1946) mixture rule is still another approximation for calculating the scattering by small coated spheres. This can be written

$$Q_{sca} \cong \frac{8}{3} v^4 \left| \left(\frac{m_1^2 - 1}{m_1^2 + 2} \right) q^3 + \left(\frac{m_2^2 - 1}{m_2^2 + 2} \right) (1 - q^3) \right|^2 \tag{5.2.6}$$

It does not hold nearly so well as (5.2.1) but it can be improved by developing a relation analogous to (5.2.2). However, even this is inferior to those results already given above.

c. Variation with v. The variation of Q_{sca} with v for a fixed value of q is quite similar to the case for homogeneous spheres and is illustrated in Fig. 5.4 for $q = 0.6$. The results for homogeneous spheres comprised of the core and coating materials are also plotted. The abscissa is the phase shift of the central ray through the sphere.

$$\rho_{12} = 2v[(m_1 - 1) + (m_2 - m_1)q] \tag{5.2.7}$$

Whereas the major oscillations of the scattering efficiency for homogeneous spheres are more or less superimposed upon each other when plotted against the phase shift parameter, coated spheres do not follow the same pattern. Indeed, the third maximum for the homogeneous cases ($\rho_{12} \cong 17.5$) corresponds precisely to the second minimum for this particular coated sphere ($q = 0.6$).

5.2.2 Complex Refractive Index

Fenn and Oser (1962, 1965) have carried out extensive calculations for coated spheres having an absorptive core. This was intended to provide a model for the absorption and scattering of sunlight by water-coated carbon particles in order to assess the effect such particles might have on the energy balance in the atmosphere. The core was assumed to have a refractive index[1] of $1.59 - 0.66i$ and that of the shell was taken as 1.33. The calculations include Q_{ext}, Q_{sca}, i_1, and i_2 for values of v ranging from 0.1 to 200 and for q at 0, 0.1, 0.2, 0.5, 0.67, 0.825, 0.911, 1.0. Only a small part of these results are presented in their report.

The effect on the scattering of the dielectric coating is illustrated in Figs. 5.5 to 5.8 where Q_{ext}, Q_{abs}, and Q_{sca} are plotted against the total size parameter v. Sufficient points were obtained only to determine the gross features of these curves. In Fig. 5.5 the radius of the core is only 0.2 that of the total particle. The core has very little effect. The value of Q_{abs} is small. Both Q_{ext} and Q_{sca} behave very much as for the purely dielectric sphere. When the ratio of the inner to total radius is $q = 0.5$ (Fig. 5.6), there is an increase of Q_{abs} and a corresponding separation of Q_{ext} and Q_{sca}, although the main structural features of the curves are preserved. In Fig. 5.7 where $q = 0.825$, Q_{abs} and Q_{sca} are comparable in magnitude. Q_{sca} still exhibits the oscillations

[1] Hodkinson (1964) has pointed out that this value for carbon at 4910 Å is incorrect even though it has gained wide currency. Probably $m = 1.95 - 0.66i$ was intended since the original measurements of Senftleben and Benedict (1918, 1919) range from $1.90-0.68i$ at 4360 Å to $2.00 - 0.66i$ at 6230 Å. Later measurements [*vide* Hodkinson (1964)] give similar values. McDonald (1962b) has calculated some extinction coefficients using the original refractive index data of Senftleben and Benedict. However, Fenn and Oser's results are still of interest because they show trends over a wide range of size parameters.

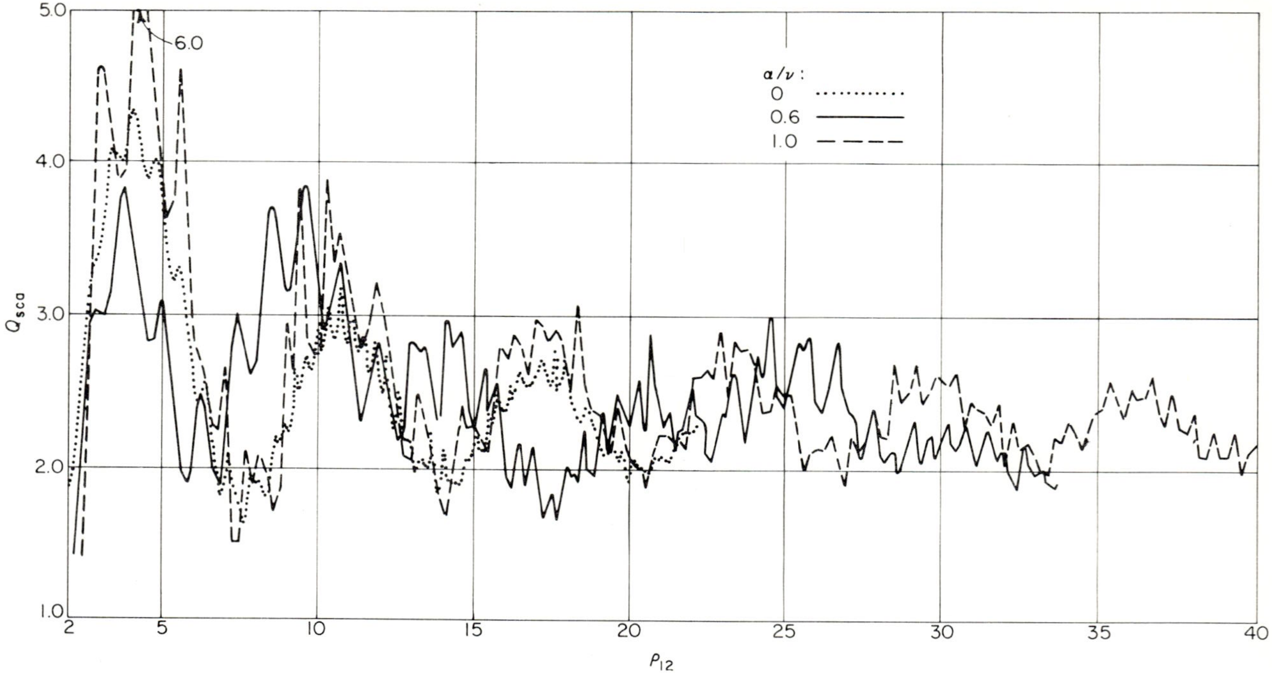

FIG. 5.4. Variation of scattering efficiency with phase shift parameter ρ_{12} for homogeneous spheres with $m_1 = 2.1050$ and $m_2 = 1.4821$, compared with coated sphere with $\alpha/\nu = 0.6$ (Kerker *et al.*, 1962).

characteristic of scattering by dielectric spheres. However, in the limit $q = 1$, which corresponds to a homogeneous sphere with the refractive index of the highly absorbing core material, this secondary structure is no longer apparent.

Interestingly, when $v > 10$, a thin film of dielectric may increase the absorption, Q_{abs}, beyond the value for a sphere of the same size consisting only of the absorbing material. This is illustrated in Fig. 5.9 where Q_{abs} relative to the value for the particle consisting only of the absorbing material is plotted against $1/q$ for various values of v, the total particle size. The absorption increases somewhat with $1/q$ and then drops sharply.

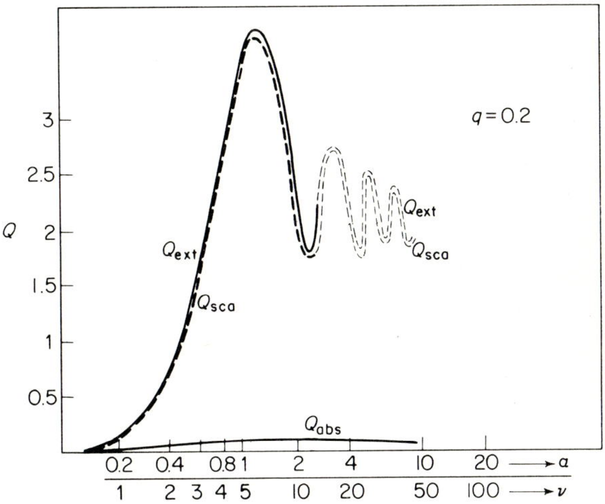

FIG. 5.5. Efficiencies for extinction, scattering, and absorption (Q_{ext}, Q_{sca}, Q_{abs}) for a coated sphere with an absorbing core ($m_1 = 1.59 - 0.66i$, $m_2 = 1.33$) plotted against size for $q = 0.2$ (Fenn and Oser, 1965).

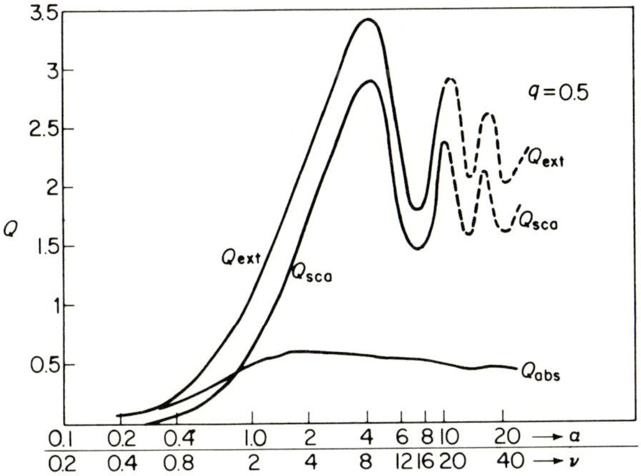

FIG. 5.6. Same as Fig. 5.5 with $q = 0.5$.

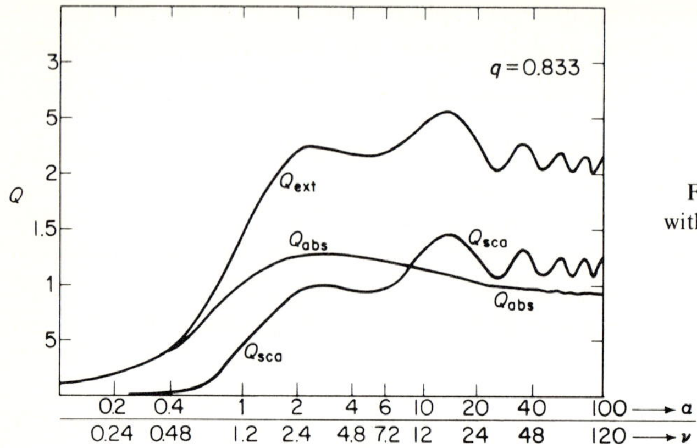

FIG. 5.7. Same as Fig. 5.5 with $q = 0.833$.

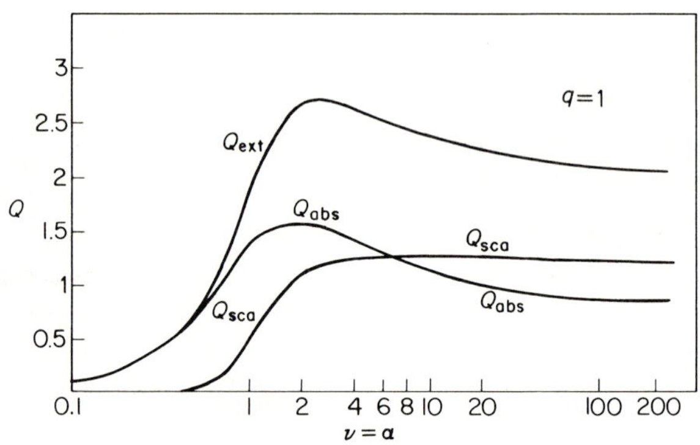

FIG. 5.8. Same as Fig. 5.5 with $q = 1.0$, which corresponds to an absorbing homogeneous sphere with $m_1 = 1.59 - 0.66i$.

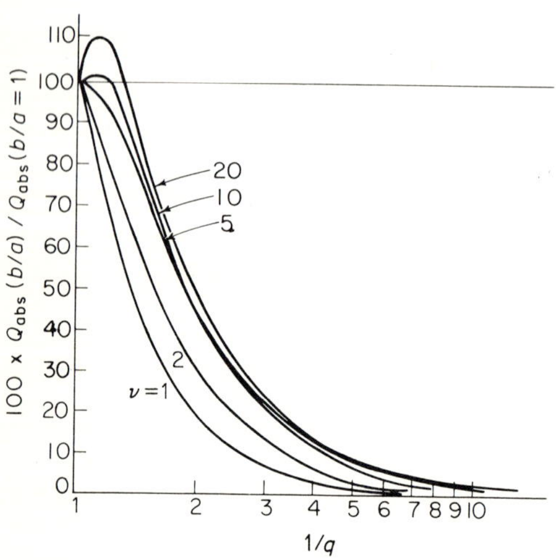

FIG. 5.9. Ratio of Q_{abs} for a coated sphere with an absorbing core ($m_1 = 1.59 - 0.66i$, $m_2 = 1.33$) to that for a homogeneous absorbing sphere of the same size plotted against $(1/q) = b/a$ for $\nu = 1$, 2, 5, 10, and 20 (Fenn and Oser, 1965).

This effect can also be noted if Q_{abs} is observed for a fixed absorbing core size with increasing coating thickness. There is a striking enhancement of the absorption as the coating thickness increases. This then decreases somewhat and levels off. This enhancement of absorption by the dielectric coating can be attributed to a focusing of the incident radiation onto the absorbing core by the dielectric coating.

The efficiency factors for a concentric sphere with a nonabsorbing core ($m = 1.5$) and an absorbing shell ($m = 1.95\text{–}0.66i$) have been calculated by Pilat (1967) for $\alpha/v = 0, 0.5, 0.99$, and 1.0 up to values of $v = 10.0$. For small values of v the efficiency for extinction was increased very markedly by the presence of the absorbing shell compared to the value for a homogeneous dielectric sphere of the same size.

Morriss and Collins (1964) have calculated the extinction efficiency of water suspensions of particles consisting of gold cores encased in a silver coating. The core diameter was 60 Å and the coating thicknesses were 5, 10, 20, 70, and 220 Å. The refractive indices corresponded to the visible wavelength range 300 to 700 mμ.

For homogeneous particles consisting either of gold or silver, the extinction plotted against wavelength exhibited a sharp peak which was not related in a simple way to the dispersion of the refractive index. Indeed, this is the effect that led Mie to develop the theory of scattering by spheres in the first place. The results for gold spheres with $a = 60$ Å are shown in Fig. 5.10. The effect on a 60 Å gold sphere of a 20 Å layer of silver is shown in Fig. 5.11. As the coating thickens, the gold core has a decreasing effect and the spectrum resembles that of a pure silver sphere of the same size.

5.3 Numerical Results for Coated Spheres; Backscatter

Other numerical studies have been concerned with backscatter. The choice of refractive indices has been determined by three physical problems of major interest—radar echoes from (1) dielectric coated metals, (2) water coated hailstones, and (3) plasma coated metals and dielectrics.

5.3.1 Totally Reflecting Core, Dielectric Coating

The effect of a dielectric coating on the backscatter of a totally reflecting sphere is of great practical interest, particularly since the plasma sheath surrounding a metallic re-entry vehicle behaves as such a dielectric. The coating may increase the backscatter over that of the single sphere by an order of magnitude or decrease it to practically zero. Earlier calculations for small spheres ($a/\lambda \leqslant 0.24$) by Scharfman (1954) had already shown

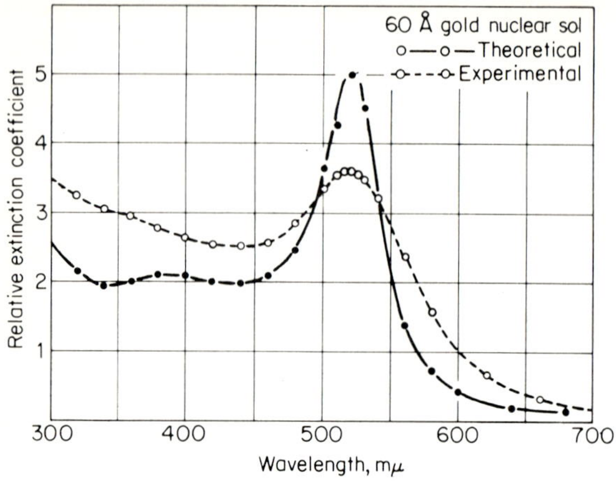

FIG. 5.10. Relative extinction for spheres 60 Å in diameter and complex refractive index corresponding to that for gold compared with experimental results for gold spheres having an average particle diameter of 59 Å (Morriss and Collins, 1964).

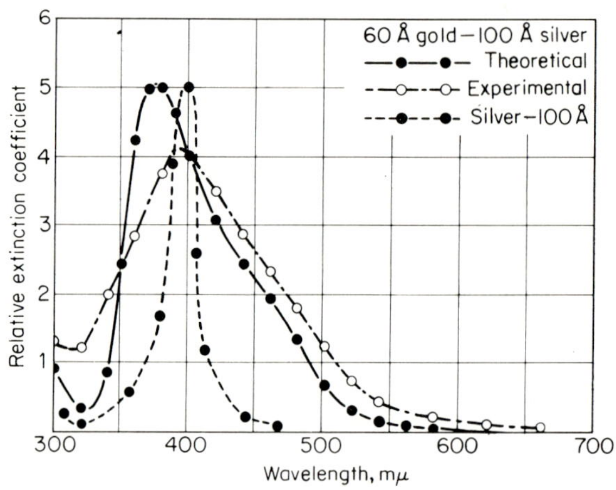

FIG. 5.11. Relative extinction for spheres with a 60 Å diameter gold core and a 20 Å silver coating (full line) compared with experimental results for 95 Å diameter silver coated gold spheres having 59 Å diameter cores (dash-dot line). The dotted curve is calculated for 100 Å silver spheres (Morriss and Collins, 1964).

increases up to 100%. Murphy (1965) has explored the effect on the backscatter of varying the electron density of the plasma coating.

Rheinstein (1964) has made extensive calculations for $m_2{}^2 = 2.56$, 4.00, and 6.00 and for core to total sphere ratios $q = 0.90$ and 0.95. The radius of the sphere relative to the wavelength, b/λ, was varied from 0.02 to 10.0 in steps of 0.02. The results for $\varepsilon = m^2 = 6.0$ and $q = 0.1$ are shown in Figs. 5.12(*a–c*). It is evident that the results are considerably different from those for the single perfectly conducting sphere (Figs. 4.27 and 4.28).

As the size parameter increases, groups of resonance-like peaks appear. These peaks occur in groups separated by intervals where the gain remains close to unity, which is the limiting value of a perfectly reflecting sphere. With

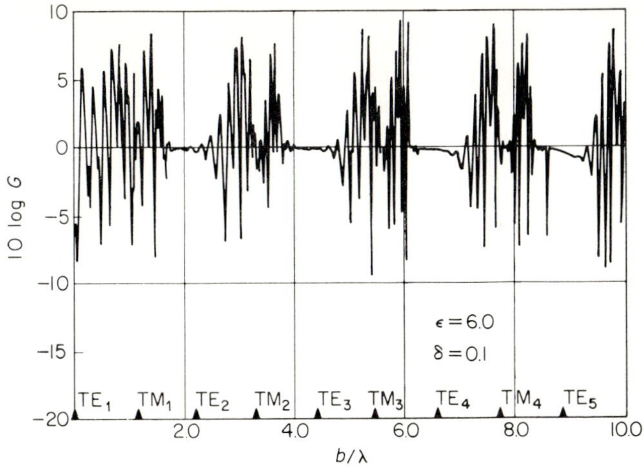

FIG. 5.12a

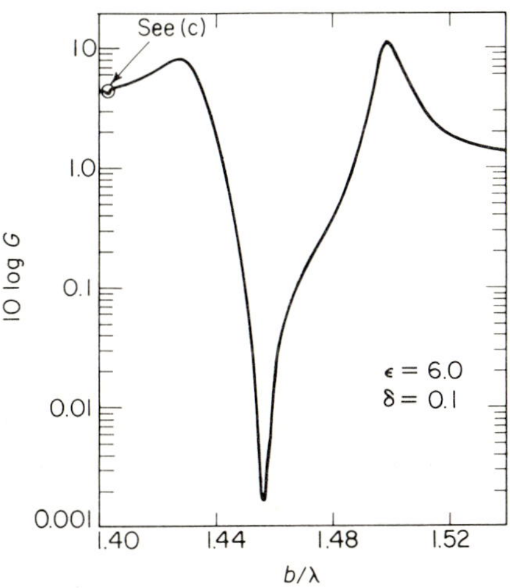

FIG. 5.12b

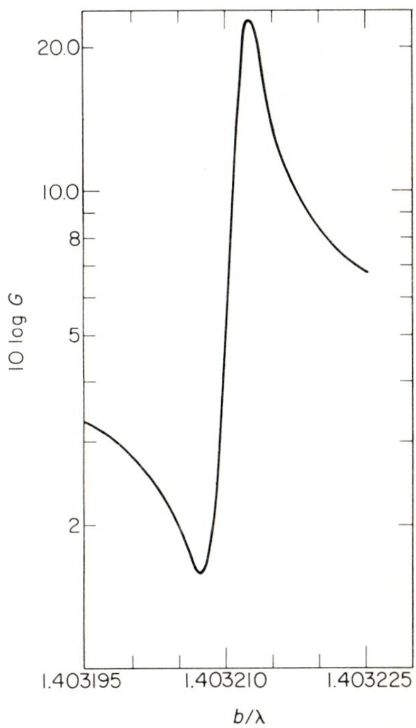

FIG. 5.12c

FIG 5.12(a). Backscatter gain for totally reflecting sphere coated with dielectric spherical shell ($m^2 = \varepsilon = 6.0$; $a/b = 0.9$) plotted against b/λ. Wedges indicate conditions for which dielectric coating behaves as waveguide for rays injected into the dielectric. (b) Scale expansion for part of (a) to show fine structure. (c) Further scale expansion to show fine structure or wriggle is encircled in (b) (Rheinstein, 1964).

higher dielectric constant, the abscissa tends to become compressed, i.e., there are more groups of resonances for a given interval of b/λ; decreasing the thickness of the coating tends to expand the abscissa. In some instances the interval $b/\lambda = 0.02$ may not be small enough to resolve all of the structure. An example of this is illustrated in Figs. 5.12(b, c) which are plotted at quite small intervals of b/λ. In Fig. 5.12(b) the minimum is nearly four orders of magnitude lower than the adjacent peaks. When the wiggle in the upper left of this curve is resolved, it gives rise to the fantastic effect in Fig. 5.12(c) where an order of magnitude change in the gain occurs when b/λ increases by 0.000008.

These resonance effects can be explained qualitatively by surface waves. The backscatter curve for a perfectly reflecting sphere has been interpreted as due to the interference between the rays traveling as a surface wave and the specularly reflected wave. The former is attenuated rapidly by being sprayed off the surface so that only those rays making one trip around the sphere make a significant contribution. Even for one trip there is considerable attenuation which becomes correspondingly greater the larger the sphere. In this way the interference between the surface wave and the axial wave accounts for the damping of the sinusoidal variation of the backscatter gain with increasing size.

For the dielectric coated sphere, the surface wave contribution may be approximately the same as for the uncoated object so that for sufficiently large spheres the scattering is mainly due to the specular reflection. This would account for those intervals between the groups of resonances in the backscatter vs. size curve where the gain is close to unity. However, the surface wave may travel around the object with very little attenuation whenever

$$2(1 - q)(b/\lambda)(m_2^{\,2} - 1)^{1/2} \geqslant n - 1 \qquad \text{or} \qquad n - \tfrac{1}{2} \tag{5.3.1}$$

Here n is an integer. Each case corresponds to the TE or TM modes, respectively, or to the two polarized components G_2 and G_1. These conditions are precisely those for injection of the rays into the dielectric, so that after reflection at the totally reflecting surface, they will be incident upon the underside of the dielectric surface at the angle for total reflection, i.e., the rays become trapped in the dielectric coating which now behaves as a wave guide. For certain combinations of the physical parameters, the waves which have circumnavigated the sphere a half-integral number of times will combine constructively so as to give rise to a scattered wave comparable to, or even larger than, the amplitude of the specular return. Interference effects between these partial waves and the specular return may cause the observed resonances. The surface wave analysis indicates that certain resonances may have fantastically narrow widths, which is certainly borne out by Fig. 5.12(c). The arrows in Fig. 5.12(a) are placed at values of b/λ where (5.3.1) applies. Apparently this condition does determine when each group of resonances will appear.

If the coating material is permitted to become absorptive, the backscatter gain no longer exhibits the above sharp resonances with increasing size (Rheinstein, 1965). This suggests that the trapped surface waves in the coating are attenuated before they escape.

5.3.2 Dielectric Core, Absorbing Coating

Another case of considerable interest is that of an ice core ($m_1 = 1.78 - 0.0024i$) surrounded by a water coating (for $\lambda = 3.21$ cm, $m_2 = 7.14 - 2.89i$; for $\lambda = 4.67$ cm, $m_2 = 7.95 - 2.20i$; for $\lambda = 10$ cm, $m_2 = 8.99 - 1.47i$). The absorption index of the ice core is sufficiently small so that, except at extremely large sizes, it behaves as if it were a nonabsorbing dielectric. On the other hand, both the real and the imaginary parts of the refractive index of the water coating are high. Accordingly, water is highly reflective. This case, then, is almost the inverse of the dielectric-coated, perfectly reflecting sphere considered in the previous section.

Herman and Battan (1961b) have calculated the backscatter at each of the above three wavelengths for 12 sphere radii up to 5 cm and for several values of q [see also Kerker *et al.* (1951)]. The curves showing the variation of the scattering efficiency as one proceeds from a homogeneous sphere composed of the core material ($q = 1$) to one composed of the coating material ($q = 0$) are similar to those obtained in the dielectric case (Figs. 5.2, 5.3). Thus for small sizes the transition proceeds smoothly (Fig. 5.13). By the time $q = 0.8$, the particle behaves essentially as if it consists only of the coating material. For larger sizes, the transition is irregular. Large dielectric spheres backscatter more intensely than perfectly reflecting spheres of the same size. Accordingly, the main effect of the coating is to decrease the backscatter. However, because of the oscillating character of the curve, there are some thicknesses of coating for which the backscatter will be enhanced over that for the homogeneous ice sphere.

Atlas and Glover (1963) have suggested, for thin coatings of a highly reflecting material, that the main feature of the backscatter may be explained as an interference among a small number of geometrical optics rays. The approach is similar to that utilized by them for the slightly absorbing homogeneous sphere. In this case, there will be two front and two rear axial rays whose amplitudes will be determined by the appropriate Fresnel factors and the divergence. Because of the high reflectivity at the two front surfaces, the axial rays reflected from them will predominate and for a thin coating they will be only slightly out of phase. Provided the film is sufficiently thin, the geometry of the glory ray will be essentially the same as for the homogeneous case. The main features of the backscatter then will be due to the interference between the front axial rays and the glory rays. For spheres of intermediate size, the contribution by the surface waves must also be considered.

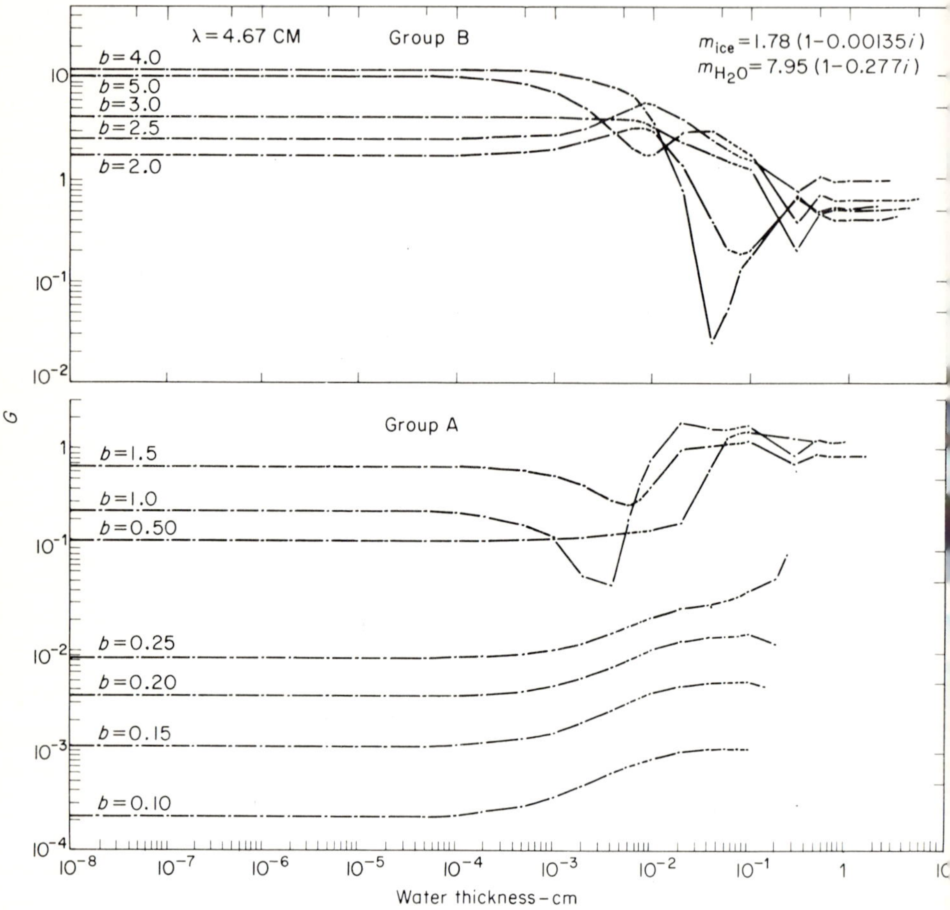

FIG. 5.13. Backscatter gain at $\lambda = 4.67$ cm for water coated ice spheres ($m_1 = 1.78 - 0.0024i$; $m_2 = 7.95 - 2.20i$) of various radii b as a function of the thickness of the water shell (Herman and Battan, 1961b).

5.3.3 PLASMA-COATED SPHERES; RAY OPTICS APPROXIMATION

Peters and co-workers have considered plasma-coated, perfectly reflecting spheres (Peters and Green, 1961; Peters and Swarner, 1961; Swarner and Peters, 1963). Their interest was stimulated by the discovery soon after the launching of the first man-made earth satellite, that a body traveling at high velocity through the ionosphere may acquire either a positive or negative charge and thereby attract a shell of particles of opposite charge. A body which is positively charged, and hence acquires a shell of electrons, may be considered to be surrounded by an equivalent dielectric shell whose refractive index is less than unity. If the shell consists of positive ions, the refractive index relative to the ambient atmosphere is greater than unity.

These workers have utilized a ray optics approach and have compared their results with the exact theory. Although their interest has been primarily in the backscatter, the method is also applicable to the angular scattering.

Two principal components of the scattered field are considered—scattering by an equivalent conducting sphere having a modified radius because of the lens action of the dielectric shell and scattering by reflection at the air-dielectric interface. These components are added vectorially to obtain the approximate field of the scattered wave. Other forms of interaction such as multiply reflected internal rays and surface waves are not considered. Good agreement is obtained with the exact theory for α from 0.12 to 1.75, for v up to 12.5, and for m_2 from 0.3 to 1.4.

The first part of the approximation is obtained by considering only the equivalent conducting sphere. This is illustrated in Fig. 5.14. Here the refractive index of the shell is greater than unity so that there is a focusing of incident energy upon the conducting core. The contribution of the core to the scattering is assumed to be equal to that of a sphere whose radius, a', is determined by the limiting rays tangential to the core. By Snell's law this is

$$a' = m_2 a \tag{5.3.2}$$

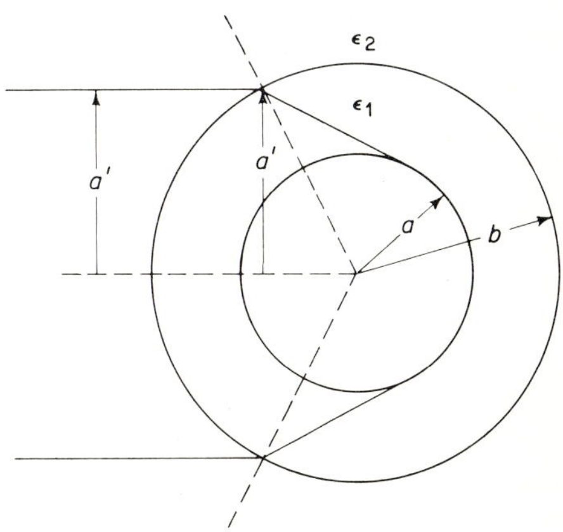

FIG. 5.14. Illustration of the change in effective radius of a conducting sphere due to focusing of energy by a dielectric shell. (Note that in this figure media 2 and 3 are designated by ε_1 and ε_2.)

For sufficiently thin shells and $m_2 > 1$, this may lead to a radius greater than that of the outer radius of the dielectric shell. Whenever this condition is obtained, the addition of a simple exponential term has been found to provide an empirical correction which is reasonably valid leading to

$$a' = a\{m_2 + (1 - m_2)\exp[-(2\pi/m_2)(b - a)/\lambda]\} \tag{5.3.3}$$

If the refractive index is less than unity, the effective radius a' is smaller than that of the core, so that the simple expression (5.3.2) is always used.

An example of the results obtained when only this equivalent conducting sphere approximation is considered is shown in Fig. 5.15 for a conducting core with $a/\lambda = 0.15$, surrounded by a dielectric shell with $m_2^2 = 0.75$. The approximation is accurate to within 3 dB for a shell radius up to 5 times that of the core radius. Up to this size, the conducting core makes the major contribution to the scattering. For larger sizes it continues to give the average value of the backscatter efficiency. However, in order to account for the large oscillations that appear, it is necessary to include another component to the scattering. This will be the contribution by reflection at the interface between the external medium and the dielectric.

The second component of the approximate solution is obtained by assuming that the contribution of the dielectric shell to the scattering is equivalent to that of a perfectly conducting sphere multiplied by the square of the Fresnel reflection coefficient at normal incidence. The size of this sphere is the same as the coated sphere. Thus the backscatter gain is

$$G = KG_m = \left(\frac{m_2 - 1}{m_2 + 1}\right)^2 G_m \tag{5.3.4}$$

where G_m is the backscatter gain of the perfectly conducting sphere. An example of this approximation is given in Fig. 5.16. This is for a core of $a/\lambda = 0.05$ and $m_2^2 = 1.25$. The solid curve represents the approximation and the dots were obtained from the exact theory. The approximation is accurate to within 3 dB provided $b/\lambda > 0.4$. The equivalent sphere approximation which is also shown in the same figure predicts the correct backscatter efficiency to within 3 dB for a shell radius up to $b/\lambda = 0.125$. For intermediate shell radii, where both components are comparable in magnitude, a combined approximation must be used.

The combined approximation is illustrated in Fig. 5.17. In this case $a/\lambda = 0.05$ for the core and the refractive index of the dielectric shell is $m_2^2 = 0.75$. The equivalent sphere approximation yields useful results for shells with $b/\lambda \leqslant 0.1$, while the dielectric sphere approximation is valid for $b/\lambda \geqslant 0.25$. Between these limits, there is a very deep null which is not even suggested by either of the above approximations alone. However when the field scattered by the equivalent conducting sphere and that scattered by the

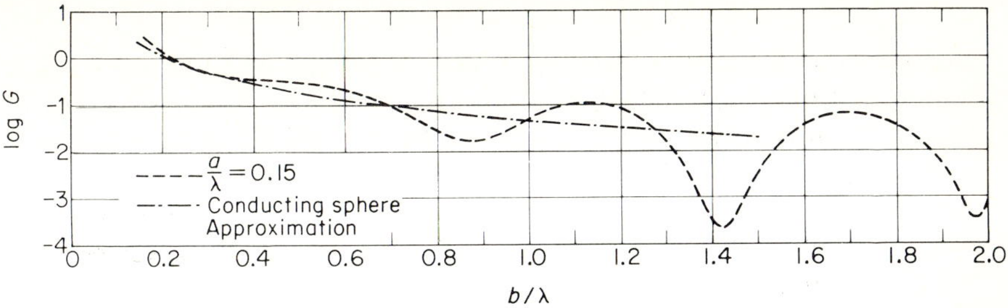

FIG. 5.15. Backscatter gain obtained by the equivalent conducting sphere approximation compared with the exact solution for a conducting sphere with $a/\lambda = 0.15$ and dielectric shell with $m_2^2 = \varepsilon = 0.75$ (Swarner and Peters, 1963).

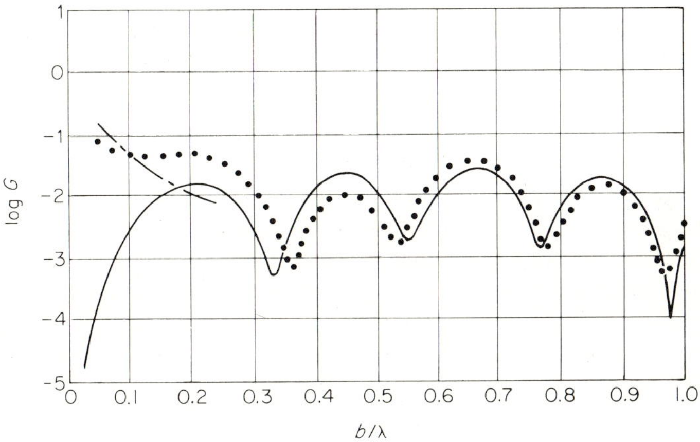

FIG. 5.16. Backscatter gain obtained by the equivalent dielectric sphere approximation (dots) compared with the exact solution (full line) for a conducting sphere with $a/\lambda = 0.05$ and a dielectric shell with $m_2^2 = \varepsilon = 1.25$ (Swarner and Peters, 1963).

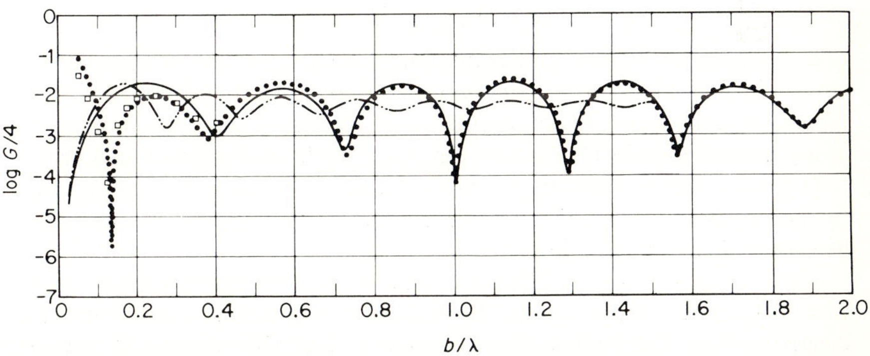

FIG. 5.17. Backscatter gain obtained by the equivalent conducting sphere approximation (dash-dot line), by the equivalent dielectric sphere approximation (full line), by the combined approximation (squares), compared with the exact solutions (heavy dots) for a conducting sphere with $a/\lambda = 0.05$ and a dielectric shell $m_2^2 = \varepsilon = 0.75$ (Swarner and Peters, 1963).

homogeneous dielectric sphere are added as phasors and the scattering is computed from the resultant field, the combined approximation curve of Fig. 5.17 results, which is in very good agreement with the exact solution.

In many cases it is sufficient to superpose the fields corresponding to the equivalent conducting sphere and the dielectric sphere directly by vector addition of the appropriate amplitude functions. However, in general there is a change in phase of the equivalent conducting sphere component due to the change in electrical path length of the rays upon passing through the dielectric shell. This is given by

$$\phi = \phi_0 + 720(m - 1)b/\lambda \tag{5.3.5}$$

where ϕ_0 is the phase (with respect to the origin) of the field scattered by a conducting sphere of radius a' in free space and the following term is the correction for the path length through the dielectric. When the radius of the equivalent conducting sphere is modified in the thin shell region by (5.3.3), a somewhat different phase correction must be used.

In addition to the phase correction, there is an amplitude correction which must be applied to the field component associated with the conducting core. This arises because of reflection at the interface between the external medium and the dielectric. As a result, only part of the energy incident upon the dielectric shell is transmitted through the shell to be reflected by the inner conducting sphere. Likewise only part of this energy will be transmitted across the interface on the way out of the shell. Consequently the amplitude function for the equivalent conducting sphere must be reduced by a factor which is the product of the transmission coefficients for normal incidence (2.4.7) on entering and leaving the dielectric shell:

$$T = \left(\frac{2}{m_2 + 1}\right)\left(\frac{2m_2}{m_2 + 1}\right) = \frac{4m_2}{(m_2 + 1)^2} \tag{5.3.6}$$

The application of this approximation to the angular scattering is illustrated in Fig. 5.18 where once again remarkable agreement is obtained between the exact and approximate results. The agreement becomes poorer for larger dielectric shells ($b/\lambda = 1.3$), especially in the backward directions.

The same approximation may also be applied to calculate the scattering by dielectric-coated, infinite circular cylinders illuminated at perpendicular incidence with comparable results.

These successes have led Peters *et al.* (1962) to apply this method to the case of scattering by a conducting sphere with a nonconcentric spherical dielectric shell. The configuration is given in Fig. 5.19 where a' is now the diameter of the equivalent conducting sphere. In order to check the approximation, a conducting sphere with a nonconcentric spherical dielectric shell with $m_2{}^2 = 1.8$, $a/\lambda = 0.143$, $b/\lambda = 0.9$, and an offset of $c/\lambda = 0.45$ was constructed and its backscatter measured on an indoor X-band CW radar facility (Swarner and Peters, 1963). The approximate backscatter efficiency gave excellent agreement with the experimental pattern.

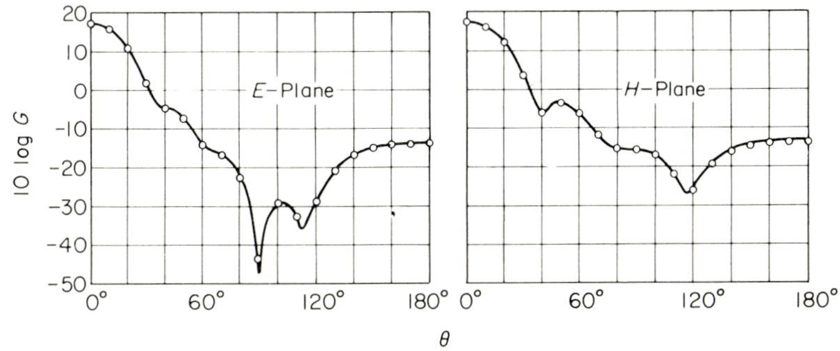

FIG. 5.18. Approximate angular gain G_1 and G_2 (circles) compared with exact values (smooth curve) for conducting core ($a/\lambda = 0.10$) and dielectric coating ($m^2 = \varepsilon = 0.75$; $b/\lambda = 1.15$). E-plane corresponds to G_1 (Swarner and Peters, 1963).

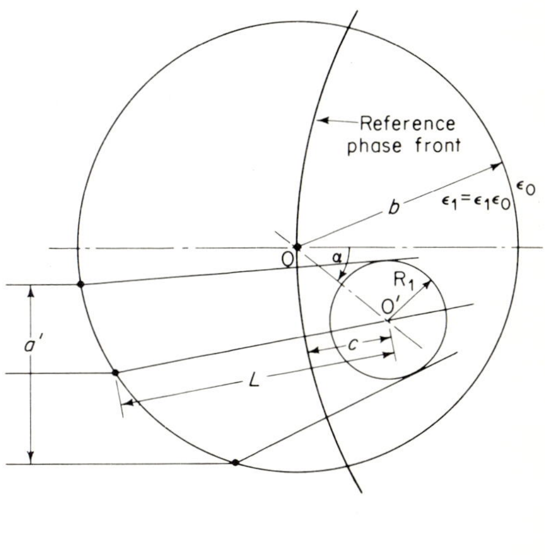

FIG. 5.19. Geometry for a conducting sphere encased in a nonconcentric dielectric sphere (Swarner and Peters, 1963).

5.3.4 Dielectric Bubble

A bubble is a special case of a coated sphere for which the core and external medium have the same refractive indices. In Fig. 5.20, the backscattering cross section of water bubbles in air at a wavelength of 3.21 cm is plotted against bubble diameter for several film thicknesses (Battan and Herman,

1961) using the exact theory of the coated sphere. Calculations with Andreasen's (1957a) equations for a thin walled bubble are not available for comparison with these results. For such thin walled bubbles ($q > 0.99$), the backscatter increases monotonically with wall thickness. The strikingly regular oscillating character of these curves with peaks four orders of magnitude greater than the troughs suggests a simple interference effect.

Peters and Thomas (1962) have calculated the backscatter on the assumption that this arises from the interference of the axial ray reflected from the front of the bubble with that transmitted through the bubble and reflected by the rear. The final expression for the backscatter efficiency is

$$G = |R|^2 \cdot |1 + T^2 e^{i(4v+\pi)}|^2 \tag{5.3.7}$$

where R is the reflection coefficient of a thin film [Stratton (1941) p. 513] of thickness $(b - a)$

$$R = \frac{r\{1 - \exp[-i2k_2(b - a)]\}}{1 - r^2 \exp[-i2k_2(b - a)]} \tag{5.3.8}$$

and T is the corresponding transmission coefficient

$$T = \frac{(1 - r^2)\exp[-i(k_2 - k_3)(b - a)]}{1 - r^2 \exp[-i2k_2(b - a)]} \tag{5.3.9}$$

Here r is the usual Fresnel coefficient at perpendicular incidence (2.4.7) for a single surface and k_2 and k_3 are the propagation constants in the bubble and in the medium. In calculating the phase shift for the rear axial ray, in addition to its propagation through the interior of the bubble, care must be taken to include a phase shift of π to allow for the crossing of a caustic in each direction (Keller and Kay, 1954).

The calculations based upon (5.3.7) agree quite well with the exact calculations. The minima and maxima occur precisely at bubble diameters corresponding to $n\lambda/2$ and $(2n - 1)\lambda/4$ which is what would be predicted for destructive and constructive interference between the axial rays, provided the caustic is taken into consideration.

a. Thin-Walled Dielectric Bubble. The limiting case of a thin-walled bubble has been treated by Andreasen (1957a). The actual coating is replaced by a shell of infinitesimal thickness δ and dielectric constant ε', chosen so that

$$(\delta\varepsilon')_{\lim \delta \to 0} = (\kappa_e - 1)\varepsilon_0 d \tag{5.3.10}$$

where κ_e and d are the relative dielectric constant (2.1.8) and thickness, respectively, of the coating, while ε_0 is the dielectric constant of the free space. Andreasen's treatment assumes that the external medium and the core consist of free space. For the case when the media on both sides of the shell

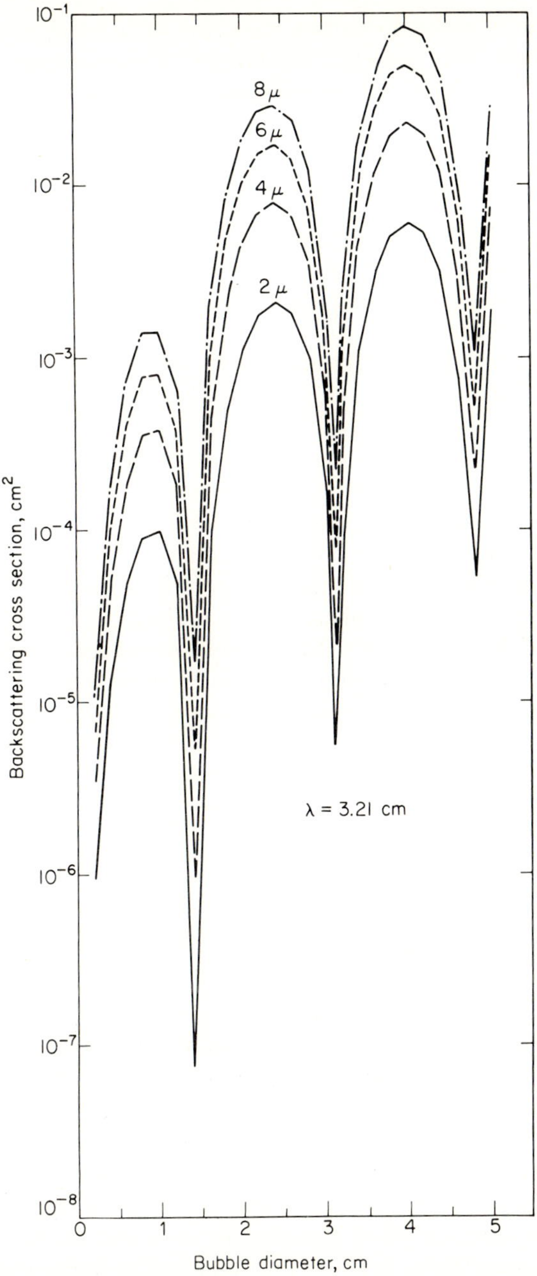

FIG. 5.20. Backscattering cross section $[(\pi a^2/4)G]$ of bubbles plotted against bubble diameter for the indicated bubble thicknesses at $\lambda = 3.21$ cm (Battan and Herman, 1961).

are the same, κ_e must be divided by the value of the corresponding quantity outside the shell. The boundary conditions of the tangential components of the fields at the infinitesimally thin coatings become (Andreasen, 1957b)

$$\begin{aligned} \mathbf{n} \times (\mathbf{H}_3 - \mathbf{H}_2) &= i[(\kappa_e - 1)k_0 d/\zeta_0](\mathbf{n} \times \mathbf{E}_2) \times \mathbf{n} \\ &= i[(\kappa_e - 1)k_0 d/\zeta_0](\mathbf{n} \times \mathbf{E}_3) \times \mathbf{n} \end{aligned} \tag{5.3.11}$$

Here subscripts 2 and 3 denote the media inside and outside the shell respectively, $\mathbf{n}$ is the unit vector normal to the shell and oriented outwards, while ζ_0 and k_0 are the specific impedance and the propagation constant, respectively, of free space.

The assumptions under which these boundary conditions are valid are

$$k_0 d \ll 1, \qquad (\kappa_e - 1)k_0 d \ll 1, \qquad \text{and} \qquad \kappa_e \gg 1 \tag{5.3.12}$$

The latter condition only becomes stringent for spheres much smaller than the wavelength. When the sphere is approximately the same size as the wavelength or greater, the dielectric constant of the shell need not be large in order to use these simplified boundary conditions.

With only one boundary, the analysis is greatly simplified. Indeed, formally it is very similar to the single sphere problem. The incident field, the scattered field, and the field within the sphere may be expanded precisely as for the single sphere case and the scattering coefficients become

$$a_n = -\frac{(\kappa_e - 1)k_0 d[\psi_n'(\alpha)]^2}{1 + i(\kappa_e - 1)k_0 d\psi_n'(\alpha)\zeta_n'(\alpha)} \tag{5.3.13}$$

$$b_n = -\frac{(\kappa_e - 1)k_0 d[\psi_n(\alpha)]^2}{1 + i(\kappa_e - 1)k_0 d\psi_n(\alpha)\zeta_n(\alpha)} \tag{5.3.14}$$

5.4 Multilayered Spheres

The extension of the theory of the coated sphere to one consisting of many layers of concentric shells is straightforward. The potential functions for the incident wave, for the scattered wave, and for the wave within the core are written as before [(5.1.1) to (5.1.4), (5.1.7), (5.1.8)]. For each of the shells, the form of the potential function is the same as for the coating [(5.1.5), (5.1.6)] using in each case the appropriate values of the propagation constant. For a system of l layers (Fig. 5.21), the boundaries will be designated as $j = 1, 2, 3 \ldots l$, the 1st boundary being between the core and the next inner layer, and the lth boundary being between the outer shell and the external medium. The relative refractive indices are $m_1, m_2 \ldots m_l$, the radii are $a_1, a_2 \ldots a_l$, and the size parameters are $\alpha_1, \alpha_2 \ldots \alpha_l$. The coefficients in the expansion of the potential functions within the sphere will be $A_n^{(1)}$, $A_n^{(2)} \ldots A_n^{(l)}$, $a_n^{(2)}$, $a_n^{(3)} \ldots a_n^{(l)}$, for the electric wave (TM mode) and $B_n^{(1)}$, $B_n^{(2)} \ldots B_n^{(l)}$, $b_n^{(2)}$ $b_n^{(3)} \ldots b_n^{(l)}$ for the magnetic wave (TE mode). The scattering coefficients will, as usual, be denoted by a_n and b_n. In order that the signs preceding these constants in the expressions below be positive, the potential functions in the odd numbered layers are written with a negative sign as in (5.1.7) and (5.1.8).

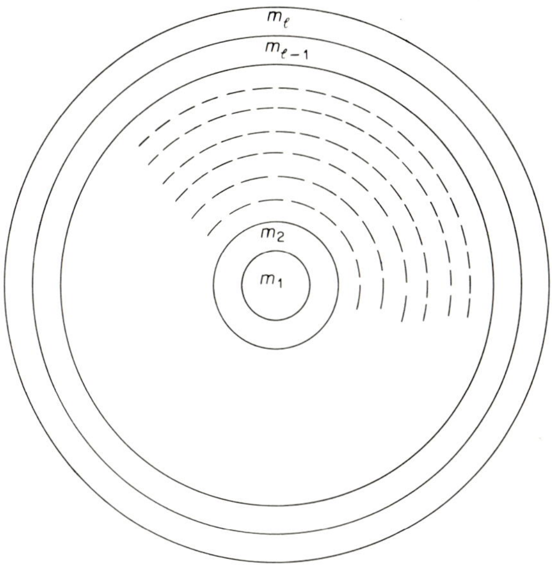

FIG. 5.21. Geometry for concentric spheres consisting of core with $(l-1)$ concentric spherical shells.

The sign for the incident wave is opposite to that for the scattered wave. The four l-families of equations obtained upon application of the boundary conditions can be written as

$$\begin{aligned}(1/m_j)[A_n^{(j)}\psi_n'(m_j\alpha_j) &+ \delta_0^{(1)} a_n^{(j)}\chi_n'(m_j\alpha_j)] \\ &+ \delta_0^{(l)}(1/m_{j+1})[A_n^{j+1}\psi_n'(m_{j+1}\alpha_j) + a_n^{j+1}\chi_n'(m_{j+1}\alpha_j)] \\ &= \delta_1{}^{l}[\psi_n'(\alpha_l) - a_n\zeta_n'(\alpha_l)]\end{aligned} \tag{5.4.1}$$

$$\begin{aligned}[A_n^{(j)}\psi_n(m_j\alpha_j) &+ \delta_0^{(1)} a_n^{(j)}\chi_n(m_j\alpha_j)] \\ &+ \delta_0{}^{l}[A_n^{(j+1)}\psi_n(m_{j+1}\alpha_j) + a_n^{(j+1)}\chi_n(m_{j+1}\alpha_j)] \\ &= \delta_1^{(l)}[\psi_n(\alpha_l) - a_n\zeta_n(\alpha_l)]\end{aligned} \tag{5.4.2}$$

$$\begin{aligned}(1/m_j)[B_n^{(j)}\psi_n'(m_j\alpha_j) &+ \delta_0^{(1)}a_n^{(j)}\chi_n'(m_j\alpha_j)]\\ &+ \delta_0^{(l)}(1/m_{j+1})[B_n^{(j+1)}\psi_n'(m_{j+1}\alpha_j) + b_n^{(j+1)}\chi_n'(m_{j+1}\alpha_j)]\\ &= \delta_1^{(l)}[\psi_n'(\alpha_l) - b_n\zeta_n'(\alpha_l)]\end{aligned} \tag{5.4.3}$$

$$\begin{aligned}(1/m_j^2)[B_n^{(j)}\psi_n(m_j\alpha_j) &+ \delta_0^{(1)}a_n^{(j)}\chi_n(m_j\alpha_j)]\\ &+ \delta_0^{(l)}(1/m_{j+1}^2)[B_n^{(j+1)}\psi_n(m_{j+1}\alpha_j) + b_n^{(j+1)}\chi_n(m_{j+1}\alpha_j)]\\ &= \delta_1^{(l)}[\psi_n(\alpha_l) - b_n\zeta_n(\alpha_l)]\end{aligned} \tag{5.4.4}$$

The delta functions are

$$\delta_0^{(1)} = 0 \quad \text{when} \quad j = 1; \qquad \delta_0^{(1)} = 1 \quad \text{when} \quad j \neq 1 \tag{5.4.5}$$

$$\delta_0^{(l)} = 0 \quad \text{when} \quad j = l; \qquad \delta_0^{(l)} = 1 \quad \text{when} \quad j \neq l \tag{5.4.6}$$

$$\delta_1^{(l)} = 1 \quad \text{when} \quad j = l; \qquad \delta_1^{(l)} = 0 \quad \text{when} \quad j \neq l \tag{5.4.7}$$

These equations reduce to (3.3.42) to (3.3.45) for a homogeneous sphere ($l = 1$) and to (5.1.17) to (5.1.24) for a coated sphere ($l = 2$). The solution for a_n written in determinantal form is given by (5.4.8). The solution for b_n can be obtained by raising the "m's" from the even numbered rows of the determinants to the odd numbered rows immediately above. The various scattering quantities and the absorption may be obtained from a_n and b_n with the formulas given in Chapter 3 for the homogeneous sphere.

$$
a_n = \frac{\begin{vmatrix}
\psi_n'(m_1\alpha_1) & \psi_n'(m_2\alpha_1) & \chi_n'(m_2\alpha_1) & 0 & 0 & \cdots & 0 & 0 & 0 \\
m_1\psi_n(m_1\alpha_1) & m_2\psi_n(m_2\alpha_1) & m_2\chi_n(m_2\alpha_1) & 0 & 0 & \cdots & 0 & 0 & 0 \\
0 & \psi_n'(m_2\alpha_2) & \chi_n'(m_2\alpha_2) & \psi_n'(m_3\alpha_2) & \chi_n'(m_3\alpha_2) & \cdots & 0 & 0 & 0 \\
0 & m_2\psi_n(m_2\alpha_2) & m_2\chi_n(m_2\alpha_2) & m_3\psi_n(m_3\alpha_2) & m_3\chi_n(m_3\alpha_2) & \cdots & 0 & 0 & 0 \\
0 & 0 & 0 & \psi_n'(m_3\alpha_3) & \chi_n'(m_3\alpha_3) & \cdots & 0 & 0 & 0 \\
0 & 0 & 0 & m_3\psi_n(m_3\alpha_3) & m_3\chi_n(m_3\alpha_3) & \cdots & 0 & 0 & 0 \\
0 & 0 & 0 & 0 & 0 & \cdots & 0 & 0 & 0 \\
0 & 0 & 0 & 0 & 0 & \cdots & 0 & 0 & 0 \\
 & & & & & & & & \\
0 & 0 & 0 & 0 & 0 & \cdots & \psi_n'(m_l\alpha_l) & \chi_n'(m\,\alpha_l) & \psi_n'(\alpha_l) \\
0 & 0 & 0 & 0 & 0 & \cdots & m_l\psi_n(m_l\alpha_l) & m_l\chi_n(m_l\alpha_l) & \psi_n(\alpha_l)
\end{vmatrix}}{\begin{vmatrix}\text{same as above except that } \psi_n'(\alpha_l) \text{ and } \psi_n(\alpha_l) \text{ are replaced by } \zeta_n'(\alpha_l) \text{ and } \zeta_n(\alpha_l)\end{vmatrix}} \tag{5.4.8}
$$

5.5 Spherically Symmetrical Lenses

Some of the interest in scattering by multilayered spheres as well as by spheres with a continuously variable but radially symmetric refractive index is related to the application of such structures to microwave technology. According to geometric optics, radially symmetric spheres can behave as perfectly focusing lenses when they have an appropriate variation of refractive index. Radially symmetrical cylinders behave in much the same fashion so that each of the lenses described below will also have its cylindrical counterpart. The exact theory of scattering by multilayered spheres which was developed in the previous section and also the theory of scattering by spheres with continuously variable refractive index which will be considered later offer the possibility of a precise determination of the limits of validity of these geometrical optics devices.

Spherical lenses of this type have a venerable history. A classical example is the Maxwell "fish-eye" for which the refractive index variation is (Born and Wolf, 1959)

$$n(r) = n_0/[1 + (r/e)^2] \tag{5.5.1}$$

where n_0 and e are constants, and r is the distance from the center of symmetry to an arbitrary point. In such a medium, all rays emanating from this point will be brought to focus elsewhere in the medium [Born and Wolf (1959) p. 146]. When the medium is a finite sphere, with r the radial distance relative to the radius of the sphere ($0 < r < 1$), this behaves as a lens which brings the rays emanating from a point source on the surface, P_0, to focus at the opposite surface point, P. The ray paths are traced in Fig. 5.22.

A Luneberg lens (Morgan, 1958; Kay, 1959) is a spherically symmetric structure with a variable refractive index, which will form perfect geometrical images of two given concentric spheres on each other. Figures 5.23 and 5.24 illustrate schematic Luneberg lenses, respectively, with two external foci and with one external and one internal focus. When one of the foci is taken on the surface of the lens and the other is at infinity, the refractive index profile takes on the following simple form:

$$n(r) = (2 - r^2)^{1/2} \tag{5.5.2}$$

with the ray path shown in Fig. 5.25. Here, a plane parallel incident beam is brought to a focus at P or the rays from a source at P may emerge parallel at the opposite surface. If the Luneberg lens is now fitted with a spherical cap reflector, it becomes a Luneberg reflector. This will behave as a perfect backscatter focusing device, returning all of the incident energy into the backward direction except for that lost at the front surface reflection. In a higher-order Luneberg lens or reflector (Huynen, 1958), the rays reverse

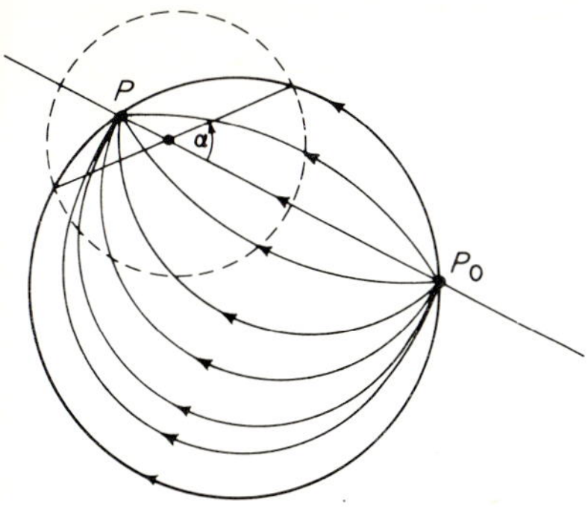

FIG. 5.22. Spherical Maxwell fish-eye. All rays emanating from P_0 travel along the indicated paths and are brought to a focus at P.

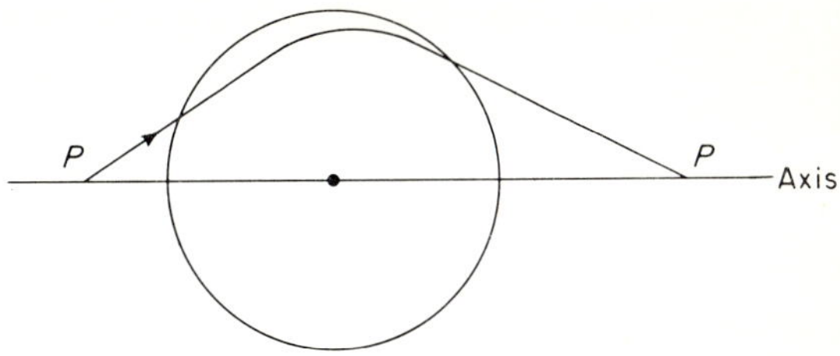

FIG. 5.23. Luneberg lens with two external foci.

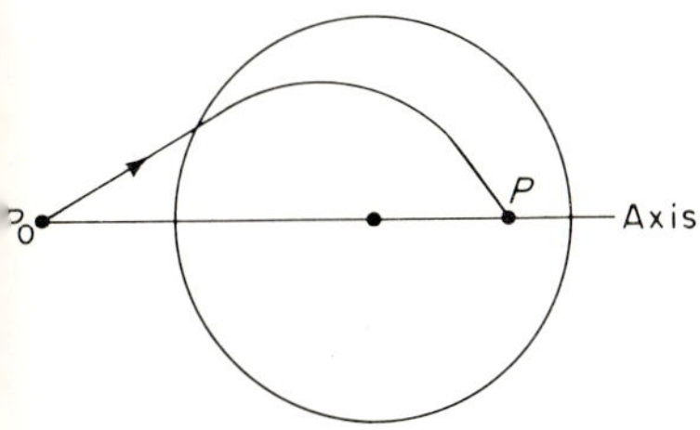

FIG. 5.24. Luneberg lens with one external focus and one internal focus.

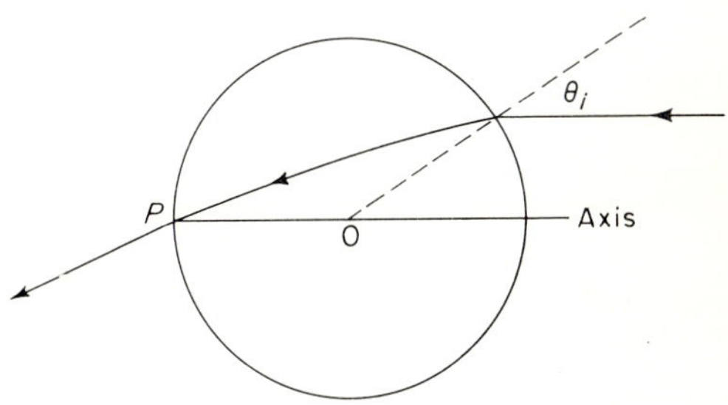

FIG. 5.25. Luneberg lens with one focus at infinity and one on the surface of the lens.

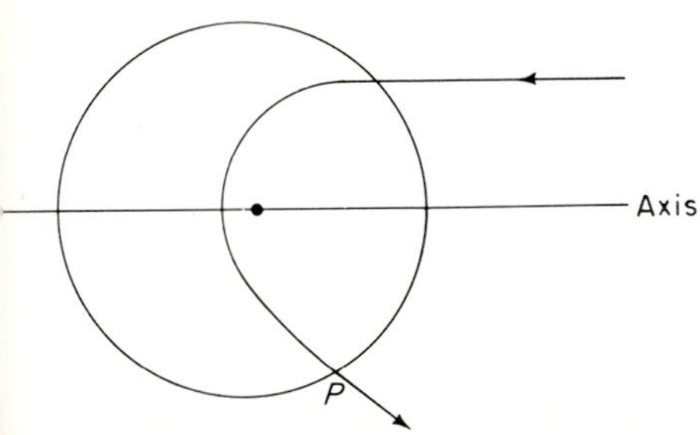

FIG. 5.26. Higher-order Luneberg lens with one focus at infinity and the other on the near-side surface of the lens.

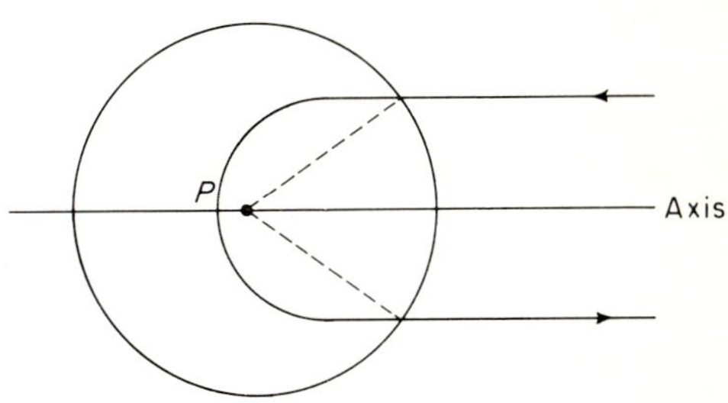

FIG. 5.27. Eaton lens.

within the sphere before coming to a focus (Fig. 5.26). The free space lens performs the inverse function as the Luneberg reflector. The rays emerge from it undeviated.

An interesting variant is the isotropic or Eaton lens first proposed as a perfect backscatter device which obviates the need for a metallic reflector. The refractive index variation is

$$n(r) = \left(\frac{2 - r}{r}\right)^{1/2} \tag{5.5.3}$$

The elliptical path of a ray through an Eaton lens emerges in the backward direction as shown in Fig. 5.27. Actually, this lens is only effective in the cylindrical case. For the spherical Eaton lens, it is necessary to consider polarization. When this is done, the integrated electric vector over the aperture of the lens for a wave traveling in the backward direction is zero (Kay, 1956). Complete destructive interference takes place and the backscatter cross section is zero.

Luneberg (1944) developed the theory of such devices as an academic exercise in classical optics. Workers in optics had no applications for such devices nor could they construct them. However radar workers immediately recognized their utility (Rudduck and Walter, 1962) and the early history of the lens after its conception has been the history of its fabrication.

In practice Luneberg and Eaton lenses are approximated by multilayered spheres fabricated from two identical hemispheres each consisting of a stack of hemispherical half shells. These lenses may vary from 3 to 48 in in diameter with up to 50 shells. The theory for the multilayered sphere is able to describe, exactly, the scattering properties of such devices, so that the effect of the number of layers and their spacing can be completely elucidated. With the general theory for a continuously variable refractive index which will be treated shortly, the actual performance of theoretical models such as those corresponding to (5.5.1) to (5.5.3) can be considered and a comparison with the layered cases can be made.

5.5.1 Numerical Results

Rheinstein (1962)[2] has calculated the scattering for a "practical" Eaton lens consisting of discrete, spherical layers. This practical lens cannot have a point singularity at the origin as indicated by (5.5.3). Rheinstein's model for a sphere of radius a consists of a core with radius $0.1a$ surrounded by $0.9m$ concentric spherical shells of thickness a/m. This is termed an "m-layer" case even though the number of regions, including the core, is $1 + 0.9\,m$.

[2] Some computations for a 5-layer and 10-layer approximation to the Eaton and Luneberg lenses have been made by Mikulski and Murphy (1963).

The refractive index of each shell is given by (5.5.3) for the Eaton lens where the value of r at the midpoint of each shell is used. Two cases were chosen for the core: $m = \sqrt{39}$ and $m = \infty$. Figure 5.28 shows the continuous variation of m according to (5.5.3) and the step-wise variation for the "20-layer" case with a dielectric core. In what follows we will refer to an m-layer Eaton lens.

In Fig. 5.29 the backscatter of an 80-layer Eaton lens with a dielectric and with a perfectly reflecting core are compared with that from a perfectly reflecting sphere of the same radius. The shaded area gives the limiting envelope of the backscatter for the perfectly reflecting sphere which undergoes a damped oscillation about the value, $G = 1$. The main point is that the backscatter of the Eaton lens is of the same order of magnitude as the perfectly reflecting sphere although the oscillations are considerably greater. This does not conform to the geometrical optics prediction of zero backscatter for the continuous variation.

Rheinstein has extended these calculations out to $\alpha = 60$. In addition, for selected regions he has taken closer intervals of α (0.1) in order to better observe the fine structure. The general pattern of Fig. 5.29 is maintained throughout this region.

The effect of the number of layers is shown in Fig. 5.30. There is a tendency for the backscatter cross section to converge to a limiting value as the number of layers is increased, especially when the number of layers is larger than α.

These results permit substantiation of the localization principle for large spheres which had been considered earlier. According to this, the scattering coefficients, a_n and b_n of order n correspond to a ray incident on the sphere at a distance

$$p = (n + 1/2)\lambda/2\pi \tag{5.5.4}$$

from the central ray. Let us assume that these rays traverse an elliptical path into the sphere and then out again in the backward direction as depicted earlier in Fig. 5.27 for the Eaton lens. This suggests that only those rays which impinge upon the core will be affected by the composition of the core. For a core whose radius relative to the total radius of the sphere is q, only those rays for which the impact parameter, p, is

$$p < a[1 - (1 - q)^2]^{1/2} \tag{5.5.5}$$

will strike the core.

Combining this with (5.5.4) we obtain for $q = 0.1$ that

$$n = 0.44\alpha - \tfrac{1}{2} \tag{5.5.6}$$

Accordingly, scattering coefficients of order less than this should be affected by the refractive index of the core while scattering coefficients of higher order should not. Computations carried out by Rheinstein show that for $\alpha > 20$ this was actually the case.

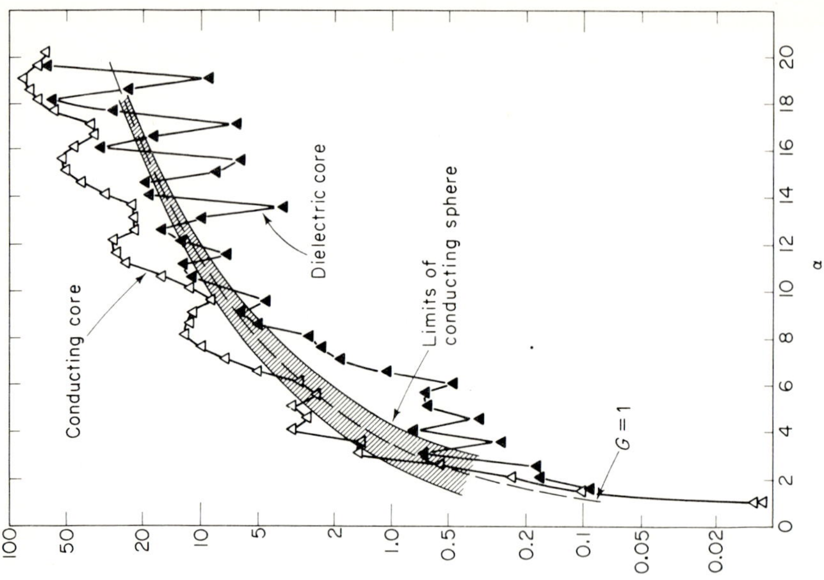

FIG. 5.29. Backscatter gain ($\alpha^2 G/4\pi$) of 80-layer Eaton lenses

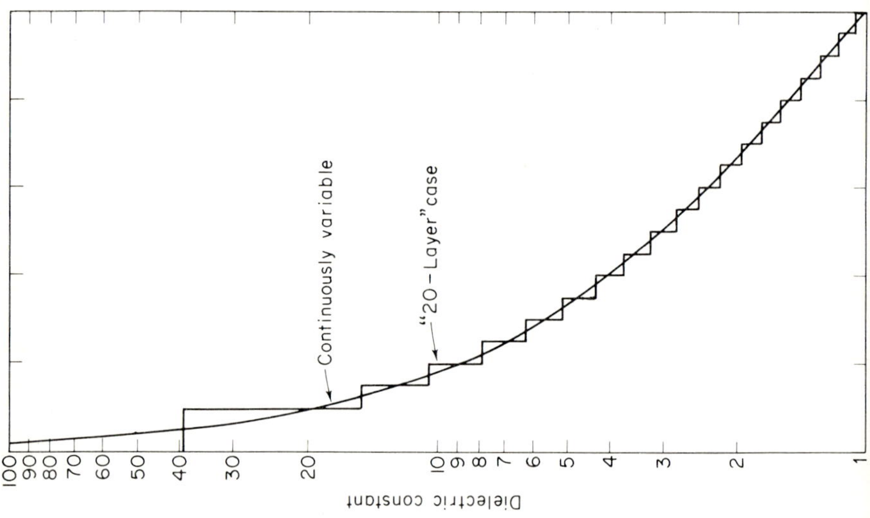

FIG. 5.28. Variation of dielectric constant ($\varepsilon = m^2$) with radial distance for a continuously variable and for a 20-layer Eaton lens (Rheinstein, 1962).

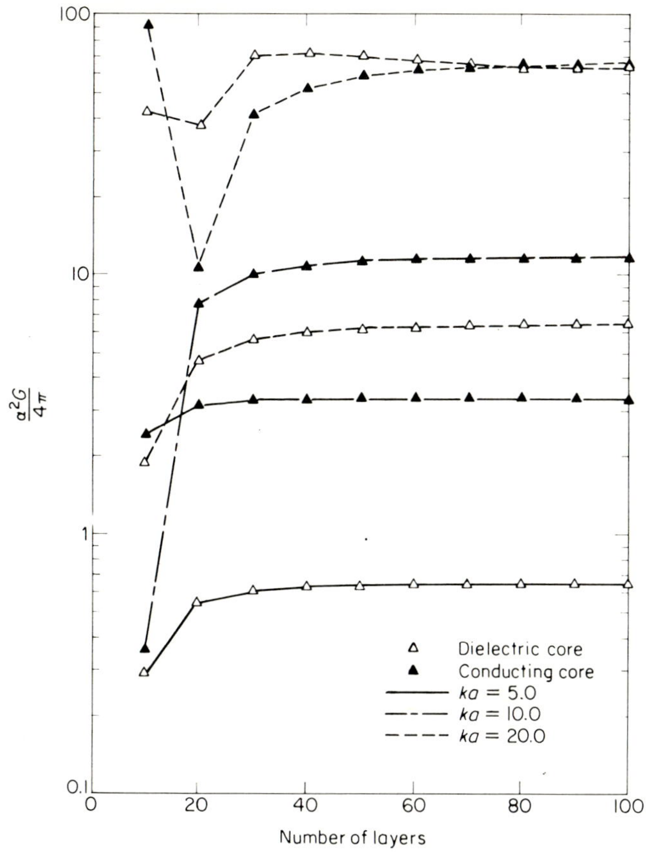

FIG. 5.30. Effect of number of layers upon the backscatter gain of layered Eaton lenses with and without a conducting core for $ka = \alpha = 5$, 10, and 20 (Rheinstein, 1962).

Finally in Fig. 5.31 and 5.32 the entire radiation pattern is shown for Eaton lenses with and without a perfectly reflecting core and $\alpha = 20$ and 59. The upper half of each pattern is for the horizontal component, i_2, while the lower half is for the vertical component, i_1. The data for these diagrams were computed at 1° intervals which was insufficient to elucidate all of the fine structure. The first of these is an 80-layer lens. In the other, the spacing of the layers is selected somewhat differently. For a dielectric core, the general character of the patterns was the same although the detailed structure was somewhat different. These patterns should be compared with those for the perfectly conducting spheres of the same size shown earlier. The front lobes, due primarily to Fraunhofer diffraction, are quite similar for both the homogeneous and the layered spheres. However, whereas the scattering

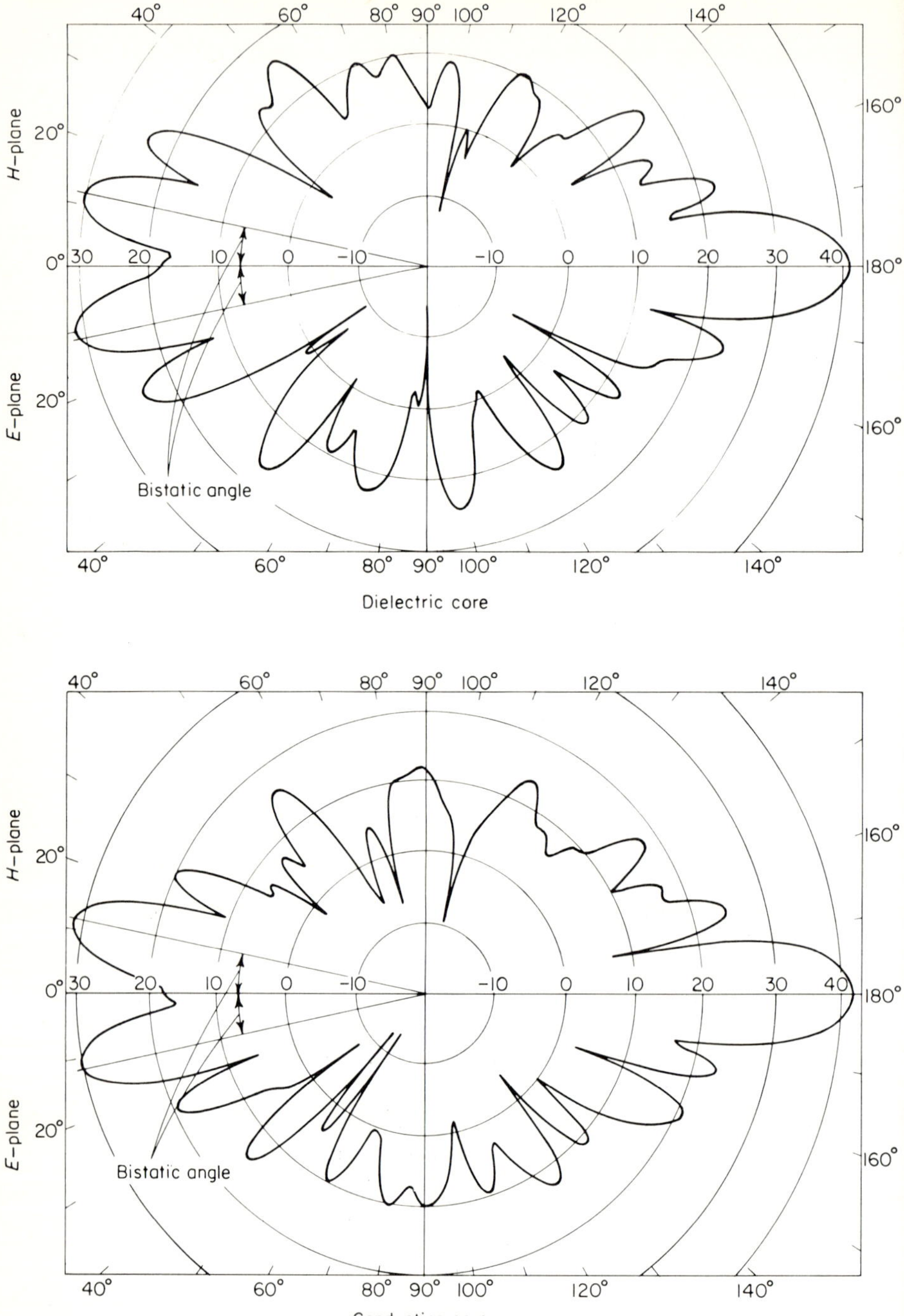

FIG. 5.31. Radiation patterns for 80-layer Eaton lens with (top) and without (bottom) a conducting core for $\alpha = 20$. Upper half of each pattern is for G_2, lower half is for G_1: marked radial distances are $10 \log(\alpha^2 G/4\pi)$ (Rheinstein, 1962).

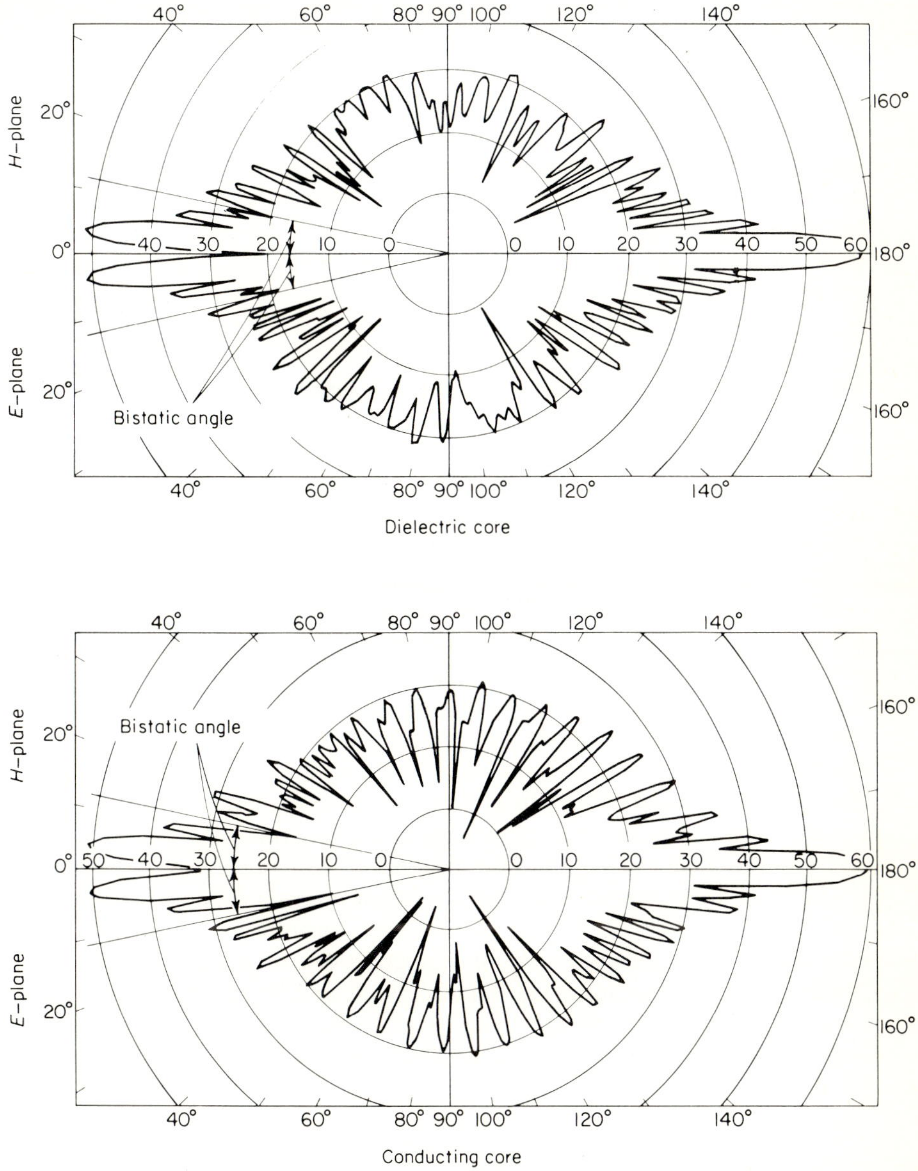

FIG. 5.32. Same as in Fig. 5.31 for $\alpha = 59$ (Rheinstein, 1962).

from the conducting sphere is uniform in the lateral and backward directions, the pattern for the Eaton lens is oscillatory. Unlike a homogeneous dielectric sphere, these objects do not exhibit extreme angular dissymetry. Except for the extreme forward lobe, the scattering in the forward direction is not particularly more intense than that in the backward direction. Although, as already noted, the backscatter is comparable to that for a perfectly reflecting sphere, there are two huge lobes directly adjacent to the direct backscatter. Accordingly if all the backward angles within 15° of the backscatter for $\alpha = 20$ and within $\pm 5°$ for $\alpha = 59°$ are considered, the total energy scattered within these backward cones is considerably greater for the Eaton lenses than for the perfectly reflecting sphere.

5.6 Spheres with Continuously Variable Refractive Index

We will now consider the scattering by a sphere with a continuously variable refractive index. The approach will be the same as that used for the homogeneous sphere or for the layered sphere consisting of discrete homogeneous regions in which the refractive index is constant.

The field is decomposed into two partial fields, each of which can be described by a scalar wave equation. The method utilizes separation of the variables and representation of the fields in terms of an arbitrary expansion of products of three wave functions, each of which is a solution of one of the three separated linear ordinary differential equations. This approach is restricted to certain conditions on the coordinate system (Bromwich, 1919; Friedman, 1962) which both spherical and cylindrical coordinates satisfy. The only other restriction is that there be radial symmetry, i.e., that the refractive index depend only upon the distance from the center of the sphere, and also the usual assumption that the sphere be isotropic.

The basic formalism was developed some time ago in connection with the problem of the propagation of radio waves through an inhomogeneous atmosphere by Luneberg (1948), Bremmer (1949), Friedman (1951, 1962), Marcuvitz (1951), Nomura and Takaku (1955), and Wait (1962). Our treatment will follow that of Wyatt (1962, 1964a). Other recent expositions of the scattering problem have been given by Tai (1958a, b), Garbacz (1962a, b), and Levine and Kerker (1963).

5.6.1 Inhomogeneous Sphere in a Homogeneous Medium

The theory follows the familiar lines already considered for the homogeneous sphere. The field vectors may be derived from the two Debye potentials, each representing a TM (electric) and a TE (magnetic) mode.

These potentials satisfy the scalar wave equation. Once again, the problem consists of two parts. First, the appropriate forms of the scalar wave equation must be solved in the spherical coordinate system. The wave inside the particle and the wave outside the particle, consisting of the incident plus the scattered wave, can then be expanded in terms of these solutions. The second part of the problem consists in matching the waves inside and outside the particle at the boundary in order that the coefficients in the expansion of the scattered wave may be determined. The determination of these scattering coefficients constitutes the solution since they can then be inserted into the formulas of Chapter 3 to obtain the physically interesting quantities.

The scattering problem is almost identical to the problem of the propagation of radio waves around an earth surrounded by an inhomogeneous atmosphere and it is in this connection that it was first discussed. There are two differences. In the latter case, it is the region external to the sphere rather than the sphere itself that is stratified. Also, for the propagation problem, the source is usually chosen to be a dipole located near the surface of the sphere rather than at infinity (plane wave).

The potential functions for the homogeneous region are identical to those given earlier. Thus, the incident and scattered waves in the homogeneous dielectric surrounding the sphere may be expressed by (3.3.32) to (3.3.35). For the radially stratified medium, the electric Debye potential satisfies

$$\nabla^2 \pi_1 - \frac{2}{k_1 r}\frac{\partial k_1}{\partial r}\frac{\partial}{\partial r}(r\pi_1) + k_1{}^2 \pi_1 = 0 \tag{5.6.1}$$

where the propagation constant within the sphere, k_1, is a function of the radial distance, r. When k_1 is constant, this reduces to the usual form of the scalar wave equation given by (3.3.18). The magnetic Debye potential satisfies the simpler wave equation

$$\nabla_1{}^2 \pi_2 + k_1{}^2 \pi_2 = 0 \tag{5.6.2}$$

The difference in the form of these equations satisfied by the electric and magnetic potentials arises from the fact that whereas k_1 is a function of r, we assume that the magnetic permeability is constant and equal to its value in free space. In the general case, the equations for the electric and magnetic potentials are completely symmetrical (Garbacz, 1962a, b).

These equations are readily solved, using the method of separation of the variables,

$$\pi_1 = R_1(r)\Theta(\theta)\Phi(\phi) \tag{5.6.3}$$

and

$$\pi_2 = R_2(r)\Theta(\theta)\Phi(\phi) \tag{5.6.4}$$

The resulting equations which depend upon the angles θ and ϕ are identical to the homogeneous case, and are given by (3.3.23) and (3.3.24) for both the electric and the magnetic waves. The radial part of (5.6.1) satisfies the differential equation

$$\frac{d^2W_n(r)}{dr^2} - \frac{2}{k_1}\frac{dk_1}{dr}\frac{dW_n(r)}{dr} + \left[k_1{}^2 - \frac{n(n+1)}{r^2}\right]W_n(r) = 0 \tag{5.6.5}$$

where the solution of this equation is a new function

$$W_n(r) = rR_1(r) \tag{5.6.6}$$

For the magnetic wave, the radial equation is

$$\frac{d^2G_n(r)}{dr^2} + \left[k_1{}^2 - \frac{n(n+1)}{r^2}\right]G_n(r) = 0 \tag{5.6.7}$$

where

$$G_n(r) = rR_2(r) \tag{5.6.8}$$

This differs from the usual Bessel equation because of the variable nature of k_1. Of course, when k_1 is constant, both equations are identical and reduce to (3.3.22).

Since the radial equations are second-order differential equations, there are two independent solutions for each of them, so that it is now necessary to deal with four separate functions. At present, explicit analytical solutions exist only for certain functional forms of the propagation constant. For the homogeneous case, these degenerate into only two solutions, the familiar Ricatti–Bessel functions, ψ_n and χ_n. However, even when the solutions cannot be expressed in terms of known functions, this need not be a deterrent as both Wyatt (1962) and Garbacz (1962a, b) have emphasized. With the aid of a computer, it is usually feasible to solve the differential equations directly using numerical methods. Even for homogeneous spheres which involve only the well-known Ricatti–Bessel functions, the computations are sufficiently formidable to require recourse to high-speed digital computers in order to obtain numerical results. Accordingly, numerical solution of the radial equations is hardly an impediment.

For the wave within the inhomogeneous sphere

$$r\pi_1{}' = \frac{1}{k_1{}^2}\sum_{n=1}^{\infty} i^{n-1}\frac{2n+1}{n(n+1)}c_n W_n^{(1)}(r)P_n^{(1)}(\cos\theta)\cos\phi \tag{5.6.9}$$

$$r\pi_2{}' = \frac{1}{k_1{}^2}\sum_{n=1}^{\infty} i^{n-1}\frac{2n+1}{n(n+1)}d_n G_n^{(1)}(r)P_n^{(1)}(\cos\theta)\sin\phi \tag{5.6.10}$$

where $W_n^{(1)}(r)$ and $G_n^{(1)}(r)$ are those solutions of (5.6.5) and (5.6.7) which are well behaved at the origin. It is convenient to introduce a new variable $\rho = k_2 r$ into these equations which are then written

$$\frac{d^2 W_n^{(1)}(\rho)}{d\rho^2} - \frac{2}{m}\frac{dm}{d\rho}\frac{dW_n^{(1)}(\rho)}{d\rho} + \left[m^2 - \frac{n(n+1)}{\rho^2}\right] W_n^{(1)}(\rho) = 0 \quad (5.6.11)$$

and

$$\frac{d^2 G_n^{(1)}(\rho)}{d\rho^2} + \left[m^2 - \frac{n(n+1)}{\rho^2}\right] G_n^{(1)}(\rho) = 0 \quad (5.6.12)$$

where m is the refractive index of the particle relative to that of the external homogeneous medium. The remainder of the procedure follows as before. The boundary conditions are the same as for the homogeneous case [(3.3.38) to (3.3.41)] and when applied at the surface of the sphere $r = a$, they lead to four linear equations involving the four coefficients, a_n, b_n, c_n, and d_n. It is the former coefficients in which we are interested. These are

$$a_n = \frac{\psi_n(\alpha) W_n^{(1)\prime}(\alpha) - m^2 W_n^{(1)}(\alpha)\psi_n'(\alpha)}{\zeta_n(\alpha) W_n^{(1)\prime}(\alpha) - m^2 W_n^{(1)}(\alpha)\zeta_n'(\alpha)} \quad (5.6.13)$$

$$b_n = \frac{\psi_n(\alpha) G_n^{(1)\prime}(\alpha) - G_n^{(1)}(\alpha)\psi_n'(\alpha)}{\zeta_n(\alpha) G_n^{(1)\prime}(\alpha) - G_n^{(1)}(\alpha)\zeta_n'(\alpha)} \quad (5.6.14)$$

where α has its usual meaning.[3] When the sphere is homogeneous so that m is constant,

$$G_n^{(1)}(\alpha) = W_n^{(1)}(\alpha) = \psi_n(\beta) \quad (5.6.15)$$

$$G_n^{(1)\prime}(\alpha) = W_n^{(1)\prime}(\alpha) = m\psi_n'(\beta) \quad (5.6.16)$$

and the coefficients reduce properly to the values given by (3.3.48) and (3.3.49). The various scattering quantities of physical interest can now be obtained from a_n and b_n in the usual way.

5.6.2 Homogeneous Sphere in an Inhomogeneous Medium

Sayasov (1961) has considered the inverse case of a homogeneous spherical particle imbedded in an inhomogeneous medium. Although his treatment is limited to a perfectly conducting sphere, extension to the dielectric case follows readily. The system continues to be centrally symmetrical, so that the inhomogeneous medium is characterized by a refractive index which is radially dependent. The further assumption is made that at a sufficient

[3] An error in the expressions for a_n and b_n originally given by Wyatt (1962, 1963a, b) has been corrected in later papers (1964a, b, 1965b).

distance from the sphere, the refractive index approaches a constant value (i.e., the relative refractive index approaches unity).

Now the expansion for the potential function within the sphere can be written exactly as for the entirely homogeneous case (3.3.36) and (3.3.37). For the incident and scattered waves

$$r\pi_1{}^i = \frac{1}{k_2{}^2} \sum_{n=1}^{\infty} i^{n-1} \frac{2n+1}{n(n+1)} W_n^{(1)}(k_2 r) P_n^{(1)}(\cos\theta)\cos\phi \tag{5.6.17}$$

$$r\pi_1{}^i = \frac{1}{k_2{}^2} \sum_{n=1}^{\infty} i^{n-1} \frac{2n+1}{n(n+1)} G_n^{(1)}(k_2 r) P_n^{(1)}(\cos\theta)\sin\phi \tag{5.6.18}$$

$$r\pi_1{}^s = -\frac{1}{k_2{}^2} \sum_{n=1}^{\infty} i^{n-1} \frac{2n+1}{n(n+1)} a_n W_n^{(2)}(k_2 r) P_n^{(1)}(\cos\theta)\cos\phi \tag{5.6.19}$$

$$r\pi_1{}^s = -\frac{1}{k_2{}^2} \sum_{n=1}^{\infty} i^{n-1} \frac{2n+1}{n(n+1)} b_n G_n^{(2)}(k_2 r) P_n^{(1)}(\cos\theta)\sin\phi \tag{5.6.20}$$

where $W_n^{(2)}(k_2 r)$ and $G_n^{(2)}(k_2 r)$ are those solutions of the radial equations (5.6.5) and (5.6.7) which are well behaved at infinity. Solution for the scattering coefficients follows in the usual way leading to

$$a_n = \frac{W_n^{(1)}(\alpha)\psi_n'(\beta) - m_1{}^2\psi_n(\beta)W_n^{(1)\prime}(\alpha)}{W_n^{(2)}(\alpha)\psi_n'(\beta) - m_1{}^2\psi_n(\beta)W_n^{(2)\prime}(\alpha)} \tag{5.6.21}$$

$$b_n = \frac{G_n^{(1)}(\alpha)\psi_n'(\beta) - \psi_n(\beta)G_n^{(1)\prime}(\alpha)}{G_n^{(2)}(\alpha)\psi_n'(\beta) - \psi_n(\beta)G_n^{(2)\prime}(\alpha)} \tag{5.6.22}$$

When the medium is homogeneous, the radial functions reduce to Ricatti–Bessel functions and the scattering coefficients reduce properly. For the case of an infinitely conducting sphere

$$a_n = [W_n^{(1)\prime}(\alpha)]/[W_n^{(2)\prime}(\alpha)] \tag{5.6.23}$$

$$b_n = [G_n^{(1)}(\alpha)]/[G_n^{(2)}(\alpha)] \tag{5.6.24}$$

Since the external functions reduce to the same asymptotic forms as the Hankel functions, viz.

$$W_n^{(2)} = G_n^{(2)} = \exp[i(kr - [(n+1)/2]\pi)] \qquad r \to \infty \tag{5.6.25}$$

the above scattering coefficients may be used directly in the usual far-field expressions, provided that the medium does not absorb.

The extension to an inhomogeneous sphere imbedded in an inhomogeneous medium follows in a straightforward manner.

5.6.3 Alternative Formulation as a Spherical Transmission Line Problem

There is a direct analogy between the formalism of transmission line theory used by electrical engineers and the theory of electromagnetic waves which permits the mathematical and numerical techniques developed for the analysis of electrical circuits to be applied directly to electromagnetic wave phenomena (Ramo and Whinnery, 1960). For a wave traveling along the z-axis

E_x	corresponds to the line voltage, V
H_y	corresponds to the current in the line, I
μ	corresponds to the induction per unit length, L
ε	corresponds to the capacitance per unit length, C
$\eta = (\mu/\varepsilon)^{1/2}$	corresponds to $Z_0 = (L/C)^{1/2}$ where Z_0 is called the characteristic impedance [cf. (2.2.5)].

Thus, at any plane z, the field or wave impedance

$$Z_0(z) = E_x(z)/H_y(z) \tag{5.6.26}$$

The boundary conditions requiring the continuity of the tangential components of E and H across surfaces of discontinuity in the absence of free charges and currents correspond to the requirement that V and I be continuous at the junction between the two transmission lines. The reflection coefficient at a plane boundary is equivalent to the impedance mismatch ratio for transmission lines.

Garbacz (1961, 1962a, b) and Wait (1963) have suggested that numerical solution of scattering by stratified spheres, either consisting of concentric shells or with a continuously variable refractive index or some combination of these, may be facilitated by considering the analogy with a nonuniform spherical transmission line. The scattering coefficients now take on the very simple form

$$a_n = \frac{\psi_n(\alpha) - i[Y_n^{(e)}(\alpha)/Y_0]\psi_n'(\alpha)}{\zeta_n(\alpha) - i[Y_n^{(e)}(\alpha)/Y_0]\zeta_n'(\alpha)} \tag{5.6.27}$$

$$b_n = \frac{\psi_n(\alpha) - i[Z_n^{(m)}(\alpha)/Z_0]\psi_n'(\alpha)}{\zeta_n(\alpha) - i[Z_n^{(m)}(\alpha)/Z_0]\zeta_n'(\alpha)} \tag{5.6.28}$$

for any kind of radial stratification—layered or continuous. At this point, the burden of computation lies in the evaluation of $Z_n^{(m)}(\alpha)/Z_0$ and $Y_n^{(e)}(\alpha)/Y_0$, the TE normalized modal surface impedances and the TM normalized modal surface admittances. Each mode can be treated as the input impedance or admittance of a transmission network. These quantities are defined by

$$Y_n^{(e)}(\rho) = \frac{H_{n\phi}^{(e)}}{E_{n\theta}^{(e)}} = -\frac{H_{n\theta}^{(e)}}{E_{n\phi}^{(e)}} \tag{5.6.29}$$

$$Z_n^{(m)}(\rho) = \frac{E_{n\theta}^{(m)}}{H_{n\phi}^{(m)}} = -\frac{E_{n\phi}^{(m)}}{H_{n\theta}^{(m)}} \tag{5.6.30}$$

where the nth field component corresponds to the nth term in the expansion of the TM and TE waves. In this way the impedances and admittances can be linked, through the boundary conditions, to the radial parts of the Debye potentials as follows:

$$\frac{Y_n^{(e)}(\rho)}{Y_0} = -i\frac{W_n(\rho)}{(1/\kappa_e)W_n'(\rho)} \tag{5.6.31}$$

$$\frac{Z_n^{(m)}(\rho)}{Z_0} = -\frac{G_n(\rho)}{(1/\kappa_m)G_n'(\rho)} \tag{5.6.32}$$

where κ_e and κ_m are the relative permittivity and relative permeability, respectively (2.1.8). The problem is now reduced to that of a spherical transmission line. The reader is referred to Garbacz (1961, 1962a, b) and Wait (1963) for further details. A similar treatment has been developed by Sazonov and Frolov (1965). Wait and Jackson (1965) have utilized this approach to obtain extensive numerical results for spheres with various types of coatings.

5.6.4 Analytical Solutions for the Radial Equations

The scattering coefficients for an isotropic continuously variable inhomogeneous sphere having radial symmetry are given by (5.6.13) and (5.6.14). These involve the functions $W_n(\alpha)$ and $G_n(\alpha)$ which are solutions of the radial equations (5.6.11) and (5.6.12). There are a number of cases for which these reduce to equations whose solutions are more or less standard functions and these will now be considered.

a. Power Law Variation of Refractive Index. Nomura and Takaku (1955) have investigated the solution of the radial equations when the refractive index has a power law dependence on the radial distance. They were concerned with the propagation of radio waves around the earth in a stratified atmosphere. However, their determination of the appropriate functions which describe the field in the inhomogeneous region can be used directly in the scattering problem. It only becomes necessary to select those of the two independent sets of solutions of the second order differential equations which are appropriate to the particular region of space.

The power law dependence of the refractive index relative to that of the medium can be written

$$m = A\rho^p \tag{5.6.33}$$

where A and p are constants which may, in general, be complex and ρ is a measure of the radial distance, r,

$$\rho = (2\pi/\lambda)r \tag{5.6.34}$$

When this explicit form of m is substituted in (5.6.11) and (5.6.12), the resulting equations have as their solutions (Kamke, 1948)

$$W_n(\rho) = \rho^{p+1}(\pi/2\rho)^{1/2} J_{\pm\nu}(X_\rho) \tag{5.6.35}$$

$$G_n(\rho) = \rho(\pi/2\rho)^{1/2} J_{\pm\mu}(X_\rho) \tag{5.6.36}$$

where J_ν, etc. are Bessel functions of the first kind. The independent solutions are comprised of the Bessel functions of either positive or negative orders. The arguments are defined by

$$X_\rho = [A/(p+1)]\rho^{p+1} \tag{5.6.37}$$

and the orders by

$$\nu = [1/(p+1)][n(n+1) + (p+\tfrac{1}{2})^2]^{1/2} \tag{5.6.38}$$

$$\mu = (2n+1)/2(p+1) \tag{5.6.39}$$

The order of the Bessel functions may be complex, and unlike the homogeneous case, they do not bear a simple relation to the particular partial wave denoted by n. When $p = 0$, the usual half integral order Bessel and Neumann functions are obtained.

The scattering coefficients for a sphere can now be obtained directly if those radial functions and their derivatives which are well behaved at the origin are used in (5.6.13) and (5.6.14), viz. those with positive orders. Levine and Kerker (1963) have developed the expressions corresponding to a coated sphere in which the refractive index of the coating follows a power law variation. In this case both solutions—those with positive and with negative orders—must be used to construct the Debye potential in the inhomogeneous region.[4] This can serve conveniently as a model for spheres with a diffuse surface.

The power law for the refractive index (5.6.33) may be expressed as

$$m_2 = A\rho^p = (Ak_3{}^p)r^p = \xi(r) + i\zeta(r) \tag{5.6.40}$$

[4] An error in the original formulation of Levine and Kerker (1963) which propagated Wyatt's (1962) original error has been corrected in later presentations (Olaofe and Levine, 1967; Kerker *et al.*, 1966d).

and further

$$Ak_3{}^p = Pe^{i\Phi} \qquad \text{and} \qquad p = s + it \tag{5.6.41}$$

where the real and imaginary parts of the refractive index are

$$\xi(r) = Pr^s \cos(\Phi + t \ln r)$$

$$\zeta(r) = Pr^s \sin(\Phi + t \ln r)$$

The variation of the refractive index is fixed by choosing the four real constants, P, Φ, s, and t and this can be done a number of ways. For example, the values of $\xi(r)$ and $\zeta(r)$ can be chosen quite arbitrarily at any two points within the inhomogeneous region, but once chosen the value of p and hence of the shape of the refractive index profile will have been determined. Alternatively, the complex refractive index can be assigned at one position and the power, p, specified.

b. Luneberg Lens. Another inhomogeneous sphere for which an analytical solution has been obtained is the spherical Luneberg lens with refractive index variation given by

$$m = [2 - (r/a)^2]^{1/2} \tag{5.6.42}$$

Following Jasik's treatment (Jasik, 1954) for the two-dimensional (cylindrical) case, Tai (1958a) obtained a solution in terms of an entirely new function not related to other well-known functions. Tai has called it a "generalized" confluent hypergeometric function. He gives the series and recurrence relations for it.

Tai (1958b, 1963) has also discussed the wave functions pertaining to Maxwell's fisheye (5.5.1) and also a bilinearly stratified medium which can be represented by

$$m^2(r) = m^2(\infty)[r + (m^2(0)/m^2(\infty))\ r_2/(r + r_2)] \tag{5.6.43}$$

where $m(\infty)$ and $m(0)$ are the refractive indices at $r = 0$ and ∞. At $r = r_2$, the value of m is equal to the mean of these extreme values. One may therefore choose values of these three parameters to simulate a monotonically increasing or decreasing function of $m(r)$ which has a finite asymptotic value at $r = \infty$.

Gould and Burman (1964) have shown that for a refractive index profile

$$m = (a + br^2 + c/r^2)^{1/2} \tag{5.6.44}$$

the radial equation is of the form of Whittaker's confluent hypergeometric equation whose solutions are known as Whittaker functions. For a profile of the form

$$m = (a + br)^{-1} \tag{5.6.45}$$

both the TM and the TE partial waves satisfy the same radial equation just as for the case of a homogeneous medium. This equation is the hypergeometric equation.

A number of workers (Lynch, 1963; Arnush, 1964; Margulies and Scarf, 1964) have used a different approach from the previous authors in order to study configurations having a dielectric constant profile given by

$$\varepsilon = 1 - \gamma/k^2r^2 \tag{5.6.46}$$

This corresponds to a cold dilute plasma such as occurs when a continuous source expands into a vacuum. The model corresponds approximately to the ionization around a flare, star, etc.

The technique has been to cast the scalar wave equation, which is satisfied by the two Debye potentials, into the form of the Schroedinger wave equation which represents the scattering of particles by a central field of force. Differences from the wave-mechanical case arise in the far-field asymptotic forms of the potential functions and in the boundary conditions. The solution is obtained with the aid of the WKB approximation.

There is still an alternative approach to the scattering by a plasma source expanding into a vacuum. If the spherical source can be discriminated from the more tenuous surrounding region which extends out to infinite distances, the system corresponds to a sphere imbedded in an inhomogeneous medium such as has already been discussed in Section 5.6.2. Only the single boundary condition at the surface of the plasma source need be considered if the propagation of the incident and scattered waves in the inhomogeneous region is adequately described by Sayasov's (1961) treatment.

Olaofe and Levine (1967) have developed solutions corresponding to the Cauchy and parabolic distributions. These refractive index profiles are generalizations of the Maxwell fisheye and spherical Luneberg lens, respectively. The Cauchy distribution is

$$m = n/(1 + \varepsilon r^2/a^2) \tag{5.6.47}$$

where a is the radius and n and ε are real, positive constants. If $n = 2$ and $\varepsilon = 1$, then this reduces to the Maxwell fisheye. The parabolic distribution is

$$m^2 = n^2(1 - 2\varepsilon r^2/a^2) \tag{5.6.48}$$

which reduces to the Luneberg lens if $n^2 = 2$ and $\varepsilon = \frac{1}{4}$.

A still more specialized variation of the refractive index has been specified by Negi (1962b). Of the three optical (electrical) constants ε_1, μ_1, and σ_1, the two permeabilities maintain a constant value within the sphere, but the conductivity obeys the following power law variation:

$$\sigma_1 = \sigma'(r/a)^p \tag{5.6.49}$$

where σ' is the conductivity at the surface of the sphere of radius a. This conductivity distribution is one which is actually encountered when prospecting for ore bodies with electromagnetic waves.

5.6.5 Numerical Results

a. Luneberg and Eaton Lenses. Garbacz (1962a) has investigated scattering by a Luneberg lens and a modified Eaton–Lippman lens, each with the size parameter $\alpha = 5.0$. The modification in the latter consisted in its having a conducting core ($m = \infty$) with radius equal to one-tenth the total lens radius. These calculations were carried out by first analyzing the problem as an analogous transmission line and then using the computational techniques devised for such problems. The results are presented in Figs. 5.33 and 5.34 for the size parameter $\alpha = 5$. Certainly for this size, neither of these lenses performs as a perfect backscatter device. Indeed, the backscatter efficiency of the Luneberg lens is exceedingly small ($G = 0.047$). On the other hand, the backscatter of the modified Eaton–Lippman lens has the considerably larger backscatter efficiency, $G = 1.732$. Although somewhat greater than for an all-metal sphere, it hardly offers an advantage as a backscatter device over a dielectric sphere with refractive index in the range $1.5 < m < 2.0$. It is interesting to note how well this agrees with Rheinstein's (1962) calculations for a multilayered modified Eaton–Lippman lens. He obtained $G = 1.711$ for a 50-layered lens and $G = 1.727$ for the 100-layered case.

b. Sphere with a Diffuse Surface. Wyatt (1962, 1963a, b, 1964c) has explored numerically the effect of a "diffuse" surface upon the scattering.[5] He utilized a numerical solution of the radial differential equations, (5.6.11) and (5.6.12), and for this reason he was at liberty to choose any form for the refractive index profile. His choice of a profile is shown in Fig. 5.35 where the real part of the refractive index is schematically depicted as a function of radius, r. The imaginary part may vary independently in a similar fashion.

The diffuse region extends from a to b. Beyond these points the refractive index is uniform, corresponding to that of the core and of the medium, respectively. The half fall-off radius is ρ_0 and the surface thickness T represents the 0.9 to 0.1 fall-off distance. The transition region is described by a Green–Wyatt function (Green, 1959) for which

$$a = \rho_0 - 0.99868T \tag{5.6.50}$$

$$b = \rho_0 + 0.99868T \tag{5.6.51}$$

[5] Fortunately, the errors introduced into Wyatt's analysis had no effect upon his computations because of the analytical properties of the refractive index profile which he chose.

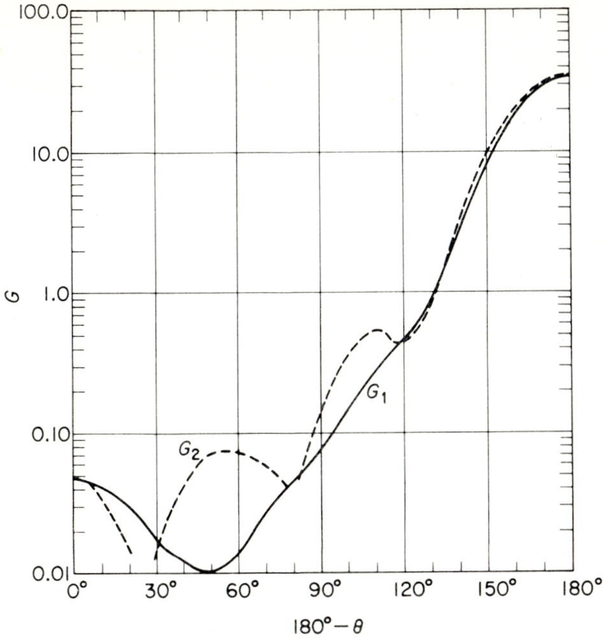

FIG. 5.33. Angular gain for Luneberg lens plotted against $180 - \theta$; $m^2 = 2 - (r/a)^2$ (Garbacz, 1962a, b).

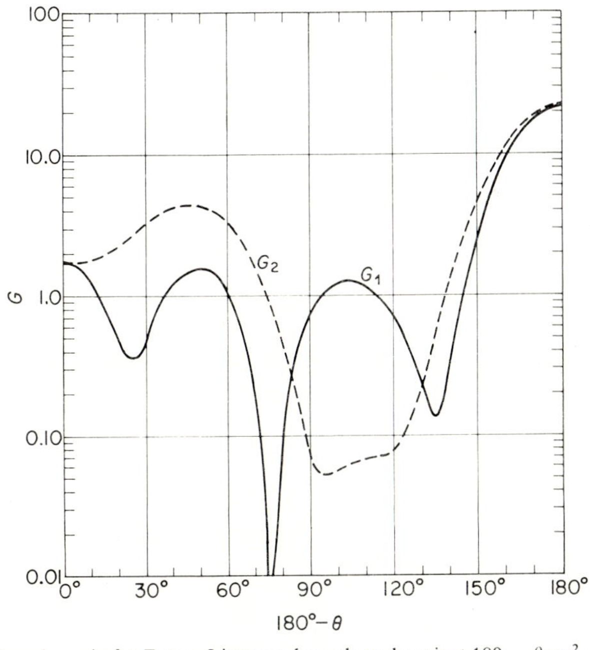

FIG. 5.34. Angular gain for Eaton–Lippman lens plotted against $180 - \theta$; $m^2 = (2 - r/a)/(r/a)$ (Garbacz, 1962a, b).

When $a = 0$, this form factor provides for a continuous variation from the outside of the particle to the center. With a nonzero value of a, it yields a coated sphere with a continuously variable coating, since from point a inwards, the sphere is actually uniform. In the notation of the coated sphere, a and b correspond to the radii of the core and the total sphere. Further complexity can be introduced by a distortion parameter (Wyatt, 1963a) illustrated in Fig. 5.36 by the dashed lines. This form factor is particularly well suited for numerical integration since its first and second derivatives are continuous throughout the range of definition, and yet the function itself vanishes for a finite argument (at the surface of the particle, $r = b$).

A few numerical examples have been studied by Wyatt. The results for the efficiency factors of a highly absorbing system (inner refractive index, $m_i = 1.67 - 0.734i$) are presented in Table 5.5. These have been normalized to the same geometrical cross section, viz. $\pi\rho_0^2$ (ρ_0 = half fall-off radius), even though the actual geometrical cross sections are $\pi(\rho_0 + T)^2$. The size parameter is $\alpha_0 = 2\pi\rho_0/\lambda$. The efficiency for absorption increases quite significantly as the particle becomes more diffuse. The scattering efficiency also increases with diffuseness, but not to the same extent. For comparison reference may be made to the efficiencies of homogeneous spheres of comparable absorptivity in Figs. 4.9, 4.10 and 4.11.

TABLE 5.5

EFFICIENCIES FOR EXTINCTION, SCATTERING, AND ABSORPTION[a]

T	$\alpha_0 = 1$			$\alpha_0 = 5.0$			$\alpha_0 = 10.0$		
	Q_{ext}	Q_{sca}	Q_{abs}	Q_{ext}	Q_{sca}	Q_{abs}	Q_{ext}	Q_{sca}	Q_{abs}
0.5	2.319	0.622	1.697	3.085	1.335	1.750	3.408	1.537	1.971
0.1	2.038	0.566	1.472	2.741	1.256	1.485	2.468	1.239	1.329
0.01	1.985	0.560	1.425	2.592	1.272	1.320	2.401	1.279	1.122
0.001	1.980	0.560	1.420	2.579	1.269	1.310	2.395	1.277	1.118
0	1.979	0.560	1.420	2.577	1.269	1.308	2.395	1.277	1.118

[a] Results are for diffuse spheres relative to the geometrical cross section, $\pi\rho_0^2$. Inner refractive index $m_i = 1.67 - 0.734i$.

A similar case has been studied by Bisbing (1966) for a spherical plasma having an exponential distribution of the electron density. Thus the refractive index, following (4.6.1), is given by

$$m^2 = \varepsilon = 1 - \frac{\omega_p^2 e^{-r/A}}{\omega(\omega + i\nu)} \tag{5.6.52}$$

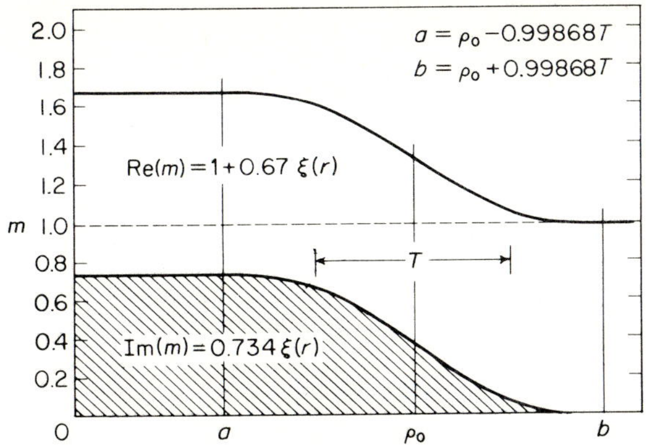

FIG. 5.35. Green–Wyatt form factor for the radial profile of a diffuse spherical shell (Wyatt, 1962).

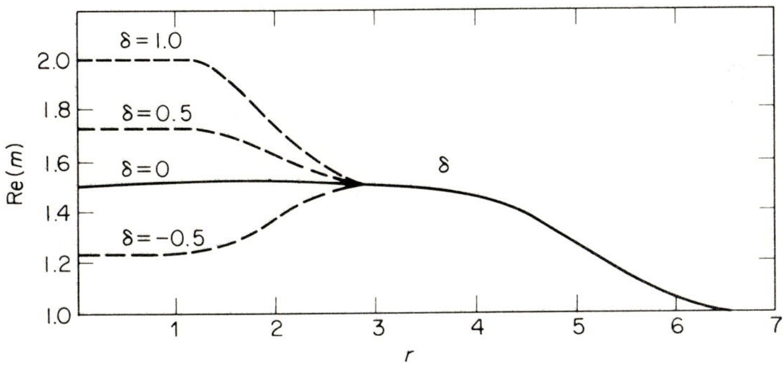

FIG. 5.36. Effect of distortion parameter δ in modifying radial profile given by Green–Wyatt form factor (Wyatt, 1963a, b).

In this case the obstacle is diffuse, lacking a definite boundary. The solution was effected with the aid of the spherical transmission line theory of Section 5.6.3. Computations up to $kA = 1.6$ were carried out for $\omega_p/\omega = 0.1, 1, 3$ and 10 and for $\nu/\omega = 0.1$, 1, and 10.

Buley (1967) has treated radially varying plasma spheres for which the assumption of zero collision frequency is made. Although a classical plasma might be expected to have an effective dielectric constant which can be represented by a real number, it can be shown that absorption takes place even in the limit of zero collision frequency.

c. Coated Sphere with Variable Coating. Kerker *et al.* (1966d) have considered a homogeneous sphere surrounded by an outer concentric spherical shell in which the refractive index is either constant or varies radially according to a power law. The refractive index profiles for which they carried out calculations are depicted in Fig. 5.37. In Case I, the core has $m_1 = 2.105$ and the coating has $m_2 = 1.482$, corresponding to the aerosols of silver chloride coated with linolenic acid which had been studied experimentally (Espenscheid *et al.*, 1965) and for which extensive computation had already been obtained (Kerker *et al.*, 1962). In Case II, the refractive index drops discontinuously at the surface of the core to 1.482 and then falls off through the shell to that of the external medium, $m = 1$, following a power of -3.74.

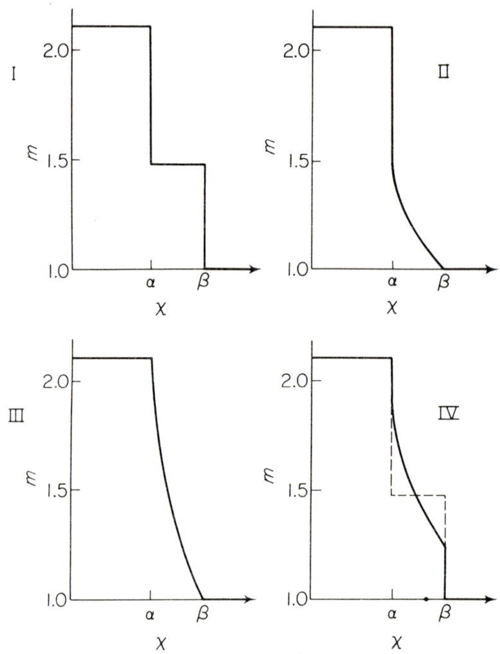

FIG. 5.37. Refractive index profiles for concentric shell (Kerker *et al.* 1966d). Abscissa is $\chi = kr$.

In Case III, the refractive index varies continuously from the value at the surface of the core to the external medium following a power -7.07. In Case IV, the power law variation (-3.42) is such that the areas enclosed by the dashed lines above and below $m = 1.482$ are equal, varying from $m = 1.78$ at the surface of the core to $m = 1.23$ at the surface of the diffuse coating. The average value of the refractive index in the core is about the same for Cases I, III, and IV. The particle size was chosen as $\alpha = 2\pi a/\lambda = 4.5$, $v = 2\pi b/\lambda = 5.0$.

The intensity functions for Cases III and IV agreed very closely with those for Case I over the entire angular range. For Case II, for which the average refractive index in the core is different, the scattering is appreciably different.

These results indicate, for a smooth variation in the refractive index of the shell, that the scattering may not be sensitive to the form of the variation. Indeed, for the two power law profiles corresponding to Cases III and IV, the scattering could have been calculated by utilizing a homogeneous concentric shell with the average refractive index of the coating. By the same token, for such a case, electromagnetic scattering will not provide a diagnostic tool sufficiently sensitive to elucidate the form of the refractive index profile. An ability to provide such a diagnosis would be of interest to space scientists concerned with radio wave scattering by meteors or space vehicles sheathed in a plasma of radially varying properties and by chemists dealing with light scattered by colloidal particles or macromolecules surrounded by a diffuse layer of ions or adsorbate. It is reasonable to expect that the same conclusions would follow for scatterers with cylindrical symmetry.

Olaofe and Levine (1967) have also calculated the scattering for concentric spheres with a homogeneous core and a coating for which the variable refractive index follows a power law. In addition they treated spherically symmetric inhomogeneous spheres corresponding to a Cauchy distribution (5.6.47) and a parabolic distribution (5.6.48). The size parameters were for $2\pi b/\lambda = 5$ and 10 and a common mean refractive index by volume, namely

$$\overline{m} = (1/V) \int_V m \, dV \tag{5.6.53}$$

where V is the volume of the sphere. Most of the examples chosen corresponded to $\overline{m} = 1.0738$ with a refractive index of 2 at the center of the sphere.

These authors also found for a concentric sphere with a variable coating that the scattering was not sensitive to the form of the variation in the shell whenever there was a constant amount of refractive material ($\overline{m} = 1.0738$). So in Fig. 5.38, the three similar scattering patterns are for a uniform sphere with $m = 1.0738$ (curve a) and size parameter 5; a concentric sphere with core refractive index 1.083, core size parameter 4.615, and an inverse first power law variation in the shell, curve b; a concentric sphere with core refractive index 1.078, core size parameter 4.84, and an inverse square power law variation in the shell.

The Cauchy and parabolic distributions having the same mean refractive index represent a considerably more drastic variation in the radial profile. The scattering for three cases is shown in Fig. 5.39. Curve D is for a Cauchy distribution with parameters $n = 1.20$ and $\varepsilon = 0.2$ while curves E and F correspond to parabolic distributions with parameters $n = 1.20$ and $\varepsilon = 0.153$ and 0.164, respectively. There is the same amount of refractive material contained in D and F but for F the refractive index at the outer boundary drops to 0.97. For curve E, the refractive index drops from 1.2 at the center to 1.0 at the boundary as does the sphere corresponding to curve D, but the total amount of refractive material is somewhat greater than in the other two cases, viz. $\bar{m} = 1.084$ compared to 1.074.

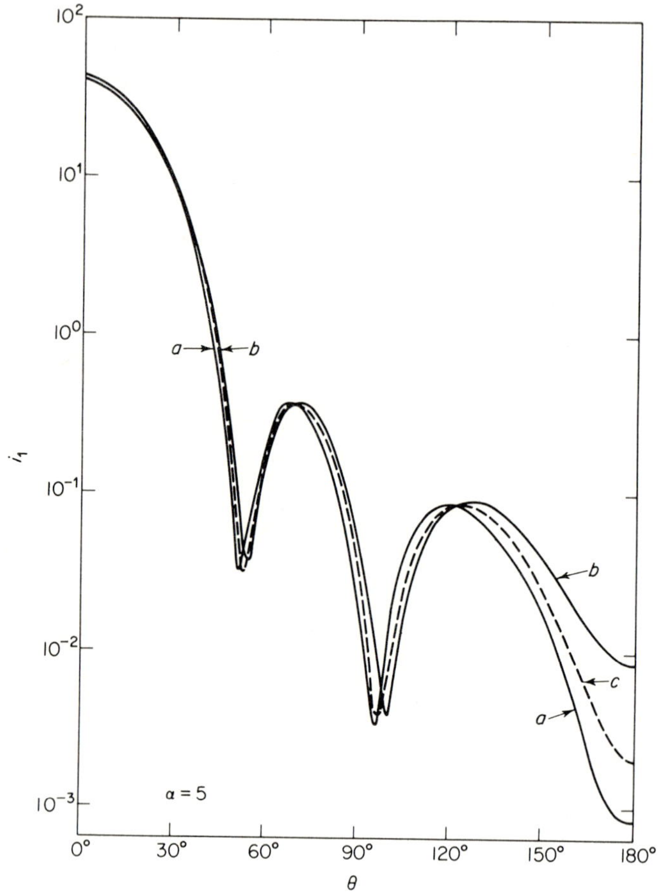

FIG. 5.38. Angular intensity function i_1 plotted against θ for $\alpha = 5.0$; a is uniform sphere with $m = 1.074$; b is uniform core with $m = 1.083$ surrounded by shell with refractive index given by power law a/r; c is uniform core with $m = 1.078$ surrounded by shell with refractive index given by power law $1.01(a/r)^2$ (Olaofe and Levine, 1967).

For these cases, the forward scattering differs only slightly from the variable shell concentric sphere model. However the backscattering of each of these distributions is considerably different from that of the concentric spheres and these results differ slightly among each other.

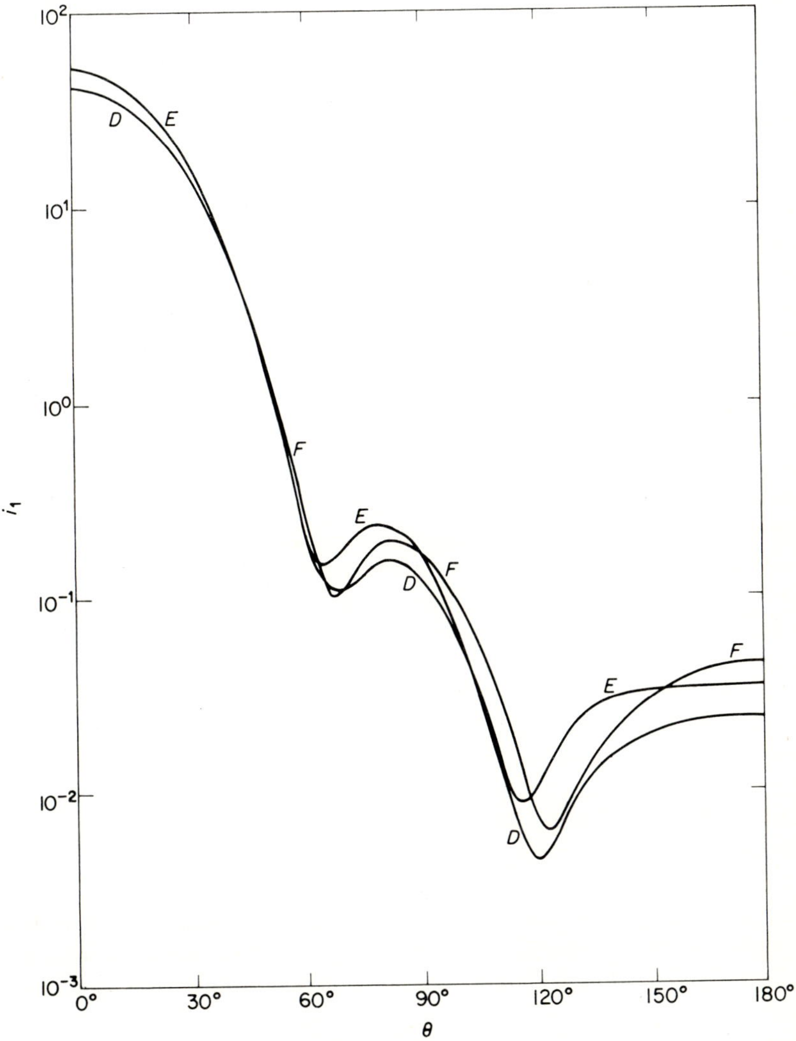

FIG. 5.39. Angular intensity function i_1 plotted against θ for $\alpha = 5.0$; D is for a Cauchy distribution $n = 1.2$, $\varepsilon = 0.2$; E and F are for parabolic distributions $n = 1.2$ and $\varepsilon = 0.153$ and 0.164, respectively; D and F have the same total refracting material (Olaofe and Levine, 1967).

5.7 Neighboring Spheres

The problem to be considered here, while not involving stratified spheres, can also be solved exactly by extending the procedure followed for single spheres. When two particles are close, their secondary fields may affect each other sufficiently so that the primary scattering process is altered and the cumulative effect cannot be obtained by simple addition of the results for single spheres. This interaction is different from that due to interference of the scattered waves or from multiple scattering.

For neighboring spheres, the scattering coefficients are still obtained by matching the fields interior and exterior to the boundary of the particle. However, now the exterior field must include the field of the neighboring particle (or particles) in addition to the incident field and the scattered field. This problem was first investigated by Trinks (1935) but seems to have been overlooked by recent workers interested in the experimental aspects of this configuration [e.g., Churchill *et al.* (1960), Kumagai and Angelakos (1961)].

Trinks developed explicit expressions for small (Rayleigh) particles in the two cases where the direction of the incident wave is parallel (end illumination) and perpendicular (broadside illumination) to the line joining the centers of the two spheres, and this was extended by Germogenova (1963) to include the partial modes due to the electric quadrupole term a_2 and the magnetic dipole term b_1. Levine and Olaofe (1968) have noted errors in the results of Trinks and of Germogenova and have themselves outlined a general solution for two equal spheres. Their results for the dipolar cases (a_1 and b_1) agree with corresponding results obtained by Twersky (1967), who has developed a completely general theory of scattering by an arbitrary configuration of arbitrary scatterers. The case of two perfectly conducting spheres of unequal radii has been treated by Liang and Lo (1967).

The results, even for the limiting case of two small spheres, is expressed by a rather intricate system of equations. Here we will only present the simpler expressions obtained for nose-on and broadside incidence.

The geometry is shown in Fig. 5.40. Two identical spheres each of radius a are fixed with their centers at O_1 and O_2, the center O_1 being the origin of the Cartesian coordinate system x', y', z'. The separation distance is

$$O_1O_2 = R_0 \geqslant 2a \tag{5.7.1}$$

Dimensionless parameters are defined in the usual way as

$$R = kR_0, \qquad \alpha = ka \tag{5.7.2}$$

The incident wave propagates in the Oz' direction with its electric vector polarized parallel to Ox'. It is convenient to consider a second Cartesian frame O_1xyz described by the three Eulerian angles γ, β, ψ with the condition that $\beta = 0$. This implies that O_1z' is in the O_1xy plane.

The condition for nose-on incidence is $\gamma = 0$ and that for broadside incidence is $\gamma = \pi/2$. Then ψ which is the angle between the direction of polarization and the line of the centers O_1O_2 takes on the values 0 and $\pi/2$.

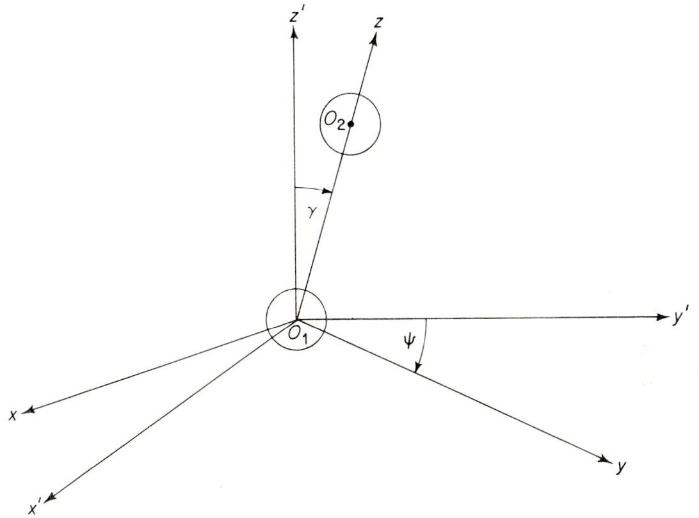

FIG. 5.40. Geometry for scattering by neighboring spheres with centers at O_1 and O_2. Incident wave propagates along z', polarized along x'.

We will be interested in the angular intensity functions $i_1(\gamma, \psi)$ and $i_2(\gamma, \psi)$ where the subscripts 1 and 2 denote incident radiation polarized perpendicular to and parallel to the scattering plane. For broadside incidence we can designate four cases $[i_1(\pi/2, 0),\ i_2(\pi/2, 0),\ i_1(\pi/2, \pi/2),\ i_2(\pi/2, \pi/2)]$ illustrated by Fig. 5.41. The scattering angle, as usual, is designated by θ.

Levine and Olaofe give the following results

$$i_1(\pi/2, 0) = 4|{}_1E_{10}|^2 \tag{5.7.3}$$

$$i_2(\pi/2, 0) = 4|{}_1E_{10}|^2 \cos^2\theta \cos^2(\tfrac{1}{2}R \sin\theta) \tag{5.7.4}$$

$$i_1(\pi/2, \pi/2) = 4|{}_1E_{11}|^2 \cos^2(\tfrac{1}{2}R \sin\theta) \tag{5.7.5}$$

$$i_2(\pi/2, \pi/2) = 4|{}_1E_{11}|^2 \cos^2\theta \tag{5.7.6}$$

where

$${}_1E_{10} = \tfrac{3}{2}a_1(1 + 3a_1\zeta_1(R)/R^2)^{-1} \tag{5.7.7}$$

$${}_1E_{11} = -(3i/2)a_1(1 + \tfrac{3}{2}a_1\zeta_1'(R)/R)^{-1} \tag{5.7.8}$$

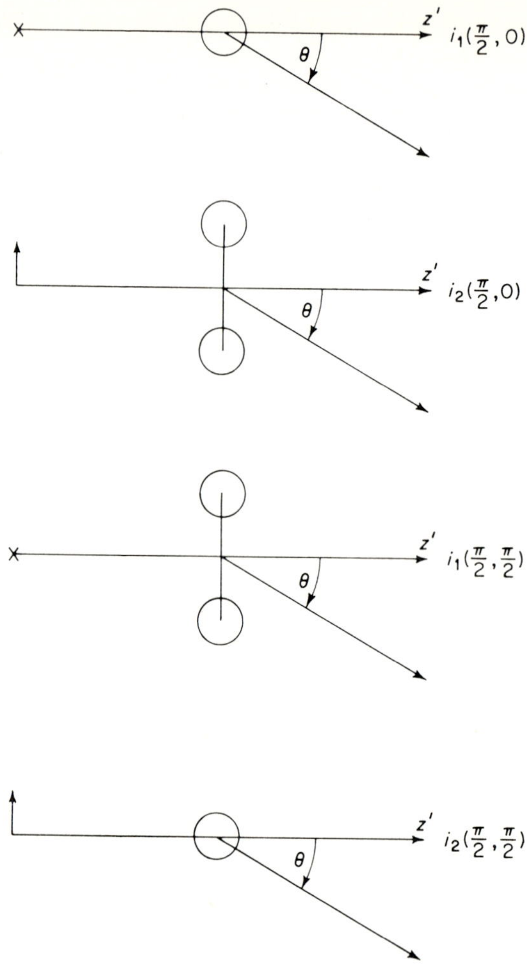

FIG. 5.41. Four cases for broadside illumination of a pair of spheres. The scattering angle θ is in the plane of the paper; × and ↑ denote that the incident beam is polarized perpendicular and parallel to the plane of the paper, respectively; the top and bottom drawings show the spheres above and below the plane of the paper; the middle drawings show them in the plane.

Here $\zeta_1(R)$ and $\zeta_1'(R)$ are the Ricatti–Bessel function and its first derivative defined by (3.3.27), while a_1 is the scattering coefficient corresponding to the electric dipole. It is not surprising to find that $i_1(\pi/2, 0)$ has no dependence on scattering angle. Here both the direction of polarization of the incident wave and the line between the centers of the spheres are perpendicular to the scattering plane.

For large separations ($R \to \infty$) the following results are obtained:

$$i_1(\pi/2, 0) = 9|a_1|^2 = 4i_1 \tag{5.7.9}$$

$$i_2(\pi/2, 0) = 4i_1 \cos^2 \theta \cos^2(\tfrac{1}{2}R \sin \theta) \tag{5.7.10}$$

$$i_1(\pi/2, \pi/2) = 4i_1 \cos^2 \theta(\tfrac{1}{2}R \sin \theta) \tag{5.7.11}$$

$$i_2(\pi/2, \pi/2) = 4i_1 \cos^2 \theta \tag{5.7.12}$$

These correspond precisely to the interference by two single Rayleigh scatterers for which the intensity functions are i_1 and $i_1 \cos^2 \theta$. The cases $i_1(\pi/2, 0)$ and $i_2(\pi/2, \pi/2)$ are for the cases where the two spheres are above and below the scattering plane on the line through the origin. The optical path to the observer is the same for each particle and the field interferes constructively.

At end-on illumination ($\gamma = 0, \psi = 0$)

$$i_1(0, 0) = |{}_1E_{11} + {}_2E_{11} \exp(-iR \cos \theta)|^2 \tag{5.7.13}$$

$$i_2(0, 0) = |{}_1E_{11} + {}_2E_{11} \exp(-iR \cos \theta)|^2 \cos^2 \theta \tag{5.7.14}$$

where

$${}_1E_{11} = -\tfrac{3}{2}a_1(1 - \tfrac{3}{2}a_1\zeta_1'(R)\, e^{iR}/R)/D_{11} \tag{5.7.15}$$

$${}_2E_{11} = -\tfrac{3}{2}a_1(e^{iR} - \tfrac{3}{2}a_1\zeta_1'(R)/R)/D_{11} \tag{5.7.16}$$

$$D_{11} = 1 - (\tfrac{3}{2}a_1\zeta_1'(R)/R)^2 \tag{5.7.17}$$

When $R \to \infty$, these equations reduce to

$$i_1(0, 0) = 4i_1 \cos^2(R \sin^2 \theta/2) \tag{5.7.18}$$

$$i_2(0, 0) = 4i_1 \cos^2(R \sin^2 \theta/2) \cos^2 \theta \tag{5.7.19}$$

It appears that, for Rayleigh scatterers, the mutual interaction of two spheres (except for interference) is negligible for spacings greater than two or three diameters. Mevel (1960) has shown experimentally for totally reflecting spheres comparable in dimension to the wavelength that the interaction is significant over a greater number of sphere diameters than for small dielectric spheres. Of course, experiments such as these in the microwave part of the spectrum can be carried out by direct manipulation of a pair of spheres with macroscopic dimensions.

In the optical region, using spheres, the problem of separating the effects of interference of the scattered waves and of multiple scattering from the mutual polarization of the adjacent particles is not simple. Doremus (1966, 1968) has proposed that the optical transmission of thin metallic films in island form provides an experiment in which only the mutual polarization is present.

When a metal is deposited onto a substrate by evaporation, small islands of the metal are formed first. These islands are approximately spherical and as deposition continues they grow, eventually merging into a continuous film. Islands are also formed if a very thin continuous metal film is evaporated. Because the particles are all in one plane, there is negligible multiple scattering and the interference effects disappear for the transmitted light. Such films can provide an example in which mutual polarization, which is dependent upon the particle-particle spacing, can be studied independently of other effects. Maxwell-Garnet's theory (1904, 1906) discussed earlier, which considers the effect of the presence of the particles upon the complex dielectric constant of the medium, can be applied to this particular situation. Instead of using the bulk dielectric constant for the colloidal particles, Doremus has introduced the free electron theory in order to explain qualitatively the observed effects. An interesting program would be to reconcile the results of this treatment with the phenomenological results described by the above equations.

CHAPTER 6

Scattering by Infinite Cylinders

6.1 Homogeneous Circular Cylinders

In order to utilize the boundary value method, the surface of the scatterer should correspond to a complete coordinate surface and the coordinate system should be one in which the wave equation is separable. The method has already been applied to the spherically symmetric case. Now we consider cylindrical symmetry.

Of the eleven separable coordinate systems, four are cylindrical: the rectangular cylinder, the circular cylinder, the elliptical cylinder, and the parabolic cylinder. The scattering problem has been solved for the circular and elliptical cylinders. Indeed, Lord Rayleigh obtained the solution for the circular cylinder at perpendicular incidence in 1881 and this particular problem was thoroughly explored in the German literature (von Ignatowsky, 1905; Seitz, 1905, 1906; von Schaeffer, 1907) even before the treatment of the sphere by Mie (1908) and Debye (1909a).

Workers in the fields of microwaves and radiowaves are often concerned with cylindrical rather than spherical structures, so that it is natural that they have studied this problem intensively (Mentzner, 1955; King and Wu, 1959; Wait, 1959). In the field of light scattering, there has not been as great an interest in cylinders as in spheres. The principal structures studied have been micron and submicron fibers which are usually sufficiently long to be treated as infinite cylinders.

6.1.1. General Theory

Our presentation is patterned after Farone and Querfeld (1965) who followed van de Hulst's (1957) notation. The circular cylindrical coordinates, r, θ, z, are depicted in Fig. 6.1. Any point in space is the intersection of the three orthogonal coordinate surfaces; the cylinder, $r =$ constant; the plane through the z-axis, $\theta =$ constant; and the plane perpendicular to the z-axis, $z =$ constant. The cylinder axis is taken to coincide with the z-axis. The simplest situation occurs when the incident plane wave of unit intensity propagates perpendicular to the z-axis. It is only this which the early workers treated and for which most of the experiments have been carried out.

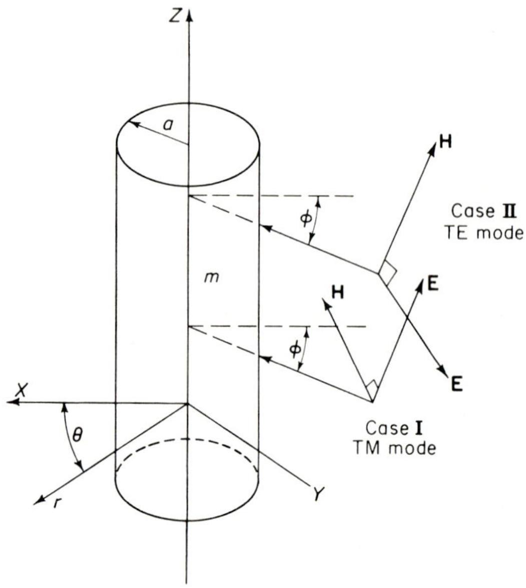

FIG. 6.1. Coordinates for scattering by cylinder with radius a. Cylindrical coordinates are z, r, θ. The cylinder axis is coincident with z. The incident plane contains the incident direction and the z-axis. The x-axis is in the incident plane; for the TM mode, **E** is in the incident plane; for the TE mode, **H** is in the incident plane; the tilt angle ϕ is the angle in the incident plane between the incident direction and the perpendicular to the z-axis

The solution for oblique incidence has been given for perfectly conducting cylinders by Mentzner (1955) and for the general case of arbitrary material constants by Wait (1955), Blank (1955), Burberg (1956), and Farone and Querfeld (1965). Van de Hulst (1957) has outlined the procedure to be followed. Earlier results obtained by Wellman (1937) seem to be incorrect.

For oblique incidence, the direction of propagation of the incident wave makes an angle ϕ (Fig. 6.1) with the normal to the z-axis. The polarized incident wave is resolved into two components, the TM mode (Case I) for which there is no z-component of the magnetic vector and the TE mode (Case II) for which there is no z-component of the electric vector. This means that for the TM mode, the magnetic vector vibrates perpendicularly to the plane formed by the incident and the z-directions, while for the TE mode it is the electric vector which vibrates perpendicularly to this plane.

Just as for spherical symmetry, the field vectors corresponding to each of these modes can be related to a scalar potential which in turn satisfies the scalar wave equation (3.3.18). In cylindrical coordinates, this is

$$\frac{1}{r}\frac{\partial}{\partial r}r\frac{\partial \pi}{\partial r} + \frac{1}{r^2}\frac{\partial^2 \pi}{\partial \theta} + \frac{\partial^2 \pi}{\partial^2 z} + k^2 \pi = 0 \tag{6.1.1}$$

The function π can be written as a product of three factors each of which is a function of only one coordinate variable, r, θ, and z and satisfies the following ordinary differential equations:

$$r\, d/dr[r\, dR(r)/dr] + [(k_0{}^2 - k^2)r^2 - n^2]R(r) = 0 \tag{6.1.2}$$

$$(d^2\Theta(\theta)/d\theta^2) + n^2\Theta(\theta) = 0 \tag{6.1.3}$$

$$(d^2 Z(z)/dz^2) + h^2 Z(z) = 0 \tag{6.1.4}$$

$$h = k_0 \sin \phi \tag{6.1.5}$$

The potential functions within the homogeneous cylinder of refractive index m must now be constructed from an appropriate superposition of

$$\pi_n = e^{i\omega t}[Z_n(r(m^2 k_0{}^2 - h^2)^{1/2})][e^{in\theta}][e^{-ihz}] \tag{6.1.6}$$

where each factor represents a solution of each of the three ordinary differential equations. Following van de Hulst (1957), the medium will be treated as if it were a vacuum so that m is the refractive index of the cylinder and k_0 is the propagation constant outside of the cylinder. Otherwise, m is the refractive index of the cylinder relative to that of the medium, $m = m_1/m_2$ and the propagation constant outside of the cylinder is k_2.

The index n takes on integral values. The radial equation (6.1.2) has two independent solutions, J_n, the integral order Bessel function, and N_n, the integral order Neumann function, each of which has been defined earlier, (3.6.2), (3.6.3), and (3.6.39). The corresponding Hankel functions are constructed in the usual way with

$$H_n^{(2)} = J_n - iN_n \tag{6.1.7}$$

The potential for the wave outside the cylinder ($r > a$) is obtained by superposition of two functions; one represents the incident wave and the other represents the scattered wave. The incident wave traveling in the direction shown in Fig. 6.1 is represented by

$$\pi^i = \exp[i\omega t - ik_0(x\cos\phi + z\sin\phi)]$$

$$= \exp[i\omega t - ihz] \sum_{n=-\infty}^{\infty} (-i)^n J_n(lr)\exp[in\theta] \qquad (6.1.8)$$

where

$$l = (k_0{}^2 - h^2)^{1/2} = k_0\cos\phi \qquad (6.1.9)$$

This wave is resolved into two components corresponding to Case I or the TM mode and to Case II or the TE mode. For spherical symmetry and also when the incident wave is perpendicular to the cylinder axis, both the internal wave and the scattered wave are of the same mode as the incident wave. In other words, incident TE waves scatter as TE waves and incident TM waves scatter as TM waves. However, for oblique incidence the incident TE waves are partially converted to TM scattered waves and similarly the incident TM waves are partially converted to TE scattered waves. In order to apply the boundary conditions, it is necessary to generate both TM and TE modes in the internal and scattered waves corresponding to each case in the incident wave. Accordingly, the potential functions are set up as follows:

Case I

$$r > a \qquad u_{\mathrm{I}} = \sum_{n=-\infty}^{\infty} F_n\{J_n(lr) - b_{n\mathrm{I}}H_n^{(2)}(lr)\} \qquad (6.1.10)$$

$$v_{\mathrm{I}} = \sum_{n=-\infty}^{\infty} F_n\{a_{n\mathrm{I}}H_n^{(2)}(lr)\} \qquad (6.1.11)$$

$$r < a \qquad u_{\mathrm{I}} = \sum_{n=-\infty}^{\infty} F_n\{d_{n\mathrm{I}}J_n(jr)\} \qquad (6.1.12)$$

$$v_{\mathrm{I}} = \sum_{n=-\infty}^{\infty} F_n\{c_{n\mathrm{I}}J_n(jr)\} \qquad (6.1.13)$$

Case II

$$r > a \qquad u_{\mathrm{II}} = \sum_{n=-\infty}^{\infty} F_n\{b_{n\mathrm{II}}H_n^{(2)}(lr)\} \qquad (6.1.14)$$

$$v_{\mathrm{II}} = \sum_{n=-\infty}^{\infty} F_n\{J_n(lr) - a_{n\mathrm{II}}H_n^{(2)}(lr)\} \qquad (6.1.15)$$

$$r < a \qquad u_{\mathrm{II}} = \sum_{n=-\infty}^{\infty} F_n\{d_{n\mathrm{II}}J_n(jr)\} \qquad (6.1.16)$$

$$v_{\mathrm{II}} = \sum_{n=-\infty}^{\infty} F_n\{c_{n\mathrm{II}}J_n(jr)\} \qquad (6.1.17)$$

where

$$F_n = (-i)^n \, e^{i(\omega t - hz + n\theta)} \tag{6.1.18}$$

and

$$j = (m^2 k_0{}^2 - h^2)^{1/2} \tag{6.1.19}$$

The boundary conditions are that the tangential components of **E** and **H** be continuous at the surface $r = a$ and this requires the continuity of the following expressions:

$$\partial v/\partial r + (inh/mk_0 r)u \tag{6.1.20}$$

$$[(m^2 k_0{}^2 - h^2)/mk_0]u \tag{6.1.21}$$

$$m \, \partial u/\partial r - (inh/k_0 r)v \tag{6.1.22}$$

$$(m^2 k_0{}^2 - h^2)v \tag{6.1.23}$$

These boundary conditions lead to eight sets of linear algebraic equations which can be solved for the scattering coefficients as

$$a_{n\mathrm{I}} = \Delta^{-1} \begin{vmatrix} -\dfrac{inh}{\alpha} J_n(la) & -\dfrac{inh}{m\alpha} J_n(ja) & -\dfrac{inh}{\alpha} H_n^{(2)}(la) & -jJ_n'(ja) \\ -\dfrac{l^2}{k_0} J_n(la) & -\dfrac{j^2}{mk_0} J_n(ja) & -\dfrac{l^2}{k_0} H_n^{(2)}(la) & 0 \\ -lJ_n'(la) & -mjJ_n'(ja) & -lH_n^{(2)\prime}(la) & \dfrac{inh}{\alpha} J_n(ja) \\ 0 & 0 & 0 & -j^2 J_n(ja) \end{vmatrix} \tag{6.1.24}$$

$$b_{n\mathrm{I}} = \Delta^{-1} \begin{vmatrix} lH_n^{(2)\prime}(la) & -\dfrac{inh}{m\alpha} J_n(ja) & -\dfrac{inh}{\alpha} J_n(la) & -jJ_n'(ja) \\ 0 & -\dfrac{j^2}{mk_0} J_n(ja) & -\dfrac{l^2}{k_0} J_n(la) & 0 \\ -\dfrac{inh}{\alpha} H_n^{(2)}(la) & -mjJ_n'(ja) & -lJ_n'(la) & \dfrac{inh}{\alpha} J_n(ja) \\ l^2 H_n^{(2)}(la) & 0 & 0 & -j^2 J_n(ja) \end{vmatrix} \tag{6.1.25}$$

$$a_{n\mathrm{II}} = \frac{\begin{vmatrix} -lJ_n'(la) & -\dfrac{inh}{m\alpha}J_n(ja) & \dfrac{inh}{\alpha}H_n^{(2)}(la) & -jJ_n'(ja) \end{vmatrix}}{} $$

$$\Delta^{-1} \begin{vmatrix} 0 & -\dfrac{j^2}{mk_0}J_n(ja) & \dfrac{l^2}{k_0}H_n^{(2)}(la) & 0 \\ \dfrac{inh}{\alpha}J_n(la) & -mjJ_n'(ja) & lH_n^{(2)\prime}(la) & \dfrac{inh}{\alpha}J_n(ja) \\ -l^2J_n(la) & 0 & 0 & -j^2J_n(ja) \end{vmatrix} \tag{6.1.26}$$

$$b_{n\mathrm{II}} =$$

$$\Delta^{-1} \begin{vmatrix} -lH_n^{(2)\prime}(la) & -\dfrac{inh}{m\alpha}J_n(ja) & -lJ_n'(la) & -jJ_n'(ja) \\ 0 & -\dfrac{j^2}{mk_0}J_n(ja) & 0 & 0 \\ \dfrac{inh}{\alpha}H_n^{(2)}(la) & -mjJ_n'(ja) & \dfrac{inh}{\alpha}J_n(la) & \dfrac{inh}{\alpha}J_n(ja) \\ -l^2H_n^{(2)}(la) & 0 & -l^2J_n(la) & -j^2J_n(ja) \end{vmatrix} \tag{6.1.27}$$

where

$$\Delta = \begin{vmatrix} lH_n^{(2)\prime}(la) & -\dfrac{inh}{m\alpha}J_n(ja) & -\dfrac{inh}{\alpha}H_n^{(2)}(la) & -jJ_n'(ja) \\ 0 & -\dfrac{j^2}{mk_0}J_n(ja) & -\dfrac{l^2}{k_0}H_n^{(2)}(la) & 0 \\ -\dfrac{inh}{\alpha}H_n^{(2)}(la) & -mjJ_n'(ja) & -lH_n^{(2)\prime}(la) & \dfrac{inh}{\alpha}J_n(ja) \\ l^2H_n^{(2)}(la) & 0 & 0 & -j^2J_n(ja) \end{vmatrix} \tag{6.1.28}$$

The far-field scattering is given by

$$\begin{aligned} I_{11} &= \frac{2}{k_0\pi r}i_{11} = \frac{2}{k_0\pi r}|T_{11}|^2 \\ &= \frac{2}{k_0\pi r}\left|b_{0\mathrm{I}} + 2\sum_{n=1}^{\infty} b_{n\mathrm{I}}\cos(n\theta)\right|^2 \end{aligned} \tag{6.1.29}$$

$$I_{12} = \frac{2}{k_0 \pi r} i_{12} = \frac{2}{k_0 \pi r} |T_{12}|^2$$

$$= \frac{2}{k_0 \pi r} \left| 2 \sum_{n=1}^{\infty} a_{n\text{I}} \sin(n\theta) \right|^2 \tag{6.1.30}$$

$$I_{21} = \frac{2}{k_0 \pi r} i_{21} = \frac{2}{k_0 \pi r} |T_{21}|^2$$

$$= \frac{2}{k_0 \pi r} \left| 2 \sum_{n=1}^{\infty} b_{n\text{II}} \sin(n\theta) \right|^2 \tag{6.1.31}$$

$$I_{22} = \frac{2}{k_0 \pi r} i_{22} = \frac{2}{k_0 \pi r} |T_{22}|^2$$

$$= \frac{2}{k_0 \pi r} \left| a_{0\text{II}} + 2 \sum_{n=1}^{\infty} a_{n\text{II}} \cos(n\theta) \right|^2 \tag{6.1.32}$$

in which I_{11} is the intensity of the TM mode of the scattered radiation where the incident radiation of unit intensity is also of the TM mode, and I_{12} is the intensity of the TE mode of the scattered radiation for the same incident radiation. When the incident radiation of unit intensity is of the TE mode, the TM and TE modes of the scattered radiation are given by I_{21} and I_{22} respectively.

As pointed out by Kerker *et al.* (1966a) the absolute values of the "cross-modes", $b_{n\text{II}}$ and $a_{n\text{I}}$ are equal and may be written as

$$b_{n\text{II}} = -a_{n\text{I}} = [nhlj^2(m^2 - 1)/ma] J_n(ja)$$
$$\cdot [J_n'(la) N_n(la) - J_n(la) N_n'(la)] \tag{6.1.33}$$

so that

$$I_{12} = I_{21} \tag{6.1.34}$$

6.1.2 Totally Reflecting Cylinder; $m = \infty$

When the cylinder is totally reflecting ($m = \infty$), the cross-modes disappear,

$$b_{n\text{II}} = a_{n\text{I}} = 0 \tag{6.1.35}$$

and the scattering coefficients simplify to (Mentzner, 1955)

$$b_n = J_n(la)/H_n^{(2)}(la) \tag{6.1.36}$$

$$a_n = J_n'(la)/H_n^{(2)\prime}(la) \tag{6.1.37}$$

where b_n corresponds to the TM mode, a_n to the TE mode.

Interest in space communication has led to the study of problems of scattering by obstacles immersed in a plasma medium. In particular, the case of a perfectly conducting cylinder has received considerable attention. A cold isotropic plasma which is assumed to be an incompressible electron fluid with equivalent numbers of cations and electrons can be characterized by a dielectric constant or refractive index so that the scattering is no different from that in any dielectric medium. In some cases, there are in addition to the usual transverse electromagnetic waves, longitudinal plasma waves. The theory of scattering of perfectly conducting cylinders immersed in such media is somewhat more complex. This has been discussed in some detail by Seshadri *et al.* (1964) for perpendicular incidence and by Wait (1965c) for oblique incidence. In addition, Rulf (1965, 1966) has discussed scattering by cylinders imbedded in a uniaxially anisotropic medium.

6.1.3 Perpendicular Incidence

Perpendicular incidence corresponds to $\phi = 0$ for which

$$h = 0; \qquad l = k_0; \qquad j = mk_0 \tag{6.1.38}$$

For this case the cross-modes reduce to zero so that

$$b_{n\text{II}} = a_{n\text{I}} = 0 \tag{6.1.39}$$

and the scattering coefficients become formally the same as for spherical symmetry except that instead of Ricatti–Bessel functions, they involve integral order Bessel functions.

$$b_n = \frac{mJ_n(\alpha)J_n'(m\alpha) - J_n(m\alpha)J_n'(\alpha)}{mH_n^{(2)}(\alpha)J_n'(m\alpha) - J_n(m\alpha)H_n^{(2)\prime}(\alpha)} \tag{6.1.40}$$

$$a_n = \frac{J_n(\alpha)J_n'(m\alpha) - mJ_n(m\alpha)J_n'(\alpha)}{H_n^{(2)}(\alpha)J_n'(m\alpha) - mJ_n(m\alpha)H_n^{(2)\prime}(\alpha)} \tag{6.1.41}$$

For a perfectly conducting cylinder at perpendicular incidence ($\phi = 0$, $m = \infty$), these equations reduce further to

$$b_n = J_n(\alpha)/H_n^{(2)}(\alpha) \tag{6.1.42}$$

$$a_n = J_n'(\alpha)/H_n^{(2)\prime}(\alpha) \tag{6.1.43}$$

It is interesting to note that the scattering coefficients for a perfectly conducting cylinder at oblique incidence are identical to those above for normal incidence with the radius reduced by the factor $\cos\phi$. The effect of tilting the perfectly conducting cylinder is to reduce its effective size.

Barakat and Levin (1964) have solved the scattering for a perfectly conducting cylindrical lamina whose cross section is a circular cylinder with a piece removed, as sketched in Fig. 6.2.

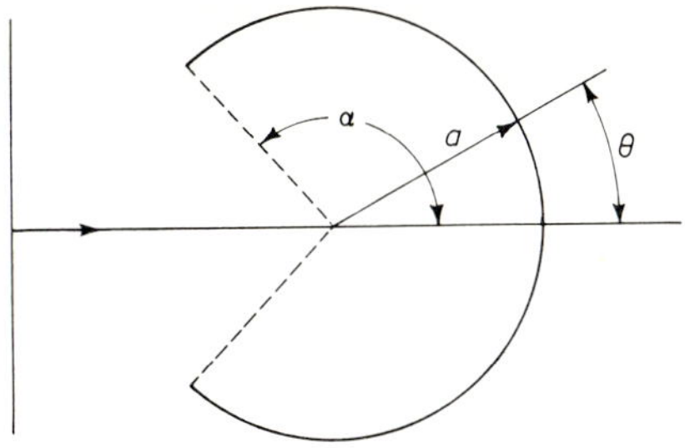

FIG. 6.2. Geometry of cylindrical lamina (Barakat and Levin, 1964).

6.1.4 PHYSICAL DISCUSSION OF THE SCATTERED LIGHT

a. Perpendicular Incidence. At perpendicular incidence, the TM mode (Case I) corresponds to a linearly polarized beam with its electric vector parallel to the cylinder axis while for the TE mode (Case II) the electric vector is perpendicular to the axis. The scattered wave is polarized in the same way as the incident wave. Just as for spherical symmetry, if the electric vector of the linearly polarized incident wave vibrates obliquely to the cylinder axis, each component of the scattered radiation is proportional to the corresponding component of the incident wave and the scattered wave will, accordingly, be elliptically polarized.

The scattered radiation propagates as a cylindrical wave. The intensity falls off as the inverse power of the radial distance, the energy flow being in the planes of constant z. Thus the scattered radiation corresponding to a particular incident ray will be observed only in that plane which is perpendicular to the cylinder's axis and which contains the incident ray. This effect can be observed when a cylinder such as a spider fiber is illuminated by a narrow parallel beam at perpendicular incidence. The fiber will appear to be brilliantly illuminated as long as the observer is in the appropriate plane. Otherwise, it will be lost to sight.

b. Oblique Incidence. The geometry of scattering for oblique incidence is shown in Fig. 6.3. Without loss of generality, the plane determined by the incident direction and the axis of the cylinder is set as $\theta = 0$. For the incident ray, the TM mode has the magnetic vector perpendicular to the plane $\theta = 0$ ($H_z = 0$) and the TE mode has the electric vector perpendicular to this plane ($E_z = 0$).

The scattered radiation is propagated along the surface of the cone with apical angle ($180° - 2\phi$) and may be viewed on the circle C, sighting towards the apex of the cone. The polarization of the scattered radiation can be visualized with the aid of Fig. 6.4 which is a section showing the z-axis and a scattered ray along an element of the conical surface. This illustration depicts the TM mode. The electric vector vibrates perpendicularly to the conical surface and to the circle C; hence the magnetic vector is tangential to the surface and perpendicular to the z-axis ($H_z = 0$). For the TE mode, the electric vector is tangential to the conical surface and perpendicular to the z-axis ($E_z = 0$). When the incident radiation is of the TM mode, the intensity

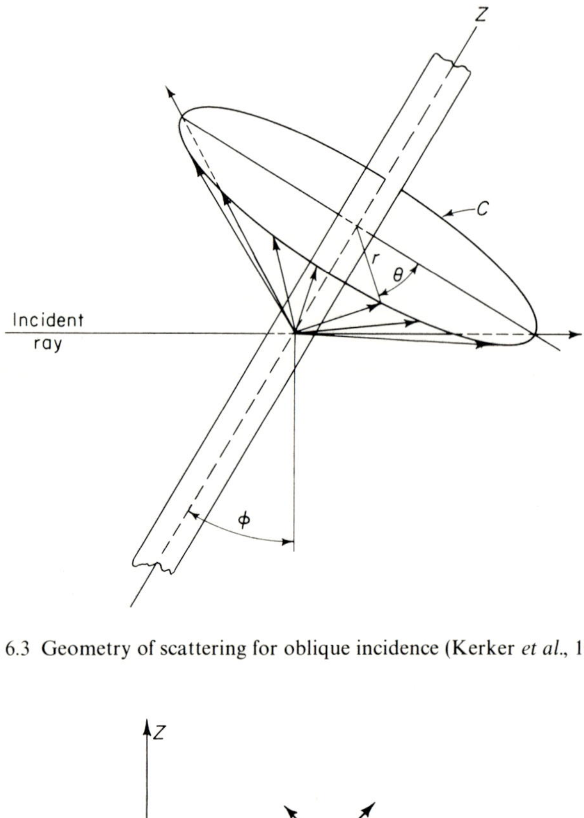

FIG. 6.3 Geometry of scattering for oblique incidence (Kerker *et al.*, 1966a).

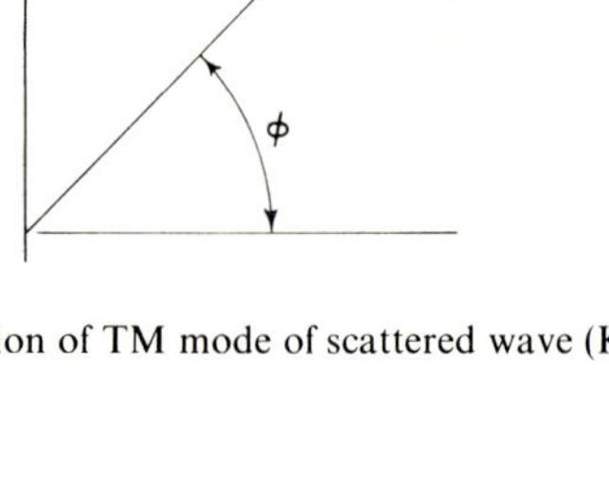

FIG. 6.4. Polarization of TM mode of scattered wave (Kerker *et al.*, 1966a).

functions corresponding to each of these polarizations are i_{11} and i_{12} respectively. For incident radiation of the TE mode, these polarizations correspond to i_{21} and i_{22}. For a perfectly conducting cylinder, there are no cross-components so that a linear polarized incident TM mode gives rise to a linearly polarized scattered TM mode, and there is a corresponding relation for an incident TE mode.

It is easy to see, as the cylinder rotates from the tilted position to one perpendicular to the incident ray, how the conical surface opens up into a plane and how this geometry becomes identical to that already described for perpendicular incidence.

6.1.5 Efficiencies for Extinction, Scattering, and Absorption

The extinction cross sections can be obtained directly from the scattering amplitudes for forward scattering with the aid of an extinction theorem analogous to that used earlier for finite particles (van de Hulst, 1949; Twersky, 1954). This is

$$C_{\text{ext}} = (2\lambda/\pi)\,\text{Re}\,T(0) \tag{6.1.44}$$

where the cross section is for an incident beam of unit intensity scattered by a unit length of cylinder. This leads to the following efficiencies for extinction per unit cross section area of the incident beam.

$$Q_{11\,\text{ext}} = (C_{11\,\text{ext}}/2a) = (2/\alpha)\,\text{Re}\left\{b_{o\text{I}} + 2\sum_{n=1}^{\infty} b_{n\text{I}}\right\} \tag{6.1.45}$$

$$Q_{12\,\text{ext}} = (C_{12\,\text{ext}}/2a) = (2/\alpha)\,\text{Re}\left\{2\sum_{n=1}^{\infty} a_{n\text{I}}\right\} \tag{6.1.46}$$

$$Q_{21\,\text{ext}} = (C_{21\,\text{ext}}/2a) = (2/\alpha)\,\text{Re}\left\{2\sum_{n=1}^{\infty} b_{n\text{II}}\right\} \tag{6.1.47}$$

$$Q_{22\,\text{ext}} = (C_{22\,\text{ext}}/2a) = (2/\alpha)\,\text{Re}\left\{a_{o\text{II}} + 2\sum_{n=1}^{\infty} a_{n\text{II}}\right\} \tag{6.1.48}$$

The scattering cross sections are obtained by integration of the scattering intensities over all values of the angle θ and this leads to

$$Q_{11\,\text{sca}} = (2/\alpha)\left\{|b_{o\text{I}}|^2 + 2\sum_{n=1}^{\infty} |b_{n\text{I}}|^2\right\} \tag{6.1.49}$$

$$Q_{12\,\text{sca}} = (2/\alpha)\left\{2\sum_{n=1}^{\infty} |a_{n\text{I}}|^2\right\} \tag{6.1.50}$$

$$Q_{21\,\text{sca}} = (2/\alpha)\left\{2 \sum_{n=1}^{\infty} |b_{n\text{II}}|^2\right\} \tag{6.1.51}$$

$$Q_{22\,\text{sca}} = (2/\alpha)\left\{|a_{\text{oII}}|^2 + 2 \sum_{n=1}^{\infty} |a_{n\text{II}}|^2\right\} \tag{6.1.52}$$

The corresponding efficiencies for absorption are obtained in the usual way by difference between the efficiencies for extinction and scattering. In general,

$$Q_{12\,\text{ext}} = Q_{21\,\text{ext}}; \qquad Q_{12\,\text{sca}} = Q_{21\,\text{sca}} \tag{6.1.53}$$

Also for perpendicular incidence and for perfectly reflecting materials it is obvious that

$$Q_{12\,\text{ext}} = Q_{21\,\text{ext}} = Q_{12\,\text{sca}} = Q_{21\,\text{sca}} = 0 \tag{6.1.54}$$

If the cross sections and efficiencies are defined relative to the incident beam without regard to the way the scattered energy is distributed between the TE and the TM modes, then

$$Q_{1\,\text{ext}} = Q_{11\,\text{ext}} + Q_{12\,\text{ext}}; \qquad Q_{1\,\text{sca}} = Q_{11\,\text{sca}} + Q_{12\,\text{sca}} \tag{6.1.55}$$

and

$$Q_{2\,\text{ext}} = Q_{21\,\text{ext}} + Q_{22\,\text{ext}}; \qquad Q_{2\,\text{sca}} = Q_{21\,\text{sca}} + Q_{22\,\text{sca}} \tag{6.1.56}$$

6.1.6 Small Dielectric Cylinders

When the diameter of the cylinder is sufficiently small relative to the wavelength, the Bessel functions can be expanded and the scattering coefficients can be expressed in powers of α in a manner similar to the one used for spheres. Wait (1965b) has given the expressions for oblique incidence. At perpendicular incidence (van de Hulst, 1957)

$$\beta_0 = (\pi\alpha^2/4)(m^2 - 1) \tag{6.1.57}$$

$$\alpha_0 = \beta_1 = (\pi\alpha^4/32)(m^2 - 1) \tag{6.1.58}$$

$$\alpha_1 = \frac{\pi\alpha^2}{4}\left(\frac{m^2 - 1}{m^2 + 1}\right) \tag{6.1.59}$$

where the phase angles α_n and β_n are defined by (4.1.1) and (4.1.2).

In the limit, only β_0 need be considered for the TM mode. However for the TE mode, it is the coefficient α_1 rather than α_0 which predominates. This leads to

$$Q_{1\,\text{sca}} = (\pi^2\alpha^3/8)(m^2 - 1)^2 \tag{6.1.60}$$

$$Q_{2\,\text{sca}} = \frac{\pi^2\alpha^3}{4}\left(\frac{m^2 - 1}{m^2 + 1}\right)^2 \tag{6.1.61}$$

$$I_1 = \frac{\pi\alpha^4}{8k_2 r}(m^2 - 1)^2 = \frac{\pi^4 a^4}{\lambda^3 r}(m^2 - 1)^2 \tag{6.1.62}$$

and provided θ is not too close to 90°

$$I_2 = \frac{\pi\alpha^4}{2k_2 r}\left(\frac{m^2 - 1}{m^2 + 1}\right)^2 \cos^2\theta = \frac{4\pi^4 a^4}{\lambda^3 r}\left(\frac{m^2 - 1}{m^2 + 1}\right)^2 \cos^2\theta \tag{6.1.63}$$

The latter formula is only applicable for θ sufficiently different from 90° so that the first term in the expression

$$I_2 = (2/\pi k_2 r)|T_2|^2 = (2/\pi k_2 r)|a_0 + 2a_1 \cos\theta + \cdots|^2 \tag{6.1.64}$$

may be neglected. For $\theta = 90°$, it is the second term which must be dropped, resulting in some residual scattering given by

$$I_2 = (\pi\alpha^8/512k_2 r)(m^2 - 1)^2 = (\pi^8 a^8/4\lambda^7 r)(m^2 - 1)^2 \tag{6.1.65}$$

Otherwise it is necessary to utilize (6.1.64), considering that for a sufficiently small value of α,

$$a_n \cong \alpha_n{}^2 + \alpha_n i \tag{6.1.66}$$

There are some interesting similarities as well as some differences between this limiting case and Rayleigh scattering for spheres. In both cases the TM mode is independent of the scattering angle, and provided that it is not too close to 90°, the intensity of the TE mode is proportional to $\cos^2\theta$. However, for the cylinder (provided that θ is not close to 90°), the ratio of the intensities of the two components is

$$\rho = I_2/I_1 = 4\cos^2\theta/(m^2 + 1)^2 \tag{6.1.67}$$

for the thin cylinder, compared to

$$\rho = I_2/I_1 = \cos^2\theta \tag{6.1.68}$$

for the small sphere. This leads to the result that both the backscatter and forward-scatter will be partially polarized even when the incident radiation is completely unpolarized (natural).

The intensity of the scattered radiation varies as the fourth power of the cylinder radius and inversely as the cube of the wavelength. For spherical Rayleigh scatterers, the intensity is proportional to the sixth and the inverse fourth powers, respectively. The dependence on the radius follows in turn from the dependence of the polarizability upon the volume. For the infinite cylinder, each element of volume radiates in the form of a cylindrical wave spreading out as a disk. Contributions from the rest of the cylinder, above and below the particular element, interfere with each other so that there is no explicit dependence upon this direction. Accordingly, the polarizability varies with the cross-sectional area or a^2 and the intensity with a^4. The inverse third power of the wavelength follows from dimensional considerations.

When the thin cylinder has a complex refractive index, it is understood that the modulus of the quantities in the parenthesis is to be taken. The efficiencies for extinction are given by

$$Q_{1\,\text{ext}} = -\frac{\pi\alpha}{2}\,\text{Im}(m^2 - 1) \tag{6.1.69}$$

$$Q_{2\,\text{ext}} = -\pi\alpha\,\text{Im}\left(\frac{m^2 - 1}{m^2 + 1}\right) \tag{6.1.70}$$

6.1.7 Angle Integration

In the usual experimental situations, the intensity is observed over a finite angle of observation, i.e.,

$$\overline{i_1} = \left[\int_{\theta_1}^{\theta_2} i_1\,d\theta\right]\bigg/(\theta_2 - \theta_1) \tag{6.1.71}$$

This can be integrated in closed form (Farone, 1964) and is given by

$$\begin{aligned}(\theta_2 - \theta_1)\overline{i_1} = \Bigg[&(\text{Re}\,b_0{}^2 + \text{Im}\,b_0{}^2)\\ &+4\left(\text{Re}\,b_0 \sum_{n=1}^{\infty} \frac{\text{Re}\,b_n \sin(n\theta)}{n} + \text{Im}\,b_0 \sum_{n=1}^{\infty} \frac{\text{Im}\,b_n \sin(n\theta)}{n}\right)\\ &+4 \sum_{n=1}^{\infty} (\text{Re}\,b_n{}^2 + \text{Im}\,b_n{}^2)\left(\frac{\theta}{2} + \frac{\sin(2n\theta)}{4n}\right)\\ &+8 \sum_{m=n+1}^{\infty} \sum_{n=1}^{\infty} (\text{Re}\,b_n\,\text{Re}\,b_m + \text{Im}\,b_n\,\text{Im}\,b_m)\\ &\cdot\left(\frac{\sin[(m-n)\theta]}{2(m-n)} + \frac{\sin[(m+n)\theta]}{2(m+n)}\right)\Bigg]_{\theta_1}^{\theta_2}\end{aligned} \tag{6.1.72}$$

The corresponding quantity for i_2 can be obtained by replacing b_n by a_n in the above expression.

Farone has explored this effect by calculating i_1 and i_2 for $\theta_2 - \theta_1 = 0$, 1.20, 2.25, and 5° for $\alpha = 5$, 10, 15, and 20, and for $m = 1.46$ and 1.47. It can become quite significant at larger values of α, particularly in the region of sharp extrema of the angular patterns.

6.2 Radially Stratified Cylinders

6.2.1 General Case

Scattering by radially stratified cylinders can be treated in a manner which is formally similar to that for spheres so that we will not treat this in great detail. The problem is to determine the scattering coefficients a_{nI}, a_{nII}, b_{nI}, b_{nII} and then to insert these in the appropriate equations such as (6.1.29) through (6.1.32). In each of the inner homogeneous regions, the radial part of the potential functions must be expressed as a linear superposition of integral order Bessel and Neumann functions. The boundary conditions are applied at each interface.

The case for perpendicular incidence was formulated by Thilo (1920) and explicit expressions were developed by Adey (1956a, b) and by Kerker and Matijević (1961a). The expressions for the scattering coefficients, a_n and b_n, will not be written out here because they can be obtained directly from those for the sphere by replacing ψ_n, ψ_n', χ_n, χ_n', ζ_n, and ζ_n' in (5.1.27) and (5.1.28) by J_n, J_n', N_n, N_n', H_n, and H_n', where these latter quantities represent the integral order Bessel, Neumann, and Hankel functions and their derivatives. This can also be extended to an arbitrary number of coaxial cylinders by a similar substitution in (5.4.8). Samaddar (1970) has shown that, at oblique incidence, there is no rigorous simple solution and that Farone and Querfeld's (1966) result is in error.

6.2.2 Dielectric Coated Conducting Cylinder

The special case of a dielectric-coated, perfectly conducting cylinder has been studied in considerable detail in much the same manner as the spherical analog consisting of a dielectric-coated, perfectly conducting sphere. Adey (1956a, b) and Tang (1957) have written down the expression for b_n. This, as well as the other scattering coefficient a_n, can be obtained directly from the more general solution for which both the inner cylinder and the concentric sheath are dielectrics. In order to obtain these results expressed in the notation used here, it is only necessary to substitute, as above, the appropriate integral order Bessel functions for the Ricatti–Bessel functions in the expressions for the spherical analog, (5.1.45) and (5.1.46). Similarly Wilhelmsson's (1963) solution for the same structure at oblique incidence can also be obtained by reduction of the above general solution. Chen and Cheng

(1964b) have explored the effects obtained when the dielectric coating is comprised of a compressible plasma which can support acoustic or plasma waves in addition to electromagnetic waves.

Kodis (1959, 1961, 1963) has considered the special case of the high frequency limit for the dielectric-coated cylinder for which the slow convergence of the conventional series representation presents considerable computational difficulties. By carrying out an asymptotic evaluation of the integral equation formulation of the scattered field (Mentzner, 1955), Kodis has obtained an expression in which each term is associated with an optical ray. The first term corresponds to an axial ray reflected from the outer dielectric boundary. A second group of terms describes the axial rays which emerge after having undergone s reflections at the conducting cylinder and $s - 1$ internal reflections at the outer dielectric surface.

The last group of terms corresponds to the nonaxial rays incident on the cylinder at larger angles. These may travel around the dielectric sleeve p times, undergoing s reflections at the conducting cylinder and $s - 1$ internal reflections at the outer dielectric interface before emerging in the backward direction. A sketch for a ray with $p = 1$ and $s = 5$ is shown in Fig. 6.5. The angles ϕ_i, ϕ_n, ϕ_r correspond to the angles of incidence, refraction, and internal reflection. They satisfy the equation

$$\phi_i + s(\phi_r - \phi_n) = p\pi \tag{6.2.1}$$

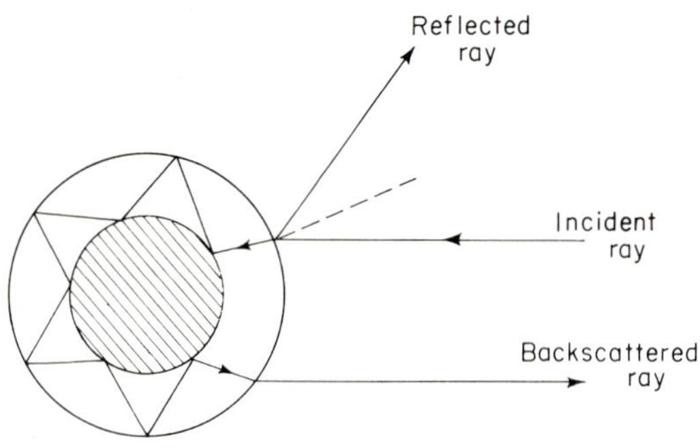

FIG. 6.5. Optical analog for the backscattered ray from a dielectric coated, totally reflecting cylinder which has circled the cylinder once after undergoing four internal reflections at the outer dielectric interface.

Following an observation of Kodis (1961), Helstrom (1963) has considered the case of a dielectric-coated cylinder with radius large compared to the wavelength but with a very thin dielectric coating. In this case, the part of the backscattered wave corresponding to the rays traveling through the dielectric material is very similar to that obtained for a bare cylinder so that the composite cylinder may be considered as a perturbation of the simpler case. Since the nonspecular part of the field arises from the passage around the cylinder of the creeping or surface waves, the analysis points up the transition from propagation around the cylinder by reflection back and forth between the metal and dielectric surfaces to propagation by creeping waves.

The cylindrical structure depicted in Fig. 6.6 has been treated by Plonus (1960). Here there is a region of free space (II) between the perfectly reflecting cylindrical core and the coaxial cylindrical sheath consisting of a dielectric material (III), all of which is surrounded by free space (IV). The exact expression for this case can be obtained directly from the analogous expressions for spherical symmetry by substitution of the integral order Bessel functions in place of the half integral order Bessel functions. In this instance, the appropriate expression corresponding to (5.4.8) with four refractive regions must be used. The relative refractive indices of regions II and IV are set equal to unity and that of region I is made infinite. Under the assumption that the surrounding shell was thin, Plonus was able to simplify the scattering coefficient into the form

$$a_n = [J_n'(\alpha)/H_n'(\alpha)](1 + \delta P_2) \tag{6.2.2}$$

$$b_n = [J_n(\alpha)/H_n(\alpha)](1 - \delta P_1) \tag{6.2.3}$$

where the first term represents the scattering coefficient for the perfectly reflecting cylindrical core alone, δ is the thickness of the shell, and P_1 and P_2 are relatively simple perturbation terms.

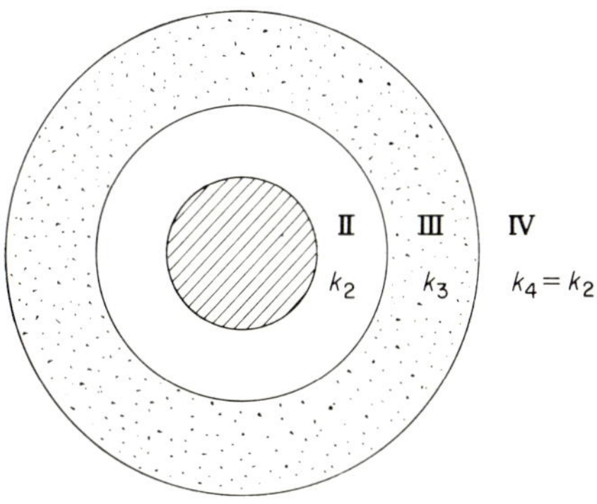

FIG. 6.6. Concentric cylindrical structure consisting of annular free space region (II) between totally reflecting cylindrical core and coaxial dielectric cylindrical sheath (III).

Plonus was motivated to test the opinion that the scattering cross section of a perfectly conducting sphere or cylinder with a dielectric shell spaced a resonant distance from the core could be increased markedly above that of the core alone. However, calculations using the exact expressions showed that although there were sharp peaks and dips as the spacing of the shell was varied, the scattering over a band of reasonable width was not significantly changed. The formulas for the spherical case were reported somewhat later and included the limiting formulas for a thin shell (Plonus, 1961).

6.3 Variable Refractive Index

Just as for spherical symmetry, the scattering functions for cylinders with a variable refractive index require evaluation of appropriate wave functions. Work in this area has been carried out (a) in order to utilize scattering techniques in the study of meteor trails which consist of long columns of ionized gas, (b) in order to predict the requirements for communicating with high-speed, re-entry vehicles of cylindrical shape which are surrounded by a sheath of ionized gas, and (c) in order to utilize microwave scattering from plasma columns as a laboratory diagnostic tool (Jones and Wooding, 1965). In each case, the ionized gas or plasma can be treated as a medium with an equivalent dielectric constant or refractive index. It is reasonable to expect that density gradients in these plasmas will modify the scattering characteristics from that expected for a homogeneous model so that the corresponding refractive index gradients must be taken into consideration. In all cases, it is assumed that radial symmetry is preserved. A rather unique example of scattering by an inhomogeneous cylinder has been treated by Negi (1962a) with a view to using long wavelength radiation in geophysical prospecting of elongated ore bodies. The model consists of a cylinder with a radial power law variation of its conductivity but with a dielectric constant and a magnetic permeability which are only negligibly different from a vacuum.

The two radial wave functions in the inhomogeneous region from which the TM and the TE modes are constructed must satisfy, respectively, the differential equations (Keitel, 1955)

$$\frac{d^2P_n(r)}{dr^2} + \frac{1}{r}\left[\frac{dP_n(r)}{dr}\right] + \left[k_0{}^2m^2 - \left(\frac{n}{r}\right)^2\right]P_n(r) = 0 \tag{6.3.1}$$

and

$$\frac{d^2T_n(r)}{dr^2} + \left[\frac{1}{r} - \frac{2d\ln m}{dr}\right]\frac{dT_n(r)}{dr} + \left[k_0{}^2m^2 - \left(\frac{n}{r}\right)^2\right]T_n(r) = 0 \tag{6.3.2}$$

where m, the refractive index relative to that of the external medium may vary with r. The potential functions may be constructed in the usual way and the scattering coefficients evaluated from the appropriate boundary conditions. The various scattering functions may then be constructed from these scattering coefficients. Thus for a circular cylinder at perpendicular incidence

$$a_n = \frac{k_0 m_a{}^2 J_n'(\alpha) T_n^{(1)}(a) - J_n(\alpha) T_n^{(1)'}(a)}{k_0 m_a{}^2 H_n^{(1)'}(\alpha) T_n^{(1)}(a) - H_n^{(1)}(\alpha) T_n^{(1)'}(a)} \tag{6.3.3}$$

$$b_n = -\frac{k_0 J_n'(\alpha) P_n^{(1)}(a) - J_n(\alpha) P_n^{(1)'}(a)}{k_0 H_n^{(1)'}(\alpha) P_n^{(1)}(a) - H_n^{(1)}(\alpha) P_n^{(1)'}(a)} \tag{6.3.4}$$

where α, a, and k_0 have their usual significance as the size parameter, radius, and propagation constant, and where m_a is the relative refractive index at a obtained by approaching the boundary of the cylinder from the interior. The functions $T_n^{(1)}(r)$ and $P_n^{(1)}(r)$ are those solutions of (6.3.1) and (6.3.2) which are well behaved at the origin.

Explicit solutions have been obtained for a number of specific radial profiles. For

$$m = (a + br)^{-1} \tag{6.3.5}$$

Burman (1965) has shown that (6.3.2) reduces to the hypergeometric equation. When the refractive index has a power law variation

$$k_0 m = ar^b \tag{6.3.6}$$

the solutions of (6.3.1) and (6.3.2) are

$$P_n(r) = Z_v\{[a/(b+1)]r^{b+1}\} \tag{6.3.7}$$

where

$$v = n/(b+1) \tag{6.3.8}$$

and Z_v is any Bessel function of order v and

$$T_n(r) = k_0 m Z_v\{[a/(b+1)]r^{b+1}\} \tag{6.3.9}$$

where

$$v^2 = (n^2 + b^2)/(b+1)^2 \tag{6.3.10}$$

This includes the special case considered earlier by Yeh and Kaprielian (1963) for which $b = -\frac{1}{2}$.

When $b = -1$, the arguments of the above Bessel functions become infinite. Instead of these Bessel functions, the differential equations, (6.3.1) and (6.3.2), have for their solutions

$$P_n(r) = r^{\pm p} \tag{6.3.11}$$

$$T_n(r) = k_0 m r^{\pm q} \tag{6.3.12}$$

where

$$p = (n^2 - a^2)^{1/2} \tag{6.3.13}$$

and

$$q = (n^2 - a^2 + 1)^{1/2} \tag{6.3.14}$$

Thus for the radial profile

$$k_0 m = a/r \tag{6.3.15}$$

each of the fields in the inhomogeneous region are expressible in terms of simple algebraic functions. Explicit expressions have been given by Burman (1966) for the backscattering cross section of a perfectly reflecting cylinder with an inhomogeneous dielectric sheath having this radial profile.

The refractive index profile corresponding to the cylindrical Luneberg lens has been studied by Jasik (1954) for the TE mode and by Tai (1956a) for the TM mode. The refractive index variation is given by

$$m^2 = \varepsilon = 2 - (r/a)^2 \tag{6.3.16}$$

The radial equations can be reduced to the standard form of the confluent hypergeometric equation which Tai (1956b) had also encountered in the theory of the spherical Luneberg lens.

Feinstein (1951) has treated a linear variation in dielectric constant

$$m^2 = \varepsilon = a + br \tag{6.3.17}$$

He obtained explicit expressions for $P_n(r)$ and $T_n(r)$ by assuming a power series solution.

The extension to compound cylindrical structures follows by methods analogous to those already described. For example Yeh and Kaprielian (1963) have written out the scattering coefficients from a perfectly reflecting cylinder coated with an inhomogeneous dielectric sheath and have also given the approximation for a very thin sheath. In the former case, the scattering coefficients may be obtained from the analogous expressions for a perfectly reflecting sphere encased in a concentric dielectric shell by replacing $\psi_n(v)$ and $\zeta_n(v)$ and their derivatives with $J_n(v)$ and $H_n^{(2)}(v)$ respectively and their derivatives, and also by replacing $\psi_n(m_2\alpha)$, $\psi_n(m_2 v)$, $\chi_n(m_2\alpha)$, $\chi_n(m_2 v)$ and their derivatives with $T_n^{(1)}(m_\alpha\alpha)$, $T_n^{(1)}(m_v v)$, $T_n^{(2)}(m_\alpha\alpha)$, $T_n^{(2)}(m_v v)$ respectively and their derivatives for a_n, and with $P_n^{(2)}(m_\alpha\alpha)$, $P_n^{(2)}(m_v v)$, $P_n^{(1)}(m_\alpha\alpha)$, $P_n^{(1)}(m_v v)$ respectively and their derivatives for b_n. The functions $T_n^{(2)}(r)$ and $P_n^{(2)}(r)$ designate those solutions of the radial equation which have a singularity at the origin. The refractive indices m_α and m_v correspond to those values at the inner edges of the boundaries $r = a$ and b, respectively. Of particular interest is the thin sheath approximation in which the refractive index in the sheath has a power law variation (6.3.6) with the exponent $b = -\frac{1}{2}$. Then

$$b_n = \frac{J_n(v) + (m_\alpha \alpha - m_v v) J'_n(v)}{H_n^{(2)}(v) + (m_\alpha \alpha - m_v v) H_n^{(2)\prime}(v)} \tag{6.3.18}$$

for a thin inhomogeneous sheath, compared to

$$b_n = \frac{J_n(v) + m(\alpha - v) J'_n(v)}{H_n^{(2)}(v) + m(\alpha - v) H_n^{(2)\prime}(v)} \tag{6.3.19}$$

for a homogeneous sheath. Apparently the cylinder with the thin inhomogeneous sheath scatters the same as one with a homogeneous sheath with relative refractive index

$$m = (m_\alpha \alpha - m_v v)/(\alpha - v) \tag{6.3.20}$$

The same conclusion follows for the TE mode.

Rusch (1964) has considered a perfectly reflecting cylinder covered by an inhomogeneous coating in which the refractive index is described by a parabolic distribution

$$m^2 = \varepsilon = A(r - a)(r - b) \tag{6.3.21}$$

This gives rise, depending upon the selection of constants, to the three types of radial dependence shown in Fig. 6.7. In Fig. 6.7a, the parabolic variation

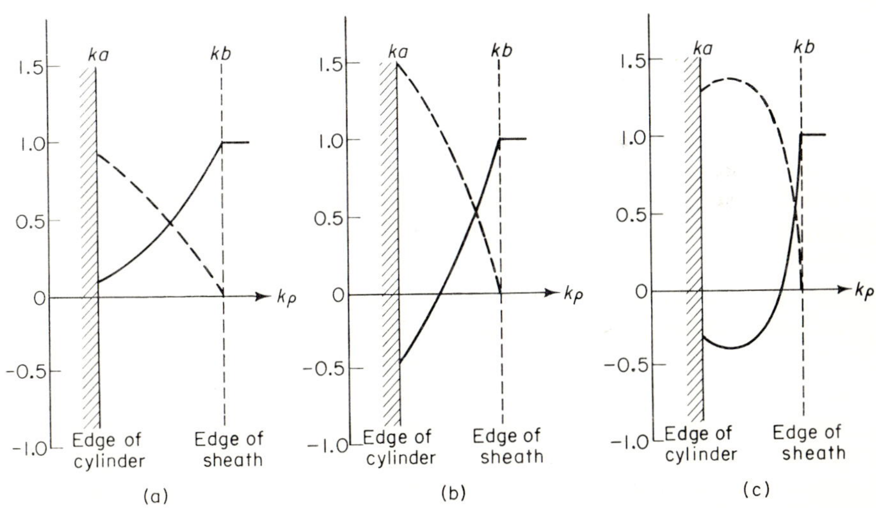

FIG. 6.7. Radial dependence of the dielectric constant, as described by Eq. (6.3.21), in a dielectric coating which covers a totally reflecting cylinder. The full curves are for the dielectric constant; the dashed curves are for the electron density of the corresponding plasmas (Rusch, 1964).

is monotonic, varying from a fractional refractive index at the conducting core to unity at the edge of the sheath. This corresponds to a "cold" plasma in which the electron density is sufficiently low so that the electron collision frequency $\omega_g = 0$. It follows from (4.6.1) that the equivalent dielectric constant is always less than unity and is given by

$$m^2 = \varepsilon = 1 - (n_e e^2/\varepsilon_0 M_e \omega^2) \tag{6.3.22}$$

In Fig. 6.7(b) the parabolic variation is still monotonic but drops to negative values of the dielectric constant. The negative region has been termed an overdense plasma medium. Finally in Fig. 6.7(c), there is a parabolic minimum. The dashed curves sketch the corresponding radial variations of electron density in each of these cases. Of course, by appropriate choice of the constant A these curves can be raised so that $m > 1$, corresponding to ordinary dielectric media.

The solutions to the radial equations for the parabolic distribution have been obtained by Rusch (1964) by expanding the radial function as an infinite power series about $r = a$. He then carried out calculations of the angular intensity functions for an inner cylinder with $\alpha = 5$ and a total cylinder with $v = 6$ for inhomogeneous sheaths corresponding to the three patterns in Fig. 6.7. The scattering patterns which he obtained were rather complex and there was no apparent correlation evident between these patterns and the refractive index distribution with the sheath.

6.4 Anisotropic Cylinders

With the current interest in space science, particular attention has been paid to scattering by long cylindrical plasmas confined in a uniform magnetic field directed along the cylinder axis. Such plasmas, which are termed gyroelectric, correspond to actual meteor trails in the upper atmosphere. They are anisotropic media with a macroscopic dielectric constant in the form of a dyadic whose elements are functions of the density of electrons and ions and the frequency of collision between them.

Both perpendicular incidence (Platzman and Ozaki,[1] 1960; Wait, 1961) and oblique incidence (Wilhelmsson, 1962) have been treated. Lee *et al.* (1965) have applied a geometrical optics approach to the gyroelectric cylinder and have compared their results at perpendicular incidence with the exact treatment. The extension to a cylindrically layered plasma follows in the usual way. The case of a perfectly conducting cylindrical core encased in a cylindrical sheath of gyroelectric medium has been discussed in detail by

[1] Samaddar (1963) has pointed out that there may be some printing errors in the results of Platzman and Ozaki (1960).

Ohba (1963), Seshadri (1964), Chen and Cheng (1964b), Yeh (1964b), and Mayhan and Schultz (1967) for perpendicular incidence and by Samaddar (1963) for oblique incidence. Rusch and Yeh (1967) have treated a perfectly conducting cylinder with a plasma sheath which is both anisotropic and inhomogeneous.

The dielectric constant of a gyroelectric medium is

$$(\varepsilon) = \begin{vmatrix} \varepsilon' & -iq_1 & 0 \\ iq_1 & \varepsilon' & 0 \\ 0 & 0 & \varepsilon'' \end{vmatrix} \tag{6.4.1}$$

where

$$D = (\varepsilon)E \tag{6.4.2}$$

or

$$D_x = \varepsilon' E_x - iq_1 E_y \tag{6.4.3}$$

$$D_y = iq_1 E_x + \varepsilon' E_y \tag{6.4.4}$$

$$D_z = \varepsilon'' E_z \tag{6.4.5}$$

The dyadic elements are related to the plasma properties by

$$\varepsilon' = 1 + \frac{\omega_p{}^2(1 - i\omega_g/\omega)}{\omega_c{}^2 - \omega^2(1 - i\omega_g/\omega)^2} \tag{6.4.6}$$

$$-q_1 = \omega_p{}^2(\omega_c/\omega)/[\omega_c{}^2 - \omega^2(1 - i\omega_g/\omega)^2] \tag{6.4.7}$$

$$\varepsilon'' = 1 - \omega_p{}^2/\omega^2(1 - i\omega_g/\omega) \tag{6.4.8}$$

where the cyclotron frequency of the electrons in a field of uniform dc magnetic induction B_0 is

$$\omega_c = eB_0/M_e \tag{6.4.9}$$

For perpendicular incidence and parallel polarization (electric vector vibrating parallel to the scattering plane or perpendicular to the cylinder axis), the scattering coefficient is (Platzman and Ozaki, 1960; Wait, 1961)

$$a_n = \left[\frac{\left(\dfrac{M}{N}\right)^{1/2} \dfrac{J_n'(\beta a)}{J_n(\beta a)} - \dfrac{nK}{\alpha} - \dfrac{J_n'(\alpha)}{J_n(\alpha)}}{\left(\dfrac{M}{N}\right)^{1/2} \dfrac{J_n'(\beta a)}{J_n(\beta a)} - \dfrac{nK}{\alpha} - \dfrac{H_n^{(2)\prime}(\alpha)}{H_n{}^2(\alpha)}} \right] \frac{J_n(\alpha)}{H_n^{(2)}(\alpha)} \tag{6.4.10}$$

where

$$M = \varepsilon'\varepsilon_0/((\varepsilon')^2 - q_1{}^2) \tag{6.4.11}$$

$$K = -q_1\varepsilon_0/((\varepsilon')^2 - q_1{}^2) \tag{6.4.12}$$

$$N = \mu_0/\mu \tag{6.4.13}$$

$$\beta = k_0/(MN)^{1/2} \tag{6.4.14}$$

When the magnetic field is removed so that the medium becomes isotropic, $q_1 = 0$ and $\varepsilon' = \varepsilon''$, Eq. (6.4.10) reduces appropriately to the usual expression (6.1.41). When the incident field is polarized with the electric vector vibrating parallel to the cylinder axis, the scattering coefficient, b_n, for the gyroelectric case is identical with the isotropic case.

There is a symmetric case in which the constant magnetic field produces a magnetic anisotropy formally analogous to the above. This occurs for certain ferromagnetic materials called ferrites of which magnetite is an example occurring in nature. Ferrites may be characterized as being gyromagnetic with the property that in a magnetic field, the magnetic permeability is the tensor (μ) such that

$$B = (\mu)H \tag{6.4.15}$$

$$(\mu) = \begin{vmatrix} \mu' & -iq_2 & 0 \\ iq_2 & \mu' & 0 \\ 0 & 0 & \mu' \end{vmatrix} \tag{6.4.16}$$

In view of the symmetry of Maxwell's equations, the equations for the gyromagnetic case can be obtained from the gyroelectric case by a straightforward transformation (Epstein, 1956). Tai and Chow (1959) have investigated a single ferrite cylinder at perpendicular incidence and Chow (1960) has extended this to two coaxial ferrite cylinders. The special case when the outer sheath is gyromagnetic and when the core is perfectly conducting has also been treated.

Cylinders with anisotropic conductivity have been studied by Kelly and Russek (1960). The two examples considered consist of cylinders which conduct only in a direction parallel to the axis or only circumferentially. Such unidirectional conductors, which may be of potential value in microwave or infrared physics, can be realized physically by imbedding fine wires in a nonconducting matrix for the first case or by a tightly wound helix in the second. The conductivity is assumed to be infinite.

The boundary conditions on a unidirectional conducting surface differ from those on a perfect isotropic conductor in that the surface is now able to sustain a tangential electric field perpendicular to the direction of conductivity. This component is now required to be continuous across the surface. Also the tangential component of the magnetic field parallel to the direction of conductivity must be continuous across the surface. With these boundary conditions, it is a straightforward matter to obtain the scattering coefficients, the efficiency for extinction, and the intensity functions. This has been effected by Kelly and Russek for oblique incidence.

A particularly striking result with the anisotropically conducting cylinder is that in the paraxial case the TE mode of the incident wave makes no contribution to the scattering, while the contribution of the TM mode is identical to that for a perfect isotropic conductor. This means that for perpendicular incidence there will be no scattering if the radiation is polarized with the electric vector perpendicular to the axis ($i_2 = 0$) so that the cylinder will be perfectly transparent to such radiation. The other polarized component will be scattered in the same manner as for a homogeneous conductor. Accordingly, the scattered radiation will be linearly polarized in all directions no matter what the state of polarization of the incident radiation might be. Hence such a cylinder might be used as a microwave or infrared polarizer.

Another illustration of the kind of effects that may be obtained is shown in Fig. 6.8. This is a polar plot of the angular gain at oblique incidence of 45° for a very fine cylinder, e.g., $\alpha = 0.1$. It should be recalled that the scattered energy is traveling along rays which make a 45° angle with the cylinder axis. The angle of observation is θ, the usual cylindrical coordinate. The

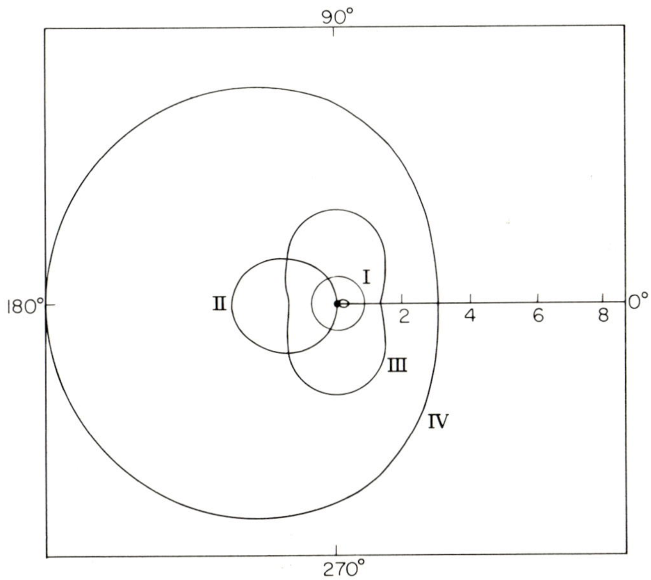

FIG. 6.8. Polar plot of angular gain at oblique incidence of 45° for a small anisotropic perfectly conducting cylinder ($\alpha = 0.1$). Curves I and II are for the TM and TE modes of an isotropic cylinder. Curves III and IV are for the TM and TE modes of a circumferentially conducting cylinder (Kelly and Russek, 1960).

planes of constant z are illuminated by the scattered radiation traveling along the surfaces of the infinite family of 45° cones emanating from along the cylinder axis. Each circle of illumination in the plane of constant z originates from a lower point on the cylinder axis.

Curve 1 is for the TM polarization for either an isotropic or a paraxial conducting cylinder and curve 2 is for the TE polarization for an isotropic cylinder. Curves 3 and 4 give the radiation patterns for the circumferentially conducting cylinder for the TM and TE modes respectively.

Chen (1966) has considered still another example of a cylinder with anisotropic conductivity, namely, the unidirectionally conducting helical sheath. By this is meant a surface on an infinite cylinder which is perfectly conducting in the direction of a helix winding around the cylinder with a particular pitch and perfectly insulating perpendicular to this direction. Physically this can be constructed by an infinite number of fine parallel wires closely wound around a circular cylinder. This degenerates to the cases considered by Kelly and Russek (1960) when the helical pitch is either zero or $\pi/2$.

The boundary conditions are that the electric field component parallel to the helical direction is zero at the surface of the sheath whereas the perpendicular component is continuous across the surface. For the magnetic field it is the parallel component which is continuous. In the case considered by Chen, the medium inside the sheath is the same as the outer medium, but there is no reason why the treatment cannot be extended to include arbitrary media. Due to the anisotropic nature of the boundary, both the TM and TE modes are present in the scattered radiation when the incident beam is exclusively of one or another mode, even at perpendicular incidence.

For a TM incident wave, the scattering coefficients are

$$b_{n\mathrm{I}} = \frac{\phi^2[J_n(\alpha)]^2}{[J_n'(\alpha)H_n^{(2)\prime}(\alpha) + \phi^2 J_n(\alpha)H_n^{(2)}(\alpha)]} \tag{6.4.17}$$

$$a_{n\mathrm{I}} = \frac{\phi J_n(\alpha)H_n^{(2)}(\alpha)}{[J_n'(\alpha)H_n^{(2)\prime}(\alpha) + \phi^2 J_n(\alpha)H_n^{2}(\alpha)]} \tag{6.4.18}$$

where

$$\phi = \tan \Psi \tag{6.4.19}$$

and Ψ is the pitch angle for the helix as measured from the scattering plane. For $\Psi = \pi/2$, the helical sheath is conducting along the z-axis. Then $a_{n\mathrm{I}} = 0$ and $b_{n\mathrm{I}}$ reduces to the expression for the scattering of the TM mode by a perfectly conducting cylinder (6.1.42). In the same manner, when $\Psi = 0$, both of the above coefficients reduce to zero and there will be no scattered radiation. Such a sheath would be completely transparent to the incident radiation. Chen has carried out a detailed numerical study over the range $\alpha = 0$ to 10 and for the whole range of pitch angles.

6.5 The Scattering Functions for Cylinders

6.5.1 Numerical Results

Published computations of scattering functions for cylinders are not nearly as extensive as those for spheres. For homogeneous cylinders at perpendicular incidence, the major tabulations are by Libelo (1962a, b) and by Farone *et al.* (1963) for dielectric media and by Libelo (1962c) for absorbing cylinders. These as well as several other less extensive results are listed in Table 6.1. Table 6.2 lists results for compound cylinders and Table 6.3 gives results obtained at oblique incidence.

TABLE 6.1

Tabulated or Graphed Scattering Functions for Homogeneous Cylinders at Perpendicular Incidence[a]

Ref.	m	α	Tabulated (T) or graphed (G) quantities
Rayleigh (1918a)	1.5	0.4(0.4)2.4	i_1 and i_2 for $\theta = 0°(30°)180°$ (T)
Kodis (1952)	∞	3.1, 10	
Froese and Wait (1954)	1.6	2.4	i_1 and i_2 for various near-field positions, near forward direction (G)
	1.6	6.3	i_1 for various near-field positions, near forward direction (G)
	$\sqrt{3.2}$	2.5	i_1 for various near-field positions, near forward direction (G)
Adey (1955)	∞	2, 3, 4, 5.97	T_1 at $\theta = 0°$ for near-field to $kr = 50$ (G)
	∞	2, 3, 4	T_1 at $\theta = 180°$ for near-field to $kr = 25$ (G)
	∞	2, 3, 4, 5.97, 8	Re T_1 at $\theta = 90°$ for near-field to $kr = 45$ (G)
Adey (1956b)	1.60	2, 3, 4, 4.25, 4.5, 5,6	T_1 at $\theta = 0°$ for near-field to $kr = 50$ (G)
	1.60	2, 3, 4	T_1 at $\theta = 180°$ for near-field to $kr = 25$ (G)
van de Hulst (1957)	1.5	0 to 6.5	Q_1, Q_2 (G)
	1.25	0 to 4.0	Q_1, Q_2 (G)
	∞	0.2(0.2)1.0 (0.5)5.0, 9.5, 10.0	T_1 and T_2 at $\theta = 0°$ (T)
	$\sqrt{2}(1 - i)$	0.4, 0.6, 0.8, 1, 1.4, 2, 3, 4	a_n, b_n, and T_1, and T_2 at $\theta = 0°$ (T)
Adey (1958)	1.60, 2.00	0 to 10	Q_1 (G)

TABLE 6.1—(*Continued*)

Ref.	m	α	Tabulated (T) or graphed (G) quantities
Larkin and Churchill (1959)	1.5, 2.0, 2.5	0.5, 1.0, 1.5, 2.0, 3.0, 4.0	a_n, b_n, Q_1, Q_2 (T)
Libelo (1962a)	1.15, 1.30,	0.1(0.2)13.5	a_n, b_n (T)
	1.20, 1.25	0.1(0.2)13.3	a_n, b_n (T)
Libelo (1962b)	1.40, 1.50,	0.1(0.15)13.15	a_n, b_n (T)
	1.60	0.1(0.1)8.2; 8.6(0.6)11.0; 8.7(0.6)11.1; 8.8(0.6)10.6; 11.8(0.2)12.8	a_n, b_n (T)
	2.0	0.1(0.06)5.68; 6.0(0.2)8.0; 8.2(0.4)10.2	a_n, b_n (T)
	3.0	0.1(0.03)3.52; 4.0(0.2)5.4	a_n, b_n (T)
Libelo (1962c)	$1.4 - 0.05i$	0.1(0.075)8.35; 9.1(0.06)13.9; 9.175(0.6) 13.975	a_n, b_n (T)
	$1.414 - 1.414i$	0.1(0.075)7.9; 8.425(0.6) 13.875; 8.5(0.6)13.9	a_n, b_n (T)
	$1.5 - 3.10i$	0.1(0.06)7.0; 7.6(0.54)8.14; 7.66(0.54)8.20; 7.72(0.54)8.26; 8.8(0.6)13.6; 8.86(0.6)13.66	a_n, b_n (T)
Farone *et al.* (1963)	1.02, 1.06	0.1(0.1)5.0	a_n, b_n (T)
	1.40, 1.46, 1.47, 1.48, 1.60	0.1(0.1)10.0	i_1, i_2 at 0°(2.5°)180°; Q_1, Q_2 (T)
	1.42	0.1(0.1)8.0	i_1, i_2 at 0°(2.5°)180°; Q_1, Q_2 (T)
	1.44, 1.52, 1.56	0.1(0.1)5.0	i_1, i_2 at 0°(2.5°)180°; Q_1, Q_2 (T) Graphs of i_1, i_2 at $\theta = 0$, 45, 90, 135, and 180° up to $\alpha = 5.0$ and for Q_1 and Q_2 at $m = 1.40$, 1.47, and 1.60 up to $\alpha = 10.0$ are given. Tabular material obtained on microfilm, footnote p. 58, *loc. cit.*

TABLE 6.1—(*Continued*)

Ref.	m	α	Tabulated (T) or graphed (G) quantities
Evans *et al.* (1964) Evans (1963)	1.5, 2.0, 2.5	0.10(0.02)6.0	Q_1, Q_2 and backward component of scattered flux, B_1, B_2 (T)
Thomas (1964)	$m = \infty, 2$	$a/\lambda = 0.0002$ to 1	$G_1(180°)$, $G_2(180°)$ (G)
Agdur *et al.* (1963)	m for Ag at visible wavelengths	$a = 0.05$ to $0.4\ \mu$	$Q_{1\,\text{abs}}$, $Q_{2\,\text{abs}}$, $Q_{1\,\text{ext}}$, $Q_{2\,\text{ext}}$ (G)
Samaddar (1963)	m corresponds to various gyroelectric media	$\alpha = 1.0$	$i_1(\theta)$ vs. θ (G)

[a] Miscellaneous results in the older literature by von Schaefer (1907, 1909), von Schaefer and Grossman (1910), Seitz (1905, 1906), and Pfenninger (1927) are not included.

TABLE 6.2

TABULATED OR GRAPHED SCATTERING FUNCTIONS FOR COMPOUND CYLINDERS AT PERPENDICULAR INCIDENCE[a]

Ref.	Refractive indices	Size parameters	Tabulated (T) or graphed (G) quantity
Adey (1956a)	$m_1 = \infty$, $m_2 = 1.60$	$\alpha = 1.25$, $v = 2.5, 3.4$	$(i_1)^{1/2}$ at $\theta = 0°$ for the near field up to $kr = 50$ and at $\theta = 90, 180°$ up to $kr = 25$ (G)
Tang (1957)	$m_1 = \infty$, $m_2 = 2.54$	$\alpha = 0.942, 2, 3.5$; $v =$ up to 6.7	G_1 (G)
	$m_1 = \infty$, $m_2 = \sqrt{6}$	$\alpha = 0.942, 1.302$; $v =$ up to 6.7	G_1 (G)
Plonus (1960)	$m_1 = \infty$, $m_2 = 1$, $m_3 = 10$	$\alpha = 12$, $v = 14.89$ to 16.00 $\mu = 14.93$ to 16.04	G_2 (G, T)
Yeh and Kaprielian (1963)	$m_1 = \infty$; $m_2 = (0.5a/r)^{1/2}$ $(2a/r)^{1/2}$, $(5a/r)^{1/2}$	$\alpha = 0.5$, v up to 10	G_1 (G)
	$m_1 = \infty$; $m_2 = (3a/r)^{1/2}$, $(5a/r)^{1/2}$, $(10a/r)^{1/2}$	$\alpha = 3$, v up to 10	G_1 (G)

TABLE 6.2—(*Continued*)

Ref.	Refractive indices	Size parameters	Tabulated (T) or graphed (G) quantity
Swarner and Peters (1963)	$m_1 = \infty$, $m_2 = (0.75)^{1/2}$	$\alpha/2\pi = 0.275$ $v/2\pi = 0.27$ to 0.97	G_1, G_2 (G)
	$m_1 = \infty$, $m_2 = (0.75)^{1/2}$	$\alpha/2\pi = 0.5$ $v/2\pi = 0.5$ to 1.33	G_1, G_2 (G)
	$m_1 = \infty$, $m_2 = (1.5)^{1/2}$	$\alpha/2\pi = 0.15$ $v/2\pi = 0.15$ to 0.73	G_1, G_2 (G)
Evans *et al.* (1964) Evans (1963)	$m_1 = 1$, $m_2 = 1.5, 2.0, 2.5$	$v = 0.10(0.02)6.0$ $\alpha/v = 0, 0.01, 0.10, 0.50, 0.90, 0.99$	Q_1, Q_2, backward components of scattered flux B_1 and B_2 (T)
Kelly and Russek (1960)	∞	0 to 3.0	Q_1 and Q_2 for cylinders with anisotropic conductivity at oblique incidence of 45° (G)
	∞	0.1	i_1 and i_2 for same as above
Wilhelmsson (1963)	$m_1 = \infty$, $m_2 = 1.59$	$\alpha = 0.0942, 0.942, 2, 3.5$ v up to 10	$G_1(180°)$ (G)
	$m_1 = \infty$, $m_2 = 2.45$	$\alpha = 0.942, 1.302$ v up to 10	$G_1(180°)$ (G)
	$m_1 = \infty$, $m_2 = 1.79$	$\alpha = 4$ v up to 10	$G_1(180°)$ (G)
Thomas (1964)	$m_1 = \infty$, $m_2 = 2$	$b/\lambda = 0.00002$ to 4, $a/b = 0.9$	$G_1(180°)$, $G_2(180°)$, (G)
Yeh (1963)	$m_1 = \infty$ m_2 is for various gyroelectric media	$\alpha = 2$, $v = 3$	$\lvert E_\theta \rvert$ vs. θ (G)
Rusch (1964)	$m_1 = \infty$, m_2 is for various radially inhomogeneous sheaths	$\alpha = 5$, $v = 6$	$\lvert E_\theta \rvert$ vs. θ (G)
Wilhelmsson (1962)	$m_1 = \infty$, $m_2 = 0.628$	$\alpha = 0.15, 1.5, 3.188, 5.58$ v up to 16	$G_1(180°)$ (G)
	$m_1 = \infty$ $m_2 = 0.408$	$\alpha = 2.308, 3.188$ v up to 25	

[a] Key:
Refractive indices from inside: $m_1, m_2, m_3, m_4, \ldots$
Radii from inside: $a, b, c, \ldots$
Size parameters from inside: $\alpha, v, \mu, \ldots$
G_1 without an angular designation is the backscatter efficiency

TABLE 6.3

TABULATED OR GRAPHED SCATTERING FUNCTIONS FOR CYLINDERS AT OBLIQUE INCIDENCE

Ref.	Refractive index	Size parameter	Tabulated (T) or graphed (G) quantity
Kerker *et al.* (1966a)	1.46	$\alpha = 0.1$	i_{11}, i_{12} for $\phi = 15(15)75, 85°$ (T)
		$\alpha = 5.0$	i_{11} for $\phi = 25, 70, 85°$ (T)
			i_{22} for $\phi = 40, 50, 70, 85°$ (T)
			$i_{12} = i_{21}$ for $\phi = 5, 40, 65, 75, 85°$ (T)
		$\alpha = 10.0$	i_{12}, i_{22} for $\phi = 5, 50, 60, 85°$ (T)
			i_{11} for $\phi = 5, 50, 60, 80°$ (T)
Lind and Greenberg (1966)	1.6	$\alpha = 0$ to 4.0	Q_1 and Q_2 for $\phi = 0(5)85, 89°$ (G)
Thomas (1964)	$m = \infty, 2$	$a/\lambda = 0.0002$ to 1	$G_1(180°), G_2(180°); \phi = 84°$ (G)
	$m_1 = \infty, m_2 = 2$	$b/\lambda = 0.0002$ to 1	$G_1(180°), G_2(180°); \phi = 84°$ (G)
	$m_1 = \infty$, $m_2 = 1.6, 2.0$	$b/\lambda = 0.78$	$G_1(180°), G_2(180°); \phi = 2°$ to $82°$ (G)
	$m_2 = 1$ to 20	$b/\lambda = 0.702$	$G_1(180°); \phi = 45°$ (G)
		$b/\lambda = 0.78$	$G_2(180°); \phi = 45°$ (G)
		$a/b = 0.9$	

The variation of the various scattering functions with the parameters such as m, α, θ, and ϕ will not be reviewed in as great detail as was done for spheres. The general trends for cylinders and for spheres are quite similar.

6.5.2 TOTALLY REFLECTING CYLINDERS, $m = \infty$

The angular gain for cylinders is defined by

$$G_1 = 4i_1/\pi\alpha; \qquad G_2 = 4i_2/\pi\alpha \tag{6.5.1}$$

The backscatter gain for the perfectly reflecting cylinder at perpendicular incidence is shown as a function of α in Fig. 6.9. The behavior of the TE mode, for which the electric vibration is perpendicular to the cylinder axis, is quite similar to that for the sphere. The gain rises steeply to a maximum value and then, with increasing α, undergoes a damped sinusoidal oscillation about the geometrical optics limit of unity. Although the period of this oscillation ($\Delta\alpha = 1.18$) is approximately the same as for the sphere, it is more highly damped, suggesting that, for the cylinder, the contribution of the surface wave to the backscatter is relatively less important than that of the specularly reflected axial ray.

The TM mode behaves quite differently. In this case, the surface wave is much more highly attenuated and does not make an important contribution

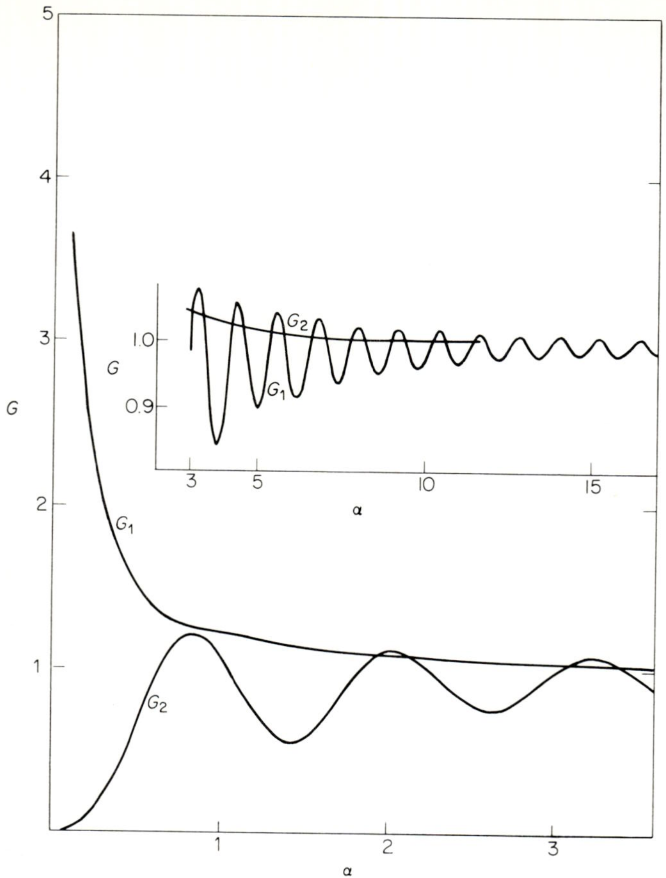

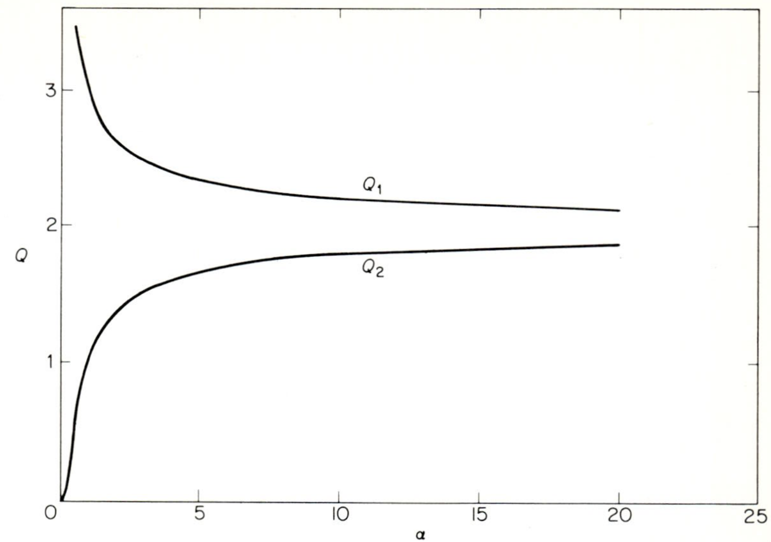

FIG. 6.10 (above). Scattering efficiencies (Q_1 and Q_2) of a totally reflecting cylinder at perpendicular incidence plotted against α.

FIG. 6.9. (left). Backscatter gain of a totally reflecting cylinder ($m = \infty$) at perpendicular incidence plotted against α.

to the backscatter. The backscatter gain drops smoothly from infinity for $\alpha = 0$ and approaches unity asymptotically. For a complete discussion in terms of the surface wave, the reader is referred to the work of Franz and Depperman (1952).

The scattering efficiencies $Q_{1\,\text{sca}}$ and $Q_{2\,\text{sca}}$ for $m = \infty$ are plotted in Fig. 6.10 over the range $\alpha = 0$ to 20. Unlike the sphere, there are no oscillations in the curves for either mode. They each approach the geometrical optics limit of 2 smoothly; the value for the TM mode drops from infinity while that for the TE mode rises from zero.

The angular variation of the gain for various values of α is shown in Figs. 6.11 and 6.12 for G_1 and G_2 respectively. The intense forward lobes are due primarily to the diffracted wave. As α becomes larger than 2, there is the development of the characteristic oscillations with angle which then damps out with still further increase of α.

When α is small, the angular gain and the scattering efficiency approach infinity for the TM mode and zero for the TE mode. That this follows analytically from the exact expressions can be seen if the Bessel functions are expanded in a power series in α, retaining only the dominant terms in much the same manner as for spheres (cf. Section 3.9). For the TM mode, the dominant coefficient is b_0 which is given by

$$b_0 = \frac{1 + i(2l/\pi)}{1 + (2l/\pi)^2} \cong T_1 \tag{6.5.2}$$

where

$$l = \gamma + \ln \alpha/2 \tag{6.5.3}$$

and γ is Euler's constant. The scattering efficiency can then be written

$$Q_{1\,\text{sca}} = Q_{1\,\text{ext}} = (2/\alpha)[1 + (2l/\pi)^2]^{-1} \tag{6.5.4}$$

and the angular intensity is

$$I_1 = (2/\pi r k_0)|b_0|^2 = (2/\pi r k_0)[1 + (2l/\pi)^2]^{-1} \tag{6.5.5}$$

These expressions are valid to within 5% up to $\alpha = 0.4$. Unlike small perfectly reflecting spheres (4.4.3), the scattering is isotropic and there is no simple power law dependence upon α, nor, in turn, upon the radius or wavelength. Indeed the wavelength dependence is now even qualitatively different from both small spheres and from small dielectric cylinders since the longer wavelengths are preferentially scattered. In Fig. 6.13, the variation of Q_1 with α is plotted using the small cylinder limiting expression (6.5.4). The slope of this log-log plot gives the dependence of Q_1 upon the power of α, a, and $1/\lambda$. In the limit of infinitesimally small particles, the scattering is directly proportional to the wavelength; in the range plotted it varies from $Q_1 \sim \lambda^{0.55}$ at $\alpha = 2 \times 10^{-2}$ to $Q_1 \sim \lambda^{0.85}$ at 2×10^{-6}.

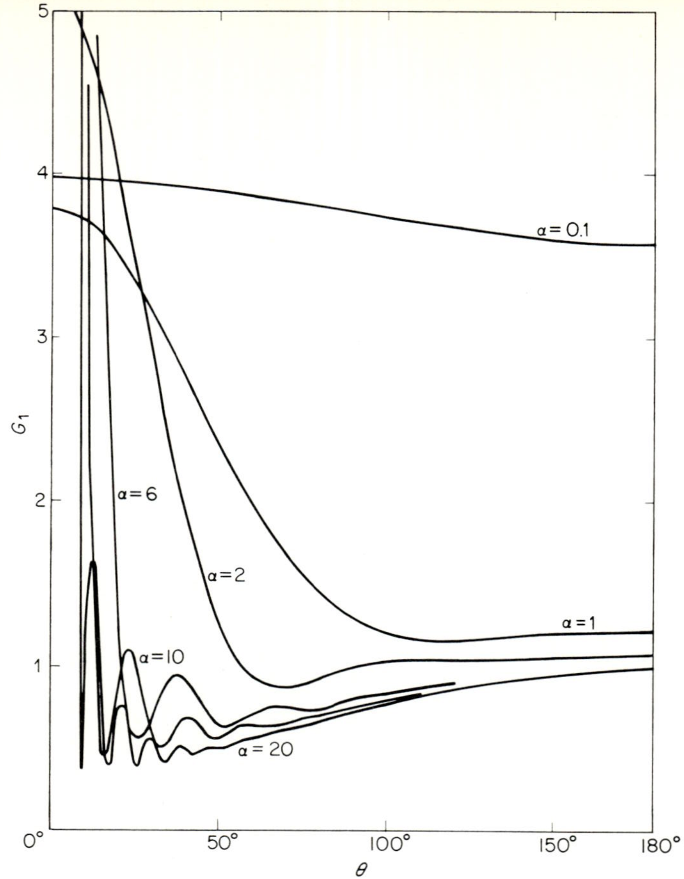

FIG. 6.11. Angular gain (TM mode; G_1) of a totally reflecting cylinder at perpendicular incidence plotted against θ for various values of α.

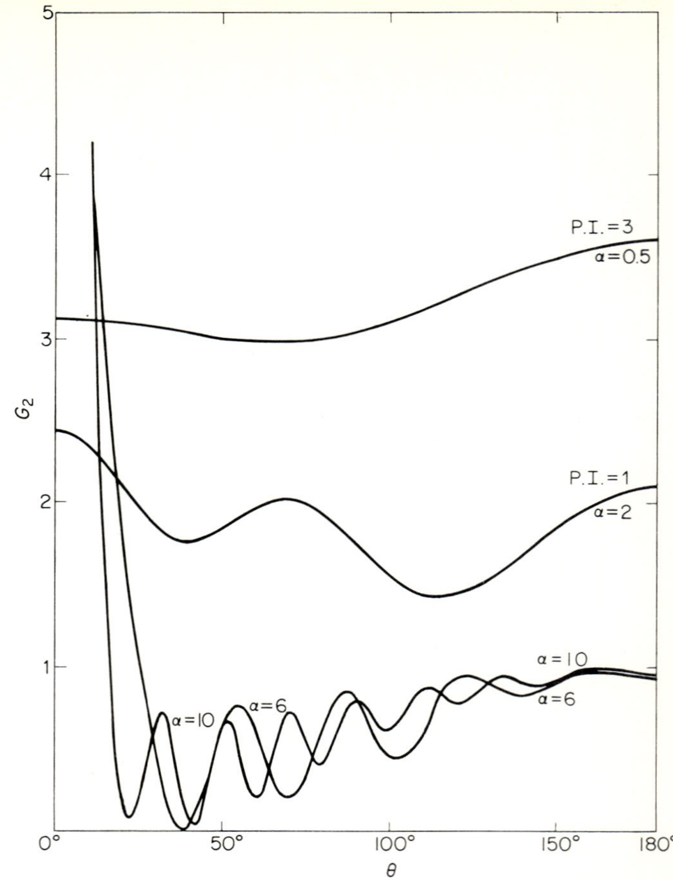

FIG. 6.12. Angular gain (TE mode; G_2) of a totally reflecting cylinder at perpendicular incidence plotted against θ for various values of α. P.I. designates a plotting increment.

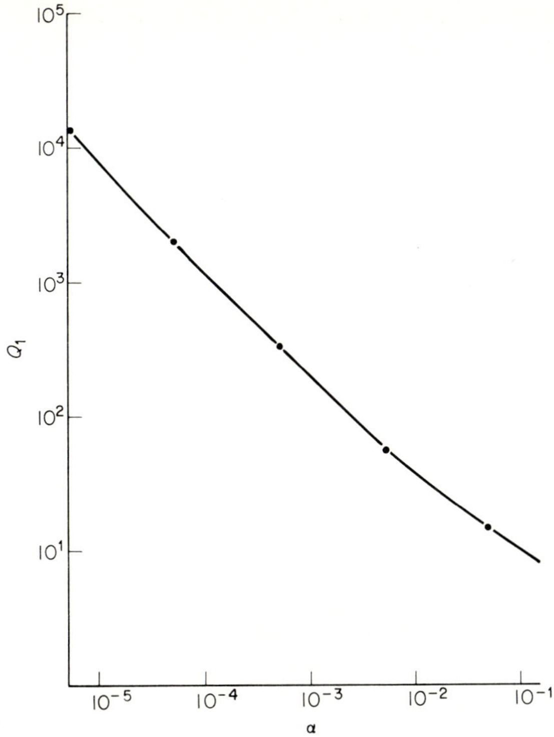

FIG. 6.13. Scattering efficiency (Q_1 or TM mode) of totally reflecting cylinder, according to small cylinder limit (Eq. 6.5.4), plotted against α.

For the TE mode both a_0 and a_1 make comparable contributions in the limiting case of small α. These are given by

$$a_0 = \frac{\pi\alpha^4}{16} - i\frac{\pi\alpha^2}{4} \tag{6.5.6}$$

$$a_1 = \frac{\pi\alpha^4}{16} + i\frac{\pi\alpha^2}{4} \tag{6.5.7}$$

which leads to

$$Q_{2\,\mathrm{ext}} = 3\pi^2\alpha^3/8 \tag{6.5.8}$$

and

$$I_2 = (2/\pi r k_0)|a_0 + 2a_1 \cos\theta|^2 \simeq (\pi\alpha^4/2rk_0)(\cos\theta - \tfrac{1}{2})^2 \tag{6.5.9}$$

The dependence upon α, and hence on a and λ, is the same for the limiting case of small dielectric cylinders considered in (6.1.60) and (6.1.61). The angular dependence of the intensity of the TE mode is the same as for small perfectly reflecting spheres.

6.5.3 Dielectric Media

a. Scattering Efficiency. The variation of the scattering efficiency with the generalized size parameter $\rho = 2\alpha(m - 1)$ for each of the polarized modes is depicted in Figs. 6.14 to 6.16 for $m = 1.48$, 2.0, and 2.5. The results for spheres with $m = 1.486$ are also shown on Fig. 6.14.

The method of anomalous diffraction ($m \to 1$ and $\alpha \to \infty$) may also be applied to infinite cylinders (van de Hulst, 1957) leading to

$$Q(\rho) = \pi\zeta_1(\rho) \tag{6.5.10}$$

where the first-order Struve function is defined by

$$\zeta_1(\rho) = (2\rho/\pi)\int_0^{2\pi} \sin(\rho\cos\alpha)\sin^2\gamma\,d\gamma \tag{6.5.11}$$

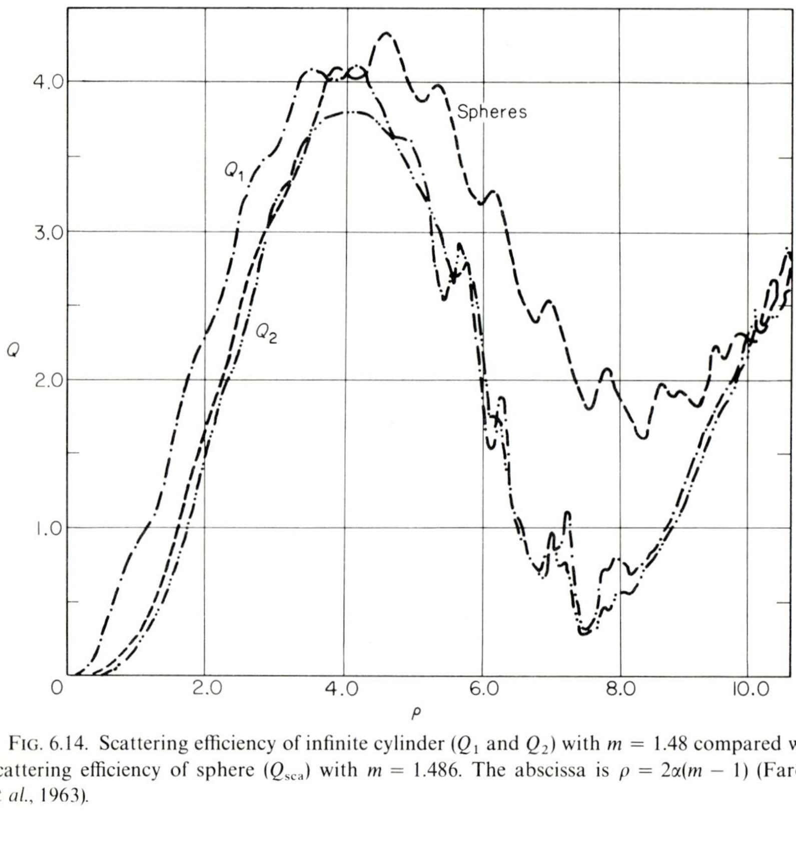

FIG. 6.14. Scattering efficiency of infinite cylinder (Q_1 and Q_2) with $m = 1.48$ compared with scattering efficiency of sphere (Q_{sca}) with $m = 1.486$. The abscissa is $\rho = 2\alpha(m - 1)$ (Farone *et al.*, 1963).

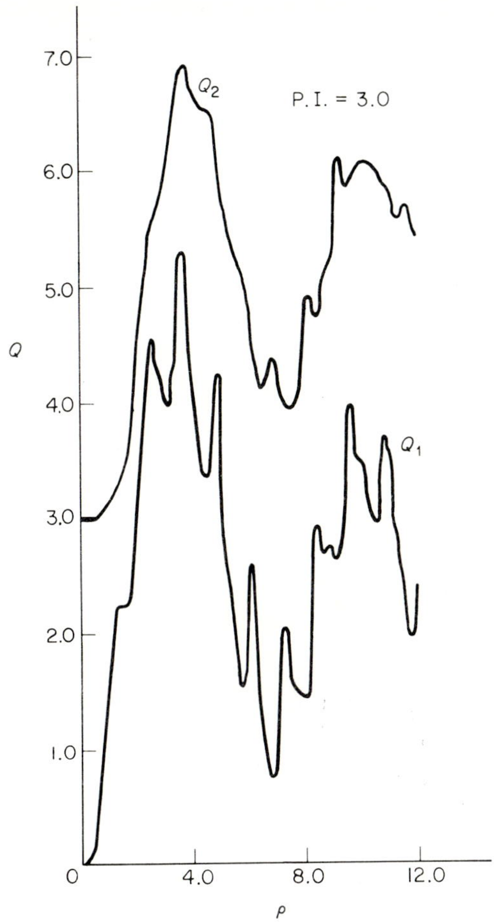

FIG. 6.15. Scattering efficiency of infinite cylinder (Q_1 and Q_2) with $m = 2.0$. P.I. is the plotting increment. The abscissa is $\rho = 2\alpha(m - 1)$.

This case is compared in Fig. 6.16 with the scattering efficiency of dielectric cylinders having $m = 2.50$. Just as for spheres, the efficiency rises from zero to a maximum and then oscillates roughly about the limiting value of 2. The principal maxima and minima fall at about the same values of ρ.

b. Angular Functions. The angular intensity functions also exhibit patterns similar to those for spheres both for the variation with angle at a given value of α and for variation with α at a given angle. The backscatter is illustrated in Figs. 6.17 and 6.18 up to $\alpha = 5$. Over this range the oscillations are regularly spaced, the period being smaller for the larger values of the refractive index. The amplitude of the oscillation increases with α, being greater for larger refractive indices. The corresponding curves for the lateral scattering are less regular and for $\alpha > 5$ exhibit such erratic fine structure that they can only be represented accurately by using an expanded scale with very small increments of α ($\Delta\alpha \sim 0.01$).

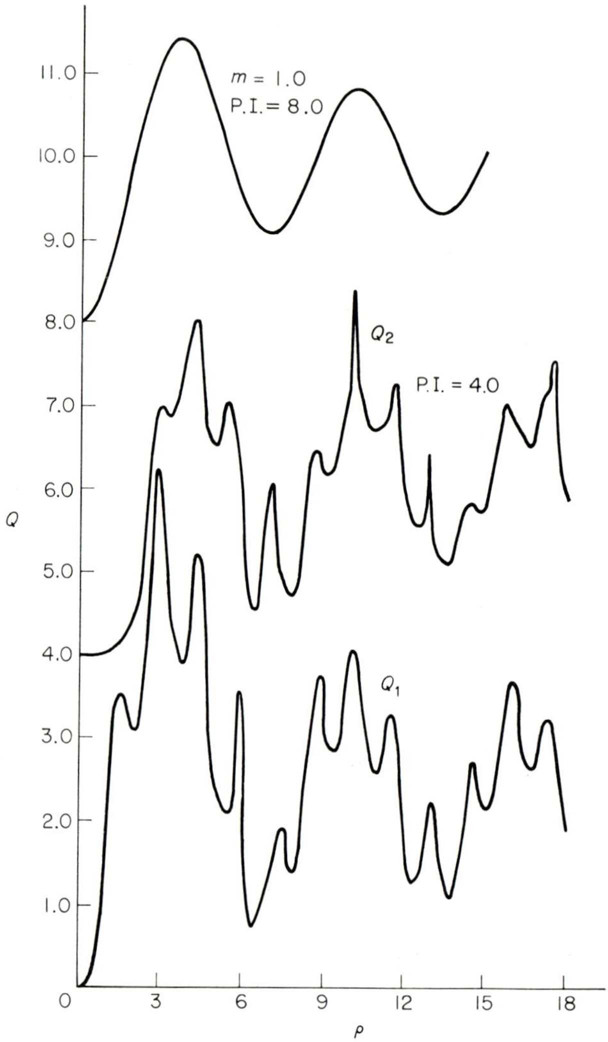

FIG. 6.16. Scattering efficiency of infinite cylinder (Q_1 and Q_2) with $m = 2.5$. P.I. is the plotting increment. The abscissa is $\rho = 2\alpha(m - 1)$. The results obtained with the anomalous diffraction theory ($m = 1$, Eq. 6.5.10) are also shown.

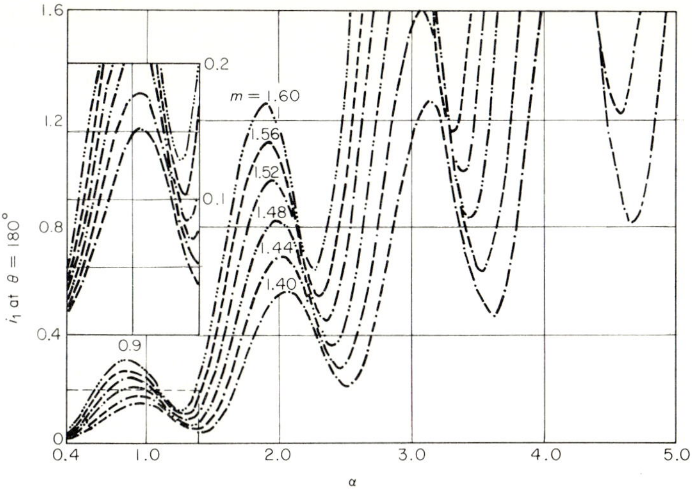

FIG. 6.17. Variation of angular scattering function i_1 with α at $\theta = 180°$ for $m = 1.40$, 1.44, 1.48, 1.52, 1.56, and 1.60. The values of the maxima near $\alpha = 3.0$ are 1.25, 1.60, 1.97, 2.32, 2.76, and 3.08, respectively, and those near $\alpha = 4.0$ are 2.06, 2.70, 3.36, 4.04, 4.68, and 5.28, respectively (Farone *et al.*, 1963).

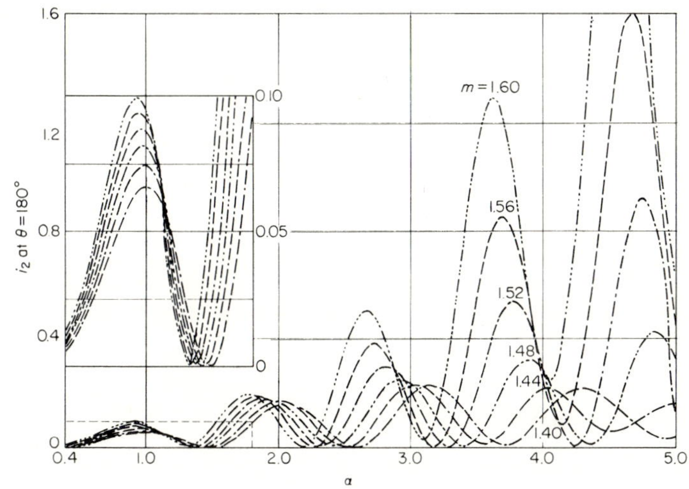

FIG. 6.18. Variation of angular scattering function i_2 with α at $\theta = 180°$ for $m = 1.40$, 1.44, 1.48, 1.52, 1.56, and 1.60. The values of the minima near $\alpha = 1.5$ are 4.25×10^{-4}, 2.78×10^{-3}, 8.67×10^{-4}, 2.03×10^{-3}, 5.98×10^{-3}, and 3.71×10^{-3}, respectively; near $\alpha = 2.5$ are 2.04×10^{-4}, 9.85×10^{-4}, 8.84×10^{-3}, 8.76×10^{-3}, 6.92×10^{-3}, and 1.74×10^{-3}, respectively; near $\alpha = 3.5$ are 5.28×10^{-3}, 1.65×10^{-2}, 9.01×10^{-3}, 4.13×10^{-3}, 6.67×10^{-3}, and 2.77×10^{-2}, respectively; and near $\alpha = 4.2$ are 0.163, 6.06×10^{-2}, 1.75×10^{-2}, 7.27×10^{-3}, 8.19×10^{-2}, and 0.266, respectively. The maximum at $\alpha = 4.5$ and $m = 1.60$ is 2.54 (Farone *et al.*, 1963).

In Fig. 6.19, the angular locations of the maxima and minima of i_1 for $m = 1.46$ are illustrated. Here too, just as for spheres, the positions of the extrema migrate towards the forward direction and the number of extrema increases with increasing α. These positions, at least for sufficiently small values of m, can be generalized on the parameter

$$\alpha \sin \theta/2 = k_i(m) \qquad \text{or} \qquad K_i(m) \tag{6.5.12}$$

where the constants have the same significance described earlier (cf. Eq. 4.4.13).

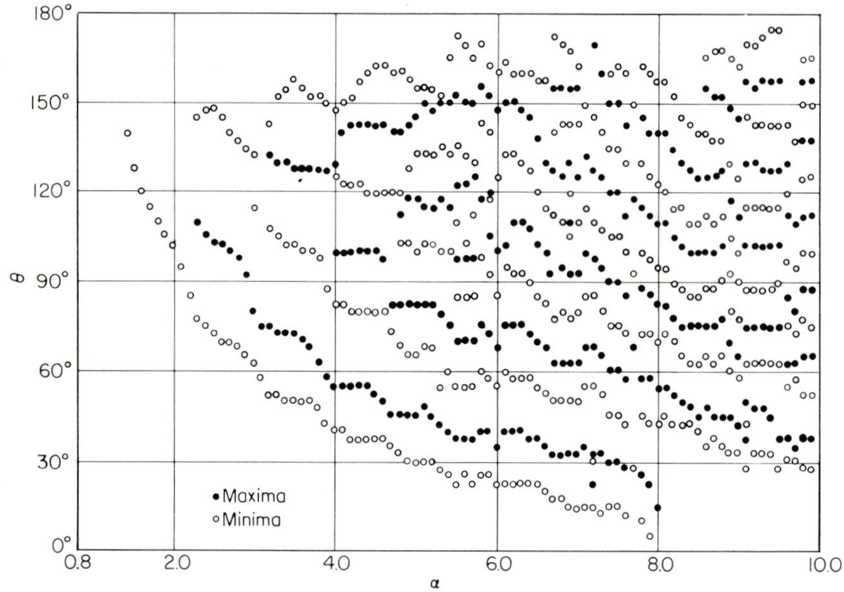

FIG. 6.19. Angular locations of the maxima (●) and minima (○) of i_1 for $m = 1.46$ (Farone *et al.*, 1963).

6.5.4 ABSORBING MEDIA

The effect of absorption is illustrated in Fig. 6.20(a–d) where Q_1 is plotted against α for $m = 1.40$, $1.40 - 0.05i$, $1.40 - 0.20i$, $1.40 - 0.50i$, and $1.40 - 2.00i$. Separate curves are shown for the extinction, for the scattering, and for the absorption. The fine structure of the nonabsorbing case is washed out when the absorption index is as high as 0.05. With increasing absorption, the absorption curve goes through a flat maximum and then drops slowly with increasing size. This maximum sharpens with still increasing absorption [Fig. 6.20(c)] and moves to smaller values of the size parameter. For

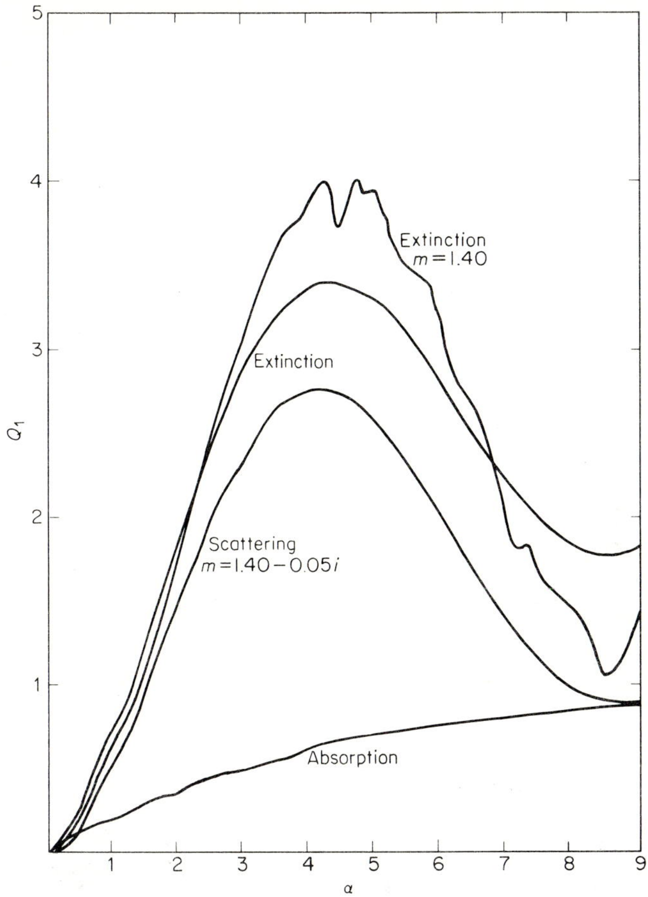

FIG. 6.20a. Extinction efficiency Q_1 for $m = 1.40$ and the extinction, scattering and absorption efficiencies for $m = 1.40 - 0.05i$.

$m = 1.40 - 2.00i$ [Fig. 6.20(d)], the sharp maximum occurs at $\alpha < 0.2$ which is the smallest value for which these calculations have been carried out.

6.5.5 Compound Cylinders

Evans *et al.* (1964) have computed scattering functions for hollow cylinders over the range of parameters listed in Table 6.2. They have also included the single cylinder which is the special case where the inner radius is zero. The effect of hollowing out the cylinder is to increase the interval between the maxima and minima in the Q versus α curve. This is shown in Fig. 6.21.

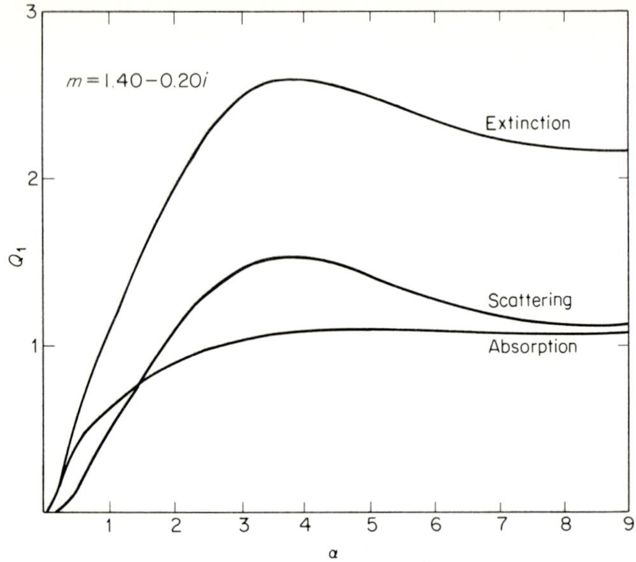

FIG. 6.20b. Same as Fig. 6.20a for $m = 1.40 - 0.20i$.

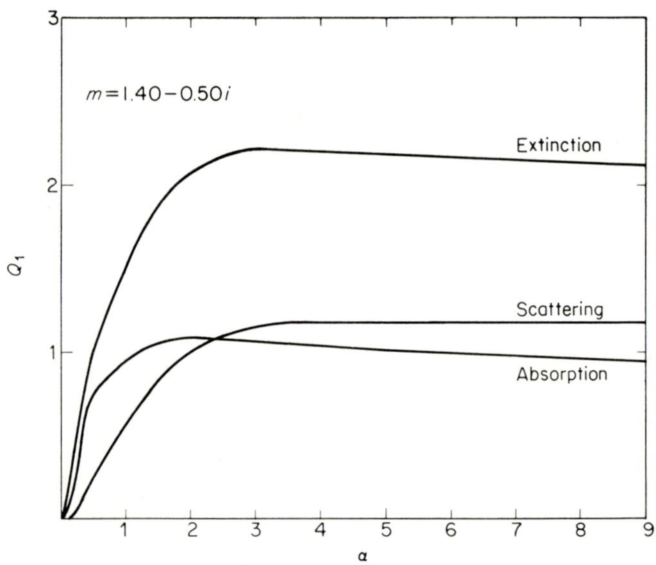

FIG. 6.20c. Same as Fig. 6.20a for $m = 1.40 - 0.50i$.

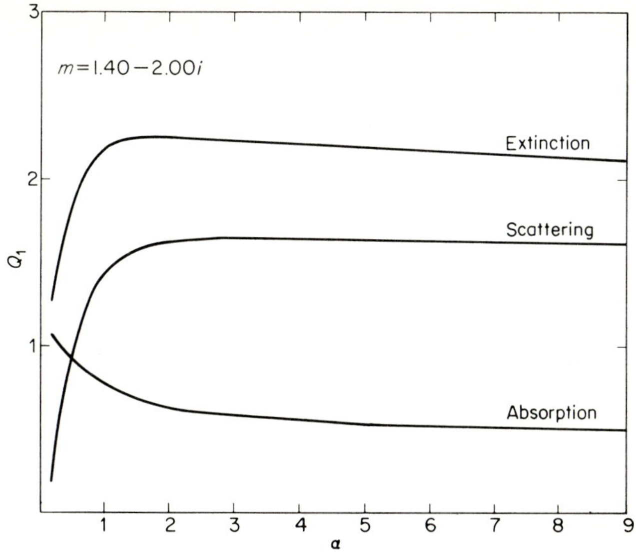

FIG. 6.20d. Same as Fig. 6.20a for $m = 1.40 - 2.00i$.

Other computations cited in Table 6.2 illustrate the influence of a dielectric sheath on a perfectly reflected cylinder. Swarner and Peters (1963) have shown that the superposition approximation discussed in connection with dielectric coatings on perfectly reflecting spherical cores is also applicable to cylinders, subject to similar restrictions. This makes it possible to avoid the lengthier computations for compound cylinders by utilizing those for equivalent perfectly conducting and dielectric cylinders.

For more highly refractive coatings than those considered by Swarner and Peters, the behavior of the dielectric-coated cylinder does not appear to present a pattern which can be predicted from a simple model. If the radius of the inner cylinder is small, the maximum backscattering occurs when the dielectric layer is approximately a multiple of one-half of a wavelength in thickness. This is an effect that would be expected from simple interference of the axial rays. A typical curve of the variation of backscatter with coating thickness is shown in Fig. 6.22. This illustrates the remarkable ability of the coating to either enhance or obscure the scattering. The sensitivity to the precise nature of the coating is illustrated in Fig. 6.23 which is a comparison between the backscattering of a cylinder coated with an inhomogeneous dielectric sheath and one coated with a homogeneous sheath of the same average refractive index. In this case, merely altering the refractive index profile changes the structure of the backscattering curve.

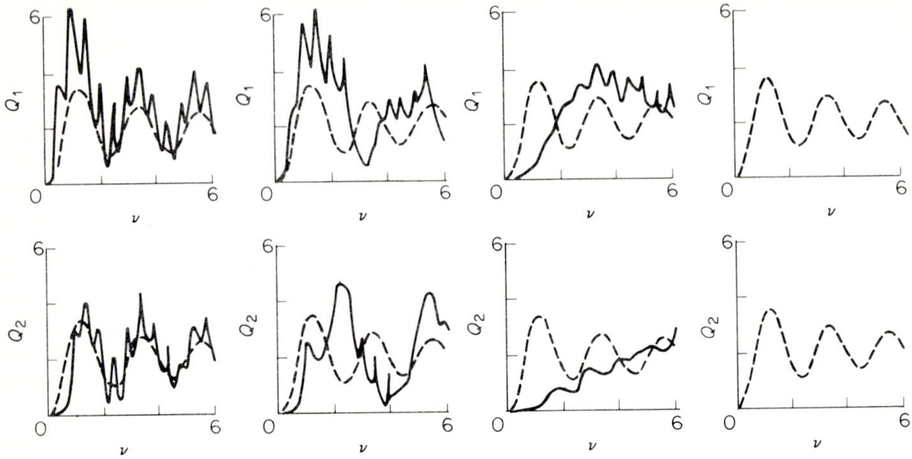

FIG. 6.21. Scattering efficiencies Q_1 and Q_2 against $2\pi b/\lambda$ for infinitely long hollow cylinders with $m = 2.5$. The ratio of the inner diameter to the outer diameter, a/b, varies from left to right; $a/b = 0, 0.5, 0.90, 0.99$. The dashed line is the approximation for $m = 1$ given by Eq. (6.5.10) (Evans *et al.*, 1964).

Shiobara (1966) has suggested that the phase shift which the electromagnetic wave experiences upon passing through a compound cylinder consisting of coaxial sheaths is simply the sum of the phase shifts produced by each of the component cylinders. However, Jones and Farone (1967) have shown that this is not the case except under very special conditions.

6.5.6 Oblique Incidence

It has already been observed that for perfectly reflecting cylinders the scattering coefficients at oblique incidence [(6.1.36) and (6.1.37)] are of the same form as at perpendicular incidence. The argument of the Bessel functions is $\alpha \cos \phi$ so that the intensity scattered by a particular cylinder at oblique incidence is precisely that of a cylinder at perpendicular incidence whose size is reduced by the factor $\cos \phi$.

The angular scattering function for dielectric cylinders ($m = 1.46$) at oblique incidence has been computed by Kerker *et al.* (1966a) for $\alpha = 0.1$, 5.0, and 10.0. The smallest of these sizes ($\alpha = 0.1$), corresponding to the dielectric needle approximation, will be discussed first.

The mode described by i_{22} is particularly simple. It reduces, at perpendicular incidence, to the TE mode i_2, which for small cylinders follows a $\cos^2 \theta$ dependence except at scattering angles close to 90° (6.1.63). When a small cylinder such as this is tilted, there is scarcely any variation in i_{22} with tilt angle.

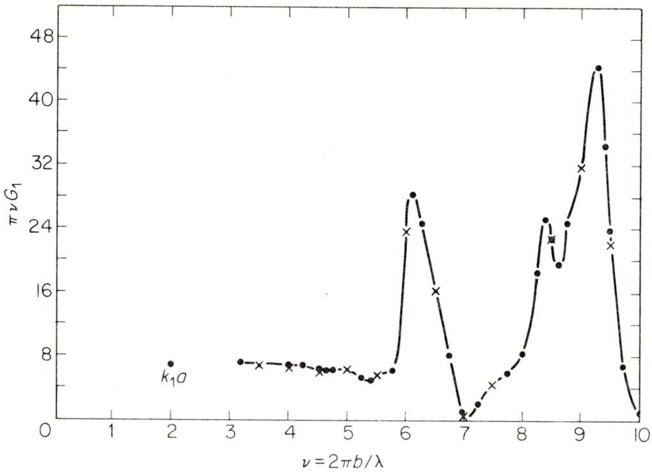

FIG. 6.22. Backscattering of a dielectric coated totally reflecting cylinder plotted as $\pi\nu G_1$ against $\nu = 2\pi b/\lambda$ where b is the cylinder radius. The size of the inner core is constant at $\alpha = (2\pi a/\lambda) = 2$. For the coating $\varepsilon = m^2 = 2.54$, the wavelength λ is 3.14 cm. Crosses are theoretical, dots are experimental (Tang, 1957).

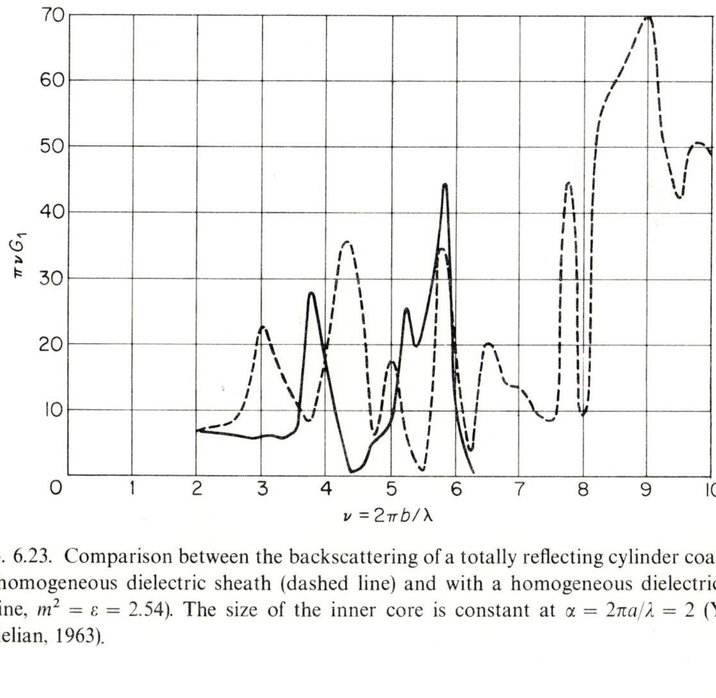

FIG. 6.23. Comparison between the backscattering of a totally reflecting cylinder coated with an inhomogeneous dielectric sheath (dashed line) and with a homogeneous dielectric sheath (full line, $m^2 = \varepsilon = 2.54$). The size of the inner core is constant at $\alpha = 2\pi a/\lambda = 2$ (Yeh and Kaprielian, 1963).

Fig. 6.24 shows how i_{11} varies with θ for various tilt angles from $\phi = 0°$ to 85°. For perpendicular incidence ($\theta = 0°$), i_{11} is constant as predicted by the dielectric needle approximation (6.1.62). However, as the tilt angle is increased, the intensity varies with scattering angles and for angles close to $\phi = 90°$, it behaves much the same as i_{22}, following the same $\cos^2 \theta$ dependence.

The scattering pattern for i_{12}, which is identically equal to i_{21}, is shown in Fig. 6.25. This cross component disappears at $\theta = 0°$ and 180° and at $\phi = 0°$. At other angles, it follows an approximately ($\sin^2 \theta \cos^2 \phi$)-dependence upon scattering angle and tilt angle.

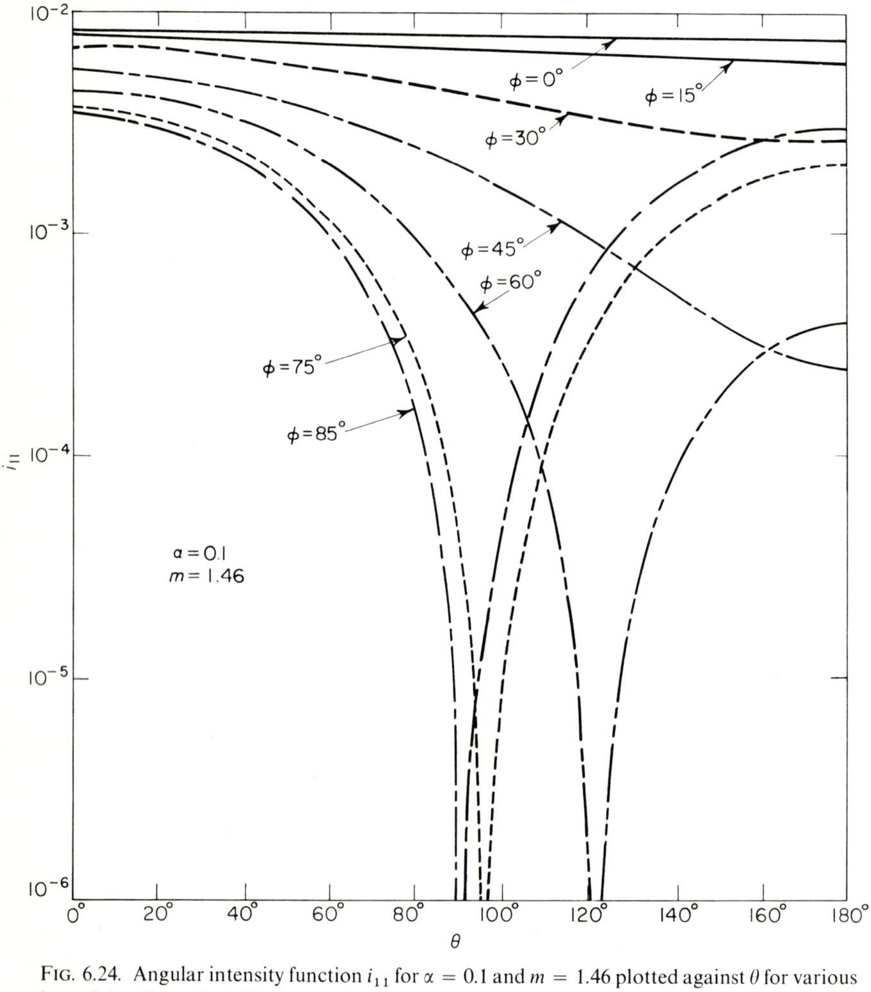

FIG. 6.24. Angular intensity function i_{11} for $\alpha = 0.1$ and $m = 1.46$ plotted against θ for various values of the tilt angle ϕ (Kerker *et al.*, 1966a).

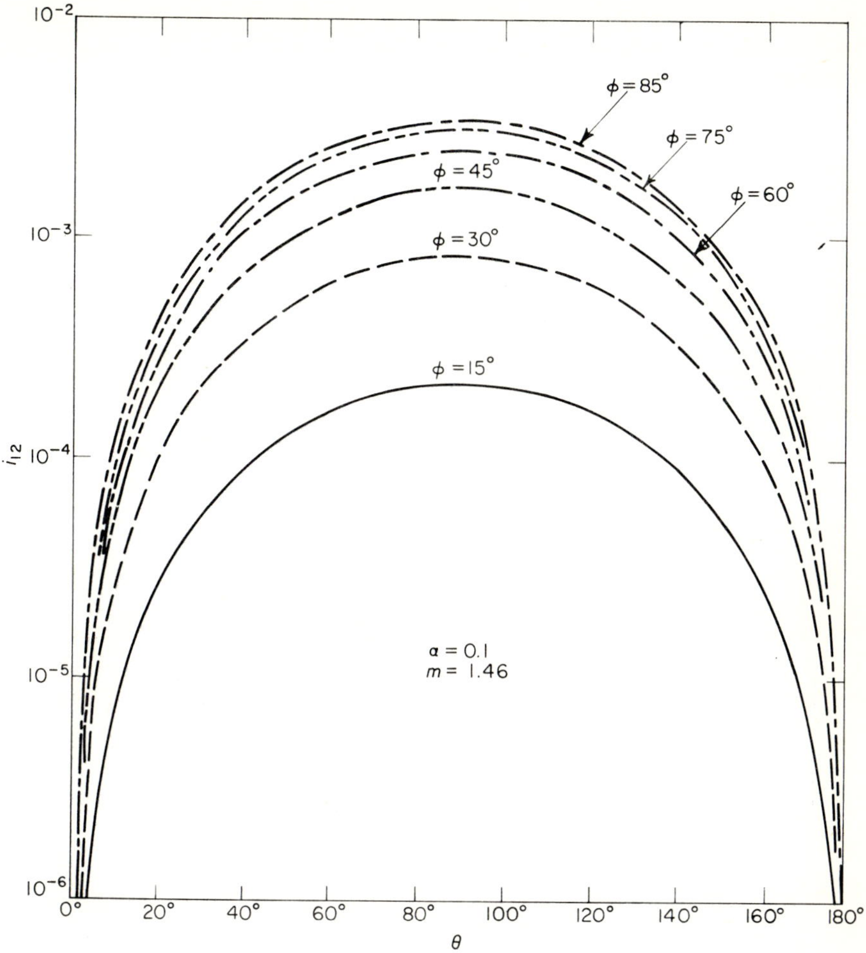

FIG. 6.25. Angular intensity function i_{12} for $\alpha = 0.1$ and $m = 1.46$ plotted against θ for various values of the tilt angle ϕ (Kerker *et al.*, 1966a).

Results for $\alpha = 10.0$ are shown in Figs. 6.26 and 6.27 where the angular variation of i_{11} and of i_{12} (or i_{21}) is plotted for various tilt angles. The typical oscillations near perpendicular incidence become damped as the cylinder is tilted. As the tilt angle approaches grazing incidence, these patterns also approach the dielectric needle approximation. For i_{11} and i_{22}, they become similar to that shown in Fig. 6.24 for large values of ϕ. Also the pattern for the cross components near $\phi = 90°$ resembles that shown in Fig. 6.25 for small α. Thus, at high obliquity the scattering by a dielectric cylinder resembles that of a dielectric needle at perpendicular incidence. In this

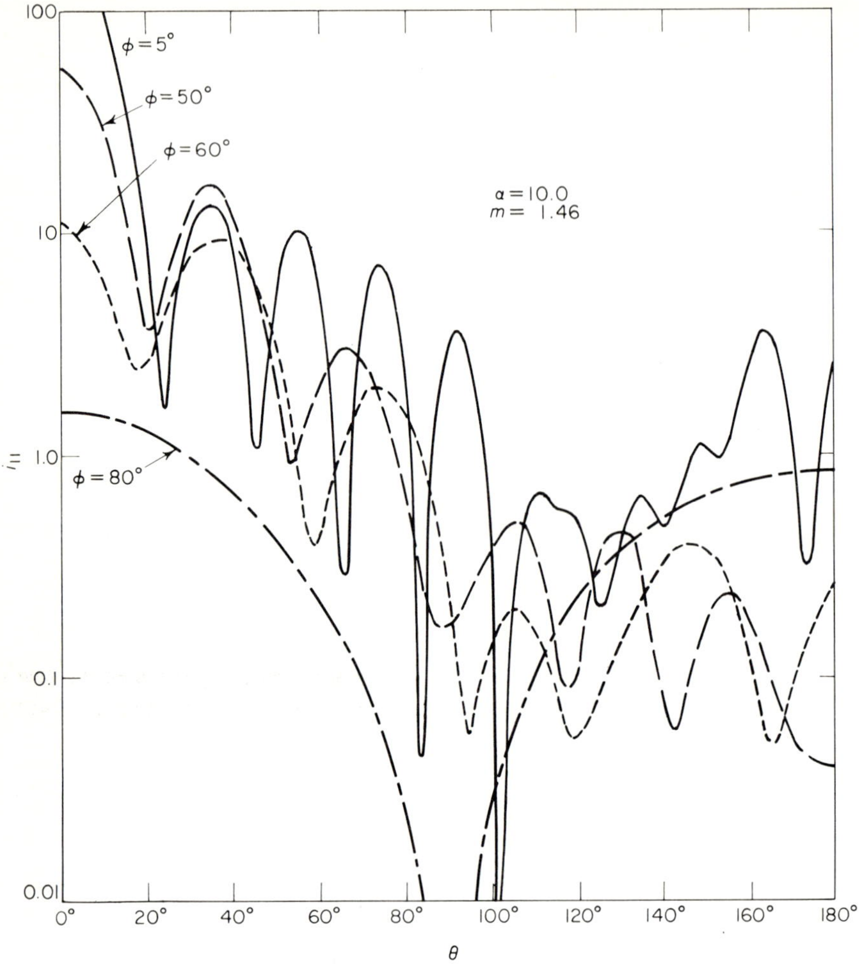

FIG. 6.26. Angular intensity function i_{11} for $\alpha = 10.0$ and $m = 1.46$ plotted against θ for various values of the tilt angle ϕ (Kerker *et al.*, 1966a).

respect, the scattering by a dielectric cylinder is similar to that by a cylinder composed of a perfectly reflecting medium.

Lind and Greenberg (1966) have calculated the scattering efficiencies up to $\alpha = 4.7$ for tilted cylinders with $m = 1.6$. The general trend over the range of the first maximum is shown for $(Q_{11} + Q_{12})$ in Fig. 6.28. The results for a TE incident wave $(Q_{21} + Q_{22})$ are quite similar.

The outstanding feature here is the appearance of two discrete sharp peaks at approximately $\alpha = 0.7$ and 3.2 for $\phi = 89°$ in contrast to the broad

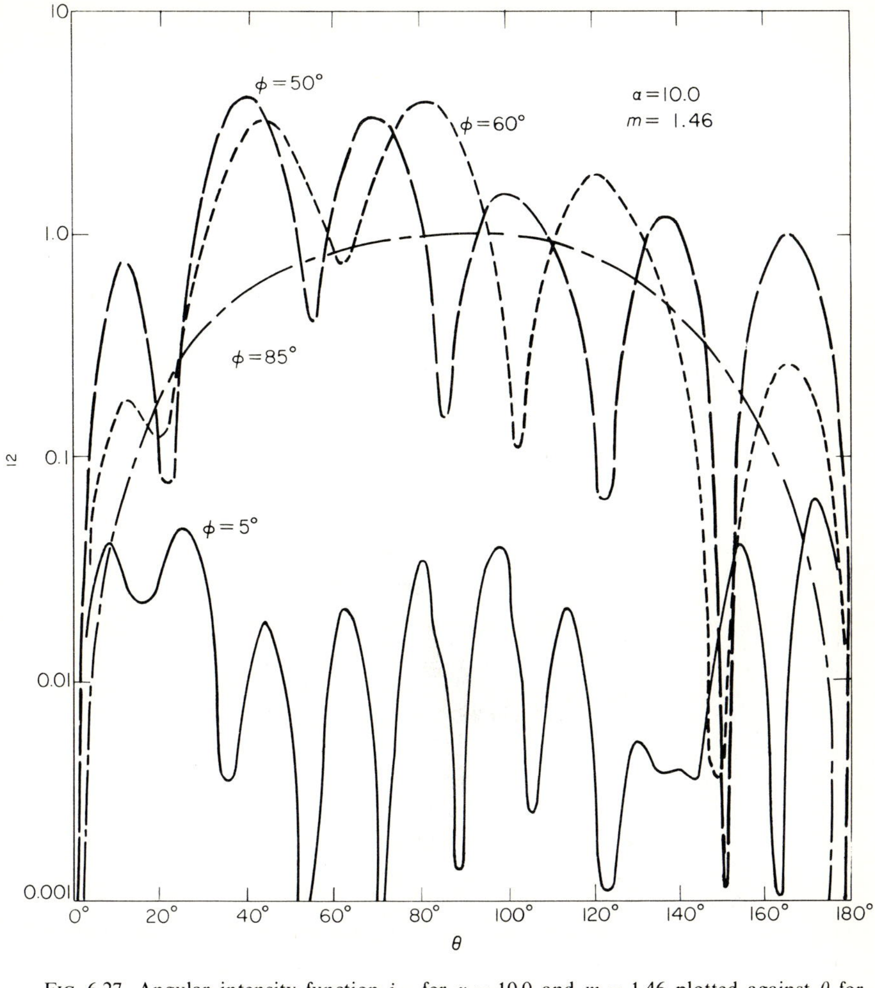

FIG. 6.27. Angular intensity function i_{12} for $\alpha = 10.0$ and $m = 1.46$ plotted against θ for various values of the tilt angle ϕ (Kerker *et al.*, 1966a).

rippled maximum at tilt angles closer to perpendicular incidence. Actually, this broad maximum is due to the superposition of many resonance peaks, each associated with a partial wave. We have already discussed, in the case of the sphere, how the scattering efficiency may be resolved into partial wave contributions as given by (4.2.15). These contributions are associated with the n terms in the series expansions of the scattering functions, such as those given by (6.1.45) to (6.1.48). Each of these partial waves, when plotted as a function of α, exhibits a pattern of resonances which is particularly

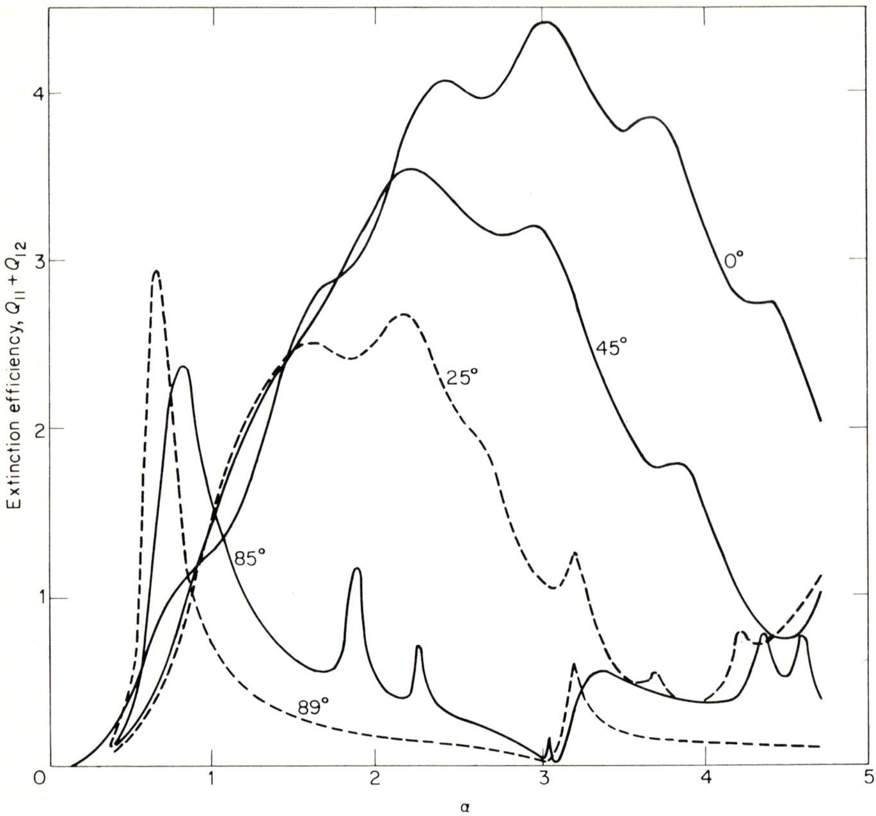

FIG. 6.28. Extinction efficiency $Q_{11} + Q_{12}$ plotted against α for various tilt angles, ϕ; $m = 1.6$ (Lind and Greenberg, 1966).

sharp for small values of the index, n. Analytically, these resonances result from a sharp minimum in the denominator of the corresponding scattering coefficient. Physically, they correspond to the resonant excitation of a particular electric or magnetic multipole.

The total scattering efficiency is obtained by superposition of the partial waves with the broadband at obliquities near perpendicular incidence, indicating that resonances from several partial waves are making significant contributions to the total effect. However, for ϕ close to 90°, only the first partial wave makes a significant contribution to the scattering. Both resonances shown in Fig. 6.28 are derived from this. This is consistent with the finding that the angular scattering patterns at high values of ϕ resemble the dielectric needle approximation which, in turn, is derived from the first partial wave.

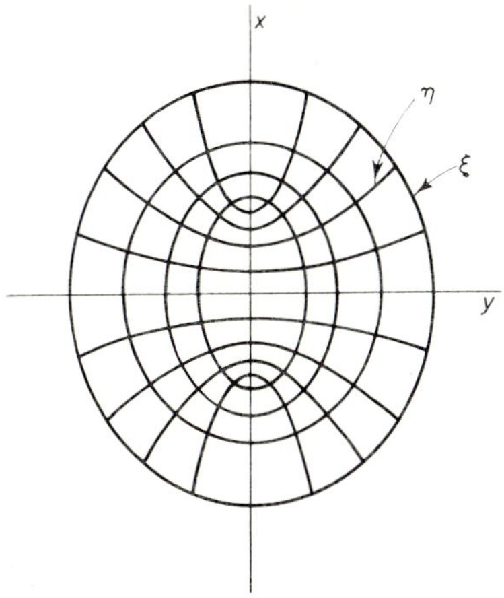

FIG. 6.29. Coordinates of the elliptic cylinder. Curves of constant ξ and η give the loci of an ellipse and of a hyperbola, respectively.

6.6 Noncircular Cylinders

6.6.1 ELLIPTICAL CYLINDERS

Scattering by an elliptical cylinder can be set up as a boundary value problem along the lines already used for the sphere and the circular cylinder. Elliptical cylindrical coordinates (ξ, η, z), shown in Fig. 6.29, are used. These are defined by

$$x = q \cosh \xi \cos \eta \tag{6.6.1}$$

$$y = q \sinh \xi \sin \eta \tag{6.6.2}$$

$$z = z \tag{6.6.3}$$

when $0 \leqslant \xi < \infty$; $0 \leqslant \eta \leqslant 2\pi$. The distance between the two foci of the elliptical cross section is $2q$. The surfaces of constant ξ are confocal elliptical cylinders and those of constant η are confocal hyperbolic cylinders.

The wave equation is separable in this coordinate system, having solutions of the form $R(\xi)\Theta(\eta)e^{-ihz}$ where R and Θ are solutions of the modified Mathieu differential equations given by

$$(d^2R/d\xi^2) = (c - 2\gamma^2 \cosh 2\xi)R = 0 \tag{6.6.4}$$

$$d^2\Theta/d\eta^2 = (c - 2\gamma^2 \cos 2\eta)\Theta = 0 \tag{6.6.5}$$

c being the separation constant and

$$\gamma^2 = k^2q^2/4 \tag{6.6.6}$$

The field of the plane incident wave can be expressed as an expansion in these functions and the scattered and internal fields are expanded similarly with unknown expansion coefficients. These are determined via the boundary conditions applied at the boundary surface $\xi = \xi_0$.

The solution for the perfectly reflecting cylinder was given as early as 1908 by Sieger and then again by Morse and Rubinstein (1938). Some numerical results have been presented by Barakat (1963) who has also extended the theory to include incident cylindrical waves (Barakat, 1965). Burke and Twersky (1964) have given a power series expansion valid when the linear dimensions of the cylindrical cross section are small compared to the wavelength and have also given some numerical results for this case [see also Burke *et al.* (1964)]. Germey (1964) has given the solution for two perfectly reflecting elliptic cylinders whose axes are parallel to each other but whose cross sections are arbitrarily oriented with respect to each other and to the perpendicularly incident plane wave. The approach is similar to that already discussed in the case of two scattering neighboring spheres in Section 5.7. Germey (1966) has also carried out the exercise of studying scattering by a perfectly reflecting elliptic cylinder adjacent to a parallel dielectric circular cylinder.

The dielectric cylinder has been solved for both perpendicular (Yeh, 1963) and oblique incidence (Yeh, 1964a, b). This becomes a cylinder of very small thickness or a "ribbon" in the limiting case $\xi_0 \to 0$.

The results for the dielectric elliptic cylinder are considerably more complicated than for either the circular dielectric cylinder or the perfectly conducting elliptic cylinder in that each expansion coefficient of the scattered or transmitted wave is coupled to all coefficients of the series expansion for the incident wave. Yeh (1963, 1965) has presented the results of calculations of $m = 2.0$ for a variety of sizes, shapes, and angles of incidence.

6.6.2 Arbitrary Cross-Sectional Shape

We have already discussed the "point matching" method for calculating scattering by nonspherical particles or by noncircular infinite cylinders. Richmond (1965, 1966) has described still another technique by which the scattered fields of dielectric cylinders of arbitrary cross-sectional shape may be determined accurately. It involves first obtaining the internal electric field intensity. This can be interpreted as an equivalent current density or "polarization current" which in turn generates the scattered field.

The total electric field intensity throughout all of space is represented as the sum of the incident field and a scattered field. When the electric vector of the incident wave is parallel or perpendicular to the cylinder axis, the scattered wave will be polarized in the same sense. Each increment of the scattered field may be generated by an equivalent electric current filament in the cylinder. Each such elemental current contributes to the scattered field. Accordingly, if the total field inside the cylinder is known, the scattered field can be calculated. Richmond has used this procedure to obtain solutions which lead to numerical results for circular shells in agreement with the boundary value solution.

6.7 Experimental Results

Analysis of the radiation scattered by cylindrical plasmas in the microwave region has been proposed as a diagnostic tool to study electron densities in plasma columns in the laboratory and some experiments have been undertaken [e.g. Stern (1963), Shiobara (1965), Jones and Wooding (1965, 1966), Faugeras (1966)]. However there have, as yet, been no definitive results. In the optical region, following exploratory work with a spider fiber (Matijević *et al.*, 1961), a comprehensive study with submicron glass and fused silica fibers has been reported by Farone and Kerker (1966).

In Fig. 6.30 the circles show typical experimental results for a silica fiber, illuminated at perpendicular incidence with unpolarized light of wavelength 546 mμ. These are compared with theoretical calculations shown as smooth curves for six combinations of the refractive index, m, and the size parameter, α. The ordinate is the ratio of the intensity of the TE mode (electric vector vibrating perpendicular to the cylinder axis) to that of the TM mode. For monochromatic radiation, this polarization ratio is, accordingly, given by

$$\rho = i_2/i_1 \tag{6.7.1}$$

The results are striking. The best fit between the experiments and theory is for $m = 1.46$ and $\alpha = 4.00$ as depicted in the lower center area of Fig. 6.30. When $\alpha = 3.98$ or 4.02, the theoretical results differ from the experiments as shown. Similarly, there are differences between theory and experiment for $\alpha = 4.00$, but with $m = 1.45$ or 1.47. In effect, as these diagrams show, this represents a determination of the size of this fiber to an accuracy of about 0.25%. This is remarkably accurate in comparison with size determination in this range by electron microscopy, which can hardly be effected to an accuracy of better than about 5%. Additional data obtained at $\lambda = 436$ mμ led to a value of $\alpha = 5.00$. The size parameters $\alpha = 4.00$ and 5.00 obtained at the respective wavelengths $\lambda = 546$ mμ and 436 mμ correspond to

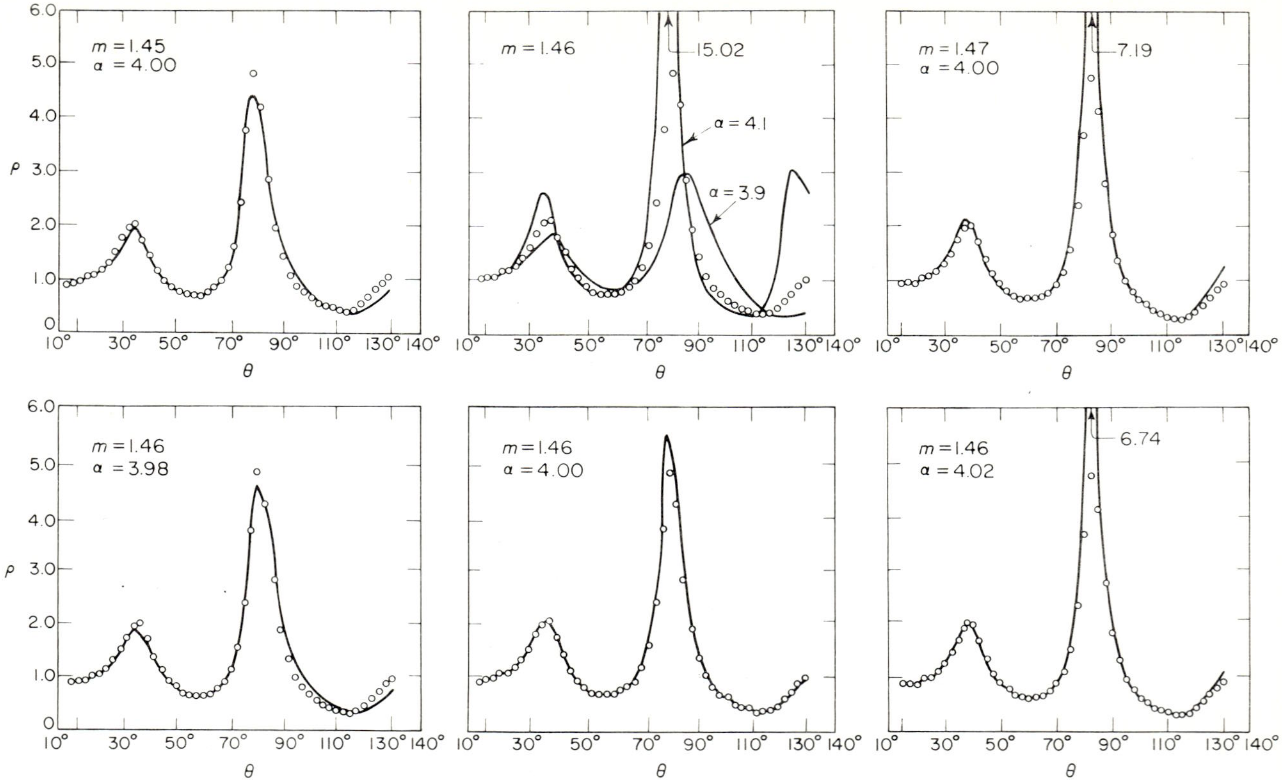

FIG. 6.30. Polarization ratio ($\rho = I_2/I_1$) plotted against θ for a Pyrex cylinder illuminated at perpendicular incidence. Comparison of experimental results (circles) and theory (smooth curves) for a silica fiber at $\lambda = 0.546\ \mu$ (Farone and Kerker, 1966).

cylinder radii of 348 and 347 mμ respectively. These experiments constitute both a verification of the scattering theory and a justification of the use of scattering as a diagnostic tool under these experimental conditions.

Additional experiments have been carried out by Lundberg (1969) using a helium-neon laser with much larger glass cylinders (6.27 to 49.9 μ in diameter). For these sizes, there are a larger number of oscillations in the angular scattering curves. It was possible to fit the experimental data with theoretical curves over short ranges of angle. The inability to represent the experimental data over all scattering angles with a single set of parameters may arise from the sensitivity of the scattering at these large values of α to the parameters m and α, coupled with geometrical and optical imperfections in the fibers.

Cooke and Kerker (1968, 1969) have obtained experimental results at oblique incidence using a specially constructed apparatus. This is illustrated in Figs. 6.31 and 6.32, which show the polarization ratios plotted against

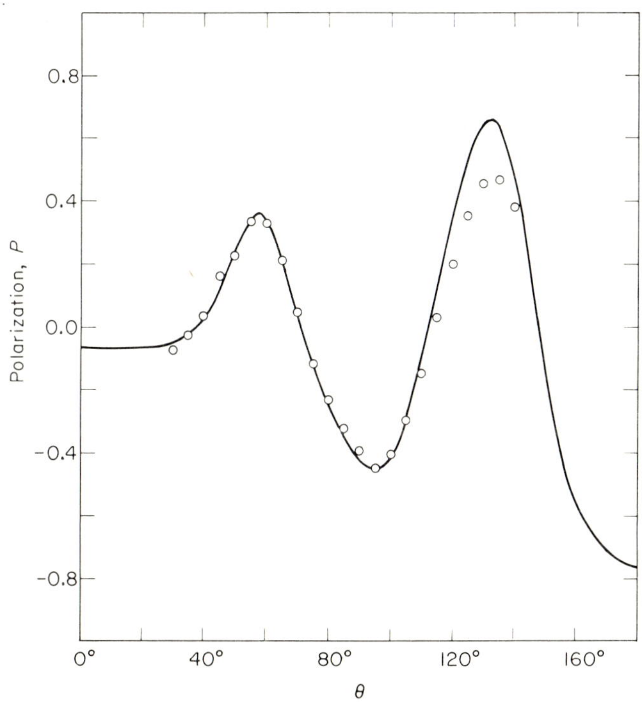

FIG. 6.31. Polarization $[P = (I_2 - I_1)/(I_2 + I_1)]$ plotted against θ for a Pyrex cylinder illuminated at perpendicular incidence. Comparison of experimental results (circles at $\lambda = 0.546\ \mu$) and theory for $\alpha = 3.02$ (smooth curve) (Cooke and Kerker, 1968).

angle of observation for $\phi = 0°$ and $\phi = 30°$, respectively. The two sets of data can be fitted to theoretical results corresponding to $\alpha = 3.02$ and 3.01, respectively. The agreement was not as good for larger angles ($\phi = 45°$ and 60°), probably because of imperfections in the glass fibers.

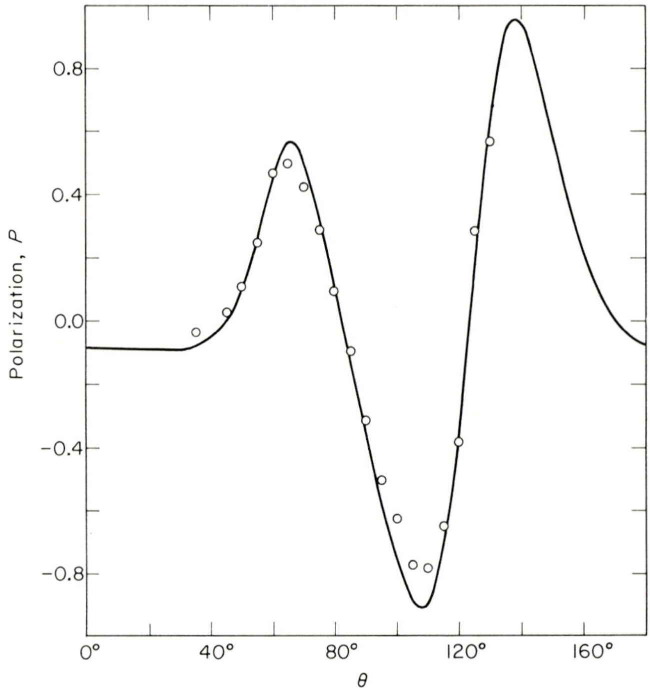

FIG. 6.32. Polarization $[P = (I_2 - I_1)/(I_2 + I_1)]$ plotted against θ for the same Pyrex cylinder as shown in Fig. 6.31 illuminated at oblique incidence ($\phi = 30°$). The theoretical curve which best goes through these points (smooth curve) is for $\alpha = 3.01$ (Cooke and Kerker, 1968).

CHAPTER 7

Analysis of Particle Size

7.1 Introduction

This chapter turns from theory to more practical matters. We have seen how the scattering of electromagnetic radiation by spheres and by infinitely long cylinders depends upon particle size, wavelength, and refractive index or, in the case of inhomogeneous particles, upon the radial profile of the refractive index. Now we are concerned with the inverse problem of determining the particle size from scattering data, especially the size of colloidal spheres. High polymer molecules which can be treated either as Rayleigh scatterers or Rayleigh–Debye scatterers will be considered later. The few optical experiments with long cylinders have already been described in the last chapter.

Light scattering offers advantages over some of the other techniques for particle size analysis such as optical and electron microscopy, sedimentation, centrifugation, filtration, diffusion, etc., because the systems under study can be observed *in situ* without significant perturbation. The method is absolute in the sense that the theory permits reduction of the data directly to the final desired results without the need of secondary schemes for calibration. The number of particles under simultaneous observation is usually sufficiently large so that in the case where there is a distribution of sizes, representative sampling is obtained automatically. Measurements are almost instantaneous and can be recorded continuously so that rate processes may be followed. Furthermore, a variety of model systems has been available with which to develop and test the various laboratory and analytical techniques. Indeed, much of the work heretofore has been of an exploratory nature with such model systems. One might expect that eventually the method will find wider use in research and industry with systems more commonly encountered in practice.

One of the deterrents to the application of light scattering to the size analysis of spherical colloidal particles has been the complexity of the numerical computation of the scattering functions. Even the extensive publication of tables of scattering functions hardly provided sufficient values to cover the range of parameters normally encountered. Yet, continued tabulation could not go on indefinitely, since the sheer bulk was self-defeating. The tables themselves became impossibly unwieldy.

The solution to this dilemma has been the availability of increasingly fast high-speed digital computers with larger memory banks. These can be used not only to calculate the specific values of the scattering functions as needed, but also to apply them directly in the reduction of the experimental data. The scattering functions can then either be stored on punched cards or magnetic tape or calculated anew for each analysis and then discarded. In any case, they remain within the computer center, intermediate between the experimental data and the final result, so that the analyst may never see them.

A second difficulty in the use of light scattering arises because of the multivaluedness of the scattering data with respect to the particle size, so that it becomes important to consider the uniqueness of a particular solution. Mirelés (1966) has shown in the case of cylinders that for each particular size there is a unique angular scattering pattern corresponding to each refractive index. However, for a polydisperse system there can still be many size distributions that lead to a particular set of data, at least within the experimental uncertainty. This is frequently a serious problem and places severe limitations on the kinds of measurements that can be utilized, as well as on the range of sizes and the width of the size distributions that are amenable to accurate treatment.

In particle size analysis, it is usually assumed that the scattering by an array of particles is incoherent so that the scattering functions corresponding to an isolated particle may be used and also that there is no significant multiple scattering. The cumulative effect is obtained by adding the intensity scattered by each particle as if it were present alone. In order for these conditions to apply, the particles must be randomly positioned in space and the system must be sufficiently dilute.

When the particles are very close to each other, their polarization fields interact, and it is necessary to match the electromagnetic fields at both particle boundaries simultaneously. This results in different scattering functions from those of the isolated particle. We have already referred to Trinks' (1935) solution for two adjacent spheres, as well as to the recent work of Olaofe and Levine (1967).

Even when the particles are more than several diameters apart so that mutual polarization is absent, some of the radiation scattered by a particular particle will be incident upon a second particle, which then rescatters it.

This is multiple scattering. It is usually possible to dilute the system sufficiently in order to avoid both mutual polarization and multiple scattering. The practice is either to dilute until the scattering is directly proportional to the particle concentration or to extrapolate data obtained at several concentrations to infinite dilution. Then, these results can be interpreted in terms of single particle scattering.

It is rather surprising that there has been so little experimental work in the opposite direction. It should be possible to study mutual polarization and multiple scattering by carrying out experiments at high concentration for a system whose single scattering behavior is well understood (Churchill *et al.*, 1960; Woodward, 1964; Smart *et al.*, 1965).

Whenever the particles are not randomly positioned, the interference of the scattered radiation among the individual wavelets is such that the intensities do not add. Since this usually arises from long-range interactions between the particles, it can also often be minimized by working at high dilution. On the other hand, for systems such as pure substances at the critical point, solutions of polyelectrolytes, plasmas, etc., the deviation from incoherent scattering can provide a powerful tool for studying such interactions.

7.1.1 Verification of the Theory

Many of the early light-scattering workers declared that one of their main aims was to check the validity of the theory. Thus, the pioneering report of La Mer and Sinclair (1943) on light scattering by aerosols was entitled "Verification of the Mie Theory," and as late as 1960, Gucker reported that the object of his work, extending from the period of World War II, in developing highly sophisticated light-scattering instrumentation for aerosols, was that this "enables a quantitative test of various scattering theories" (Gucker and Rowell, 1960). The literature is replete with similar references to experimental verification of the theory.

What is sought in all of this work is to establish a concordance between the predictions of theory and experimental observations. However, the establishment of such concordance has a dual function. It not only constitutes a test of the theory, but just as importantly, and sometimes even more importantly, it may serve to confirm assumptions about the physical constitution of the scatterer or to justify the design and execution of the experiments. This is particularly true when the underlying physical laws are as soundly established as Maxwell's theory of electromagnetism. Then, concordance between theory and experiment develops confidence in the experimental data so that these may be interpreted to obtain physical information about the system under investigation—in this case, the size, geometry, and optical properties of the scatterers. It is precisely this information which is often the unknown factor in colloidal systems.

On the other hand, with electromagnetic waves in the millimeter and centimeter range (microwaves), objects with dimensions comparable to the wavelength can be easily fabricated and manipulated. For this reason, scattering of microwaves by cylinders and spheres provides the best opportunity to test the theory directly. The results of some of this microwave work will be reviewed, and then some recent light-scattering work, which constitutes an example of concordance, will be considered.

a. Microwave Scattering. Aden (1950, 1951) measured the backscatter from both perfectly conducting (aluminum) and partially conducting (water) spheres at $\lambda = 16.230$ cm. He utilized the standing-wave method rather than direct radar intensity measurements, which are not as practical as a laboratory method. The interaction between the incident and scattered fields along the lines between the scatterer and the source causes standing waves to be set up in space. These can be described in terms of a reflection coefficient which, in turn, is related in a simple way to the backscatter efficiency.

In Aden's work, the scattering object was a hemisphere exposed over a large-image screen, which by image theory can be interpreted as equivalent to a complete sphere in free space. The 18 metal hemispheres ($m = \infty$), ranging from 0.500 to 3.875 in. in radii, were carefully machined from aluminum. Their corresponding values of α were from 0.49 to 3.81. The water was contained in hemispherical shells of styrofoam mounted on aluminum disks. Styrofoam is a sufficiently low dielectric material so that it is effectively transparent to microwaves. The radii of the thirty hemispherical containers ranged from 0.75 to 6.00 in. The refractive index of the water at 20° was taken to be $m = 9.01 - 0.43i$.

The experimental results for the backscatter efficiency, together with the theoretical curves, are shown in Fig. 7.1 for the aluminum and Fig. 7.2 for the water. The phase shift is shown in Fig. 7.3. The agreement is well within the experimental error so that, inasmuch as these experiments are capable of doing so, they have verified the theory.

Other experiments with spheres confirm these results. For example, Atlas *et al.* (1960) obtained radar echoes from aluminum and plexiglas ($m = 1.60$) spheres suspended from a tethered balloon at a range of 1.5 miles. The total scattering efficiency from lucite spheres ($m = 1.603$), as well as the lateral (bistatic) scattering, has been determined by Greenberg, *et al.* (1961) with an outside accuracy of 3%. Mevel (1960) has done similar work with conducting spheres. Among the measurements on long (infinite) cylinders consisting of perfectly reflecting, absorbing, and dielectric materials, mention

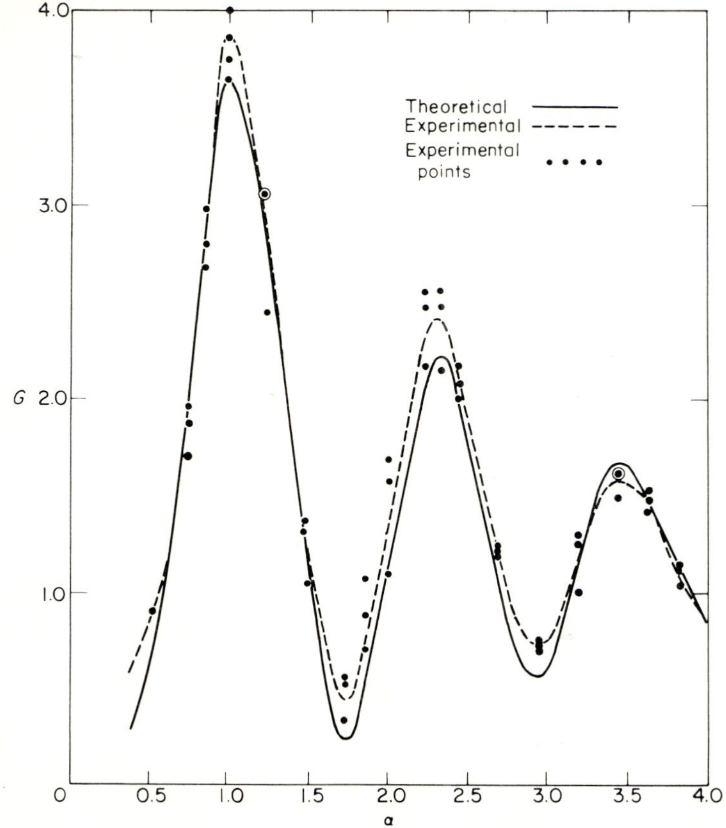

FIG. 7.1. Backscatter gain for aluminum spheres at $\lambda = 16.23$ cm; comparison of theory and experiment (Aden, 1951).

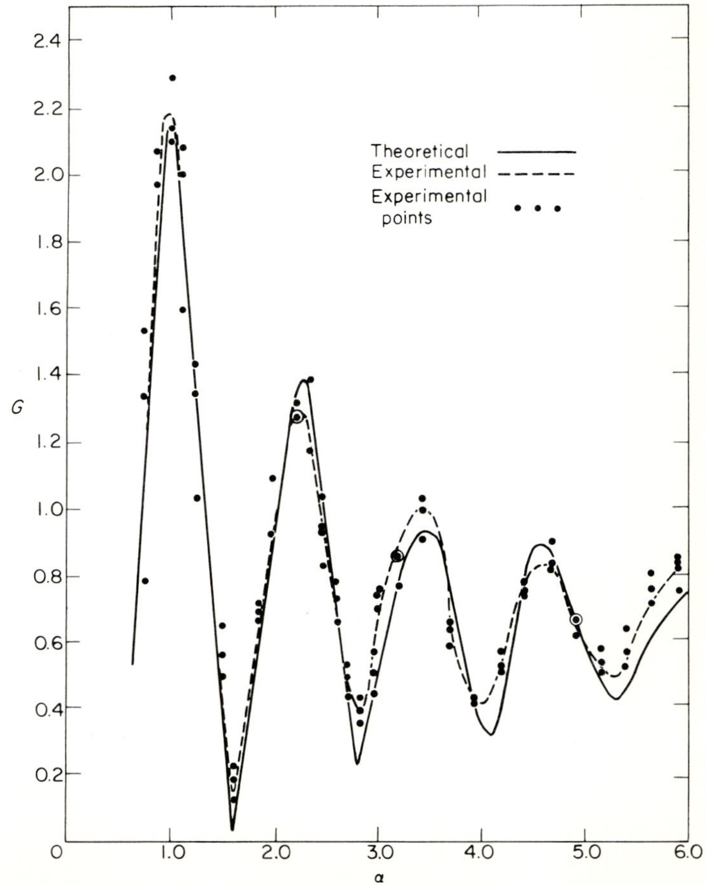

FIG. 7.2. Backscatter gain for water spheres at $\lambda = 16.23$ cm; comparison of theory and experiment (Aden, 1951).

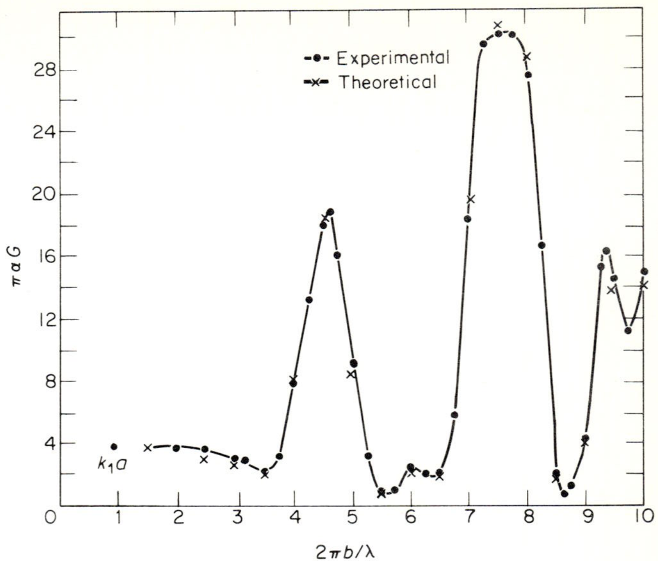

FIG. 7.4.(top) Backscatter gain (plotted as $\pi\alpha G$) of a dielectric coated conducting cylinder $\alpha = 0.942$, $\lambda = \pi$, $m_2{}^2 = 2.54$; comparison of experiment and theory. Total cylinder radius is b (Tang, 1957).

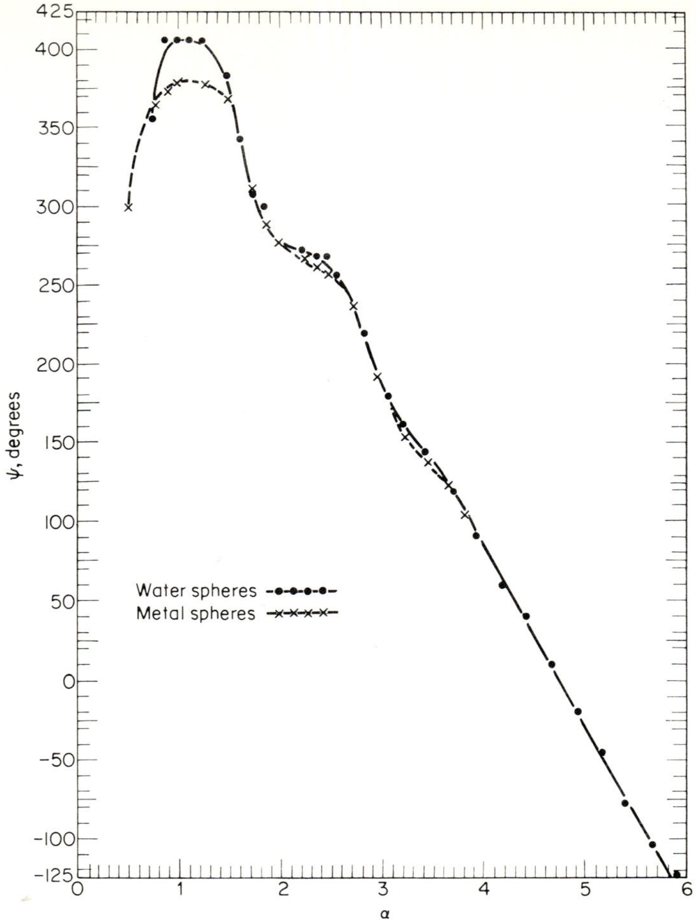

FIG. 7.3.(left) Phase shift for aluminum and water spheres at $\lambda = 16.32$ cm: comparison of theory and experiment (Aden, 1951).

may be made of the near-field scattering in the forward direction by Kodis (1952), Wiles and McLay (1954), Subbarao and McLay (1956), McLay and Subbarao (1956), and Adey (1955, 1956b). Comparison of measurements with theory has also been carried out for dielectric-coated, perfectly reflecting cylinders by Adey (1956a) and by Tang (1957). Tang estimates that the absolute accuracy of his measurements is at least 1%. A typical example of the remarkable agreement between his experiments and the theory is illustrated in Fig. 7.4.

b. Light Scattering. Kratohvil and Smart (1965) have shown that the absolute scattering intensity for a dispersion of polystyrene spheres can be accurately represented by theory. The sphericity of the particles in this particular suspension had long been established by electron microscopy and their optical isotropy was verified by polarization measurements. The particle concentration was determined gravimetrically. Because of the experimental uncertainties in the electron microscopy, this technique only provided a close first approximation to the size distribution. The final theoretical calculations of the light scattering were obtained by adjusting the parameters in the size distribution until there was an optimum fit between the measured and calculated values.

The results for a particular latex, plotted as the angular intensity functions i_1 and i_2 versus the scattering angle, are shown in Figs. 7.5 and 7.6. The circles are the experimental points for $\lambda = 0.546\ \mu$, and the curve represents the theoretical values corresponding to a size distribution of the form given by (7.5.16). The squares and triangles correspond to theoretical calculations for monodisperse systems with $\alpha = 9.8$ and 10.0, respectively.

7.2. Colloids with Narrow Size Distributions

A dispersion is said to be monodisperse when all of the particles are of the same size. Although such systems are rare, a variety of so-called monodisperse colloids have been prepared, and these can then be used to test and develop techniques for particle size analysis. Actually, as these techniques have become more refined, it has been possible to determine the distribution of sizes in systems which had previously been considered monodisperse. For this reason, we speak here of colloids with narrow size distributions. This section constitutes a digression from the actual problem of scattering in order to describe the preparation of several such colloids which have served as model systems in scattering studies.

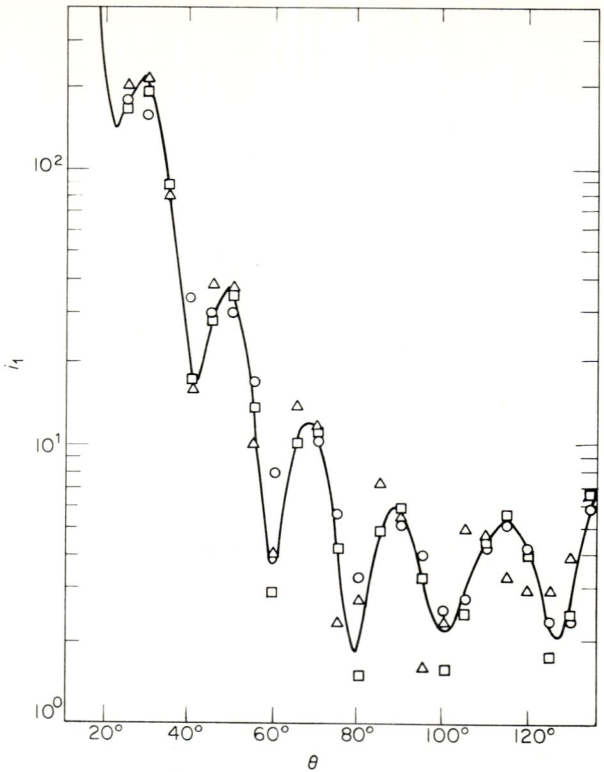

FIG. 7.5. Perpendicular component of angular scattering for a suspension of polystyrene spheres (circles) compared with theory for $\alpha = 9.8$ (squares), $\alpha = 10.0$ (triangles), and with a size distribution (Eq. 7.5.16) for which $\alpha_M = 9.8$, $\sigma_0 = 0.02$ (smooth curve) (Kratohvil and Smart, 1965).

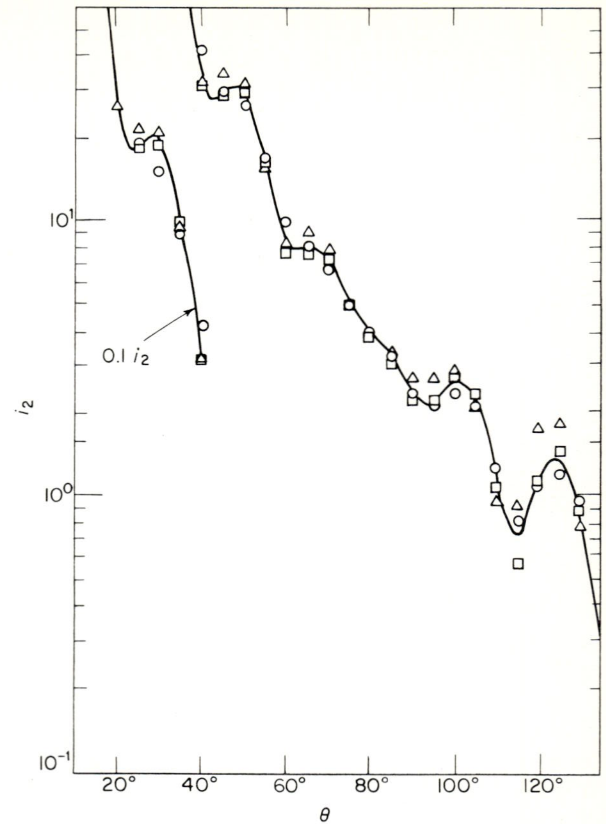

FIG. 7.6. Same as Fig. 7.5 for the parallel component.

7.2.1 AEROSOLS

Sinclair and La Mer (1949) developed a method for the production of aerosols consisting of submicron and micron-size liquid droplets of very uniform particle size which involves condensation from a mixture of vapor and nuclei in a temperature gradient. Their original generator is shown in Fig. 7.7.

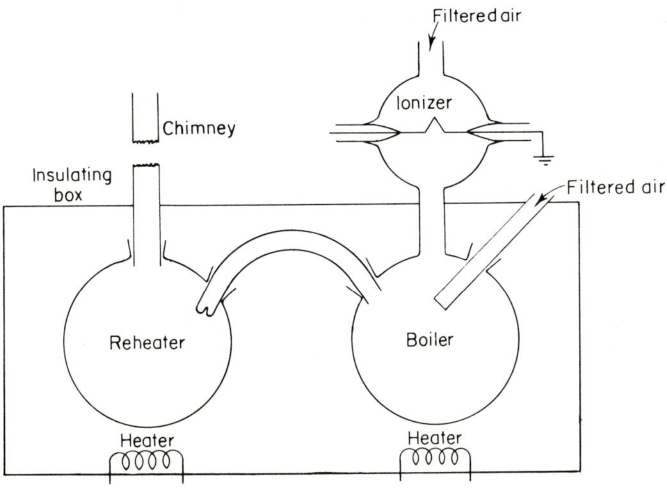

FIG. 7.7. Homogeneous liquid aerosol generator of Sinclair and La Mer (1949).

The substance from which the aerosol is to be formed is contained in the flask designated as the "boiler," which is generally maintained at about 100 to 200°C. The condensation nuclei may be formed by a high-voltage electric spark as shown in the figure, by evaporation of sodium chloride from an electrically heated coil of wire, by dehydration of an aerosol produced by atomizing a solution of sodium chloride (Wilson and La Mer, 1948), or by a volatile nucleating agent such as the oxides of nitrogen (Kerker *et al.*, 1955), or H_2SO_4 (La Mer and Gordieyeff, 1950). The "reheater" is maintained at about 300°C and serves to evaporate any spray carried over from the boiler and also to permit better mixing of nuclei and vapors. The "chimney" is an air condenser in which the aerosol is formed by gradual condensation upon the nuclei along the temperature gradient. Two filtered air streams enter the boiler, one through the nucleating chamber, and the other is bubbled directly through the hot liquid. The concentration of nuclei is controlled by the relative flow rates of these streams, the total rate of flow usually being from 1 to 4 l./min. The aerosol leaving the chimney may be diluted in order to reduce the rate of coagulation and to minimize multiple scattering.

This generator is capable of producing aerosols in which the standard deviation of the particle radius is about 10 to 20% of the average radius, which, in turn, may vary typically from 1 to 2 μ down to 0.01 μ and even less. The intensity of scattering for the smallest particles ($r < 0.05\ \mu$) is so low that they do not give a noticeable Tyndall beam. However, these particles can be grown by exposing them to the vapors of a solvent with which they are miscible. Thus, sulfuric acid particles will grow in the atmosphere of a more dilute solution of sulfuric acid by absorption of water vapor until the particles and the ambient solution have equilibrated. The growth ratio can be calculated and the size of the grown particles determined by one of the light-scattering techniques to be considered below (La Mer *et al.*, 1950; Coutarel *et al.*, 1967).

In general, the size of the aerosol particles increases with boiler temperature and decreases with nuclei concentration. The aerosol concentrations are typically about 10^5 to 10^6 particles per cc. Uniform aerosols of this type can be produced from a variety of high-boiling liquids, such as oleic acid, triphenyl phosphate, dibutyl phthalate, sulfuric acid, etc. This generator can also be used for the preparation of mercury aerosols or of aerosols of volatile solids such as sulfur (Kerker *et al.*, 1955). Wachtel and La Mer (1962) have prepared emulsions by passing oil aerosols from this generator through water containing an emulsifier.

A number of other aerosol generators which operate by condensation of vapor have been designed. For example, the generator of Rapaport and Weinstock (1955) uses a nebulizer for primary formation of an aerosol, which is subsequently mixed with nuclei and then evaporated on passing through a heated tube. An aerosol of uniform particle size is finally produced upon condensation of the vapor. This generator permits higher flow rates (up to 10 l./min.) than that of Sinclair and La Mer and also provides higher mass concentrations. Improvements in this generator which give better stability and reproducibility of the output have been described by Preining (1962) and, particularly, by Liu *et al.* (1966). The latter workers have found that the particle size of the emergent aerosol can be precisely controlled by using a solution of the aerosol-forming material in a volatile solvent such as ethyl alcohol. Then, the radius of the particles varies with the one-third power of the concentration.

A schematic diagram of an aerosol generator used by Matijević *et al.* (1962) is shown in Fig. 7.8. The nuclei, consisting of a salt such as sodium chloride or silver chloride, are picked up from a boat (4) contained in a combustion tube (3) by a stream of filtered helium (13) and then passed through a flask (8) containing the liquid from which the aerosol is to be made. This flask, which serves as the boiler, is maintained at constant temperature by an oil-bath thermostat (7). The aerosol is formed by condensation onto the nuclei upon emergence of the mixture from the boiler. A variety of by-passes and valves permits dilution of the nuclei and of the aerosol. This generator has the advantage of better temperature control of the nucleator and boiler and can be operated more reproducibly than the La Mer–Sinclair generator.

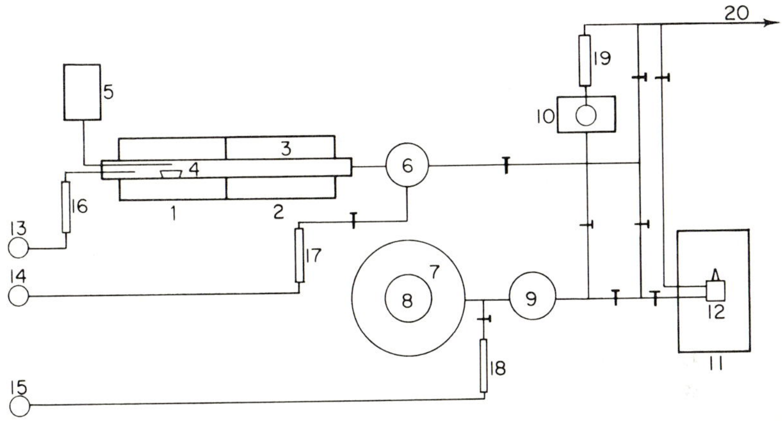

FIG. 7.8. Homogeneous liquid aerosol generator of Matijevic *et al.* (1962). (1) Boiler furnace for AgCl, (2) reheater furnace for AgCl, (3) combustion tube, (4) combustion boat with AgCl, (5) thermocouple and potentiometer, (6) and (9) viewing chambers, (7) oil bath, (8) linolenic acid evaporator, (10) thermopositor collector, (11) light-scattering instrument, (12) light-scattering cell for aerosols, (13)–(15) helium sources, (16)–(19) flowmeters, (20) exhaust line.

Many of the nuclei described above are themselves aerosols consisting of small solid particles, and their size may be increased to the micron range. Thus, when the furnace of Fig. 7.8 is operated at a sufficiently high temperature, the nucleator will generate an aerosol whose particles may be as large as several microns in diameter (Matijevic *et al.*, 1960). By growing these particles successively in a series of furnaces such as shown in Fig. 7.9 (Espenscheid *et al.*, 1964b), aerosols consisting of solid spheres of NaCl or AgCl of quite uniform size are obtained. These particles consist either of spherical aggregates of very small microcrystals or are glass-like since, unlike single crystal aerosol particles, they do not exhibit diffraction lines either in an electron or X-ray beam. Accordingly, they provide still another model system for carrying out light-scattering studies. Espenscheid *et al.* (1965) have combined the features of the generator for solid particles with that for liquid particles to prepare aerosols consisting of spherical cores of AgCl coated by a concentric shell of linolenic acid. In effect, the large AgCl particles act as nuclei for the linolenic acid. These particles are particularly convenient to serve as a model for scattering by concentric spheres because their refractive indices are quite different ($m_1 = 2.105$ for AgCl; $m_2 = 1.482$ for linolenic acid) and also because linolenic acid can be fixed for electron microscope observation so that the size can be determined independently by this latter technique. Still another generator has been designed by Jacobsen *et al.* (1967) to operate at furnace temperatures as high as 1500°C. This was used to obtain aerosols consisting of amorphous spheres of vanadium pentoxide.

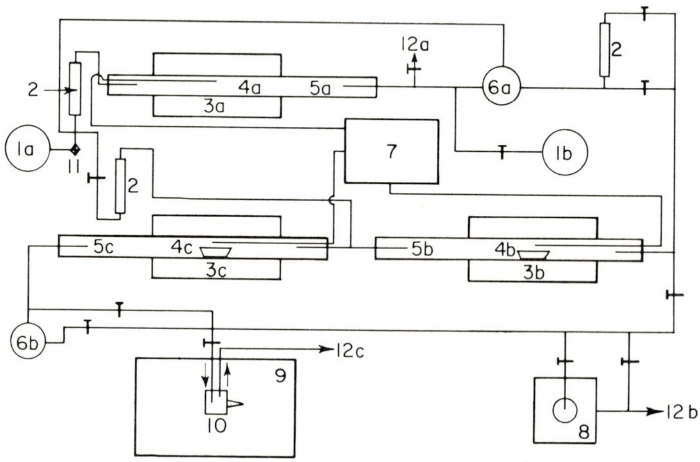

FIG. 7.9. Homogeneous solid aerosol generator of Espenscheid *et al.* (1964b). (1a, b) helium sources, (2) flow meters, (3a–c) combustion furnaces, (4a–c) combustion boats, (5a–c) combustion tubes, (6a, b) viewing chambers, (7) potentiometer, (8) thermopositor, (9) light-scattering photometer, (10) light-scattering cell, (11) gas filter, and (12a–c) exhausts.

Aerosols of uniform particle size can also be prepared by nebulizing a suspension of a monodisperse hydrosol with subsequent dehydration to give an aerosol of the dried suspended particles (O'Konski, 1955). The hydrosol must be sufficiently dilute so that there is only a small probability of having two colloidal particles per liquid droplet. There are available a number of monodisperse synthetic latexes which are useful for this purpose, among the most common of which are polystyrene and poly (vinyl toluene). These may be prepared by emulsifying the monomers in water with the aid of a detergent and then polymerizing the monomer with a catalyst. The latex formed consists of particles which, within the limits of electron microscope resolution, are perfect spheres and when grown under appropriate conditions (Vanderhoff *et al.*, 1956) are of uniform size. They have been prepared with particle radii ranging from about 0.01 to nearly 1.0 μ and with standard deviations, as determined electron microscopically, in the neighborhood of 1 to 5% (Bradford and Vanderhoff, 1955, 1963). Langer and Pierrard (1963) have pointed out that, in addition to the latex particles, there are frequently smaller residual particles formed by the stabilizer associated with the latex.

Stern *et al.* (1959) have used a similar nebulizer-dehydration technique to form aerosols of monodisperse spheres consisting of T3E coliphage and Type 3 poliomyelitic virus. The particle radii and standard deviations of the resultant aerosols were 35 ± 1.4 mμ and 26 ± 1.3 mμ, respectively, as determined by electron microscope count.

In the above examples, the particles which comprise the aerosol are already contained in the colloidal dispersion, which is then atomized. If a liquid solution containing a nonvolatile solute is nebulized, solid aerosol particles will form after evaporation. Then, the distribution of sizes is determined by the drop-size distribution of the spray and the concentration of the solution. Although most nebulizers produce a rather broad distribution of droplet sizes, this tends to be truncated at the lower end. It is possible to truncate the upper end of the aerosol as well by passing it through an impactor consisting of a metal plate containing many fine holes. Such nebulizer-impactor combinations are capable of generating quite monodisperse aerosols.

Atomization may also be accomplished by feeding a liquid onto the center of a sharp-edged rapidly rotating disk from which the liquid is spun off as a mixture of primary droplets and a considerably larger number of smaller satellites. The satellites may be separated out to give a uniform aerosol. Disk speeds of about 70,000 to 85,000 rpm are generally used, although some workers have gone as high as 150,000 rpm. Whitby *et al.* (1965) have reviewed some of the extensive literature on both the nebulizer-impactor and the spinning-disk aerosol generators.

7.2.2 Hydrosols

The most convenient hydrosol suspensions for exploratory light-scattering studies are the polymer latexes which have just been described since they consist of isotropic spheres with a very narrow distribution of sizes. Furthermore, various sizes can be prepared, which can then be mixed to simulate a variety of size distributions. The latex is stored as a concentrate which can be diluted sufficiently to eliminate multiple scattering, or, alternatively, when the scattering is studied as a function of concentration, it can provide a system for the experimental study of multiple scattering. The particle concentration for a latex of a given size can be determined from the mass concentration and the density of the latex particles. Pugh and Heller (1957) have reported the density of polystyrene particles to be 1.057 ± 0.002, and of poly (vinyl toluene) to be 1.026 ± 0.001. The refractive indices of these materials at $\lambda = 5461$ Å are 1.602 and 1.586, respectively (Heller and Pugh, 1957). Somewhat lower values have been proposed more recently for polystyrene (Smart and Willis, 1967). The Cauchy dispersion formula for the polystyrene is (Bateman *et al.*, 1959)

$$n = 1.5683 + 10.087 \times 10^{-11}/\lambda_0^2 \tag{7.2.1}$$

In order to obtain the relative refractive index in aqueous suspension, the formula for water must also be used.

$$n = 1.324 + 3.046 \times 10^{-11}/\lambda_0^2 \tag{7.2.2}$$

Here, λ_0 is the vacuum wavelength in centimeters. This gives a refractive index relative to water which is close to 1.20 so that values of the angular intensity functions and of the scattering efficiency can be obtained by interpolation from tables prepared by Pangonis *et al.* (1957), Pangonis and Heller (1960), and by Denman *et al.* (1963).

The hydrosol with which La Mer and his co-workers carried out their extensive light-scattering studies consists of a suspension of sulfur in water (La Mer and Barnes, 1946; Barnes and La Mer, 1946). When illuminated with white light, the scattered light appears brilliantly colored, the particular colors varying with the angle of observation. This effect, which has been termed high-order Tyndall spectrum (HOTS), may also be observed with other colloids of narrow size distribution. The theory will be considered later in this chapter.

The La Mer sulfur sols are prepared by rapid mixing of sodium thiosulfate with mineral acids to give a final concentration of about 0.002 M of each component. The principal reaction is

$$H^+ + S_2O_3^{=} \rightarrow HSO_3^- + S \tag{7.2.3}$$

although there are also various side reactions involving the formation of polythionates (Zaiser and La Mer, 1948; Smellie and La Mer, 1954; Kerker, 1951). The reaction proceeds sufficiently slowly at these concentrations so that colloidal sulfur does not precipitate until after about 30 min. After the initial precipitation, the continually generated sulfur diffuses to the existing particles, which then grow uniformly. The variation of the size distribution with time can be followed by light scattering (Kerker *et al.*, 1963a; Rowell *et al.*, 1968b). The growth may be stopped at any stage by destruction of the unreacted sodium thiosulfate through the addition of iodine.

Silica sols have also provided model systems for light scattering studies. They may be prepared by addition of acid to solutions of sodium silicate or, alternatively, by passing these solutions through appropriate ion exchange resins which remove hydroxyl ions. The tetrafunctional monosilicic acid,

$Si(OH)_4$, then polymerizes, forming a sol with approximately spherical particles ranging from 35 to 100 Å in radius (Dezelić *et al.*, 1960). The most widely used sol has been distributed by the Dupont Company for experimental studies under the trade name "Ludox."

Although hydrosols of lanthanum and lead iodate have not been widely used (Herak *et al.*, 1958), they could provide an extremely useful model system. The particles are quite spherical and have a narrow distribution of sizes. These sols are prepared by direct mixing of dilute solutions of lead nitrate with potassium iodate (0.001 to 0.003 N for both components), and of lanthanum nitrate (0.001 N) with iodic acid (0.03 N). Unfortunately, they are rather unstable and must be utilized shortly after formation.

7.3 Average Size from Transmission

The transmission methods described in this section are strictly valid only for monodisperse systems. Otherwise, only a particular moment of the size distribution will be obtained. Although this is only one of the parameters that describes the size distribution, it may be adequate for most purposes if the distribution is sufficiently narrow. Methods which are based upon a two-parameter distribution function will be treated later in this chapter.

The transmission to monochromatic radiation is defined by

$$T = I_l/I_0 = \exp(-NC_{\text{sca}}l) = \exp(-\tau l) \tag{3.2.29}$$

where N is the number of particles per unit volume, C_{sca} is the particle scattering cross section, l is the path length, and τ is the turbidity. For partially absorbing particles, C_{ext} would be used in place of C_{sca}. The medium is transparent, the system is sufficiently dilute so that multiple scattering may be neglected, and the particles are randomly positioned so that the intensity of scattering is simply an additive property of the number of particles. The latter condition implies the absence of particle interactions, which might result in some ordering of the system.

7.3.1 Rayleigh Scatterers

For dielectric particles which are sufficiently small so that Rayleigh's law holds,

$$C_{\text{sca}} = \frac{24\pi^3 V^2}{\lambda^4}\left(\frac{n^2 - 1}{n^2 + 2}\right)^2 \tag{3.2.24}$$

where n is the relative refractive index, λ is the wavelength in the medium, and V is the volume of the spherical particles. The turbidity, which represents the total scattering per unit volume or the loss in intensity by the primary beam as it traverses a unit length of the scattering volume, is given by

$$\tau = NC_{\text{sca}} \tag{7.3.1}$$

In general, the turbidity is obtained directly from transmission measurements with the aid of (3.2.29). However, for Rayleigh scatterers it may be evaluated alternatively from the intensity scattered at any particular angle

$$R_\theta = r^2 N I_{\text{u}} = (3/16\pi)\tau(1 + \cos^2 \theta) \tag{3.2.31}$$

where R_θ, the Rayleigh ratio at scattering angle θ, is the intensity per unit solid angle scattered by a unit volume in the direction θ when illuminated with an unpolarized incident beam of unit intensity.

It is often convenient to express the turbidity, relative to the volume fraction, ϕ, of the scattering material, by

$$\left(\frac{\tau}{\phi}\right) = 24\pi^3 \left(\frac{n^2 - 1}{n^2 + 2}\right)^2 \frac{V}{\lambda^4} \tag{7.3.2}$$

or, alternatively, relative to the weight fraction, ω', by

$$\left(\frac{\tau}{\omega'}\right) = 24\pi^3 \left(\frac{n^2 - 1}{n^2 + 2}\right)^2 \left(\frac{\rho_{12}}{\rho_2}\right) \frac{V}{\lambda^4} \tag{7.3.3}$$

where ρ_{12} and ρ_2 are the densities of the solution and the spheres, respectively. Here,

$$\phi = (\rho_{12}/\rho_2)\omega' = c/\rho_2 \tag{7.3.4}$$

where c is the concentration expressed as grams of solute per milliliter. When the concentration is expressed as c and the particle size as the molecular weight, M, the turbidity becomes (Heller, 1959a, b)

$$\left(\frac{\tau}{c}\right) = \frac{24\pi^3}{N_A} \left(\frac{n^2 - 1}{n^2 + 2}\right)^2 \frac{M}{\rho_2{}^2 \lambda^4} \tag{7.3.5}$$

where N_{A} is Avogadro's number. The quantity (τ/c) is called the specific turbidity. If the refractive index of the particles and the appropriate concentrations and densities are known, then measurement of the turbidity or of the Rayleigh ratio permits determination of the particle size or molecular weight from one of the above formulations.

The turbidity must be evaluated at a sufficiently low concentration where multiple scattering is not significant. Ideally, the data should be obtained at several low concentrations and extrapolated to zero. Also, the particles must be randomly positioned in space in order that the scattering from the individual wavelets be completely incoherent. If there are interparticle forces which tend to arrange the system in some order, this approach cannot be used.

When there is a distribution of particle size, it is the volume average volume

$$V_v = \sum N_i V_i^2 / \sum N_i V_i \tag{7.3.6}$$

and the weight-average molecular weight

$$M_w = \sum N_i M_i^2 / \sum N_i M_i \tag{7.3.7}$$

which are determined by this method. Here, N_i is the number of particles per unit volume having a volume V_i or a molecular weight M_i. The particle radius which is determined by this method has been termed the turbidity average (Beyer, 1959) and is given by

$$a_\tau = (\sum N_i a_i^6 / \sum N_i a_i^3)^{1/3} = [(3/4\pi) V_v]^{1/3} \tag{7.3.8}$$

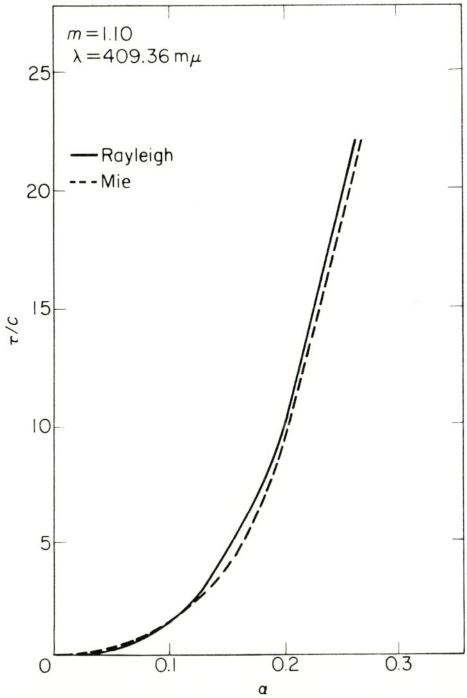

FIG. 7.10. The specific turbidity (τ/c) calculated using the Rayleigh theory (full line) and the exact theory (dashed line) when $m = 1.10$, $\lambda = 0.4094\,\mu$, and $\rho_{12} = \rho_2$ (Dezelić and Kratohvil, 1960b).

Despite the simplicity of this formulation, it has not been exploited frequently. Dezelić and Kratohvil (1960a, b; K. and D., 1962) determined the turbidity average particle radius of four samples of Ludox colloidal silica and compared the results with the average size as determined by the electron microscope. The transmission data were obtained with a conventional spectrophotometer, which was modified to avoid errors due to stray light and multiple scattering. The curves in Fig. 7.10 were used to obtain the size parameter, α, from the specific turbidity. The full line is for the Rayleigh formula, whereas the dotted line is based upon the exact calculations for spheres. The relative refractive index was 1.100. These curves were calculated for the "standard case" of $\tau_{0,s} = 546.1$ mμ and $\rho_2 = \rho_1$. Therefore, the experimental value of the specific turbidity must be converted to the standard value in order to apply these curves.

$$(\tau/\omega')_s = (\tau/\omega') \cdot (\rho_2/\rho_1) \cdot (\lambda_0/\lambda_{0,s}) \tag{7.3.9}$$

Some of the results are given in Table 7.1. The agreement with the electron microscope values is well within the experimental limits of the method. The radius determined from the electron micrographs is the turbidity average radius as determined from the experimental histograms.

TABLE 7.1

COMPARISON OF PARTICLE SIZE OF LUDOX DETERMINED BY TRANSMISSION WITH THAT OBTAINED BY ELECTRON MICROSCOPY[a]

λ_0(mμ)	m	Sample I	$a_w = 11.7$ mμ	Sample II	$a_w = 10.0$ mμ
		$(\tau/c)_0$	a_τ	$(\tau/c)_0$	a_τ
578	1.100	2.67	11.7	1.80	10.3
546	1.100	3.27	11.6	2.29	10.3
436	1.101	7.94	11.4	5.81	10.3
405	1.101	10.40	11.3	7.94	10.3
366	1.101	16.98	11.6	12.20	10.4
			av. 11.5		av. 10.3

[a] The values for a_τ have been corrected by the factor 1.013 [see Dezelić and Kratohvil (1962)].

7.3.2 SPECIFIC TURBIDITY METHOD

When the particles are sufficiently large so that Rayleigh's law of scattering is invalid, it is no longer possible to relate the particle size to the turbidity through simple equations such as (7.3.2) to (7.3.5), and, instead, the general expression for the scattering cross section must be used. The specific turbidity is then given by

$$(\tau/c) = (3\pi/2\lambda\rho_2)(Q_{\text{sca}}/\alpha) = (3/4\rho_2)(Q_{\text{sca}}/a) \tag{7.3.10}$$

and this will exhibit oscillations with particle size similar to those of Q_{sca}. In Fig. 7.11, the specific turbidity for $m = 1.10, 1.20, 1.30$ is plotted up to

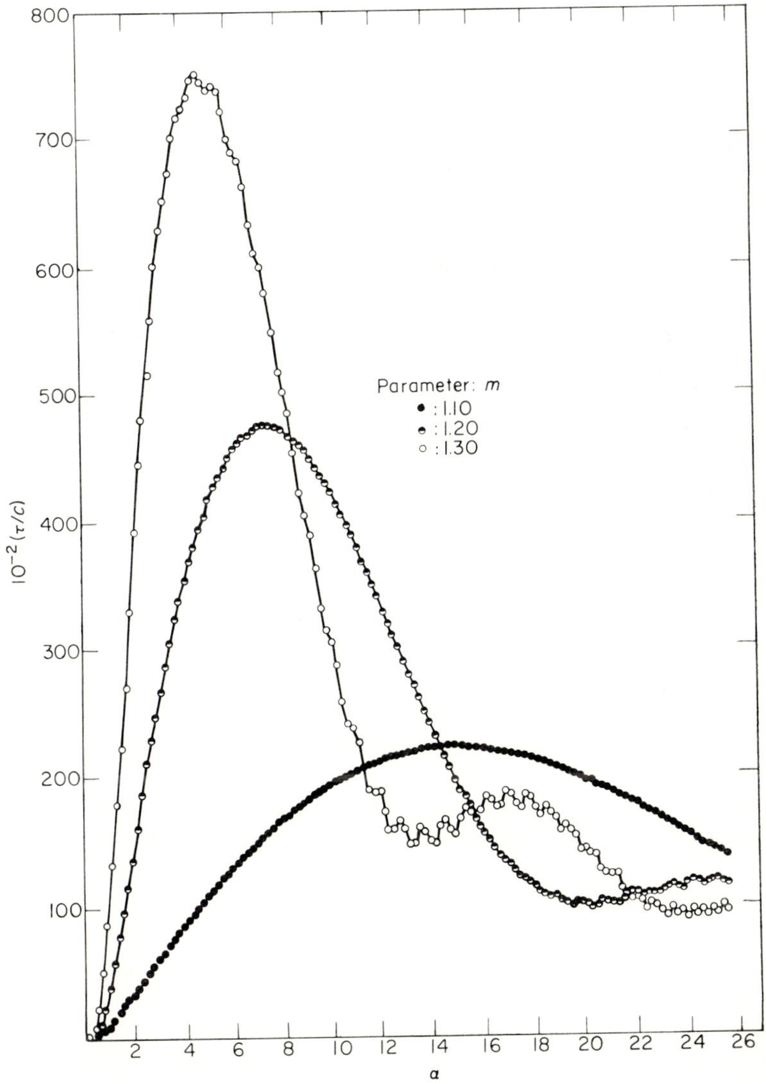

FIG. 7.11. The specific turbidity (τ/c) for $m = 1.10, 1.20$, and 1.30 plotted up to $\alpha = 26$ (Heller and McCarty, 1958).

$\alpha = 26$ for green light in water. This covers the range of colloidal sizes—up to radii of 2 μ—which are most likely to be investigated by light scattering. For this range the multivaluedness is reduced to the possibility of two values. Measurements at two wavelengths allow one to decide unambiguously whether one is operating on the ascending or descending branch of these curves, and, by comparison of the experimental results with the theoretical curve, the size parameter can be obtained.

Samples of polystyrene and poly (vinyl toluene) with a narrow range of sizes have provided a convenient system for testing the practicality of using the specific turbidity for particle size analysis. The work of Tabibian *et al.* (1956), carried out at one wavelength ($\lambda_0 = 546$ mμ), has been extended by Dezelić and Kratohvil (1961) to the four principal lines of the mercury spectrum (405, 436, 546, and 578 mμ); by Hodkinson (1963b) to three mercury lines (365, 436, and 546 mμ); by Maron *et al.* (1963c) to fourteen wavelengths from 350 to 1000 mμ; and by Bateman *et al.* (1959) to eight wavelengths from 370 to 950 mμ. Hermanie and van der Waarden (1966) have used the specific turbidity method to estimate the average drop size of water hazes in hydrocarbon liquids for which $m < 1$.

Bateman *et al.* paid particular attention to incorporating the dispersion of the refractive index of the polystyrene into the theoretical scattering functions and also to interpolation procedures for obtaining the precise values of these from the existing tabulations. The coefficient of variation of the radii, determined at the eight wavelengths for each particular latex, ranged from 2 to 4%. Such internal consistency over this broad range of wavelength is well within the limits that can be expected, when one considers that in addition to the experimental errors there were uncertainties in the values of the scattering functions themselves which were obtained by interpolation between exact values. In addition, there is some uncertainty in the value of the refractive index appropriate for such colloidal particles. Smart and Willis (1967) have proposed values slightly lower than those of Bateman. The radii obtained were, in all cases, in good agreement with the electron microscope values and, in the case of the results of Bateman *et al.*, within 2% of the cited values. These small discrepancies may not have been entirely due to errors in the light scattering, since they are well within the range of errors normally expected for electron microscopy.

According to Walton and Hlabse (1963), the specific turbidity at two wavelengths leads to good size estimates for irregularly shaped particles such as those found in barium sulfate hydrosols, even when the calculations are based upon the theory of scattering by spheres.

Metz (1963) has proposed a technique based upon turbidity measurements at two wavelengths which can be utilized in the range where the scattering efficiency is a multivalued function of the size parameter α. A particular refinement consists in using for the theoretical curve, against which the experimental data are to be compared, a plot of log Q_{sca} versus log ρ where

$$\rho = 2\alpha(m - 1) \tag{4.2.1}$$

We have already seen (Section 4.2.1*b*) for small values of $(m - 1)$ that this gives a universal curve independent of m. For higher values of m, the main features of the curve are independent of the refractive index so that this plot minimizes errors due to the neglect of the dispersion of the refractive index with wavelength.

The procedure is to plot on transparent or translucent paper the values of $\log(Q_{sca}/a)$ and $\log(\rho/a)$ as the ordinate and abscissa, respectively, for two wavelengths. These quantities are obtained from the experimental values of the turbidity, τ, and the wavelength in the medium, λ, from

$$(Q_{sca}/a) = (4\rho_2/3\omega'\rho_{12})\tau \tag{7.3.11}$$

$$(\rho/a) = (4\pi/\lambda)(m - 1) \tag{7.3.12}$$

where a is the radius, ρ_2 and ρ_{12} are the densities of the particles and dispersion, respectively, and ω' is the weight fraction of the particles. The two experimental points on the transparent paper are now displaced over the theoretical curve of log Q_{sca} vs. log ρ until the points line up with the curve. The value of log a may then be obtained directly from the difference in the numerical value of each coordinate. Metz utilized this technique for the size analysis of monodisperse silver bromide suspensions.

Meehan and Miller (1968) have used specific turbidity measurements to calculate the complex refractive index of silver bromide hydrosols at ultraviolet wavelengths. The particle size was first determined from measurements at visible wavelengths where the refractive index was presumably better known.

a. Effect of Polydispersity. Since there is usually a finite range of sizes present, there is no simple method of comparing the size determined by the electron microscope with the "average" size obtained by light scattering. Indeed, not only will different light-scattering methods give different averages, but each particular distribution of sizes will have its own characteristic average. The so-called turbidity average defined earlier applies only to Rayleigh scatterers. We can set

$$C_{sca} = K\alpha^{y} \tag{7.3.13}$$

or

$$Q_{sca} = K'\alpha^{y-2} \tag{7.3.14}$$

where the exponent y is equal to 6 for Rayleigh scatterers. Then

$$\tau = K\sum N_i\alpha_i^{y_i} \tag{7.3.15}$$

and the specific turbidity

$$(\tau/c) = K''(\sum N_i\alpha_i^{y_i}/\sum N_i\alpha_i^{3}) = K''\overline{\alpha^{y-3}} \tag{7.3.16}$$

where $\overline{\alpha^{y-3}}$ corresponds to the $(y - 3)$rd moment of the distribution. The value of y at a given value of α for a particular refractive index m can be obtained from a plot of $\log(Q_{sca}/\alpha)$ vs. $\log \alpha$. Since

$$d \ln(Q_{sca}/\alpha) = (y - 3)d \ln \alpha \tag{7.3.17}$$

the slope of such a plot is $y - 3$. Meehan and Beattie (1960) have explored the value of y for $m = 1.30$ and have found that, from the Rayleigh range to $\alpha = 14$, it decreases from 6 to 0 and then increases again to 2. The curve of Q_{sca} vs. α for $m = 1.30$ is shown in Fig. 7.12 with the corresponding values of y.

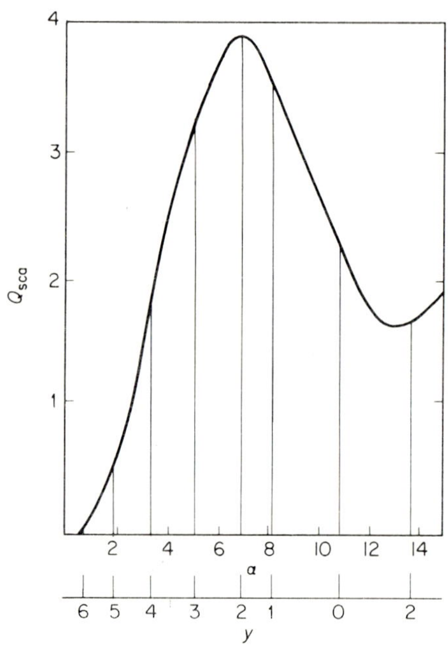

FIG. 7.12. Values of the exponent y (Eq. 7.3.13) corresponding to the values of α for spheres with $m = 1.30$ (Meehan and Beattie, 1960).

The effect of polydispersity upon the average diameter determined by the specific turbidity method has been explored empirically by Maron *et al.*

(1963c). They studied a large number of polydisperse butadiene-styrene latexes for which electron microscope data were available including systems with bimodal size distributions. Measurements were made at 14 wavelengths between 3500 and 10,000 Å. In the absence of data on the dispersion of the refractive index, constant values of $m = 1.15$ and 1.17 were used for the relative refractive indices of the 70:30 and 50:50 butadiene-styrene polymers, respectively.

Despite neglect of the dispersion of the refractive index, consistent particle diameters were obtained for each sample over the entire wavelength range. This is illustrated in Table 7.2 for various samples of these latexes and for one polystyrene latex. This test may not be too critical since the largest average value of α is only 4 and most of the values are sufficiently small so that the y-exponent defined by (7.3.11) is close to the Rayleigh limiting value $y = 6$ for which the turbidity average radius would be obtained. That these are actually turbidity averages is borne out by Table 7.2, where the results obtained from the specific turbidity are compared with both the number averages and the turbidity averages as determined from electron microscope counts.

TABLE 7.2

COMPARISON OF RADII OBTAINED BY TRANSMISSION MEASUREMENTS WITH THE NUMBER AND TURBIDITY AVERAGES OBTAINED FROM ELECTRON MICROSCOPE COUNTS

Butadiene-styrene 50:50			Butadiene-styrene 70:30			Polystyrene		
$a_{S.T.}$[a]	$a_{N.A.}$	a_τ	$a_{S.T.}$	$a_{N.A.}$	a_τ	$a_{S.T.}$	$a_{N.A.}$	a_τ
0.040	0.020	0.037	0.057	0.018	0.057	0.131	0.129	0.130
0.038	0.031	0.037	0.052	0.023	0.062			
0.037	0.023	0.042	0.065	0.022	0.067			
0.041	0.027	0.042	0.123	0.037	0.123			
0.046	0.029	0.045	0.184	0.034	0.196			
0.047	0.025	0.047						
0.138	0.027	0.138						

[a] Key: $a_{S.T.}$–radius from specific turbidity, $a_{N.A.}$–number average radius from electron microscopy, a_τ–turbidity average radius from electron microscopy.

Dobbins and Jizmagian (1966) have noted, for a wide variety of monomodal size distribution functions $[p(a)\, da]$ with radii comparable to and larger than the wavelength, that the specific turbidity can be represented by

$$\tau/c = (3/4\rho_2)\, \overline{Q}_{sca}/a_{32} \tag{7.3.18}$$

where a_{32} is the volume surface mean radius

$$a_{32} = \int_0^\infty p(a)a^3\,da \bigg/ \int_0^\infty p(a)a^2\,da \tag{7.3.19}$$

and the mean scattering efficiency is

$$\overline{Q}_{\mathrm{sca}} = \int_0^\infty Q_{\mathrm{sca}}p(a)a^2\,da \bigg/ \int_0^\infty p(a)a^2\,da \tag{7.3.20}$$

They have constructed curves of $\overline{Q}_{\mathrm{sca}}/a_{32}$ *vs.* a_{32} which are independent of the particular form of the distribution so that from a single measurement of the turbidity and concentration, this quantity can be calculated, and hence the volume-surface mean radius can be determined. When the concentration is not known, it becomes necessary to measure the turbidity at two widely separated wavelengths.

7.3.3 DQ Method

As early as 1931, Teorell proposed a method for particle size determination based upon the ratio of the turbidity at two wavelengths termed the "dispersion-quotient"

$$\mathrm{DQ} = \tau_1\lambda_2{}^2/\tau_2\lambda_1{}^2 \tag{7.3.21}$$

For Rayleigh scatterers,

$$\mathrm{DQ} = (\lambda_2/\lambda_1)^6 \tag{7.3.22}$$

provided the dispersion of the refractive index can be neglected. For larger particles, DQ decreases monotonically with size until the size range where the scattering efficiency becomes oscillatory with α is reached. Thus, up to this range, the DQ-value can be used as a monotonic measure of size.

In the absence of sufficient appropriate theoretical values of the scattering cross section, earlier workers (Evva, 1952) developed empirical curves for the variation of DQ-values with particle radius. However, with the availability of the many computed functions, it is now possible to prepare such curves from existing tabulations. As an example, a typical DQ-curve is shown in Fig. 7.13 for $\lambda = 0.405\,\mu$ and $0.546\,\mu$ and relative refractive indices of 1.214 and 1.200, respectively, corresponding to polystyrene latex. Dezelić *et al.* (1963) and Sakurada *et al.* (1964) have used this curve to determine the particle size of various polymer latexes with good results.

Despite the simplicity with which the experimental data can be obtained and analyzed, this method has hardly been exploited. Unlike methods based upon absolute determination of the turbidity, it does not require that the concentration be known. Also, because it is less sensitive to concentration effects, it is not necessary to reduce the concentration to as low values as in other methods.

Evva (1953a, b, 1954) has explored, theoretically, the dependence of the DQ-curves upon both the refractive index and the degree of polydispersion.

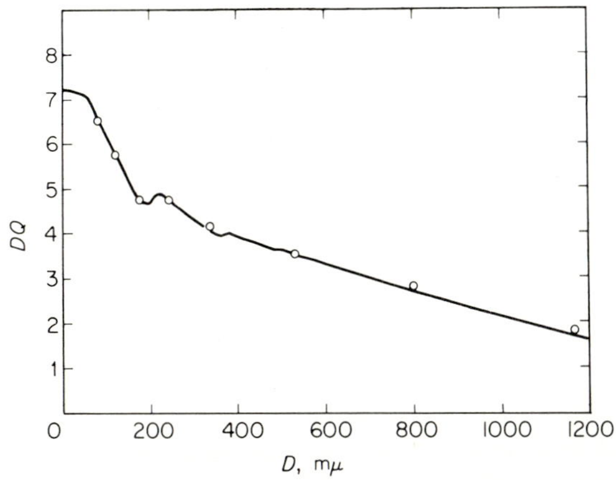

FIG. 7.13. Dispersion quotient DQ plotted against diameter D for $\lambda = 0.405\ \mu$ and $0.546\ \mu$ and for $m = 1.214$ and 1.200. Circles are experimental values obtained with polystyrene latexes (Dezelić *et al.* 1963).

7.3.4 TURBIDITY MAXIMUM

La Mer and his co-workers have utilized a somewhat different technique for determining the particle size of a monodisperse system of spherical particles from transmission data. This depends upon comparison of the first maximum in the theoretical curve of Q_{sca} vs. α with the corresponding maximum in the experimental curve of optical density vs. wavelength (La Mer and Sinclair, 1943; Barnes and La Mer, 1946). The experimental data are obtained spectrophotometrically in the usual way. If the dispersion of the refractive index is neglected, the maximum in each curve corresponds to the same value of α. Thus, from the wavelength in the medium $(\lambda)_{expt}$ at which the experimental maximum occurs and the value $(\alpha)_{theor}$ at which the theoretical maximum occurs, the particle radius can be calculated by

$$a = (\lambda)_{expt}(\alpha)_{theor}/2\pi \qquad (7.3.23)$$

Furthermore, it follows from (3.2.29) that the optical density is

$$\text{O.D.} = -\log T = Q_{sca}\pi a^2 Nl/2.3 \qquad (7.3.24)$$

so that

$$Na^2 = 2.3\text{O.D.}/Q_{\text{sca}}\pi l \tag{7.3.25}$$

Thus, the concentration N can be determined if a has already been obtained with the aid of (7.3.23).

Only that part of the theoretical curve in the neighborhood of the characteristic maximum need be considered. In order to include the effect of dispersion, the refractive index appropriate to the characteristic wavelength $(\lambda)_{\text{expt}}$ should be used in the calculation of Q_{sca}. Bateman *et al.* (1959) have obtained good agreement between the particle size of monodisperse polystyrene latexes obtained by this method and that obtained by electron microscopy.

This technique is limited to a size range for which a characteristic feature, such as a particular maximum or minimum in the transmission curve, occurs over the accessible range of wavelengths. Since only the range of the optical wavelengths is usually available, the experimental curve for any particular particle size may correspond to a range of α considerably smaller than that in which the first maximum lies. It is often possible to remedy the situation by generating new functions for which the maximum occurs at a smaller value of α (La Mer *et al.*, 1950). One such function may be obtained by multiplying the optical density by the wavelength,

$$\lambda(\text{O.D.}) = (Q_{\text{sca}}/\alpha)2\pi^2 a^3 Nl/2.3 \tag{7.3.26}$$

which leads to

$$Na^3 = 2.3\lambda(\text{O.D.})/(Q_{\text{sca}}/\alpha)2\pi^2 l \tag{7.3.27}$$

The new function (Q_{sca}/α) has a maximum at $\alpha = 4.5$ for $m = 1.33$, compared to $\alpha = 6$ for Q_{sca}. Further multiplication by λ shifts the maximum to still smaller values; i.e., Q_{sca}/α^2 has its maximum at $\alpha = 2.5$. It is sometimes feasible to multiply by fractional powers of λ. In general,

$$Na^{p+2} = 2.3\lambda^p(\text{O.D.})/(Q_{\text{sca}}/\alpha^p)(2\pi)^p\pi l \tag{7.3.28}$$

Thus, by plotting $\lambda^p(\text{O.D.})$ vs. λ and $(Q_{\text{sca}}/\alpha^p)$ vs. α and comparing the corresponding characteristic points on the two curves, a and Na^{p+2} can be determined.

The maximum tends to broaden with increasing p so that it becomes more difficult to identify the values of $(\alpha)_{\text{theor}}$ and $(\lambda)_{\text{expt}}$ corresponding to the points with a common α, and this results in a loss of accuracy in determining the radius from (7.3.23). On the other hand, this broadening enables one to identify even more easily the ordinates, $\lambda^p(\text{O.D.})$ and $(Q_{\text{sca}}/\alpha^p)$, at the same α. Thus, it is sometimes feasible to utilize two values of p when determining both N and a. For example, if characteristic points are chosen corresponding to a particular value of p, then both N and a can be determined from the following:

$$a = (Na^{p+2})/(Na^{p+1}) \tag{7.3.29}$$

and

$$N = Na^p/a^p \tag{7.3.30}$$

Since Na^{p+2}, Na^{p+1}, and Na^p are each evaluated from the ordinates of the characteristic points, viz. λ^p(O.D.) and (Q_{sca}/α^p), the need for a precise evaluation of $(\lambda)_{expt}$ and $(\alpha)_{theor}$ is obviated. Then (7.3.28) is used to obtain N. Inn (1951) has demonstrated the feasibility of this procedure.

In practice, it may turn out for a particular system over a limited range of particle sizes that the wavelength at which the maximum in the optical density curve occurs is directly proportional to the particle radius, so that

$$a = C(\lambda)_{expt} \tag{7.3.31}$$

Watillon and van Grunderbeeck (1956) have found this to be the case for monodisperse selenium hydrosols. They evaluated the proportionality constant by comparison of the light scattering results with electron microscope measurements. Then, the location of the wavelength at which the maximum in absorption occurred sufficed to determine the particle size.

Another interesting study of the problem of particle size analysis by location of the wavelength at which the maximum in the turbidity curve occurs has been carried out by DeVore and Pfund (1947). The turbidity of suspensions of powders consisting of uniform size but irregularly shaped particles was measured over the range of infrared wavelengths from 0.8 to 3.0 μ. The powders were suspended in liquid media covering a broad range of refractive index. These suspensions exhibited the typical turbidity maximum characteristic of spherical particles. Because of the irregular shape of the particles, it was not possible to use the method described above which requires computation of the theoretical functions. The analysis was carried out on the basis of a universal scattering curve generalized upon the parameter (Bailey, 1946),

$$\frac{a}{\lambda}\left(\frac{n^2 - 1}{n^2 + 2}\right) \tag{7.3.32}$$

rather than upon the phase shift parameter ρ. The maximum in Q_{sca} occurs at the value of 0.10 for Bailey's parameter. The sizes obtained by DeVore and Pfund, by comparison of the wavelength at which the maximum in the experimental turbidity occurs, were in excellent agreement with the sizes obtained by light microscopy.

A second effect which they noted was that for each particular solvent the wavelength at which the maximum occurred was linearly related to the refractive index of the solvent. This is illustrated in Fig. 7.14. Furthermore, extrapolation to zero wavelength gave the refractive index of the particle in vacuum. This interesting effect provides a method for the determination of the refractive index of such particles.

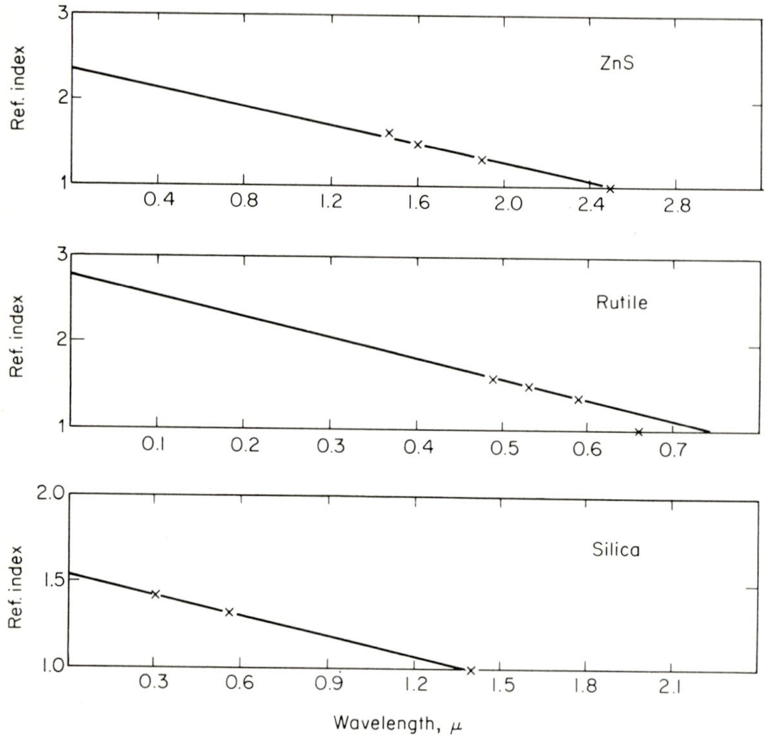

FIG. 7.14. The relative refractive index plotted against the wavelength at which the maximum in the absorption curve occurs for zinc sulfide, rutile, and silica suspensions in various solvents (DeVore and Pfund, 1947).

7.3.5 WAVELENGTH DEPENDENCE OF TURBIDITY

The dependence of the turbidity upon the wavelength has long been recognized as the source of the color imparted to white light upon transmission through a particulate medium (Kiessling, 1884a, b), and this color effect will be considered in greater detail later. Our present concern is with attempts to utilize the wavelength dependence to determine the particle size of a monodisperse dispersion. The wavelength effect is influenced most directly through the variation of Q_{sca} with α and, more subtly, by the dispersion of the refractive index with wavelength.

At the time that light scattering was being exploited for particle size analysis during World War II, Heller and Vassy (1943, 1946) and La Mer (1948) pointed out that the particle size could be characterized by the wavelength dependence of either the scattering or the turbidity. For the specific turbidity, this can be expressed by

$$(\tau/\omega') = k\lambda_0^{-g} = k'\lambda^{-g'} \tag{7.3.33}$$

where ω' is the weight fraction of spheres in the suspension, and the wavelength exponents, g and g', are respectively either in terms of the vacuum wavelength, λ_0, or the wavelength in the medium λ. For Rayleigh scatterers, g will be 4, and with increasing particle size this exponent will decrease, eventually assuming negative values. As long as it is single-valued, it may be used for determining particle sizes, and both Heller *et al.* (1946) and Bateman *et al.* (1959) have applied the method to polystyrene latexes with some success. This has been seriously limited by the lack of theoretical values of the exponents; however, the recent work of Heller *et al.* (1962) and of Heller (1964) has provided numerical data for refractive indices up to $m = 1.30$.

The experimental quantity which is determined directly from the transmission data is

$$-\left[\frac{d\ln(\tau/\omega')}{d\ln\lambda_0}\right] = g - \frac{d\ln k}{d\ln\lambda_0} = g_0 \tag{7.3.34}$$

and may be obtained from the slope of a plot of $\ln(\tau/\omega')$ vs. $\ln\lambda_0$. When the dispersion of k with wavelength is small, g_0 may be used directly instead of g. Indeed, the experimental work cited above was carried out under this approximation. For the most accurate work it is necessary to evaluate g as a function of α and m from the scattering theory and also to evaluate the second term in (7.3.34) in order to relate the experimental quantity g_0 to g.

For Rayleigh scatterers

$$k_R = \frac{\rho_{12}}{\rho_2} 32\pi^4 a^3 \left(\frac{n^2-1}{n^2+2}\right)^2 \tag{7.3.35}$$

$$g_0 = 4\left[1 - \left(\frac{\lambda_0}{n_2}\right)\left(\frac{dn_2}{d\lambda_0}\right)\right] - \left[\frac{12n}{(n^2-1)(n^2+2)}\right]\frac{\lambda_0}{n_2}\left[\frac{dn_1}{d\lambda_0} - n\frac{dn_2}{d\lambda_0}\right] \tag{7.3.36}$$

where n_1 and n_2 are the refractive indices of the spheres and the medium respectively, and $n = n_1/n_2$. When $n_1 > n_2$, the dispersion will usually be such that g_0 will almost invariably be greater than 4.0. This seems to have been overlooked in many discussions of atmospheric scattering which have reported deviations from the inverse fourth-power law.

For spheres of arbitrary size, (7.3.4) and (7.3.10) lead to

$$k = \frac{3}{2a}\frac{\rho_{12}}{\rho_2}\frac{\lambda_0{}^g}{\alpha^2}\sum \tag{7.3.37}$$

where Σ is an abbreviation for

$$\sum_{n=1}^{\infty}(2n+1)|a_n|^2 + |b_n|^2 \tag{7.3.38}$$

derived from the general expression for Q_{sca} in (3.3.81). The quantity g_0 can now be expressed as

$$g_0 = gF_1 - PF_2 \tag{7.3.39}$$

$$g = \alpha\frac{\partial \ln \sum}{\partial \alpha} - 2 = \alpha\frac{\partial \ln(\tau/\omega')}{\partial \alpha} + 1 \tag{7.3.40}$$

$$F_1 = 1 - \frac{d \ln n_2}{d \ln \lambda_0} \tag{7.3.41}$$

$$F_2 = \frac{dn}{d \ln \lambda_0} \tag{7.3.42}$$

$$P = \frac{d \ln \sum}{dn} = \frac{d \ln(\tau/\omega')}{dn} \tag{7.3.43}$$

Equation (7.3.39) provides the basis for comparison of g_0 as obtained experimentally (7.3.34) and g as obtained from theory (7.3.40). The factors F_1 and F_2 must be determined experimentally from the dispersion of n and n_2. The wavelength exponent, g, and the quantity, P, are best evaluated on the basis of theory from Σ rather than from the variation of the specific turbidity with α and with n. The numerical problems in evaluating g from the tabulated values of Σ are discussed by Heller *et al.* (1962), who computed this for $\alpha = 0.2(0.2)25.0$ and $n = 1.05(0.05)1.30$. Their results are illustrated in Fig. 7.15.

A striking feature of the variation of the wavelength exponent with α is the short term oscillation, especially for $n \geqslant 1.15$. This raises the problem of multivaluedness in making comparisons between the experimental values as determined from g_0. However, if the system is slightly heterodisperse, the resulting curves become smooth and are not sensitive to the degree of heterodispersion. For example, Fig. 7.16 gives g for a rectangular distribution with half-widths from 0.40 to 0.45 in α. These curves follow very closely the "backbone" of the oscillating curves and should be useful for analyzing slightly heterodisperse systems. In these cases, the value obtained is reasonably close to the weight average of α for the system investigated. Experimental applications of this method have not yet been reported.

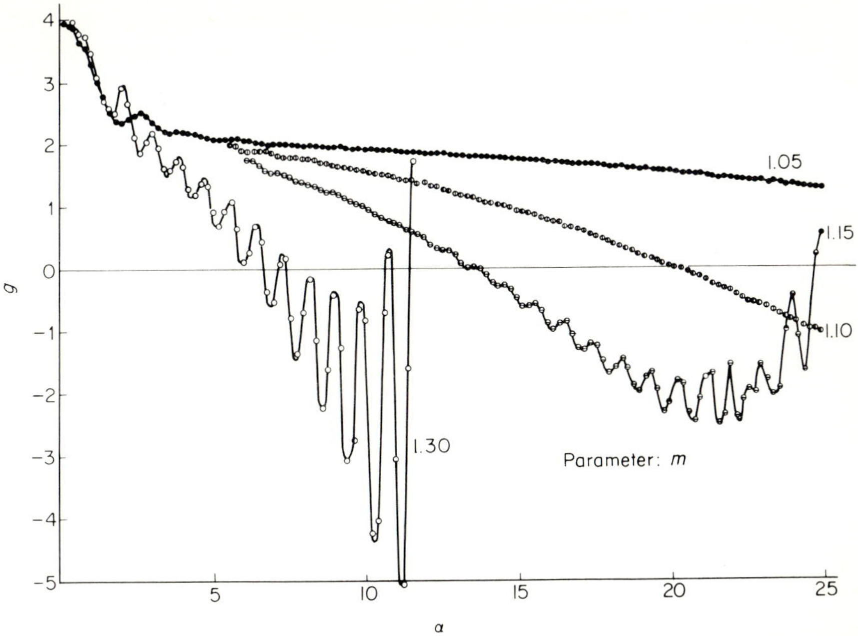

FIG. 7.15. The variation of the wavelength exponent g with α (Heller *et al.*, 1962).

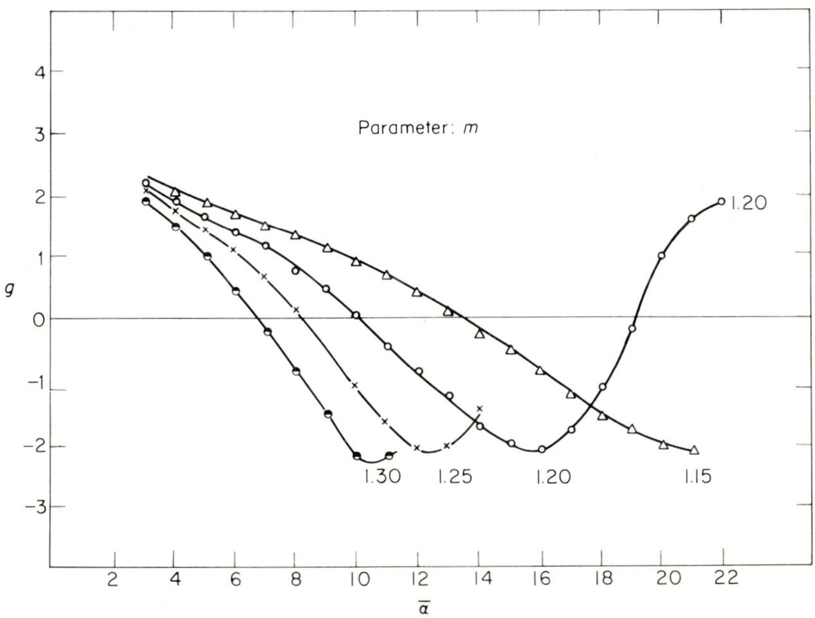

FIG. 7.16. The variation of the wavelength exponent g for a slightly heterodisperse system. The abscissa is the mean value of α for a rectangular distribution (Heller *et al.*, 1962).

Heller (1964) has also developed relations for sizes just larger than Rayleigh scatterers using the Stevenson (1953a, b) extension of the Rayleigh equation

$$\left(\frac{\tau}{\omega'}\right) = 24\pi^3 \frac{\rho_{12}}{\rho_2} \frac{V}{\lambda^4} \left(\frac{n^2 - 1}{n^2 + 2}\right)^2 \left[1 + \frac{6}{5}\left(\frac{n^2 - 2}{n^2 + 2}\right)\alpha^2\right] \quad (7.3.44)$$

This equation gives excellent agreement with the general theory, provided $n \leqslant 1.35$ and $\alpha \leqslant 0.8$. The wavelength exponents can be represented in closed form by

$$g = 4 + \frac{12(n^2 - 2)\alpha^2}{5(n^2 + 2) + 6(n^2 - 2)\alpha^2} \quad (7.3.45)$$

or, since α is small, this is approximated reasonably well by

$$g = 4 + \frac{12(n - 2)\alpha^2}{5(n^2 + 2)} \quad (7.3.46)$$

Then,

$$g_0 = nF_1 - RF_2 \quad (7.3.47)$$

where F_1 and F_2 are defined above, and

$$R = \frac{12n}{(n^2 - 1)(n^2 + 2)} + \frac{48\alpha^2 n}{5(n^2 + 2) - 6(n^2 - 2)\alpha^2} \quad (7.3.48)$$

The Rayleigh case is obtained by letting $\alpha = 0$. Equations (7.3.45) and (7.3.46) may be rearranged, leading to

$$\alpha^2 = \frac{5(g - 4)(n^2 + 2)}{6(6 - g)(n^2 - 2)} \quad (7.3.49)$$

and

$$\alpha^2 = \frac{5(g - 4)(n^2 + 2)}{12(n^2 - 2)} \quad (7.3.50)$$

The validity of these approximate expressions for g has been checked by comparison with the exact values, and the results are summarized by the error contour chart in Fig. 7.17. The region to the left of the curves represents that part of the $n\alpha$-domain in which (7.3.45) and (7.3.46) give values of g to within 5% of the exact values and in which α, as calculated by (7.3.49) or (7.3.50), is within 25% of the correct value.

Bhatnagar and Heller (1964) have also discussed the wavelength exponent of the scattering intensity at 90°, defined by

$$g_{90} = \frac{d \ln(r^2 I/I_0 \omega')}{d \ln \lambda_0} = \frac{d \ln R_{90}/\omega'}{d \ln \lambda_0} = \frac{d \ln(i_1 + i_2)}{d \ln \lambda_0} \quad (7.3.51)$$

where R_{90} is the Rayleigh ratio at 90° and ω' is the weight fraction of spheres. They found that the nearly monotonic decline of g_{90}, from the limiting value of 4.0 to its first minimum, extends over an appreciably larger range of α than in the case of the wavelength exponent of the turbidity. Here, the α values at the minimum were 2.9 and 2.5 at $n = 1.05$ and 1.30, respectively, as compared with 2.0 and 1.8 for the turbidity. Also, there was a greater drop so that in this case the wavelength exponent was a more sensitive measure of particle size. Because of this, coupled with the fact that for small values of α experimental measurements of the turbidity are frequently more difficult than angular scattering, it follows that it is advantageous to work with the angular measurements, at least up to the first minimum in the g_{90} versus α curve. On the other hand, at values of α beyond the first minimum, g_{90} exhibits extremely steep maxima and minima, which persist even when the system is somewhat heterodisperse. For this reason the wavelength exponent of the turbidity is more useful in this range.

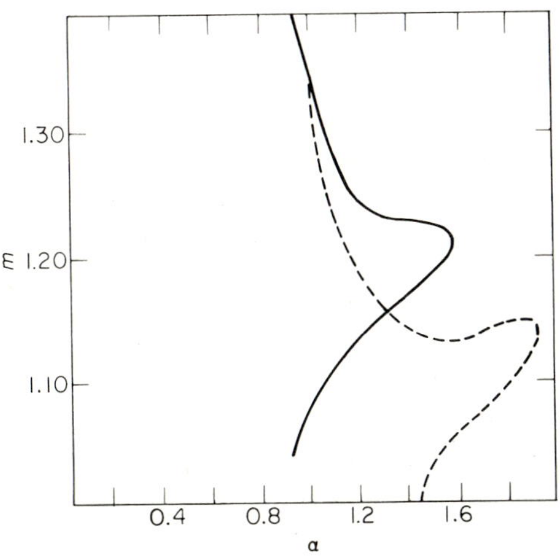

FIG. 7.17. The 5% error contour for Eqs. (7.3.45) (full line) and (7.3.46) (dashed line) (Heller, 1964).

7.4 Average Size from Angular Variation of Scattering

The intensity of light scattered at a particular angle increases rapidly with particle size up to a maximum value at about $\alpha \cong 2$ and then oscillates in a complicated fashion as the size increases further. It follows then, unless the particles are known to be smaller than the size at which the intensity becomes oscillatory, that a single measurement of the scattered intensity will not suffice to determine the particle size. Tabibian and Heller (1958) have discussed the limitations of determining the particle size in polystyrene latexes from a single measurement of the intensity scattered at 90°.

We have already seen that the scattering by a sphere varies in a complicated oscillatory fashion with angle of observation and that the particular pattern depends upon both the size parameter α and the relative refractive index. The pattern is sufficiently unique so that it can serve as a measure of particle size. Indeed, merely the angular location of the maxima and minima in the scattering often suffices to determine the size. Furthermore, because the angular positions of the extrema are hardly sensitive to small departures from monodispersity, the system need not be perfectly monodisperse. In addition to the intensity, the angular variation of the polarization or of the phase angle of the scattered monochromatic light can be used to determine the particle size. The angular distribution of the color of the scattered light for incident white light and its use in particle size analysis will be discussed in Section 7.6.2.

7.4.1 Maxima and Minima in the Scattered Intensity

Dandliker (1950) has compared the location of the first minimum of the scattered intensity from the forward direction, as obtained from theory, with the experimental value in order to estimate the particle size in a polystyrene latex of narrow size distribution. He restricted his considerations to the perpendicularly polarized component. Dezelić and Kratohvil (1961) extended this work by measuring the angular location of each of the accessible minima and maxima of the scattered intensity for several monodisperse polystyrene latexes. The particle size was then determined by comparing these results with the location of the extrema from the calculations of the angular intensity functions for the appropriate refractive index. The results were in good agreement with the sizes obtained by electron microscopy and by a number of other procedures for evaluating the light scattering.

In this work, Dandliker had suggested that the angular location of the first minimum might be expected to follow the expression

$$(2a/\lambda)\sin(\theta/2) = k \tag{7.4.1}$$

which is based upon the Rayleigh–Debye theory. This theory, which will be treated in the next chapter, is applicable to particles with a relative refractive index close to unity and is increasingly valid in the forward directions.

The above expression has been generalized to higher minima and maxima by Maron and co-workers (Maron and Elder, 1963a; Maron *et al.*, 1963b; Pierce and Maron, 1964) and utilized successfully for particle size determination in polystyrene latexes. They proposed a generalization of Dandliker's equation to

$$(2a/\lambda)\sin(\theta/2) = k_i \quad \text{or} \quad K_i \tag{7.4.2}$$

where k_i is a parameter corresponding to the angle at which the ith intensity minimum occurs, and K_i corresponds to the ith intensity maximum.

These parameters, which are dependent upon the refractive index, may be obtained from the scattering theory for spheres. Values for the first four pairs of extrema up to $m = 1.60$ are given in Table 7.3 (Kerker *et al.*, 1964a). They are based upon scattering functions computed for 1° intervals of θ and are averaged from parameters obtained for each of the values of α listed in Table 7.4. The individually computed parameters were constant to within 2% of the average except for those values placed in brackets for which the constancy was to within 10%. Figures 7.18 and 7.19 are error contour charts which summarize the range of validity of (7.4.2). The region to the left of

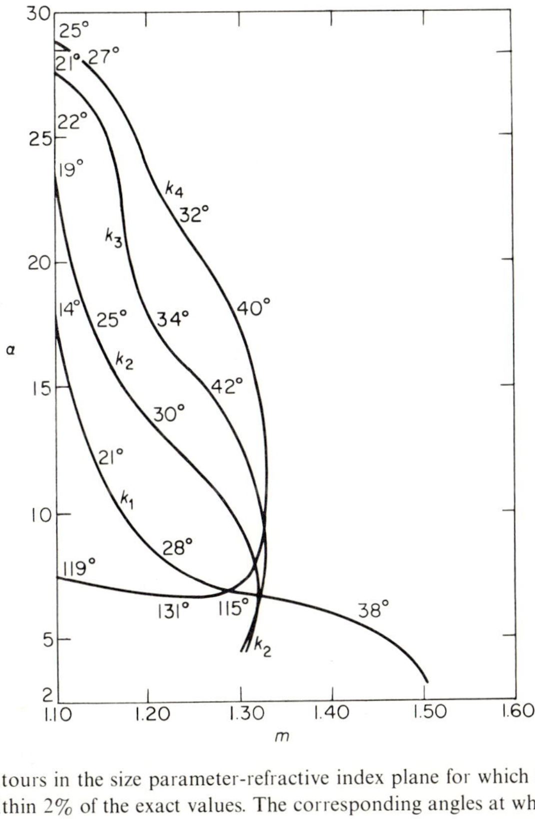

FIG. 7.18. Contours in the size parameter-refractive index plane for which k_1, k_2, k_3, and k_4 (Eq. 7.4.2) are within 2% of the exact values. The corresponding angles at which the extremum in the angular gain occurs is indicated along the curves (Kerker *et al.*, 1964a).

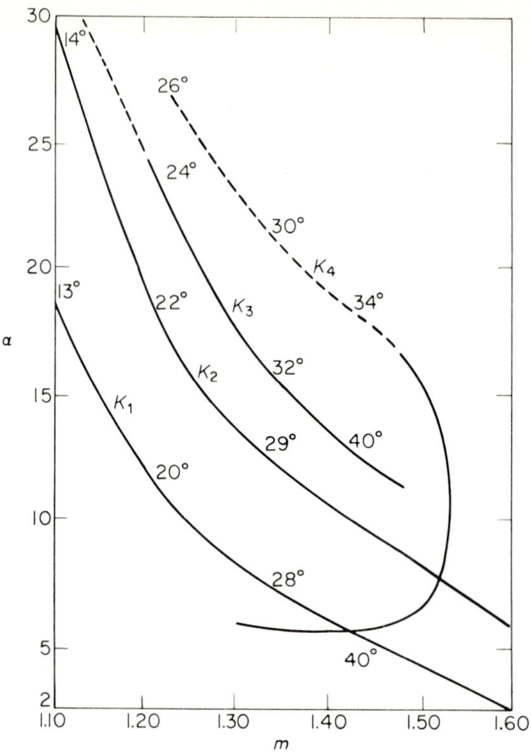

FIG. 7.19. Same as Fig. 7.18 for K_1, K_2, K_3, and K_4 (Kerker *et al.*, 1964a).

each contour delimits the $m\alpha$-domain where the individual values of k_i and K_i agree to within 2% of the values given in Table 7.3.

TABLE 7.3

VALUES OF THE PARAMETERS k_i AND K_i FOR THE FIRST FOUR INTENSITY MAXIMA AND MINIMA

m	k_1	k_2	k_3	k_4	K_1	K_2	K_3	K_4
1.00	0.715	1.23	1.74	2.25	0.917	1.45	1.96	2.47
1.02	0.706	1.218	1.708	2.11	0.904	1.427	1.927	(2.23)
1.10	0.671	1.158	1.633	2.10	0.864	1.363	1.848	2.32
1.14	0.664	1.137	1.585	2.04	0.843	1.336	1.799	2.25
1.18	0.646	1.109	1.568	2.00	0.825	1.298	1.761	2.20
1.22	0.631	1.085	1.521	1.97	0.805	1.269	1.713	2.15
1.30	0.599	1.047	1.493	1.87	0.770	1.228	1.692	2.03
1.34	0.587	(0.99)	(1.39)	(1.78)	0.756	(1.19)	(1.61)	(1.95)
1.44	0.572	(0.99)	(1.35)	(1.66)	0.755	(1.19)	(1.50)	(2.14)
1.50	0.566	(0.94)	(1.28)	—	(0.68)	(1.17)	(1.42)	—
1.60	(0.51)	(0.91)	—	—	—	(1.02)	—	—

TABLE 7.4

VALUES OF REFRACTIVE INDEX, m, AND SIZE PARAMETER, α, FOR WHICH SCATTERING COEFFICIENTS, a_n AND b_n, HAVE BEEN COMPUTED FOR USE IN TABLE 7.3

m	α	m	α
1.02	2(1)16	1.30	2(1)10; 12(3)21
1.10	2(1)16; 18(3)39	1.34	2(1)10; 12(3)21
1.14	2(1)12; 15(3)39	1.44	2(1)9; 12(3)18
1.18	2(1)12; 15(3)24	1.50	2(1)10
1.22	2(1)10; 12,14,16,18(3)27	1.60	3(2)9

For each value of m, there is a lower limit of α at which a minimum first appears in the radiation pattern. As α increases, this minimum shifts to the forward direction ($\theta = 0$), and successive maxima and minima appear at higher values of θ. It is the regularity of this appearance and migration of extrema that is the basis of (7.4.2). Accordingly, there is a lower limit in α at which each extremum occurs, e.g., the first minimum does not occur until $\alpha = 2$; the fourth maximum until $\alpha = 6$. These lower limits are not indicated on the error contour charts in order to avoid making them overly complicated. In some cases k_i and K_i are not constant in the region of the lowest α values and this is indicated by a bending back of the contour lines.

As α increases, the first minimum and maximum flatten out and disappear. When this occurs, the original index of the other extrema is still maintained. What had been the ith extremum from the forward direction actually becomes the $(i - 1)$th extremum. However the index is retained as well as that of the associated value of k_i or K_i.

Experimentally, there is a problem in assigning the proper index to the various extrema both because of the effect just described and because of the fact that the lower extrema may occur at values of θ which are too low to be experimentally accessible. Pierce and Maron (1964) have pointed out that the ratios of the k_i and K_i are, to a close approximation, independent of m and are characteristic of the particular values of the index i. The ratios of parameters for successive maxima-minima are given in Table 7.5, where they are compared with the values obtained earlier by Pierce and Maron using the Rayleigh–Debye equation. The agreement between their results and those from the exact calculations (Kerker *et al.*, 1964a) is excellent. These ratios can be used to identify particular extrema. Thus, if the parameter ratio for the first two experimentally encountered minima is 1.408 rather than 1.718, the parameters must be counted as k_2 and k_3 rather than k_1 and k_2.

TABLE 7.5
PARAMETER RATIOS

	Exact theory	Pierce and Maron		Exact theory	Pierce and Maron
$K_1/k_1 =$	1.280	1.283	$k_2/k_1 =$	1.718	1.720
$k_2/K_1 =$	1.347		$K_2/K_1 =$	1.577	1.578
$K_2/k_2 =$	1.174	1.176	$k_3/k_2 =$	1.408	1.410
$k_3/K_2 =$	1.199		$K_3/K_2 =$	1.354	1.354
$K_3/k_3 =$	1.128	1.130	$k_4/k_3 =$	1.282	1.292
$k_4/K_3 =$	1.138		$K_4/K_3 =$	1.252	1.259
$K_4/k_4 =$	1.101	1.102			

Kenyon and La Mer (1949) studied the angular distribution of the ratio of the intensity of the perpendicular component at two wavelengths. In Fig. 7.20 the smooth curves are the loci of the maxima of such a ratio plotted in the $\alpha\theta$-domain. These curves are for the ratio of the intensity at $\lambda = 0.436\,\mu$ to that at $\lambda = 0.365\,\mu$ for $m = 1.51$. Thus, at $\alpha = 5.0$, maxima would occur at 20, 38, 55, 115, 128, 143, and 157°. The circles correspond to values of the maximum of this ratio obtained experimentally for various sulfur hydrosols.

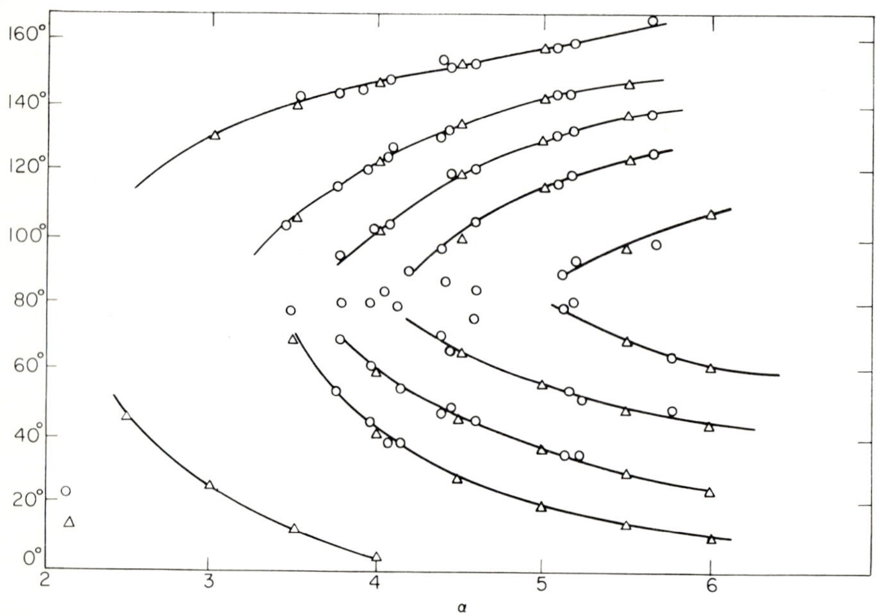

FIG. 7.20. Loci of the various maxima of the ratio of I_1 at $\lambda = 0.436\,\mu$ to the value at $0.365\,\mu$ for $m = 1.51$ (Kenyon and La Mer, 1949).

The procedure in determining the size from the extrema of the intensity ratio at two wavelengths is to place the set of experimental points on the θ-coordinate and then shift this along the α-coordinate until it intersects with the theoretical curves at a common value of α. This value serves to define the average size.

The dissymmetry method is still another technique for the analysis of dispersions of narrow size distribution. This utilizes the ratio of the intensity of the perpendicular component at two scattering angles symmetrical about 90° such as $\theta = 45$ and 135°. It has been widely used primarily in connection with systems obeying the Rayleigh–Debye theory and will be discussed later in that connection. Maron *et al.* (1964) have examined its applicability to spheres whose relative refractive index is so large that the general theory must be utilized. They have concluded that here its usefulness is limited to sizes up to $\alpha \cong 1.8$ since beyond this it becomes multivalued.

7.4.2 Maxima and Minima in the Polarization Ratio

A similar procedure to that of Kenyon and La Mer (1949) was used by Kerker and La Mer (1950) based upon the polarization ratio $\rho(\theta)$ which is the ratio of the intensity of the parallel to the perpendicular component of the scattered light for unpolarized incident light.

The polarization ratio increases monotonically from its value of $\cos^2 \theta$ for small particles up to a maximum value at approximately $\alpha = 2.5$ (Sinclair and La Mer, 1949; Kerker, 1950). If *a priori* knowledge about the system provides assurance that the particles are within this size range, a single measurement of $\rho(\theta)$ at one angle provides a measure of particle size. Heller and Tabibian (1962) have suggested that in the multivalued regime the particle size might be estimated from measurements of the polarization ratio at just 90° if values were obtained at several wavelengths. Graessley and Zufall (1964) have noted that for sufficiently broad distributions, $\rho(90°)$ retains its monotonic character to sizes somewhat larger than $\alpha = 2.5$.

Maron *et al.* (1963a) have called attention to the sensitivity of $\rho(\theta)$ to possible intrinsic optical anisotropy of the particles and have proposed that the particles of polystyrene latex exhibit such anisotropy. However, Wallace and Kratohvil (1967) have shown that the evidence for such anisotropy is insubstantial.

For larger sizes, the polarization ratio oscillates with size and for each particular size the variation with angle of observation is also oscillatory. The angular location of the maxima and minima permits construction of a diagram similar to Fig. 7.20. This can then be used to determine particle size in the same manner as for the ratio of the intensities at two wavelengths (Kerker and La Mer, 1950).

Maron *et al.* (1964) have noted, in analogy with (7.4.2), that for the polarization ratio

$$(2a/\lambda)\sin(\theta/2) = l_i \quad \text{or} \quad L_i \tag{7.4.3}$$

where l_i and L_i are parameters which determine the scattering angle at which the ith minimum and maximum, respectively, of the polarization ratio occur. Calculated values of these parameters for refractive indices up to $m = 1.25$ are given in Table 7.6.

TABLE 7.6
CALCULATED PARAMETERS l_i AND L_i

m	l_1	l_2	l_3	L_1	L_2
1.05	0.605	1.10	1.22	0.681	1.15
1.10	0.551	0.989	1.30	0.680	1.17
1.15	0.506	0.949	1.38	0.661	1.12
1.20	0.446	0.879	1.37	0.630	1.09
1.25	0.412	0.848	1.27	0.611	1.06

7.4.3 MAXIMA AND MINIMA IN THE PHASE DIFFERENCE

If the incident light is linearly polarized with the electric vector oriented obliquely to the scattering plane, the scattered light will be elliptically polarized. The amplitudes of the parallel and perpendicular components of the scattered light are proportional to the amplitudes of the corresponding incident components. The phase difference δ between these amplitudes is independent of the orientation of the incident light and a function only of the refractive index of the particle and the size parameter α. The phase difference exhibits the same kind of oscillatory variation with angle and with size as does the intensity so that a diagram depicting the loci of the extrema of the phase angle in the $\alpha\theta$-plane can be constructed (Kerker and La Mer, 1950) in the same manner as for Fig. 7.20. This can also be used to determine particle size if the angular positions of the extrema of the phase difference are determined experimentally.

The method for measuring phase difference developed by Kerker and La Mer (1950) utilizes a polarizer in the incident beam and a quarterwave plate and analyzer in the scattered beam. With the polarizer oriented 45° from the scattering plane, the following quantities are measured:

- I_3, intensity of light scattered with analyzer oriented 45° from the scattering plane; quarterwave plate out of beam.
- I_4, intensity of light scattered with analyzer oriented 135° from the scattering plane, quarterwave plate out of beam.

I_5, same as I_3 with quarterwave plate in the beam.
I_6, same as I_4 with quarterwave plate in the beam.

Then,

$$\tan \delta = (I_5 - I_6)/(I_3 - I_4) \tag{7.4.4}$$

and δ can be calculated from (3.3.63).

7.4.4 Shape of Forward Scattering Lobe

Hodkinson (1966) has noted that the angular distribution of the scattered intensity within the central portion of the forward lobe can provide a useful measure of particle size, particularly when the refractive index is not known. This latter condition is a consequence of the forward lobe being due primarily to Fraunhofer diffraction which is independent of the optical properties of the particle since it arises from the light passing near the particle rather than from rays undergoing reflection and refraction. The angular intensity function is given by

$$i_1 + i_2 = 2\alpha^2[J_1{}^2(\alpha \sin \theta)/\sin^2 \theta] \tag{7.4.5}$$

Figure 7.21 shows how the ratio of the intensities scattered at selected pairs of angles varies with α according to the above diffraction formula. It thus seems practical to determine the size of particles from $\alpha = 1$ to 18 without requiring measurements at scattering angles smaller than 5° or comparison of intensities differing by more than a factor of ten. Actually, the upper limit is set only by the difficulty of measuring intensities at angles close to the forward direction. The ability to measure at $\theta = 1°$ would increase this limit to about $\alpha = 100$.

7.5 Particle Size Distribution

7.5.1 Distribution Functions

We now turn to the problem of determining the distribution of particle sizes by light scattering when the spheres comprising the dispersion are no longer of nearly the same size. The object of such an investigation is to determine a distribution function $p(a)$ such that

$$P(a) = \int_a^{a+\Delta a} p(a)\, da \tag{7.5.1}$$

gives the fraction of particles with radii between a and $a + \Delta a$. The distribution function is normalized so that the integral over all values of the radius

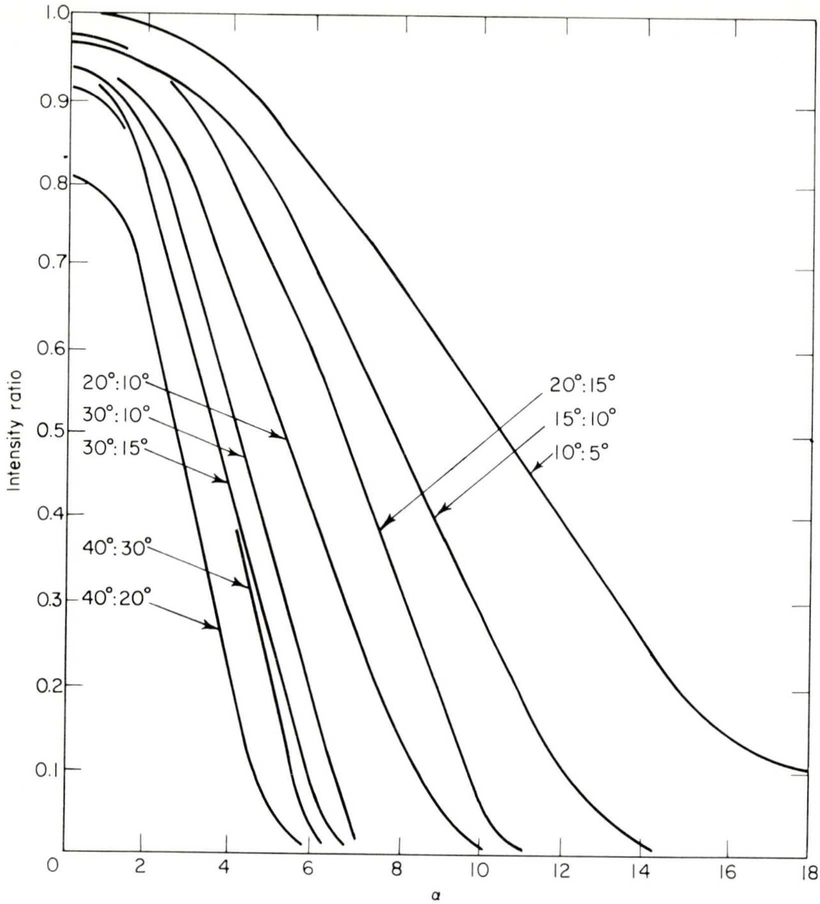

FIG. 7.21. Variation of the intensity ratios at the indicated angle according to Fraunhofer diffraction plotted against α (Hodkinson, 1966).

is unity. It may be represented in a variety of ways, e.g., in tabular form, as a histogram, graphically, or as an analytic function. The latter form is usual and, in order to avoid computational labor, analytic functions involving only two parameters are generally used. The success of this approach hinges upon selection of the form of an appropriate two-parameter distribution function that approximates the actual distribution. There is no *a priori* reason for assuming that this can always be done.

A complete discussion of the various distribution functions which have been proposed to describe different particulate systems will not be attempted here. Instead, we will consider only those distributions which have been employed widely in light scattering by recent workers.

Undoubtedly, the best known distribution function is the normal distribution (Hald, 1962)

$$p(a) = \frac{1}{(2\pi)^{1/2}\sigma} \exp\left[-\frac{(a - \bar{a})^2}{2\sigma^2} \right] \tag{7.5.2}$$

where the parameters $\bar{a}$ and σ, called the mean and the standard deviation, are defined by

$$\bar{a} = \int_{-\infty}^{\infty} ap(a)\, da \tag{7.5.3}$$

and

$$\sigma^2 = \int_{-\infty}^{\infty} (a - \bar{a})^2 p(a)\, da \tag{7.5.4}$$

Here, the exponential factor in the distribution describes a Gaussian curve, and the pre-exponential factor normalizes the expression. Because of its symmetry, two other parameters of interest, the modal value, a_M, and the median value, a_m, are identical with the mean. The mode is the value of a at the maximum frequency; the median is the value below which 50% of the population falls.

Obviously, a normal distribution cannot represent a distribution of particle sizes with highest precision because it admits negative values of a. In addition, unlike the normal distribution which is symmetrical, naturally occurring populations are frequently skewed. A satisfactory representation of many such populations is the logarithmic normal distribution

$$p(a) = \frac{1}{(2\pi)^{1/2}\sigma_g a} \exp\left[-\frac{(\ln a - \ln a_m)^2}{2\sigma_g{}^2} \right] \tag{7.5.5}$$

In this distribution it is $\log a$ rather than a which is normally distributed, so that

$$\log a_m = \int_{-\infty}^{\infty} \ln a\, p(a)\, da \tag{7.5.6}$$

and

$$\sigma_g{}^2 = \int_{-\infty}^{\infty} (\ln a - \ln a_m)^2 p(a)\, da \tag{7.5.7}$$

Thus, $\log a_m$ is the mean value of $\log a$ and a_m is, in this case, both the median and the geometric mean value of a;

$$a_g = a_m = (a_1 \cdot a_2 \cdot a_3 \ldots a_n)^{1/n} \tag{7.5.8}$$

The second parameter of the distribution, σ_g, is the standard deviation of $\log a$ and is termed the geometric mean standard deviation. The mode, the median, and the mean are related to the geometric mean by

$$\ln a_M = \ln a_g - \sigma_g^2 \tag{7.5.9}$$

$$\ln a_m = \ln a_g \tag{7.5.10}$$

$$\ln \bar{a} = \ln a_g + 0.5\sigma_g^2 \tag{7.5.11}$$

Figure 7.22 shows three frequency curves for a logarithmic normal distribution having the same values of a_m, but different σ_g. The skewness of the distribution depends upon the logarithmic standard deviation; indeed, for sufficiently small values of σ_g, there is so little skewness that the frequency curve can be closely approximated by a normal distribution. As a general rule for practical work, both a and $\log a$ can be considered to be normally distributed as long as $\sigma_g < 0.14$. Another feature of this distribution is the movement of the modal value of a towards smaller values as σ_g increases for a fixed value of a_m.

In Fig. 7.23, the cumulative curves for the above distributions have been plotted on logarithmic probability paper, i.e., the radius is plotted as the abscissa on a logarithmic scale, and the ordinate is

$$P(a) = \int_0^a p(a)\, da \tag{7.5.12}$$

Then, the cumulative distribution curves form straight lines. By plotting a sample population on such a graph, it is possible to determine whether the population corresponds to a logarithmic normal distribution and, if so, the parameters a_g and σ_g can be picked out. In some cases, where the simple logarithmic normal distribution does not seem to fit the data, the system may be heterogeneous, i.e., it consists of two or more simple distributions. The treatment of such populations is discussed in detail by Kottler (1952).

In a logarithmic normal distribution, 95% of the population is contained within the interval of $\log a_g \pm 1.96\,\sigma_g$, which corresponds in turn, to

$$a_g/(1 + \delta) \leqslant a \leqslant a_g(1 + \delta) \tag{7.5.13}$$

where

$$1 + \delta = 10^{1.96\sigma_g} \tag{7.5.14}$$

The ratio of this upper limit to the median is equal to the ratio of the median to the lower limit. Each of these intervals contains 47.5% of the population.

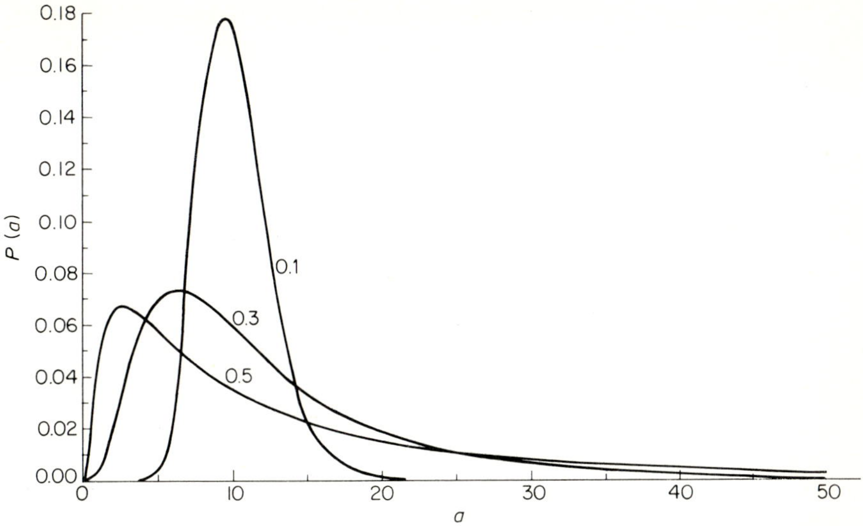

FIG. 7.22. Three logarithmic normal distribution curves with $a_m = 10$ and $\sigma_g = 0.1, 0.3$, and 0.5 (Hald, 1962).

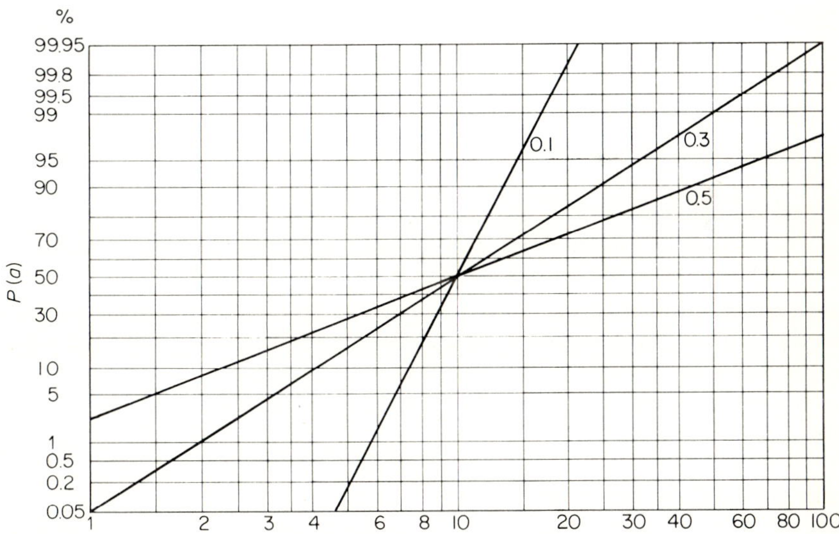

FIG. 7.23. Cumulative curves for the distributions of Fig. 7.22 plotted on log probability scales (Hald, 1962).

Sometimes, the logarithmic normal distribution is represented as

$$p(a) = \frac{1}{(2\pi)^{1/2}\sigma_g} \exp\left[-\frac{(\ln a - \ln a_m)^2}{2\sigma_g^2} \right] \tag{7.5.15}$$

which differs from (7.5.5) by the factor $1/a$. This gives the logarithmic normal distribution correctly if it is plotted against a on semilogarithmic paper and if, in obtaining the population density, it is integrated with respect to $\log a$. On the other hand, if (7.5.15) is integrated with respect to a, as indicated by (7.5.1), it leads to a new distribution function given by

$$p(a) = \exp\left[-\frac{(\ln a - \ln a_M)^2}{2\sigma_0^2} \right] \Big/ \left[\int_0^\infty \exp\left[-\frac{(\ln a - \ln a_M)^2}{2\sigma_0^2} \right] da \right.$$

$$= \exp\left[-\frac{(\ln a - \ln a_M)^2}{2\sigma_0^2} \right] \Big/ (2\pi)^{1/2}\sigma_0 a_M \exp[\sigma_0^2/2] \tag{7.5.16}$$

This has been called a zeroth order logarithmic distribution (ZOLD) function by Espenscheid *et al.* (1964a), who have described some of its properties. Although earlier work from the author's laboratory had referred to the logarithmic normal distribution, in point of fact, the ZOLD represented by (7.5.16) had been utilized throughout. A similar error has been made by Meehan and Beattie (1960) and also by Gledhill (1962). Evva (1953a, b) has also used a somewhat similar distribution which he called a logarithmic normal distribution, but which he correctly recognized had the modal rather than the geometric mean size as a parameter in the distribution function.

The ZOLD is defined by two parameters, a_M, which is the modal value of a, and σ_0, which is a measure of the width and skewness of the distribution. The latter is related to the standard deviation in a manner which will be described below.

The frequency function for this logarithmic distribution is plotted in Fig. 7.24 for $a_M = 3.0$ and $\sigma_0 = 0.1$, 0.2, 0.3, and 0.5. Perhaps the most interesting feature of these curves is that the modal value of a remains constant as the value of σ_0 is varied. It offers the possibility of exploring the effect of changing the breadth of the distribution for a constant mode.

The relation between the modal value of a and the mean value is

$$\ln \bar{a} = \ln a_M + 1.5\,\sigma_0^2 \tag{7.5.17}$$

The standard deviation is given by

$$\sigma = a_M[\exp(4\sigma_0^2) - \exp(3\sigma_0^2)]^{1/2} \tag{7.5.18}$$

$$\sigma = a_M\sigma_0\left(1 + \frac{7\sigma_0^2}{2!} + \frac{37\sigma_0^4}{3!} + \cdots\right)^{1/2} \tag{7.5.19}$$

so that for $\sigma_0 \ll 1$,

$$\sigma \cong a_M \sigma_0 \tag{7.5.20}$$

Thus, for a sufficiently narrow distribution, σ_0 approximates the coefficient of variation

$$\sigma_0 \cong C = \sigma/a \cong \sigma/a_M \qquad (\sigma_0 \ll 1) \tag{7.5.21}$$

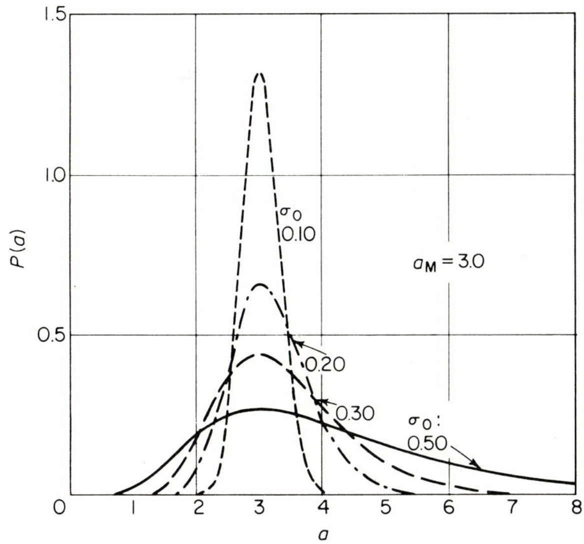

FIG. 7.24. ZOLD distribution with $a_M = 3.0$ and $\sigma_0 = 0.10, 0.20, 0.30$, and 0.50 (Espenscheid *et al.*, 1964a).

Actually, both the logarithmic normal distribution and the ZOLD are special cases of a general family of logarithmically skewed distributions represented by

$$p_n(a) = C_n a^n \exp[-(\ln a - \ln a_n)^2/2\sigma_n{}^2] \tag{7.5.22}$$

where C_n is the normalizing factor. The quantity a_n locates one of the moments of the distribution (median, mean, mode, etc.) and σ_n is some measure of the breadth of the distribution. The normalizing factor is obtained from the definite integral

$$\begin{aligned} C_n^{-1} &= \int_0^\infty a^n \exp[-(\ln a - \ln a_n)^2/2\sigma_n{}^2]\, da \\ &= (2\pi)^{1/2}\sigma_n a_n^{n+1} \exp[(n+1)^2\sigma_n{}^2/2] \end{aligned} \tag{7.5.23}$$

It is immediately apparent that when $n = -1$, (7.5.22) reduces to the logarithmic normal distribution, and when $n = 0$, it becomes the ZOLD. For these distributions, as pointed out above, a_{-1} and a_0 correspond to the median and modal values of the radius, respectively. For each value of the index n, a_n will correspond to a particular moment of the distribution which can be maintained invariant as the breadth parameter σ_n is changed. For $n = -\frac{3}{2}, -2, -\frac{5}{2}$,

$$a_n = \int_0^\infty a^m p_n(a)\, da \tag{7.5.24}$$

where $m = 1, 2$, and 3, respectively, so that a_n corresponds to first, second, or third moments of the distribution, respectively. Holding these values constant while varying σ_n corresponds to a constant mean radius, constant surface area, or constant volume, respectively.

Still another type of distribution has been adopted by Heller and his collaborators (Stevenson *et al.*, 1961) in connection with their determination of particle size from the spectra of the scattering and the turbidity. This is defined by

$$p(a) = c(a - a_0) \exp -[(a - a_0)/s]^3, \qquad a \geqslant a_0 \tag{7.5.25}$$

$$= 0, \qquad a < a_0 \tag{7.5.26}$$

where c is the normalization constant, a_0 is the radius of the smallest particle present in consequential numbers, while s determines the modal radius, a_M, the half width w, and the "half-spread" $(a_M - a_0)$, through the relations

$$w = 0.9015s \tag{7.5.27}$$

$$a_M - a_0 = 3^{-1/3}s \tag{7.5.28}$$

The half-width is the distance between the two points of the distribution curve at which the frequency is one-half the value at the modal radius. When treating light-scattering data for spheres, it will be more convenient to utilize the parameter α so that distribution will be expressed as

$$p(\alpha) = c(\alpha - p) \exp - [(\alpha - p)/q]^3, \qquad \alpha \geqslant p \tag{7.5.29}$$

$$= 0, \qquad \alpha < p \tag{7.5.30}$$

where the quantities p and q are defined by

$$p = 2\pi a_0/\lambda \tag{7.5.31}$$

$$q = 2\pi s/\lambda \tag{7.5.32}$$

In Fig. 7.25, $p(\alpha)$ is plotted against α for $p = 1.84$ and $q = 7.97$. This distribution coincides very closely with those found empirically for emulsions.

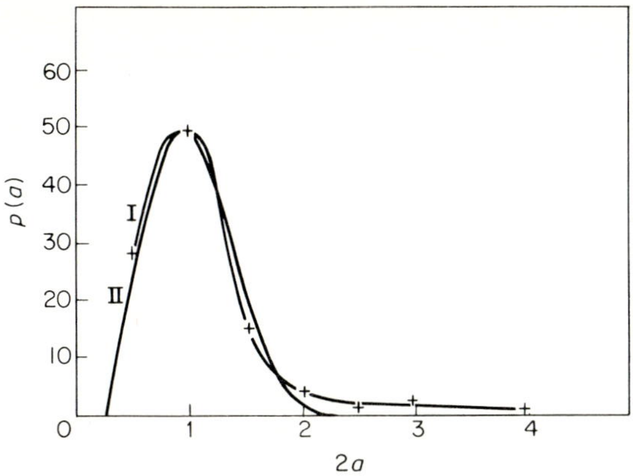

FIG. 7.25. Distribution curve according to Eq. (7.5.25) for $a = 0.12$, $s = 0.52$ (II) compared with the experimental values (I) for an ethylbenzene-water emulsion (Stevenson *et al.*, 1961).

Nakagaki and Shimoyama (1964) have discussed the effect of a three-parameter distribution function upon the light scattering at 45, 90, and 135°. One of the parameters characterizing this distribution was the modal value of the size parameter α_M. The branch of the distribution curve for $\alpha < \alpha_M$ is a Gaussian function with a standard deviation α_1, whereas the other branch for $\alpha > \alpha_M$ is a Gaussian function with a different standard deviation, α_2. Although, in principle, it should be possible to determine all three distribution parameters from light scattering experiments, these authors concluded that the experimental accuracy required was excessively high, and they recommended using a two-parameter distribution.

Hilbig (1966) has used a Maxwell distribution defined by

$$p(y) = \frac{2}{\Gamma[(n-2)/2]} y^{n-3} \exp(-y^2) \tag{7.5.33}$$

where $y = a/c$ and a is the particle radius, c and n are distribution parameters, and Γ is the gamma function. This has the advantage that for the case of anomalous diffraction, the scattering efficiency can be integrated over this function in closed form, leading to an analytical expression involving confluent hypergeometric functions.

7.5.2 Polarization Method

This method for evaluating the distribution of sphere sizes (Kerker *et al.*, 1964c) utilizes the polarization of the scattered radiation obtained with monochromatic light at various angles of observation. The intensity of the components of the scattered light whose electric vector vibrates perpendicular and parallel to the plane of observation, $I_1(\theta)$ and $I_2(\theta)$, is measured at a number of angles; e.g., the above workers used nineteen angles from $\theta = 40$ to 130° at 5° intervals. The polarization ratio defined by

$$\rho(\theta) = \frac{I_2(\theta)}{I_1(\theta)} = \int p(\alpha) i_2(\theta, \alpha)\, d\alpha \Big/ \int p(\alpha) i_1(\theta, \alpha)\, d\alpha \tag{7.5.34}$$

is used in the analysis. Because this is a ratio, it is sufficient to use instrument readings rather than absolute intensities. The distribution function utilized in this work was the ZOLD (7.5.16), which is described by the modal value of the size parameter, α_M, and the ZOLD standard deviation σ_0. The problem is to determine α_M and σ_0 from the experimental data. This is done by comparison of the experimental values of $\rho(\theta)$ with theoretical calculations corresponding to the refractive index of the system at the wavelength under investigation.

Theoretical values of $\rho(\theta)$ were computed by the above workers for an extensive array of combinations of α_M, σ_0, m, and θ; viz. $\alpha_M = 1.9$ to 15.0 in steps of $\Delta\alpha_M = 0.1$, $\sigma_0 = 0.005$ to 0.155 in steps of $\Delta\sigma_0 = 0.005$ and from 0.16 to 0.30 in steps of $\Delta\sigma_0 = 0.01$; $\theta = 30$ to 130° in steps of $\Delta\theta = 5°$; and $m = 1.43$, 1.51, and 2.074. The refractive indices were selected to correspond to values for octanoic acid aerosols, sulfur hydrosols or sodium chloride aerosols, and silver chloride aerosols. These very extensive calculations comprised 382,536 cases. The integrals in (7.5.34) were solved by numerical integration, using intervals of 0.1 in α over a sufficiently broad range so that the integrand reached a limit. The results were stored on punched cards. Wallace and Kratohvil (1968) have reviewed recent extensions of the computer routine.

The procedure used to determine the parameters of the size distribution involved comparison of these stored values of $\rho(\theta)$ with the experimental values at each of nineteen angles of observation. "Solutions" to the problem consisted of pairs of α_M and σ_0 values for which agreement was obtained between the experimental and theoretical values of $\rho(\theta)$ at all angles, or nearly all angles, to within a prescribed tolerance. The final best solution was selected from those values for which the mean square deviation between the experimental and theoretical values was a minimum.

The effect upon $\rho(\theta)$ of changing the logarithmic standard deviation, while keeping the modal value of the size parameter fixed, is shown in Figs. 7.26 and 7.27 for $m = 1.43$ and for $\alpha_M = 2.0$ and 5.0, respectively. The curves in each figure correspond to six values of σ_0 ranging from 0.100 to 0.300. The distribution curves for the two extreme cases are in the insets, where the radii plotted along the abscissa correspond to light of wavelength 0.546μ ($a = \lambda\alpha/2\pi$).

For $\sigma_0 = 0.100$, these curves show the typical oscillations characteristic of monodisperse systems. The frequency of the oscillations increases as the

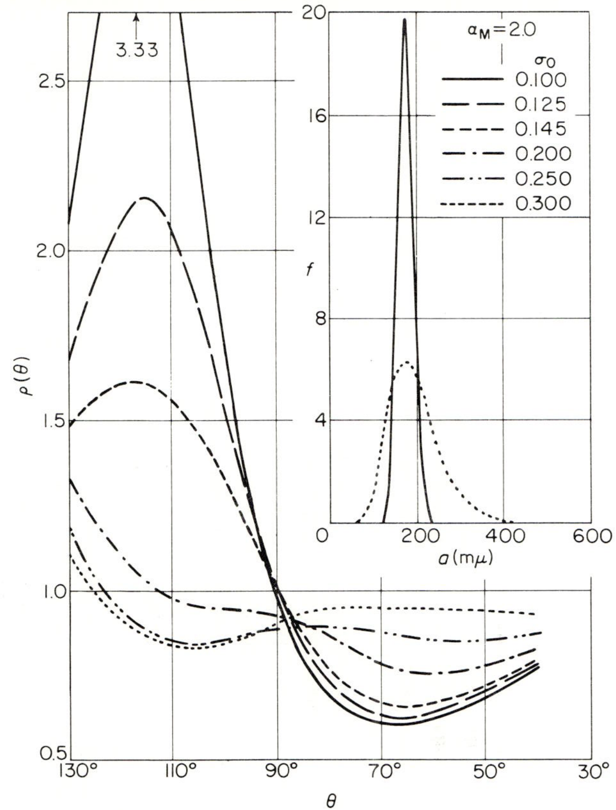

FIG. 7.26. Polarization ratio $\rho(\theta)$ plotted against θ for $\alpha_M = 2.0$ and $\sigma_0 = 0.100, 0.125, 0.145, 0.200, 0.250$, and 0.300. Inset shows the distribution curves for $\sigma_0 = 0.100$ and 0.250 when $\lambda = 0.546\ \mu$ (Kerker *et al.*, 1964c).

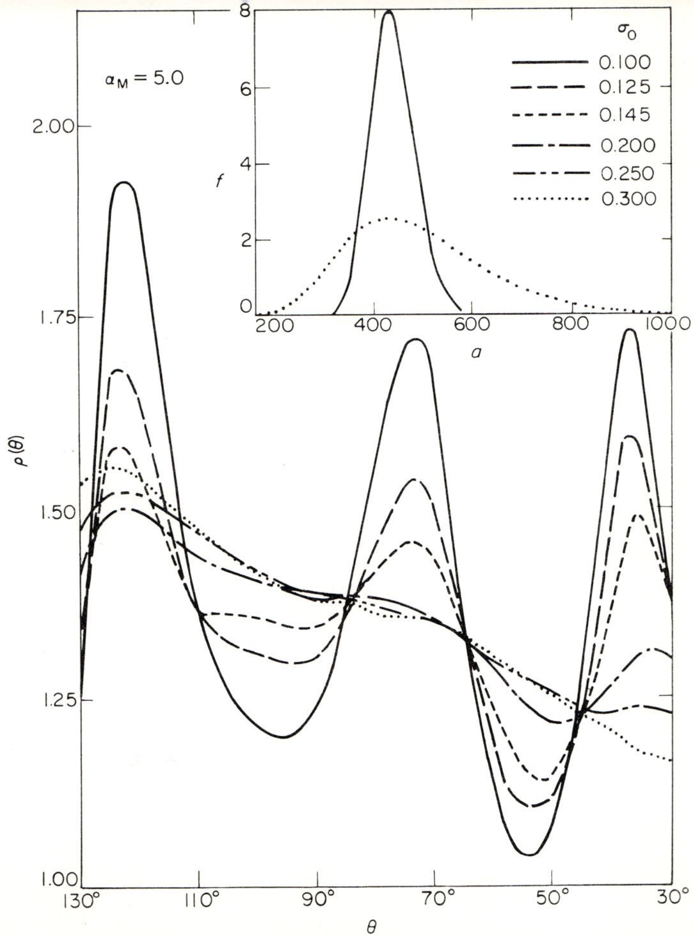

FIG. 7.27. Same as Fig. 7.26 for $\alpha_M = 5.0$.

size becomes larger. However, this structured character is obliterated as the distribution of sizes becomes broader so that for $\alpha_M = 2.0$ and $\sigma_0 = 0.300$, the scattered light is nearly completely unpolarized in all directions when the incident light is natural. Since it is the structure in these curves which permits the precise determination of the size distribution from light scattering data, it would appear that $\sigma_0 = 0.300$ may represent the upper limit of the polydispersion for which the present method is applicable (Wallace and Kratohvil, 1967).

As these distributions become broader and the polarization ratio curves become flatter, the positions of the extrema remain at nearly the same angular locations. This is a very useful property. Earlier, it had been noted that the location of these extrema in the angular distribution of the polarization ratio could be used to obtain the size in a monodisperse system. In view of the lack of sensitivity of these angular positions to the width of the distribution, that method will continue to give a correct value for the modal radius even when there is appreciable deviation from monodispersity.

In Figs. 7.26 and 7.27, the variation of $\rho(\theta)$ with θ has been represented by smooth curves drawn through the theoretical points determined at every 5°. Actually, the method described here does not involve such constructions, but utilizes a direct point by point numerical comparison of $\rho(\theta)$ at each of the nineteen values of θ. On the other hand, for those methods which are dependent upon angular location of the extrema of $\rho(\theta)$, it is important that the angular functions be determined at sufficiently close intervals to delineate the extrema, especially when $\alpha > 5$. In Fig. 7.28, $\rho(\theta)$ has been calculated every 0.5° over the interval $\theta = 130$ to 150° for $\alpha_M = 6.7$, $\sigma_0 = 0$, and $m = 1.51$. It is obvious that the maximum at 137.2° could hardly be located if the computations were available at only 10° intervals and could be located only with considerable uncertainty for data at 5° intervals.

Fortunately, the polarization ratio is not highly sensitive to refractive index so that a small difference between the value of m of the experimental system and that used for the theoretical calculation does not affect the results significantly. The polarization ratio for $\alpha_M = 5.5$ and $\sigma_0 = 0.025$ and for $m = 1.50$, 1.51, and 1.52 is depicted in Fig. 7.29. These results were obtained at 10° intervals, and the calculated points were connected by straight lines. It is obvious that small changes of m (i.e., $m = \pm 0.01$) are only significant at very high values of $\rho(\theta)$. In this particular case, higher values of $\rho(\theta)$ are obtained at 70° and 100° for $m = 1.50$, but there is no general pattern. In other cases, the higher values may be associated with either $m = 1.51$ or $m = 1.52$.

The sensitivity of $\rho(\theta)$ to both α_M and σ_0 is illustrated in Table 7.7, where values of $\rho(\theta)$ are given at 10° intervals at a fixed value of $\sigma_0 = 0.04$ for $\alpha_M = 8.2$, 8.3, 8.4, and 8.5, and a fixed value of $\alpha_M = 8.0$ for $\sigma_0 = 0.005$,

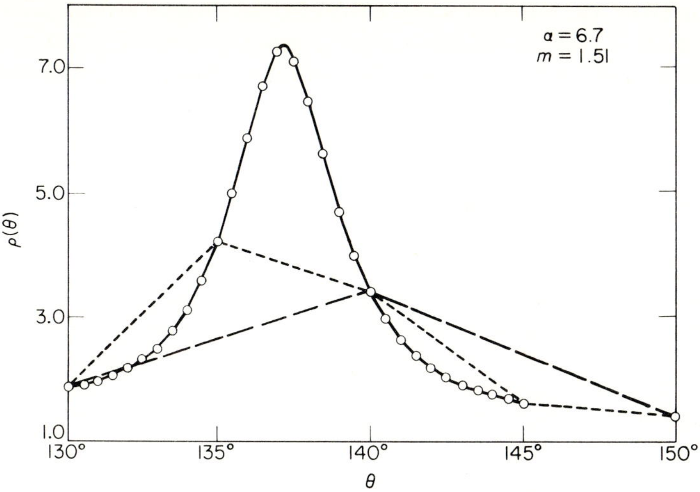

FIG. 7.28. Plot of $\rho(\theta)$ vs. θ for $\alpha = 6.7$, $m = 1.51$, illustrating the necessity for close angular intervals in order to determine the detailed shape of the curve (Kerker *et al.*, 1964c).

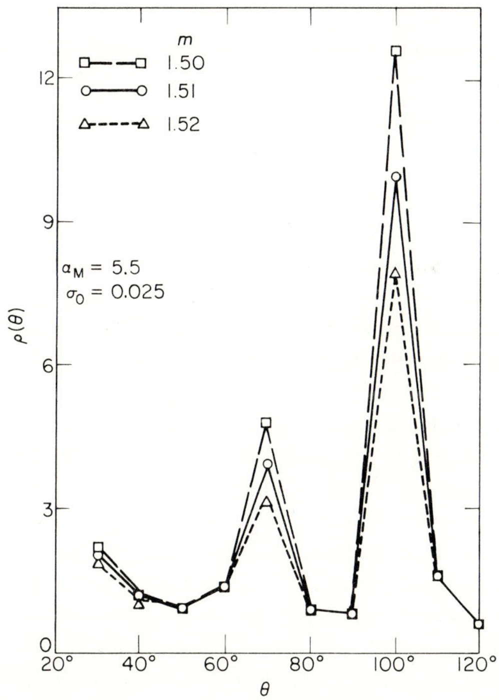

FIG. 7.29. Influence of refractive index upon polarization ratio for $\alpha_M = 5.5$, $\sigma_0 = 0.025$, and $m = 1.50$, 1.51, and 1.52 (Kerker *et al.*, 1963a).

0.025, and 0.05. Usually, $\rho(\theta)$ is less sensitive to the changes in these parameters at the forward angles, but there is no general pattern of behavior. Thus, whereas at 100° and $\sigma_0 = 0.04$, $\rho(\theta)$ increases 2.3-fold as α_M changes from 8.2 to 8.5, there is a comparable decrease in $\rho(\theta)$ at 110° for the same change in α_M. Extreme effects are also obtained at a fixed value of $\alpha_M = 8.0$ as σ_0 changes. Indeed, it is this sensitivity to these parameters that makes $\rho(\theta)$ so useful in the precise determination of narrow size distributions.

TABLE 7.7

COMPARISON OF $\rho(\theta)$ AT CONSTANT σ_0 FOR NEIGHBORING VALUES OF α_M AND AT CONSTANT α_M FOR NEIGHBORING VALUES OF σ_0

θ	$\sigma_0 = 0.04$				$\alpha_M = 8.0$		
	$\alpha_M = 8.2$	$\alpha_M = 8.3$	$\alpha_M = 8.4$	$\alpha_M = 8.5$	$\sigma_0 = 0.005$	$\sigma_0 = 0.025$	$\sigma_0 = 0.050$
30	0.87	0.89	0.92	0.95	0.89	—	—
40	1.20	1.13	1.05	0.94	0.53	1.04	1.34
50	0.85	0.85	0.87	0.91	0.87	0.89	0.90
60	1.99	2.22	2.27	2.07	0.72	1.24	1.51
70	0.89	0.87	0.78	0.80	1.81	1.46	1.23
80	1.91	2.50	3.10	3.40	0.47	0.89	1.23
90	1.27	0.96	0.80	0.76	52.8	6.39	2.13
100	1.63	2.17	2.89	3.74	0.28	0.69	1.20
110	2.24	1.57	1.18	0.99	9.71	8.04	2.80
120	1.74	1.98	2.31	2.69	0.56	1.13	1.67

A typical result obtained with an octanoic aerosol of narrow size distribution is shown in Fig. 7.30 (Matijević *et al.*, 1964) where the experimental data, plotted as circular points, are compared with four theoretical curves which best agree with them. The distribution parameters corresponding to each of these curves are shown on the figure, as well as ΣD^2, the sum of the squares of the percentage difference between the experimental and theoretical values of $\rho(\theta)$ at each of nineteen angles. There is little doubt that $\alpha_M = 2.9$, $\sigma_0 = 0.160$ best represent these data.

Since the size parameter α depends on wavelength, the value of α_M obtained for a particular aerosol should change when the wavelength used in the experiments is changed. This is illustrated in Fig. 7.31 (Espenscheid *et al.*, 1964b) which gives the experimental light-scattering results obtained with a sodium chloride aerosol using two wavelengths: $\lambda = 436$ and $546\ m\mu$. The theoretical results which best agree with these data correspond to $\alpha_M = 2.5$, $\sigma_0 = 0.115$, and $\alpha_M = 2.0$, $\sigma_0 = 0.130$, respectively. The size frequency distributions are shown in the inset. While the scattering curves are quite different, the frequencies calculated from these two sets of data are in excellent agreement.

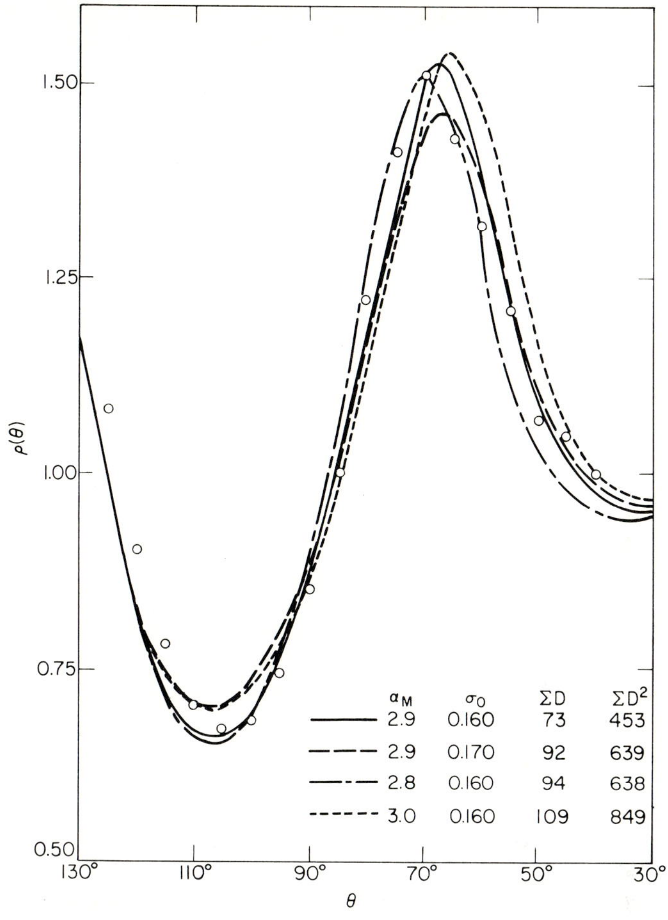

FIG. 7.30. Experimental values of $\rho(\theta)$ obtained from an octanoic acid aerosol compared with the theoretical values for the four sets of distribution parameters which fit best (Matijevic *et al.*, 1964).

An even more stringent test of the accuracy of the method has been provided by experiments with vanadium pentoxide aerosols (Jacobsen *et al.*, 1966). Experiments were conducted at six wavelengths in the interval $\lambda = 406$ to 689 mμ. A special feature of vanadium pentoxide is the very great dispersion of its refractive index over the visible spectrum (Jacobsen and Kerker, 1967). At the lower wavelengths, the optical absorption is exceedingly high, comparable to that for metals, whereas at the red end of the spectrum there is very little absorption, so that it behaves as a dielectric. Thus, when the wavelength is changed, the scattering changes both because

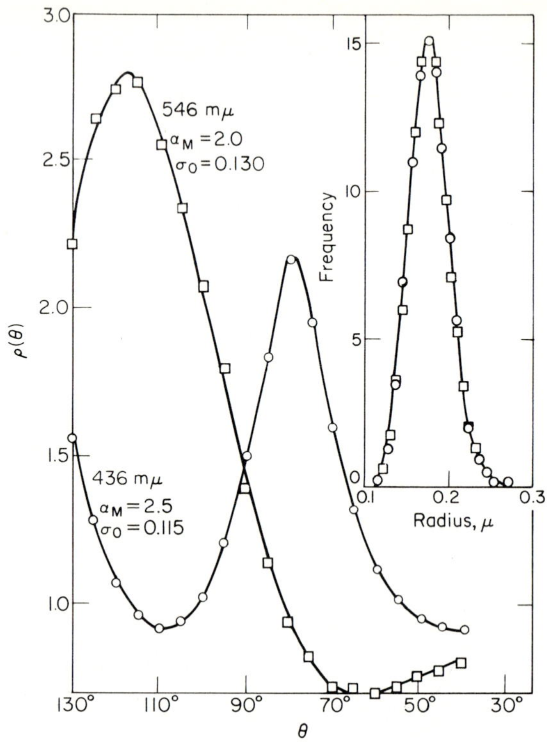

FIG. 7.31. Values of $\rho(\theta)$ measured at $\lambda = 0.546\,\mu$ and $0.436\,\mu$ for the same sodium chloride aerosol. Inset shows resulting frequency distributions obtained from each set of data (Espenscheid *et al.*, 1964b).

of the effect upon the "optical size," α, and upon the refractive index. A typical result is shown in Fig. 7.32 for $\lambda = 0.406, 0.436, 0.546, 0.578\,\mu$. In this case, the experimental quantity plotted is the angular distribution of the degree of polarization P rather than the polarization ratio $\rho(\theta)$, where

$$P = \frac{i_2 - i_1}{i_2 + i_1} = \frac{\rho(\theta) - 1}{\rho(\theta) + 1} \tag{7.5.35}$$

The various points plotted on the inset correspond to the size distributions determined from the data at each of the four wavelengths and it is apparent that they agree with each other and with the histogram which was obtained from electron microscope measurements.

There is an advantage to using P rather than $\rho(\theta)$ in the graphical representation of polarization. Because the interval of $\rho(\theta)$ over which the perpendicularly polarized component predominates only varies between 0 and 1,

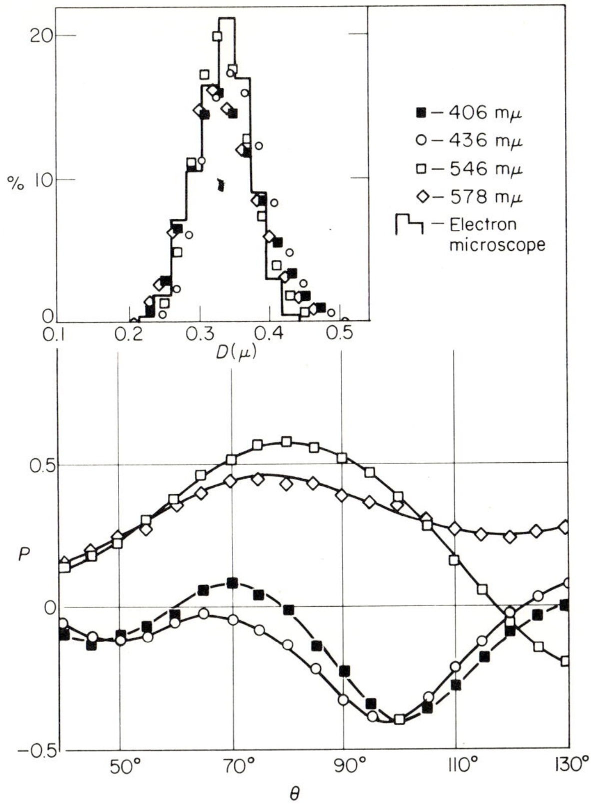

FIG. 7.32. Polarization of vanadium pentoxide aerosols at four wavelengths. Inset shows resulting frequency distributions compared with an electron microscope histogram (Jacobsen *et al.*, 1967).

while the interval over which the horizontal component predominates ranges from 1 to infinity, the maxima in $\rho(\theta)$ are exaggerated compared to the minima. On the other hand, these two intervals span equal ranges of P; viz., -1 to 0 when i_1 predominates, and 0 to 1 when i_2 predominates. The effect is illustrated in Fig. 7.33. The two polarization ratio curves, as shown in part A, appear quite different. Curve I appears more "structured" than Curve II, and yet these curves actually differ only in that i_1 and i_2 have been interchanged. When these data are plotted as P against θ, as in part B of Fig. 7.33, the similarity of the two results is apparent.

For any iterative matching technique, such as used here, one must raise the question of the uniqueness of the solution. Granted that there is excellent concordance between experimental and calculated results, and also that there

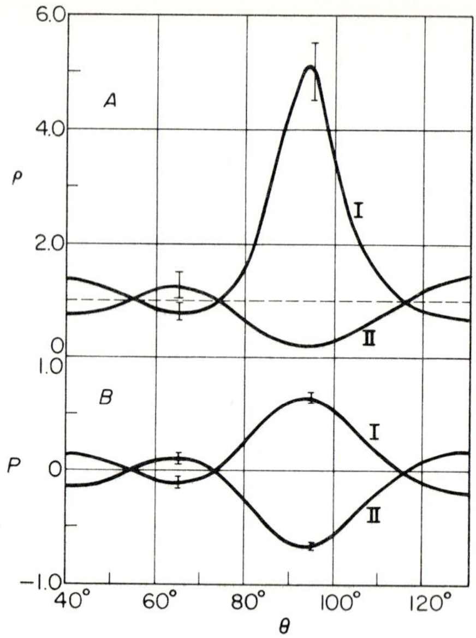

FIG. 7.33. Comparison of data plotted as $\rho(\theta)$ with the same data plotted as P. Curves I and II differ by interchange of I_1 and I_2 (Jacobsen *et al.*, 1967).

is internal consistency among results obtained at several different wavelengths, might there not be other size distributions that would fit these data equally well?

An error contour map for the aerosol corresponding to Fig. 7.32 is shown in Fig. 7.34 in order to illustrate this point. This map is for $\lambda = 0.578\ \mu$. The contour lines in the $\alpha_M\sigma_0$-domain represent loci of equal error between experiment and calculations as measured by the sum of the squares of the deviations ΣD^2. The lines designated 1, 2, 3, etc., correspond to $\Sigma D^2 = 0.1$, 0.2, 0.3, etc. The topography in Fig. 7.34 is a deep well with steep sides. The precision of the solution at the bottom of the well is depicted by the shaded area within the lowest contour line. Since this is the only well, this solution is unique over the domain shown.

Considerable experience with this method has been accumulated in the author's laboratory in connection with particle size analysis of La Mer sulfur sols (Kerker *et al.*, 1963a; Rowell *et al.*, 1968a), octanoic acid aerosols (Matijević *et al.*, 1964), sodium chloride aerosols (Espenscheid *et al.*, 1964b), silver chloride aerosols (Espenscheid *et al.*, 1964b), vanadium pentoxide aerosols (Jacobsen *et al.*, 1967), polystyrene latexes (Kratohvil and Smart, 1965), and sulfuric acid aerosols (Coutarel *et al.*, 1967). In general, a unique solution was always obtained.

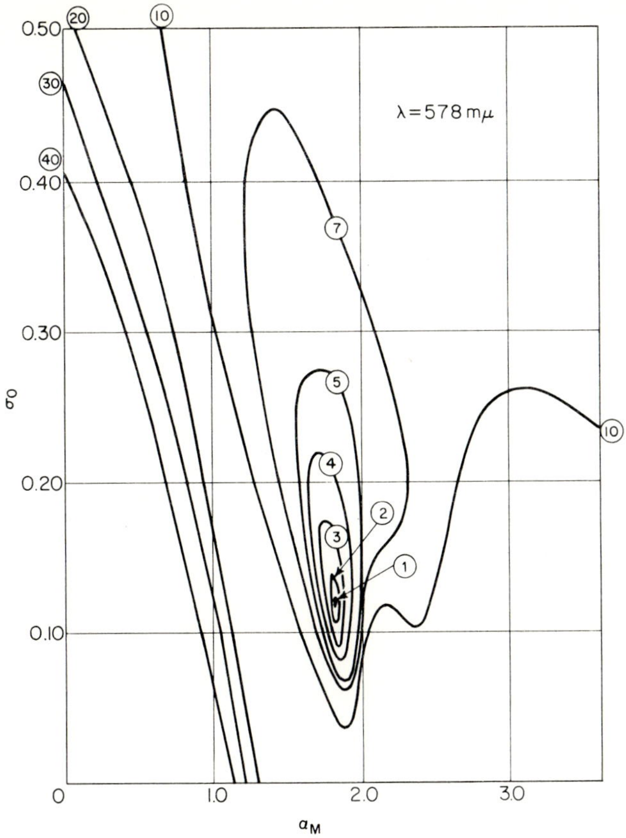

FIG. 7.34. Error contour map for the results in Fig. 7.32 at $\lambda = 0.578\ \mu$ (Jacobsen *et al.*, 1967).

A unimodal distribution is normally anticipated, and when this is not too broad any of the two-parameter distribution functions heretofore considered provide an adequate qualitative description. However, there are often naturally occurring distributions of particles that may have two or more maxima in the frequency curve. For example, if an initially unimodally-distributed system of spherical droplets undergoes coalescence as a result of Brownian coagulation, the frequency curve will, after a time, exhibit a second maximum at the size corresponding to a doublet. The bottom curve in Fig. 7.35, drawn with a full line, is the frequency at the half-lifetime for a distribution initially characterized by $a_M = 0.25\ \mu$ and $\sigma_0 = 0.10$ (Willis *et al.*, 1967). The polarization ratio for this coagulated suspension, computed for $m = 1.50$ and $\lambda = 0.546\ \mu$, is shown in the upper curve. Now, if these polarization ratio data are used as input for the method described above, based upon

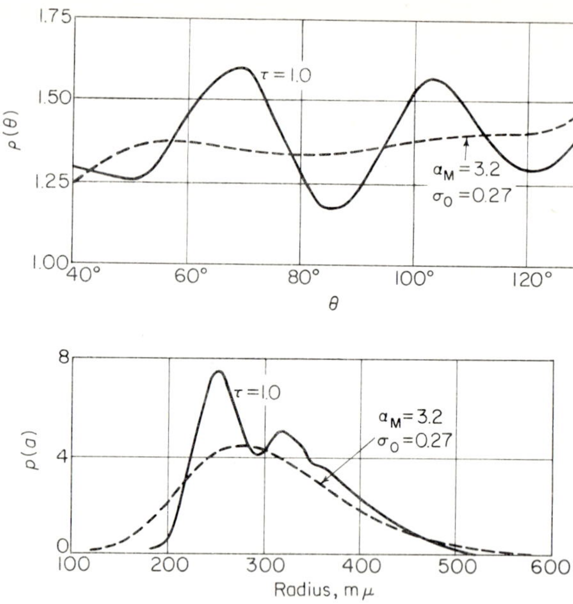

FIG. 7.35. Full line, lower curve: frequency distribution at $\tau = 1$ for initial distribution $a_M = 0.25\ \mu$, $\sigma_0 = 0.10$. Upper curve: full line is polarization ratio calculated for $\tau = 1$; dashed line is polarization ratio for unimodal distribution which fit this best. Dashed line, lower curve: frequency distribution of the unimodal distribution (Willis *et al.*, 1967). τ is half lifetime.

the assumption of a ZOLD, the result is the frequency distribution shown as a dashed curve. Obviously, it cannot depict the bimodal character of the actual distribution, but for some purposes it might not be considered too bad a representation of it. One might anticipate that when the peaks are sharper and further removed from each other, the assumption of a unimodal model would result in a failure of the method by an inability to find a distribution leading to the experimental data, and, certainly, even if a "solution" were obtained, it would necessarily be erroneous.

The size distribution of an aerosol consisting of coated spheres has been determined by Espenscheid *et al.* (1965). The spherical cores were of silver chloride, and the concentric spherical shells were of linolenic acid. These materials had refractive indices of 2.105 and 1.482, respectively, for which extensive theoretical functions had been computed (Kerker *et al.*, 1962), based upon the theory of Aden and Kerker (1951).

For a system of concentric spheres with size parameters α and ν, there are two distinct distribution functions, viz. $p(\alpha)$, which defines the distribution of the inner spheres, and $g(\nu)$, which is the distribution function for the concentric spheres and is determined by the manner in which the material of the spherical shell is distributed upon the cores. The intensity function, i_1, corresponding to these distributions is

$$i_1 = \int\int p(\alpha)g(\nu)i_1(\alpha, \nu)\,d\alpha\,d\nu \tag{7.5.36}$$

In order to avoid the extremely long calculation when both $p(\alpha)$ and $g(\nu)$ are independent distribution functions, two special models were considered. In both cases the inner spheres were assumed to follow a ZOLD. On the one hand, the spherical shell of every particle was assumed to have a constant thickness parameter β so that

$$i_1 = \int p(\alpha)i_1(\alpha, \nu)\,d\alpha \tag{7.5.37}$$

$$\nu = \alpha + \beta \tag{7.5.38}$$

On the other hand, it was assumed that the total volume of material in each spherical shell is the same, independent of the core size. If this volume of material corresponds to an equivalent sphere with size parameter γ, then

$$\nu = (\alpha^3 + \gamma^3)^{1/3} \tag{7.5.39}$$

is the appropriate value to use in (7.5.37).

In Figs. 7.36 and 7.37 some computed results, corresponding to each of these models, are illustrated. These are both for cores having $\alpha_M = 2.0$ and $\sigma_0 = 0.10$ and 0.095, respectively. The values of β and γ have been chosen so that the corresponding curves in each figure will have approximately the same value of ν_M. It is obvious that the light scattering is quite sensitive to a small change in coating thickness. On the other hand, it is not very sensitive to whether the coating material is distributed according to one model or the other.

An experimental result is shown in Fig. 7.38. The square and circular points respectively give the polarization ratio for a silver chloride aerosol prior to and after being coated with linolenic acid. The analysis of the data for the core aerosol resulted in $\alpha_M = 1.9$ and $\sigma_0 = 0.095$. In attempting to reduce the data for the coated aerosol, both the constant thickness model and the constant volume model were used. Starting with the size distribution, already deduced for the core, various values of β and γ were used until an optimum fit between experimental and calculated results was obtained. In this process, the original core size distribution was also perturbed somewhat, leading to values of α_M, σ_0, and ν_M listed in the upper left corner of the figure. The theoretical polarization values, which are shown as a dashed and a dotted curve, correspond quite well with the experimental curve, and the size distributions obtained agree quite well with each other, as well as with the electron microscope size histograms obtained for both the cores and the coated aerosol, as shown in the inset.

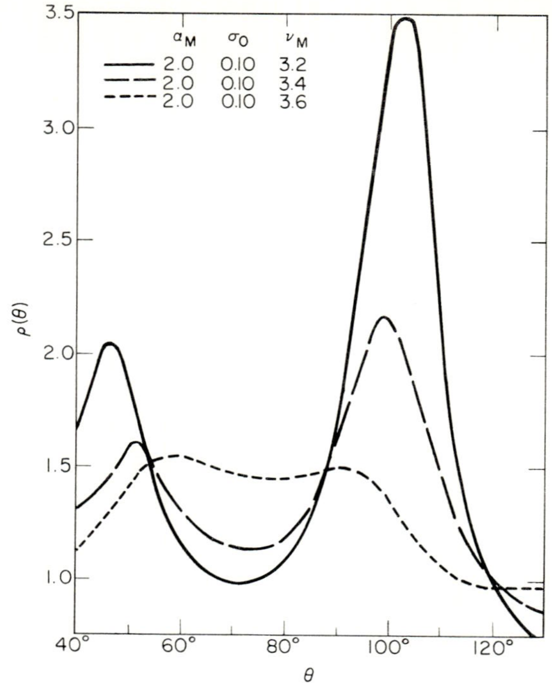

FIG. 7.36. Computed polarization ratio for an aerosol consisting of concentric spheres, assuming constant thickness of coating: $m_1 = 2.1050$, $m_2 = 1.4821$, $\alpha_M = 2.0$, $\sigma_0 = 0.10$, thickness of the coating $\beta = 1.2$, 1.4, and 1.6, respectively (Espenscheid *et al.*, 1965).

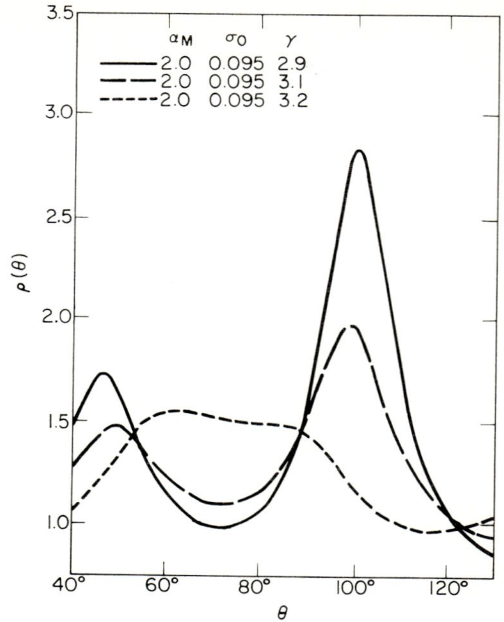

FIG. 7.37. Computed polarization ratio for an aerosol consisting of concentric spheres, assuming constant volume of coating: $m_1 = 2.1050$, $m_2 = 1.4821$, $\alpha_M = 2.0$, $\sigma_0 = 0.095$, thickness of coating $\gamma = 2.9$, 3.1, and 3.2, respectively (Espenscheid *et al.*, 1965).

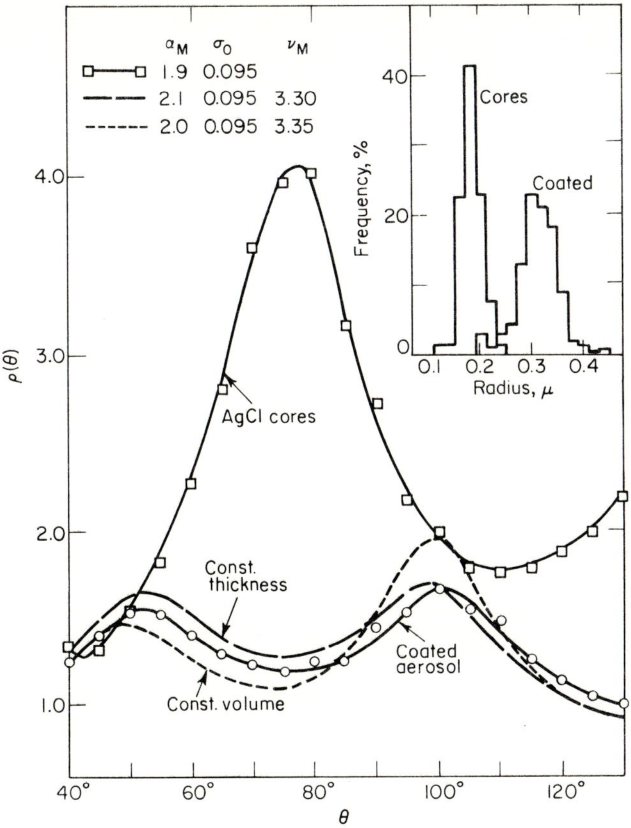

FIG. 7.38. A comparison of the experimental values of $\rho(\theta)$ for AgCl cores (squares) with the same aerosol after being coated with linolenic acid (circles). Computed results: AgCl cores, $\alpha_M = 1.9$, $\sigma_0 = 0.095$; coated aerosol: $\alpha_M = 2.1$, $\sigma_0 = 0.095$, $\nu_M = 3.30$ (constant-thickness model—dashed curve) or $\alpha_M = 2.0$, $\sigma_0 = 0.095$, $\nu_M = 3.35$ (constant-volume model—dotted line). Wavelength 546 mμ. The inset shows the frequency histograms, obtained by electron microscopy of the AgCl cores and of the coated aerosol (Espenscheid *et al.*, 1965).

7.5.3 METHODS BASED UPON THE SPECTRA OF THE SCATTERED LIGHT OR TRANSMITTED LIGHT

a. Scattering Ratio. A method for the analysis of particle size distributions developed by Stevenson *et al.* (1961) utilizes the variation with wavelength of the polarization ratio at a particular angle of observation. These workers specify that the quantity to be measured, which they call the scattering ratio, is the ratio of the intensity of two beams scattered when a polarizing prism, placed in the incident beam, transmits the parallel and perpendicular components, respectively. For spherical isotropic particles, provided there is no multiple scattering, it makes no difference whether the polarizing prisms are placed in the incident beam, in the scattered beam, or in both, since there are no depolarized components. The scattering ratio is given by

$$\rho(p, q) = \frac{I_2(p, q)}{I_1(p, q)} = \frac{\int_p^\infty i_2(\alpha)(\alpha - p)\exp(-[(\alpha - p)/q]^3)\, d\alpha}{\int_p^\infty i_1(\alpha)(\alpha - p)\exp(-[(\alpha - p)/q]^3)\, d\alpha} \tag{7.5.40}$$

where p and q are the two parameters that determine size distribution. This particular distribution has already been discussed [(7.5.25) to (7.5.32)].

Extensive computations of $I_1(p, q)$, $I_2(p, q)$, and $\rho(p, q)$ have been carried out for angle of observation $\theta = 90°$ and the following ranges of m, q, and p:

$$m = 1.05(0.05)1.30$$

$$q = 0.0(0.2)12.8$$

$$p = 0.0(0.2)25.2 \qquad \text{for} \qquad q = 0.2$$

$$p = 0.0(0.2)24.8 \qquad \text{for} \qquad q = 0.4$$

$$\vdots \qquad\qquad\qquad\qquad \vdots$$

$$p = 0.0 \qquad\qquad \text{for} \qquad q = 12.8$$

The smallest value of the size parameter present in the distribution is given by p (7.5.31), and the spread of the size distribution by q (7.5.32). The values above were selected so that the upper size limit of intensity functions that need be considered is within the range of previously tabulated values (Pangonis *et al.*, 1957). The tabulated results include values of $\rho(p, q)$ and of $I_1(p, q)$ (Stevenson and Heller, 1961; Yajnik *et al.*, 1968).

Figures 7.39 and 7.40 illustrate how $\rho(p, q)$ varies with the distribution parameters for $m = 1.20$. In Fig. 7.39, the scattering ratio is plotted against p for $q = 0.0$ (monodisperse case), 1.0, and 4.0. The oscillations characteristic of the monodisperse case become progressively smaller with increasing polydispersion, resulting in a nearly linear curve for $q = 4.0$. Figure 7.40 is a plot of the scattering ratio against q for fixed values of p. Here, the scattering ratio drops linearly with decreasing polydispersion and then goes through a minimum, provided the size parameter for the smallest size $p > 2$.

We now consider the manner of treating the experimental data. It is convenient to normalize the data by using the wavelength in water of the green line of mercury spectrum as a reference wavelength $\lambda_R = 0.4094\ \mu$ ($\lambda_0 = 0.5461\ \mu$), and to define

$$p_R = 2\pi a_0/\lambda_R = p(\lambda/\lambda_R) \tag{7.5.41}$$

$$q_R = 2\pi s/\lambda_R = q(\lambda/\lambda_R) \tag{7.5.42}$$

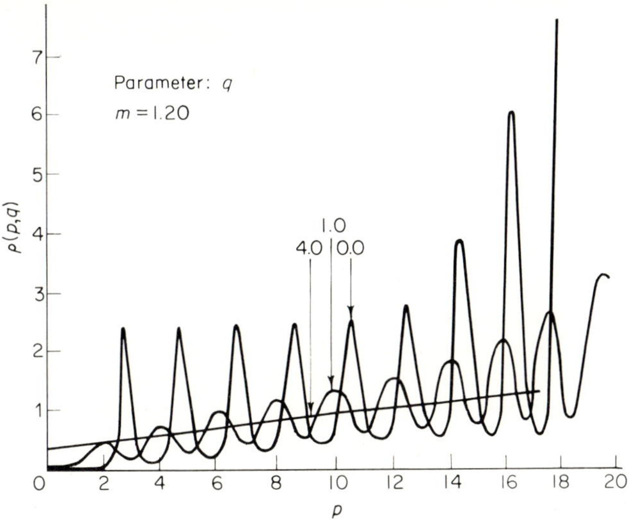

FIG. 7.39. Variation of the scattering ratio $\rho(p, q)$ with parameter p for various values of parameter q at $m = 1.20$ (Stevenson *et al.*, 1961).

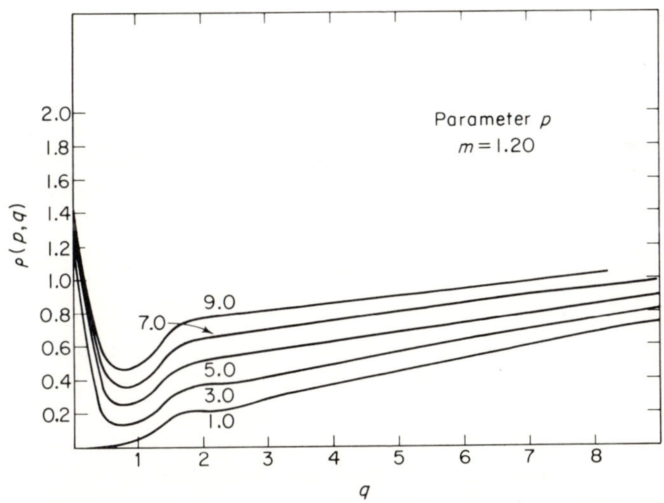

FIG. 7.40. Variation of scattering ratio $\rho(p, q)$ with parameter q for various values of parameter p (Stevenson *et al.*, 1961).

Then, for any distribution defined by p_R, q_R, there will be a value of the polarization ratio corresponding to each wavelength which constitutes the spectrum of the polarization ratio. Two sets of such spectra are shown in Figs. 7.41 and 7.42 for $p_R = 2.8$ and 8.2, respectively. By comparing the experimental spectrum with the various theoretical values, one may arrive at an estimate of the standard parameters, p_R and q_R, and through (7.5.41) and (7.5.42), those corresponding to the distribution parameters, a_0 and s. In this treatment, the dispersion of the relative refractive index with wavelength has been neglected.

Heller and Wallach (1963) have utilized this method to determine the size distribution in a polystyrene latex. A polydisperse system was obtained by mixing, selectively, eighteen relatively monodisperse latexes so as to approximate a distribution of the form given by (7.5.25). The histogram of this mixture was determined by an electron microscopic examination and the spectrum of the polarization ratio at 90° was obtained over the range of wavelengths, $\lambda_0 = 0.451$ to $0.597\ \mu$. The experimental data (dots) and the theoretical curves which fit them most closely are shown in Fig. 7.43. By interpolation among these curves, the best fit turns out to be $p_R = 6.84$, $q_R = 1.49$. The distribution curve corresponding to these parameters is compared in Fig. 7.44 with the histogram determined with the electron microscope. The modal radius, as obtained by the two methods, differs by 5%, the half-width by 22%, the "half-spread" by 2%, and the smallest particle radius by 6%. These results are well within the range of experimental error attributable to the electron microscope analysis.

In addition, these authors investigated two preparations having a Gaussian distribution and a negatively skewed distribution, respectively. In the former case, the distribution parameters that led to the best fit with the experimental ρ-spectrum yielded a positively skewed size distribution, which agreed quite well with the Gaussian-shaped histogram obtained with the electron microscope. However, with the system having the negatively skewed distribution of particle sizes, there was no theoretical spectrum that satisfied the experimental one in satisfactory approximation over a sufficiently extensive spectral range.

If a satisfactory fit between the theoretical and experimental ρ-spectra cannot be obtained, this may indicate that the form of the actual distribution is different from that which was assumed. Heller and Wallach (1963) have proposed, in such a case, to represent the distribution function as a heterogeneous one, i.e., as the superposition of two or more of the simple distributions. However, this would require the evaluation of five independent parameters and it is highly unlikely that these could be fit uniquely to a real set of experimental data.

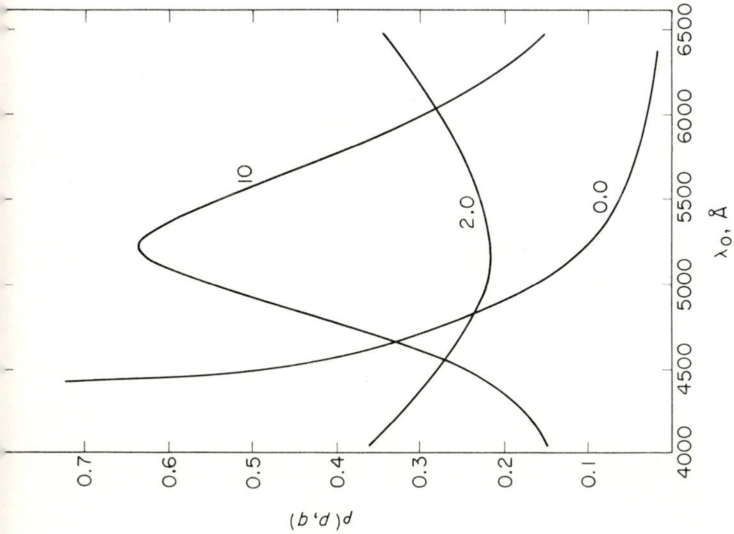

FIG. 7.41. (above) Spectra of the scattering ratio for $p_R = 2.8$ and $q_R = 0$, 2.0, and 10 at $m = 1.20$ (Stevenson *et al.*, 1961).

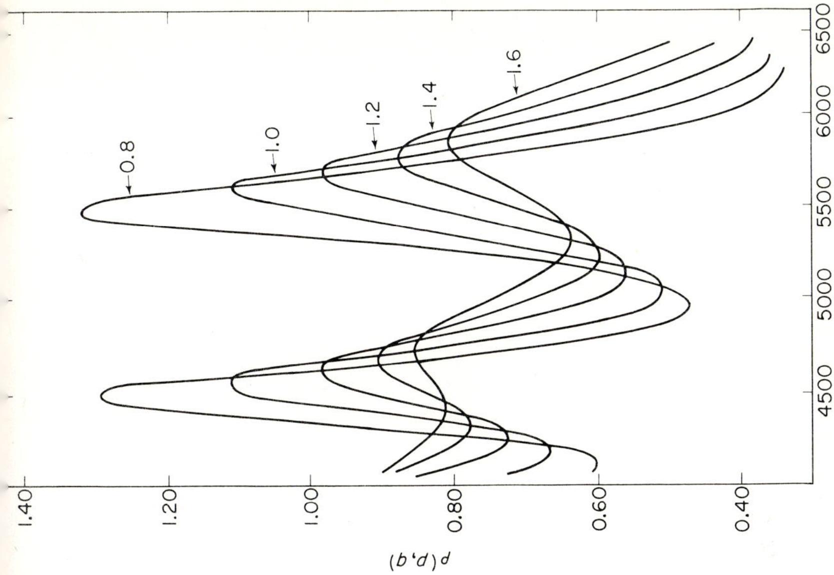

FIG. 7.42. (right) Same as Fig. 7.41 for $p_R = 8.2$ and the indicated values of q_R.

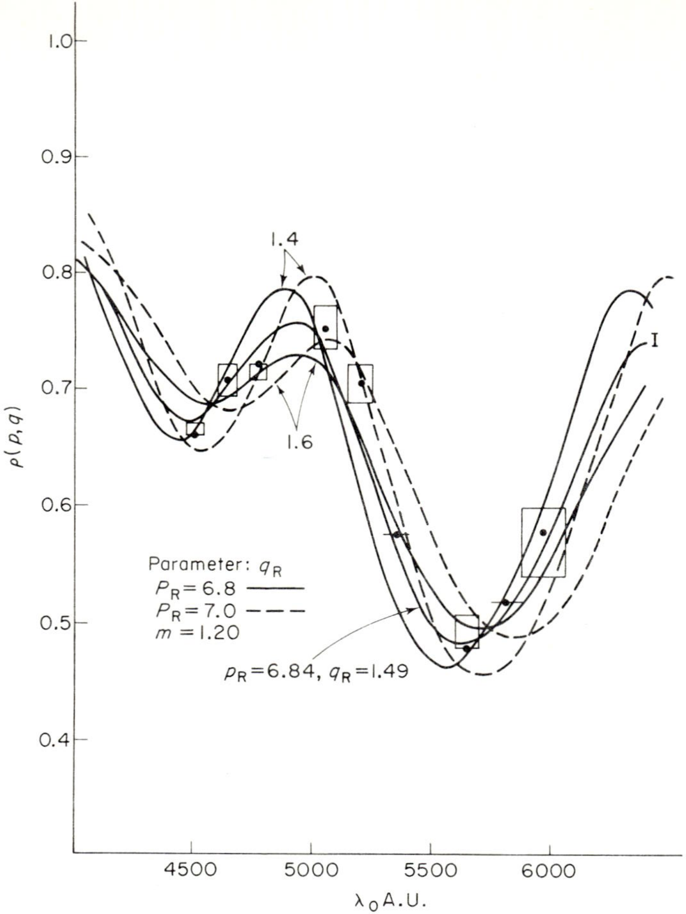

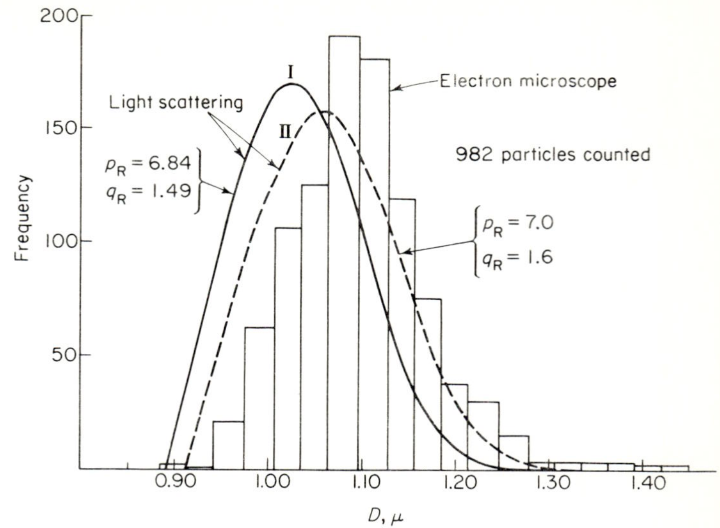

FIG. 7.44. (above) Comparison of size distribution obtained by light scattering (curve I) with the electron microscope histogram. Curve II represents the extreme distribution considered in the analysis, giving the spread of the results by light scattering (Heller and Wallach, 1963).

FIG. 7.43. (left) Spectrum of polarization ratio for a heterodisperse polystyrene latex (dots) compared with various theoretical curves. Heavy curve for $p_R = 6.84$ and $q_R = 1.49$ fits the experimental points best (Heller and Wallach, 1963).

Heller and Wallach (1964) have also explored the sensitivity of this method for very narrow distributions. They analyzed the polarization spectra of two polystyrene latexes which are normally referred to as monodisperse. The distribution curves determined from the light scattering data for these samples agree quite well with the histograms obtained from the electron microscope counts (Figs. 7.45 and 7.46). The resolution of this light scattering method is such that it clearly enables one to determine the essential features of the size distribution in such systems. Kratohvil and Smart (1965) found that the polarization ratio method also was able to resolve the detailed features of such narrow distributions.

b. Turbidity Spectra. There is no *a priori* reason why the polarization ratio must be selected at 90°, and more information could be obtained by working at additional angles. Alternatively, other scattering properties can be used. For example, Wallach *et al.* (1961) have worked out a procedure analogous to the above for evaluating the turbidity spectra. These spectra can be obtained with conventional spectrophotometric instruments, and their analysis requires far less computational work than the polarization spectra. For a polydisperse system, the turbidity is given by

$$\tau = (\lambda^2/2\pi) \int_{a_0}^{\infty} \Sigma_{\text{sca}} p(a)\, da \tag{7.5.43}$$

where in this case the distribution function is normalized to give the number of particles per unit volume, and the cross section for scattering is given by

$$C_{\text{sca}} = (\lambda^2/2\pi)\Sigma_{\text{sca}} = (\lambda^2/2\pi) \sum_{n=1}^{\infty} (2n + 1)\{|a_n|^2 + |b_n|^2\} \tag{7.5.44}$$

It is convenient to utilize the dimensionless quantity

$$\frac{\lambda\tau}{\phi} = \frac{3\lambda^3}{8\pi^2} \int_{a_0}^{\infty} \Sigma_{\text{sca}} p(a)\, da \bigg/ \int_{a_0}^{\infty} a^3 p(a)\, da \tag{7.5.45}$$

where ϕ, the volume fraction of scattering material, is given by

$$\phi = \int_{a_0}^{\infty} (4\pi a^3/3) p(a)\, da \tag{7.5.46}$$

and ϕ is related to the weight fraction, ω', and the concentration expressed as grams of solute per milliliter of solution, c, through (7.3.4). Upon replacing a, a_0, and s in (7.5.45) by α, p, and q, this becomes

$$\frac{\lambda\tau}{\phi} = \frac{3\pi \displaystyle\int_{p}^{\infty} \Sigma_{\text{sca}}(\alpha - p) \exp(-[(\alpha - p)/q]^3)\, d\alpha}{\displaystyle\int_{p}^{\infty} \alpha^3(\alpha - p) \exp(-[(\alpha - p)/q]^3)\, d\alpha} \tag{7.5.47}$$

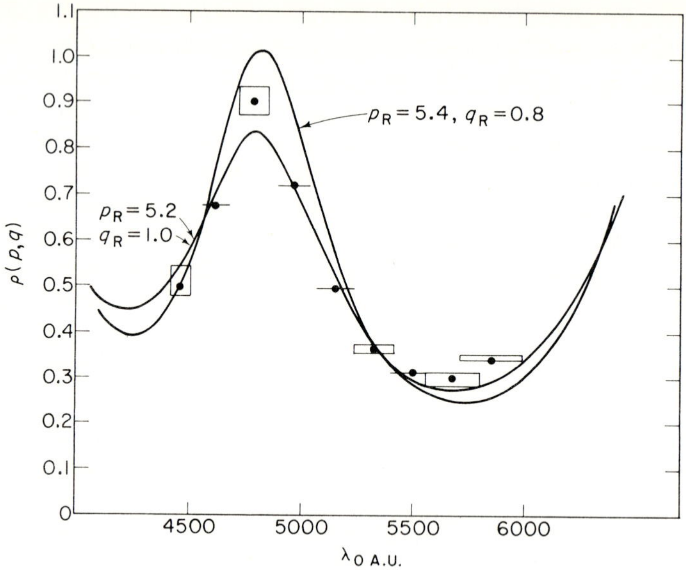

FIG. 7.45. Spectrum of polarization ratio for a "monodisperse" polystyrene latex (dots) compared with theoretical curves (Heller and Wallach, 1964).

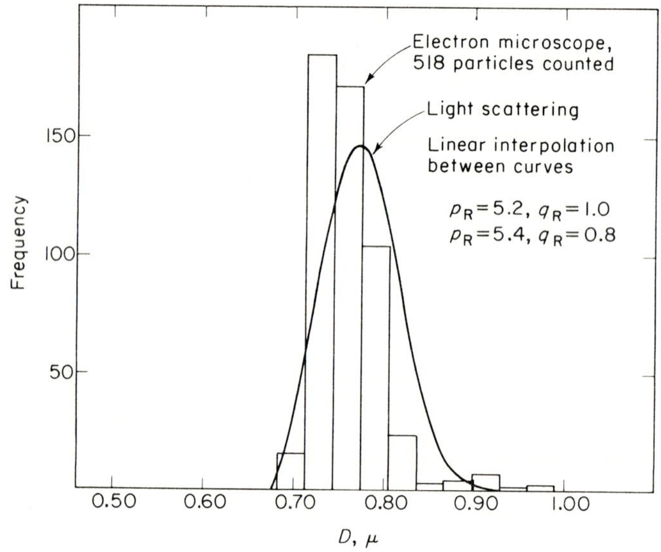

FIG. 7.46. Comparison of size distribution obtained by light scattering (data of Fig. 7.45) with the electron microscope histogram (Heller and Wallach, 1964).

Values of this quantity have been tabulated by Wallach *et al.* (1961) for $m = 1.05(0.05)1.30$ over a convenient range of values of p and q.

Again, it was found convenient to standardize with respect to the green mercury line. Both the experimental values of $\lambda\tau/\phi$ and the theoretical values were standardized with respect to the value at λ_R. Such normalized theoretical spectra are illustrated in Fig. 7.47. In this figure there are six curves, corresponding to six sets of values, of the parameters p_R and q_R. For each of these, the value of $\lambda\tau/\phi$, relative to its value at λ_R, is plotted against wavelength over the visible range. Comparison of the corresponding experimental turbidity spectra with these results can provide a relatively rapid estimation of the size distribution. If the answer obtained here is concordant with that obtained from the polarization spectrum, this provides additional reassurance of the reliability of the method.

Wallach and Heller (1964) have tested this method with precisely the same three latex systems used in connection with the polarization spectra method. A comparison of the size distribution curve obtained for the Gaussian distribution by the two methods is shown in Fig. 7.48, where the histograms obtained by electron microscopy are also given. There is excellent agreement among all three results.

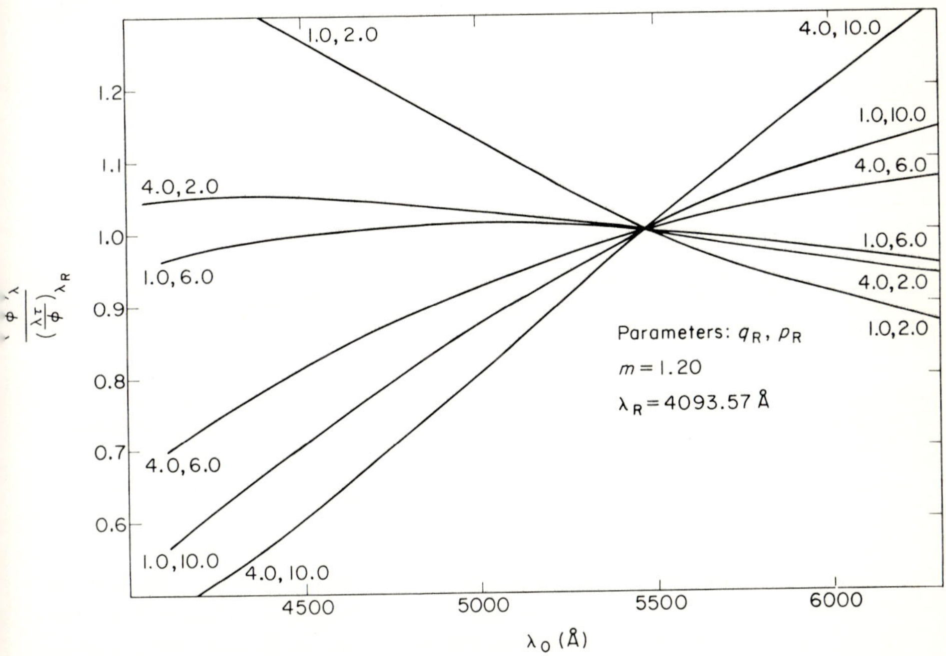

FIG. 7.47. Normalized theoretical spectra of $\lambda\tau/\phi$ for six sets of values of q_R and P_R (Wallach *et al.*, 1961).

These authors have discussed the relative advantages of each of these two light scattering methods. In addition to the obvious advantage that turbidity spectra may be obtained with conventional spectrophotometers requiring only relatively simple modifications, the theoretical curves can be derived and constructed rapidly. Furthermore, provided $a_0 < 0.5\ \mu$, the sensitivity of the turbidity spectra is preserved even at relatively high degrees of dispersion (a half width of 0.5 μ). On the other hand, the polarization spectra are much more sensitive to changes in the smallest size, a_0. Also, the turbidity spectra become quite insensitive to the particle size distribution as the particles become large compared to the wavelength ($a_0 \cong 4\ \mu$), since then the scattering efficiency approaches the limiting value of 2. The polarization spectra method is not subject to this restriction, provided that the degree of heterodispersion is moderate. Of course, both methods will be completely insensitive for determination of particle size distribution if all of the particles are so small that they behave as Rayleigh scatterers.

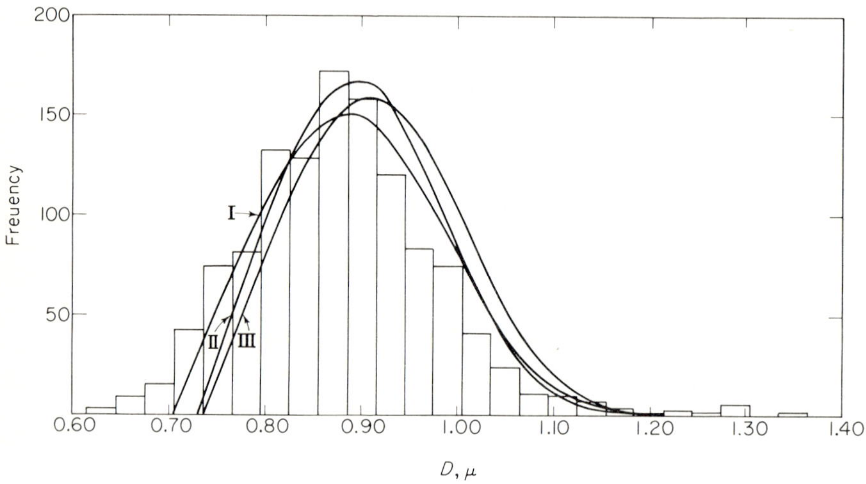

FIG. 7.48. Particle size distribution derived from turbidity spectra (curves I and II) and that from the scattering ratio (curve III) compared with the electron microscope histogram (Wallach and Heller, 1964).

Maxim *et al.* (1966) have examined the possibility of multivaluedness in the turbidity spectra technique and have found for the particular latex system which they investigated that this was a serious problem. Because a single turbidimetric determination specified a family of alternative distributions, they recommended combining this with other techniques for measuring various moments of the distribution. This, indeed, is what the polarization ratio method, discussed in Section 7.5.2, does. Sufficient different moments of the distribution are obtained implicitly from the polarization measurements at 19° angles of observation. It would appear, then, if the scattering ratio method were extended to additional angles other than 90°, that the problem of multivaluedness could be eliminated, provided the distribution does not become too broad.

7.5.4 Other Turbidimetric Methods

A number of workers have utilized a method which may be considered to be a somewhat less sophisticated version of the turbidity spectra method of Wallach *et al.* (1961). The particles supposedly follow a two-parameter, logarithmic distribution so that the turbidity may be calculated from an expression equivalent to (7.5.43). This calculation is carried out over a suitable range of values of the two parameters. The turbidity of the dispersion is measured at only two to four wavelengths. Each of these measurements defines a family of values for each of the distribution parameters, and the point of intersection of the loci corresponding to each set uniquely determines the parameters for the system.

Wales (1962) studied polyisoprene latexes with radii ranging up to several microns in radius, taking measurements at four wavelengths in the near infrared region of 0.6 to 1.1 μ. He utilized a logarithmic normal distribution on a weight basis. There was considerable discrepancy between the results obtained by this method and those from a Coulter counter, which was attributed to failure to obtain adequate monochromaticity in the turbidity measurements and other errors of measurement, both in the turbidity and the Coulter counter experiments.

Meehan and Beattie (1960) obtained measurements at only two wavelengths ($\lambda = 0.450$ and $0.800\ \mu$), working with silver bromide hydrosols. Although they claimed to have used a logarithmic normal distribution, in actual fact, their distribution is the ZOLD described earlier (7.5.16). There was moderately good agreement between the size distributions obtained turbidimetrically and by electron microscopy, despite the considerable uncertainties in the latter technique (10 to 20% estimated error for these particles). Because the turbidimetric measurements were obtained at only two wavelengths, there was no internal check on the consistency of these results.

Still another application of this method was carried out by Mailliet and Pouradier (1961) using silver chloride hydrosols as well as silver chloride suspended in a solution of 90% glycol–10% water. The latter system gives a different relative refractive index, and thereby provides a more stringent test of the consistency of the method. Measurements with the hydrosols were carried out at three wavelengths ($\lambda = 0.440$, 0.572, and $0.700\ \mu$ and

$m = 1.58$, 1.55, and 1.536, respectively), and at $\lambda = 0.589$ ($m = 1.45$) with the glycol-water solution. The logarithmic normal distribution was assumed to represent the particles, and this was confirmed by electron microscopy.

Some results are shown in Fig. 7.49, where the loci corresponding to the size parameter and the breadth parameter of the distribution are shown at each of the three wavelengths for three sols. For these selected examples, the curves intersect at a unique point, thereby determining the distribution. It is interesting to note that for the distributions with the smallest particles, although the size parameter is well established, there is considerable uncertainty in the breadth parameter. This is because these particles are nearly Rayleigh scatterers for which the turbidity depends only upon the average size (turbidity average) and is independent of the breadth of the distribution.

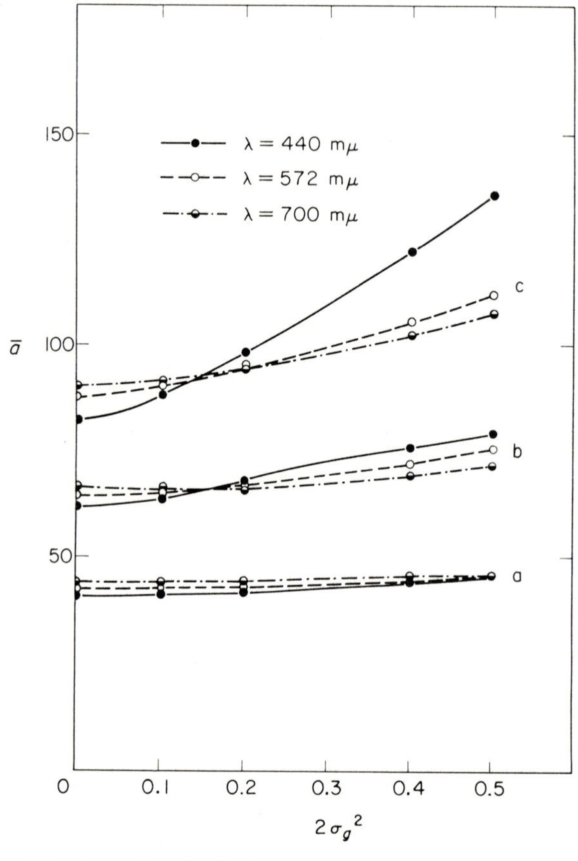

FIG. 7.49. Loci for those size distributions satisfying the turbidimetric data at $\lambda = 0.440$, 0.572 and 0.700 μ for three silver chloride hydrosols (a, b, c). The points of intersection determine the size distribution in terms of the mean radius $\bar{a}$ and σ_g (Mailliet and Pouradier, 1961).

In other cases, the three curves did not intersect at a single point, thereby introducing considerable uncertainty in the determination. In general, these workers found that the size parameter obtained was in good agreement with that from electron microscopy, usually to within 10%. However, the determination of the breadth parameter was considerably less precise. Of course, in these experiments an error is introduced because it is unlikely that either the silver chloride or silver bromide particles are perfectly spherical.

A somewhat different method based upon transmission measurements has been described by Gledhill (1962). He also assumed a two-parameter distribution, which he termed a logarithmic normal distribution, but which, in fact, was similar to the ZOLD discussed above (7.5.16). Two experimental quantities for the estimation of the distribution parameters were selected to be the specific turbidity at a particular wavelength and the wavelength dependence at that wavelength, g_0, defined earlier (7.3.34). These are related to the distribution in the usual way by

$$(\tau/\phi) = \int_0^\infty (\tau/\phi)_\alpha p(\alpha)\, d\alpha \tag{7.5.48}$$

$$(g_0\tau/\phi) = \int_0^\infty (g_0\tau/\phi)_\alpha p(\alpha)\, d\alpha \tag{7.5.49}$$

The quantities on the left are measured, and the integrals on the right-hand side are evaluated for various values of the distribution parameters until concordance is obtained between the experimental and theoretical quantities. The method was tested at $\lambda = 0.546\ \mu$ for suspensions of spherical particles in the submicron range having a rather broad distribution of sizes. Good agreement was obtained between the distribution parameters obtained by this method and those from electron microscopy.

Dettmar *et al.* (1963) have proposed still another technique for treating transmission data with a view to obtaining the particle size distribution [see also Lode *et al.* (1962)]. They computed the scattering cross section for a sample containing a unit volume of particular material using

$$s = \frac{3}{4} \frac{\int_0^\infty a^2 Q_{\text{sca}} p(a)\, da}{\int_0^\infty a^3 p(a)\, da} \tag{7.5.50}$$

where $p(a)$ is the logarithmic normal distribution function defined by (7.5.5). This was done for the seven refractive indices, $m = 1.23, 1.60, 1.78, 2.06, 2.37, 2.54$, and 2.74, corresponding to the various pigment materials in which they were interested. Values of the geometric mean standard deviation ranged from 1.1 to 2.0 in steps of 0.1, and the geometric mean radii were selected so that the distribution would be encompassed by the values for which they had carried out calculations, viz., from $\alpha = 0.1$ to 20.0 for the first four refractive indices, and from $\alpha = 0.1$ to 10.0 for the last three. The computed results are presented as plots of $a_m \cdot s$ against λ/a_m, such as that shown in Fig. 7.50 for

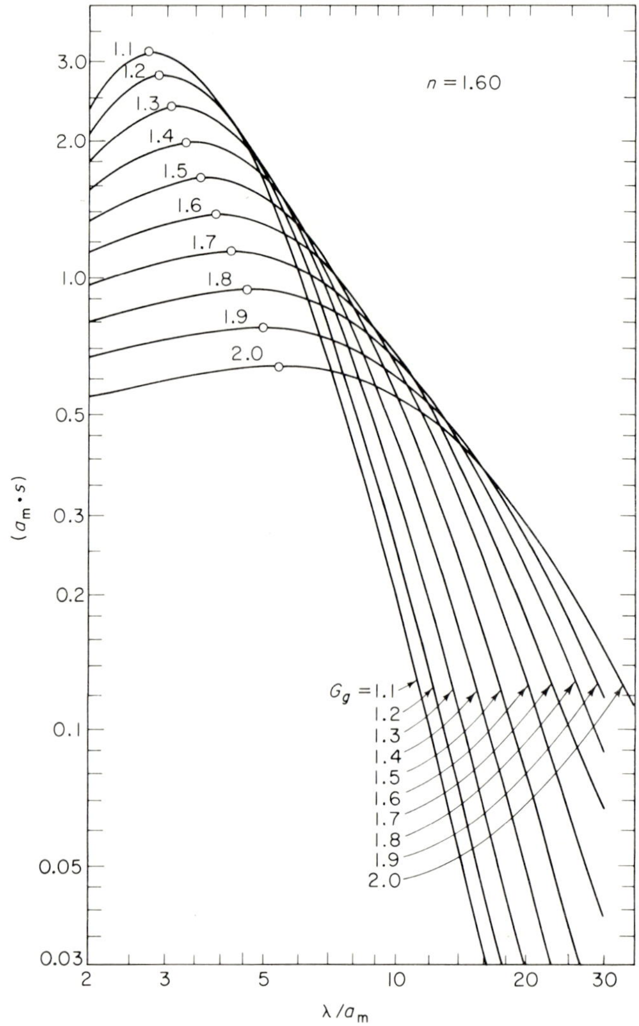

FIG. 7.50. Plot of $(a_m \cdot s)$ against λ/a_m for various values of σ_g where s is defined by Eq. (7.5.50) for $m = 1.60$ (Dettmar *et al.*, 1963).

$m = 1.60$. The parameter for each separate curve is the geometric mean standard deviation, σ_g.

In order to determine the distribution parameters of a particular experimental sample, a plot of s versus wavelength is superimposed upon Fig. 7.50. This may be done by plotting the experimental data on transparent graph paper, using logarithmic grids, and then moving this over the theoretical curves. The value of s and the turbidity are related by

$$s = \tau/\phi \tag{7.5.51}$$

Although the experimental curve will be displaced on the log–log plot, its shape should correspond to one of the theoretical curves, provided the assumption of a logarithmic normal distribution is correct, and also provided that the system conforms to the requirements of the scattering theory.

The value of σ_g is obtained directly by the matching procedure. The value of a_m is obtained, in turn, by comparison of the abscissas λ/a_m and λ of corresponding points on the matched curves.

The results obtained by this technique, with various pigments, have been compared with electron microscope results by Weber (1963). The agreement was moderately good, considering that the colloidal particles deviate from the required spherical shape and, in some cases, from the logarithmic normal law. However, the various suspensions tested covered only a narrow range of size distributions ($a_m = 0.12$ to $0.25\ \mu$ and $\sigma_g = 1.2$ to 1.8), so that the usefulness of this method over a broader range of size distributions remains to be investigated.

Walstra (1968) has also used a curve-matching technique in which the specific turbidity over the entire visible wavelength range ($\lambda_0 = 0.42$ to $0.75\ \mu$) is matched with various theoretical curves. This gave results for emulsions (milk, oil in water) in agreement with Coulter counter and direct microscope counts.

Watillon and Dauchot (1968) have used a direct curve matching technique with selenium sols which is particularly interesting because these sols consist of highly colored particles. Extensive calculations of the extinction efficiency were carried out (Dauchot and Watillon, 1967) using the complex refractive index over the wavelength range $\lambda = 0.240$ to $1.100\ \mu$. The absorption index of selenium rises very rapidly below about $0.650\ \mu$ to a maximum of about 1.3 at about $0.260\ \mu$. The selenium sols had quite narrow size distributions with average radii ranging from approximately 0.020 to $0.250\ \mu$.

These selenium sols were sufficiently monodisperse so that a first approximation could be obtained by comparison of the location of the maxima in the experimental extinction curve with theoretical curves for monodisperse systems. Then the entire experimental curves were compared with a set of calculated extinction curves corresponding to average sizes close to the approximate size and various values of the standard deviation. A normal distribution was used. Excellent concordance was obtained between the experimental curve and the best theoretical curve. The results agreed with the size distribution of the same system obtained by direct electron microscopic and ultramicroscopic counting.

Napper and Ottewill (1963a, c, d, 1964) have examined in detail the deviations obtained between experimental values of scattering by certain nonspherical particles and the results calculated, on the assumption that these particles scatter as equivalent spheres. The systems observed were silver bromide hydrosols, consisting of either cubes or octahedra and having a narrow distribution of particle size (Ottewill and Woodbridge, 1961). This was determined by electron microscopy. In calculating the various light-scattering quantities from the histograms for the cubes, the radius of the equivalent sphere was taken to be half of the face diagonal. For the octahedra, this distance was apparently the distance from the center to a corner.

For the octahedral particles, there was reasonably good agreement between experimental and calculated values of the angular intensity. However, the scattering by the cubes exhibited considerable discrepancy from the theoretical calculations for equivalent spheres. In both cases, there was quite good agreement between the experimental and computed values of the turbidity. Indeed, the agreement in this latter case was such as to indicate justification of the various turbidimetric methods described in this section for particles which deviate somewhat from spherical shape.

7.5.5 Scattering Combined with Differential Settling

Each of the techniques for determination of the particle size distribution described above required an *a priori* assumption of the form of the distribution. The frequency curve was determined from an analytical expression involving two parameters: e.g., one parameter gave a measure of average size; the other measured the width of the distribution. It was sometimes also necessary to specify the total amount of particulate material. A second limitation was the restriction to distributions which were not too broad; otherwise, the characteristic patterns in the light-scattering curves were washed out by the superposition of the many patterns characteristic of each of the varied sizes.

Gumprecht and Sliepcevich (1953a) have described a method based upon transmission measurements combined with differential settling which does not suffer from these limitations. The only requirement is that the particles be sufficiently dense relative to the ambient medium and sufficiently large so that differential settling measurements can be obtained.

At the start of the experiment, the particles are uniformly distributed throughout the container. These authors note that for most instruments encountered in practice, the transmission is given by

$$\ln(I_0/I) = -\ln T = \pi l \int RQ_{\text{sca}} a^2 p(a)\, da \tag{7.5.52}$$

R [see (7.5.58)] is a correction factor arising from the interception by the detector of the light scattered within a narrow forward cone (Gumprecht and Sliepcevich, 1953b), in addition to the directly transmitted light.

As tranquil settling takes place, the particles will drop through the light beam, as depicted schematically in Fig. 7.51. The uppermost layer of particles of each size range will fall through the beam at characteristic times and at each of these times there will be a corresponding decrease in the scattering and an increase in the transmission. The particular size corresponding to each time can be calculated from Stokes's law

$$a = \left(\frac{9h\mu}{2g(\rho_1 - \rho_2)t} \right)^{1/2} = \frac{k}{\sqrt{t}} \tag{7.5.53}$$

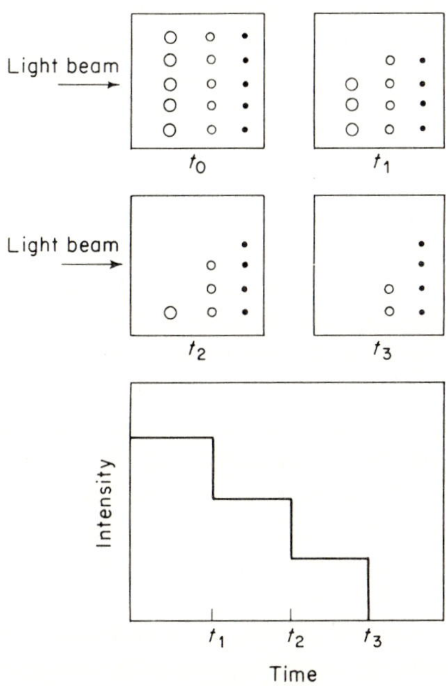

FIG. 7.51. The upper four boxes schematically illustrate tranquil settling at times t_0, t_1, t_2, and t_3. The curve below shows how the scattered intensity would decay with time (Kerker *et al.*, 1955).

where h is the height from the top of the container to the light beam, μ is the viscosity of the settling medium, g is the acceleration of gravity, and $(\rho_1 - \rho_2)$ is the difference in density of the dispersed particles and the surrounding medium. When the particle size is comparable with the mean free path of the gas molecules, the Cunningham correction to this formula must be applied (Davies, 1945).

The time rate of change of the optical transmission due to the settling is given by

$$-d(\ln I)/dt = -d(\ln T)/dt = \pi l R Q_{\text{sca}} a^2 p(a)\, da/dt \tag{7.5.54}$$

where $p(a)\,da$ is the number of particles in the size range between a and $a + da$ which irreplaceably drop out of the light beam during the time interval between t and $t + dt$. By means of (7.5.53), a can be eliminated from the above, leading to

$$p(a) = \frac{2t^{5/2}}{\pi l k^3 R Q_{\text{sca}}} \frac{d(\ln I)}{dt} \tag{7.5.55}$$

This determines the number of particles of each particular size a. Thus, the frequency function defining the distribution is obtained in terms of the slope of the transmission curve at a characteristic time. This, in turn, is related to the particular radius through (7.5.53) and, of course, R and Q_{sca} also correspond to the same radius. In this way the frequency function of the distribution can be determined from the time rate of decay of the intensity without any *a priori* assumptions regarding the form of the distribution or of the concentration of the particles.

Some results obtained by Gumprecht and Sliepcevich (1953a) are shown in Fig. 7.52 for a kerosene aerosol produced by atomization. These particles are sufficiently large to give an appreciable settling rate and considerably larger than either the aerosols or hydrosols used in the work described previously. Indeed, it is doubtful whether a system with such large particles and such a broad distribution could be analyzed by the earlier methods.

A somewhat different treatment of the transmittance-settling data has been proposed by Rose (1953). The curve of ln $1/T$, taken at the various times, is plotted against the particle radius corresponding to the size falling through the beam as calculated from (7.5.53). A typical example is shown in Fig. 7.53. Provided that R and Q_{sca} are constant ($R \to 1$ for a sufficiently small solid angle of acceptance by the detector, and $Q_{\text{sca}} \to 2$ for large values of α), the shaded area is proportional to the size frequency on a weight basis. However, the conditions of constancy for both parameters are difficult to attain since, as α becomes larger, the solid angle of acceptance necessary to maintain the value of R close to unity becomes correspondingly smaller. In order to correct for this, the instrument is calibrated by evaluating the quantity

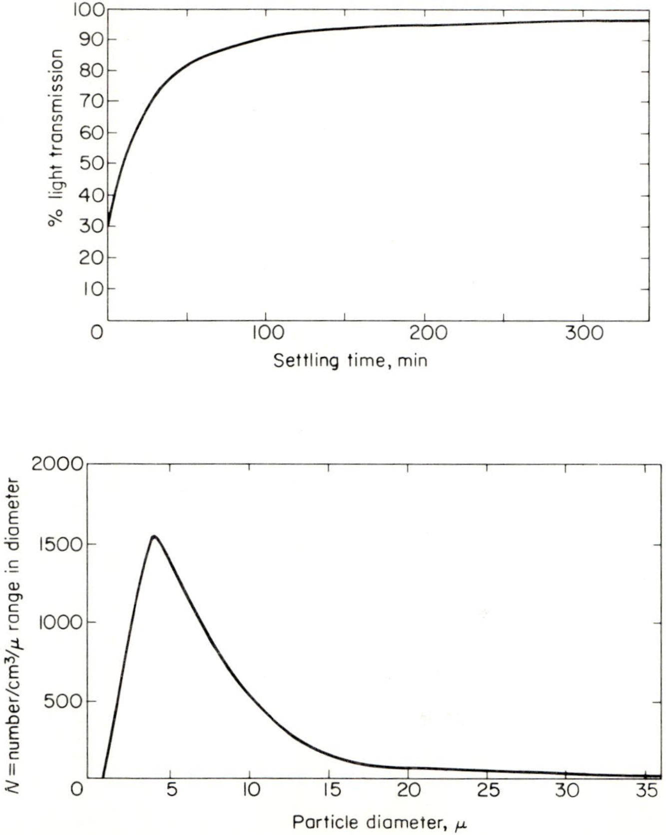

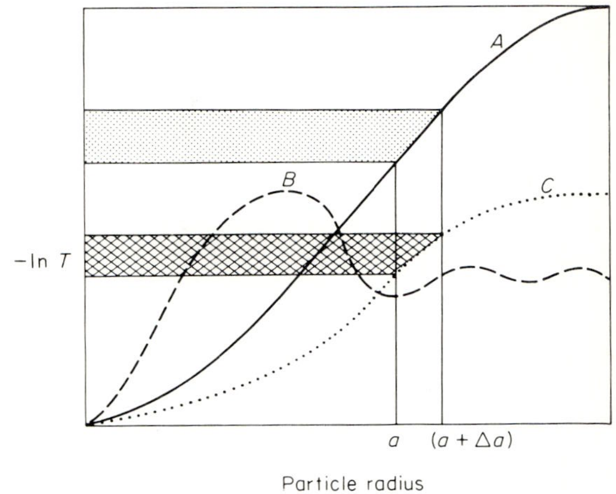

FIG. 7.53. (above) Particle size from transmittance of a settling dispersion.

FIG. 7.52. (left) Variation of optical transmission of a kerosene aerosol with time of settling (upper curve). Size distribution obtained from these data (lower curve) (Gumprecht and Sliepcevich, 1953a).

$R \times Q_{sca}$ as a function of particle radius and a corrected transmittance curve is obtained by dividing each ordinate of the curve by the corresponding value of $R \times Q_{sca}$. This reduced transmittance curve can then be used to obtain the size frequency curve. An advantage of this method is that such a graphical integration procedure is often more accurate than is the graphical differentiation required in the procedure used by Gumprecht and Sliepcevich. Kaye and Allen (1965) have explored some of the problems involved in the calibration procedure.

Instead of measuring the transmittance, the differential settling method can also be adapted to utilize scattering measurements at any particular angle of observation. This has been done by Kerker *et al.* (1955) in connection with mercury and sulfur aerosols comprised of particles with submicron radii. Their results were compared with independent determinations of the size distribution obtained by electron microscopy. Because they were unable to control the convection currents during the sedimentation, especially for the longer times necessary to permit the smallest particles ($r \sim 0.1\ \mu$) to settle through the beam, they were only able to determine the upper end of the distribution.

Turbidimetric-sedimentation measurements, similar to those described above, are widely used in industrial work to analyze powders and pigments in the size range from 0.1 to 50 μ in diameter, and yet, much of this work, as carried out in practice and as reported in the literature, may not be reliable. Michaels (1958) has presented a summary of results obtained by six industrial laboratories on four standard different-sized tungsten metal powders. There was a wide divergence among these results.

7.5.6 Measurement of the Forward Scattering Cone

Chin *et al.* (1955a, b) have proposed still another technique based upon the angular variation of the intensity of forward scattered light at very small angles and at a single wavelength. This does not require a second kind of measurement such as differential settling, and yet it is applicable to broad distributions for which it is unnecessary to assume an *a priori* form.

It follows from (3.3.78) that the total efficiency for scattering is given by

$$Q_{sca} = (1/\alpha^2) \int_0^{\pi} (i_1 + i_2) \sin\theta \, d\theta \tag{7.5.56}$$

However, when a detector used to measure the transmission also views some of the forward scattered light, the effective efficiency is reduced by the fraction of light scattered through the cone of half-angle, ω, so that

$$Q_{eff} = Q_{sca} - (1/\alpha^2) \int_0^{\omega} (i_1 + i_2) \sin\theta \, d\theta \tag{7.5.57}$$

The parameter R, which has already been used in connection with the transmission-differential settling method, is defined by

$$R = Q_{\text{eff}}/Q_{\text{sca}} \tag{7.5.58}$$

so that $R \cdot Q_{\text{sca}}$ gives the effective efficiency for scattering as measured by the particular optical system. The value of R may be calculated readily from the theoretical equations. Gumprecht and Sliepcevich (1953b) have shown that it may be approximated for large spheres ($\alpha \geqslant 20$) by

$$R = \tfrac{1}{2}[1 + J_0{}^2(\alpha\omega) + J_1{}^2(\alpha\omega)] \tag{7.5.59}$$

provided

$$(\alpha\omega) < 30° \tag{7.5.60}$$

The J's are the Bessel functions of orders zero and one. In this expression Q_{sca} is approximated by the value 2 which is the large sphere limit. The amount of light scattered into the forward cone was obtained from diffraction theory.

Substitution of (7.5.59) into (7.5.52) now results in an integral equation which can be inverted by means of a Mellin transform. After some development, the following expression is obtained:

$$Q_{\text{sca}}p(\alpha)\alpha^2 = 2\pi \int_0^\infty M(\omega)[-J_1(\alpha\omega)N_1(\alpha\omega)(\alpha\omega)]\,d\omega \tag{7.5.61}$$

where

$$M(\omega) = \frac{4}{l\lambda^2}\frac{d}{d\omega}\left(\omega^2 \frac{d(\ln I)}{d\omega}\right) \tag{7.5.62}$$

The quantities $J_1(\alpha\omega)$ and $N_1(\alpha\omega)$ are the Bessel and Neumann functions of order one. Provided that experimental arrangements for varying ω in a known way can be devised, $M(\omega)$ may be calculated from the angular variation of the flux contained in the forward scattered cone. Then, $Q_{\text{sca}}p(\alpha)\alpha^2$ may be calculated from (7.5.61) for various values of α by graphical integration. The integrand converges quite rapidly at low values of ω. The value of the number distribution, $p(\alpha)$, may be evaluated for each particular value of α.

Chin *et al.* (1955b) have described an apparatus for obtaining the requisite data and have tested the method with a dispersion of Pyrex glass beads having a known distribution as obtained from microscopic counting. Some experimental results are shown as a plot of τ vs. ω in Fig. 7.54 for two runs, having the same relative size distribution, but different concentrations. The comparison of the calculated size distribution with the histogram obtained by microscopic count is shown in Fig. 7.55.

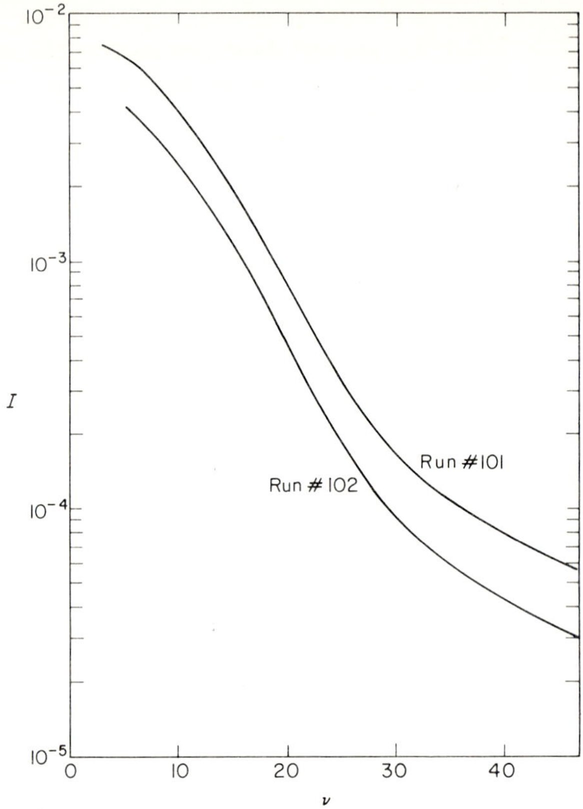

FIG. 7.54. Scattering intensity (relative scale) plotted against the quantity ν which is proportional to the angle θ for two concentrations of a dispersion of glass spheres in water (Chin *et al.*, 1955a, b).

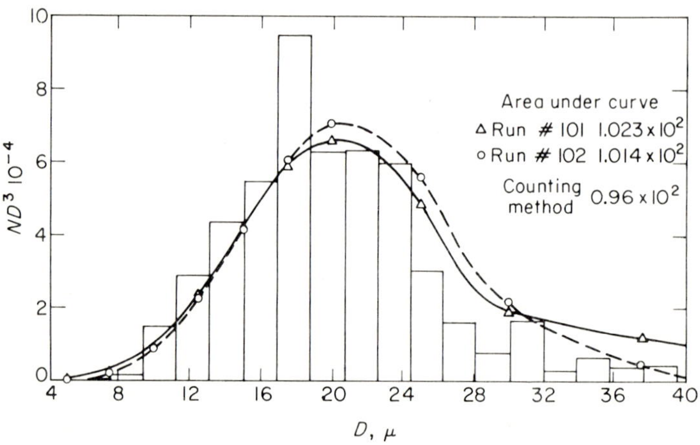

FIG. 7.55. Size distribution obtained from data of Fig. 7.54 plotted as ND^3 versus D, where N is the number of particles per unit volume having a diameter D, compared to the histogram obtained by microscopic count (Chin *et al.*, 1955a, b).

The agreement with the microscopic counts is quite satisfactory. These particles are rather large and the distribution broad compared to those amenable to treatment by either the polarization ratio method or the method based on the spectra of the polarization ratio or turbidity. Chin *et al.* have proposed the design of a commercial high-speed measuring and computing unit for the determination of particle size distribution using this technique.

More recently, Walstra (1965) has obtained agreement between experimental values of the extinction and calculations based upon the above diffraction theory as the angle of acceptance was varied from 0.013 to 0.026 rad.

A similar method for the analysis of the particle size distribution from the angular variation of the forward scattering has been developed independently by Shifrin, using a slightly different point of view. For unit incident intensity, the intensity of light scattered at very small angles by a polydispersion of spheres is given by diffraction theory as

$$I(\theta) = \int_0^\infty a^2[J_1{}^2(\alpha\theta)/\theta^2]\,p(a)\,da \tag{7.5.63}$$

Shifrin *et al.* (1966a, b) have developed asymptotic formulas useful in approximating the above integral for a gamma size distribution function [see (8.3.36)] and have published extensive numerical tables of the integral.

The inversion of the above integral equation has been accomplished with the aid of an appropriate transform, and the size distribution may be expressed as (Shifrin, 1957)

$$p(\alpha) = -(2/\alpha^2)\int_0^\infty \alpha\theta J_1(\alpha\theta)N_1(\alpha\theta)\psi(\theta)\,d\theta \tag{7.5.64}$$

where

$$\psi(\theta) = (d/d\theta)\{I(\theta)[(2\pi/\lambda)J_1(\theta)]^3\} \tag{7.5.65}$$

The stability of this inversion, for the case of a gamma size distribution, has been explored with regard to the range of the parameters (particle radius and scattering angle) and also with regard to the accuracy of the experimental scattering data. The computational errors do not exceed 10% for particle radii in the range 2 to 50 μ, provided $\theta < 8°$. Shifrin and Golikov (1961) have constructed a laboratory apparatus to obtain the necessary forward-angle scattering data ($\theta < 2°$) and have found excellent agreement between the size distributions of dispersions of water and of poly(methyl methacrylate) powders obtained by this method and that from direct microscope counts.

7.5.7 Absolute Intensity

The work of Kratohvil and Smart (1965) in obtaining concordance between measured values of the Rayleigh ratios ($V_v(\theta)$, $H_h(\theta)$) and theoretical calculations was cited earlier in connection with the verification of the theory (Section 7.1.2*b*). Such measurements, obtained with a system having a finite but not too broad size distribution, can also be used to determine the parameters of the distribution, including the concentration. The working equations are

$$V_v(\theta) = (N\lambda^2/4\pi^2)\int_0^\infty i_1 p(\alpha)\,d\alpha \tag{7.5.66}$$

$$H_h(\theta) = (N\lambda^2/4\pi^2)\int_0^\infty i_2 p(\alpha)\,d\alpha \tag{7.5.67}$$

The particle size distribution can be determined by comparing the polarization ratio (7.5.34) or the degree of polarization (7.5.35) with theoretical calculation or the comparison can be made directly using the Rayleigh ratios. If the absolute value of the Rayleigh ratio is known, the number concentration N can be determined directly from either of the above equations at any of the angles of observation.

This method for determining particle concentration has been applied to La Mer sulfur sols and to polystyrene latexes by Rowell *et al.* (1968b). Excellent agreement was obtained for the latter system with values of N obtained by direct analysis.

7.6 Color Effects

7.6.1 Introduction

The color effects of light scattering are among nature's most beautiful phenomena—the blue of the sky, the sunset, the rainbow, the corona, and to some extent, the hues of the sea are due to light scattering. Rayleigh showed that the color of the sky arose from scattering by small particles. We have already seen that the intensity of the light scattered by such particles varies with the inverse fourth power of the wavelength and that the wavelength exponent increases as the particles become larger, eventually even taking on positive values. The wavelength exponent also varies with the refractive index. It is the wavelength exponent which determines the color of both the transmitted and scattered light.

This effect is manifested in nature by the disk of the sun and the moon taking on various hues, such as blues or greens, whenever an exceptional amount of colloidal dust is generated, for example, by volcanic activity (Kiessling, 1884a) or forest fires (Penndorf, 1953). In the laboratory, the color of light transmitted by clouds of water droplets and its dependence upon the particle size had been investigated many years ago in connection with cloud chamber experiments designed to study nucleation of water clouds (Kiessling, 1884b; Aitken, 1892; Wilson, 1897; Mecke, 1920) and with experiments on steam jets (Barus, 1902, 1908; Ruedy, 1944). It was in order to explain the brilliant colors of metal hydrosols that Mie (1908) developed his general theory of scattering by spheres.

An early report that colors other than the Tyndall blue were to be observed in the laterally scattered light is contained in a paper by Rayleigh (1881). Upon adding a small quantity of sulfuric acid to a dilute solution of sodium thiosulfate, a colloidal suspension of small sulfur particles was formed. At first the scattered light appeared blue, but as the particles increased in size upon the continued generation of sulfur, this blue became much "richer," and then various colors were observed in different directions. Rayleigh was interested in the polarization so he used approximately monochromatic light in order to eliminate the color effects, since "the observation is easier than with white light, by the uniformity of the colour scattered in various directions."

Similar observations were made by Keen and Porter (1913). However, it was Ray (1921) who first described these colors in some detail. He noted how the scattered light varied in intensity and color in different directions and how, if observed through a Nicol prism, the succession of colors with angle of observation was also dependent upon the particular polarization viewed. Ray pointed out that the colors in the scattered light arise because the scattered intensities at any given angle and particle size are different for each wavelength. This follows because the angular intensity functions vary in a very sensitive way with the size parameter α which in turn is related to wavelength through $\alpha = 2\pi a/\lambda$.

La Mer and Barnes (1946) perfected a technique for preparing sulfur sols which exhibit these color phenomena most vividly. Actually, although their first published description of these brilliant spectral colors was of these sulfur sols, the effect had been studied in La Mer's laboratory for some time in connection with secret work on aerosols during the war years (Sinclair and La Mer, 1949). These colors were designated as "higher-order Tyndall spectra," H.O.T.S. It was pointed out that the brilliance of the colors depended upon the degree of monodispersity of the dispersion. Since each particle of a particular size generates a characteristic band of colors, the presence of a broad distribution of sizes leads to a washout of the colors upon superposition of the various H.O.T.S.

We will discuss first the technique developed in La Mer's laboratory for estimation of the particle size by analysis of the H.O.T.S. This is particularly useful because it provides a very simple and rapid visual method for size analysis. Following this, the relation between color and the spectral distribution of a light source will be reviewed in order to provide a basis for a more quantitative discussion of the color effects of scattering by both monodisperse and heterodisperse systems of spheres and cylinders.

7.6.2 Particle Size from H.O.T.S.

La Mer's method is based upon the observation that red and green bands predominate in the scattered light and that as particle size increases, there is a shift in the positions and the numbers of these bands. Accordingly, if their angular location can be calculated theoretically, experimental observations will then permit an estimate of the particular size.

Johnson and La Mer (1947) determined the positions of the orders (centers of the bands) by computing the ratio of the intensity of the red scattered light to the green scattered light for each particular value of α and then determining the angles at which this ratio was a maximum (red order) or a minimum (green order). The assumption was made that the white light could be approximated by two monochromatic components of equal intensity, viz. a red component for which $\lambda_R = 6290$ Å and a green component for which $\lambda_G = 5240$ Å, the ratio of these two wavelengths having been selected at 1.2. Therefore, for a particular size, a, the intensity of red light is given by

$$I_R(\theta) = \lambda_R{}^2 i_1(\alpha_R, \theta)/4\pi^2 r^2 \tag{7.6.1}$$

and of the green light by

$$I_G(\theta) = \lambda_G{}^2 i_1(\alpha_G, \theta)/4\pi^2 r^2 \tag{7.6.2}$$

where λ_G and λ_R are the wavelengths in the medium, and

$$\alpha_G = 2\pi a/\lambda_G; \qquad \alpha_R = 2\pi a/\lambda_R \tag{7.6.3}$$

In this case only the vertically polarized component is considered.

A typical result, neglecting the dispersion of refractive index with wavelength, is shown in Fig. 7.56. The maxima locate positions of red orders, and the minima locate positions of green orders. When the positions of each order are plotted against particle radius, the curves shown in Fig. 7.57 are obtained. The particle size can now be determined by locating the orders experimentally and comparing these with Fig. 7.57. For example, red orders at 49, 77, 118, and 148° would indicate a particle size of 0.35 μ. Berner (1965) has presented the ratio of the red to green intensities for $m = 1.33$ as a function of size and scattering angle in the form of a contour diagram, the ridges representing the red orders and the valleys representing the green orders.

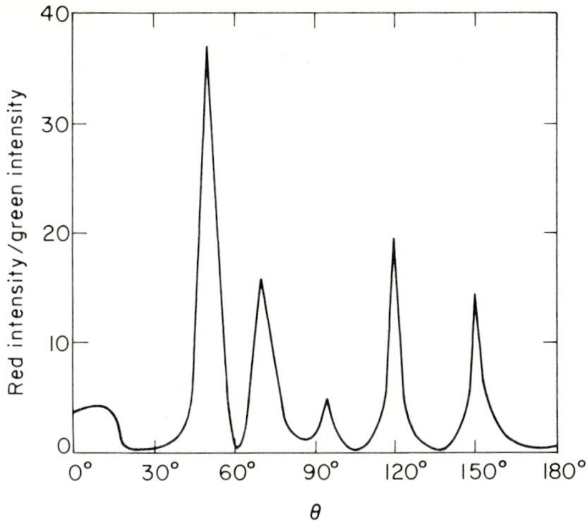

FIG. 7.56. Ratio of intensity of red ($\lambda = 0.629\ \mu$) to green ($\lambda = 0.524\ \mu$) scattered light against θ for $m = 1.44$ and $\alpha_{\text{red}} = 5$ (Johnson and La Mer, 1947).

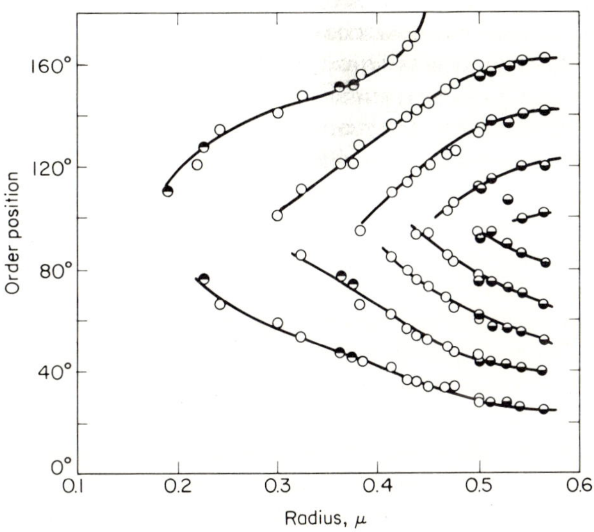

FIG. 7.57. Angular positions of red orders against radius for $m = 1.44$ (Johnson and La Mer, 1947).

Johnson and La Mer found good agreement between the particle size in sulfur hydrosols obtained from the H.O.T.S. and that from sedimentation experiments. Also, the experimental positions of the orders for two monodispersed polystyrene latexes (La Mer and Plesner, 1957) agreed with those calculated by this method (Kerker, 1958). Although Petro (1960) found that the particle size of La Mer sulfur sols determined by the H.O.T.S. disagreed with that from a Coulter counter, the latter instrument is known to be precarious for the determination of submicron sizes. Kerker *et al.* (1963) found that the modal sizes obtained using the polarization ratio method were in good agreement with those from the H.O.T.S.

Maron and Elder (1963b) have studied the possibility of determining particle size from the H.O.T.S. by location of the angular position of only the first red and green orders. Kitani (1956, 1960) had earlier proposed a somewhat similar approach. If r_1 and g_1 represent the angular positions of the first red and green orders (from the forward direction), then for monodisperse butadiene-styrene latexes for which $m = 1.17$

$$a \sin(r_1/2) = 1150 \tag{7.6.4}$$

$$a \sin(g_1/2) = 1560 \tag{7.6.5}$$

and, for monodisperse polystyrene latexes ($m = 1.20$)

$$a \sin(r_1/2) = 1290 \tag{7.6.6}$$

These formulas were developed by calibration of light scattering results against electron microscope values for twenty butadiene-styrene latexes ranging from $a = 0.130$ to $0.505\ \mu$ and for six polystyrene latexes ranging from $a = 0.170$ to $0.586\ \mu$. Pierce and Maron (1964) have proposed a technique for estimating the location of the red and green orders at larger angles (r_i, g_i where $i > 1$) where either r_1 or g_1 is known.

Napper and Ottewill (1963a, b) have observed H.O.T.S. with silver bromide hydrosols consisting of narrow distributions of either cubical or octahedral particles. In the absence of a theory of scattering for such nonspherical shapes, they correlated the observed positions of the orders with computations for "equivalent" spheres, taking into account the detailed size distribution as determined with the electron microscope. In most cases, they obtained good agreement between the positions of the orders observed with those computed. For the octahedra, the corner-to-corner cross-sectional distance was taken as the diameter of the equivalent sphere, while for the cubes the best results were obtained when this diameter was taken to be the length of the face diagonal.

There has been some discussion in the literature regarding the influence of polydispersity on the appearance of the H.O.T.S. In particular, an experiment by Johnson and La Mer (1947) has led to some confusion on this point. Two of the La Mer sulfur sols designated as A and B were found to have radii of 0.467 and 0.476 μ, respectively. Sol A had a red order at 136°, while sol B exhibited a green order in the same region. The ratio of the red to green intensities, as determined experimentally by these authors, is plotted in Fig. 7.58. This ratio is also plotted for a mixture of 90% *A* and 10% *B*, as well as a 50–50 mixture of these two components. In the former case, there is a measurable effect due to the presence of 10% of a particle which is 2% larger, and for the 50–50 mixture the order is practically obliterated.

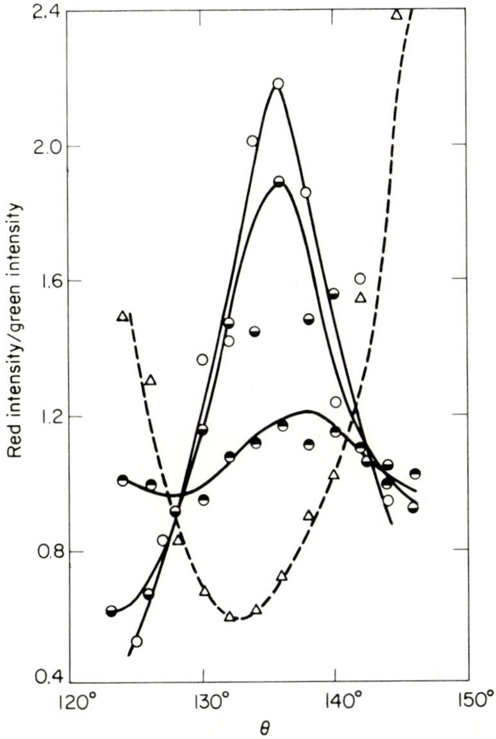

FIG. 7.58. Red-green intensity ratios for the light scattered by homogeneous sulfur sols and mixtures: Sol *A*, 0.467 μ radius; Sol *B*, 0.476 μ radius; ○, 100% *A*; ◒, 90% *A*, 10% *B*; ◓, 50% *A*, 50% B; Δ, 100% *B* (Johnson and La Mer, 1947).

Although it has been suggested from this experiment that a 2% spread in particle sizes will obliterate the H.O.T.S., this is far from the case. The example is chosen at an angle where there is a fortuitous cancellation of the color effect. At other angles, the color may actually be enhanced, rather than annulled, by the superposition of orders. When two sols exhibiting H.O.T.S. containing widely differing sizes are mixed in equal proportions, the mixture will still usually exhibit H.O.T.S. Qualitative calculations (Kerker and La Mer, 1950) show that H.O.T.S. will be observed even for a rectangular distribution spread from $a = 0.216$ to $0.252\ \mu$.

7.6.3 Color Theory

The next section will consider some exact calculations of the color effects obtained when white light is scattered by dispersions of spheres and cylinders. Before proceeding to this, some elements of color theory will be reviewed here.

Color may be defined as that which distinguishes the perception of one kind of light from another. The fundamental law of normal color perception is that almost all colors can be produced by a combination of three differently colored lights called primaries which are most conveniently, but not necessarily, chosen from the middle and near each end of the visible spectrum; i.e., red, green, and blue. The amounts of the three primaries necessary to produce a particular color are called the tristimulus values.

Physically, the light is completely defined by a spectral distribution curve in which the radiant flux is plotted as a function of wavelength. The psychophysical problem is to determine the color sensation corresponding to each particular curve.

The psychological attributes of color sensation are hue, brightness, and saturation. Hue is the quality of sensation by which an observer becomes aware of major differences in wavelength. Light, such as white, black, and the grays, which does not give rise to a color sensation, is said to be hueless or achromatic. The achromatic colors are characterized only by brightness, which is the second attribute of color. Finally, colors may differ from each other in saturation. This is sensed when different colors of the same hue and brightness are compared. Thus, saturation may be defined as the degree to which a chromatic sensation differs from an achromatic sensation of the same brightness.

There are three physical characteristics of color which have been settled upon to parallel the psychological attributes. They are dominant wavelength, luminance, and purity, corresponding in a general way to hue, brightness, and saturation. The dominant wavelength is the wavelength which is most intense in the spectral distribution, or at least close to it. Luminance is derived from the energetic quantity, radiance, through the luminosity. Luminosity shows the capacity of energies of various wavelengths to evoke for an observer visual sensations of equal brightness. Purity is the extent to which the dominant wavelength appears to predominate.

The dominant wavelength and purity are lumped together as chromaticity. Operationally, the dominant wavelength is the wavelength of spectrally pure light which would have to be mixed with an appropriate amount of achromatic or "white" light in order to match the chromaticity of a sample. The purity is the proportion of the spectrally pure component in the mixture.

The convention used to describe the color of a light utilizes three standard primaries adopted by the International Commission on Illumination in 1931, which correspond to narrow bands of red, blue, and green light. Any color can be matched by a mixture of these three primaries, the amounts of each being given by the tristimulus values, X, Y, and Z. The procedure for obtaining the tristimulus values utilizes the relations

$$X = K \int_0^\infty \overline{x_\lambda} P_\lambda \, d\lambda \tag{7.6.7}$$

$$Y = K \int_0^\infty \overline{y_\lambda} P_\lambda \, d\lambda \tag{7.6.8}$$

$$Z = K \int_0^\infty \overline{z_\lambda} P_\lambda \, d\lambda \tag{7.6.9}$$

where P_λ is the spectral energy distribution of the light, λ is the wavelength, and $\overline{x_\lambda}$, $\overline{y_\lambda}$, and $\overline{z_\lambda}$ are the tristimulus values of a specific equal energy source as given in standard references [e.g., C.C.O.S.A. (1963)].

When the color is to be described in terms of the dominant wavelength and the purity, i.e., the chromaticity, it is only necessary to specify two parameters which may be chosen as the relative tristimulus values or chromaticity coordinates

$$x = X/(X + Y + Z) \tag{7.6.10}$$

and

$$y = Y/(X + Y + Z) \tag{7.6.11}$$

These can be employed as coordinates for the representation of the chromaticity by the position of a point in a plane diagram. The x and y values are plotted as abscissa and ordinate, respectively. Figure 7.59 depicts such a chromaticity diagram.

The curved line known as the spectrum locus represents the locus of spectrally pure colors. These are characterized by the indicated relative tristimulus values and may be matched by an appropriate mixture of standard stimuli, as well as by the spectral colors. Thus, what is perceived as a spectral color may actually be a mixture. The wavelength markings indicate the dominant wavelength of such mixtures. The region bounded by the spectrum locus and the straight line joining its ends encloses the domain of real colors, within which each point corresponds to a color.

The achromatic point, P, may represent the chromaticity of a flat source which radiates equal amounts of energy in equal intervals of wavelength throughout the spectrum, or, alternatively, it is possible to locate other practical "white" light sources in adjacent regions of the diagram which may also represent a source point or achromatic point. Just as for the spectrum locus, although each source point has a unique pair of relative tristimulus values and is represented by a particular point in the chromaticity diagram, each such point may correspond to an infinite number of physical stimuli with distinct spectral distributions, but similar color responses.

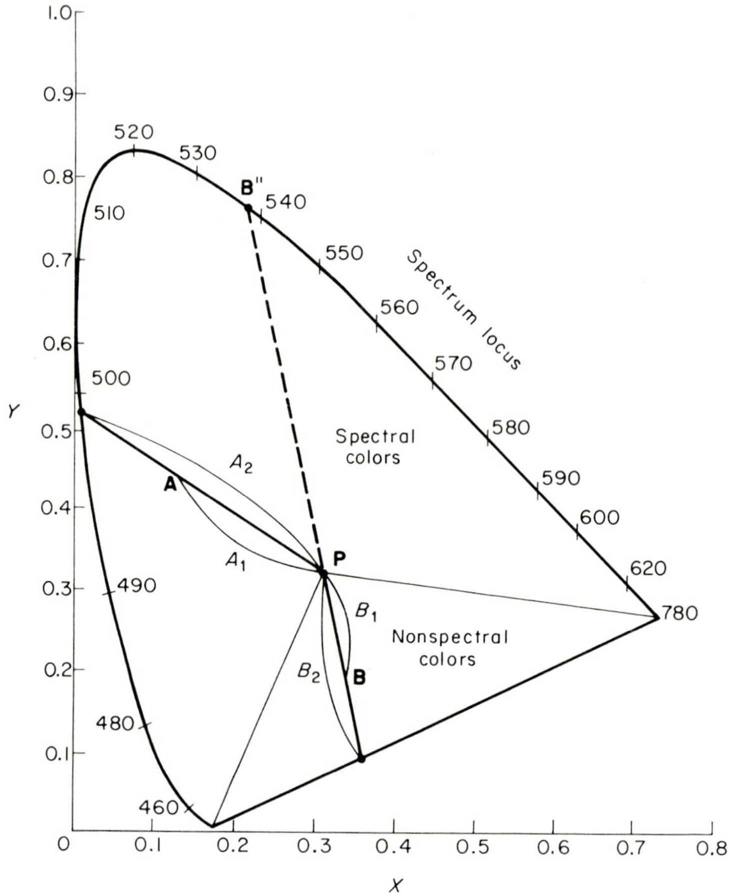

FIG. 7.59. Chromaticity diagram.

In order to describe any color in terms of the dominant wavelength and purity, its position in the chromaticity diagram is related to an achromatic point and a spectrum color. The chromaticity of a point such as A on Fig. 7.59 is determined by extending the segment PA until it intersects the spectrum locus, in this case at $\lambda = 499$ mμ. Then, the dominant wavelength of the color indicated by point A is 499 mμ, and the purity is A_1/A_2. What this means is that the chromaticity of point A can be matched by mixing A_1 parts of spectrally pure radiation of 499 mμ with $(A_2 - A_1)$ parts of radiation emitted by source P. Obviously, the chromaticity is quite arbitrarily defined relative to the particular "achromatic" point which serves as a reference.

When the sample is located at a point such as B, which is within the lower triangular region, the extension of PB intersects the straight line joining the extremities of the spectrum locus. Then, the dominant wavelength is complementary to the wavelength obtained by extending PB to the spectrum locus, in this case at $\lambda = 538$ mμ. This complementary wavelength is designated $\lambda = 538$c mμ. The purity is obtained in the same manner from the ratio of the length B_1 to the total length B_2. It signifies that light of chromaticity B may be synthesized by adding B_1 parts of $\lambda = 538$c mμ to $(B_2 - B_1)$ parts of the achromatic source, P.

The complementary color, $\lambda = 538$c mμ, when mixed with its complement $\lambda = 538$ mμ, will give some color along the extension of line PB, depending upon the proportions mixed. Thus, if an amount of $\lambda = 538$c mμ, proportional to the length of the extension of PB to the straight line, is mixed with an amount of $\lambda = 538$ mμ proportional to PB'', the chromaticity of the resultant point is represented by point P.

Although the colors along the lower straight line do not correspond to spectral colors, they do vary continuously from violet to red. They merge directly without intermediate hues, unlike the physical spectrum which proceeds through the blues, greens, and yellows. Thus, this interval joins the two ends of the wavelength spectrum, so that, psychologically, it becomes a closed circuit of hues.

7.6.4 Theory of H.O.T.S.

We now consider the scattering of white light by a sphere or infinite cylinder. The intensity of the perpendicular component is

$$I_1 = (1/4\pi^2 r^2) \int_0^\infty \lambda^2 P_\lambda i_1(m, a, \lambda)\, d\lambda \tag{7.6.12}$$

for spheres, and

$$I_1 = (1/\pi^2 r) \int_0^\infty \lambda P_\lambda i_1(m, a, \lambda)\, d\lambda \tag{7.6.13}$$

for cylinders. The dependence of i_1 upon α is shown explicitly in terms of a and λ in order to emphasize the λ-dependence. In addition, the relative refractive index will vary with wavelength, depending upon the dispersion of both the particle and the medium. P_λ is the normalized spectral radiant flux of the incident light, $P_\lambda d\lambda$ being the fraction of incident energy contained in the spectral band between λ and $\lambda + d\lambda$.

The expressions under the integral in each of the above equations give the spectral distribution of the scattered light. The tristimulus values for spheres are obtained from

$$X_1 = (1/4\pi^2 r^2) \int_0^\infty \lambda^2 \overline{x_\lambda} P_\lambda i_1(m, a, \lambda)\, d\lambda \tag{7.6.14}$$

$$Y_1 = (1/4\pi^2 r^2) \int_0^\infty \lambda^2 \overline{y_\lambda} P_\lambda i_1(m, a, \lambda)\, d\lambda \tag{7.6.15}$$

$$Z_1 = (1/4\pi^2 r^2) \int_0^\infty \lambda^2 \overline{z_\lambda} P_\lambda i_1(m, a, \lambda)\, d\lambda \tag{7.6.16}$$

and the corresponding values for cylinders follow in the same manner. The chromaticity coordinates can then be calculated from (7.6.10) and (7.6.11). These, in turn, can be interpreted in terms of dominant wavelength and purity with the aid of a chromaticity diagram such as Fig. 7.59.

Extensive calculations of this type have been made over the range of parameters listed in Table 7.8 [(Kerker *et al.*, 1966b; Kerker *et al.*, 1966c). The detailed numerical results are deposited with the American Documentation Institute as cited in these two references.] These were carried out at every 2.5° over the entire angular range for two light sources corresponding to noon sunlight and a 3200°K black-body radiator. The computed results consisted of the dominant wavelength, the purity, the luminous flux, as determined from (7.6.16), and the intensity, as determined from (7.6.12) or (7.6.13). Results were obtained for unpolarized incident light as well as for each of the polarized components.

TABLE 7.8

PARAMETERS FOR WHICH TRISTIMULUS VALUES HAVE BEEN OBTAINED

System	Radius, μ	Refractive index
Single spheres	0.2(0.1)0.8(0.2)1.2	1.31, 1.46
Single spheres	0.1, 0.2, 0.4	For V_2O_5
Single cylinders	0.4(0.1)1.4	1.31, 1.46
Distribution of spheres	$a_M = 0.4$; $\sigma_0 = 0.01, 0.02, 0.05, 0.07, 0.10(0.02)0.18$	For water at 18°C

A typical example of the angular variation of the visible intensity is shown in Fig. 7.60. This is for the perpendicularly polarized component of the visible intensity of a 3200°K source scattered by spheres with $m = 1.46$ and

$a = 0.4$, 0.8, and $1.2\,\mu$. It represents the scattering as averaged over the spectral distribution of the source weighted by the visual response of a standard observer. The characteristic oscillations which are obtained when monochromatic light is used are considerably damped out in the lateral directions, although there are still well-defined lobes both in the forward and backward directions.

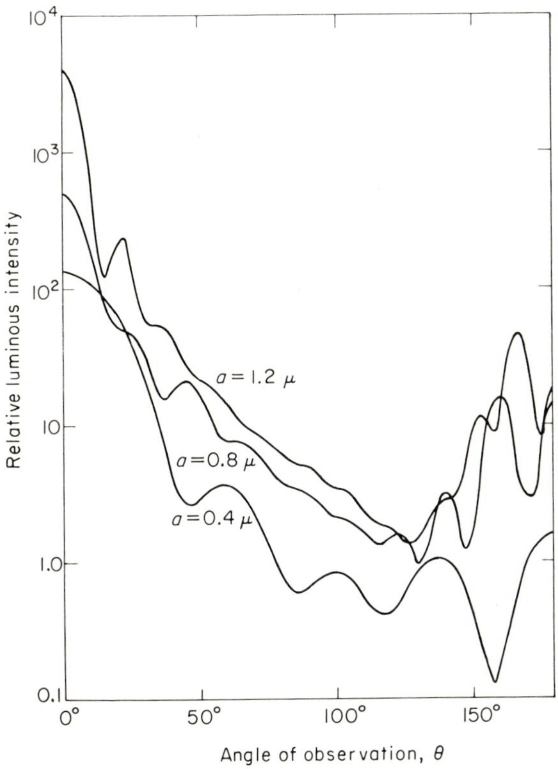

FIG. 7.60. Relative luminous intensity of the perpendicular component of the scattered light for $m = 1.46$, $a = 0.4$, 0.8, $1.2\,\mu$ and a 3200°K source (Kerker *et al.*, 1966c).

The color of the scattered light for $m = 1.46$ and $a = 0.4$, 0.6, $1.0\,\mu$ is given in Table 7.9. The hues are designated by the notation of Table 7.10. The purity has been designated by prefixing a letter before the capital letter indicating the hue as follows: p (pale) for purity 0 to 0.249; no prefix for purity 0.250 to 0.499; b (bright) for purity 0.500 to 0.749; and br (brilliant) for purity $\geqslant 0.750$.

TABLE 7.9
COLORS OF THE PERPENDICULAR COMPONENT OF THE LIGHT SCATTERED BY A SPHERE; $m = 1.46$, 3200°K SOURCE

θ	Radius (microns)			θ	Radius (microns)			θ	Radius (microns)		
	0.4	0.6	1.0		0.4	0.6	1.0		0.4	0.6	1.0
0	Y	bY	G	62.5	bG	bG	Y	125	bRO	bY	Y
2.5				65	brY	bY	bO	127.5	brO	brYG	bY
5			↓	67.5		bY	bO	130	brOY	bG	bY
7.5			bY	70	↓	bO	OY	132.5	brY	G	O
10		↓	bY	72.5	bY	O	Y	135	bY	B	pY
12.5		brOY	brY	75	bO	RO	G	137.5	Y	P	pY
15		brO	bY	77.5	R	O	Y	140	G	bP	O
17.5	↓	brO	R	80	bR	O	Y	142.5	pG	bO	bO
20	bY	bR	P	82.5		G	bO	145	pB	brOY	bY
22.5		bR	B	85	↓		bO	147.5	B	brY	bY
25	↓	P	B	87.5	R		Y	150	B	bY	G
27.5	brY	pB	G	90	pR	↓		152.5	bB	U	B
30		pG	G	92.5	Y	Y		155	bB	pG	P
32.5	↓	G	Y	95	G	Y		157.5	B	pG	bO
35	brO	bY	O	97.5	bG	O		160	Y	B	brY
37.5	bO	brY	R	100	brG			162.5	brY	B	bY
40	R	brY	R	102.5	bG	↓		165	bY	pB	Y
42.5	R	pY	O	105	bG	bY	↓	167.5	bY	R	pG
45	P	bO	G	107.5	G	bY	bY	170	Y	brY	B
47.5	pP	R	G	110	pG	bG		172.5		bY	pB
50	pB	R	Y	112.5	pO	Y		175	↓	Y	bY
52.5	pGB	pR	Y	115	R	O	↓	177.5	pY		
55	pG	pG	bO	117.5	bR	R	bO	180	pY	↓	↓
57.5	G	pG	bO	120		O	O				
60	G	G	OY	122.5	↓	bY	Y				

TABLE 7.10
NOTATION FOR DOMINANT WAVELENGTHS

Dominant wavelength (mμ)	Notation	Hue
400–450	V	Violet
451–500	B	Blue
501–570	G	Green
571–590	Y	Yellow
591–610	O	Orange
611–700	R	Red
<515 C	R	Red
>515 C	P	Purple

The H.O.T.S. phenomenon is clearly apparent in the sequence of colors observed as the scattering angle varies from 0 to 180°. The red and green "orders" are designated as occurring at about the midpoint of each band of red or green scattered light. There does not appear to be any correlation between the angular positions of the extrema in the visible intensity and the angular location of the H.O.T.S. Accordingly, it is not strictly correct to refer to a spectrally pure order as bright or brilliant if by this is also meant that the order has a high visible intensity.

The angular positions of the orders exhibited by the perpendicular component are shown in Table 7.11 for $m = 1.46$ and the 3200°K source. The counting is from the forward direction; thus, in this sense, R-3 and G-3 designate the angular positions of the centers of the third red and green bands.

TABLE 7.11

ANGULAR LOCATION OF RED AND GREEN ORDERS (PERPENDICULAR COMPONENT)
$m = 146$; 3200°K Source

	Radius, a (μ)							
Order	0.2	0.3	0.4	0.5	0.6	0.8	1.0	1.2
R-1								
G-1	25	7.5				B		Y
R-2	92.5	57.5	40	30	20	G		O
G-2	140	77.5	57.5	42.5	32.5	17.5	2.5	
R-3		105	77.5	60	50	32.5	17.5	12.5
G-3		150	95	72.5	60	40	27.5	17.5
R-4			115	102.5	75	50(O)	37.5	27.5
G-4			140	117.5	85	62.5	45	32.5
R-5				135	100(O)	77.5(O)	55(O)	42.5(O)
G-5				115(B)	110	87.5	65	50.0
R-6				165	117.5	100(O)		
G-6					130	110		
R-7					142.5(O)	117.5(O)		
G-7					157.5	127.5	Y	Y
R-8					167.5	132.5(O)	O	O
G-8						142.5		
R-9						150		
G-9						162.5	150	
R-10							157.5(O)	
G-10							170	155
R-11								160
G-11								170

The number of orders increases with increasing radius up to about $a = 0.8\ \mu$ in accordance with the approximate calculations and the observations of Sinclair and La Mer (1949). Indeed, their rough rule of thumb that the particle size in microns is approximately equal to one-tenth of the number of red or green orders is borne out.

For particles larger than about $a = 0.8\ \mu$, the proliferation of orders does not continue, but instead the light scattered into the lateral directions becomes uniformly yellow (e.g., for $m = 1.46$ and $a = 1.2$, the scattered light is yellow over the range 50 to 155°). This "wash out" of the H.O.T.S. takes place at a somewhat larger radius for $m = 1.31$. Infinite cylinders behave in very much the same manner, showing an initial increase in the number of orders with increasing radius and then a "wash out" to almost uniformly yellow scattered light when a reaches about $1.4\ \mu$. This limits the phenomenon of H.O.T.S., even for absolutely monodisperse systems, to the range of sizes where the radius is comparable in magnitude to the wavelength of light.

The "wash out" of the orders is pertinent to the consideration of H.O.T.S. as a criterion of monodispersity. Sinclair and La Mer reasoned that with increasing polydispersity the H.O.T.S. would be washed out as a result of the superposition of light of different colors at each particular scattering angle, and they proposed that the purity of the colors in the H.O.T.S. was a measure of the monodispersity. While this may be the case if the system is restricted to a particular range of sizes, it is not generally true. Part of a calculation (Kerker *et al.*, 1966b) for a zeroth order logarithmic distribution (7.5.16) having a modal radius, $a_M = 0.4\ \mu$ and various breadth parameters varying from $\sigma_0 = 0.01$ to 0.18 is shown in Table 7.12. For the narrow distribution corresponding to $\sigma_0 = 0.01$, as shown in the second column, there is a red order at 82.5° and a green order at about 102.5 to 105°. The "wash out" of orders with increasing polydispersity occurs slowly so that these orders still persist when the breadth parameter is as large as $\sigma_0 = 0.07$. Thus, such a system, which is not monodisperse, exhibits H.O.T.S. On the other hand, a perfectly monodisperse system, with $a = 1.2\ \mu$, is uniformly yellow over this angular range. Ultimately, as the distribution with the $0.4\ \mu$ modal radius becomes still broader, the scattered light appears uniformly yellow. This is probably due to the effect of the larger sizes in the distribution which scatter yellow light. These particles predominate because they scatter more intensely.

Calculations were also carried out for a highly absorbing material having a dispersion comparable to vanadium pentoxide. Such material in the bulk is colored a very deep red. It is only slightly absorbing in the red region of the spectrum, but in the blue its index of absorption is very high, comparable to that of metals. Yet, the scattered light exhibits the full range of colors. For $a = 0.1$, the scattering is uniformly blue or purple, constituting the transition region from Rayleigh scattering. When $a = 0.2$, there is a broad orange-red band in the backward directions and a broad green one in the forward directions, and at $a = 0.4$, the number of bands increases.

TABLE 7.12
VARIATION OF DOMINANT WAVELENGTH AND PURITY WITH INCREASING BREADTH OF DISTRIBUTION (σ_0) FOR WATER SPHERES WITH VERTICAL POLARIZATION ($a_M = 0.4\ \mu$, SOURCE 3200°K)

θ	σ_0								
	0.01	0.02	0.05	0.07	0.10	0.12	0.14	0.16	0.18
77.5	595	596	596	595	592	590	587	586	584
	0.628	0.610	0.564	0.526	0.469	0.437	0.423	0.406	0.403
80.0	604	606	605	602	595	591	588	586	584
	0.536	0.514	0.476	0.452	0.420	0.406	0.398	0.395	0.396
82.5	620	625	620	612	597	591	588	585	584
	0.436	0.416	0.386	0.370	0.374	0.378	0.379	0.388	0.389
85.0	496C	496C	660	618	597	590	587	585	584
	0.357	0.355	0.302	0.307	0.333	0.353	0.366	0.377	0.382
87.5	497C	497C	496C	612	593	588	585	584	583
	0.323	0.313	0.238	0.259	0.308	0.336	0.359	0.372	0.381
97.5	505	508	527	551	570	576	579	581	582
	0.252	0.238	0.192	0.249	0.316	0.338	0.353	0.363	0.371
100.0	505	508	521	544	567	574	578	580	581
	0.354	0.324	0.225	0.250	0.312	0.333	0.349	0.360	0.370
102.5	507	508	520	542	566	574	578	580	581
	0.351	0.334	0.225	0.237	0.299	0.327	0.343	0.355	0.365
105.0	510	512	525	546	568	574	578	580	581
	0.274	0.257	0.196	0.216	0.284	0.311	0.332	0.348	0.359
107.5	523	526	542	557	571	576	579	580	581
	0.185	0.183	0.179	0.203	0.262	0.296	0.320	0.339	0.353
110.0	559	560	566	572	578	579	580	581	581
	0.249	0.239	0.210	0.213	0.252	0.282	0.308	0.329	0.345

Greenberg and Roark (1966) have carried out calculations of the color of star light scattered by reflection nebulas in order to determine whether there are major color differences between dielectric and metallic particles. Their conclusion, in agreement with the above finding, was that the dominant quality of the interstellar grains in the determination of the nebular colors is the size, and not the optical properties.

There may be special interest in the forward- and back-scatter for which the visible intensity is plotted over the range of radius from $a = 0.1$ to 1.2 in Figs. 7.61 and 7.62. For each calculated point which is plotted as a circle, the color of the forward- or back-scattered light is shown.

We have already seen how the angular variation of the polarization of the scattered monochromatic light may be used as a diagnostic tool for particle size analysis. In fact, measurement of the polarization for incident white light at two or three angles near the backward direction may also serve to

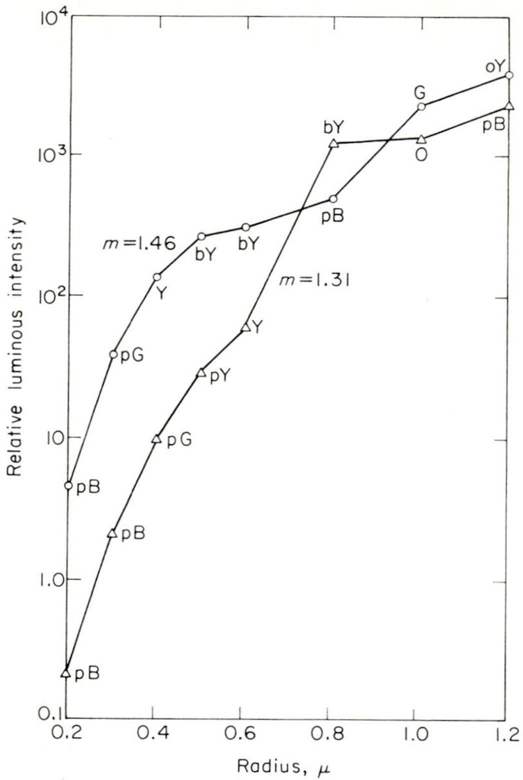

FIG. 7.61. Relative luminous intensity and color of the forward scattered light ($\theta = 0$) against particle radius for $m = 1.31$ and 1.46 and a 3200°K source (Kerker *et al.*, 1966c).

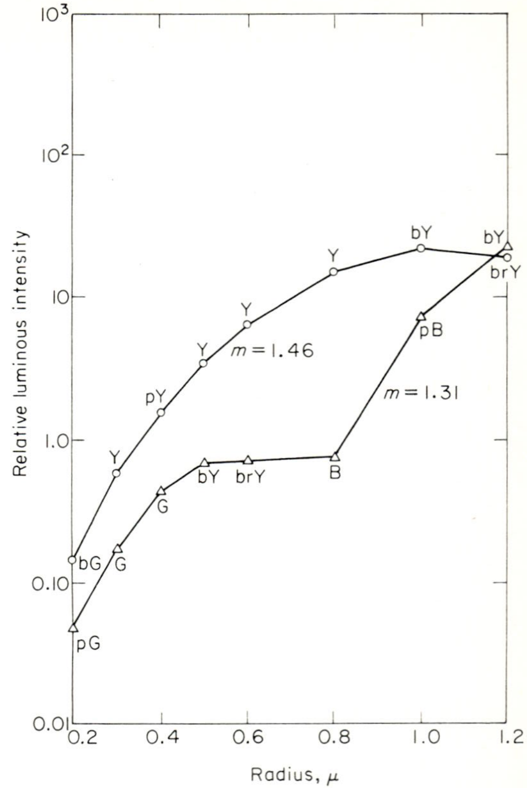

FIG. 7.62. Same as Fig. 7.61 for the backscattered light, $\theta = 180°$ (Kerker *et al.*, 1966c).

determine the particle radius. This is illustrated for $a = 0.2$ to 1.2 and $m = 1.46$ in Fig. 7.63 (Kerker *et al.* 1966c) in which the variation of the polarization with size is shown at the three angles, $\theta = 150$, 160, and 170°. The polarization at these three angles is sufficiently different so that its measurement at each of them unambiguously permits estimation of the particle size. This method cannot be used as successfully at smaller angles because there is not such a sharp variation of polarization with particle radius.

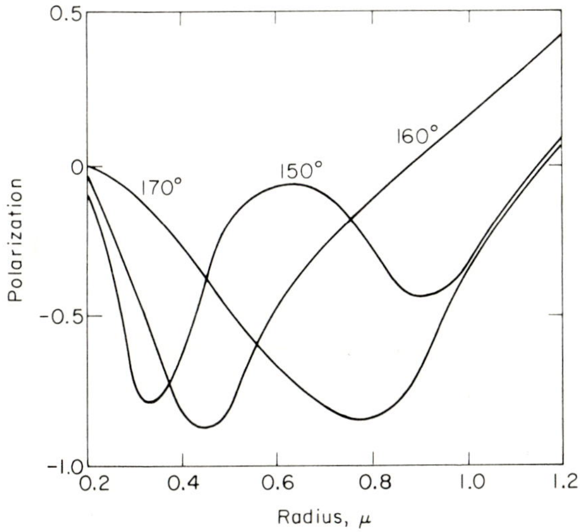

FIG. 7.63. Polarization, $P = (I_2 - I_1)/(I_2 + I_1)$, of the scattered light as a function of radius at $\theta = 150$, 160, and 170° for a 3200°K source (Kerker *et al.*, 1966c).

CHAPTER 8

Rayleigh–Debye Scattering

8.1 General Theory

In the same paper in which he developed the theory of scattering by small particles, Rayleigh (1881) also presented an approximate theory for particles of any shape and size having a small relative refractive index. He applied this to homogeneous spheres, spherical shells, radially inhomogeneous spheres, as well as to infinite cylinders. The full formulas for spheres were developed in later papers (Rayleigh, 1910, 1914, 1918b). Further contributions were made by Debye in 1915.

In 1925, Gans rederived the scattering formula for a homogeneous sphere. In order to distinguish it from Rayleigh scattering, which is restricted to particles small compared to the wavelength, this approach has generally been termed Rayleigh–Gans scattering. Actually Gans' contribution to this method was hardly significant and it seems more appropriate to call it Rayleigh–Debye scattering. The wave mechanical analog is known as the Born (Schiff, 1955) approximation.

The physical basis of Rayleigh–Debye scattering is depicted in Fig. 8.1 where a particle of arbitrary shape is subdivided into volume elements, two of which are designated A and B. Each element is treated as a Rayleigh scatterer excited by the incident field, which is assumed to be unperturbed by the presence of the rest of the particle. For incident radiation polarized perpendicular to the scattering plane, the amplitude function for each volume element is given by

$$dS_1(\theta) = i\frac{3}{4\pi}k^3\left(\frac{m^2-1}{m^2+2}\right) e^{i\delta}\, dV \tag{8.1.1}$$

where $e^{i\delta}$ relates the phase of each elemental wavelet at the position of the observer to a common reference such as the plane P. Thus, at observation points such as (1) or (2), far removed from the particle, the resultant amplitude function which arises from interference of each of the wavelets is obtained by vector summation leading to

$$S_1(\theta) = i\frac{3}{4\pi}k^3\left(\frac{m^2-1}{m^2+2}\right)\int e^{i\delta}\,dV \qquad (8.1.2)$$

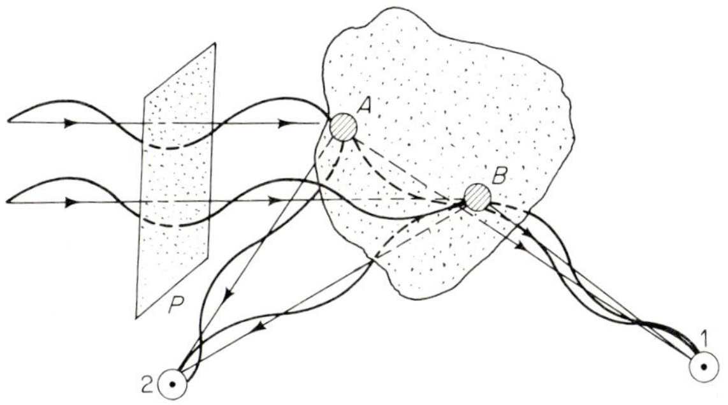

FIG. 8.1. Mutual interference between wavelets emanating from points A and B.

The fundamental approximation in the Rayleigh–Debye approach is that the "phase shift" corresponding to any point in the particle be negligible, i.e., that

$$2ka(m-1) \ll 1 \qquad (8.1.3)$$

where a is the longest dimension through the particle. For this reason neither the particle size nor the relative refractive index can become too large. It is a consequence of this approximation that the phase of each wavelet in (8.1.1) is determined only by the position of each volume element and is independent of the material properties of the particle.

With m sufficiently close to unity, the expression containing the refractive index in (8.1.2) can be simplified to give

$$S_1(\theta) = (i/2\pi)k^3(m-1)\int e^{i\delta}\,dV \qquad (8.1.4)$$

and provided m is real, the scattered intensity is

$$I_1 = (k^4V^2/4\pi^2r^2)(m-1)^2P(\theta) \qquad (8.1.5)$$

where

$$P(\theta) = (1/V^2)\left|\int e^{i\delta}\,dV\right|^2 \qquad (8.1.6)$$

When the incident radiation is polarized parallel to the scattering plane, the scattered radiation is obtained in the usual way as for a Rayleigh scatterer

$$I_2 = (k^4 V^2/4\pi^2 r^2)(m - 1)^2 \cos^2\theta \, P(\theta) \tag{8.1.7}$$

This assumes no depolarization of the scattered radiation, even for non-spherical particles. Furthermore, irrespective of the particle size, there is no phase retardation between the perpendicular and parallel components.

The quantity $P(\theta)$ is known as the form factor. That part of each of the above expressions which precedes it is the intensity scattered by a small sphere with volume V, so that $P(\theta)$ represents the modification of the intensity due to the finite size of the particle and to its deviation from sphericity.

The phase δ of each scattered wavelet can be described in terms of Fig. 8.2. The two wavelets scattered in the direction θ at the origin O and at the arbitrary point A have each traveled the distances NOP and RAQ. It is readily shown by geometry that the phase of the scattered ray at Q lags behind that at P by

$$\delta = hs \tag{8.1.8}$$

where

$$h = 2k \sin(\theta/2) = (4\pi/\lambda)\sin(\theta/2) \tag{8.1.9}$$

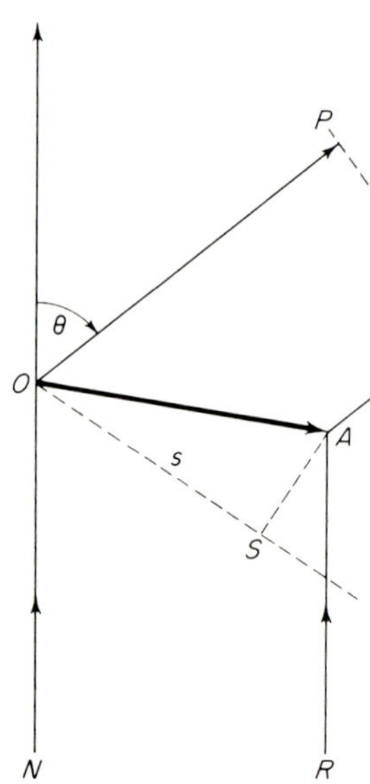

FIG. 8.2. Geometry for Rayleigh–Debye scattering. OS is the bisectrix of angle NOP.

In the construction, the base line of the triangle OSA bisects the angle NOP. Actually, all points in the plane through A, which is perpendicular to the bisectrix OS, have the same phase shift, so that, following Rayleigh (1918b), the integration in (8.1.6) may be performed from slice to slice along OS. The phase corresponding to each slice is weighted by the area bounded by the particle, leading to

$$P(\theta) = (1/V^2)\left|\int Se^{ihs}\,ds\right|^2 \tag{8.1.10}$$

where the limits of the integration are taken through the particle along s, and the area of the slice is S. In the forward direction where $\theta = 0$, this reduces to unity. This is because each of the individual wavelets is precisely in phase, still assuming that the phase shift of a ray is not altered by passing through the particle.

Premilat and Horn (1966) have proposed that the same theory can be used for weakly absorbing particles by taking the modulus of the complex numbers obtained from the above expressions when the complex refractive index is inserted. Thus, the Rayleigh scattering by each volume element would be given by (3.2.8). This implies that as the wave travels through the particle there is no significant reduction of the intensity nor is there any appreciable alteration of the phase due to the presence of the particle.

8.1.1 Spherical Symmetry

The form factor has been evaluated for a variety of configurations. As already indicated, Rayleigh gave the result for a homogeneous sphere with radius a as

$$P(\theta) = [(3/u^3)(\sin u - u\cos u)]^2 = (9\pi/2u^3)J_{\frac{3}{2}}{}^2(u) \tag{8.1.11}$$

where

$$u = ha = ka2\sin(\theta/2) = 2\alpha\sin(\theta/2) \tag{8.1.12}$$

and $J_{\frac{3}{2}}(u)$ represents the three-halves order Bessel function.

The scattering efficiency is obtained by integrating the intensity over the whole spherical surface and dividing the result by πa^2. This is

$$Q_{\text{sca}} = r^2/a^2\int_0^\pi (I_1 + I_2)\sin\theta\,d\theta \tag{8.1.13}$$

$$Q_{\text{sca}} = (m-1)^2(4/9)\alpha^4\int_0^\pi P(\theta)(1+\cos^2\theta)\sin\theta\,d\theta \tag{8.1.14}$$

The integral has been evaluated for a sphere by Rayleigh (1914), who obtained

$$Q_{\text{sca}} = (m-1)^2\left\{\frac{5}{2} + 2\alpha^2 - \frac{\sin 4\alpha}{4\alpha} - \frac{7}{16\alpha^2}(1-\cos 4\alpha) + \left(\frac{1}{2\alpha^2} - 2\right)(\gamma + \log 4\alpha - Ci[4\alpha])\right\} \tag{8.1.15}$$

where γ is Euler's constant (0.577) and the cosine integral is

$$Ci(x) = -\int_x^\infty (\cos u)/u\, du \tag{8.1.16}$$

Ryde (1931) has shown that when α is very much less than unity, the above reduces to

$$Q_{\text{sca}} = (32/27)\alpha^4(m-1)^2 \tag{8.1.17}$$

the same as that for a Rayleigh scatterer with a refractive index close to that of the medium. On the other hand, if α is very large, one obtains

$$Q_{\text{sca}} = 2(m-1)^2\alpha^2 \tag{8.1.18}$$

Rayleigh (1918b) also considered a sphere with a radially inhomogeneous refractive index. Using the method of slices, he derived the amplitude function of an infinitesimal spherical shell having a radius s and an infinitesimal thickness ds, as

$$dS_1(\theta) = (2i/3)k^3(m-1)s^2[\sin(hs)]/hs\, ds \tag{8.1.19}$$

Integration for a homogeneous sphere, i.e., constant m, leads to (8.1.11). For a spherical shell with inner and outer radii a and b, respectively,

$$S_1(\theta) = (i/2\pi)k^3(m-1)[V_bG_b(\theta) - V_aG_a(\theta)] \tag{8.1.20}$$

Here V_a and V_b are the volumes, and $G_a(\theta)$ and $G_b(\theta)$ are the square roots of the form factor corresponding to spheres with refractive index m and with radii a and b, respectively. The effective form factor for the spherical shell is

$$P(\theta) = [G_b(\theta) - (a/b)^3G_a(\theta)]^2 \tag{8.1.21}$$

provided V_b, the volume of a sphere with radius b, is used in (8.1.5).

Kerker *et al.* (1962) have considered the case where there are two concentric regions, each of which is optically homogeneous, i.e., a sphere of one medium encased in a spherical shell of a second medium. The relative refractive indices of the inner region with radius a and the outer spherical shell with radius b are m_1 and m_2, respectively. Then

$$S_1(\theta) = (i/2\pi)k^3[(m_2-1)V_bG_b(\theta) + (m_1-m_2)V_aG_a(\theta)] \tag{8.1.22}$$

This expression reduces to that for a spherical shell when $m_1 = 1$ and to the appropriate single sphere case when either of the following conditions apply:

$$m_1 = m_2; \qquad m_2 = 1; \qquad a = b; \qquad a = 0 \tag{8.1.23}$$

A form factor can be written for this case, such that

$$P(\theta) = \left[G_b(\theta) + \left(\frac{m_1 - m_2}{m_2 - 1} \right) \left(\frac{a}{b} \right)^3 G_a(\theta) \right]^2 \tag{8.1.24}$$

where again the total volume V_b and m_2 are used in (8.1.5). This treatment can be extended to any arbitrary number of homogeneous shells.

The scattering efficiency for the concentric spheres is obtained by integrating (8.1.14). This leads to

$$Q_{\text{sca}} = (m_2 - 1)^2 \phi(v) + (m_1 - m_2)^2 (a/b)^2 \phi(\alpha)$$
$$+ 2(m_2 - 1)(m_1 - m_2)(a/b)\phi(\alpha, v) \tag{8.1.25}$$

where

$$\phi(\alpha, v) = \tfrac{4}{9}\alpha^2 v^2 \int_0^\pi G_a(\theta) G_b(\theta)(1 + \cos^2\theta) \sin\theta \, d\theta \tag{8.1.26}$$

and $\phi(\alpha)$ and $\phi(v)$ are the corresponding expressions evaluated in (8.1.15). For $\phi(\alpha, v)$, involving the cross terms, the evaluation can be carried out by numerical integration. The size parameter v is $2\pi b/\lambda$. Oster and Riley (1952*a*, b) have derived the analogous formulas for the case of small angle X-ray scattering.

When the refractive index of the sphere is no longer constant, but varies in the radial direction, integration of (8.1.1) must be carried out appropriately. Rayleigh (1918b) has presented explicit results for the case where the internal structure is radially periodic with $(m - 1)$ proportional to $\sin(2ks \sin\theta/2)$, and Oster and Riley (1952a) have examined the case where $(m - 1)$ is proportional to $\cos^2(\pi js/a)$. The integer j gives the number of concentric shells. More recently, Albini (1962) has considered the influence of several other functional forms of the refractive index profile upon the backscattering cross section.

Sometimes these expressions are formulated in terms of the polarizability rather than the refractive index. When the sphere is radially symmetric its polarizability is given by

$$\alpha'_v = \int_0^a 4\pi s^2 \alpha'(s) \, ds \tag{8.1.27}$$

where, according to the Lorentz–Lorenz equation, the polarizability per unit volume is related to the refractive index by

$$\alpha'(s) = \frac{3(m^2 - 1)V}{4\pi(m^2 + 2)} = \frac{(m - 1)V}{2\pi} \tag{8.1.28}$$

This leads to

$$I_1 = \frac{k^4}{r^2}\left[\int_0^a 4\pi\alpha'(s)s^2\frac{\sin(hs)}{hs}\,ds\right]^2 \tag{8.1.29}$$

A formula similar to this was first given by Ehrenberg and Schafer (1932) in connection with X-ray scattering. In particular, when the polarizability has a Gaussian distribution such that

$$\alpha'(s) = \alpha_0' \exp[-(s/a)^2] \tag{8.1.30}$$

the resultant intensity is (Peterlin, 1951)

$$I_u = [k^4(\alpha_v')^2/2r^2](1 + \cos^2\theta)\exp(-u^2/2) \tag{8.1.31}$$

An interesting application of scattering by spheres with a variable refractive index or polarizability arises in certain glasses and alloys in which precipitation or phase separation occurs. Maurer (1962), confirming the results of Russian workers, observed a considerable excess of backward light scattering over forward scattering for phase separated glass. This prevalence of backscattering is quite unusual. Furthermore, Gurevich (1955) reported a dependence of the turbidity on the inverse eighth power of the wavelength rather than the more usual inverse fourth power. Walker and Guinier (1953) found an unusual effect in the small angle X-ray scattering by an aluminum-silver alloy in which the forward scattering mounts rapidly with increasing angle, reaches a peak, and then decreases more slowly.

It has been proposed that in each of these systems there is a growing particle of a new phase surrounded by a region in which there is a depletion of the diffusing species. The system is examined at times sufficiently early so that the depleted regions do not overlap.

The refractive index profile in the neighborhood of a precipitated particle is shown qualitatively in Fig. 8.3 for the case that the new phase has a higher refractive index than the medium. The refractive index is uniform out to the surface of the particle at $s = a$. In the depleted region, the refractive index increases out to a distance b where it is not significantly different from that of the original system. The integrals in (8.1.29) can now be separated for each of the two regions. For example,

$$\int_0^a 4\pi s^2\alpha_a'(s)\frac{\sin(hs)}{hs}\,ds + \int_a^b 4\pi s^2\alpha_{ab}'(s)\frac{\sin(hs)}{hs}\,ds \tag{8.1.32}$$

The first term is simply the usual expression for a uniform sphere of radius a as given by (8.1.11). The second term can be evaluated numerically if the polarizability profile for the depleted region ab is known.

Goldstein (1963a, b) has calculated the polarizability profile for such a diffusion controlled precipitation process in region ab and has found that the observed wavelength dependence and the prevalence of backscattering can indeed be satisfactorily explained as due to the interference between the particle and its diffusion field. Further experimental confirmation of this model for phase-separated glasses has been obtained by Hammel and Ohlberg (1965) and Hammel *et al.* (1968).

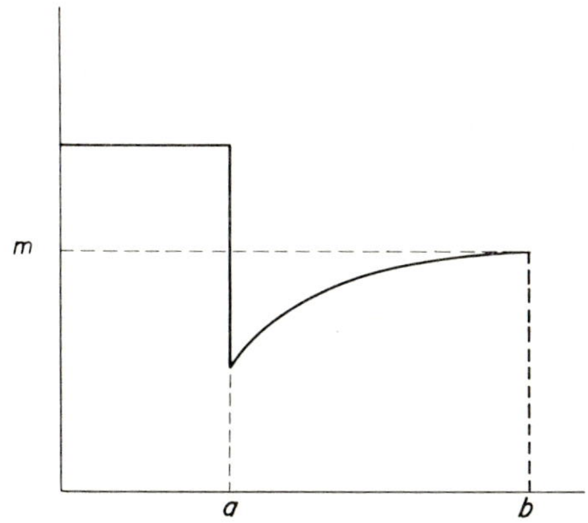

FIG. 8.3. Refractive index profile in the neighborhood of a freshly precipitated particle.

There is an alternative point of view; namely, that the precipitated particles are not randomly located so that the scattering is not incoherent, and the interference among the particles may account for the anomalous scattering effects. It is possible, using transform techniques to be discussed in Section 8.4, to describe an appropriate arrangement of the particles in a statistical way with the aid of a correlation function, and this has been done by Voishvillo (1961). However, this does not provide the same detailed insight into the physical processes occurring in glass as the depletion model. Also it cannot account for the observed wavelength effects.

8.1.2 Macromolecules; Discrete Scattering Elements

Macromolecules whose dimensions are comparable to or larger than the wavelength of light can be treated as Rayleigh–Debye scatterers when their effective index of refraction is close to that of the solvents in which they are suspended. The macromolecule may be envisaged as an assembly of discrete scattering centers comprised of the monomer units, each of which is an elemental Rayleigh scatterer. The phase relations among these elemental wavelets are determined by the geometry of the macromolecule. It is the interference among these wavelets which determines the form factor for each particular configuration. Figure 8.4 depicts a linear, randomly coiled molecule.

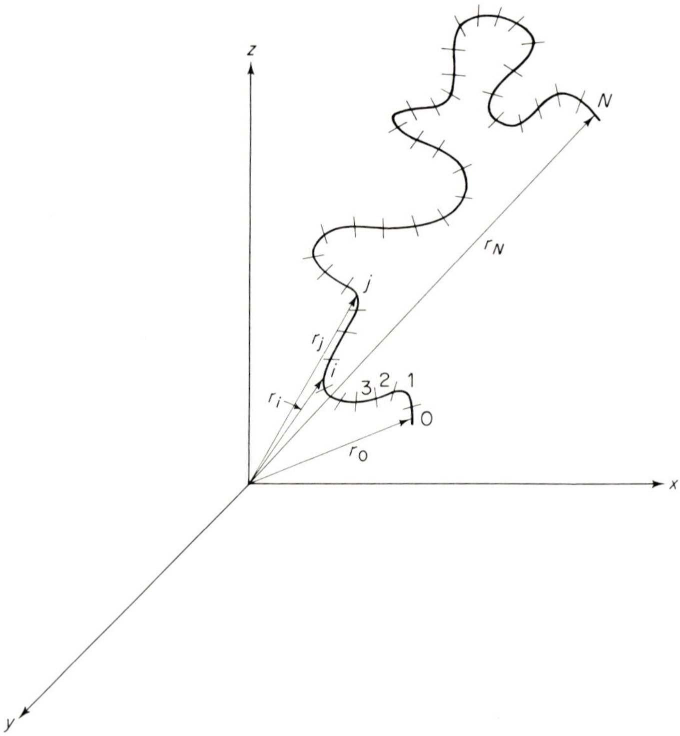

FIG. 8.4. Scattering elements of a large coiled polymer molecule.

Debye (1915) formulated the theory for an aggregate of discrete elements rather than for a continuum in connection with the scattering of X-rays by molecules in a gas. The intensity for an assembly of N scattering elements can be expressed as

$$I_1 = \frac{\lambda^2}{4\pi^2 r^2}\left\{\frac{1}{N^2}\sum_i^N \sum_j^N S_i S_j \cos(hs_{ij})\right\} \tag{8.1.33}$$

where S_i and S_j are the amplitude functions for the elements i and j and s_{ij} is the distance between their centers. If each of the scattering elements is identical and if the molecule assumes all random orientations, then it is possible to derive the following average value of the form factor:

$$P(\theta) = \frac{1}{N^2} \sum_{i}^{N} \sum_{j}^{N} \frac{\sin hs_{ij}}{hs_{ij}} \tag{8.1.34}$$

This expression has been evaluated for a number of configurations which are encountered in macromolecular chemistry.

For a spherical distribution of segments or particles, Debye (1930) has given

$$P(\theta) = \frac{\int_0^\infty 4\pi s^2 G(s) \frac{\sin hs}{hs}\, ds}{\int_0^\infty 4\pi^2 s^2 G(s)\, ds} \tag{8.1.35}$$

where $G(s)\,ds$ is the probability density of scattering material within the spherical shell having radii s and $s + ds$. This may be compared to the expression (8.1.19) in which the weighting factor is $(m - 1)$ or to (8.1.29) where it is the polarizability $\alpha'(s)$. With macromolecules we focus attention upon the distribution of monomeric segments in the polymer chain, assuming that each of these has the same refractive index.

Perhaps the case of greatest interest in macromolecular chemistry is the spherical distribution of matter which occurs in the randomly coiled high polymer chain. Here, each link usually makes a definite angle with the preceding one and is free to rotate with respect to a plane determined by the preceding two links. This is a much studied problem in polymer statistics (Tanford, 1961). If no restrictions are imposed upon the orientation of any link, the conformation of the coil corresponds to a three-dimensional random walk for which one obtains the Gaussian radial distribution function

$$G(s)\, ds = 4\pi s^2 \,(3/2\pi \overline{s^2})^{3/2} \exp[-\tfrac{3}{2}(s^2/\overline{s^2})]\, ds \tag{8.1.36}$$

where $G(s)\,ds$ is the probability that one end-point of the chain is found at a distance between s and $s + ds$ from the other end and $\overline{s^2}$ is the mean square of the end-to-end distance of the chain. Debye (1947) has derived the following form factor for this case:

$$P(\theta) = (2/w^2)(e^{-w} + w - 1) \tag{8.1.37}$$

Here

$$w = h^2\overline{s^2}/6 = h^2\overline{R_g{}^2} \tag{8.1.38}$$

The radius of gyration is defined in terms of an assembly of mass elements, m_i, each located a distance s_i from the center of mass by

$$R_g^2 = \sum m_i s_i^2 / \sum m_i \tag{8.1.39}$$

In (8.1.38) the mean square value of the radius of gyration over all statistical configurations is used.

The modification of the form factor when the chain segments are stiff so that the Gaussian distribution does not apply has been obtained by Benoit and Goldstein (1953), Benoit and Doty (1953), Miyake (1960), and Peterlin (1963) for various non-Gaussian distributions.

There are many possible models corresponding realistically to macromolecules for which a form factor can be calculated. As an illustration, one might consider the detailed theoretical study of scattering from "comb-like" branched molecules carried out by Casassa and Berry (1966). A schematic diagram of the comb model is shown in Fig. 8.5. The backbone is a random

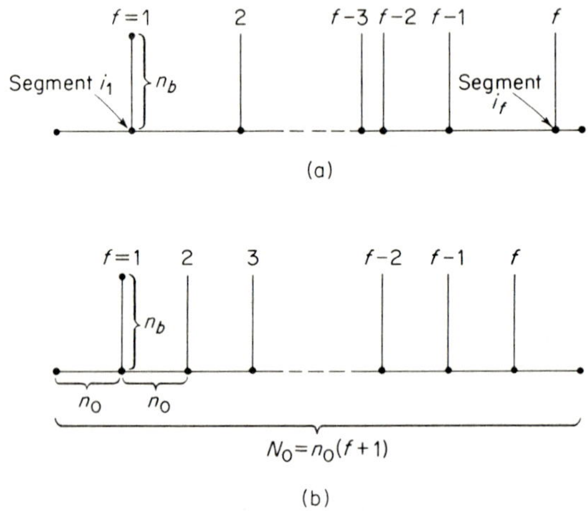

FIG. 8.5. Schematic diagrams of comb molecules constructed from a backbone chain of N_0 chain segments and f branches, each containing n_b segments: (a) an irregular comb; (b) a symmetrical comb (Casassa and Berry, 1966).

coil consisting of N_0 uniform segments. Each molecule contains f branches, each branch having n_b segments. In the "symmetrical comb" model, the branches are attached equidistantly along the backbone chain so that each molecule is identical. In the "irregular comb" model, the branches are distributed randomly along the chain so that, although the molecules are still homogeneous in mass, there is a distribution of structures superimposed upon the distribution of random configurations. Finally, a system of "heterogeneous combs" is generated, still consisting of the uniform branches and main chains. However, the total number of branches and of main chains are now imagined to couple randomly, resulting in a distribution of molecules with varying numbers of branches. Each molecular size is heterogeneous with regard to the placement of the branches on the chains. Closed form expressions for $P(\theta)$ for each of these models have been obtained by Casassa and Berry (1966), who also discuss the implications for discriminating among these models by light scattering experiments. Dautzenberg and Ruscher (1967) have calculated $P(\theta)$ for other branched configurations including the so-called cascade and star models.

8.1.3 Cylinders

Neugebauer (1943) has derived for the form factor for a thin rod of length L

$$P(\theta) = (1/z)\int_0^{2z} \frac{\sin w}{w}\,dw - \left(\frac{\sin z}{z}\right)^2 \tag{8.1.40}$$

where

$$z = (2\pi L/\lambda)\sin(\theta/2) \tag{8.1.41}$$

Rayleigh (1881) first treated the infinite circular cylinder at perpendicular incidence for which the form factor is

$$P_1(\theta) = G_1{}^2(\theta) = (2J_1(u)/u)^2 \tag{8.1.42}$$

and the amplitude function is (Montroll and Greenberg, 1954)

$$T_1(\theta) = (\pi\alpha^2/2)(m - 1)G_1(\theta) \tag{8.1.43}$$

This holds for the electric vector polarized parallel to the cylinder axis. The perpendicular case is obtained from the above by multiplication by $\cos^2\theta$. For a homogeneous cylindrical shell

$$T_1(\theta) = [\pi(kb)^2/2]\,(m - 1)[G_b(\theta) - (a/b)^2 G_a(\theta)] \tag{8.1.44}$$

where a and b are the inner and outer radii of the cylindrical shell (Oster and Riley, 1952b). For an infinitely thin cylindrical shell the form factor is

$$P_1(\theta) = J_0{}^2(u) \tag{8.1.45}$$

For an assembly of N identical long cylinders in parallel orientation and located in fixed positions relative to each other, the form factor is

$$P(\theta) = \frac{1}{N^2}\sum_i^N \sum_j^N J_0(hs_{ij}) \tag{8.1.46}$$

where s_{ij} is the interparticle distance separating nearest neighbors. This is the two-dimensional analog of (8.1.33). As examples, one may consider two parallel cylinders for which

$$P(\theta) = \tfrac{1}{4}[2 + 2\,J_0(hs)] \tag{8.1.47}$$

or seven cylinders in a central hexagonal arrangement for which

$$P(\theta) = (1/49)[7 + 24J_0(hs) + 6J_0(2\,hs) + 12J_0(\sqrt{3}\,hs)] \tag{8.1.48}$$

where s is the distance between nearest neighbors. The form factor now represents the factor by which the scattered intensity is changed due to the interparticle interference from what would have been obtained by simple addition of the intensities scattered by the individual particles.

Montroll and Greenberg (1952) and Albini and Nagelberg (1962) have treated the case of oblique incidence, and the latter have compared their results with the limiting expressions derived from Wait's (1955) exact theory. Midzuno (1961a) has derived the Rayleigh–Debye expression for scattering at perpendicular incidence from a nonuniform anisotropic plasma. The anisotropy corresponds to the gyromagnetic medium which occurs when a plasma is in a uniform magnetic field. Analytical solutions are obtained for some special forms of the radial distribution of electron density. The results have also been carried out when the incident radiation is a spherical wave (Midzuno, 1961b).

The form factor for a randomly oriented thin disk was derived by Kratky and Porod (1949) as

$$P(\theta) = (2/u^2)[1 - (2/u)J_1(u)] \tag{8.1.49}$$

Here $u = ha$ where a is the radius of the thin disk. Frequently, practical systems containing thin disks in which there is interest, such as solutions of polymer single crystals, are anisotropic. Then four separate form factors must be considered corresponding to the horizontal (parallel) and vertical (perpendicular) components of the scattered light when the incident light is polarized vertically (H_v and V_v) and the corresponding quantities for incident light polarized horizontally (V_h, H_h). If the polarizabilities in the directions parallel and perpendicular to the surfaces of the disk are α_1' and α_2' and the polarizability of the medium is α_s', the anisotropy can be defined by

$$\delta = (\alpha_1' - \alpha_2')/3(\alpha_0' - \alpha_s') \tag{8.1.50}$$

where

$$\alpha_0' = (2\alpha_1' + \alpha_2')/3 \tag{8.1.51}$$

The various form factors have been derived by Picot *et al.* (1968) as

V_v $$P(\theta) = (2/u^2)\{I_1 + \delta(3I_2 - 2I_1) + \delta^2[I_1 - 3I_2 + (27/8)I_3]\} \tag{8.1.52}$$

$V_h = H_v$ $$P(\theta) = (9\delta^2/u^2)\{I_2 - I_3 + (\sin^2 \theta/2)[(5/4)I_3 - I_2]\} \tag{8.1.53}$$

H_h $$\begin{aligned} P(\theta) = (2/u^2)\{&I_1 \cos^2 \theta - \delta \cos \theta(2 - \sin^2 \theta/2)(3I_2 - 2I_1) \\ &+ \delta^2[(2 - \sin^2 \theta/2)^2(I_1 - 3I_2) \\ &+ 9I_3[1 - \sin^2 \theta/2 + (3/8)\sin^4 \theta/2]\} \end{aligned} \tag{8.1.54}$$

where

$$I_n = \int_0^\pi J_1{}^2(u \sin \gamma)(\sin \gamma)^{2n-3}\, d\gamma \tag{8.1.55}$$

8.1.4 Range of Validity of the Rayleigh–Debye Theory

The range of validity of the Rayleigh–Debye theory can be tested by comparison of exact calculations for spheres and cylinders with those obtained with the approximate formulas. The most extensive testing of this sort has been carried out by Kerker *et al.* (1963b), Farone *et al.* (1963), and Heller (1963).

The results for spheres are presented as error contour charts in the $m\alpha$-domain in Figs. 8.6 through 8.9. In each chart, region I delimits the range of values of α and m for which the exact theory and the Rayleigh–Debye theory agree to within 10%. In region II the deviations are no greater than 100%, whereas in region III they are greater than this. There are small islands in regions II and III where the agreement between the two theories is better than 10% and 100%, respectively. The quantities tested are Q_{sca} in Fig. 8.6, and i_1 at $\theta = 10$, 20, and 45° in Figs. 8.7, 8.8, and 8.9, respectively.

The Rayleigh–Debye approximation is best for small values of m and for small angles (forward direction). Thus, for $m = 1.10$, the 10% contour line falls at $\alpha = 10.0$ when $\theta = 10°$, at $\alpha = 7.5$ when $\theta = 20°$, and at $\alpha = 4.0$ when $\theta = 45°$. The agreement rapidly becomes poorer as θ increases beyond 45°. For Q_{sca}, the 10% contour line falls at $\alpha = 9.2$.

The failure to obtain 10% agreement at the smallest sizes when $m > 1.25$ is not due to any limitation of the Rayleigh–Debye theory but rather to the numerical approximation usually made concerning the refractive index, viz.

$$(m^2 - 1)/(m^2 + 2) = \tfrac{2}{3}(m - 1) \qquad \text{as} \qquad m \to 1 \tag{8.1.56}$$

The error due to this approximation is precisely 10% when $m = 1.25$. Had it not been introduced, region I would have included the area just above the abscissa presently designated as region II. The present peninsular strip of region I extended towards the right of each diagram is probably due to the cross-over from the positive deviations introduced by (8.1.56) to negative deviations due to differences between the exact and approximate theories.

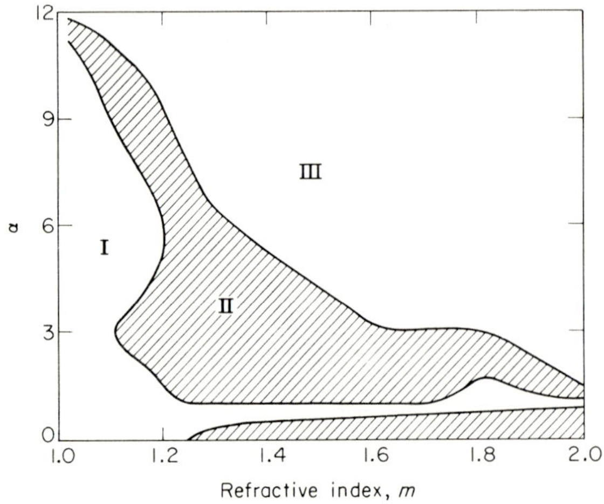

FIG. 8.6. Error contour chart for Q_{sca}. In region I the Rayleigh–Debye theory is correct to within at least 10%; in region II to within at least 100%; in region III, the error is greater than 100% except in small islands (Kerker *et al.*, 1963b).

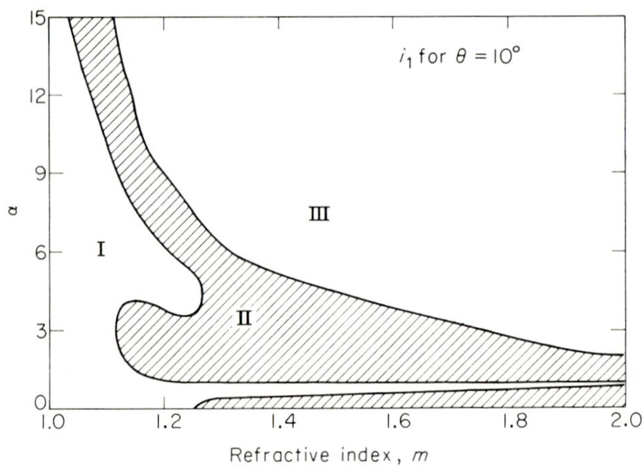

FIG. 8.7. Same as Fig. 8.6 for i_1 at $\theta = 10°$.

The results for the horizontal component of the intensity function, i_2, are quite similar at 10 and 20°. However, at 45° the deviations are considerably greater, especially at larger values of α.

For the cylinder, the trends are quite similar except that the Rayleigh–Debye approach is not valid for as broad a range of either α or m.

Farone and Robinson (1968) have carried out a similar study of the range of validity of the anomalous diffraction approximation for scattering by spheres. This approximation is successful over an even larger range of parameters than the Rayleigh–Debye theory, particularly at larger sizes.

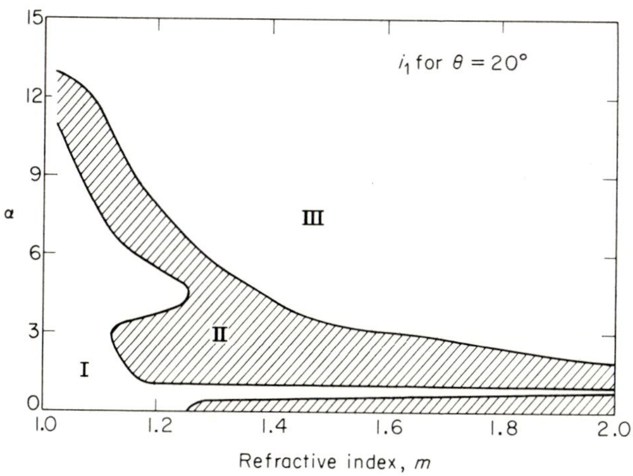

FIG. 8.8. Same as Fig. 8.6 for i_1 at $\theta = 20°$.

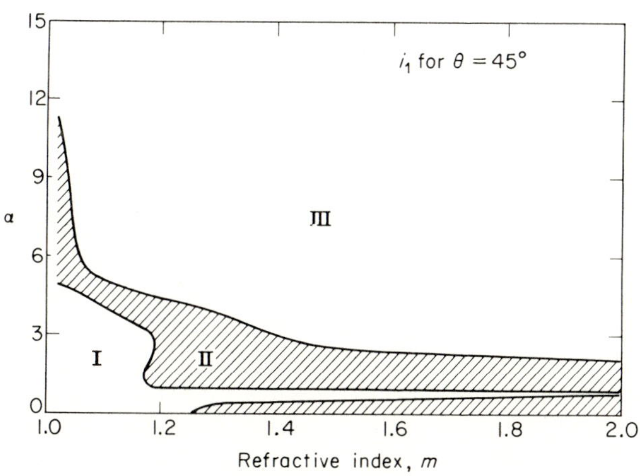

FIG. 8.9. Same as Fig. 8.6 for i_1 at $\theta = 45°$.

8.1.5 Integral Equation Formulation

As an alternative to the boundary value solution to the wave equation such as that first obtained for a cylinder by Rayleigh (1881) or for a sphere by Lorenz (1890, 1898b), the scattering problem can be formulated as an integral equation. This formulation may be derived, without approximation, directly from Maxwell's equations (Saxon, 1955), and it automatically incorporates all of the boundary conditions. There are no restrictions upon the particle shape nor upon the optical properties of the particle. Indeed, the particle need be neither homogeneous nor isotropic but may be described by a completely general refractive index.

For a plane incident wave, the electric field of the scattered light is given by

$$\mathbf{E}_s = -\frac{\exp(ik_0 r)}{r}\frac{k_0^2}{4\pi}\int_v \exp(-ik_0\mathbf{s}\cdot\mathbf{r})[m^2(r)-1][\mathbf{s}\times\mathbf{s}\times\mathbf{E}(r)]\,dV \qquad (8.1.57)$$

where k_0 is the propagation constant of the wave in the medium, r is the distance of the point of observation from the particle along the direction of the unit vector $\mathbf{s}$, $\mathbf{r}$ the position vector of any point, $m(r)$ is the relative refractive index which will be assumed to be real, and $\mathbf{E}(r)$ is the time independent part of the electric field within the particle. The volume integration is carried out inside the particle. The magnetic inductive capacity is assumed to have the same value inside and outside.

The magnetic field of the scattered light may be obtained from the electric field in the usual way and the intensity function formulated with the aid of Poynting's theorem. Alternatively, Saxon (1955) has expressed the magnetic field in terms of the internal magnetic field, $\mathbf{H}(r)$.

$$\mathbf{H}_s = -\frac{\exp(ik_0 r)}{r}\frac{k_0^2}{4\pi}\int_v \exp[-ik_0\mathbf{s}\cdot\mathbf{r}][m^2(r)-1][\mathbf{s}\times\mathbf{s}\times\mathbf{H}(r)]\,dV + \frac{ik_0}{4\pi}\int_v \exp[-ik_0\mathbf{s}\cdot\mathbf{r}]\,\nabla m^2(r)\times\mathbf{E}(r)\,dV \qquad (8.1.58)$$

The interesting feature of this is that the scattering amplitudes may be expressed in terms of a relatively simple looking integral of the internal field over the scattering object. However, before the integral can be evaluated, the internal field must be known and this can only be obtained from the solution of the boundary value problem. It would appear then that the full circle has been turned to little avail.

However, the utility of the integral equation formulation lies in the opportunities which it affords for approximate solutions. Indeed, the Rayleigh–Debye theory corresponds precisely to this formulation when the internal field is replaced by the incident field. Other, possibly more realistic, trial fields might be considered. For example, Ikeda (1963) has expanded the internal field as a Taylor series of the following form:

$$\mathbf{E}(r) = \sum_{n=0}^{\infty} (m^{-2} - 1)^n \mathbf{E}_n(r) \tag{8.1.59}$$

where $\mathbf{E}_n(r)$ may be expressed explicitly in terms of the incident field.

Saxon (1955) has examined several simpler representations of the internal field obtained from elementary ray optics considerations. The Rayleigh–Debye approximation may be modified so as to assume that the internal wave undergoes a phase shift equal to that of the axial ray. In another approach, termed the W.K.B. (Wentzel–Kramers–Brillouin) approximation, the internal wave continues to propagate rectilinearly with no change in polarization. Unlike the modified Rayleigh–Debye approximation, the phase is not assumed constant over the wavefront, but is modified by the discontinuous change in the refractive index as each ray enters the spherical interface. However, there is no bending of the rays upon penetrating into the sphere. A somewhat more sophisticated approximation modifies the W.K.B. approximation to include bending of the incident rays so that both the phase shift and the refraction upon entering the particle are considered.

An alternative also considered by Saxon is to estimate the interior field from the exact solution. The Debye potential for the field inside the particle is given by (3.3.36) and (3.3.37). The coefficients c_n and d_n may be obtained directly as solutions of (3.3.42) to (3.3.45). A simplified expression is obtained for the case of large particles by using the asymptotic forms of the spherical Bessel functions for large arguments. The final results are similar to those of the modified Born approximation except that in addition to the forward traveling wave there is also a wave traveling in the backward direction corresponding to a wave reflected at the back surface of the particle. The addition of reflected and refracted waves offers many other possibilities for approximations that still remain to be explored.

The results of the W.K.B. approximation (parallel internal field) have been examined critically by Deirmendjian (1957, 1959) for the case of spherical scatterers, and test computations have been compared with the exact theory. The expressions for the scattering efficiency and for the forward scatter are identical with those obtained earlier for the case of anomalous diffraction (Section 4.2), but these two approximate treatments differ as the scattering angle moves away from the forward direction. Then the superiority of the W.K.B. approach becomes apparent. It shows better agreement with the exact theory, both with respect to the angular intensity pattern and with respect to polarization. Indeed, Deirmendjian (1959) has found it quite useful in describing the main diffraction peak in the forward scattered light of atmospheric aerosols which gives rise to the phenomenon of the solar aureole.

8.2 Size and Shape of Particles and Macromolecules

We have already seen how the angular variation of the scattering can be utilized to determine the particle size. A number of special techniques developed for the Rayleigh–Debye approximation will now be considered. These need not be restricted to spherical or cylindrical symmetry.

8.2.1 Dissymmetry Method

The dissymmetry is defined as the ratio of the intensity scattered at two angles symmetrical about 90°. Thus,

$$z(\theta) = i_1(\theta)/i_1(180° - \theta) \tag{8.2.1}$$

For a particular shape, $z(45°)$ increases rapidly with increasing particle size. This is illustrated in Fig. 8.10 where the 45° dissymmetry for spheres, random

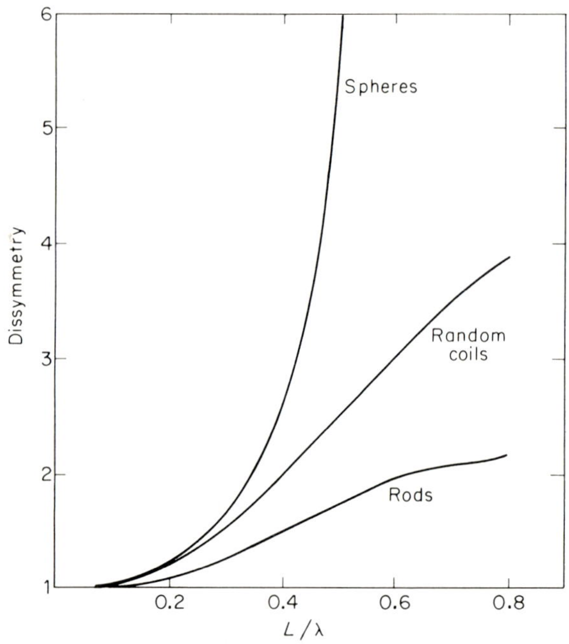

FIG. 8.10. Dissymmetry at 45° relative to 135° plotted against the relative characteristic length L/λ for spheres, random coils, and rods.

coils, and rods as calculated from (8.1.11), (8.1.37), and (8.1.40) is plotted against L/λ. The characteristic dimension L designates the diameter of the sphere, the root mean square distance between the ends of the coil, and the length of the rod, respectively. Provided the particular shape of the particle is known and the size is within the indicated range, a single measurement of the dissymmetry can be used to determine the dimension L. Obviously, this method will be most useful when the size is just beyond the range of validity of simple Rayleigh scattering. Jennings and Jerrard (1965a) have compared results obtained for rubber latex samples when the dissymmetry was calculated both by the Rayleigh–Debye equation and by the exact theory for spheres. The agreement was within 5% for values of α up to 1.6. In this case, $m = 1.15$.

8.2.2 Radius of Gyration

A very powerful method for determining particle size utilizes the angular variation of the scattering close to the forward direction. By expanding the function $\sin hs_{ij}$ in (8.1.34) into a power series and dropping higher-order terms under the limiting condition of small θ, Guinier (1939) has shown that

$$P(\theta)_{\theta \to 0} = 1 - (h^2 R_g{}^2/3) = 1 - (16\pi^2/3\lambda^2) R_g{}^2 \sin^2(\theta/2) \qquad (8.2.2)$$

where R_g is the radius of gyration of the particle (8.1.39). This leads to a measure of the size of the particle independent of any assumptions regarding its shape since the radius of gyration can be obtained by plotting $P(\theta)$ vs. $\sin^2(\theta/2)$ and calculating R_g from the initial slope according to

$$R_g{}^2 = (3\lambda^2/16\pi^2)(-\text{slope}) \qquad (8.2.3)$$

Here, λ is the wavelength in the medium.

The value of $P(\theta)$ itself can be determined from the scattered intensity with the aid of (8.1.5) if the number of particles in the scattering region and the particle volume are known. Otherwise, recourse may be had to the treatment based upon concentration fluctuations (Chapter 9). In this case, it is necessary to know the concentration of the dispersed phase and the differential refractive index. Both of these quantities may be readily obtained experimentally without any prior knowledge of particle size, particle number, or particle shape.

It will be shown that for Rayleigh scatterers, the fluctuation treatment leads to

$$\frac{Kc}{R_\theta} = \frac{1}{M} + 2Bc + 3Cc^2 + \cdots \qquad (8.2.4)$$

where

$$K = (2\pi^2 n^2/\lambda_0{}^4 N_A)(\partial n/\partial c)^2_{T,p}(1 + \cos^2\theta) \tag{8.2.5}$$

and c is the concentration in g/ml, R_θ is the Rayleigh ratio for unpolarized incident light at the scattering angle θ, M is the molecular weight, n is the refractive index of the dispersion (the particles plus the medium), λ_0 is the wavelength *in vacuo*, N_A is Avogadro's number, and B, C, etc., are virial coefficients. The concentration dependent terms involving the virial coefficients arise whenever the particles are no longer randomly positioned in space so that their scattered intensities are no longer additive. For a Rayleigh–Debye scatterer, the form factor $P(\theta)$ is introduced.

$$\frac{Kc}{R_\theta} = \frac{1}{MP(\theta)} + \frac{2Bc}{P(\theta)} + \cdots \tag{8.2.6}$$

In the limit of zero concentration and at sufficiently low angles so that (8.2.2) is valid,

$$Kc/R_\theta = (1/M)[1 + (16\pi^2/3\lambda^2)R_g{}^2 \sin^2(\theta/2)] \tag{8.2.7}$$

Here the reciprocal of $P(\theta)$ has been approximated by

$$\frac{1}{P(\theta)} = \frac{1}{1 - (h^2 R_g{}^2/3)} = 1 + \frac{h^2 R_g{}^2}{3} \tag{8.2.8}$$

On the other hand, in the limit of zero angle where $P(\theta) = 1$ and at sufficiently low concentrations so that only the second virial coefficient need be considered

$$Kc/R_\theta = (1/M) + 2Bc \tag{8.2.9}$$

Finally, in the limit of both zero angle and zero concentration

$$Kc/R_\theta = 1/M \tag{8.2.10}$$

Thus, M can be evaluated from (8.2.10), B can be evaluated from the slope of a plot of Kc/R_θ vs. c with the aid of (8.2.9) and knowledge of M, and R_g can be evaluated from the slope of a plot of Kc/R_θ vs. $\sin^2(\theta/2)$ with the aid of (8.2.7). Each of these objectives can be achieved with a Zimm plot in which Kc/R_θ is plotted vs. $\sin^2(\theta/2) + gc$ where g is arbitrarily chosen to provide a convenient spread of the data. A typical example is shown in Fig. 8.11. The curve drawn through the circles represents the results at various values of θ extrapolated to $c = 0$ and thus corresponds to (8.2.7). Similarly, the curve through the triangles corresponds to (8.2.9). It is the locus of the experimental points at various values of c extrapolated to $\theta = 0$. Boël (1966) has studied the effects of the choice of the value of g upon the precision of the results.

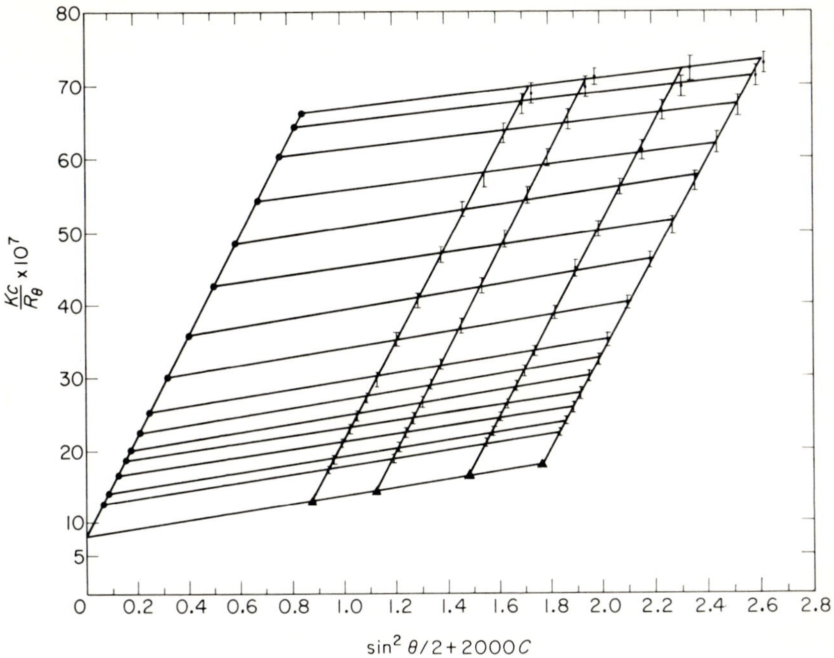

FIG. 8.11. Zimm plot for nitrocellulose in acetone at 25°C (Benoit *et al.*, 1954).

Nakagaki (1966) has calculated the corrections to both the molecular weight and to the radius of gyration which are required when the spherical particles no longer correspond to the requirements of the Rayleigh–Debye theory. The initial slope, from which R_g is obtained, is hardly affected provided $m \leqslant 1.30$, $\alpha \leqslant 7.0$, and $ha \leqslant 1.0$. However, the molecular weight obtained by extrapolation of the Zimm plot must be corrected by the factor

$$F = 1 - 0.155Q + bQ^{1.5} \tag{8.2.11}$$

where

$$b = 0.0115 + 0.00310(m - 1.01) \tag{8.2.12}$$

and

$$Q = (8\lambda\rho_2/3c)R_u(0) \tag{8.2.13}$$

provided

$$Q \leqslant 0.01068/b^2 \tag{8.2.14}$$

Here ρ_2 is the density of the particles.

If, in addition to the radius of gyration, the molecular weight and density or the particle volume are known, it may be possible to predict the particle shape. For a random coil

$$R_g = (\alpha^2\beta^2/6M_0)^{1/2}M^{1/2} = c'M^{1/2} \tag{8.2.15}$$

for a sphere with radius a

$$R_g = (3/4\pi\rho N_A)^{1/3}(3/5)^{1/2}M^{1/3} = c''M^{1/3} \tag{8.2.16}$$

and for a rod with fixed radius a and variable length L

$$R_g = (\sqrt{12}\,\pi a^2\rho N_A)^{-1}M = c'''M \tag{8.2.17}$$

where α and β are molecular parameters defined by Tanford (1961, p. 168), ρ is the density, N_A is Avogadro's number. For example, for $\rho = 1$ and $M = 10^6$, the radius of gyration is 56 Å for a sphere. However, for a rod with a diameter of 25 Å, the radius of gyration would be 980 Å. Thus, a determination of both the radius of gyration and the molecular weight provides a very sensitive method of distinguishing among possible shapes. An example is the use of this method by Eliezer and Silberberg (1967) to follow the changes in the shape of multichain macromolecules upon mild heat treatment. Luzzati *et al.* (1961) estimated the pitch of the α-helix in a polypeptide (polyglutamate) from the dependence of the radius of gyration upon the molecular weight. In another interesting study, Eisenberg and Felsenfeld (1967) studied the variation of the radius of gyration of a synthetic polynucleotide over a considerable range of temperature. They dissolved the polymer at each temperature in a "theta" solvent for which the long range interactions due to polymer–solvent interactions and polymer-polymer interactions are compensated. In such a solvent which has a zero value for the second virial coefficient B in Eq. (8.2.9), the change of the radius of gyration can be attributed to local structure and the hindered rotations about the chemical bonds.

A more complete analysis of the expansion of (8.2.2) (Debye, 1947; Geiduschek and Holtzer, 1958) shows that when higher terms are retained, a more general expression can be developed, leading to

$$P(\theta) = 1 - \frac{h^2}{3!}[s]_2 + \frac{h^4}{5!}[s]_4 - \frac{h^6}{7!^6}[s]_6 + \cdots \tag{8.2.18}$$

where

$$[s]_{2j} = \int_v s^{2j}G(s)\,ds \Big/ \int_v G(s)\,ds \tag{8.2.19}$$

is the $2j$th moment of the distribution of mass within the particle or macromolecule and $G(s)$ is once again the probability density of scattering material. The first of these moments $[s]_2$ is just twice the radius of gyration. Thus, (8.2.18) connects the particle scattering factor with the particle shape in a completely general way.

8.2.3 Copolymers

Copolymers are macromolecules formed by linking two different monomers which will be designated A and B. If the arrangement of monomers is random such as $ABBAABAB\ldots$, the preparation is called a statistical copolymer; if the monomers are arranged in a definite order with large blocks such as $AAA\ldots AABBB\ldots BB\ldots AAAA\ldots$ successively following each other, it is called a block copolymer. Here, we will only consider copolymers whose molecular configuration is that of a random coil.

In studying the light scattering by such a system in solvents of different refractive index, it is found that the apparent molecular weight (Tremblay *et al.*, 1952) and radius of gyration calculated from a Zimm plot are quite dependent upon the refractive index of the solvent. This phenomenon is a consequence of each kind of monomer segment having a different refractive index relative to the solvent and hence a different polarizability. The theory has been derived by Stockmayer *et al.* (1955) and by Bushuk and Benoit (1958a, b). Its validity has been investigated experimentally by these workers and by Krause (1961) who have shown that it is possible to learn a great deal about the composition and internal structure of copolymers by carrying out light scattering measurements in several solvents.

The form factor can be written as a sum of three terms

$$P(\theta) = (1/v^2)[v_A{}^2x^2P_A + v_B{}^2(1 - x^2)P_B + 2v_Av_Bx(1 - x)Q_{AB}] \qquad (8.2.20)$$

where x is the mass fractional composition of component A. The macromolecules may show heterogeneity of chain composition in addition to polydispersity of molecular weight. It is assumed that

$$v = v_Ax + v_B(1 - x) \qquad (8.2.21)$$

where v, v_A, v_B are the refractive index increments, dn/dc, for the copolymer solution and for the two homopolymer solutions corresponding to A and B. The latter equation is a statement of the fact that the specific refractive index increment of a polymer solution depends only on the weight fraction of each monomer component (x and $1 - x$) and not upon the molecular weight nor on the manner in which the monomer is distributed in the macromolecule. The quantity v is incorporated in the factor K in (8.2.4) and the expressions that follow it.

P_A and P_B represent form factors corresponding to the pair interference between the A segments alone or the B segments alone. If the molecule consisted of only one or the other of these segments, these quantities would represent the corresponding form factors. Q_{AB} is the contribution to the form factor which results from the pairwise interference between A segments and B segments. Benoit and Wippler (1960) have given explicit formulas for these quantities for the case that the copolymer is a Gaussian chain. The spiral case for which the segments A comprise the backbone and polymers of segments B comprise the side chains has also been considered.

If the angular light scattering data are formulated as a Zimm plot and treated in the usual way, they will lead to an apparent molecular weight M_{ap} and an apparent radius of gyration $R_g{}^*$. The former is given by Bushuk and Benoit (1958a, b):

$$M_{ap} = (1/v^2)[xM_w{}^A v_A{}^2 + (1 - x)M_w{}^B v_B{}^2 + 2v_A v_B M_w^{AB}] \tag{8.2.22}$$

The partial weight average molecular weight of component A is defined by

$$M_w{}^A = \sum_i \sum_j c_{ij}^A M_i{}^A \Big/ \sum_i \sum_j c_{ij} \tag{8.2.23}$$

where c_{ij} is the weight concentration of component A in a macromolecule having a molecular weight of $M_i{}^A$ in component A and a mass fractional composition x_j. There is a similar definition for $M_w{}^B$. The cross term is defined by

$$M_w^{AB} = \sum n_{ij} M_i{}^A M_i{}^B \Big/ \sum n_{ij}(M_i{}^A + M_i{}^B) \tag{8.2.24}$$

where n_{ij} is the number of macromolecules having a molecular weight $(M_i{}^A + M_i{}^B)$ and composition x_j. By carrying out experiments at three different solvent refractive indices, it is possible to determine $M_w{}^A$, $M_w{}^B$, and M_w^{AB}, thereby characterizing the composition of the copolymer.

An alternative form for expressing (8.2.22) is

$$M_{ap} = M_w + 2P\left(\frac{v_A - v_B}{v}\right) + Q\left(\frac{v_A - v_B}{v}\right)^2 \tag{8.2.25}$$

in which

$$P = \sum_i \sum_j c_{ij} M_i (x_j - x) \Big/ \sum_i \sum_j c_{ij} \tag{8.2.26}$$

and

$$Q = \sum_i \sum_j c_{ij} M_i (x_j - x)^2 \Big/ \sum \sum c_{ij} \tag{8.2.27}$$

A plot of M_{ap} against $(v_A - v_B)/v$ as in Fig. 8.12 leads to the values of M_w, $M_w{}^A$, $M_w{}^B$, P, and Q as indicated on the diagram. Benoit (1963, 1966) has discussed in detail how these parameters characterize the copolymer composition and in some cases lead to information about the mechanism of formation.

The initial slope of the angular part of a Zimm plot leads to an apparent radius of gyration which (Benoit and Wippler, 1960), for a copolymer formed of molecules having the same molecular weight and composition, is

$$(R_g{}^2)^* = \overline{R_A{}^2}\left(\frac{x\nu_A}{\nu}\right)^2 + \overline{R_B{}^2}\left(\frac{(1-x)\nu_B}{\nu}\right)^2 + \frac{2x(1-x)\nu_A\nu_B}{\nu^2}\,\overline{L^2} \qquad (8.2.28)$$

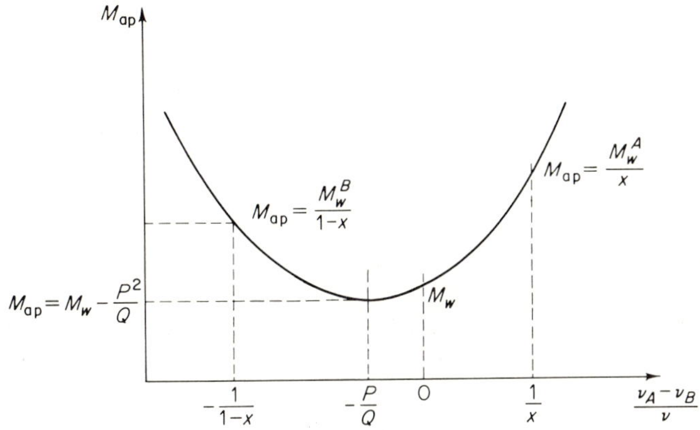

FIG. 8.12. The various parameters obtained by a plot of M_{ap} against $(\nu_A - \nu_B)/\nu$ for a copolymer solution (Benoit, 1963).

Here, $\overline{R_A{}^2}$ is the mean square radius of gyration of part A about its center of mass, $\overline{R_B{}^2}$ is the same for part B, and $\overline{L^2}$ is the mean square distance between the centers of gravity of parts A and B. These three quantities can be determined if measurements are carried out in three solvents with different refractive indices.

8.2.4 EFFECT OF SHAPE ON THE SCATTERING CURVE

The radius of gyration can be determined precisely without the assumption of any supplementary hypothesis. Indeed, the scattering curve for particles having the same radius of gyration but different shapes will coincide at very small angles. This will follow the Guinier law (8.2.2) which may also be expressed as

$$P(\theta)_{\theta\to 0} = \exp(-h^2 R_g{}^2/3) \qquad (8.2.29)$$

Beyond some point the curves for different shapes will diverge, and this offers the possibility of determining the particle shape by comparison of an experimental curve with families of theoretical curves. For example, in Fig. 8.13, the scattering intensity at small angles is plotted against hR_g for prolate spheroids. The lower curve for which the eccentricity is unity corresponds to a sphere and the upper curves to increasingly elongated ellipsoids of revolution. The abscissa has been chosen so that each particle has the same radius of gyration. At scattering angles immediately beyond the Guinier range, the curve drops off less steeply as the particle gets stretched out from spherical shape.

Figure 8.14 concentrates on the behavior of the scattering curve where the first maxima develop. As the sphere is stretched out to a prolate spheroid with an axial ratio of 2, the maxima shift toward larger angles and become considerably flattened. Mittelbach (1964) has discussed these trends in great detail for ellipsoids, prisms, cylinders, and plates having a wide variety of forms.

In practice it is very difficult to deduce information about the particle shape from the scattering curve because large changes in the former have only subtle effects on the latter. Conversely, particles of very different shapes can have very nearly the same scattering curves.

These difficulties are compounded when, as is usually the case in practice, there is both a distribution of particle shapes and particle sizes. These two effects influence the scattering in a qualitatively similar way so that it is difficult to disentangle them. Thus, the scattering by a dispersion of ellipsoids of the same size and shape is equivalent to the scattering of a polydispersion of spheres. Mittelbach (1965) has carried through a numerical exercise in which Roess' (1946) equations were used to obtain the integral transform inversion for a polydispersion of spheres (see Section 8.3.3). Alternatively, these data could have been interpreted in terms of a monodispersion of ellipsoids. Indeed, it is even possible to have equivalent scattering by a polydispersion of ellipsoids and a polydispersion of spheres.

Schmidt (1958) has made calculations of the predicted small angle scattering from polydisperse systems of ellipsoidal particles. These have been extended by Mittelbach (1965) who has explored the separate influences of anisotropy and polydispersity on the scattering curves.

8.2.5 Turbidimetric Titration of Polymers

An interesting application of the Rayleigh–Debye theory is to the determination of the concentration of a polymer component in solution by turbidimetric titration (Beattie, 1965; Beattie and Jung, 1968). The addition of a precipitant to a polymer solution may form a new phase consisting of

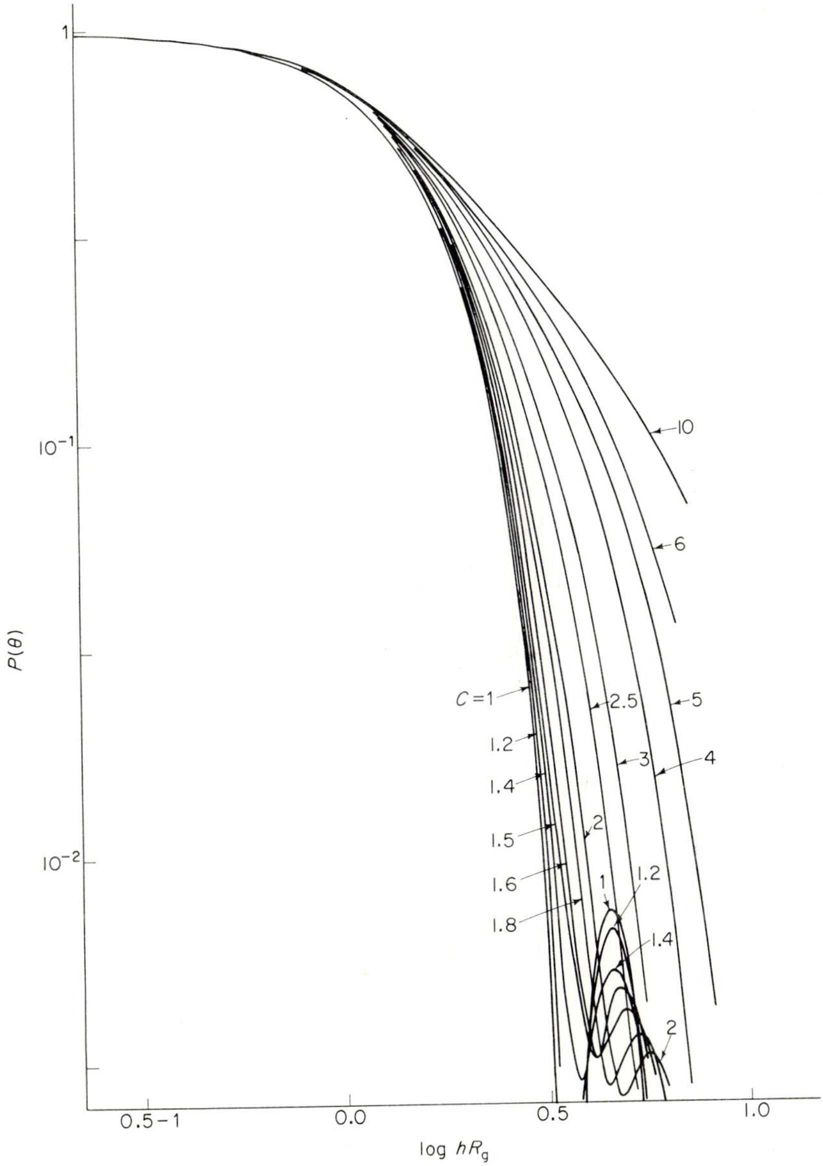

FIG. 8.13. Variation of $P(\theta)$ with hR_g at low angles for various prolate spheroids having the indicated eccentricities (Mittelbach, 1964).

small droplets composed of the solvent, the precipitant, and most of the polymer. The system becomes turbid and the problem is to calculate the concentration of polymer from either the light transmission or the angular light scattering. The turbidity may continue to change slowly with time after the initial precipitation due to the coalescence of the particles and the consequent increase in the average particle size. With continued particle growth, the turbidity increases to a maximum value, decreases to a minimum, and continues to oscillate between maxima and minima.

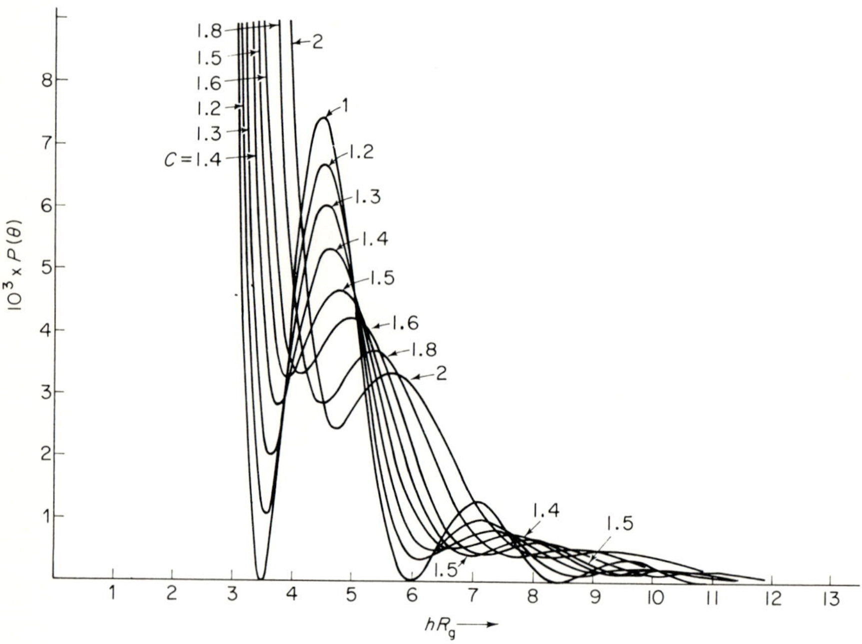

FIG. 8.14. Same as Fig. 8.13 for larger values of hR_g.

A precipitant is chosen which has the same refractive index as the solvent. In this way the refractive index of the precipitated droplet depends only upon the concentration of the polymer in the droplet. Furthermore, the refractive index of this droplet, m, may be related to the volume fraction of polymer in that phase by a mixing rule such as the Gladstone–Dale equation

$$(m - 1)\mu_0 = (\mu - \mu_0)\phi \tag{8.2.30}$$

where μ is the refractive index of the dry polymer, μ_0 is the refractive index of the solvent and precipitant, and ϕ is the volume fraction of polymer in the precipitated droplets.

If the spherical particles scatter light according to the Rayleigh–Debye theory, the concentration of polymer (g/ml) in the precipitated phase is

$$c = \frac{3V_{\theta,v}\lambda\rho \sin^3(\theta/2)}{(\mu - \mu_0)(m - 1)U(u)} \tag{8.2.31}$$

where $V_{\theta,v}$ is the vertical component of the Rayleigh ratio at scattering angle θ, λ is the wavelength, ρ is the density of pure polymer, and

$$U(u) = u^3P(\theta)/4 \tag{8.2.32}$$

where $P(\theta)$ is the form factor for a spherical particle with $u = ha$.

All the quantities on the right side of (8.2.31) except $U(u)$ are measurable. This will vary with u in an oscillatory fashion and the product of the measurable quantities in the numerator of (8.2.31) will vary with u in the same way. At the first maximum in the curve of $U(u)$ vs. u, this quantity takes on the value 0.974, so that

$$c = [2.92\rho/(\mu - \mu_0)(m - 1)][V_{\theta,v}\lambda \sin^3(\theta/2)]_{\max} \tag{8.2.33}$$

where the Rayleigh ratio is evaluated at the time at which coalescence has proceeded to give the first maximum. If the particle size is not changing, it may be possible to maximize the quantity in the square bracket by varying the angle and wavelength. A limited degree of polydispersity will not appreciably affect the results.

The initial concentration in the polymer solution can now be calculated since the particle size is known from the value of u at the maximum and the concentration of droplets can be obtained from the numerical value of $V_{\theta,v}$.

8.3 Polydisperse Systems

For a polydisperse polymer solution at infinite dilution

$$\begin{aligned} R_\theta &= K \int_0^\infty c_n M_n P_n(\theta) p(n)\, dn \\ &= K(N/N_A) \int_0^\infty M_n{}^2 P_n(\theta) p(n)\, dn \end{aligned} \tag{8.3.1}$$

where n is the degree of polymerization or number of monomer segments in the molecule, $p(n)\,dn$ is the fraction of polymer molecules with degree of polymerization n, and c_n, M_n, and $P_n(\theta)$ are the particle concentration (g/ml), molecular weight, and form factor associated with each molecular size. Also, N is the number of molecules per milliliter and N_A is Avogadro's number.

Zimm (1948b) has explored the consequences of polydispersity upon the Rayleigh ratio of a dispersion of randomly coiled macromolecules having a distribution of molecular weights equal to that expected in certain types of polymerization reactions for which (8.3.1) can be integrated, while Goldstein (1953), Holtzer (1955), and Casassa (1955) have done the same for a dispersion of rod-like molecules. Debye *et al.* (1963a) have carried out the integration for a gamma distribution such as that described later by (8.3.36). The effect of the polydispersity is to introduce curvature into the Zimm plot obtained when the reciprocal of the Rayleigh ratio is plotted against $\sin^2(\theta/2)$ at infinite dilution in accordance with (8.2.7). In principle the shape of this plot should furnish information about the dispersion of molecular weights. However, it appears quite hopeless to obtain data of sufficient precision to provide quantitative results.

In the combined limit of infinite dilution and of small angle, the form factor goes to unity and the above reduces to

$$\begin{aligned} R_\theta &= K(N/N_\mathrm{A})\int_0^\infty M_n{}^2 p(n)\,dn \\ &= Kc\int_0^\infty M_n{}^2 p(n)\,dn \Big/ \int_0^\infty M_n p(n)\,dn \\ &= KcM_\mathrm{w} \end{aligned} \tag{8.3.2}$$

where M_w is the weight average molecular weight. This is precisely what is expected for an array of Rayleigh scatterers.

On the other hand, the Zimm plot leads to the so-called z-average value of the radius of gyration. This can be seen when each value of $P_n(\theta)$ in (8.3.1) is expanded according to (8.2.2) (Ehrlich and Doty, 1954), leading to

$$\begin{aligned} P(\theta) &= 1 - \left\{(h^2/3)\int_0^\infty R_\mathrm{g}{}^2 M_n{}^2 p(n)\,dn \Big/ \int_0^\infty M_n{}^2 p(n)\,dn\right\} \\ &= 1 - (h^2/3)\langle R_\mathrm{g}{}^2\rangle_z \end{aligned} \tag{8.3.3}$$

If the functional relation between the radius of gyration and the molecular weight is known, then it may be possible to obtain an additional moment of the molecular weight distribution. For the three most commonly encountered configurations, the radius of gyration is proportional to the one-half power, one-third power, and first power of the molecular weight, respectively [(8.2.15), (8.2.16), (8.2.17)]. This leads to the following relations between the z-average value of the radius of gyration and the molecular weight.

For random coils:

$$\langle R_g^{\,2} \rangle_z = c' \int_0^\infty M_n^{\,3} p(n)\, dn \Big/ \int_0^\infty M_n^{\,2} p(n)\, dn = c' M_z \tag{8.3.4}$$

For spheres:

$$\langle R_g^{\,2} \rangle_z = c'' \int_0^\infty M_n^{8/3} p(n)\, dn \Big/ \int_0^\infty M_n^{\,2} p(n)\, dn = c'' M_s^{2/3} \tag{8.3.5}$$

For rods:

$$\langle R_z^{\,2} \rangle_z = c''' \int_0^\infty M_n^{\,4} p(n)\, dn \Big/ \int_0^\infty M_n^{\,2} p(n)\, dn = c''' M_R^{\,2} \tag{8.3.6}$$

If all the molecules are known to be either random coils, spheres, or rods, the above expressions may be used to obtain the additional moment of the molecular weight distribution corresponding to M_z, M_s, or M_R as defined in the above equations.

Burnett *et al.* (1958) followed this procedure to obtain the two moments for the size distribution of a polystyrene latex. They expressed this size as particle diameter rather than molecular weight and they utilized a discrete rather than a continuous size distribution for which the Rayleigh ratio can be written

$$R_\theta = \frac{c\pi^3 \rho n^2}{3\lambda_0^{\,4}} \left(\frac{\partial n}{\partial c}\right)^2 D_w^{\,3} \left(1 - \frac{4\pi^2}{5\lambda^2} D_s^{\,2} \sin^2(\theta/2)\right) \tag{8.3.7}$$

where the weight average value of the cube of the diameter obtained from the extrapolation to zero angle is

$$D_w^{\,3} = \sum n_i D_i^{\,6} \Big/ \sum n_i D_i^{\,3} \tag{8.3.8}$$

and the other moment obtained from the limiting slope of the Rayleigh ratio vs. $\sin^2(\theta/2)$ is

$$D_s^{\,2} = \sum n_i D_i^{\,8} \Big/ \sum n_i D_i^{\,6} \tag{8.3.9}$$

and λ_0 and λ are the wavelengths *in vacuo* and in the medium, respectively. These workers obtained good agreement between these two averages obtained in the above manner with those computed from the distribution obtained by electron microscopy.

Wesslau (1963) has also compared the electron microscope size of several polystyrene latexes with values from the limiting slope of the Zimm plot. The light scattering experiments were carried out both in an aqueous medium in which the relative refractive index of the polystyrene is 1.20 and in a 60% aqueous solution of sucrose in which this was only 1.10. Whereas in water there were considerable deviations from the electron microscope results for radii greater than 0.20 μ, the Rayleigh–Debye treatment of the light scattering data gave excellent results for radii up to at least 0.40 μ when the data were obtained in the sucrose solution. Such improved agreement at lower relative refractive index is precisely what would be expected in view of the stated limitation of the Rayleigh–Debye theory to values of the relative refractive index close to unity.

8.3.1 Aggregates of Spheres

Wippler *et al.* (1959) have developed equations for scattering by multiplets of spheres using the Rayleigh–Debye theory. Their initial purpose was to determine whether the deviation between the particle size obtained by light scattering and by electron microscopy in poly(vinyl chloride) latexes might not be due to the existence of such multiplets. Having ascertained that this was the case, they were then able to use the relations that had been developed in order to follow the kinetics of coagulation of this system (Benoit *et al.*, 1962).

If the individual particles, whether they be singlet spheres or aggregates of spheres, scatter independently, then the Rayleigh ratio for the system will be given by

$$R_\theta = R_{s\theta}\left[1 + \sum_{i=2}^{\infty} (v_i/v)A_i\right] \tag{8.3.10}$$

where $R_{s\theta}$ is the Rayleigh ratio that would be obtained were all of the v spheres in the form of singlets. Each of the terms in the summation corrects for the effect of having v_i particles of the ith configuration and A_i is a configurational factor. Values of this latter quantity are tabulated for some simple configurations in Table 8.1. Oster and Riley (1952a, b) have also given the form factors for a number of aggregates of spherical particles.

8.3.2 Asymptotic Behavior at Large Angles

Benoit (1953a, b, 1963) has shown that for a randomly coiled polymer there is a useful asymptotic result at large values of the parameter w [see also Benoit *et al.* (1954)]. This depends upon the disappearance of the exponential term in (8.1.37) leading to

$$P(\theta) = (2/w) - (2/w^2) \tag{8.3.11}$$

TABLE 8.1
CONFIGURATIONAL FACTORS FOR AGGREGATES OF SPHERES
[A_i IN EQ. (8.3.10)]

Configuration	Configurational factor
Doublet	$\frac{\sin 2u}{u}$
Triplet	
Triangle	$\frac{3 \sin 2u}{u}$
Linear	$\frac{2 \sin 2u}{u} + \frac{\sin 4u}{2u}$
Quadruplet	
Tetrahedron	$\frac{6 \sin 2u}{u}$
Square	$\frac{4 \sin 2u}{u} + \frac{2 \sin 2\sqrt{(2)}u}{\sqrt{(2)}u}$
Opposite spheres in contact	$\frac{5 \sin 2u}{u} + \frac{\sin 2\sqrt{(3)}u}{\sqrt{(3)}u}$
Linear	$\frac{3 \sin 2u}{u} + \frac{\sin 4u}{u} + \frac{\sin 6u}{3u}$

When this expression is used, the intercept in the Zimm plot leads to twice the number average molecular weight $2M_n$, and the slope leads to the number average value of the square of the radius of gyration.

$$\langle R_g{}^2 \rangle_n = \frac{3\lambda^2 \text{ (slope of asymptote)}}{8\pi^2 \text{ (intercept of asymptote)}} \tag{8.3.12}$$

All the information that can be analyzed from a light-scattering diagram of a system of linear but polydisperse Gaussian chains is illustrated in Fig. 8.15. Thus, by analyzing both the small and large angle asymptotes of the scattering curve, several different moments of the size distribution can be observed. Loucheux *et al.* (1958) have extended the treatment to polymer chains which do not follow Gaussian statistics. There is also a large-scale asymptote for rods that leads to the number average values of the molecular weight and the radius of gyration. However, Reichmann (1959) has pointed out that the evaluation of M_n of rods by this technique may be extremely inaccurate in practice.

Benoit points out that, since light-scattering measurements are usually carried out in the range from $\theta = 30$ to $135°$ and for wavelengths (in the medium) of from 0.300 to 0.400 μ, when $\langle R_g{}^2\rangle_z^{1/2} < 800$ Å, one is limited to the small-angle case. On the other hand, for $\langle R_g{}^2\rangle_z^{1/2} > 3000$ Å, only the large-angle asymptote is accessible. At intermediate sizes, the curvature in the curve of Kc/R_θ vs. $\sin^2 \theta/2$ will be such as to make it impossible to achieve any of the information depicted in Fig. 8.15. Prud'homme and Sicotte (1968) have emphasized that within the experimental errors the Zimm plot may appear to be linear even for values of the radius of gyration which are larger than the range of validity of the small-angle case. Furthermore, if measurements are carried out only to a scattering angle of 30°, with $\lambda_0 = 0.546\ \mu$ for polystyrene in toluene solutions, the error in $\langle R_g{}^2\rangle_z^{1/2}$ is about 5% for a 300 Å radius of gyration, 10% for 400 Å, and 15% for 500 Å. Mijnlieff and Coumou (1968) have noted that at large angles, Eq. (8.2.2) is no longer valid and that the angular variation of $P(\theta)$ may depend strongly upon the specific shape of the particles.

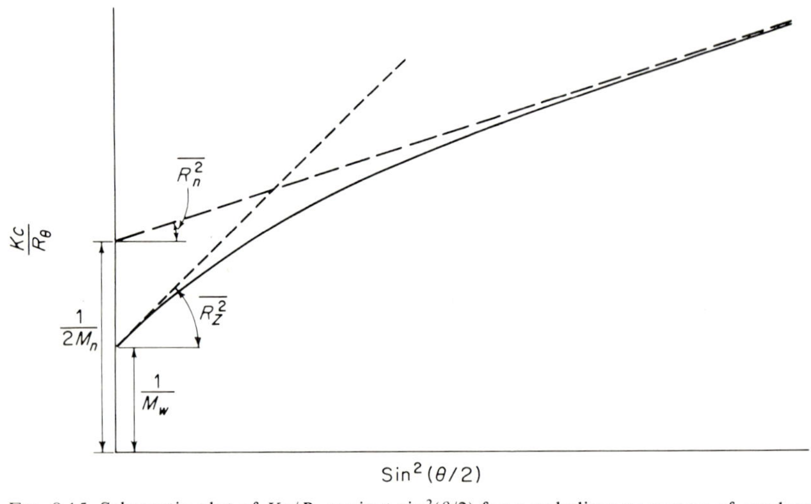

FIG. 8.15. Schematic plot of Kc/R_θ against $\sin^2(\theta/2)$ for a polydisperse system of randomly coiled macromolecules. The initial intercept and slope give the weight average molecular weight (M_w) and the z-average radius of gyration $\langle R_g{}^2\rangle_z$. The asymptotic intercept and slope give the number average molecular weight (M_n) and the number average radius of gyration $\langle R_g{}^2\rangle_n$ (Benoit *et al.*, 1954).

Carpenter (1966) has determined the molecular weight distribution by direct comparison of the experimental angular scattering with theoretical calculations rather than by obtaining two moments of the distribution from the asymptotes at high and low scattering angles. A particular form for the molecular weight distribution such as a log-normal distribution is assumed and the form factor obtained experimentally over the accessible angular range is compared with that calculated for various values of the distribution parameters consistent with the value of M_w obtained from the small angle of extrapolation. The accessible range of $P(\theta)$ was extended by carrying the measurements out at three wavelengths ($\lambda = 546, 436, 365$ mμ) as well as over the angular range 30 to 150°.

8.3.3 ALGEBRAIC INVERSION OF SCATTERING DATA

The "inverse" scattering problem is this: Given certain experimental scattering data, determine the particle size distribution or the particle structure.

In determining the particle structure, it is customary to treat either a single particle or a monodisperse array of identical particles. The structure may be described by the variation of the refractive index in the particle whose shape and volume are presumably known. Accordingly, the problem can be formulated in terms of inversion of the following equation:

$$S_1(\theta) = (ik^3/4\pi) \int (m^2 - 1)\, e^{i\delta}\, dV \tag{8.3.13}$$

The structure factor $(m^2 - 1)$ is to be expressed in terms of the amplitude function $S_1(\theta)$.

Hart and Gray (1964) have developed a solution based upon expansion of the structure function in associated Legendre polynomials. However, they have not considered in detail the effects of experimental and computational errors on the accuracy of the final result so that the practical limitations of their technique remain to be assessed.

In determining the particle size distribution, the shape and optical constants are known or values for them are assumed, and it is also assumed that there are no multiple scattering or interparticle interference effects. We have seen earlier how it is possible, when dealing with a collection of homogeneous spheres, to assume a plausible size distribution and then to modify the values of the parameters of the distribution in order to bring the calculated scattering pattern into agreement with that which is found experimentally. The use of a high-speed computer renders this method practical, but it must be used with caution because it may not give a unique result.

On the other hand, the determination of the particle size distribution may be obtained without *a priori* assumptions about the distribution by solution of the following integral equation

$$I(y) = \int_0^\infty I(\alpha, y)\, p(\alpha)\, d\alpha \tag{8.3.14}$$

$I(y)$ is an intensity determined experimentally, $I(\alpha, y)$ is this quantity as known from theory for a single particle of size α, and $p(\alpha)$ is the size distribution frequency function. The inversion of this equation may be effected for certain simple cases such as Rayleigh–Debye scattering, anomalous diffraction, or Fraunhofer diffraction. Even when a formal solution is obtained, this may often be "unstable," i.e., small errors in the measurements or in the computations may lead to large errors in the size distribution or even to nonsensical results such as negative values. One of the principal goals of research in this area is to determine the cause of such instabilities and to attempt to remove them.

a. Small Angle X-Ray Scattering; Rayleigh–Debye Theory. Much of the practical work to obtain the size distribution of polydisperse systems from an integral transform of the scattered intensity has been carried out in the realm of small angle X-ray scattering. This is described precisely by Rayleigh–Debye theory. Roess (1946) and Riseman (1952) developed detailed equations for a system composed of spherical particles and Luzzati (1957) extended these to particles with the same form but different sizes. It will be convenient to introduce, first of all, the notation used by workers in the X-ray field and to relate it to the optical notation.

Small angle X-ray scattering may be attributed to the action of free electrons whose polarizability can be expressed as

$$\alpha' = -e^2/m_e k^2 c^2 \tag{8.3.15}$$

where e is the electronic charge, m_e is the mass of the electron, k is the propagation constant, and c is the velocity of light. For X-rays, these latter two quantities do not deviate sensibly in the presence of matter from their *in vacuo* values. This leads to the Thomson equation for scattering by a single electron:

$$\frac{I_e}{I_o} = \frac{e^4}{2r^2 m_e^2 c^4}(1 + \cos^2 \theta) = \frac{7.90 \times 10^{-26}}{2r^2}(1 + \cos^2 \theta) \tag{8.3.16}$$

The effective refractive index of an electron gas with electron density ρ_e is

$$m^2 = 1 - (4\pi e^2 \rho_e / m_e k^2 c^2) \tag{8.3.17}$$

The small angle X-ray scattering by a small particle is practically independent of the "short-range order" of the atoms and can be considered to arise from an electron gas having the exterior form and dimensions of the particle and an electron density determined by the distribution of matter in the particle. For a particle of uniform electron density

$$I = I_e N_e^2 P_a(\theta) = I_e V^2 \rho_e^2 P_a(\theta) = I_e F_a^2(h) \tag{8.3.18}$$

where N_e is the number of free electrons in the particle, V is the volume, $P_a(\theta)$ is the form factor of a particle of characteristic length a, and $F_a^2(h)$ is the intensity factor.

We will now describe a procedure followed by Schmidt for determining the particle size distribution from the low angle X-ray scattering intensity (Letcher and Schmidt, 1966; Brill, 1967; Schmidt *et al.*, 1968; Schmidt and Brill, 1967). It utilizes an integral transform of the experimental data and does not presuppose any particular form for the distribution.

The point of departure is a characteristic or correlation function $\gamma_0(s)$ introduced by Porod (1951, 1952) which represents the probability that a point at a distance s in an arbitrary direction from an origin within the particle will itself be within the particle. If the particle possesses a center of symmetry, the origin is taken as this center. This function is related to the intensity factor by

$$F_D^2(h) = \rho_e^2 V \int_0^\infty \gamma_0(s/D)[(\sin hs)/hs]4\pi s^2\, ds \tag{8.3.19}$$

It contains all the information about particle size and shape which can be obtained from scattering experiments. The particles are assumed to be of the same geometrical form and to be randomly positioned so that they scatter independently. An average intensity factor may be defined by

$$\overline{F^2(h)} = \int_0^\infty p(D)\overline{F_D^2(h)}\, dD \tag{8.3.20}$$

where $p(D)$ is the normalized size distribution function and $\overline{F_D^2(h)}$ is the intensity factor for a particle of characteristic length D as averaged over random orientations. Now, a new characteristic function may be defined (Letcher and Schmidt, 1966) such that

$$\overline{F^2(h)} = \rho_e^2 \overline{V} \int_0^\infty \gamma(s)(\sin hs/hs)4\pi s^2\, ds \tag{8.3.21}$$

where the volume of a particle with characteristic length D is $V_0 D^3$ and

$$\gamma(s) = \frac{V_0}{\overline{V}} \int_0^\infty \gamma_0(s/D)D^3 p(D)\, dD \tag{8.3.22}$$

The average particle volume $\overline{V}$ is

$$\overline{V} = V_0 \int_0^\infty p(D)D^3\, dD \tag{8.3.23}$$

The characteristic function can be obtained by Fourier inversion of the intensity factor

$$\gamma(s) = (1/2\pi^2\rho_e{}^2\overline{V}) \int_0^\infty (\sin hs/hs)h^2\overline{F^2(h)}\, dh \tag{8.3.24}$$

and the size distribution can in turn be obtained from

$$p(D) = -\frac{2\overline{V}}{\pi}\frac{d}{ds}\left[\frac{\gamma''(s)}{s}\right]_{s=D} \tag{8.3.25}$$

where the double prime designates the second derivative with respect to the argument s. If the particles are spheres so that D is the diameter

$$\gamma_0(s/D) = 1 - (3s/2D) + (s^3/2D^3) \tag{8.3.26}$$

then

$$p(D) = (1/\pi^3\rho_e{}^2D^2) \int_0^\infty [h^4\overline{F^2(h)} - c_4]\alpha(hD)\, dh \tag{8.3.27}$$

where

$$\alpha(hD) = \left[\cos hD\left(1 - \frac{8}{h^2D^2}\right) - \frac{4\sin hD}{hD}\left(1 - \frac{2}{h^2D^2}\right)\right] \tag{8.3.28}$$

and

$$c_4 = \lim_{h\to\infty} h^4\overline{F^2(h)} \tag{8.3.29}$$

For large values of h, the following asymptotic expression may be valid:

$$h^4\overline{F^2(h)} \simeq c_4 + (c_6/h^2) + (c_8/h^4) \tag{8.3.30}$$

The values of the constants must be determined from the experimental scattering curve and a final expression for the evaluation of the size distribution takes the form

$$p(D) = (1/\pi^3\rho_e{}^2D^2)\left[\int_0^{h_m} [h^4\overline{F^2(h)} - c_4]\alpha(hD)\, dh + c_6\int_{h_m}^\infty (\alpha(hD)/h^2)\, dh + c_8\int_{h_m}^\infty (\alpha(hD)/h^4)\, dh\right] \tag{8.3.31}$$

where h_m corresponds to the angle beyond which the asymptotic expression (8.3.30) is valid. Best results are not necessarily obtained by choosing h_m as the largest scattering angle at which data can be recorded, since in this angular range the intensity may be so low that the data are not as reliable as data for somewhat smaller angles.

A number of further observations are in order. The outer part of the scattering curve must be proportional to h^{-4} in order that the above integrals converge. The results for small sizes are strongly dependent upon the asymptotic form of $\overline{F^2(h)}$ so that there are stringent requirements on the accuracy of the constants c_4, c_6, and c_8. The X-ray data must be available at sufficiently large angles to permit an accurate estimation of these constants. If the data do not fit well in the asymptotic region where these constants are presumed to apply, small errors in the intensity data at these angles may result in incorrect values for the constants used to describe the scattering curve in the extrapolated region $h > h_m$. These errors are relatively more important for the smaller radii in the distribution.

Preparatory to applying the theory to experimental scattering curves, Schmidt and his associates tested it by calculating the size distributions from theoretical curves corresponding to assumed distributions. Random errors were introduced into these hypothetical data in order to determine the sensitivity of the calculation to experimental error. The main practical problem turned out to be the accurate calculation of the constants c_4, c_6, and c_8. Experiments were carried out with gold and silica sols containing spherical particles with average diameters of the order of 100 to 200 Å. Figure 8.16 illustrates the excellent agreement obtained between a histogram

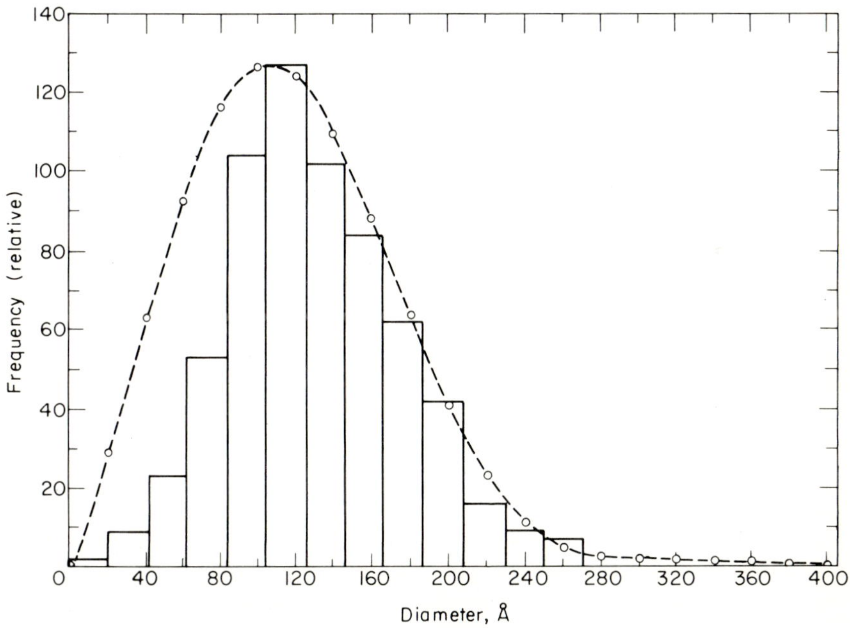

FIG. 8.16. Diameter distribution of a silica hydrosol obtained by small angle X-ray scattering compared with electron microscope histogram (Brill *et al.*, 1968).

for colloidal silica as determined with the electron microscope and the size distribution curve obtained by inversion of the forward angle X-ray scattering (Brill *et al.*, 1968).

b. Transmission by Spheres Exhibiting Anomalous Diffraction. Shifrin and Perelman (1965a) have calculated the spectral transmission of polydispersions of spheres which scatter according to the law of anomalous diffraction. This is applicable to particles which are both "soft" ($m \to 1$) and are large relative to the wavelength. Throughout this section, we will retain the notation of these authors in order to facilitate cross reference to their detailed work, but will present in square brackets the comparable notation used elsewhere in this book.

The average scattering cross section $[C_{\mathrm{sca}}]$ is defined by

$$g\left(\frac{x}{2}\right) = \int_0^\infty K\left(\frac{x}{2}a\right) m(a)\, da \tag{8.3.32}$$

where $K(x/2)a)$ is one-half of the scattering efficiency $[Q_{\mathrm{sca}}]$ and is given by

$$\begin{aligned} K(xa/2) &= \tfrac{1}{2}Q_{\mathrm{sca}} \\ &= 1 - \frac{2\sin(xa)}{xa} + \frac{2[1 - \cos(xa)]}{(xa)^2} \end{aligned} \tag{8.3.33}$$

The size distribution is represented as

$$m(a) = 2a^2 f(a)\, da \tag{8.3.34}$$

The quantity a is the ratio of the particle radius r to an arbitrarily selected mode r_0 of the size distribution. The parameter x is the phase shift of the central ray through the particle $[\rho]$ defined by

$$x = 4\pi(m - 1)r_0/\lambda \tag{8.3.35}$$

The particles are described by a gamma size distribution

$$f(a) = N{r_0}^3[\Delta^{\mu+1}/\Gamma(\mu + 1)]a^\mu e^{-\Delta a} \tag{8.3.36}$$

where N is the number of particles per unit volume, μ is a positive number which measures the breadth of the distribution, and

$$\Delta = 2\pi(m - 1)r_0 \tag{8.3.37}$$

A monodisperse system corresponds to $\mu = \infty$. Actually, the parameter Δ is a function of μ, the form of which depends upon the choice of r_0. If r_0 is chosen as the modal radius, then $\Delta = \mu$; if r_0 is the mean radius, then $\Delta = \mu + 1$; if r_0 is the mean square radius, then $\Delta = [(\mu + 2)(\mu + 1)]^{1/2}$, etc. Shifrin and Perelman have developed approximate formulas to facilitate numerical computations and have presented the results graphically for various distributions.

These authors then made a comprehensive study of the problem of determining the size distribution by direct inversion of the integral equation (8.3.32). Because the kernel of this Fredholm equation of the first kind appears as a function of the product of the variables, the equation can be dealt with rigorously by using Mellin's transform. The analysis is described in detail in a series of papers (Shifrin and Perelman, 1963, 1964a, b, 1965a, b, 1966a, b) where references to earlier work may also be found. Here, we will only present some of the final results. The size distribution may be expressed as a contour integral

$$m(a) = (1/2\pi i)\int_{c+i\infty}^{c-i\infty} [G(1-p)/L(1-p)]a^{-p}\,dp \tag{8.3.38}$$

where $G(p)$ and $L(p)$ are Mellin's transformations of functions $g(x/2)$ and $K(xa/2)$. We first suppose that beyond a particular value of the parameter x, the experimental data can be adequately represented by an expression of the form

$$g(x/2) = c_0 + (c_2/x^2); \qquad x \geqslant \tau \tag{8.3.39}$$

It is obvious that c_0 can be determined from the limiting value at $x \to \infty$ (or $\lambda \to 0$) and then c_2 is determined from

$$c_0 k + c_2 \sum_{j=1}^{k} (1/x_j^2) = \sum_{j=1}^{k} g(x_j/2) \tag{8.3.40}$$

Here, k is the number of points ($x_j > \tau$) at which the various spectral values of the cross section $g(x_j/2)$ are measured. Then, if the scattering law is given by (8.3.33),

$$m(a) = -(1/\pi)\sum_{j=1}^{m} g(x_j/2)w(ax_j)\Delta x_j + c_0\tau w_0(a) + c_2(w_2(a\tau)/\tau) \tag{8.3.41}$$

where

$$w(y) = y\sin y + \cos y - 1 \tag{8.3.42}$$

$$w_0(y) = \cos y - 2[(\sin y)/y] + 1 \tag{8.3.43}$$

$$w_2(y) = \cos y - 1 \tag{8.3.44}$$

The method has been investigated numerically by calculating the scattering cross section corresponding to a gamma distribution for which r_0 is taken as the modal value and upon which an error has been superimposed. The stability of the calculation to the errors and to the ranges of the various physical parameters has been discussed in detail (Shifrin and Perelman, 1965a, b). It was possible to obtain excellent coincidence (Shifrin and Perelman, 1964a, b) between the calculated values of $m(a)$ and the values corresponding to the initial distribution.

In practice it is important that the transmission measurements be accurate to about 1% and that the intervals Δx_j be sufficiently small. Furthermore, accuracy depends upon a successful choice of τ, and this in turn depends on utilizing a sufficiently small wavelength in the extrapolation ($\lambda \to 0$) of (8.3.39) to obtain c_0.

The range of wavelength within which the transmission data must be kept is $2\pi(m-1)r_0$ to $5\pi(m-1)r_0$ where r_0 is the modal radius. Thus, for a water aerosol with $r_0 = 0.1\ \mu$, the measurements should range from $\lambda = 0.21$ to $0.52\ \mu$, whereas for fog droplets with $r_0 = 1\ \mu$, the range is $\lambda = 2.1$ to $5.2\ \mu$. For these estimations, $m = 1.33$.

The authors claim that experimental checks with microcrystals of AgBr in gelatin ($r_0 = 0.3$) and spores ($r_0 = 2.5$) gave satisfactory agreement even though $m = 1.5$ and the particles were not perfectly spherical (Shifrin and Perelman, 1967).

In a separate numerical study, Bakhtiyarov *et al.* (1966) explored the effect of utilizing the anomalous diffraction expression (8.3.33) for the extinction. They calculated the average scattering cross section (8.3.32) using both this expression and the exact values for $m = 1.5$. The results were quite similar and when each of these sets of data were inverted, the initial frequency distribution of the particle sizes was obtained.

c. Small-Angle Light Scattering; Rayleigh–Debye Theory. Shifrin and his collaborators have developed a technique for direct inversion of the Rayleigh–Debye scattering along lines analogous to the transmission method which is outlined in the previous section (Shifrin and Perelman, 1966a, b). The stability of the calculations to experimental errors in the scattering data has been discussed by Shifrin and Perelman (1965c) and to the range of particle sizes by Shifrin and Chayanova (1966). In principle, this is similar to the small angle X-ray technique used by Schmidt *et al.* As in the last section, we will retain the author's notation.

The relation between the Rayleigh ratio for unpolarized incident light of a single particle $R_{\theta,r}$, of the particle size distribution $f(r)\,dr$, and of the polydisperse Rayleigh ratio $\bar{R}_\theta$ is

$$\bar{R}_\theta = \int_0^\infty R_{\theta,r} f(r)\,dr \tag{8.3.45}$$

where for a Rayleigh–Debye scatterer

$$R_{\theta,r} = (8\pi^4 r^6/\lambda^4)[(m^2-1)/(m^2+1)]^2(1+\cos^2\theta)P(\theta) \tag{8.3.46}$$

Here, r denotes the radius of a sphere having volume equivalent to the particle and the quantity in the square bracket has not been reduced to $\frac{2}{3}(m-1)$ as is usually done when the refractive index is close to unity. The final solution can be represented by

$$f(r) = (4/\pi r_0{}^4 a^2) \sum_{j=1}^{n} g(x_j/2)\kappa(ax_j)\,\Delta x_j + C_0 \tau \kappa_0(a\tau) + C_2(\kappa_2(a\tau)/\tau) \tag{8.3.47}$$

where r_0, r, and a are defined the same as in the previous section. The experimental data are cast into the form

$$g(x/2) = (r_0/\psi(\theta))\bar{R}_\theta \tag{8.3.48}$$

in which

$$\psi(\theta) = \frac{9}{8}\left[\frac{m^2-1}{m^2+2}\right]^2 \frac{(1+\cos^2\theta)}{(1-\cos\theta)^2} \tag{8.3.49}$$

and now

$$x = (8\pi r_0/\lambda)\sin(\theta/2), \qquad \tau = (8\pi r_0/\lambda)\sin(\theta_{max}/2) \tag{8.3.50}$$

Values of x_j correspond to scattering angles θ_j for which the scattering was measured, $\theta = \theta_{max}$ being the largest of such angles. The coefficients C_0 and C_2 are determined from the relationship

$$g(x/2) \simeq C_0 + (C_2/x^2), \qquad x > \tau \tag{8.3.51}$$

which approximates the "tail" of the scattering function. The other functions are defined by

$$\kappa(y) = [1 - (8/y^2)]\cos y + [(8/y^3) - (4/y)]\sin y + \tfrac{1}{3} \tag{8.3.52}$$

$$\kappa_0(y) = [\kappa(y) - w_0(y)]/2 \tag{8.3.53}$$

and

$$\kappa_2(y) = [\kappa(y) - w_2(y)]/4 \tag{8.3.54}$$

The previous section should also be consulted for the definitions of w_0 and w_2.

One condition for the success of this technique is that

$$\tau > (1.6 - 2.0)x_0 \tag{8.3.55}$$

where x_0 is the modal value of x. This means that beyond this value $g(x/2)$ must decrease monotonically, and this is equivalent to the requirement that all of the peculiarities of the scattering pattern must be concentrated in the interval $(0, \tau)$. Of course, close to $\theta = 0°$, the form factor goes to unity and the scattering approaches Rayleigh scattering. Therefore, the size parameter

α must be large enough so that the scattering pattern will differ sufficiently from a Rayleigh pattern within an intermediate range of values. On the other hand, the particles may not be so large that the Rayleigh–Debye approximation is not valid. Shifrin (1965) has found that reliable inversion can be obtained at $\lambda = 0.55\ \mu$ for a gamma distribution corresponding to mean radii listed in Table 8.2 if the data are obtained within the angular intervals θ_{min} and θ_{max} as specified in the table.

TABLE 8.2
LIMITING ANGLES FOR USE OF (8.3.47)

$\lambda = 0.55\ \mu$	Gamma distribution ($\mu = 2$)					
Mean radius (microns)	0.2	0.4	0.6	1	2	5
θ_{max} (degrees)	130	55	35	20	10	5
θ_{min} (degrees)	30	15	10	6	3	1

8.4 Inhomogeneous Media

In treating the scattering by a particle, we usually assume that the medium of which the particle is comprised is optically homogeneous. However, extended continuous media themselves scatter light, indicating that they contain inhomogeneities. Thus, the turbidity of gases is due to incoherent scattering by the molecules which form a random array of Rayleigh scatterers. For liquids, the scattering may be thought of as arising from density fluctuations resulting from the thermal motions of the molecules. Inhomogeneities, normally present in both crystalline and amorphous solids, will also scatter light.

The problem of characterizing the inhomogeneities in solids and of relating these to the scattered intensity has been treated by Debye and Bueche (1949) and by Porod (1951, 1952) using the Rayleigh–Debye theory. Their approach has been utilized both for small angle X-ray scattering and light scattering. Porod's papers and the book by Guinier and Fournet (1955) review the fundamentals quite comprehensively, using X-ray notation. The method was originally proposed by Zernike and Prins (1927) and developed by Debye and Menke (1930) for determining the spatial distribution of atoms in a liquid such as mercury.

The angular variation of the intensity provides information about the product of the scattering angle and a linear dimension of the particle divided by the wavelength, *ha*. The information about the linear dimension obtained by X-ray scattering is normally of a different order of magnitude from that obtained by light scattering because of the large difference in wavelength between X-rays and light. However, the power of the X-ray technique is enhanced by the ability, with modern devices, to obtain scattering data at angles to within fractions of minutes of arc from the forward direction. Therefore, by going to very small scattering angles, the accessible structural features obtained by X-ray scattering can be made to overlap with those obtained by light scattering.

8.4.1 The Correlation Function

In what follows, we will use the optical notation, treating first the case where the inhomogeneity of the medium is due to a continuous variation of the dielectric constant or refractive index rather than to the presence of discrete particles. The macroscopic properties of the solid appear to be uniform. However, the dielectric constant varies from point to point and may be represented as the sum of the average value $\bar{\varepsilon}$ and a local variation η_A so that

$$\varepsilon_A = \bar{\varepsilon} + \eta_A \tag{8.4.1}$$

Two aspects of this variation will be considered; namely, (1) the mean square value $\overline{\eta^2}$ and (2) the average extension of the inhomogeneities as measured by a so-called correlation distance $\overline{\eta_A \eta_B}$. In order to visualize the correlation distance, consider the product of the fluctuations η_A and η_B at two points A and B separated by the distance s. The average of this product for all points throughout the solid will depend upon the distance s. For $s = 0$, $\overline{\eta_A \eta_B}$ is obviously equal to the mean square value of the fluctuation $\overline{\eta^2}$. For large values of s, $\overline{\eta_A \eta_B} = 0$ since in the absence of any correlation η_A and η_B will vary randomly, and there will be equal numbers of negative and positive values. A correlation function $\gamma(s)$ may be defined by

$$\overline{\eta_A \eta_B} = \gamma(s)\overline{\eta^2} \tag{8.4.2}$$

The steepness with which this function drops off from 1 to 0 is a measure of the average extension of the inhomogeneities.

The fluctuations are assumed to be small compared to the average value of the dielectric constant, so that the Rayleigh–Debye approximation is valid. Following (8.1.1), the amplitude function for any volume element is given by

$$dS_1 = (ik^3/4\pi)(\varepsilon_A - 1)\, e^{i\delta}\, dV \tag{8.4.3}$$

or with (8.4.1)

$$dS_1 = (ik^3/4\pi)(\bar{\varepsilon} + \eta_A - 1)\, e^{i\delta}\, dV \tag{8.4.4}$$

The model can be visualized with the aid of Fig. 8.1 except that the particle boundary is now extended *ad infinitum*. The elemental amplitude function is now integrated over all space by vector addition and the intensity is obtained from the square of the modulus. These operations will result in three sets of terms [Guinier and Fournet, (1955) p. 76]. The first of these corresponds to a homogeneous medium for which there is zero scattering. In the second set, the exponential is weighted by the fluctuations η, and these also average to zero. The nonzero set involves the product $\eta_A \eta_B$ whose average falls off to 0 for the interval over which these fluctuations are correlated. This results in

$$I_1 = (k_0^4/16\pi^2 r^2)\overline{\eta^2} V \int_v \gamma(s)\, e^{i\delta}\, dV \tag{8.4.5}$$

When $\gamma(s)$ is spherically symmetric, depending only upon the distance s and not on the direction in space

$$I_1 = (k_0^4/16\pi^2 r^2)\overline{\eta^2} V \int_0^\infty 4\pi s^2 \gamma(s)(\sin(hs)/hs)\, ds \tag{8.4.6}$$

The integral in the above equation is called the correlation-volume

$$w = \int_0^\infty 4\pi s^2 \gamma(s)(\sin(hs)/hs) ds \tag{8.4.7}$$

so that

$$I_1 = (k_0^4/16\pi^2 r^2) V \overline{\eta^2} w \tag{8.4.8}$$

This shows that for a given illuminated volume it is only the product $\overline{\eta^2} w$ which determines the scattering. Thus, small fluctuations coupled with long range correlations are as effective in scattering as larger fluctuations in which the extension of the fluctuations is not as great. Furthermore, the correlation volume depends upon angle, decreasing somewhat as scattering angle increases from zero. If the average extension of the inhomogeneity is so small that $s \ll \lambda$ for any distances over which $\gamma(s)$ is appreciably different from zero or if the angle θ is small, then the correlation volume reduces to

$$w_0 = \int_0^\infty 4\pi s^2 \gamma(s)\, ds \tag{8.4.9}$$

and it becomes independent of angle. In this case the scattering follows the Rayleigh law in which the scattered intensity is symmetrical about 90° and is proportional to λ^{-4}.

If the correlation function is known, the scattering can be predicted. For example, for the simple exponential function

$$\gamma(s) = \exp(-s/a) \tag{8.4.10}$$

the dissymmetry factor is

$$z(\theta) = (1 + h_1{}^2a^2)/(1 + h_2{}^2a^2)^{-2} \tag{8.4.11}$$

where h_1 and h_2 correspond to the two angles for which the dissymmetry is measured and the turbidity is

$$\tau = 2k_0{}^4\overline{\eta^2}a^3\left[\frac{(b+2)^2}{b^2(b+1)} - \frac{2(b+2)}{b^3}\log(b+1)\right] \tag{8.4.12}$$

where

$$b = 4k^2a^2 \tag{8.4.13}$$

The parameter a, sometimes called the persistence length, characterizes the average distance between fluctuations in dielectric constant. For example, Gallacher and Bettelheim (1962) have utilized dissymmetry measurements to calculate a value of a in a cross-linked polymer. This quantity provided a means of following the cross-linking process.

8.4.2 The Inverse Problem

The inverse problem consists in determining the correlation function from the angular distribution of the scattered light. A new function may be defined by

$$\Phi = R_\theta h \bigg/ \left(\frac{\pi^2 V}{2\lambda_0{}^4}\frac{(1+\cos^2\theta)}{2}\right) \tag{8.4.14}$$

where R_θ is the Rayleigh ratio at angle θ for unpolarized incident light and $(1 + \cos^2\theta)/2$ is the usual polarization factor. Using (8.4.6)

$$\Phi = 8\pi\overline{\eta^2}\int_0^\infty s\gamma(s)\sin(hs)\,ds \tag{8.4.15}$$

which can be solved by a Fourier transformation. This leads to

$$4\pi^2\overline{\eta^2}s\gamma(s) = \int_0^\infty \Phi\sin(hs)\,dh \tag{8.4.16}$$

The left-hand side of this equation is determined for each value of s by evaluating the above integral as a function of angle either graphically or numerically from the low-angle scattering data. This leads to the distribution with s of the product $\overline{\eta^2}\gamma(s)$, which is, in turn, the correlation distance $\overline{\eta_A\eta_B}$. The value of $\gamma(s)$ can be obtained explicitly by remembering that it must go to unity as s goes to zero. In this way $\gamma(s)$ can be evaluated even if the absolute values of the Rayleigh ratio are not determined.

This procedure is only practical when the extensions of inhomogeneity are sufficiently large so that a pronounced dissymmetry occurs. The reader is referred to the papers of Debye and Bueche (1949), Beebe and Marchessault (1964), and Wilkes and Marchessault (1966) for a discussion of some of the experimental and computational problems. Of course, the procedure is also applicable to scattering by X-rays, and in this case inhomogeneities considerably smaller than those encountered with visible light can be analyzed.

In practice, the integration of (8.4.16) to obtain the correlation function is carried out with some upper limit of angle, determined by the experimental arrangement, rather than infinity. The largest physically realizable angle is 180°. For X-ray scattering, the rapid decrease of the intensity with increasing angle to values which are too small to measure accurately is the limiting factor. Caulfield and Ullman (1962) have explored the effect of this cutoff upon the determination of $\gamma(s)$ by carrying out calculations assuming various cut-off values. The resultant error in the correlation function can be considerably greater than the error in the intensity itself. For a system of uniform spherical inhomogeneities, the error depends upon h^*a where a is the radius and h^* is the angular cut-off value. Accordingly, a larger value of h^* must be used, the smaller the spherical inhomogeneity.

8.4.3 Spherulites

Solid polymers may scatter light because of the presence of crystalline regions in an amorphous matrix. In some cases the crystallites are arranged in spherical aggregates called spherulites whose optical anisotropy may be characterized by a radial and a tangential refractive index. This gives rise to interesting scattering effects which in turn provide useful information about the internal structure of the polymer (Keane *et al.* 1956; Stein and Rhodes, 1960). Similar effects may be observed with dispersions of small starch granules in oil (Borch and Marchessault, 1968) or with liquid crystals (Frenkel *et al.* 1967; Stein *et al.* 1968).

The prevalent technique utilizes low angle scattering from a polymer film contained between polarizing prisms such as the arrangement in Fig. 8.17. In the case of liquid crystals, the liquid must be contained in an appropriate thin, flat, thermostated cell. Scattering at both the usual polar angle and the azimuthal angle is recorded on a photographic film. Some typical patterns are shown in Fig. 8.18.

For unpolarized light, the scattering is cylindrically symmetrical about the incident beam direction. The V_v pattern in Fig. 8.18 is obtained when both the polarizer and the analyzer are oriented in the vertical direction, whereas the V_h pattern is for crossed polarizer and analyzer.

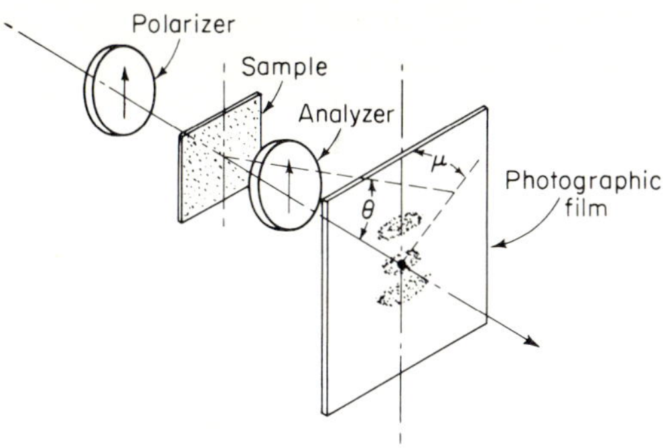

FIG. 8.17. The experimental arrangement for photographic light scattering from films (Stein and Rhodes, 1960).

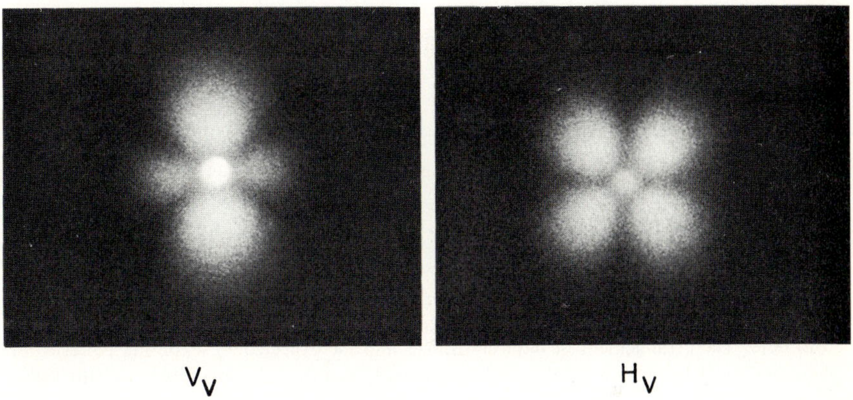

FIG. 8.18. Typical light scattering photographs for V_v polarization and H_v polarization (courtesy Professor R. S. Stein).

The qualitative features of the scattering can be understood with reference to Fig. 8.19, which illustrates two extreme possibilities for the V_v case. In (a) it is assumed that the spherulite is only polarizable in a tangential direction. It is apparent that dipoles are induced in only the equatorial regions of the spherulite so that it appears optically as an extended particle whose long dimension is perpendicular to the direction of polarization. In case (b), the spherulite is polarizable only along the radial directions and the induced dipoles are principally in the polar regions. Such a spherulite appears as a particle whose major extension is parallel to the direction of polarization. Guinier and Fournet (1955) have shown that the intensity falls off most rapidly with angle in the plane which is parallel to the direction in which the scattering particle presents its largest dimension. The observed scattering pattern for the V_v case extends along the direction of polarization so that the intensity falls off more slowly in the plane parallel to this direction. It follows that the particles are more extended in the direction perpendicular to the polarization direction and that it is the tangential component of the polarizability which mainly gives rise to the scattering.

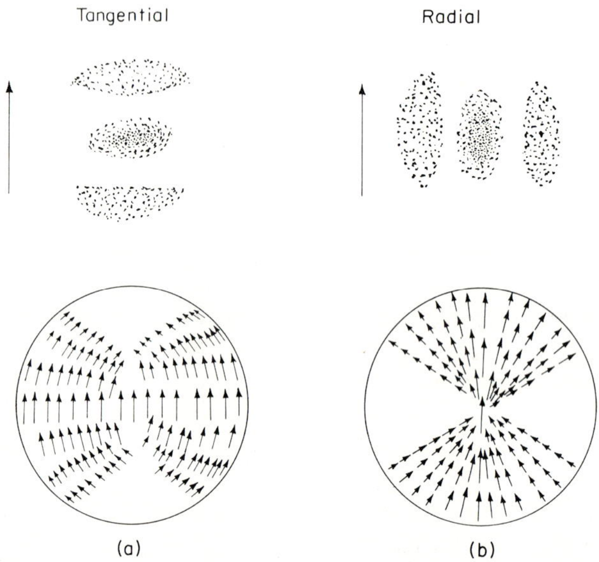

FIG. 8.19. The arrangement of induced dipoles and the expected V_v scattering patterns for spherulites which are (a) only polarizable tangentially and (b) only polarizable radially (Stein and Rhodes, 1960).

The cloverleaf type pattern obtained with crossed polaroids may also be understood in the same way. If the dipoles induced as Fig. 8.19(a) are viewed through the crossed analyzer, those along the equator will not pass through the analyzer, and the maximum contribution to the scattering will come from the dipoles in the plane oriented 45° to both the analyzer and the polarizer.

Stein and Rhodes (1960) have derived equations for the Rayleigh ratio scattered by spheres having different radial and tangential polarizabilities, α_r' and α_t'. The results are

$$\begin{aligned} V_v = {} & k^4 V^2 (3/u^3)^2 [(\alpha_t' - \alpha_s')(2 \sin u - u \cos u - Si(u)) \\ & + (\alpha_r' - \alpha_s')(Si(u) - \sin u) \\ & + (\alpha_t' - \alpha_r') \cos^2 (\theta/2) \cos^2 \mu \, (4 \sin u - u \cos u - 3Si(u))]^2 \end{aligned} \tag{8.4.17}$$

and

$$H_v = k^4 V^2 (3/u^3)^2 [(\alpha_t' - \alpha_r') \cos^2(\theta/2) \sin\mu \cos\mu \times (4 \sin u - u \cos u - 3Si(u))]^2 \tag{8.4.18}$$

where α_s' is the polarizability of the matrix and μ is the azimuthal angle as shown in Fig. 8.17. If the polarizability of the matrix is also anisotropic, the tangential component is used in the first term of V_v and the radial value in the second term. $Si(u)$ designates the integral

$$Si(u) = \int_0^u (\sin x)/x \, dx \tag{8.4.19}$$

For $\alpha_t' = \alpha_r'$ and an isotropic matrix, $H_v = 0$ and V_v reduces to the Rayleigh ratio with the Rayleigh–Debye form factor for isotropic spheres given by (8.1.11).

It is apparent that H_v arises entirely from the anisotropy of the spherulite $(\alpha_t' - \alpha_r')$ whereas only the last term of V_v arises from the anisotropy. The experimental observation in Fig. 8.18 that the V_v scattering is strongly oriented in the polarization direction indicates that the last term predominates. This means that the difference between the tangential and radial components of the polarizability of the spherulite is greater than the difference between either of them and that of the surrounding matrix. Thus, it is the anisotropy of the spherulite which is the principal cause of the low angle scattering in this case.

Polar plots of V_v and H_v for typical values of the polarizability are given in Figs. 8.20 and 8.21. The lines represent contours of constant logarithmic intensity. Further results for a variety of polarizability values have been reported by Stein and Wilson (1962).

The H_v scattering pattern can be used in a simple way to obtain the average spherulite size. From (8.4.18), it can be shown that for a fixed value of μ such as $\mu = 45°$, the scattered intensity goes through a maximum at

$$u = (4\pi a/\lambda) \sin(\theta_m/2) = 4.0 \tag{8.4.20}$$

The radius a can now be obtained directly from the measured value of θ_m and the wavelength.

A still more complicated situation may be encountered because of the so-called "ring structure" which has been observed within spherulites by microscopic examination with polarized light and by micro X-ray diffraction studies. This arises from a helicoidal orientation of the crystallites of the spherulite. At a given radius, the crystallites are all oriented in phase, but this orientation varies periodically with radial distance along the spherulite.

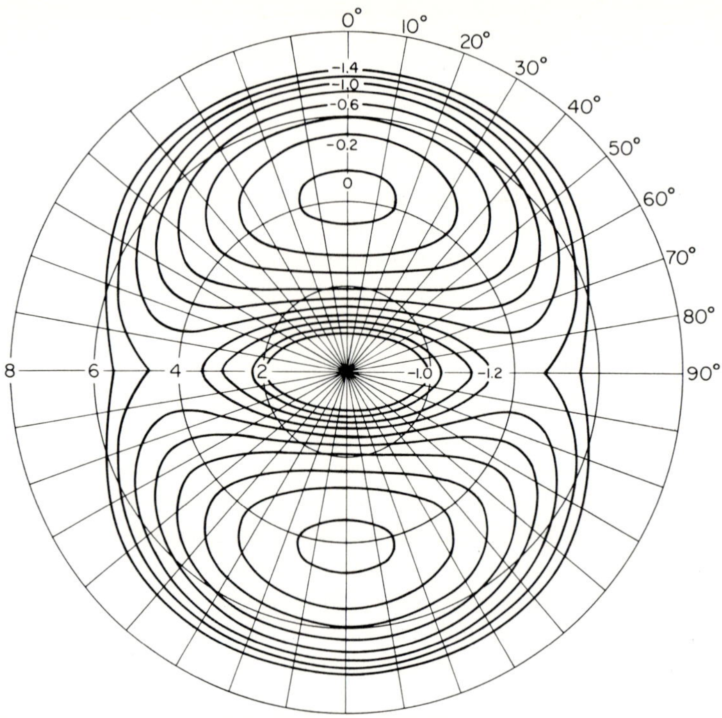

FIG. 8.20. A polar contour plot of the calculated intensity of V_v scattering for an anisotropic sphere having polarizabilities, $(\alpha_t' - \alpha_r') = 3$ (arbitrary units), $(\alpha_t' - \alpha_s') = 1$, and $(\alpha_r' - \alpha_s') = -2$ in isotropic surroundings (Stein, 1963).

This results in a periodically varying tangential polarizability but a constant value along the radial direction. The theory which has been worked out by Stein and Rhodes (1960) predicts a scattering maximum at larger angles (about 15°), which corresponds to diffraction from the internal periodicity. This effect has been observed.

The two-dimensional analog of the spherulite is an arrangement of crystallites which may be described by an anisotropic disk. The theory as developed by Clough *et al.* (1965) is capable of explaining a variety of experimental observations. For example, in Fig. 8.22 there is a comparison of the H_v scattering pattern from polyethylene (similar to that in Fig. 8.18) with that from polytetrafluoroethylene. In the latter case, the four-leaf clover pattern occurs with the scattering lobes lying in the directions of polarization, rather than at 45° to them as with polyethylene. This pattern can be attributed to anisotropic disks in which the optic axis of the crystallites is restricted to the plane of the disk and makes the angle 45° or −45° to the radial direction.

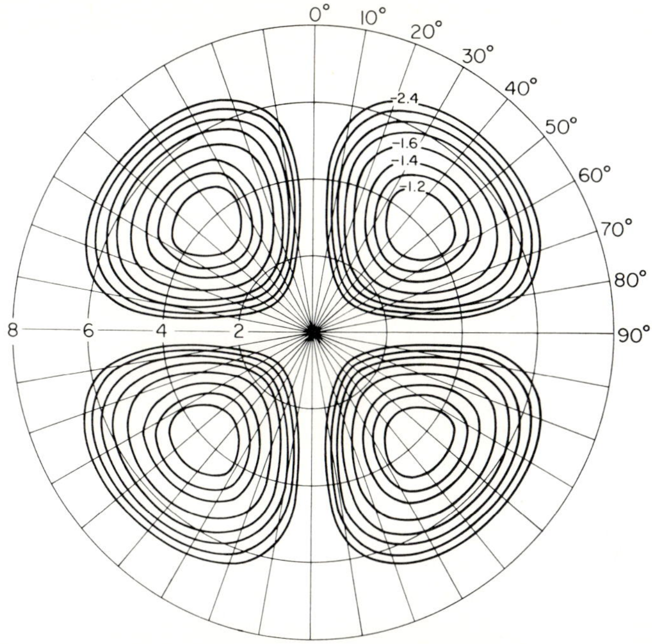

FIG. 8.21. A polar contour plot of the calculated intensity of H_v scattering for an anisotropic sphere having anisotropy $(\alpha_t - \alpha_r) = 3$ (arbitrary units) (Stein, 1963).

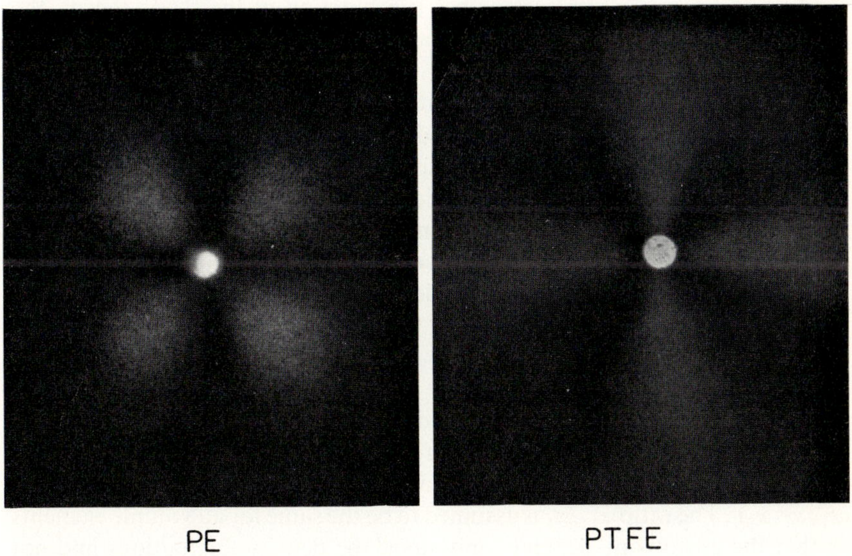

FIG. 8.22. A comparison of the H_v scattering pattern from polyethylene (PE) with that from polytetrafluoroethylene (PTFE) (Clough *et al.*, 1965).

The effect of varying the orientation of the optic axis including helicoidal variation out of the plane of the disk and the effect of distortion of the disk into other shapes have been treated in detail.

8.4.4 ANISOTROPY

Goldstein (1959, 1962, 1963b; Goldstein and Michalik, 1955) has extended the treatment of Debye and Bueche (1949) to include both the effects of heterogeneities due to a local refractive index variation and of anisotropy in the medium which may arise from elastic strains. These strains may, in turn, arise as a result of the presence of the heterogeneities and be correlated with them. The strain scattering will be detectable if the magnitude of the optical effects, as measured by the photoelastic constants of the medium and the strength of the strain field, is appreciable compared to the intrinsic refractive index of the heterogeneities.

An interesting result of the analysis is that although the presence of elastic strains changes the intensity, it does not affect the angular dependence of the perpendicular component of the Rayleigh ratio for incident light which is polarized in the same sense, V_v. This can accordingly be analyzed as though the strains were absent. However, the corresponding parallel component H_h may yield information about the strains. Whereas in the absence of strain this is equal to $V_v \cos^2 \theta$, in the presence of strain it is equal to the product of V_v and an expression that is a function of the strain tensor, the photoelastic constants, and the scattering angle, but not of the size and shape of the heterogeneities. This has provided a useful technique for studying radiation damage in glass by light scattering (Maurer, 1960).

The expressions developed by Goldstein and Michalik are general, but quite complicated and difficult to apply. Stein and Wilson (1962) have therefore proposed a somewhat simpler model for an anisotropic polymer medium. This was done after an extension of the Debye–Bueche theory in which the density fluctuations were considered to be geometrically anisometric, but optically isotropic (Norris and Stein, 1958; Bhatnagar, 1960, 1961), had proven to be inadequate to account for scattering data from certain anisotropic polymers.

Like Goldstein and Michalik, they assume that the anisotropic volume elements have cylindrical symmetry and may be described by two polarizabilities, α'_{11} defined in the principal direction and α'_1 perpendicular to this direction. The anisotropy of the ith volume element is specified by $\delta_i = (\alpha'_{11} - \alpha'_1)_i$. The ratio α'_{11}/α'_1 is assumed to be the same for all volume elements so that the anisotropy depends only upon the density fluctuations and not on fluctuations in the intrinsic anisotropy. The density fluctuations are described by the usual type of correlation function (8.4.2). In addition to density fluctuations, there will be fluctuations in the orientation of the volume elements. The orientation correlation function is given by

$$f(s) = \tfrac{1}{2}\overline{(3\cos^2\theta_{ij} - 1)} \tag{8.4.21}$$

where θ_{ij} is the angle between the principal axes of the ith and the jth volume elements. An additional quantity is defined by

$$u(s) = 1 + (\overline{\eta^2}/\overline{\alpha'}^2)\gamma(s) \tag{8.4.22}$$

where $\overline{\alpha'}$ is the average polarizability. Both $\gamma(s)$ and $f(s)$ are spherically symmetric, depending only upon the magnitude of s. Stein and Wilson derive the following equations for the various polarized components of the scattered light expressed as Rayleigh ratios:

$$\begin{aligned} V_v = (64\pi^5/\lambda^4)\Bigg[\overline{\eta^2}\int_0^\infty \gamma(s)[(\sin hs)/hs]s^2\,ds \\ + (4/45)\delta^2\int_0^\infty f(s)u(s)[(\sin hs)/hs]s^2\,ds\Bigg] \end{aligned} \tag{8.4.23}$$

$$H_v = V_h = (64\pi^5/15\lambda^4)\delta^2\int_0^\infty f(s)u(s)[(\sin hs)/hs]s^2\,ds \tag{8.4.24}$$

$$H_h = V_h\sin^2\theta + V_v\cos^2\theta \tag{8.4.25}$$

where δ is the average optical anisotropy. From (8.4.23) and (8.4.24) it follows that

$$V_v - (4/3)H_v = (64\pi^5/\lambda^4)\overline{\eta^2}\int_0^\infty \gamma(s)[(\sin hs)/hs]s^2\,ds \tag{8.4.26}$$

The density correlation function $\gamma(s)$ and $\overline{\eta^2}$ can be obtained from the following Fourier transforms:

$$\gamma(s) = (\lambda^4/32\pi^6\overline{\eta^2})\int_0^\infty (V_v - (4/3)H_v)h\sin hs\,dh \tag{8.4.27}$$

$$\gamma(0) = 1 = (\lambda^4/32\pi^6\overline{\eta^2})\int_0^\infty (V_v - (4/3)H_v)h^2\,dh \tag{8.4.28}$$

In addition, the transform of (8.4.24) gives

$$f(s)u(s) = (\lambda^4/32\pi^6\delta^2 s)\int_0^\infty H_v h\sin hs\,dh \tag{8.4.29}$$

$$u(0) = 1 + (\overline{\eta^2}/\overline{\alpha'}^2) = (4/32\pi^6\delta^2)\int_0^\infty H_v h^2\,dh \tag{8.4.30}$$

These two equations yield $f(s)$ and δ^2 since $u(s)$ is already known once $\gamma(s)$ and $\overline{\eta^2}$ have been determined.

Although the scattering from well-defined microcrystalline polymers can be understood in terms of the spherulitic structures, for less perfectly organized specimens such as shock-cooled (crystallizable) polymers, the supermolecular structure can be best described by the phenomenological quantities $\gamma(s)$ and $f(s)$ as outlined above (Keijzers *et al.*, 1965). The theory has been extended to include nonrandom orientation fluctuations in a two-dimensional solid (Stein *et al.*, 1966; Keijzers *et al.*, 1968). The case is considered where the probability of a given angle between two optic axes depends upon the separation of the scattering elements s and the angle which the optic axis makes with the interconnecting vector.

8.4.5 Two-Phase Media

In the case of some solids, the inhomogeneous substance can be considered to consist of two homogeneous phases. One of these may be cellular regions of varying and undetermined shape distributed within a homogeneous matrix. Only two dielectric constants need be considered: ε_1 in the matrix and ε_2 in the cells. If region 2 consists of holes, then ε_2 is unity. Alternatively, the matrix may be characterized by its dielectric constant relative to that in the cells, in which case the cells would have a relative dielectric constant of unity. For small angle X-ray scattering, a dielectric constant of unity corresponds to a zero value for the electron density.

The "profile" of the dielectric constant along a particular direction is shown in Fig. 8.23. The average dielectric constant is given by

$$\bar{\varepsilon} = \varepsilon_1\phi_1 + \varepsilon_2\phi_2 \tag{8.4.31}$$

where ϕ_1 and ϕ_2 are the volume fractions occupied by each phase. The local variation of the dielectric constant from the average value is now given by

$$\eta_1 = \varepsilon_1 - \bar{\varepsilon} = (\varepsilon_1 - \varepsilon_2)\phi_2 \tag{8.4.32}$$

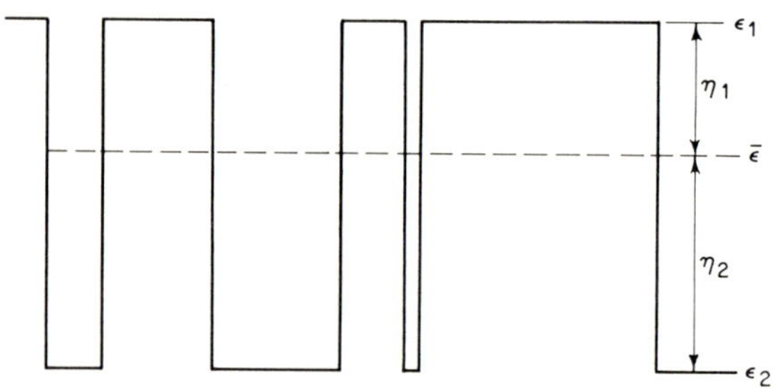

Fig. 8.23. Dielectric constant profile of a two-phase material.

and

$$\eta_2 = \varepsilon_2 - \bar{\varepsilon} = (\varepsilon_2 - \varepsilon_1)\phi_1 \tag{8.4.33}$$

and the mean square variation of the dielectric constant can be written as

$$\overline{\eta^2} = \phi_1 \eta_1{}^2 + \phi_2 \eta_2{}^2 = (\varepsilon_1 - \varepsilon_2)^2 \phi_1 \phi_2 \tag{8.4.34}$$

so that following (8.4.8)

$$I_1 = (k_0{}^4/16\pi^2 r^2)V([\varepsilon_1 - \varepsilon_2]^2 \phi_1 \phi_2)w \tag{8.4.35}$$

The correlation volume w continues to be defined by (8.4.7) and involves the correlation function which is still defined by (8.4.2).

If we now consider an arbitrary distance s within the scattering medium as a measuring rod which is thrown at random into the medium, it can be shown that the probability of "dissimilar ends" is given by

$$P_{\mathrm{D}} = 2\phi_1 \phi_2 [1 - \gamma(s)] \tag{8.4.36}$$

This is the probability that the ends of the measuring rod are in different phases or that the rod cuts through the surface between the phases. Porod (1951, 1952) has shown that the slope of the correlation function plotted vs. s at $s = 0$ (the "initial slope") is related to the specific surface S_{sp} by

$$\left[\frac{d\gamma(s)}{ds}\right]_{s=0} = -\frac{1}{4\phi_1 \phi_2} S_{\mathrm{sp}} = -\frac{1}{4\phi_1 \phi_2} \frac{S}{V} \tag{8.4.37}$$

where S is the interfacial area in a sample of volume V.

The scattering at large angles is determined primarily by the short range correlations. Porod has demonstrated a connection between the initial slope of the correlation function and the actual scattering intensity at large angles by expanding the correlation function in a Taylor series. Then upon dropping all terms higher than the linear terms, substituting in (8.4.6), and integrating by parts, a large angle asymptote between the intensity and specific surface is found:

$$I_1 = \frac{k_0{}^4 V}{4\pi^2 r^2} \frac{2\pi\varepsilon^2}{h^4} S_{\mathrm{sp}} \tag{8.4.38}$$

According to this, the intensity at large angles should be proportional to the inverse fourth power of $\sin \theta/2$ and should be proportional to the total internal surface.

In the above treatment, the radius of curvature of the interface has been assumed to be large compared to the magnitude of s. Weigel *et al.* (1965) have extended the treatment to include a finite radius of curvature of the surface R_0. In this case

$$I_1 = \frac{k_0{}^4 V}{4\pi^2 r^2}\left[\frac{2\pi\varepsilon^2}{h^4} + \frac{2\pi\varepsilon^2}{R_0 h^6}\right] S_{\text{sp}} \tag{8.4.39}$$

These authors claim that it is possible to estimate an average value of R_0 from the intermediate part of the scattering curve.

Kirste and Porod (1962) have examined the effect of retaining higher terms in the Taylor expansion of the correlation function. This results in

$$\gamma(s) = 1 + a_1 s + a_2 s^2 + a_3 s^3 + \cdots \tag{8.4.40}$$

where

$$a_1 = -(1/4\phi_1\phi_2)S_{\text{sp}}; \qquad a_2 = 0 \tag{8.4.41}$$

$$a_3 = [(1/12)\overline{k_1 k_2} + (1/32)\overline{(k_1 - k_2)^2}](1/4\phi_1\phi_2)S_{\text{sp}} \tag{8.4.42}$$

and the high angle asymptote for the intensity

$$I_1 = (k_0{}^4 V/4\pi^2 r^2)(\pi\varepsilon^2/h^4)[-2!a_1 + (4!a_3/h^2)] \tag{8.4.43}$$

The quantities k_1 and k_2 are the two principal curvatures of the surface. The values of $\overline{k_1 k_2}{}^2$ and $\overline{(k_1 - k_2)^2}$ for regularly shaped regions such as spherical shells and rotation ellipses have been calculated by Kirste and Porod.

One possibility for treating experimental data is first to multiply the intensity by h^4 and then to plot this quantity against h or $\sin \theta/2$. This should converge at high angles to an asymptotic value from which a_1, and thus the specific surface, may be calculated. Next, the asymptotic value can be substracted from the above product and then this result multiplied by h^2. If this new product is plotted against h or $\sin \theta/2$, it should approach a high angle asymptote from which a_3, and hence information about the curvature, can be obtained.

Kirste and Porod have also extended Porod's treatment to "oriented" systems by which they mean cylindrically symmetric systems. In this case the correlation function $\gamma_c(s)$ is restricted to the plane perpendicular to the cylinder axis, and the analog of the correlation volume which now has the dimensions of an area is given by

$$w_c = \int_0^\infty 2\pi s \gamma_c(s) J_0(hs)\, ds \tag{8.4.44}$$

The correlation function for a two-phase cylindrically symmetric medium may be obtained from the angular distribution of the intensity in the same manner as for the spherically symmetric system.

It has been shown by Debye *et al.* (1957) that if the cells of one phase are randomly distributed in a solid matrix of a second phase, the correlation function is exponential and of the same form as (8.4.10). It follows that

$$S_{sp} = S/V = 4\phi_1\phi_2/a \tag{8.4.45}$$

and further that

$$I_1 = \frac{k_0{}^4}{16\pi^2 r^2} \frac{8\pi a^3 \overline{\eta^2} V}{[1 + (ha)^2]^2} \tag{8.4.46}$$

or at small angles

$$I_1 = \frac{k_0{}^4}{16\pi^2 r^2} \frac{8\pi a^3 \overline{\eta^2} V}{(1 + k^2 a^2 \theta^2)^2} \tag{8.4.47}$$

For such a material, a plot of $(I_1)^{-1/2}$ vs. θ^2 at low angles should yield a straight line for which

$$a = \frac{1}{k}\left[\frac{\text{slope}}{\text{intercept}}\right]^{1/2} \tag{8.4.48}$$

Debye *et al.* (1957) found that it was necessary to add a small Gaussian correction term to the correlation function in order to obtain good agreement with the experimental results for the small angle X-ray scattering by a variety of porous catalytic materials. More recently, Chu and Tan Creti (1967) have obtained linear plots of $I^{-1/2}$ against θ^2 for the low angle X-ray scattering from an ion exchange resin, demonstrating that this porous medium has a random structure. The specific surface was calculated from (8.4.37) after the correlation function was obtained from (8.4.10) using the value of a obtained from (8.4.48). The specific surface agreed quite well with a value from gas adsorption measurements.

The significance of the parameter a in the exponential correlation function can be generalized. Porod (1951) defines a coherence length by

$$l_c = 2\int_0^\infty \gamma(r)\, dr \tag{8.4.49}$$

With an exponential correlation function, l_c is obviously twice the distance a. The coherence length has the following significance. An arbitrarily oriented straight line of length L will have a mean length of $L\phi_1$ in the matrix part of the material and a mean length of $L\phi_2$ in the cells. If now the condition is imposed that one end of L must be located in the matrix region, then the mean occupied length is increased by $\phi_1 l_c/2$.

8.4.6 Three-Phase Media

Peterlin (1965) has extended the above treatment to the case where the inhomogeneous medium consists of three phases having optical dielectric constants ε_1, ε_2, and ε_3 and volume fractions ϕ_1 and ϕ_2. Then the effective mean square variation of the optical dielectric constant is given by

$$\overline{\eta^2} = (\varepsilon_1 - \varepsilon_3)^2\phi_1(1 - \phi_1) + (\varepsilon_2 - \varepsilon_3)^2\phi_2(1 - \phi_2) - 2(\varepsilon_1 - \varepsilon_3)(\varepsilon_2 - \varepsilon_3)\phi_1\phi_2 \tag{8.4.50}$$

and the correlation function is

$$\gamma(s) = \overline{\eta^2}[(\varepsilon_1 - \varepsilon_3)^2\phi_1(1 - \phi_1)\gamma_1(s) + (\varepsilon_2 - \varepsilon_3)^2\phi_2(1 - \phi_2)\gamma_2(s) - 2(\varepsilon_1 - \varepsilon_3)(\varepsilon_2 - \varepsilon_3)\phi_1\phi_2\gamma_{12}(s)] \tag{8.4.51}$$

In this expression, $\gamma_1(s)$ and $\gamma_2(s)$ are the correlation functions that would result in the case, on the one hand, that components 2 and 3 were equal, and on the other hand, that components 1 and 3 were equal.

Various light-scattering data (Beebe and Marchessault, 1964; Beebe *et al.*, 1966; Coalson *et al.*, 1966) for semicrystalline samples of cellulose swollen by absorption of solvent have been analyzed by this theory. Such materials are thought to consist of three regions—polymer crystals, amorphous polymer, and solvent. The solvent can be exchanged among materials of varying refractive index in such a way that the special structure of the sample remains unchanged. Beebe and Marchessault exchanged the solvent from water to methanol, then to benzene, and finally to a silicone oil. The measured intensity decreased markedly with increasing refractive index of the solvent.

Such an experiment, as pointed out by Peterlin, affords the possibility of determining $\gamma_1(s)$ and $\gamma_2(s)$ by utilizing solvents (considered to be component 3) with the same refractive index as the crystalline and the amorphous polymer, respectively. Then for a solvent with still a different refractive index, one can determine $\gamma_{12}(s)$ from the above equation. From the initial slope of $\gamma_{12}(s)$ it is then possible to determine the area of the interface between components 1 and 2 in the unswollen polymer sample.

8.4.7 Particulate Systems

The roles played by the cellular fraction and the matrix in the above analysis are symmetric so that the characteristic function is the same for an object and its complementary object. A porous material consisting of a distribution of cells or holes in a solid matrix will scatter in the same way, were the holes and the solid to be interchanged. This would then comprise a particulate system. The above analysis can be useful in treating such systems, particularly densely packed powders or concentrated dispersions. In a densely packed system, the initial slope of the correlation function offers an independent technique, comparable to the BET gas adsorption method, of determining the surface area of fine powders even when there is no information about the size distribution and shapes of the particles.

The following equations are valid when the disposition of the particles is respectively spherically symmetric or cylindrically symmetric (Riley and Oster, 1951):

$$(I_i - I_n)/I_i = N\int_0^\infty 4\pi s^2[1 - g(s)](\sin hs/hs)\, ds \tag{8.4.52}$$

and

$$(I_i - I_n)/I_i = N\int_0^\infty 2\pi s[1 - g(s)]J_0(hs)\, ds \tag{8.4.53}$$

I_i is the scattering that this system of N particles per milliliter would exhibit if these were randomly positioned so that the scattering were incoherent. I_n is the actual intensity scattered by the highly concentrated dispersion of particles. The quantity $[1 - g(s)]$, which formally resembles the correlation function, contains the radial distribution function $g(s)$. This is the probability of finding a particle at a distance s from a reference point for the spherical distribution or at a perpendicular distance s from a reference axis for the cylindrical distribution. An interesting study on cellulose fibers has been carried out by Heyn (1955) in which the radial distribution function of oriented microcrystalline cylinders in the fiber was determined by small angle X-ray scattering under different conditions of swelling. The microcrystals are more randomly positioned in the swollen fibers, and it is this which made it possible to separate the effects of I_i and I_n.

One can speak about a characteristic function for a single particle designated $\gamma_0(s)$, which formally plays the same role for the particle as the correlation function does for the inhomogeneous solid. For example, for a uniform sphere of radius a [also see (8.3.26)]

$$\gamma_0(s) = 1 - (3/4)(s/a) + (1/16)(s/a)^3 \tag{8.4.54}$$

which leads to the appropriate form factor when substituted into (8.4.6) (Stein, 1963). A single particle correlation volume can be defined by

$$w_0 = \int_0^\infty 4\pi s^2 \gamma_0(s)(\sin hs/hs)\, ds \tag{8.4.55}$$

where $\gamma_0(s)$ represents the probability that a point at a distance s in an arbitrary direction from a given point in the particle will also be in the particle. In the limit of small angles, the correlation volume equals the geometrical volume

$$w_0 = \int_0^\infty 4\pi s^2 \gamma_0(s)\, ds = V \tag{8.4.56}$$

The characteristic function varies from unity for $s = 0$ and becomes zero beyond a value of s corresponding to the line of maximum length through the particle.

In complete analogy with the case for a porous medium (8.4.37), the initial slope of $\gamma_0(s)$ is proportional to the specific surface

$$\left|\frac{d\gamma_0(s)}{ds}\right|_{s=0} = -\frac{S}{4V} = -\frac{1}{\bar{l}} \tag{8.4.57}$$

This defines a characteristic length $\bar{l}$ which depends upon both the size and shape of the particle. In addition, the integral of the characteristic function for an isolated particle defines a coherence length

$$l_c = 2\int_0^\infty \gamma_0(s)\, ds \tag{8.4.58}$$

This is the average length of the lines passing through all points in the particle in all directions and terminating on its boundaries. Expressions for $\bar{l}$ and l_c for various particle forms are given in Tables 8.3 and 8.4, respectively.

TABLE 8.3
EXPRESSIONS FOR $\bar{l}$ FOR VARIOUS FORMS

Sphere, radius a	$\frac{4}{3}a$
Prolate spheroid, axes b, b, a	$\frac{8}{3}b\left[\frac{b}{a} + \frac{\arcsin[1-(b/a)^2]^{1/2}}{[1-(b/a)^2]^{1/2}}\right]^{-1} \xrightarrow[a\to\infty]{} \frac{16}{3\pi}b$
Oblate spheroid, axes b, a, a	$\frac{8}{3}a\left[\frac{a}{b} + \frac{\arcsin[(a/b)^2-1]^{1/2}}{[(a/b)^2-1]^{1/2}}\right]^{-1} \xrightarrow[a\to\infty]{} \frac{8}{3}a$
Rectangular prism, sides A, B, C	$\frac{2}{(1/A)+(1/B)+(1/C)} \xrightarrow[c\to\infty]{} \frac{2}{(1/A)+(1/B)} \xrightarrow[B,C\to\infty]{} 2A$
Cylinder, radius a, height h	$\frac{2ah}{a+h} \xrightarrow[h\to\infty]{} 2a \xrightarrow[a\to\infty]{} 2h$
Spherical shell, radii ka, a	$\frac{4}{3}a\frac{1-k^2}{1+k^2}$

TABLE 8.4
EXPRESSIONS FOR l_c FOR VARIOUS FORMS

Sphere radius a	$\frac{3}{2}a$
Prolate spheroid, axes b, b, a	$\frac{3}{2}b\frac{\arcsin[1-(b/a)^2]^{1/2}}{[1-(b/a)^2]^{1/2}} \xrightarrow[a\to\infty]{} \frac{3\pi}{4}b$
Oblate spheroid, axes b, a, a	$\frac{3}{2}a\frac{\arcsin[(a/b)^2-1]^{1/2}}{[(a/b)^2-1]^{1/2}} \xrightarrow[a\to\infty]{} \frac{3}{2}b\ln\left(\frac{2a}{b}\right) \to \infty$
Spherical shell, radii ka, a	$\frac{3}{2}a\frac{(1-k)^2}{1-k^3}\left[(1+k)^2\left(1+\frac{1}{4}\ln\frac{1+k}{1-k}\right)-\frac{k}{2}\right]$
Infinitely long cylinder, radius a	$\frac{8}{3}a$

8.5 Size versus Shape Effects

Equivalent scattering patterns may be obtained from an appropriate polydispersion of spheres, monodispersion of ellipsoids, or polydispersion of ellipsoids. Accordingly, it is very difficult to disentangle the effects of size from those of shape. A procedure has been proposed for doing this (Mittelbach and Porod, 1965; Mittelbach, 1965), and hence for determining from scattering data both the size distribution and the particle anisometry.

The size distribution of a particular linear dimension in the particle is described by two parameters, and the particles are assumed to be either prolate or oblate spheroids having a particular axial ratio. The procedure requires knowledge of the average radius of gyration, the average volume, the characteristic length, and the coherence length. The first two quantities can be obtained from the angular variation of the intensity near $\theta = 0$, using a procedure such as the Zimm plot. The characteristic length can be obtained from the initial slope of the correlation function (8.4.57) and the coherence length from its integral over all correlation distances (8.4.58). These experimentally determined quantities are appropriately averaged over the size distribution

$$\overline{R_g{}^2} = \sum n_i V_i^2 R_{gi}^2 \Big/ \sum n_i V_i^2 \tag{8.5.1}$$

$$\overline{V} = \sum n_i V_i \cdot V_i \Big/ \sum n_i V_i \tag{8.5.2}$$

$$\bar{\bar{l}} = \sum n_i O_i \bar{l}_i \Big/ \sum n_i O_i \tag{8.5.3}$$

$$\bar{l}_c = \sum n_i V_i l_{ci} \Big/ \sum n_i V_i \tag{8.5.4}$$

where n_i designates the number of particles of size i and O_i is the surface area.

For a given form, the size can be characterized by a dimensionless parameter p_i such that $R_{gi}^2 = R_{gz}^2 p_i^2$, $\bar{l}_i = \bar{l}_z p_i$, $l_{ci} = l_{cz} p_i$, $V_i = V_z p_i^3$, and $O_i = O_z p_i^2$ where the subscript z denotes the value for which $p_i = 1$. For a sphere, one might choose p_i as the ratio of the radius to the mean radius in which case $R_{gz}^2 = (3/5)\bar{a}^2$, $\bar{l}_z = (4/3)\bar{a}$, $l_{cz} = (3/2)\bar{a}$, $V_z = (4/3)\pi\bar{a}^3$, and $O_z = 4\pi\bar{a}^2$. Then when all the particles are assumed to have the same shape, any of the above quantities can be calculated for each value of p_i. It now follows that

$$\overline{l_c}/\bar{\bar{l}} = (l_{cz}/\bar{l}_z) \cdot (\overline{p^4}\,\overline{p^2}/(\overline{p^3})^2) = F \cdot P \tag{8.5.5}$$

where

$$\overline{p^m} = \sum n_i p_i^m \tag{8.5.6}$$

The ratio of $\bar{l}_c/\bar{\bar{l}}$ is the product of two factors, the first of which depends only upon the form of the particles and the other of which depends only upon the statistical distribution of the sizes.

Also, it can be shown that

$$\frac{(\overline{R_g^2})^{1/2}}{(\overline{V})^{1/3}} = \frac{R_{gz}}{(V_z)^{1/3}} \left[\frac{\overline{p^8}}{\overline{\overline{p^6}}}\right]^{1/2} \left[\frac{\overline{p^3}}{\overline{\overline{p^6}}}\right]^{1/3} = F^* \cdot P^* \tag{8.5.7}$$

where the appropriately averaged values of the volume (or molecular weight) and radius of gyration may be obtained experimentally in the usual way from the zero angle limiting value of the intensity and its initial slope. Again, the experimentally determined ratio is the product of a form factor and a polydispersity factor.

If the particle shape is sufficiently well defined to calculate F or F^*, then the polydispersity factor can be determined. Mittelbach and Porod have used the gamma distribution function

$$g(p)\,dp = \frac{c^{n+1}}{\Gamma(n+1)} p^n e^{-cp}\,dp \tag{8.5.8}$$

for which they have shown

$$P = 1 + [\sigma^2/(1 + 2\sigma^2)] \tag{8.5.9}$$

and

$$P^* = \frac{[(1 + 6\sigma^2)(1 + 7\sigma^2)]^{1/2}}{[(1 + 3\sigma^2)(1 + 4\sigma^2)(1 + 5\sigma^2)]^{1/3}} \tag{8.5.10}$$

where σ, which measures the breadth of the distribution, is given by

$$\sigma^2 = 1/(n+1) = [\overline{p^2}/(\bar{p})^2] - 1 \tag{8.5.11}$$

The distribution function is plotted in Fig. 8.24. It possesses a maximum only for $\sigma < 1$ and becomes increasingly narrow as σ approaches zero, at which point the polydispersity factors go to unity. These are plotted in Fig. 8.25. The form factors for ellipsoids of revolution are plotted against the axial ratios v in Fig. 8.26. The abscissa is either v or $1/v$ according to whether the ellipsoid is prolate ($v > 1$) or oblate ($v < 1$). The values plotted have been normalized to unity for a sphere by multiplying F by $\frac{8}{9}$ and F^* by 2.08.

It is possible now to estimate both the form and polydispersity from the simultaneous determination of $\bar{l}_c$, $\bar{\bar{l}}$, $\overline{R_g{}^2}$, and $\bar{V}$. This can be done by combining Figs. 8.25 and 8.26 as shown in Fig. 8.27. The dashed curves are the loci of σ and v calculated for the designated values of the measurable parameter $K = \frac{8}{9}\bar{l}_c/\bar{l}$. In a similar way, the full curves give those combinations of σ and v which yield designated values of the measurable parameter $K^* = 2.06(\overline{R_g{}^2})^{1/2}/(\bar{V})^{1/3}$. The intersection of the curves corresponding to the experimentally determined values of K and K^* now lead to the polydispersity σ and the eccentricity v. This still requires an *a priori* assumption that the particles are either prolate or oblate ellipsoids of revolution. The efficacy of this method still remains to be tested experimentally.

Each of the experiments which leads to the determination of the quantities defined by (8.5.1) to (8.5.4) also determines a particular moment of the size distribution. For example, remembering the respective values of R_z, V_z, O_z, $\bar{l}_z$, and l_{cz} for spheres, the following moments of the distribution of radii for a dispersion of spheres are obtained:

$$a_R = \left[\frac{\overline{a^8}}{\overline{a^6}}\right]^{1/2} = \left[\frac{5\overline{R_g{}^2}}{3}\right]^{1/2} \tag{8.5.12}$$

$$a_v = \left[\frac{\overline{a^6}}{\overline{a^3}}\right]^{1/3} = \left[\frac{3\bar{V}}{4\pi}\right]^{1/3} \tag{8.5.13}$$

$$a_l = \frac{\overline{a^3}}{\overline{a^2}} = \frac{3\bar{\bar{l}}}{4} \tag{8.5.14}$$

$$a_{l_c} = \left[\frac{\overline{a^4}}{\overline{a^3}}\right] = \frac{2\bar{l}_c}{3} \tag{8.5.15}$$

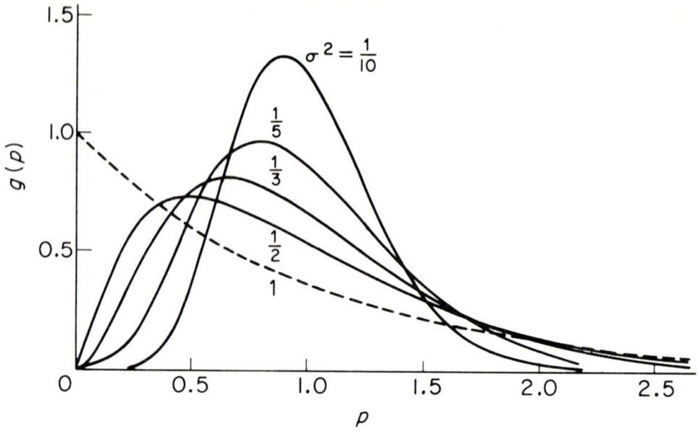

FIG. 8.24. Gamma distribution function (Eq. 8.5.8) $g(p)$ plotted against p for various values of the parameter σ^2 (Eq. 8.5.11) (Mittelbach and Porod, 1965).

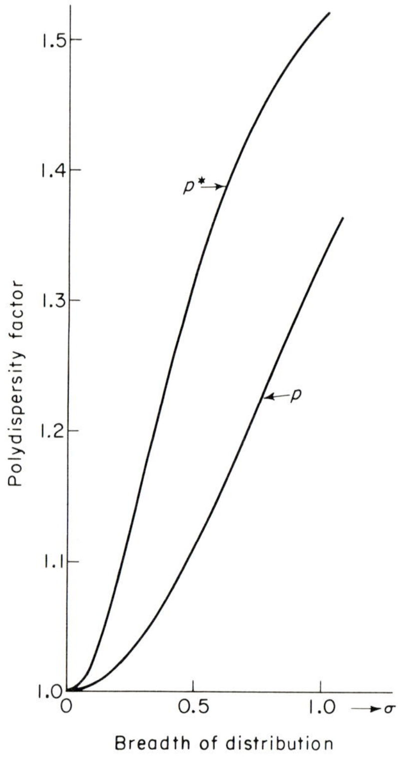

FIG. 8.25. Polydispersity factors P and P^* plotted against σ for a gamma distribution (Mittelbach and Porod, 1965).

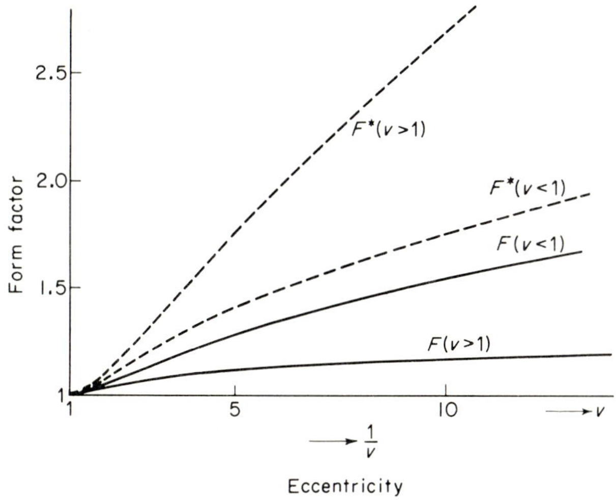

FIG. 8.26. Form factors F and F^* plotted against the spheroid axial ratio v (prolate) or $1/v$ (oblate) (Mittelbach and Porod, 1965).

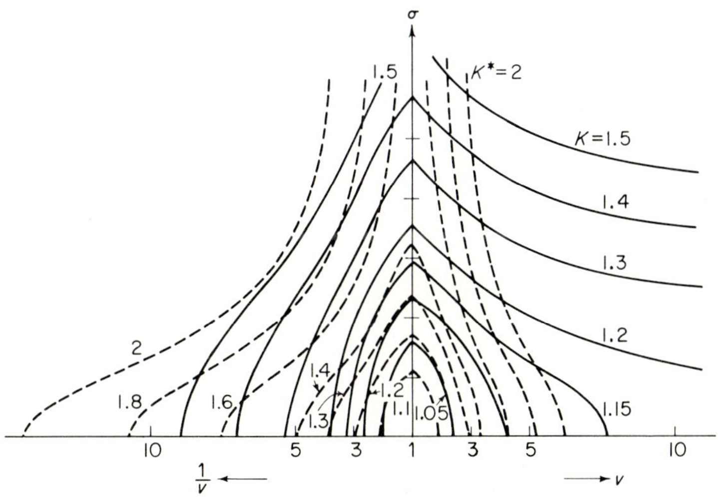

FIG. 8.27. Combination of Figs. 8.25 and 8.26 to give lines of constant F and F^* in the σv-plane. Left-hand side is for oblate spheroids; right-hand side is for prolate spheroids (Mittelbach and Porod, 1965).

TABLE 8.5
PARTICLE SCATTERING FACTORS FOR VARIOUS SHAPES

Shape	Definition of variables	$P(\theta)$	Derivation Ref.
Sphere	Radius $= a$ $u = ha$	$\left[\frac{3(\sin u - u\cos u)}{u^3}\right]^2 = \frac{9\pi}{2}\frac{J_{\frac{3}{2}}{}^2(u)}{u^3}$	Rayleigh (1914)
Concentric sphere including spherical shell	Inner radius $= a$ Outer radius $= b$ $u = ha$ $v = hb$ Refractive index Inner region $= m_1$ Outer region $= m_2$	$\frac{9\pi}{2}\left[\frac{J_{\frac{3}{2}}(v)}{v^{\frac{3}{2}}} + \left(\frac{m_1 - m_2}{m_2 - 1}\right)\left(\frac{a}{b}\right)^3\frac{J_{\frac{3}{2}}(u)}{u^{\frac{3}{2}}}\right]^2$	Kerker *et al.* (1962)
Sphere with Gaussian distribution of polarizability	Polarizability is $\alpha' = \alpha_0{}'\exp[-(s/a)^2]$ $u = ha$	$R_\theta = \frac{k^4(\alpha_v')^2}{2}(1 + \cos^2\theta)\exp[-u^2/2]$ where $\alpha_v' = \int_0^a 4\pi s^2\alpha'\,ds$	Peterlin (1951)
Random coil (Gaussian distribution of end-to-end distance)	$w = \frac{h^2\overline{s^2}}{6} = h^2\overline{R_g{}^2}$ where $\overline{s^2}$ is the mean square end-to-end distance and $\overline{R_g{}^2}$ is the mean square radius of gyration	$\frac{2}{w^2}(e^{-w} + w - 1)$	Debye (1947)

Straight chain of spheres (randomly oriented)	Radius $= a$ $u = ha$ n is number of spheres	$\frac{9\pi J_{\frac{3}{2}}{}^2(u)}{u^3}\left[\frac{n}{2} + (n-1)\frac{\sin 2u}{2u} + (n-2)\frac{\sin 4u}{4u} + \cdots \frac{\sin[2(n-1)u]}{2(n-1)u}\right]$	Kratky (1948) For other configurations of touching spheres, Kratky and Porod (1949), Oster and Riley (1952a)
Infinite circular cylinder at perpendicular incidence	Radius $= a$ $u = ha$	$\left[\frac{2J_1(u)}{u}\right]^2$	Rayleigh (1881)
Circular cylinder	Radius $= a$ Height $= l$ $u = ha$ $v = hl/2$ β is angle between axis of cylinder and bisectrix	$\frac{\pi}{2v \cos \beta}\left[J_{\frac{1}{2}}(v \cos \beta) \times \frac{2J_1(u \sin \beta)}{u \sin \beta}\right]^2$ For random orientation: $\int_0^{\pi/2} P(\theta) \sin \beta \, d\beta$	Fournet (1951) van de Hulst (1957) For hollow circular cylinder and for elliptic cylinder see Mittelbach and Porod (1961a, b)
Thin circular cylinder	v, β are same as above	$\frac{\pi}{2v \cos \beta}[J_{\frac{1}{2}}(v \cos \beta)]^2$ $\int_0^{2v} \frac{\sin t}{t} dt - \frac{\sin^2 v}{v^2}$ (randomly oriented)	van de Hulst (1957) Neugebauer (1943)

TABLE 8.5—(*Continued*)

Shape	Definition of variables	$P(\theta)$	Derivation Ref.
Thin disk	u, β are same as above	$\left[\frac{2J_1(u \sin \beta)}{u \sin \beta}\right]^2$ $\frac{2}{u^2}\left(1 - \frac{1}{u}J_1(2u)\right)$(randomly oriented)	van de Hulst (1957) Kratky and Porod (1949)
Two parallel infinite circular cylinders at perpendicular incidence	Radius $= a$ $u = ha$ σ is distance between cylinder axes $\gamma = \sigma/2a$	$\frac{J_1^{\,2}(u)}{u^2}[2 + 2J_0(2\gamma u)]$	Oster and Riley (1952b) See this reference for other arrays of parallel infinite cylinders
Ellipsoid of revolution	$u = ha\left(\cos^2 \beta + \frac{b^2}{a^2}\sin^2 \beta\right)^{1/2}$ semiaxes are a and b β is angle between figure axis and bisectrix	$\frac{9\pi}{2}\frac{J_{\frac{3}{2}}^{\,2}(u)}{u^3}$ For random orientation, $\int_0^{\pi/2} P(\theta) \cos \beta \, d\beta$	Guinier (1939) Shull and Roess (1947) Porod (1948) Malmon (1957) Schmidt and Hight (1959)
Ellipsoid	Semiaxes are a, b, c $u = hA$ $A^2 = a^2 \cos^2 \alpha + b^2 \cos^2 \beta + c^2 \cos^2 \gamma$ where α, β, γ are the direction cosines on the bisectrix	$\frac{9\pi}{2}\frac{J_{\frac{3}{2}}^{\,2}(u)}{u^3}$ See reference for random orientation	Mittelbach and Porod (1962)

Cube	l is edge of cube γ, ε are the polar angular coordinates of the bisectrix $E(u) = \frac{\sin u}{u}$	$P(\theta, \gamma, \varepsilon) = E^2\left(\frac{hl}{2}\cos\varepsilon\sin\gamma\right) \times E^2\left(\frac{hl}{2}\sin\varepsilon\sin\gamma\right) \times E^2\left(\frac{hl}{2}\cos\gamma\right)$ For random orientation, $P(\theta) = \frac{2}{\pi}\int_0^{\pi/2}\int_0^1 P(\theta, \gamma, \varepsilon)\, d(\cos\gamma)\, d\varepsilon$	Napper (1968) For square pyramid and regular octahedron see Napper (1968). For rectangular parallelepiped see Mittelbach and Porod (1962)

In a monodisperse dispersion, these quantities will all be equal. Otherwise, $a_R > a_v > a_{l_c} > a_l$, and they will be more nearly equal to each other for a narrow distribution than when the distribution is broad. Brill *et al.* (1968) have verified that this does turn out to be the case for small angle X-ray scattering by various aqueous dispersions of colloidal silica.

8.6 Table of Form Factors

The form factors for a number of regular shapes have been collected in Table 8.5. The references cited frequently also contain tabulations of computed values of these functions. Additional computations for spheres will be found in Guinier and Fournet (1955) and for polydispersions of random coils in Beattie and Booth (1960a). Beattie and Booth (1960b) have also published dissymmetries for disks, spheres, rods, and random coils. Beidl *et al.* (1957) give values for a variety of shapes.

CHAPTER 9

Scattering by Liquids

9.1 Pure Liquids

Simple molecules, being small compared with the wavelength, may be treated as Rayleigh scatterers even though they cannot be assigned a refractive index. The scattering by a random array of molecules such as those comprising a gas may be expressed in terms of the macroscopic refractive index of the gas, using (3.2.33) if the value of Avogadro's number is known. Indeed, among the first uses to which quantitative measurements of scattering were put was the calculation of Avogadro's number. When the correction for the optical anisotropy of the molecules is made, the agreement between theory and experiment is within the range of the experimental error (Bhagavantam, 1942). This confirms not only the Rayleigh theory, but also that in a gas each molecule acts as an incoherent, randomly located scattering center. The total intensity is the sum of the intensities from each molecule.

This is no longer the case for a liquid. Although the qualitative features of Rayleigh scattering, such as the wavelength dependence, the polarization, and the angular distribution of the scattered light, may still be obtained, the intensity of the scattering is one or two orders of magnitude less than that predicted for a gas-like system. This is because the molecules in a liquid are no longer completely incoherent, randomly located scatterers.

The situation for a liquid is intermediate between that for a gas and for a solid. In a perfectly crystalline solid, there would be no light scattering. All of the molecules would be fixed in a completely ordered array at small distances from each other. Although the molecules can each be thought of as centers for secondary radiation, the waves emanating from them would interfere destructively at any point of observation. This is because, except for those at the edge of the system, the molecules can be paired so that the waves from each pair traveling in a particular direction are exactly out of phase. Such a crystal behaves as a uniform medium and scatters no light.

It might be possible, at least in principle, to determine the mutual interference among the molecules of a liquid if at any given time the position of each molecule could be specified. It would be necessary to add the scattering amplitudes, taking into account all of the phase differences. In addition, because of the close packing of the molecules, it would also be necessary to consider the mutual electromagnetic interaction on the polarization of the molecules as well as the effect of multiple scattering.

On the other hand, there is a phenomenological way of looking at the scattering that lends itself to a direct statistical-thermodynamic description. A liquid is not uniform. The distribution of molecules in any given small volume element will vary with time because of the thermal motion and each particular region will differ from its neighbors. Thus, the local density of matter is constantly changing.

This local density may be considered to consist of two contributions. The major part is the macroscopic or average density and this is uniform throughout the liquid; superimposed upon this is a fluctuation from the average which varies in an irregular manner. Scattering by the uniform part will be completely eliminated by interference between paired regions just as in the perfect crystal, except in the forward direction where it will contribute to the transmitted beam. If the density fluctuations in each volume element are random, thus giving rise to incoherent scattered wavelets, each element of volume will make an additive contribution to the intensity of the transverse scattering in the same manner as the scattering by the molecules of a gas. This point of view was first developed by Smoluchowski (1907, 1912), who proposed that it could explain the phenomenon of critical opalescence. Although this does not, in fact, provide a correct model for critical opalescence, it is quite adequate for fluids away from the critical point. Later, Einstein (1910) showed how this model could be treated quantitatively.

9.1.1 Scattering due to Density Fluctuations

A fluctuation in density is accompanied by a corresponding fluctuation in the polarizability α' of a volume element, δV. This volume is chosen sufficiently large, compared with molecular dimensions, so that it may be assigned the usual phenomenological thermodynamic properties, and yet it is maintained small enough, compared with the wavelength of light, so that it can be treated as a particle for which the Rayleigh equation is valid. In this way, the fluctuation manifests itself as a region of excess polarizability $\delta\alpha'$. These regions may be likened to small Rayleigh scatterers with a dielectric constant different from that of a uniform homogeneous medium in which they are imbedded. However, they are generally not isotropic because of the anisotropy of the molecules. The effect of the anisotropy can be corrected for by the Cabannes factor which will be considered later. Accordingly, it is only the isotropic part of the scattering which is being dealt with now.

Because of the random nature of the thermal process in both space and time, the scattered wavelets from these regions may be treated as completely incoherent. The turbidity is given by

$$\tau = \frac{128\pi^5}{3\lambda^4}\left(\frac{\overline{(\delta\alpha')^2}}{\delta V}\right) \tag{9.1.1}$$

where $\overline{(\delta\alpha')^2}$ is the mean square of the excess polarizability corresponding to the volume element δV and λ is the wavelength in the uniform medium. This is obtained directly from (3.2.3), (3.2.19), and (3.2.30), remembering that the number of volume elements per unit volume is $(\delta V)^{-1}$.

There now remain two steps in order to connect the above equation with directly measurable quantities. First, the fluctuation in excess polarizability will be expressed in terms of density fluctuations; then, this will be expressed in terms of the experimental parameters.

The excess polarizability of the fluctuating volume element is related to the dielectric constant through the Lorenz–Lorentz formula [see (3.2.3)]

$$\delta\alpha' = \frac{3}{4\pi}\left(\frac{\varepsilon' - \varepsilon}{\varepsilon' + 2\varepsilon}\right)\delta V \tag{9.1.2}$$

where ε is the average dielectric constant of the medium and ε' is the particular value in the region exhibiting the fluctuation. Assuming that the magnitude of the fluctuation is small, this reduces to

$$\delta\alpha' = \frac{\delta V}{4\pi}\left(\frac{\varepsilon' - \varepsilon}{\varepsilon}\right) = \frac{\delta V}{4\pi}\frac{\delta\varepsilon}{\varepsilon} \tag{9.1.3}$$

so that

$$\tau = \frac{8\pi^3}{3\lambda^4}\frac{\overline{(\delta\varepsilon)^2}}{\varepsilon^2}\delta V \tag{9.1.4}$$

Since ε can be considered to be a function of the density ρ and one other thermodynamic variable, we may write either

$$\delta\varepsilon = (\partial\varepsilon/\partial\rho)_T\,\delta\rho + (\partial\varepsilon/\partial T)_\rho\delta T \tag{9.1.5}$$

or

$$\delta\varepsilon = (\partial\varepsilon/\partial\rho)_p\,\delta\rho + (\partial\varepsilon/\partial p)_\rho\delta p \tag{9.1.6}$$

where T and p are the absolute temperature and pressure, respectively. Because the temperature and pressure coefficients of the dielectric constant at constant density are small, the second term in these equations may usually be neglected. At this point we shall arbitrarily select the expression involving the density derivative at constant temperature rather than that at constant pressure. The reason for this choice will be discussed shortly. Then

$$\overline{(\delta\alpha')^2} = \frac{(\delta V)^2}{16\pi^2\varepsilon^2}\left(\frac{\partial\varepsilon}{\partial\rho}\right)_T^2 \overline{(\delta\rho)^2} \tag{9.1.7}$$

and

$$\tau = \frac{8\pi^3\delta V}{3\lambda_0{}^4}\left[\rho\left(\frac{\partial\varepsilon}{\partial\rho}\right)_T\right]^2 \frac{\overline{(\delta\rho)^2}}{\rho^2} \tag{9.1.8}$$

where λ_0 is the wavelength in vacuo.

Next, the mean square fluctuation in density occurring in the volume element δV must be calculated. This is done using the Boltzmann theorem. If the system is constrained to constant volume and temperature, the probability of a change from the average value of any property is related to the Helmholtz free energy δA expended in bringing about the change. This probability is proportional to

$$\exp(-\delta A/kT) \tag{9.1.9}$$

where k is the Boltzmann constant. For small deviations about the average, expansion in a Taylor series gives

$$\delta A = (\partial A/\partial\rho)_{T,V}\delta\rho + (1/2!)(\partial^2 A/\partial\rho^2)_{T,V}(\delta\rho)^2 + \cdots \tag{9.1.10}$$

The first term is zero since the coefficient $(\partial A/\partial\rho)_{T,V}$ is evaluated at the equilibrium condition where the Helmholtz free energy is a minimum under the indicated constraints. Therefore, neglecting higher terms

$$\delta A = \tfrac{1}{2}(\partial^2 A/\partial\rho^2)_{T,V}(\delta\rho)^2 \tag{9.1.11}$$

The mean square value of $\delta\rho$ can now be obtained by averaging over the probability distribution

$$\begin{aligned}\overline{(\delta\rho)^2} &= \frac{\displaystyle\int_{\delta\rho=0}^{\infty} (\delta\rho)^2 \exp[-(\partial^2 A/\partial\rho^2)_{TV}\,\delta\rho^2/2kT]\,d(\delta\rho)}{\displaystyle\int_{\delta\rho=0}^{\infty} \exp[-(\partial^2 A/\partial\rho^2)_{TV}\,\delta\rho^2/2kT]\,d(\delta\rho)} \\ &= \frac{kT}{(\partial^2 A/\partial\rho^2)_{TV}}\end{aligned} \tag{9.1.12}$$

The use of well-known thermodynamic relations leads to

$$\overline{(\delta\rho)^2} = kT\beta_T\rho^2/\delta V \tag{9.1.13}$$

and finally to

$$\tau = \frac{8\pi^3}{3\lambda_0{}^4}\left[\rho\left(\frac{\partial\varepsilon}{\partial\rho}\right)_T\right]^2 kT\beta_T \tag{9.1.14}$$

where the isothermal compressibility is

$$\beta_T = -\frac{1}{V}\left(\frac{\partial V}{\partial p}\right)_T \tag{9.1.15}$$

This equation for the turbidity is actually equivalent to the one given earlier for a perfect gas (3.2.33). The quantity in the square bracket as evaluated from the Clausius–Mossotti equation

$$P_0 = \left(\frac{\varepsilon - 1}{\varepsilon + 2}\right)\frac{M}{\rho} \tag{9.1.16}$$

is

$$[\rho(\partial\varepsilon/\partial\rho)_T] = 2(\mu - 1) \tag{9.1.17}$$

and for a perfect gas

$$\beta_T = 1/p \tag{9.1.18}$$

Here, μ is the macroscopic refractive index of the gas. Thus, as pointed out by Einstein, even for the completely incoherent scattering by a perfect gas, there is no need to make any assumptions about the discrete distribution of matter on the molecular level, provided the system can be described thermodynamically. However, these equations are only valid if the fluctuations themselves are random.

An elementary thermodynamic transformation gives

$$\rho\left(\frac{\partial\varepsilon}{\partial\rho}\right)_T = \frac{1}{\beta_T}\left(\frac{\partial\varepsilon}{\partial p}\right)_T = \frac{2n}{\beta_T}\left(\frac{\partial n}{\partial p}\right)_T \tag{9.1.19}$$

where n is the refractive index, and upon substitution in (9.1.14) the turbidity is now

$$\tau = (32\pi^3 kT/3\lambda_0{}^4\beta_T)n^2(\partial n/\partial p)_T^2 \tag{9.1.20}$$

This allows direct calculation of the turbidity from measurements of β_T and $(\partial n/\partial p)_T$.

a. Calculation of $(\partial n/\partial p)_T$ and $\rho(\partial\varepsilon/\partial\rho)_T$ from Empirical Expressions. Until recently (Waxler and Weir, 1963; Coumou *et al.*, 1964b; Reisler and Eisenberg, 1965), there have been very few reliable values of the piezo-optic coefficient, $(\partial n/\partial p)_T$, reported for pure liquids, and for this reason some effort had been expended in searching for an appropriate empirical expression from which to calculate $\rho(\partial\varepsilon/\partial\rho)_T$ or $(\partial n/\partial p)_T$. Among those expressions which have been used, we have listed in order below the Clausius–Mossotti, the Onsager (1936), the Eykman (1895), the Gladstone–Dale (1858), and the Laplace–Ramanathan (1927) equations.

$$\frac{n^2 - 1}{n^2 + 2} = \text{const} \times \rho \qquad \left[\frac{(n^2 - 1)(n^2 + 2)}{3}\right] \tag{9.1.21}$$

$$\frac{(n^2 - 1)(2n^2 + 1)}{9n^2} = \text{const} \times \rho \qquad \left[(n^2 - 1)\frac{(2n^4 + n^2)}{(2n^2 + 1)}\right] \tag{9.1.22}$$

$$\frac{n^2 - 1}{n + 0.4} = \text{const} \times \rho \qquad \left[\frac{(2n^2 + 0.8n)(n^2 - 1)}{n^2 + 0.8n + 1}\right] \tag{9.1.23}$$

$$n - 1 = \text{const} \times \rho \qquad [2n(n - 1)] \tag{9.1.24}$$

$$n^2 - 1 = \text{const} \times \rho \qquad [n^2 - 1] \tag{9.1.25}$$

The expressions in the square brackets are the corresponding values of $\rho(\partial\varepsilon/\partial\rho)_T$. When this quantity and the measured values of the isothermal compressibility and the refractive index are inserted in (9.1.14), the calculated and measured turbidities may be compared. Partington (1953) has discussed the validity of each of the above relations critically and finds that none of them gives satisfactory agreement for all liquids. Carr and Zimm (1950), Shakhparanov (1962), and Parfitt and Wood (1968) have found that the Eykman equation, even though it does not appear to have a theoretical basis, describes the available experimental data best. The latter workers have compared in Fig. 9.1 the percentage deviation between the measured value of the isotropic part of the scattering and the value calculated from each of the above equations for the six indicated liquids. All of these comparisons are based on the assumption that the available compressibility data are accurate. Meeten (1968) has proposed still another empirical equation, but the results are not significantly different from the Eykman equation.

The advantage of the Eykman equation over the others for these liquids is apparent. It is obvious that the procedure of many workers in using the Clausius–Mossotti equation directly, or even dropping the second factor, using only

$$\rho(\partial\varepsilon/\partial\rho)_T \cong (n^2 - 1) \tag{9.1.26}$$

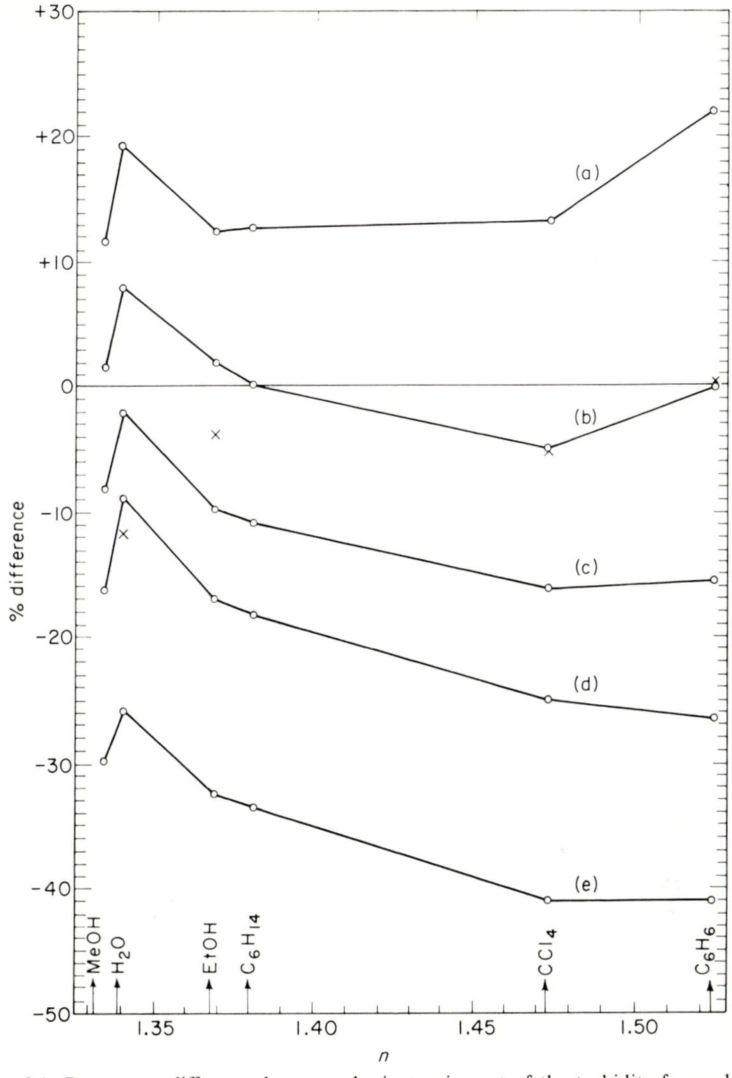

FIG. 9.1. Percentage difference between the isotropic part of the turbidity for each of the indicated liquids (plotted along the refractive index scale) and the value calculated with the aid of the (a) Clausius–Mossotti, (b) Eykman, (c) Gladstone–Dale, (d) Onsager, and (e) Laplace equations. The value obtained using Rosen's equation is designated by × (Parfitt and Wood, 1968).

cannot always be justified. Coumou *et al.* (1964b) have found from their measurements of the pressure dependence of the refractive index of a number of organic liquids that the use of the Clausius–Mossotti equation may result in scattering factors which are 10 to 20% too high. Their results at $\lambda = 0.546\ \mu$ and at 25°C are given in Table 9.1. The experimental values of $\rho(\partial\varepsilon/\partial\rho)_T$ were calculated with the aid of (9.1.19) from their measured values of $(\partial n/\partial p)_T$ and n and values of β_T from the literature. The calculated values, listed in the last column of this table, were obtained from the Clausius–Mossotti expression (9.1.21).

TABLE 9.1
EXPERIMENTAL VALUES OF $\rho(\partial\varepsilon/\partial\rho)_T$ COMPARED WITH THOSE CALCULATED FROM THE CLAUSIUS–MOSSOTTI EQUATION

	$n_{546}^{23^\circ}$	$\rho(\partial\varepsilon/\partial\rho)_T$	
		(expt)	(calc)
Benzene	1.503	1.65	1.79
Toluene	1.499	1.58	1.76
Cyclohexane	1.426	1.29	1.39
Iso-octane	1.391	1.15	1.22
n-Hexane	1.374	1.07	1.15
n-Octane	1.398	1.18	1.26
n-Decane	1.413	1.26	1.33
n-Hexadecane	1.435	1.35	1.43
Carbon tetrachloride	1.460	1.45	1.57
Carbon disulfide	1.634	2.37	2.60
Methyl ethyl ketone	1.379	1.13	1.17

Brown (1950) has proposed the use of an appropriate truncation of the formula given below, which is actually an expansion of the reciprocal of the Clausius–Mossotti formula in powers of the density. This expansion has a theoretical basis as well as considerable empirical reliability.

$$(n^2 + 2)/(n^2 - 1) = (\rho_0/\rho) - c_0' + c_1'(\rho/\rho_0) - c_2'(\rho/\rho_0)^2 + \cdots \quad (9.1.27)$$

The constants ρ_0 and c_n' can be derived from the radial distribution function which describes the structure of the liquid statistically. Earlier, Rosen (1949) had obtained an expression involving only the first two terms from an empirical analysis of both refractive index data and dielectric constant data for a large number of liquids and gases over a very extensive range of pressure. The values of $\rho(\partial\varepsilon/\partial\rho)_T$ obtained from the experimental quantities $(\partial n/\partial p)_T$, n, and β_T have been compared by Dezelić (1966) with values obtained using Rosen's equation. The results in Table 9.2 show agreement between the two sets of values to within 2%, which is considerably better than that obtained with the Clausius–Mossotti equation.

TABLE 9.2
EXPERIMENTAL VALUES OF $\rho(\partial\varepsilon/\partial\rho)_T$ COMPARED WITH THOSE CALCULATED FROM ROSEN'S EQUATION[a]

	n^a	$\rho(\partial\varepsilon/\partial\rho)_T$ (expt)	$\rho(\partial\varepsilon/\partial\rho)_T$ (calc)
Benzene	1.503	1.655	1.629
Carbon tetrachloride	1.460	1.455	1.426
Ethanol	1.360	1.026	1.014
Water	1.333	0.859	0.843

[a] Values of n are for $\lambda = 0.546\ \mu$ at 23°C.

Eisenberg (1965) has analyzed very accurate refractive index data for water over a wide range of temperature and pressure and has found that these can be described by

$$f = (n^2 - 1)/(n^2 + 2) = A\rho^{\beta} \exp(-CT) \tag{9.1.28}$$

This follows from the observation that

$$(\partial \ln f/\partial p)_T = B\beta_T \tag{9.1.29}$$

$$-(\partial \ln f/\partial T)_p = \alpha_p B + C \tag{9.1.30}$$

$$-(\partial \ln f/\partial T)_V = C \tag{9.1.31}$$

where β_T is the isothermal compressibility and α_p is the volume expansion coefficient

$$\alpha_p = (1/V)(\partial V/\partial T)_p \tag{9.1.32}$$

This leaves only one arbitrary constant A in (9.1.28). The group of empirical equations cited earlier [(9.1.21) to (9.1.25)] differ from this result by having $B = 1$ and $C = 0$ so that

$$(1/\beta_T)(\partial \ln f/\partial p)_T = -(1/\gamma)(\partial \ln f/\partial T)_p = 1 \tag{9.1.33}$$

If instead of using the factor f, which is the left-hand side of the Clausius–Mossotti equation, the Eykman or Gladstone–Dale equations are used, the corresponding expressions are nearly as accurate with somewhat different values for the constants. Reisler and Eisenberg (1965) have measured the pressure dependence of the refractive index of deuterium oxide and of methanol and found that the above equations also apply to these liquids.

b. Expansion of $\delta\varepsilon$. We now return to a fuller consideration of the alternative expansions of $\delta\varepsilon$ as a function of density and temperature (9.1.5) or as a function of density and pressure (9.1.6) (Coumou *et al.*, 1964b). In the first case

$$\overline{(\delta\varepsilon)^2} = \left(\frac{\partial\varepsilon}{\partial\rho}\right)_T^2 \overline{(\delta\rho)^2} + 2\left(\frac{\partial\varepsilon}{\partial\rho}\right)_T\left(\frac{\partial\varepsilon}{\partial T}\right)_\rho \overline{\delta\rho\,\delta T} + \left(\frac{\partial\varepsilon}{\partial T}\right)_\rho^2 \overline{(\delta T)^2} \tag{9.1.34}$$

A statistical mechanical treatment analogous to that leading to the expression for $\overline{(\delta\rho)^2}$ (9.1.13) results in (Landau and Lifschitz, 1959)

$$\overline{(\delta T)^2} = kT^2/c_v \tag{9.1.35}$$

where c_v is the heat capacity of the volume element at constant volume and

$$\overline{(\delta\rho\,\delta T)} = 0 \tag{9.1.36}$$

Further analysis leads to

$$\tau = \frac{32\pi^3 kT}{3\lambda_0{}^4\beta_T} n^2\left(\frac{\partial n}{\partial p}\right)_T^2\left[1 + \left(\frac{\beta_T}{\beta_S} - 1\right)\chi^2\right] \tag{9.1.37}$$

where β_S is the adiabatic compressibility. In the above

$$\chi = 1 + \frac{\beta_T(\partial n/\partial T)_p}{\alpha_p(\partial n/\partial p)_T} \tag{9.1.38}$$

In the case that the fluctuation in dielectric constant is expanded as a function of density and pressure

$$\overline{(\delta\varepsilon)^2} = \left(\frac{\partial\varepsilon}{\partial\rho}\right)_p^2 \overline{(\delta\rho)^2} + 2\left(\frac{\partial\varepsilon}{\partial\rho}\right)_p\left(\frac{\partial\varepsilon}{\partial p}\right)_\rho \overline{\delta\rho\,\delta p} + \left(\frac{\partial\varepsilon}{\partial p}\right)_\rho^2 \overline{(\delta p)^2} \tag{9.1.39}$$

where now

$$\overline{(\delta p)^2} = kT/\beta_S\delta V \tag{9.1.40}$$

and

$$\overline{\delta p\,\delta\rho} = kT\rho/\delta V \tag{9.1.41}$$

This leads to

$$\tau = \left(\frac{32\pi^3}{3\lambda_0{}^4}\right)\left(\frac{kT\beta_T}{\alpha_p{}^2}\right)n^2\left(\frac{\partial n}{\partial T}\right)_p^2\left[1 + 2\left(\frac{\chi}{1-\chi}\right) + \left(\frac{\beta_T}{\beta_S}\right)\left(\frac{\chi^2}{(1-\chi)^2}\right)\right] \tag{9.1.42}$$

or if the quantity in the square bracket were sufficiently close to unity

$$\tau = \left(\frac{32\pi^3}{3\lambda_0{}^4}\right)\left(\frac{kT\beta_T}{\alpha_p{}^2}\right)n^2\left(\frac{\partial n}{\partial T}\right)_p^2 \tag{9.1.43}$$

TABLE 9.3
COEFFICIENTS OF THE TEMPERATURE AND PRESSURE DEPENDENCE OF THE REFRACTIVE INDEX

	$(\partial n/\partial T)_p \times 10^5$ $(°C)^{-1}$	$(\partial n/\partial p)_T \times 10^{12}$ $(cm^2/dyne)$	$\alpha_p \times 10^3$ $(°C)^{-1}$	$\beta_T \times 10^{12}$ $(cm^2/dyne)$	χ
Benzene	63.8	52.3	1.21	95	0.046
Toluene	56.2	48.6	1.08	92	0.015
Cyclohexane	53.8	50.8	1.21	112	0.018
Iso-octane	48.7	62.9	1.19	152	0.011
n-Hexane	52.8	66.5	1.38	170	0.022
n-Octane	47.6	52.8	1.15	125	0.020
n-Decane	44.8	46.9	1.04	105	0.036
n-Hexadecane	40.6	39.1	0.90	83	0.042
Carbon tetrachloride	58.6	52.8	1.21	106	0.028
Carbon disulfide	81.6	68.2	1.19	94	0.055
Methyl ethyl ketone	51.0	44.2	1.30	108	0.041

Coumou *et al.* (1964b) have measured the refractive index dependence on temperature and pressure for a number of organic liquids. Their results are given in Table 9.3 along with values of α_p and β_T obtained from the literature and also values of χ calculated with the aid of (9.1.38). The values of χ are all in the range 0.01 to 0.05 and since the ratio β_T/β_S is approximately 1.3 for all liquids, the correction factor in (9.1.37) is of the order of 10^{-4} to 10^{-3}, which is completely negligible. Indeed, when this factor is neglected, the turbidity reduces to the expression obtained earlier (9.1.20) by dropping, in the expansion for the fluctuation of the dielectric constant (9.1.5), the second term containing the temperature coefficient. On the other hand, the second term in the bracket of (9.1.42) varies from 0.02 to 0.10 so that this cannot be neglected without introducing a significant error. Accordingly, the simpler expression (9.1.43) will only be useful when an error of the above magnitude is acceptable. Both α_p and $(\partial n/\partial T)_p$ are usually readily available from density and refractive index data obtained at various temperatures at atmospheric pressure. When the compressibility is not known, it may be useful to introduce the approximation

$$\beta_T \cong \frac{-\alpha_p(\partial n/\partial p)_T}{(\partial n/\partial T)_p} \tag{9.1.44}$$

into either equation (9.1.43) or (9.1.20), resulting in

$$\tau \cong \left(\frac{-32\pi^3}{3\lambda_0{}^4}\right)\left(\frac{kTn^2}{\alpha_p}\right)\left(\frac{\partial n}{\partial T}\right)_p\left(\frac{\partial n}{\partial p}\right)_T \tag{9.1.45}$$

c. Correction for Depolarization. It has been assumed above that the density fluctuations which give rise to the scattering can be treated by Rayleigh's theory for small isotropic spheres. This theory predicts that the polarization ratio ($\rho_u = i_2/i_1$) for unpolarized incident light will vary as $\cos^2 \theta$. Actually, the light scattered by liquids is generally depolarized because of the molecular anisotropy. There will be contributions to the parallel component of the scattered light from the perpendicular component of the incident light and a similar cross component for the perpendicular component of the scattered light. The polarization ratio no longer follows the $\cos^2 \theta$ law. For such media, the polarizability can be represented by a symmetric tensor and the scattering by such a configuration can be calculated precisely. The net effect is an enhancement of the scattering. This enhancement, called the Cabannes factor, can be directly related to the polarization ratio of the scattered light. This will be discussed in detail in Chapter 10. It will suffice at present simply to state that expressions for the turbidity must be multiplied by the Cabannes factor

$$C_\tau = \frac{6 + 3\rho_u(90)}{6 - 7\rho_u(90)} \tag{9.1.46}$$

where $\rho_u(90)$ is the ratio of the parallel to the vertical component of the scattered light at 90° when the incident light is unpolarized. The appropriate factor by which the theoretical value of the Rayleigh ratio at 90° must be multiplied is

$$C_u(90) = \frac{6 + 6\rho_u(90)}{6 - 7\rho_u(90)} \tag{9.1.47}$$

In the discussion which follows, we will frequently refer to the Rayleigh ratio at 90° rather than the turbidity, especially when data being cited from a particular reference in the literature are in this form. The relation between the Rayleigh ratio and the turbidity is given by (3.2.31). This permits any of the above expressions to be transformed from turbidity into the corresponding equation for the Rayleigh ratio.

9.1.2 Experimental Results

The first step in the experimental verification of the above theory of scattering by pure liquids is the acquisition of reliable turbidimetric data. This has by no means been a trivial matter and there has been a plethora of conflicting data since the early measurements by Martin (1913, 1926) and others. The available results for benzene, toluene, carbon tetrachloride, and carbon disulfide have been reviewed critically by Kratohvil *et al.* (1962). Earlier, the results had been bunched into two widely divergent groups of values, but the evidence in favor of the higher of the two groups has been overwhelming, so that only these are now considered to be valid.

The work on benzene has been particularly extensive because it has frequently been used as a secondary turbidimetric standard. There is a considerable range of values reported for both the Rayleigh ratio and the polarization ratio. Kratohvil *et al.* list 37 published values of the Rayleigh ratio of benzene at 0.436 μ ranging from 44 to 50 $\times 10^{-6}$ cm^{-1}. They favor $R_{90} = 46.5 \times 10^{-6}$ cm^{-1} at $\lambda = 0.436\,\mu$ and $R_{90} = 15.5 - 16.0 \times 10^{-6}$ cm^{-1} at $\lambda = 0.546\,\mu$. These agree with a number of more recent determinations (Shakhparonov, 1962; Huisman, 1964; Claesson and Ohman, 1964; Coumou *et al.*, 1964b; Smart, 1965; Dezelić and Vavra, 1966; Bello and Guzman, 1966; Parfitt and Wood, 1968).

Coumou *et al.* (1964b) have tested the theory by comparing the Cabannes factor, obtained from polarization measurements at 90° (9.1.47), with the value obtained from the ratio of the measured Rayleigh ratio to the calculated value of the isotropic Rayleigh ratio. The results, which are shown in Table 9.4, are for $\lambda_0 = 0.546\,\mu$ and room temperature (22 to 24°C). The second column of data consists of the measured values of the Rayleigh ratio at 90°, while the third column is calculated from experimental quantities using (9.1.20) and (3.2.31). The fourth column contains the Cabannes factor, which is calculated from the measured values of ρ_u listed in the first column using (9.1.47). This is to be compared with the last column which lists values of the Cabannes factor obtained from the ratio $R_{90}/R_{90}^{\text{is}}$. The two sets of values agree to within 6% which is within the experimental accuracy. More recently Parfitt and Wood (1968) have carried out a similar study with

TABLE 9.4
LIGHT SCATTERING AND CABANNES FACTOR

	$\rho_u(90)$	$R_{90} \times 10^6$ (cm^{-1})	$R_{90}^{\text{is}} \times 10^6$ (cm^{-1})	C (Eq.(9.1.47))	C $R_{90}/R_{90}^{\text{is}}$
Benzene	0.42	15.8	5.90	2.68	2.78
Toluene	0.48	18.4	5.23	3.52	3.36
Cyclohexane	0.049	4.56	4.25	1.07	1.11
Iso-octane	0.047	5.15	4.57	1.12	1.11
n-Hexane	0.073	5.32	4.46	1.19	1.17
n-Octane	0.12	4.85	3.96	1.22	1.30
n-Decane	0.15	4.95	3.79	1.30	1.39
n-Hexadecane	0.26	5.75	3.44	1.67	1.81
Carbon tetrachloride	0.042	5.38	5.09	1.05	1.09
Carbon disulfide	0.65	83.9	12.0	6.99	6.83
Methyl ethyl ketone	0.16	4.18	3.12	1.34	1.42

comparable results. Also, Shakhparonov (1963) has critically reviewed the correlation between various other turbidimetric data and the Einstein–Smoluchowski theory. An interesting study along similar lines was carried out by Lundberg *et al.* (1964) using a laser at $\lambda = 6937$ Å, a part of the spectrum not commonly employed for light-scattering studies.

Schmidt (1968) has chosen to test the theory by comparing the isothermal compressibility, calculated from (9.1.20) as corrected by the appropriate Cabannes factor, with values reported by direct measurements. The piezo-optic coefficient $(\partial n/\partial p)_T$ was obtained from the Clausius–Mossotti equation. In using this equation, an empirical factor was used to correct for the deviations found by Coumou *et al.* (1964b). The results shown in Table 9.5 are in excellent agreement except for aniline and *n*-hexane.

TABLE 9.5
ISOTHERMAL COMPRESSIBILITIES OF PURE LIQUIDS

Compound	Temp. (°K)	$\beta_T \times 10^{12}$ (scattering)	$\beta_T \times 10^{12}$ (literature)	Ref.
Benzene	298.15	96.7	96.9	*a*
	308.15	102	104.2	*a*
Carbon tetrachloride	298.15	107.2	107.7	*a, b*
2-Propanol	298.15	118.5	117 (112)	*b, c*
Dichloromethane	298.15	95.0	97.4	*c*
Dibromomethane	295.15	62.0	64.7	*c*
	300.15	64.0	68.0	*d*
Diiodomethane	293.15	47.5	42.0	*d*
Bromoform	288.15	50.0	50.0	*d*
Chloroform	303.15	108.0	108.6	*c*
Aniline	298.15	61.9	46.7	*c*
n-Hexane	313.06	229.3	183.1	*c*

[a] Holder, G. A., and Whalley, E., *Trans. Faraday Soc.* **58**, 2095 (1962).

[b] Katti, P. K., and Shil, S. K., *J. Chem. Eng. Data* **11**, 601 (1966).

[c] "Handbook of Chemistry and Physics" (C. D. Hogdman, ed.). Chemical Rubber Co., Cleveland, Ohio (1963).

[d] Jacobson, B., *Acta Chem. Scand.* **6**, 1485 (1952).

Another test of the validity of the fluctuation theory of light scattering by pure liquids has been proposed by Dezelić (1966), where temperature dependence of the theoretical and experimental Rayleigh ratios of benzene are compared. The results are given in Table 9.6. The values of $\rho(\partial\varepsilon/\partial\rho)_T$ were calculated from Rosen's equation and were used in (9.1.14) along with the compressibility data of Holder and Whalley (1962) to calculate the isotropic

Rayleigh ratio. The experimental values of the isotropic Rayleigh ratio at $\lambda_0 = 0.546\ \mu$ were obtained by Ehl *et al.* (1964).

TABLE 9.6
ISOTROPIC SCATTERING OF BENZENE AT VARIOUS TEMPERATURES

°C^t	$\rho(\partial\varepsilon/\partial\rho)_T$	$\beta_T \times 10^{12}$ (cm^2 dyne^{-1})	$R_{is}(t)/R_{is}(25°C)$ (expt)	(calc)
20	1.637	92.5	0.957	0.952
25	1.624	96.5	1.000	1.000
30	1.608	101.0	1.043	1.043
35	1.592	105.0	1.080	1.092
40	1.578	109.5	1.125	1.133
45	1.561	114.5	1.170	1.176
50	1.548	119.0	1.215	1.225
55	1.531	124.5	1.262	1.263
60	1.518	129.5	1.310	1.303

Each of the tests described above has utilized polarization data in order to account for the Cabannes factor; yet it would be interesting to calculate the turbidity independent of any light-scattering measurement. One of the possibilities is to calculate the polarization ratio from the Kerr constant using a relation derived by Gans (1923)

$$K = \frac{3\beta_T}{8\pi n^2}\rho\left(\frac{\partial\varepsilon}{\partial\rho}\right)_T \rho\left(\frac{\partial\varepsilon_s}{\partial\rho}\right)_T \frac{\rho_u}{6 - 7\rho_u} \tag{9.1.48}$$

In this equation, which is written in a form given by Stuart and Buckheim (1938), the Kerr constant is

$$K = [(n_{\parallel} - n_{\perp})/n]E^{-2} \tag{9.1.49}$$

$n_{\parallel}$ and $n_{\perp}$ denoting the refractive index when the electric vector of the light is parallel and perpendicular, respectively, to the impressed electric field E. Also, ε_s represents the static dielectric constant.

Dezelić (1966) has compared the experimental values of ρ_u for benzene and carbon disulfide with values calculated from the Kerr constant. Each of these ρ_u values was used to calculate the Rayleigh ratio, which was then compared with the experimental values. The results shown in Table 9.7 are within the experimental errors. The theory appears to be successful, although additional data carried out with greater precision and for additional liquids would be useful in providing a still more definitive test.

TABLE 9.7

COMPARISON OF POLARIZATION RATIO AND RAYLEIGH RATIO CALCULATED FROM KERR CONSTANT WITH EXPERIMENTAL VALUES

λ_0 (mμ)	ρ_u (calc)	ρ_u (expt)		$R_{90} \times 10^6$ (cm^{-1}) (calc)		$R_{90} \times 10^6$ (cm^{-1}) (expt)		
		(*a*)	(*b*)	(*d*)	(*e*)	(*f*)	(*b*)	(*g*)
Benzene								
578	0.40	0.40		11.9	12.5	12.1		12.1
546	0.40	0.40	0.41	15.5	16.4	15.4	15.8	16.0
436	0.42	0.41	0.42	43.9	46.8	44.0	45.6	46.4
405	0.42	0.41		64.2	67.7	62.9		67.3
365	0.43	0.42	0.44	110	116	103	100	106
Carbon disulfide								
		(*c*)		(*d*)	(*e*)	(*c*)		
546	0.66	0.64		89	81	86		
436	0.68	0.66		289	261	265		

[a] Massoulier (1963).
[b] Coumou (1960).
[c] Dezelić and Vavra (1966).
[d] Cabannes factor from Kerr constant.
[e] Cabannes factor from exptl ρ_u
[f] Cantow (1956).
[g] Claesson and Ohman (1964)

The turbidity of pure water is considerably lower than that of organic liquids, and for that reason it has been even more difficult to obtain reliable data for water. However, recent efforts appear to have established the value for water and to have correlated it with calculations from compressibility data and the pressure dependence of the refractive index. Kratohvil *et al.* (1965) have concluded, both on the basis of their own experiments and of a critical review of the literature, that the Rayleigh ratios for water at $\lambda = 0.436\ \mu$ and $0.546\ \mu$ are 2.6×10^{-6} cm^{-1} and 1.00×10^{-6} cm^{-1}, respectively. They obtained calculated values of 2.59×10^{-6} and 0.987×10^{-6}. Actually, the experimental results are considerably less accurate than these figures would indicate, and it should be emphasized particularly that the calculated results are quite dependent upon which values of the available polarization data one chooses to utilize.

A similar comparison of the experimental and calculated light scattering by both water and deuterium oxide has been carried out by Cohen and Eisenberg (1965). They determined the scattering relative to benzene rather than the absolute turbidity. The ratio of the flux of scattered light was calculated from

$$\frac{L}{L_B} = \frac{\beta_{T,B}}{\beta_T}\left[\frac{(\partial n/\partial p)_T}{(\partial n/\partial p)_{T,B}}\right]\frac{C(\theta)}{C(\theta)_B} \tag{9.1.50}$$

If the Rayleigh ratio for benzene is known, the value for H_2O or D_2O is given by

$$R = R_B(L/L_B)(n/n_B)^2 \tag{9.1.51}$$

where the refractive index ratio corrects for the refraction of the scattered rays upon emergence from the cell. The results are given in Table 9.8. Almost identical values for the water to benzene ratio have also been obtained by Morel (1966).

TABLE 9.8
RAYLEIGH RATIO OF H_2O AND D_2O

	$\lambda_0(\mu)$	L/L_B (calc)	L/L_B (expt)	$R_{90} \times 10^6$ (cm^{-1})
H_2O	0.546	0.0708	0.0692	0.865
H_2O	0.436	0.0676	0.0648	2.32
D_2O	0.546	0.0684	0.0680	0.843
D_2O	0.436	0.0657	0.0650	2.30

The calculated and experimental values of L/L_B are in close agreement. The Rayleigh ratio of water given in the last column of the table was obtained by adopting Coumou's (1960) value for benzene. Morel (1966) has also measured the ratio of the Rayleigh ratio of water to that of benzene at five wavelengths obtaining 0.054 (0.578 μ), 0.054 (0.546 μ), 0.048 (0.436 μ), 0.046 (0.405 μ), and 0.44 (0.366 μ). These results are almost identical to those in Table 9.8 after taking the $(n/n_B)^2$ factor into account. They are about 13% lower than the ratio of the values for water reported by Kratohvil *et al.* (1965). However, differences in the calculated values can easily amount to this because of the possible use of different reported values of β_T, $(\partial n/\partial p)_T$, and ρ_u. At least to within this accuracy, the turbidity of water appears to obey the Einstein–Smoluchowski theory.

Mysels (1964) had proposed, upon the basis of earlier experimental evidence and calculations, that the turbidity of water was significantly higher than the calculated values and that this could provide evidence for determining the "iceberg" structure of water. The term "iceberg" is used to designate an aggregate of hydrogen bonded water molecules. Litan (1968) has pointed out that any shortcomings in the assumptions upon which the fluctuation theory of light scattering is based will result in a lowering of the calculated results so that the experimental scattering may exceed but may never be lower than the fluctuation theory values. For example, although a pure liquid consists of only a single thermodynamic component, it might be comprised of more than one molecular species. The fluctuations in density of each of these might not be the same for each microscopic state having the same density. The net effect would be an increase in scattering. Utilizing a model in which water is comprised of "icy" and "liquid" molecules, Litan has estimated that the augmentation of the calculated turbidity due to this effect would be expected to be less than 1.5% so that it cannot be definitively measured by present techniques.

Cohen and Eisenberg (1965) have also measured the temperature dependence of the turbidity of benzene, H_2O, and D_2O in the range 5 to 65°C. The calculated values made use of the empirically established relation

$$(\partial n/\partial p)_T/\beta_T(n-1) = B' \tag{9.1.52}$$

where B' is a constant, independent of temperature, and only very slightly dependent upon wavelength. This leads to the following expression for the Rayleigh ratio at temperature t relative to that at 25°C:

$$\frac{R_T}{R_{25}} = \left[\frac{n_t(n_t-1)}{n_{25}(n_{25}-1)}\right]^2 \frac{\beta_{T,t}C(\theta)_t(273.2+t)}{\beta_{T,25}C(\theta)_{25}(298.2)} \tag{9.1.53}$$

where n_t, $\beta_{T,t}$, and $C(\theta)_t$ are the refractive index, isothermal compressibility, and Cabannes factor at temperature t. There was good agreement between the experimental ratios and the values calculated with this expression.

9.2 Binary Solutions

9.2.1 CONCENTRATION FLUCTUATIONS

The expression for the turbidity of a homogeneous liquid consisting of more than one component continues to be (9.1.4) provided that the volume elements are still chosen sufficiently small to be treated as Rayleigh scatterers and yet large enough so that their statistical thermodynamic behavior can be described by the usual macroscopic properties. Also, the fluctuations must be random. Einstein (1910) treated the case of a binary solution briefly and this was considered in greater detail by Raman (Raman and Ramanathan, 1923; Raman and Rao, 1923).

The fluctuation in dielectric constant may be written

$$\delta\varepsilon = (\partial\varepsilon/\partial\rho)_{T,m}\delta\rho + (\partial\varepsilon/\partial T)_{\rho,m}\delta T + (\partial\varepsilon/\partial m)_{T,\rho}\delta m \tag{9.2.1}$$

where ρ is the density of the solution and the molality m is the number of moles of solute per kilogram of solvent. Just as in the case of a pure liquid, the second term involving the temperature coefficient of the dielectric constant is assumed to be small and is neglected. Furthermore, it is customary to assume that the density and concentration fluctuations are not correlated so that upon taking the square of $\delta\varepsilon$ and averaging, the average of the cross terms reduces to zero, i.e.,

$$(\partial\varepsilon/\partial\rho)_{T,m}(\partial\varepsilon/\partial m)_{T,\rho}\overline{\delta\rho\ \delta m} = 0 \tag{9.2.2}$$

resulting in

$$\overline{(\delta\varepsilon)^2} = (\partial\varepsilon/\partial\rho)^2_{T,m}\overline{(\delta\rho)^2} + (\partial\varepsilon/\partial m)^2_{T,\rho}\overline{(\delta m)^2} \tag{9.2.3}$$

This leads to two terms in the expression for the turbidity, one of them arising from density fluctuations, τ_d, and the other arising from concentration fluctuations, τ_E.

$$\tau = \tau_d + \tau_E \tag{9.2.4}$$

The latter is sometimes called the excess turbidity.

The density fluctuations for the solution can be calculated as before, resulting in (9.1.14) where now ρ, ε, and β_T are the appropriate values for the solution. In practice, for dilute solutions, these differ only slightly from the value for the pure solvent so that its turbidity is frequently utilized for τ_d in the absence of sufficiently accurate optical and compressibility data for the solution.

The excess turbidity is

$$\tau_E = \frac{8\pi^3}{3\lambda_0^{\ 4}}\,\delta V\left(\frac{\partial\varepsilon}{\partial m}\right)^2_{T,\rho}\overline{(\delta m)^2} \tag{9.2.5}$$

The evaluation of $\overline{(\delta m)^2}$ follows a procedure analogous to that for the mean square density fluctuation. The statistical device utilized is the grand canonical ensemble in which the system is divided into small volume elements which are free to exchange molecules and energy with each other. It is then possible to calculate the distribution of molecules among the volume elements and to calculate the appropriate average; in this case, the mean square fluctuation of solute molecules expressed as $\overline{(\delta m)^2}$. Davidson (1962) utilizes a rather long thermodynamic argument in order to derive the following expression in terms of readily obtained experimental quantities:

$$\tau_E = \frac{8\pi^3}{3\lambda_0{}^4 N_A}\left(\frac{\partial \varepsilon}{\partial m}\right)^2_{T,p} \frac{RT}{10^{-3}\rho_0(\partial\mu_2/\partial m)_{T,p}} \tag{9.2.6}$$

or

$$\tau_E = \frac{H'RT}{10^{-3}\rho_0(\partial\mu_2/\partial m)_{T,p}} \tag{9.2.7}$$

where ρ_0 is the weight of solvent per unit volume of solution,

$$H' = \frac{32\pi^3 n^2}{3\lambda_0{}^4 N_A}\left(\frac{\partial n}{\partial m}\right)^2_{T,p} \tag{9.2.8}$$

and μ_2 is the Gibbs chemical potential

$$\mu_2 = (\partial A/\partial n_2)_{T,V} = (\partial G/\partial n_2)_{T,p} \tag{9.2.9}$$

with n_2 the number of moles of solute and A and G the Helmholtz and Gibbs free energies, respectively. The factor $[10^{-3}\rho_0]^{-1}$ is the volume of solution corresponding to m moles of solute or to one kilogram of solvent.

By utilizing the Gibbs–Duhem equation

$$n_1\, d\mu_1 + n_2\, d\mu_2 = 0 \tag{9.2.10}$$

the excess turbidity can be related to the chemical potential of the solvent as follows:

$$\tau_E = \frac{H'M_1RTm}{\rho_0(-\partial\mu_1/\partial m)_{T,p}} \tag{9.2.11}$$

It will be useful to express the concentration as c, the weight of solute per unit volume of solution, which is related to the molality by

$$c = \frac{\rho_0 M_2}{1000}m \qquad \text{or} \qquad \frac{\rho M_2}{1000 + mM_2}m \tag{9.2.12}$$

M_1 and M_2 are the molecular weights of the two components. Then,

$$\tau_E = \frac{HRT(\partial c/\partial m)_{T,p}}{10^{-3}\rho_0(\partial\mu_2/\partial c)_{T,p}} \tag{9.2.13}$$

and

$$\tau_E = \frac{HM_1RTc(\partial c/\partial m)_{T,p}}{\rho_0(-\partial\mu_1/\partial c)_{T,p}}\,\frac{1000}{\rho_0 M_2} \tag{9.2.14}$$

Now

$$H = \frac{32\pi^3 n^2}{3\lambda_0{}^4 N_A}\left(\frac{\partial n}{\partial c}\right)^2_{T,p} \tag{9.2.15}$$

For dilute solutions where $\rho_0 \cong \rho$ so that

$$c = (\rho M_2/1000)m \tag{9.2.16}$$

$$\tau_E = \frac{HM_2RT}{(\partial\mu_2/\partial c)_{T,p}} \tag{9.2.17}$$

and

$$\tau_E = \frac{HcM_1RT}{\rho_0(-\partial\mu_1/\partial c)_{T,p}} = \frac{Hc\overline{V_1}RT}{(-\partial\mu_1/\partial c)_{T,p}} \tag{9.2.18}$$

where $\overline{V_1}$ is the partial molal volume of the solvent. It should be emphasized that (9.2.17) and (9.2.18) are only valid at low concentration and that significant errors can arise in concentrated solutions unless the correct forms involving $(\partial c/\partial m)_{T,p}$ are used. It may be useful to set down the various forms of the lumped optical constant that will appear in the remainder of this chapter.

$$H = \frac{32\pi^3 n^2}{3\lambda_0{}^4 N_A}\left(\frac{\partial n}{\partial c}\right)^2_{T,p} = H''\left(\frac{\partial n}{\partial c}\right)^2_{T,p} = \frac{H'}{(\partial c/\partial m)^2_{T,p}} \tag{9.2.19}$$

a. The Debye Equation. Debye (1944, 1947) has shown how (9.2.18) can be developed into an expression which can be used to determine the molecular weight of the solute. This has been applied very widely, particularly to solutions of high polymers.

The chemical potential of the solvent is related simply to the osmotic pressure π, which, in turn, can be expanded as a virial equation in concentration.[1]

$$\mu_1 - \mu_1{}^0 = -\pi\overline{V_1} = -RT\overline{V_1}c(1/M + Bc + Cc^2 + \cdots) \tag{9.2.20}$$

where $\mu_1{}^0$ is the chemical potential of the pure solvent and $\overline{V_1}$ is the partial molal volume of the solvent component in the solution. For dilute solutions this is practically the same as the molar volume. Then

$$\tau_E = \frac{HcRT}{(\partial\pi/\partial c)_{T,p}} \tag{9.2.21}$$

[1] Whenever M is used without a subscript, it designates the molecular weight of the solute component.

If the above series is truncated after the second term, the van't Hoff equation is obtained, which leads to

$$\tau_E = \frac{Hc}{(1/M) + 2Bc} \tag{9.2.22}$$

or the well-known Debye equation

$$Hc/\tau_E = (1/M) + 2Bc \tag{9.2.23}$$

All of the quantities on the left-hand side can be determined experimentally. The usual technique for evaluating the molecular weight is to plot Hc/τ_E versus c. At low concentrations where the van't Hoff equation is valid, a linear plot should be obtained with an intercept equal to $1/M$ and a slope equal to twice the second virial coefficient B. Actually, the relationship between the limiting slope in the turbidity plot and the second virial coefficient obtained from osmotic pressure is somewhat more subtle. This is discussed in detail by Hill (1959) and applied to aqueous solutions of sucrose by Stigter (1960a, b, 1963). For an ideal solution, $B = 0$, and the turbidity is directly proportional to the molecular weight, viz.

$$\tau_E = HcM \tag{9.2.24}$$

Just as for scattering by pure liquids, there will be an additional contribution to the turbidity if the volume elements comprising the concentration fluctuations are not isotropic. Accordingly, each of the above expressions for the turbidity must be multiplied by the Cabannes factor (9.1.46). Otherwise, the calculated molecular weights will be too high by this factor.

In the event that the molecules or particles are comparable in size to the wavelength of the light, the double extrapolation described earlier (Fig. 8.11) of Hc/τ_E against both concentration and scattering angle can be used. This is known as the Zimm plot. Both curves should extrapolate to the value $1/M$. The second virial coefficient is obtained from the slope of the curve plotted against c and the radius of gyration is obtained from the slope of the curve plotted against $\sin^2(\theta/2)$.

Yang (1957) has proposed some alternative techniques for plotting the data which may improve the accuracy in some cases. For very large molecules such as DNA, Eigner and Doty (1965) have emphasized the importance of obtaining measurements at sufficiently low angles in order to provide an accurate extrapolation of the Zimm plot. Nakagaki (1966) has elucidated quantitatively the effect of deviations from the conditions of the Rayleigh–Debye theory when the particles are spherical. He has given a correction factor which must be applied to the intercept at $\theta = 0$ in order to obtain the correct molecular weight. In order to obtain the radius of gyration from the limiting slope of this angular plot, the extrapolation must be carried out to sufficiently small angles. Nakagaki has delineated this condition quantitatively using the exact theory for scattering by spheres.

The application of the Debye equation to the determination of the molecular weight and of the second virial coefficient has resulted in light scattering becoming one of the most important methods of studying macromolecules. Actually, even small molecules can be studied, light scattering determinations having ranged from molecular weights of less than 10^2 to as high as 10^8. In the latter case, the appropriate form factor derived from Rayleigh–Debye theory must be utilized or else the Rayleigh ratio must be extrapolated to $\theta = 0$ such as is done in the Zimm plot.

The validity of the technique has been confirmed not only by the successful determination of the molecular weight of well-defined molecules such as CS_2 (Sicotte and Rinfret, 1962), sucrose (Maron and Lou, 1955), and $H_4SiW_{12}O_{40}$ (Kerker *et al.*, 1963c) but also by the agreement between the molecular weight obtained from light scattering compared to values from other techniques such as osmotic pressure, ultracentrifugation, and viscosity. A vast body of literature has accumulated in this field, and it is not our intention here to review this nor to consider the many practical experimental difficulties. Stacey (1956) has discussed some of the practical aspects of the determination of molecular weight by light scattering, and Fishman (1957, 1958) and Kratohvil (1964, 1966) have assembled very comprehensive annotated bibliographies. There were more than 400 publications dealing with molecular weight determination by light scattering during the two year period 1964–1965.

Even though the technique has become routine, it is not without its difficulties. Some of these relate to the preparation and stability of the solutions themselves, others to the variety of instrumental errors that may be encountered, including the calibration of the instruments. For example, the Commission on Macromolecules of the International Union for Pure and Applied Chemistry distributed three samples of polystyrene of narrow molecular-weight distribution to numerous laboratories throughout the world in order to compare the results of viscosity, osmotic pressure, and light scattering measurements in a variety of solvents. The result was a very considerable variation in the reported values of the molecular weight, second virial coefficient, and angular dissymmetry (Atlas and Mark, 1961). A subsequent round-robin program has also led to disappointing results (Kratohvil, 1968).

b. Interdependence of Concentration and Density Fluctuations. As long as the density fluctuations refer to the density of the solution rather than to the partial density of the solvent component, the density and concentration fluctuations are not independent and (9.2.4) is not strictly valid. Bullough

(1960, 1963a, b) has calculated the contribution of the cross terms to the turbidity using a molecular model, and this has been verified by Sicotte (1964), using a classical statistical thermodynamic derivation. Accordingly,

$$\tau = \tau_d + \tau_E + \tau^* \tag{9.2.25}$$

where

$$\tau^* = (64\pi^3/3\lambda_0{}^4)n^2kT(\partial n/\partial p)_{T,c}c(\partial n/\partial c)_{T,p} \tag{9.2.26}$$

This term only makes a significant contribution when both components of the solution have a low molecular weight. Since light scattering is more commonly used to study high molecular weight species, it can usually be neglected. If this is done, the quantities determined with the aid of (9.2.23) are an apparent molecular weight

$$M^* = M + 2RT\frac{(\partial n/\partial p)_{T,c}}{(\partial n/\partial c)_{T,p}} + \cdots \tag{9.2.27}$$

and an apparent second virial coefficient

$$B^* = (M/M^*)^2B + \cdots \tag{9.2.28}$$

For most solutions, the molecular weight correction is unlikely to exceed 100 molecular weight units, a value which can be neglected for macromolecules.

Sicotte and Rinfret (1962) studied the scattering by two binary solutions of known low molecular weight in order to check these results. They compared the apparent molecular weights obtained with the aid of (9.2.23) with those values from

$$\frac{Hc}{\tau - (\tau_d + \tau^*)} = \frac{1}{M} + 2Bc \tag{9.2.29}$$

where τ^* is calculated from (9.2.26). For carbon disulfide dissolved in carbon tetrachloride, the apparent molecular weight was 125, whereas a value of 77 was obtained with the above equation. The correct molecular weight is 76. For diethyl ether in benzene, the corresponding results were 69 and 73, respectively, compared to a correct value of 74.

An interesting corollary of this work is a method for determining the turbidity of a pure liquid from measurements on a solution containing a known low molecular-weight solute. The above equation can be written

$$\frac{HcM}{(\tau/\tau_0) - [(\tau_d + \tau^*)/\tau_0]} = \tau_0 + 2\tau_0 MBc \tag{9.2.30}$$

where τ_0 is the turbidity of the solvent. Of course, at sufficiently low concentrations

$$\tau_d = \tau_0 \tag{9.2.31}$$

The left-hand side now involves only relative turbidities which can be determined only from the meter readings for the solution and solvent. The intercept of a plot of the left-hand side versus c gives the turbidity of the pure solvent. The method has been checked for a number of liquids including benzene. The results are internally quite consistent and, although slightly higher for benzene than some of the more recently reported values, are well within the accepted range of values.

9.2.2 Comparison of Turbidity and Activity

The relations between the excess turbidity of a binary solution and the partial derivative of the chemical potential of one component or the other (9.2.7, 11, 13, 14, 17, 18) offer the possibility of studying the thermodynamics of solutions by light scattering. The thermodynamic behavior of each component may be described by the activity which is related to the chemical potential by

$$\mu_i - \mu_i^0 = RT \ln a_i \tag{9.2.32}$$

where μ_i^0 refers to a standard state. For the solvent this is normally chosen to be the pure liquid. Alternatively, it may be convenient to express the chemical potential of the solvent in terms of the rational osmotic coefficient g_{os}

$$\mu_1 - \mu_1^0 = g_{os} RT \ln x_1 \tag{9.2.33}$$

or the practical osmotic coefficient ϕ_{os}

$$\mu_1 - \mu_1^0 = -\phi_{os} RT(M_1/1000)\sum m_i \tag{9.2.34}$$

where x_1 is the mole fraction of the solvent, M_1 is its molecular weight, and m_i represents the molality of each molecular species dissolved or each ionic species in the case of electrolytes.

This section will describe a number of attempts to correlate the chemical potential or activity obtained from the turbidity with values derived by various other experimental techniques. Steinberg and Katchalski (1963) have summarized a number of relations correlating turbidity with diffusion, sedimentation, and osmotic pressure.

a. Miscible Low Molecular Weight Liquids. Coumou and Mackor (1964) have calculated activities from measurements of the light scattering by binary mixtures of benzene with cyclohexane, neopentane, and methanol over the complete range of concentrations and have compared these with the results of vapor pressure measurements. The isotropic part of the turbidity τ_{is} was obtained by using the Cabannes factor (9.1.46) to correct the measured turbidity and the excess turbidity τ_E was obtained by subtracting from this the turbidity due to density fluctuations τ_d using equation (9.1.20). Those quantities needed for this calculation which were not available from the literature were measured directly. The effect of the cross term (9.2.26) accounting for correlations between density and concentration fluctuations was neglected.

For completely miscible components covering the entire composition range, it is convenient to express the concentration as mole fraction. Then

$$\tau_E = \frac{32\pi^3 n^2(\partial n/\partial x_2)^2_{T,p} VRTx_1}{3\lambda_0^4 N_A(\partial \mu_2/\partial x_2)_{T,p}} \tag{9.2.35}$$

where V is the molar volume of the solution. If the solution is ideal

$$\mu_2 - \mu_2^0 = RT \ln x_2 \tag{9.2.36}$$

and the ideal excess turbidity is

$$\tau_{id} = 32\pi^3 n^2(\partial n/\partial x_2)^2_{T,p} Vx_1x_2/3\lambda_0^4 N_A \tag{9.2.37}$$

For real solutions

$$\mu_2 - \mu_2^0 = RT \ln a_2 = RT \ln f_2x_2 = RT \ln x_2 + \mu_2^E \tag{9.2.38}$$

where a_2, f_2, and μ_2^E are the activity, the activity coefficient, and the excess free energy, respectively. Then

$$1 - (\tau_{id}/\tau_E) = (-\partial \ln f_1/\partial \ln x_1)_{T,p} = (-\partial \ln f_2/\partial \ln x_2)_{T,p} \tag{9.2.39}$$

or

$$1 - (\tau_{id}/\tau_E) = -(x_2/RT)(\partial \mu_2^E/\partial x_2)_{T,p} = (x_1x_2/RT)(\partial^2 G^E/\partial x_1 \partial x_2)_{T,p} \tag{9.2.40}$$

where G^E, the molar excess Gibbs free energy of the solution, is

$$G^E = x_1\mu_1^E + x_2\mu_2^E = RT(x_1 \ln f_1 + x_2 \ln f_2) \tag{9.2.41}$$

The activity coefficients can now be obtained by an appropriate graphical integration of $[1 - (\tau_{id}/\tau_E)]$ plotted as a function of $\log x_1$ and $\log x_2$, remembering that the value for each pure liquid is unity. The light scattering results for methanol-benzene mixtures are shown in Fig. 9.2 as the corresponding Rayleigh ratios, R_{id}, R_d, R_{is}, R_c. The large difference between the excess Rayleigh ratio R_c and R_{id} shows that this solution deviates considerably from ideality. The values of $\log f_1$, $\log f_2$, and G^E calculated from these light scattering results are shown in Fig. 9.3 as solid curves. The

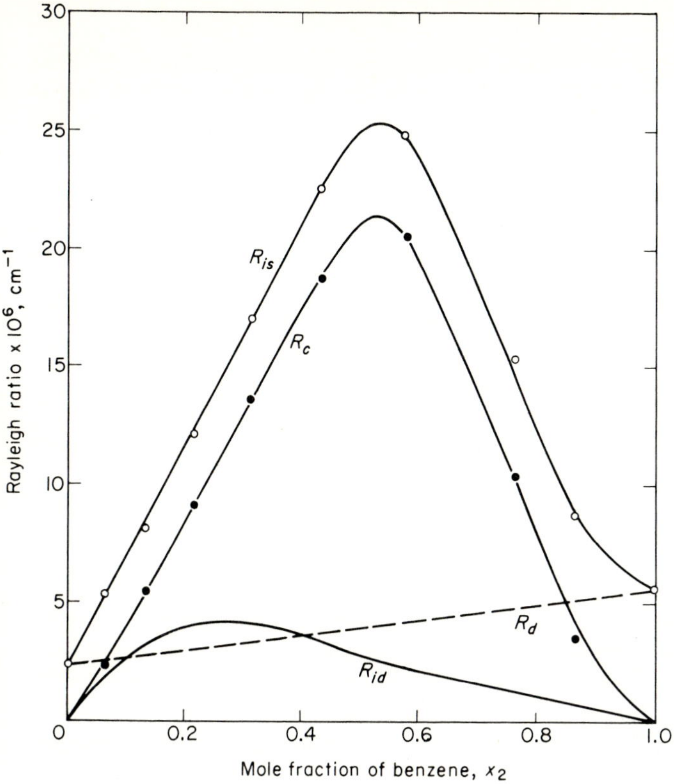

FIG. 9.2. Isotropic light scattering in methanol-benzene mixtures (Coumou and Mackor, 1964).

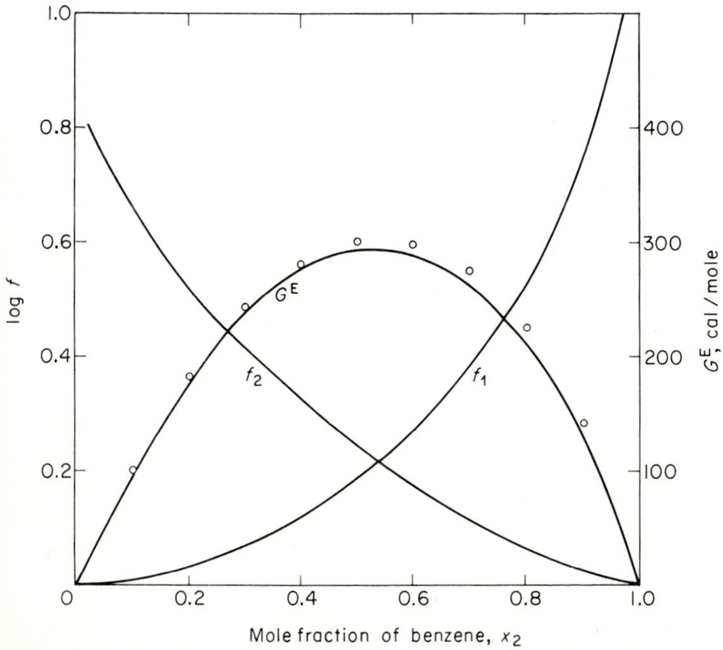

FIG. 9.3. Activity coefficients (log f) and molar excess Gibbs free energy G^E in methanol-benzene mixtures obtained by light scattering (curves) compared with values obtained from vapor pressure measurements (circles) (Coumou and Mackor, 1964).

points are values of G^E obtained from vapor pressure measurements (Scatchard and Ticknor, 1952). The agreement is excellent. Similar results were obtained for other solutions, indicating that for systems such as these, light scattering can be used to obtain accurate activity data.

Malmberg and Lippincott (1968) have measured the Rayleigh ratio and depolarization of solutions of acetone, acetonitrile, benzene, chloroform, nitrobenzene, and nitromethane in carbon tetrachloride, using a continuous laser with $\lambda_0 = 6328$ Å. They calculated the contribution of both density and concentration fluctuations to the isotropic part of the Rayleigh ratio. The system benzene-carbon tetrachloride showed nearly ideal behavior in the excess turbidity. The acetone, acetonitrile, and nitromethane solutions showed a positive molar excess Gibbs free energy, whereas the nitrobenzene and chloroform solutions exhibited negative values. Similar studies have been carried out by Schmidt and Clever (1968) for binary mixtures of dichloromethane with dibromomethane, diiodomethane, chloroform, and bromoform. There was no attempt in either of these studies to compare these light scattering results with other thermodynamic measurements.

b. Aqueous Solutions of Sucrose. Early attempts to compare the turbidity of aqueous sucrose solutions with the osmotic pressure (Debye, 1947; Halwer, 1948) appeared to be successful, but this was the result of two compensating errors—neglect of the Cabannes factor on the one hand and incorrect calibration of the instruments. Maron and Lou (1955) repeated this work with greater success. Although they also erred in the use of the wrong Cabannes factor (Prins, 1961) and of an incorrect formulation of the so-called volume correction (Kerker *et al.*, 1964b), these did not affect the final results. The measured turbidities are compared with those calculated from

$$\tau = \frac{8\pi^3(n + n_0)^2}{3\lambda_0{}^4 N_A}\left(\frac{\partial n}{\partial c}\right)^2_{T,p} \frac{cRT}{(\partial\pi/\partial c)_{T,p}} \tag{9.2.42}$$

in Fig. 9.4. In the above expression, n_0 is the refractive index of the solvent and π is the osmotic pressure. The use of $(n + n_0)^2$ rather than $4n^2$ appears to be another minor error. The agreement between the experimental values of the turbidity (points) and those calculated from the osmotic pressure (full line) is remarkably good. The intercept of the plot of Hc/τ against c leads to a molecular weight of 338, in excellent agreement with the known value of 342.

Stigter (1960a, 1963) has utilized the osmotic coefficients of aqueous sucrose solutions obtained from vapor-pressure data to calculate the turbidity according to

$$\frac{HcM}{\tau} = \left(\frac{c}{m_2'}\right)^2 \frac{[\phi_{os}' + m_2'(\partial\phi_{os}'/\partial m_2')_{T,p}]}{(\partial c/\partial m_2')^2_{T,p}} \tag{9.2.43}$$

where m_2' is the solute concentration in moles per mole of solvent and

$$\mu_2 - \mu_2{}^0 = -\phi_{os}' RTm_2' = RT \ln (p/p^0) \tag{9.2.44}$$

with p and p^0 the vapor pressures of the solution and solvent, respectively. This expression for the practical osmotic coefficient ϕ_{os}' differs somewhat from that defined earlier (9.2.34). The results are compared with Maron and Lou's turbidity data in Fig. 9.5. There is a significant discrepancy at higher concentrations which Stigter attributes to inaccuracies in the light scattering data. Mijnlieff and Zeldenrust (1965) have shown that this can be partially resolved if the turbidity data are corrected for the effect of the optical activity of this system. However, there is still a substantial difference so that additional experiments on sucrose appear to be advisable.

c. Solutions of Simple Electrolytes. There have been a number of studies in which the scattering by aqueous solutions of simple electrolytes has been compared with calculated results obtained from other measurements of the activity. Sweitzer (1927) compared his 90° scattering results for 15 electrolytes ($NaCl$, KCl, NH_4Cl, $NaNO_3$, KNO_3, NH_4NO_3, KI, Na_2SO_4, $(NH_4)_2SO_4$, $BaCl_2$, $CaCl_2$, $Ca(NO_3)_2$, $Pb(NO_3)_2$, HCl, $HC_2H_3O_2$) with calculated values for which the activity was obtained from vapor pressure measurements. The calculated and observed scattering agreed reasonably well, considering the experimental limitations. For example, the incident light from a carbon arc was filtered through a copper ammonia sulfate solution and was therefore hardly monochromatic.

Lochet (1953) carried out a similar study. He expressed turbidity due to concentration fluctuations by

$$\tau_E = (HM^2/1000 v_n)[m/\{1 + (\partial \ln \gamma_\pm)/(\partial \ln m)_{T,p}\}] \tag{9.2.45}$$

where m is the molality, v_n is the number of ions into which the electrolyte dissociates, and $\gamma_\pm$ is the mean ionic activity coefficient. The values of $\gamma_\pm$ were obtained from the treatise by Harned and Owen (1950). Lochet found that the values of the quantity in the square bracket obtained from light scattering for 25 electrolytes were always significantly higher than those obtained by direct calculations from the tabulated activity coefficients. Subramanian (1962) has brought the results of Lochet into better agreement by proposing an equation for the turbidity in which $(\partial n/\partial c)_{T,p}$ is replaced by a rather arbitrary empirical expression, and a distribution law different from the simple Boltzman expression (9.1.9) is used.

The activity coefficient tends to unity at high dilution where the above expression (9.2.45) reduces to (9.2.24) except for the factor $1/v_n$, and the molecular weight can be obtained directly from the limiting slope of the turbidity vs. the concentration. Following this procedure, Lochet obtained

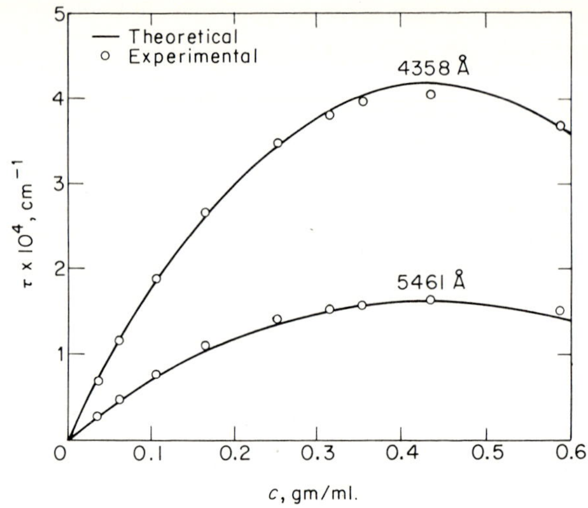

FIG. 9.4. Comparison of experimental values of the turbidity of sucrose solutions (circles) with values calculated from osmotic pressure data (curve) (Maron and Lou, 1955).

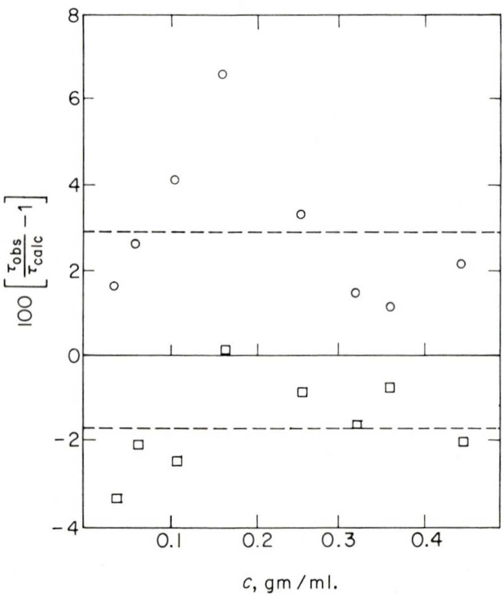

FIG. 9.5. Observed turbidity (Maron and Lou, 1955) with values calculated from Eq. (9.2.43), using vapor pressure data (Stigter, 1960a) plotted as $100[(\tau_{obs}/\tau_{calc}) - 1]$ against c. Squares are for 4358 Å; circles are for 5461 Å.

values for the molecular weights that were in excellent agreement with correct values, despite the discrepancy between the turbidity and activity data at higher concentrations.

Pethica and Smart (1966) have measured the turbidity of three simple electrolytes (KNO_3, KCl, and NaI) at $\lambda_0 = 0.436$ and $0.546\ \mu$ and compared their results with activities obtained from the vapor pressure data (Robinson and Stokes, 1955). They utilized the following expression for the isotropic part of the excess turbidity

$$Hc/\tau_E = -V_1^{-1}(\partial \ln a_1/\partial c)_{T,p} \tag{9.2.46}$$

which is obtained directly from (9.2.18) with the aid of (9.2.32). The left-hand side of this equation as obtained from light scattering by aqueous KNO_3 and KCl is compared in Fig. 9.6 with the right-hand side as calculated from vapor pressure data. The broken curve is calculated from the Debye–Hückel limiting law in order to show the behavior predicted theoretically at the lowest

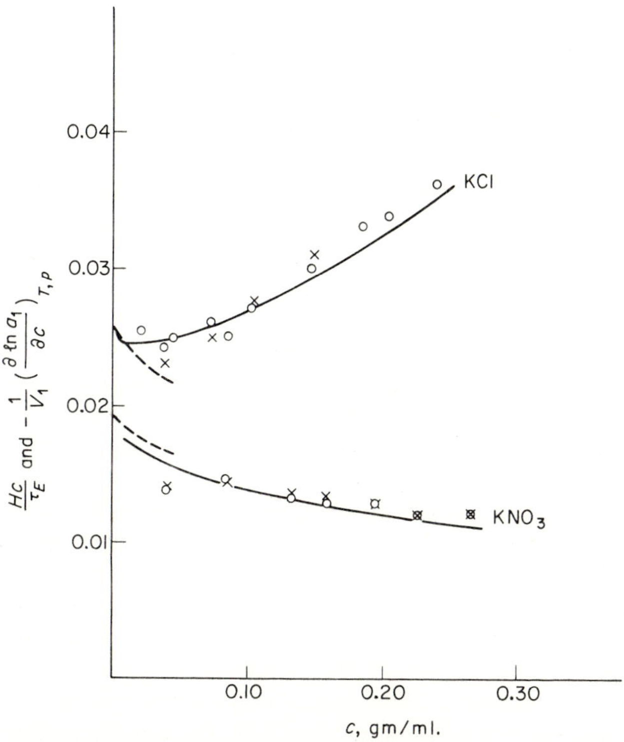

FIG. 9.6. Hc/τ_E at $0.546\ \mu$ (circles) and at $0.436\ \mu$ (crosses) compared with $-(1/V_1)(\partial \ln a_1/\partial c)_{T,p}$ obtained from vapor pressure data (curves) for KCl and KNO_3. The broken curve is calculated from the limiting Debye–Hückel theory (Pethica and Smart, 1966).

concentrations. The agreement between the light scattering and vapor pressure data is generally good although the light scattering points tend to lie above the activity curves, particularly at high concentrations. Pethica and Smart estimate that their results for KCl and KNO_3 agree with those of Sweitzer with an average difference of about 6%.

d. Heteropoly Acids. The heteropoly acids have provided still another system with which to compare activities obtained from the turbidity and those obtained by other methods. These compounds have attracted increasing interest in recent years as a bridge between the solution chemistry of simple salts and macromolecules. Among those receiving special attention have been 12-tungstophosphoric acid ($H_3PW_{12}O_{40}$), 9-tungstophosphoric acid ($H_6P_2W_{18}O_{62}$), and particularly 12-tungstosilicic acid ($H_4SiW_{12}O_{40}$). They are convenient for light scattering studies because of an unusual combination, for inorganic compounds, of a high molecular weight (hence, high scattering power) and a remarkably high solubility. Indeed, it is possible to form aqueous solutions which are 80% by weight of a heteropoly acid and which have a density greater than 3 g/ml. These compounds are also highly soluble in polar organic solvents. The dissolved species have been shown to be monodisperse by diffusion (Baker *et al.*, 1955) and equilibrium ultracentrifugation (Johnson *et al.*, 1960) measurements.

Kerker *et al.* (1958) found that a plot of Hc/τ_E against c led to correct values of the molecular weight in accordance with the Debye equation (9.2.23) when an organic solvent such as ethanol or propanol was used, but that this was not the case for aqueous solutions. This was attributed to the ionization of these compounds in water to highly charged ionic species as demonstrated by acidity (Matijević and Kerker, 1959a, b; Keller *et al.*, 1961), coagulation (Matijević and Kerker, 1958; Matijević *et al.*, 1963), and transference number (Kerker *et al.*, 1961a) measurements. Under these circumstances the simple van't Hoff equation for the activity is no longer valid. It is known that electrostatic interactions are frequently minimized in the presence of a high concentration of a neutral electrolyte, and this appears to be the case for the heteropoly acids. With the addition of a neutral salt such as KCl or NaCl, the turbidity increases (Matijević and Kerker, 1959b; Kratohvil *et al.*, 1966). Figure 9.7 depicts the results for 12-tungstosilicic acid[1a] plotted as Hc/τ_E against c. The upper curve which represents the data in pure water obviously does not extrapolate to the correct value of the reciprocal of the molecular weight designated on the curve as $1/M$. We will see in Section 9.2.3 that the extrapolated value in the pure water may go to a value other than $1/M$. Even in the presence of 0.02

[1a] Attention is called to the correction (Kerker *et al.*, 1963c) of similar data for 9-tungstophosphoric and 12-tungstophosphoric acids (Matijevic and Kerker, 1959b).

and 0.2 *M* NaCl the data extrapolate to the correct molecular weight only if obtained at very low concentrations. For 0.3 *M* NaCl and higher, the curves are quite linear and extrapolate to the correct molecular weight from concentrations which are readily accessible experimentally. At the highest salt concentration, the solution behaves nearly ideally, i.e., the curve is nearly horizontal, showing the value of the second virial coefficient to approach zero. The excess turbidity was obtained in each of these cases by subtracting the turbidity of the salt solution from the total turbidity. Actually, this is a three-component solution for which the present two-component theory is inadequate and this will be considered more fully in Section 9.3.

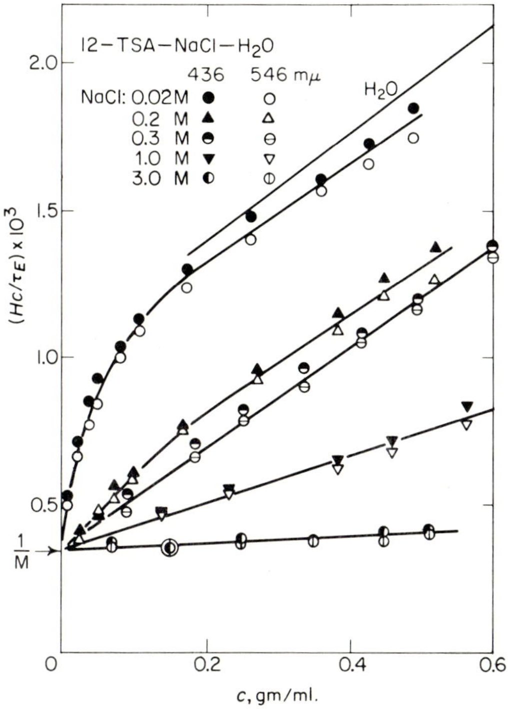

FIG. 9.7. Hc/τ_E for 12-tungstosilicic acid in NaCl solutions. The arrow indicates the reciprocal of the formula weight $M = 2875$ (Kratohvil *et al.*, 1966).

The sharp deviation from ideality of solutions of the heteropoly acids in pure water can provide still another test for the correlation between the activity as measured by light scattering and by other methods. There have been extensive studies of the turbidity and vapor pressure of 12-tungstosilicic acid by Goehring *et al.* (1959), Kronman and Timasheff (1959), Johnson *et al.*

(1960) (JKS), Kerker *et al.* (1963c) (KKOM), Kratohvil *et al.* (1966), Tyree *et al.* (1966), Johnson and Rush (1968), and Oppenheimer (1967). The data are summarized in Fig. 9.8 in which the results of KKOM are represented by the full line covering the concentration range of $c < 0.01$ to 0.5 g/ml and the results of JKS at five concentrations from 0.014 to 0.11 g/ml which are shown by diamond-shaped points. Additional data recently obtained by Oppenheimer (1967) agree with the results of KKOM except at the lower concentrations where the upturn of Hc/τ_E observed by JKS is confirmed.

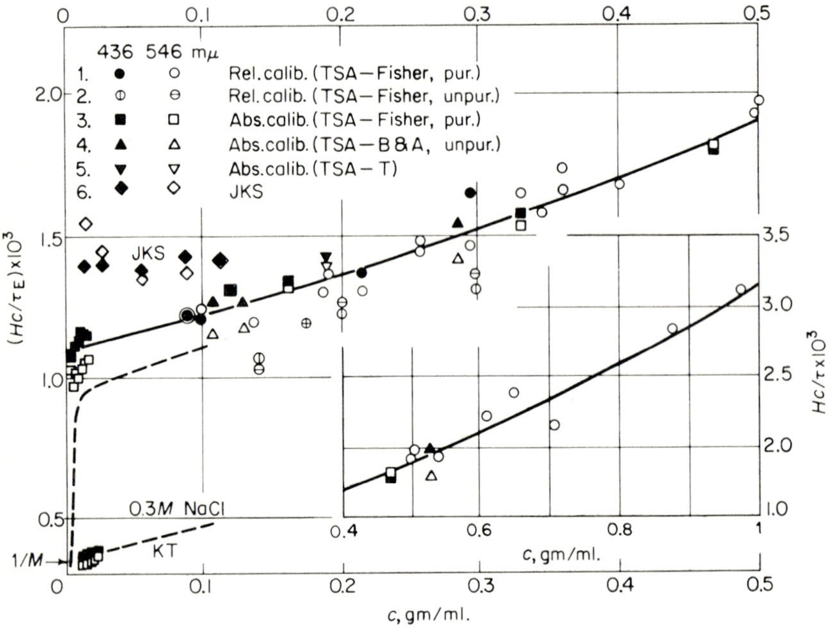

FIG. 9.8. Hc/τ_E for 12-tungstosilicic acid in water as well as in 0.3 *M* NaCl at $\lambda = 0.546\,\mu$ and $0.436\,\mu$. See original paper for significance of various experiments (Kerker *et al.*, 1963c).

Oppenheimer's values of Hc/τ_E at these concentrations are somewhat lower than those of JKS, but within the range of experimental error. A comparison of the measured turbidities of JKS and those calculated from activities obtained by equilibrium ultracentrifugation is shown in Table 9.9. The agreement is excellent, particularly considering the precariousness of the light-scattering measurements at such low values of τ_E.

Oppenheimer's comparison of the activities obtained from both turbidimetric and vapor pressure lowering data is shown in Table 9.10. These experiments were carried out in aqueous solutions as well as in methanol.

TABLE 9.9
COMPARISON OF EXPERIMENTAL TURBIDITIES WITH TURBIDITIES COMPUTED FROM EQUILIBRIUM ULTRACENTRIFUGATION ($H_4SiW_{12}O_{40}$)

m	$\tau_E \times 10^4$			
	(expt)		(calc)	
	$\lambda_0 = 0.436\ \mu$	$\lambda_0 = 0.546\ \mu$	$\lambda_0 = 0.436\ \mu$	$\lambda_0 = 0.546\ \mu$
0.0404	2.51	0.90	2.63	0.93
0.0310	1.92	0.71	1.89	0.67
0.0196	1.25	0.45	1.28	0.46
0.00973	0.60	0.21	0.62	0.22

TABLE 9.10
COMPARISON ACTIVITY OF SOLVENT FROM LIGHT SCATTERING AND FROM VAPOR PRESSURE LOWERING ($H_4SiW_{12}O_{40}$)

Conc. (g/ml)	a_1 (aqueous soln.)	
	L.S.	V.P.
0.1	0.9978	0.9974
0.2	0.9956	0.9947
0.3	0.9931	0.9919
0.4	0.9904	0.9889
0.5	0.9874	0.9855
0.6	0.9841	0.9818
0.8	0.9767	0.9731
1.0	0.9680	0.9626
1.2	0.9579	0.9481
1.4	0.9461	0.9347
1.6	0.9325	0.9168
1.8	0.9166	0.8955

The agreement in water at low concentrations is good, but the results diverge at high concentrations. This is not surprising considering the very high concentrations attained. Concentrations of $c = 0.5$ and 1.0 g/ml correspond to weight fractions of 35% and 54%, respectively. It would hardly be surprising if such concentrated solutions failed to conform to the requirements of the theory that the fluctuations be independent and have a small refractive index increment over that of the solution.

Oppenheimer (1967) also carried out a similar experiment in methanol, methanol-water mixtures, and methanol-sodium chloride mixtures. The lower curve in Fig. 9.9 is a plot of Hc/τ_E against c in methanol containing

0.1 *M* NaCl, while that above it is in pure methanol. Just as for the aqueous solution, the curve in the pure solvent turns upwards at low concentration, while that containing the supporting electrolyte extrapolates linearly to the reciprocal of the molecular weight. A comparison of the solvent activities calculated from light scattering and vapor pressure lowering is shown in Table 9.11 for solutions of 12-tungstosilicic acid in methanol. These results are similar to those in water, the agreement between the values obtained by the two techniques becoming progressively poorer with increasing concentration. If the calculated turbidities rather than the activities are compared, the relative error appears to be even greater. This may be due to the greater sensitivity to error when $\ln a_1$ is differentiated in an expression such as (9.2.46).

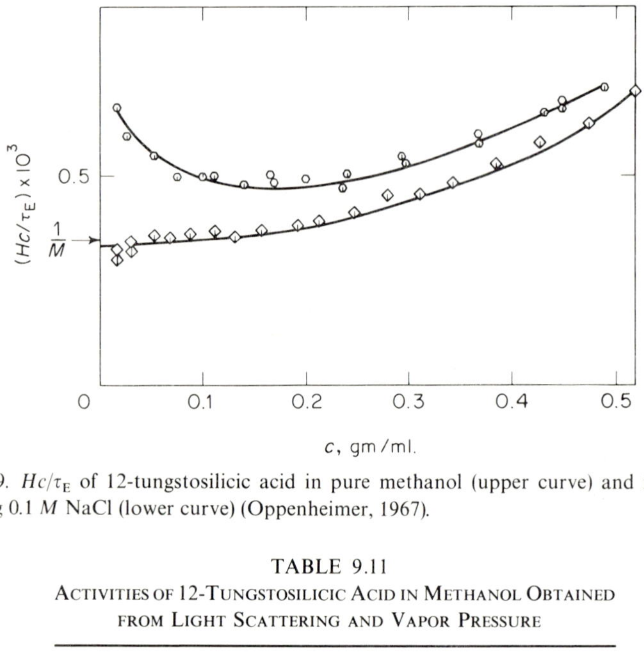

FIG. 9.9. Hc/τ_E of 12-tungstosilicic acid in pure methanol (upper curve) and in methanol containing 0.1 *M* NaCl (lower curve) (Oppenheimer, 1967).

TABLE 9.11
ACTIVITIES OF 12-TUNGSTOSILICIC ACID IN METHANOL OBTAINED FROM LIGHT SCATTERING AND VAPOR PRESSURE

Conc. (g/ml.)	a_1 L.S.	a_1 V.P.
0.05	0.9987	0.9987
0.1	0.9976	0.9973
0.2	0.9957	0.9947
0.3	0.9938	0.9918
0.4	0.9915	0.9893
0.5	0.9889	0.9862

e. Polyelectrolytes. Alexandrowicz (1959) has demonstrated the correlation between light-scattering and osmotic-pressure measurements in the case of polyelectrolytes by comparing the osmotic coefficients obtained by the two techniques. He used ionized polymethacrylic acid (PMA), polyacrylic acid (PAA), and bovine serum albumin (BSA) dissolved in water without the use of a salt. Some of the results are presented in Fig. 9.10 as the osmotic coefficient ϕ_{os} against the degree of ionization for one concentration of PAA and three concentrations of PMA. The concentrations are expressed in molality of monomer, m_m; the degree of ionization is the number of moles of free hydrogen ions per polyion ν divided by the degree of polymerization z. The osmotic

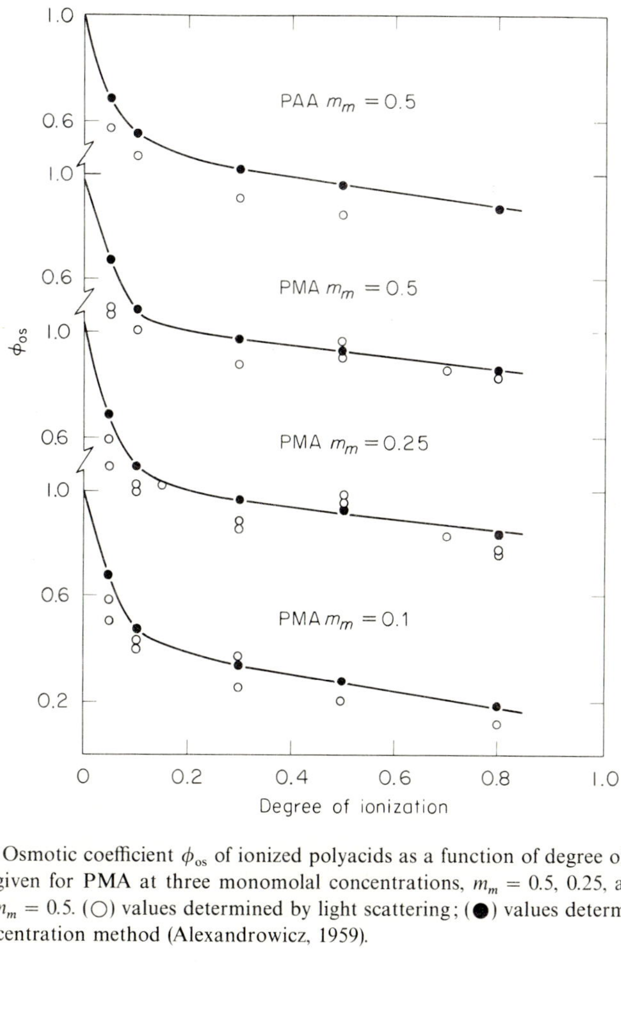

FIG. 9.10. Osmotic coefficient ϕ_{os} of ionized polyacids as a function of degree of ionization. Results are given for PMA at three monomolal concentrations, m_m = 0.5, 0.25, and 0.1, and for PAA at m_m = 0.5. (○) values determined by light scattering; (●) values determined by the osmotic concentration method (Alexandrowicz, 1959).

coefficient was calculated from the turbidity with the aid of

$$\phi_{os} = H\frac{z}{\nu + 1}\left[\left(\frac{\partial n}{\partial m_n}\right)^2_{T,p}\frac{m_m}{\tau_E} - \left(\frac{\partial \phi_{os}}{\partial \ln m_m}\right)_{T,p}\right] \tag{9.2.47}$$

and from the osmotic pressure from

$$\phi_{os} = (\pi \bar{V}_A/m_m)[55.5/(\nu + 1)RT] \tag{9.2.48}$$

The relatively small last term in the square bracket $(\partial\phi_{os}/\partial \ln m_m)_{T,p}$ in (9.2.47) was evaluated from the osmotic pressure. The agreement between the two techniques is reasonably good considering the complexity of the system and the experimental difficulties. For example, the polydispersity of the polymer samples was neglected.

f. Polymers. Solutions of polymeric nonelectrolytes such as polystyrene may behave according to a theory developed by Maron (1959) which leads to an expression for the activity given by

$$\ln a_1 = X_1 + \left[\left(\mu - \frac{\partial \mu}{\partial \phi_2}\right)\phi_1\right]\phi_2^2 \tag{9.2.49}$$

where ϕ_1 and ϕ_2 are the volume fractions of each component and X_1 is a known function of the composition and partial molal volume. The interaction parameter μ is a constant for any particular solute-solvent system. The quantity $(\mu - \partial\mu/\partial\phi_2)$ can be evaluated in dilute solution from values of the activity obtained from either light scattering or osmotic pressure, utilizing a relation such as (9.2.46) on the one hand and, on the other, with the aid of

$$-\pi\bar{V}_1 = RT \ln a_1 \tag{9.2.50}$$

which follows from (9.2.20). Maron and Nakajima (1960a) have obtained excellent agreement between the values of $(\mu - \partial\mu/\partial\phi_2)$ calculated from turbidity (Maron and Lou, 1954) and those from osmotic pressure (Maron and Nakajima, 1960b; Goldberg *et al.*, 1947) for polystyrene dissolved in toluene and in methyl ethyl ketone.

9.2.3 Solutions of Electrolytes

The theory of fluctuations as developed heretofore may not always represent an adequate approach to the problem of light scattering by solutions of electrolytes. The thermal motions of the ions will in general bring about free space charges with their attendant long-range forces, and the result of this can be that the fluctuations in neighboring volume elements will no longer be independent.

Hermans (1949) pointed out that there are three regimes. At sufficiently high dilutions, the different ionic species will fluctuate independently of each other, and they can be treated as if they were gas-like. An approximate criterion for this condition is that

$$\kappa^2 V^{2/3} < 1 \tag{9.2.51}$$

where V is the volume containing an appreciable number of ions, e.g., at least about 10, and κ is the reciprocal length encountered in the Debye–Hückel theory of strong electrolytes.

$$\kappa^2 = 4\pi \sum_j e_j{}^2 n_j \Big/ \varepsilon kT \tag{9.2.52}$$

Here e_j and n_j are the charge and number per unit volume of ions of type j, ε is the static dielectric constant, k the Boltzmann constant, and T the temperature. In order for this condition to prevail, the concentration must be less than 10^{-4} m/l. for a 1–1 electrolyte and less than 10^{-3} m/l. for a 4–1 electrolyte such as 12-tungstosilicic acid, $H_4SiW_{12}O_{40}$.

When the ions are independent of each other

$$\tau_E = \sum_j \tau_j = \sum H_j c_j M_j \tag{9.2.53}$$

where τ_j is the contribution of the jth ionic species to the turbidity, c_j and M_j are the concentration and gram ionic weight, respectively, for each of these species, and

$$H_j = (32\pi^3 n^2/3\lambda_0{}^4 N_A)(\partial n/\partial c_j)^2_{T,p} \tag{9.2.54}$$

Accordingly, this system scatters as if it consists of a single molecular species with a concentration $\sum c_j$ and apparent molecular weight

$$M' = \sum_j c_j M_j \Big/ \sum_j c_j \tag{9.2.55}$$

This assumes that each of the refractive index coefficients is about the same

$$(\partial n/\partial c_j)_{T,p} \cong (\partial n/\partial c)_{T,p} \tag{9.2.56}$$

Therefore

$$\tau_E = HcM' \tag{9.2.57}$$

If a determination of the molecular weight were made by extrapolation of Hc/τ_E to zero concentration, the apparent molecular weight would correspond to the weight average gram ionic weight of the ions. For a 1–1 electrolyte, this is

$$M' = (M_1{}^2 + M_2{}^2)/(M_1 + M_2) = (M_1{}^2 + M_2{}^2)/M \tag{9.2.58}$$

where M is the molecular weight of the electrolyte. If the two ionic species have equivalent gram ionic weights

$$M' = M/2 \tag{9.2.59}$$

On the other hand, if, as in the case of the heteropoly acids, $M_1 \ll M_2 \cong M$, then the correct molecular weight will be obtained provided the extrapolation is carried out to a sufficiently high dilution. However, sufficiently accurate light-scattering data are hardly attainable at the required high dilutions.

As the concentration increases, the electrostatic forces perturb the ions from a random distribution, so that the scattering is no longer incoherent. The ionic structure can be described in terms of an electrical double layer in which each ion is surrounded by an excess of ions of opposite charge. The resultant optical interference is accounted for by the Zernike–Prins (1927) formula which, in this case, can be written

$$R_\theta = \sum_j R_j + \sum_j \sum_k M_j (K_j K_k)^{1/2} c_j \int_0^\infty 4\pi s^2 [(\sin hs)/hs] g_{jk}(s)\, ds \tag{9.2.60}$$

The scattering is now expressed by the excess Rayleigh ratio R_θ rather than the excess turbidity. The angular variation of the scattering no longer follows the $(1 + \cos^2 \theta)$ Rayleigh law so that the simple relation between the Rayleigh ratio and turbidity given by (3.2.31) is no longer valid.[2] The parameter K is

$$K = (3/16\pi)(1 + \cos^2 \theta)H = (2\pi^2 n^2/\lambda_0{}^4 N_\mathrm{A})(\partial n/\partial c)^2_{T,p}(1 + \cos^2 \theta) \tag{9.2.61}$$

The second term of (9.2.60) accounts for the reduction of the scattered intensity from the value given by (9.2.53) as a result of the interference between pairs of ions. This is completely analogous to the interference encountered in Rayleigh–Debye scattering as discussed in the previous chapter. The radial distribution function $g_{jk}(s)$ describes the excess concentration of ions k over the average concentration c_k at a distance s from an ion j.

The Debye–Hückel limiting law can be used to obtain the radial distribution function

$$g_{jk}(s) = -(c_j e_j e_k/\varepsilon kT)[\exp(-\kappa s)/s] \tag{9.2.62}$$

[2] A number of workers including Hermans have overlooked this point and have continued to utilize expressions involving the turbidity. Since the experiments are carried out by measurement of the Rayleigh ratio which is then converted to the turbidity by (3.2.31), there will be no error in plotting this erroneous value of Hc/τ_E against c for the purpose of obtaining a molecular weight, rather than the value of Kc/R_θ against c.

which upon integration of (9.2.60) yields

$$R_\theta = \sum_j R_j - \frac{\left(\sum_j K_j^{1/2} c_j e_j\right)^2 \kappa^2}{\sum_j (c_j/M_j) e_j^2 (\kappa^2 + \sigma^2)} \tag{9.2.63}$$

In the limit of low concentration where κ becomes very small this reduces to

$$R_\theta = \sum_j R_j \tag{9.2.64}$$

which is equivalent to (9.2.53). In practice an appreciable light-scattering signal can be obtained accurately only when $\kappa \gg \sigma$ or when the above reduces to

$$R_\theta = \sum_j R_j - \frac{\left(\sum K_j^{1/2} c_j e_j\right)^2}{\sum (c_j/M_j) e_j^2} \tag{9.2.65}$$

which in turn is given exactly by

$$R_\theta = KcM/\nu_n \tag{9.2.66}$$

Here, ν_n is the number of moles of ions formed per formula weight.

When the extrapolation of Kc/R_θ against c is carried out in this range of concentration, the intercept will be ν_n/M, leading to an apparent molecular weight of M/ν_n. The limiting Debye–Hückel law is valid only at low concentrations and the use of this law in a quantitative way becomes increasingly precarious as one proceeds from uni-univalent electrolytes to higher charge types.

Pethica and Smart (1966) have reported apparent molecular weights for KNO_3, KCl and NaI, which were approximately $M/2$, in accord with what would be expected in the regime where the Debye–Hückel limiting law is valid. However, from their curves shown in Fig. 9.6, it is apparent that the experimental points below $c = 0.10$ are rather uncertain. For the two salts illustrated, $c = 0.10$ corresponds to about 1 m/l., which is well above the range of validity of the Debye–Hückel law. Morel (1966) has carried out similar experiments with NaCl in the range $c = 0.01$ to 0.1 g/ml and also claims the data agree with (9.2.66), leading to the apparent molecular weight, $M/2$. However, there is considerable uncertainty in the light-scattering data at these concentrations.

It should be noted that when the cations and anions of a 1–1 electrolyte have nearly equal masses, no distinction can be made between the Debye–Hückel range and the very low concentration range where the ions are independent of each other. The weight average gram ionic weight, which is obtained in the latter case (9.2.58), will be approximately $M/2$ which is the same as M/ν_n predicted for the Debye–Hückel case.

Figure 9.11 shows some results for 12-tungstosilicic acid at low concentrations (Oppenheimer, 1967), which indicate that Hc/τ_E bends upwards in the direction of the value ν/M as predicted for the concentration range where the Debye–Hückel theory is valid. This trend had first been noted by Johnson *et al.* (1960). However, this result should only be considered as a qualitative

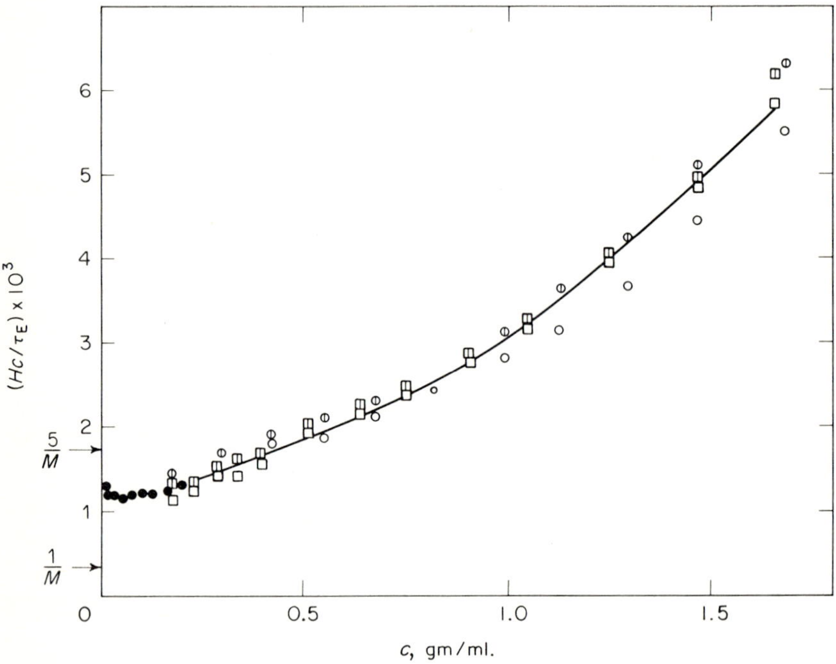

FIG. 9.11. Hc/τ_E for 12-tungstosilicic acid in water at low concentrations (Oppenheimer, 1967).

trend. The value ν_n/M for the reciprocal of the apparent molecular weight may not be the precise limiting value because the Debye–Hückel theory can hardly be considered to be precisely valid for a 1–4 electrolyte at the concentration $c = 0.01$ g/ml. At sufficiently high dilutions, the curve Hc/τ_E would be expected to turn down again towards $1/M$. This is because the hydrogen ions are so light compared to the tungstosilicate ions which comprise most of the mass of the species that the weight average ionic weight, which would be obtained in this low concentration range, is practically equal to the molecular weight.

At sufficiently high concentration the number of ions of a particular charge within the volume undergoing a concentration fluctuation is not significantly different from that of ions of the opposite charge. Then the scattering will be determined by fluctuations in concentration of the electrolyte as a whole. If the higher virial coefficients are sufficiently small, the Debye equation for determining molecular weight by extrapolation of Hc/τ_E to zero concentration may lead to the correct molecular weight in the usual fashion. Thus, Lochet (1953) obtained correct values for the molecular weight of simple electrolytes by using the Debye formula.[3] His data appear to have been obtained at relatively high concentrations, probably greater than 0.1 M/l.

We have already seen how the addition of a neutral electrolyte to aqueous solutions of heteropoly acids sufficiently screens the charge effects so that the Debye equation can be interpreted in the usual manner to obtain the molecular weight (Fig. 9.8). Similar results obtained in methanol (Oppenheimer, 1967) were shown in Fig. 9.9. The lower curve for 0.1 M NaCl in methanol exhibits the screening effect where Hc/τ_E extrapolates directly to $1/M$. The curve above this is for 12-tungstosilicic acid in pure methanol, and it exhibits the same up-turn of Hc/τ_E at low concentrations that was observed in water. A similar result is obtained in pure ethanol as shown in Fig. 9.12. However, when isopropanol is the solvent (Fig. 9.13) Hc/τ_E appears to extrapolate to $1/M$, which indicates that in this case the ionization of the 12-tungstosilicic acid may be considerably suppressed so that the system behaves as a nonelectrolyte.

a. Polyelectrolytes. Hermans (1949) has also considered the effect of the ionic double layer on the scattering by a polyelectrolyte solution. If the colloidal particle with its double layer is treated as a Rayleigh–Debye scatterer surrounded by a diffuse double layer for which the Debye–Hückel approximation is valid, then the Rayleigh ratio is

$$R_\theta = \frac{Kc}{M_p}\left[M_p - \frac{e_p \sum c_j e_j}{\sum (c_j/M_j) e_j^2}\right]^2 P(\theta) \tag{9.2.67}$$

where $P(\theta)$ is the form factor for the spherical colloidal particle, the subscripts p and j refer to the colloid and the ions in the diffuse double layer, respectively.

Similar expressions can be derived for other approximations describing the ionic atmosphere or for other configurations for the polyelectrolyte such as random coils or spheres having a Gaussian density distribution. In

[3] There is some ambiguity here because, in different parts of his paper, Lochet introduces both HcM and HcM/v as limiting relations for the turbidity. However, it appears from his discussion and conclusions that the former was used when he determined the molecular weight from the limiting slope of the plot of the turbidity against the concentration.

all cases, the effect of the ions surrounding the colloid particle is to reduce its scattering from what would be obtained in the absence of an ionic atmosphere. Hermans has calculated that in the most favorable cases this could amount to 10% of the light scattering, although it would normally be considerably less than this.

It should be emphasized that the above model assumes that each colloidal ion with its ionic atmosphere scatters independently and that these atmos-

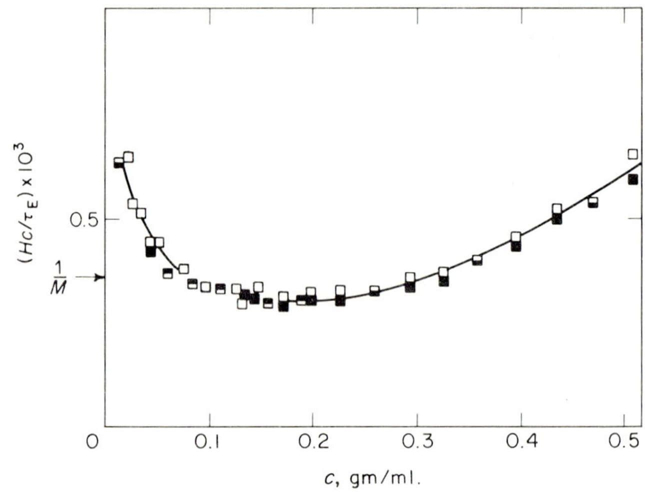

FIG. 9.12. Hc/τ_E for 12-tungstosilicic acid in ethanol (Oppenheimer, 1967).

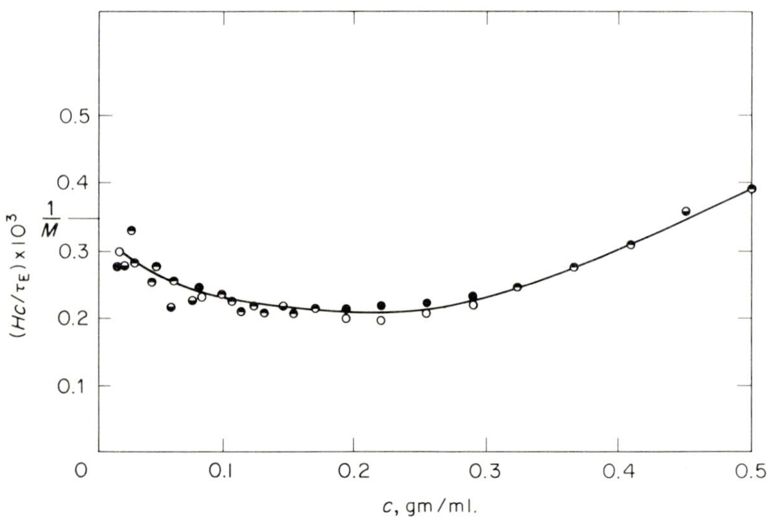

FIG. 9.13. Hc/τ_E for 12-tungstosilicic acid in isopropanol (Oppenheimer, 1967).

pheres do not overlap. This latter condition will seldom apply so that interference between separate colloidal particles may actually give rise to larger effects than those due to the internal interference between a particle and its own ionic atmosphere.

b. Charge Fluctuations in Isoionic Polyelectrolytes. The charge on a polyelectrolyte molecule is determined by the ionization of the polar groups. In the case of a polyampholyte such as a protein which contains both acidic and basic groups, the charge will vary from positive at low pH to negative at high pH. The intermediate pH of zero charge constitutes the isoelectric region. Although there is no net statistical charge in this region, there will be fluctuations from this configuration for individual molecules, so that even when the average net charge is zero, there is a nonzero mean square charge.

Kirkwood and Shumaker (1952) have shown that this charge fluctuation gives rise to an attractive force and have calculated its contribution to the concentration coefficient of the chemical potential, $(\partial \mu_2/\partial c)_{T,p}$, and in turn, with the aid of (9.2.17), its contribution to the turbidity. In addition to the linear term containing the second virial coefficient, the Debye equation (9.2.23) will now contain a term dependent upon the square root of the concentration, viz.

$$\frac{Hc}{\tau_E} = \frac{1}{M}\left\{1 - \frac{1}{2}\left(\frac{\pi N_A}{M}\right)^{1/2}\left(\frac{e^2\overline{Z^2}}{\varepsilon kT}\right)^{3/2} c^{1/2} + Bc + \ldots\right\} \tag{9.2.68}$$

where $\overline{Z^2}$ is the mean square charge fluctuation, e is the protonic charge, and ε is the static dielectric constant. According to this equation, Hc/τ_E should become proportional to $c^{1/2}$ at high dilution, and $\overline{Z^2}$ may be determined from the limiting slope of Hc/τ_E plotted as a function of $c^{1/2}$.

Values of $\overline{Z^2}$ have been determined for a number of proteins by Timasheff *et al.* (1955, 1957). Figure 9.14 shows a plot of Hc/τ_E against c for bovine serum albumin in salt-free water and in the presence of 0.001 M NaCl. The salt-free solution shows a decided upward curvature, while the curve for the solution in the presence of the salt is linear. The latter effect is quite usual and is attributed to the screening effect of the added electrolyte, which, in this case, is due to the binding of chloride ion by the protein molecules. When the salt-free data are plotted against $c^{1/2}$, the linear relation shown in Fig. 9.15 is obtained. Possible sources of error in the interpretation of these results have been discussed in detail by the authors who conclude finally that these and other results provide convincing evidence of the validity of the charge-fluctuation mechanism. The actual values of the root mean square charge fluctuation (in this example, a value of 3.5 protonic units) were in excellent agreement with the results of titration and dielectric constant measurements.

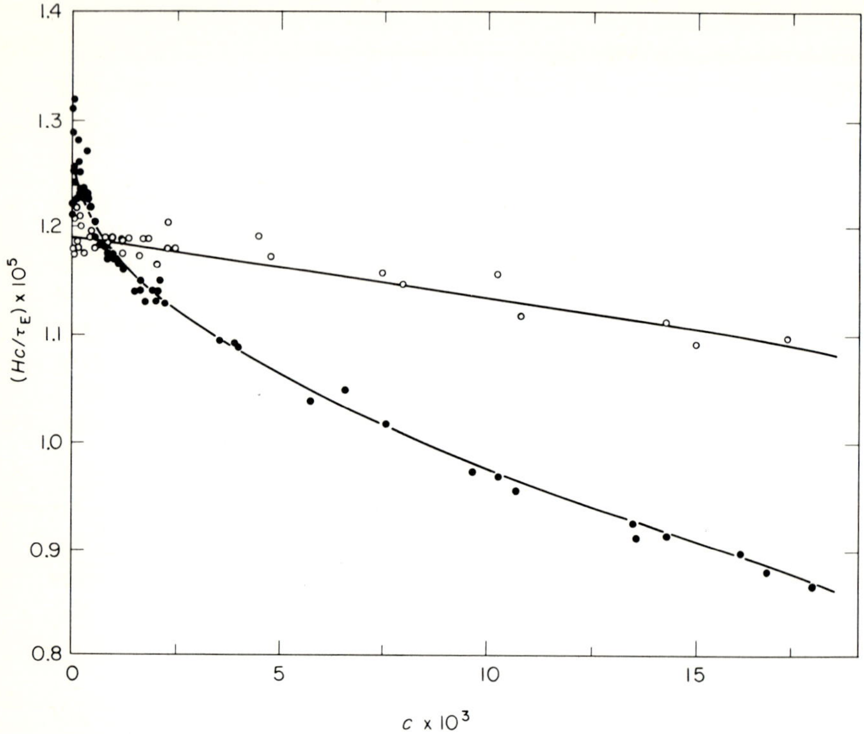

FIG. 9.14. Hc/τ_E of BSA in water (lower curve) and in 0.001 *M* NaCl (upper curve) (Timasheff *et al.*, 1955).

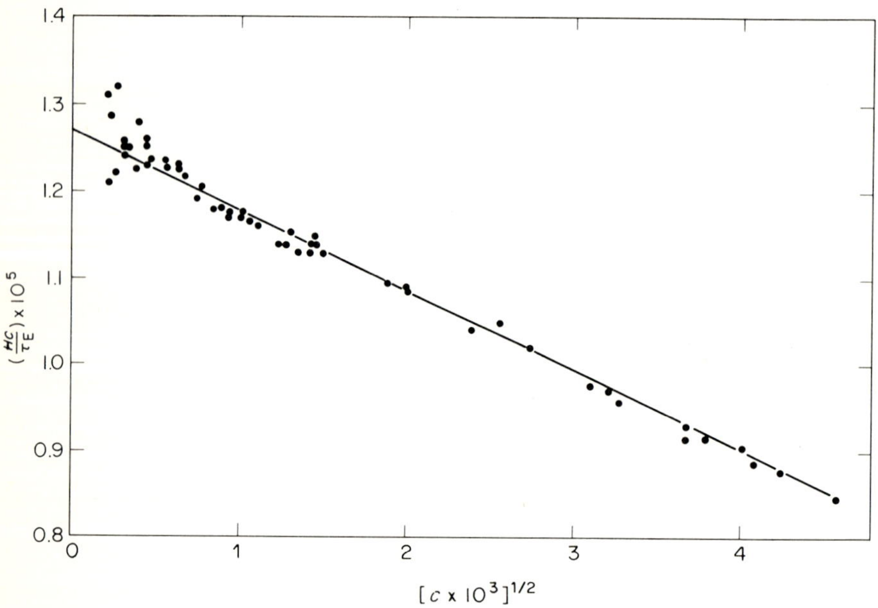

FIG. 9.15. Hc/τ_E of BSA in water plotted against $c^{1/2}$ (Timasheff *et al.*, 1955).

9.3 Multicomponent Solutions

9.3.1 THEORY

The fluctuation theory was extended to multicomponent solutions by Zernike (1915, 1918), but this was overlooked until it was utilized by Brinkman and Hermans (1949). Shortly thereafter, Kirkwood and Goldberg (1950) and Stockmayer (1950) developed the theory independently. We will follow Stockmayer's treatment.

The problem, just as for the one and two component systems, is to determine the mean square fluctuation of the optical dielectric constant or of the refractive index for insertion in (9.1.4). For the refractive index n

$$\delta n = \left(\frac{\partial n}{\partial T}\right)_{p,m} \delta T + \left(\frac{\partial n}{\partial p}\right)_{T,m} \delta p + \sum_{i=2}^{l} \left(\frac{\partial n}{\partial m_i}\right)_{T,p,m} \delta m_i \tag{9.3.1}$$

where l is the number of components. The index 1 is reserved as usual for the solvent which is present in excess and the subscript m indicates that all molalities except the one which is being differentiated are to be held constant. This can be rewritten as

$$\delta n = \left[\left(\frac{\partial n}{\partial \ln \rho}\right)_m \left(\sum_{i=2}^{l} \overline{V_i}\, \delta m_i - \delta V\right) \Big/ V\right] + \sum_{i=2}^{l} \psi_i\, \delta m_i \tag{9.3.2}$$

where $\overline{V_i}$ is the partial molal volume of the ith component and

$$\psi_i = (\partial n / \partial m_i)_{T,p,m} \tag{9.3.3}$$

What is desired now are the fluctuations of volume and of m_i in a part of the system containing a fixed quantity of component 1 and in which p, T, and the chemical potentials μ_i are maintained constant. The appropriate averages are found to be

$$\overline{(\delta V)^2} = kT\left[\beta_T V - \sum_{i=2}^{l} \overline{V_i}\left(\frac{\partial m_i}{\partial p}\right)_{T,\mu,m_1}\right] \tag{9.3.4}$$

$$\overline{(\delta m_i \delta V)} = -kT\left(\frac{\partial m_i}{\partial p}\right)_{T,\mu,m_1} \tag{9.3.5}$$

$$\overline{(\delta m_i \delta m_j)} = kT\left(\frac{\partial m_i}{\partial \mu_j}\right)_{T,p,\mu,m_1} = kT\left(\frac{\partial m_j}{\partial \mu_i}\right)_{T,p,\mu,m_1} \tag{9.3.6}$$

When these values are inserted in the expression for $\overline{(\delta n)^2}$, the result is

$$\overline{(\delta n)^2} = kT\left\{\left[\beta_T\left(\frac{\partial n}{\partial \ln \rho}\right)_m^2 \Big/ V\right] + \sum_{i=2}^{l} \sum_{j=2}^{l} \psi_i \psi_j \left(\frac{\partial m_i}{\partial \mu_j}\right)_{T,p,\mu,m_1}\right\} \tag{9.3.7}$$

The first term represents the contribution from density fluctuations and leads to the familiar formula developed earlier for pure liquids (9.1.14). The remainder of the expression is the contribution to the mean square fluctuation of the refractive index arising from fluctuations in the concentrations. This contains cross terms which in general cannot be neglected. Were these not appreciable, the diagonal terms would lead to independent contributions from each component. Before inserting (9.3.7) into the expression for the turbidity, it will prove convenient to express the coefficients in somewhat more familiar quantities with the aid of

$$(\partial m_i/\partial \mu_j)_{T,p,\mu,m_1} = A_{ij}/|a_{ij}| \tag{9.3.8}$$

where

$$a_{ij} = (\partial \mu_i/\partial m_j)_{T,p,m} = (\partial \mu_j/\partial m_i)_{T,p,m} \tag{9.3.9}$$

and A_{ij} represents the cofactor of the determinant $|a_{ij}|$. The cofactor is obtained by striking out the row and column containing a_{ij}; otherwise,

$$A_{ij} = \partial |a_{ij}|/\partial a_{ij} \tag{9.3.10}$$

Now

$$\tau_E = \frac{32\pi^3 n^2 RTV}{3\lambda_0{}^4 N_{\mathrm{A}}} \sum_{i=2}^{l} \sum_{j=2}^{l} \psi_i \psi_j \frac{A_{ij}}{|a_{ij}|} \tag{9.3.11}$$

This reduces appropriately for a binary system to (9.2.7). The volume V must be chosen to correspond to the constant quantity m_1 of the solvent component that is assumed in the evaluation of a_{ij}. For concentration expressed as molality, which is utilized above

$$V = 10^3/\rho_0 \tag{9.3.12}$$

9.3.2 Applications

a. Polymers in a Binary Solvent. Ewart *et al.* (1946) obtained what appeared to be abnormal results for the scattering of polystyrene in mixtures of benzene and methanol. Although the intercept of a Debye plot of Hc/τ_E against c in pure benzene led to correct molecular weights, varying intercepts were found for data obtained in mixtures of the two solvent components. This is illustrated in Fig. 9.16 for solvent mixtures up to 15% methanol.

These results were interpreted as due to selective adsorption of benzene by the polymer molecule with a corresponding depletion of this component from the bulk of the system. Such adsorption would affect the scattering by altering the refractive index of the polymer relative to the ambient medium.

This model leads to

$$\tau_E = \frac{32\pi^3 n^2}{3\lambda_0^4 N_A}\left[\left(\frac{\partial n}{\partial c}\right)_{T,p} + \alpha\left(\frac{\partial n}{\partial \phi}\right)_{T,p}\right]^2 cM \tag{9.3.13}$$

where ϕ is the volume fraction in the solvent mixture of the component which becomes selectively adsorbed, and the adsorption constant is

$$\alpha = -(\partial\phi_1/\partial c)_{T,p} \tag{9.3.14}$$

where ϕ_1 is the volume fraction of adsorbed solvent in the medium surrounding the polymer molecules. Strazielle and Benoit (1961) have developed a somewhat more detailed molecular model in which

$$\alpha = (X_1/M)V_1 \tag{9.3.15}$$

where V_1 is the molar volume of the solvent which is selectively adsorbed and X_1 is the number of molecules adsorbed per polymer molecule. Each of the above equations correlates the data quite well.

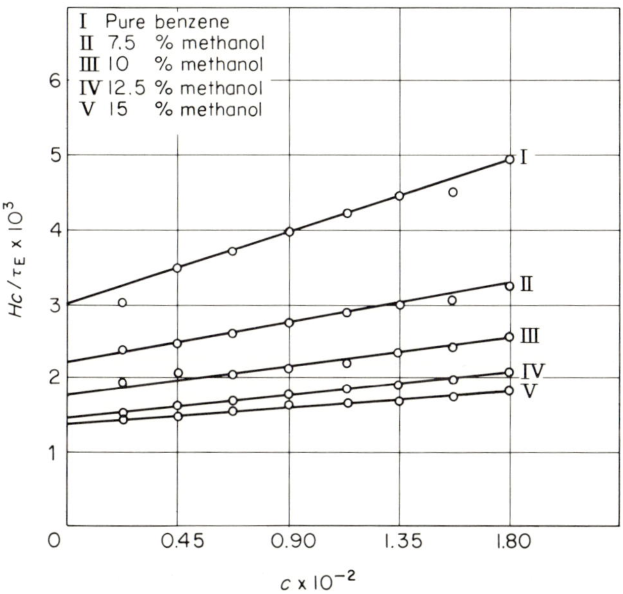

FIG. 9.16. Hc/τ_E of polystyrene in various benzene-methanol mixtures (Ewart *et al.*, 1946).

It is necessary that the solvent components have different refractive indices in order that there be an observable effect, i.e., $(\partial n/\partial\phi)_{T,p} \neq 0$. Figure 9.17 shows the results for polystyrene in dioxane-butanone mixtures for which $(\partial n/\partial\phi)_{T,p} = 0.0004$ and $(\partial n/\partial c)_{T,p} = 0.219$. This behaves as a two-component system because the selective adsorption can only be seen by the light scattering when the refractive indices of the two solvents are different. Each of the curves intercepts the concentration axis at the same point even though the slopes are quite different. Of course, in this instance there is no assurance that this behavior is due to the absence of selective absorption ($\alpha = 0$) rather than to the purely optical effect caused by the similarity of the refractive index of the two solvent components.

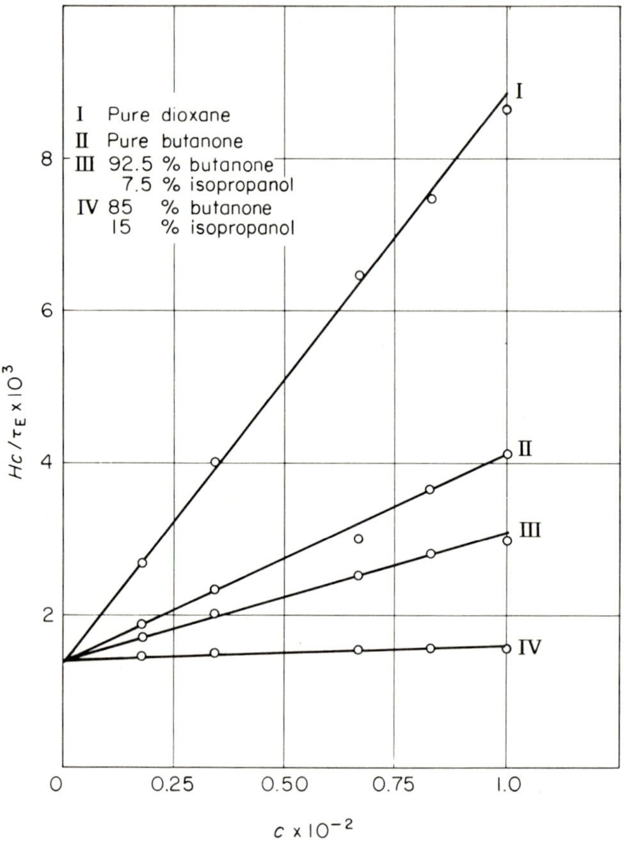

FIG. 9.17. Hc/τ_E of polystyrene in various butanone-isopropanol mixtures (Ewart *et al.*, 1946).

Stockmayer (1950) has shown how these data can also be analyzed with the aid of the multicomponent theory. The excess turbidity will now be designated as the difference between the turbidity of the three-component system and that of the two-component system of the same composition, but without the polymer. This is

$$\frac{\tau_E}{H^*V} = \frac{\psi_2{}^2 a_{33} - 2\psi_2\psi_3 a_{23} + \psi_3{}^2 a_{22}}{a_{22}a_{33} - a_{23}^2} - \frac{\psi_3{}^2}{a_{33}^0} \tag{9.3.16}$$

where

$$H^* = 32\pi^3 n^2 RT/3\lambda_0{}^4 N_A \tag{9.3.17}$$

The superscript zero denotes the solution containing no polymer. Component 1 represents the principal solvent, such as benzene in the above example, component 2 is the polymer, and component 3 is the second solvent, or in later examples component 3 may be an added low molecular weight salt. H^*V for both the two- and three-component systems are assumed above to be approximately equal.

The above expression can be cast into the same form as (9.3.13) if it is written

$$\frac{\tau_E}{H^*V} = \frac{[\psi_2 - (a_{23}/a_{33})\psi_3]^2}{a_{22} - (a_{23}^2/a_{33})} + \psi_3{}^2\left(\frac{1}{a_{33}} - \frac{1}{a_{33}^0}\right) \tag{9.3.18}$$

At low concentrations of polymer the last term becomes negligible. The quantity in the square bracket may be compared to that in the square bracket in (9.3.13) provided that a_{23}/a_{33} is proportional to the adsorption constant α. The correspondence between these is better seen when it is noted that

$$\frac{a_{23}}{a_{33}} = \frac{(\partial\mu_3/\partial m_2)_{T,p,m_3}}{(\partial\mu_3/\partial m_3)_{T,p,m_2}} = -\left(\frac{\partial m_3}{\partial m_2}\right)_{T,p,\mu_3} \tag{9.3.19}$$

This coefficient describes how the relative amount of m_3 must be decreased as polymer is added in order to keep the chemical potential of component 3 fixed. If the mechanism by which the depletion of component 3 under these constraints were preferential adsorption, this quantity corresponds precisely to the adsorption coefficient. Also a_{22} becomes inversely proportional to m_2 at low concentrations and a_{23}^2/a_{33} approaches a constant value. Accordingly, at low concentrations this approaches the modified form of the Debye equation such as (9.3.13) in which there is an additional term $[\alpha(\partial n/\partial\phi)_{T,p}]$ or $(-a_{23}/a_{33})\psi_3$ added to the refractive index increment.

Ullman and Benoit (1962) have derived the above empirical expression from an alternative formulation of the multicomponent theory given by

$$\tau_E^* = (32\pi^3 n^2 cM/3\lambda_0{}^4 N_A)(\partial n/\partial c)_{\mu_1,\mu_3}^2 \tag{9.3.20}$$

Here, τ_E^* is the excess turbidity relative to a binary solution which is in osmotic equilibrium with the ternary system, and the refractive index increment is measured while maintaining μ_1 and μ_3 constant; i.e., also in osmotic equilibrium. This equation will be discussed in some detail later in this section. It follows for dilute solutions that

$$(\partial n/\partial c)_{\mu_1,\mu_3} = (\partial n/\partial c)_{m_3} + (X_1 V_1/M)(\partial n/\partial \phi)_{m_3} \tag{9.3.21}$$

in accord with (9.3.13) and (9.3.15).

It is useful to cast (9.3.18) into a form involving somewhat more familiar experimental quantities utilizing the following notation of Scatchard (1946):

$$(\mu_i - \mu_i^0)/RT = \ln m_i + \ln \gamma_i \tag{9.3.22}$$

$$a_{ii}/RT = (1 + \beta_{ii} m_i)/m_i \tag{9.3.23}$$

$$(a_{ij}/RT) = \beta_{ij} \qquad (i \neq j) \tag{9.3.24}$$

where the interaction coefficients are defined by

$$\beta_{ij} = (\partial \ln \gamma_i/\partial m_j)_{T,p,m} \tag{9.3.25}$$

To an approximation involving only linear terms in the concentrations, the activity coefficient can be expanded as

$$\ln \gamma_i = \sum_{j=1}^{l} \beta_{ij} m_j \tag{9.3.26}$$

Then

$$\frac{\tau_E RT}{H^* V} = \frac{m_2[\psi_2 - m_3\beta_{23}\psi_3/(1 + \beta_{33}m_3)]^2}{1 + m_2[\beta_{22} - m_3\beta_{23}^2/(1 + \beta_{33}m_2)]} \tag{9.3.27}$$

Interestingly, the denominator is proportional to the derivative of the osmotic pressure just as for the simple two-component case. Also, it can be seen that when m_3 vanishes, this reduces directly to the two-component case

$$\tau_E RT/H^* V = m_2 \psi_2{}^2/(1 + m_2\beta_{22}) \tag{9.3.28}$$

which is identical with (9.2.6).

Kirkwood and Goldberg (1950) have used the data of Ewart *et al.* (1946) and an expression similar to (9.3.27) to calculate the interaction coefficients. However, these were not checked against values obtained by an independent method.

Read (1960) and Strazielle and Benoit (1961) have independently carried out studies on the system benzene-cyclohexane-polystyrene. The following expression for the adsorption constant α can be obtained from (9.3.13)

$$\alpha = \left[\frac{(\tau'/c)_0^{1/2}}{(\tau/c)_0^{1/2}} - 1\right]\frac{(\partial n/\partial c)_{T,p}}{(\partial n/\partial \phi)_{T,p}} \tag{9.3.29}$$

where the prime refers to the turbidity in the single solvent, and the subscript zero is the value in the limit of zero concentration of polystyrene. Furthermore,

$$\alpha = (V_3/V_2)\overline{V_2}\phi_1(a_{23}/a_{33}) \tag{9.3.30}$$

where V_3 and V_2 are the molar volumes of the second solvent and the polymer, $\overline{V_2}$ is the partial molal volume of the polymer, and ϕ_1 the volume fraction of the first solvent in the mixed solvent. It is the first solvent which becomes preferentially adsorbed.

The chemical potential of a ternary polymer solution as a function of the composition and of the three interaction coefficients of the separate binary systems is given by the Flory–Huggins theory. This makes it possible to express the adsorption constant α in terms of these coefficients, which can in turn be determined experimentally from vapor pressure data on the three separate binary solutions. This offers the possibility of correlating light-scattering and vapor-pressure data for such a ternary system. Read (1960) has extended the solution theory by introducing some of the higher order binary interaction coefficients and one ternary interaction coefficient which was chosen to best represent the final results. Some results are shown in Fig. 9.18 where the adsorption coefficient is plotted as a function of the volume fraction composition of the mixed solvent. Curve (a) was based on

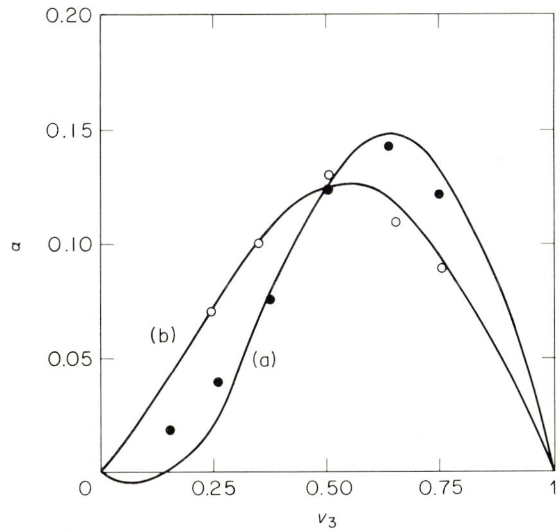

FIG. 9.18. The adsorption constant α as a function of volume fraction v_3 of cyclohexane in the mixed solvent, benzene-cyclohexane. Curve (a) is the Flory–Huggins theory; curve (b) is Read's extension. Open circles are Read's (1960) light scattering data; solid circles are the results of Strazielle and Benoit (1961).

the Flory–Huggins theory while curve (b) was obtained from Read's extension of this theory, using the fitted constant. Except for this one constant, these curves are based upon measurements of vapor pressure over the binary system.

The open circles which fit the extended theory are calculated from the light scattering data of Read. The closed circles are based upon the data of Strazielle and Benoit (1961). These fit the simple Flory–Huggins theory. The latter workers claim that Read's results are in error because the $(\partial n/\partial c)_{T,p}$ values used by him were too high. Since the curve with which these results agree contains a fitted constant, this could be spurious. It would appear that the results of Strazielle and Benoit, based only upon parameters obtained from the binary systems, successfully correlate the turbidity and vapor pressure for this system.

The case of two polymer constituents in a single solvent has been discussed by Hyde and Tanner (1968). Their system was complicated by having molecules sufficiently large so that they no longer behave as Rayleigh scatterers. It is only necessary, in such a case, to change the turbidity to the Rayleigh ratio (3.2.31) and to include the product of the form factors $P_i(\theta)P_j(\theta)$ inside the double summation of (9.3.11) (Blum and Morales, 1952). This offers the possibility of measuring the interactions between the polymer constituents, particularly those interactions which may be dependent upon changes in size and configuration. Yamakawa (1967) has investigated, theoretically, the case of a single polymer constituent, which must be described by a Rayleigh–Debye form factor, in a binary solvent.

b. Polyions in Supporting Electrolyte. We now consider the application of light scattering to the study of the hydrolytic aggregation of ions to form polyions. Simple inorganic salts frequently undergo complex chemical reactions in aqueous solution such as hydrolysis, complexing, and polymerization so that the empirical formula of the dry salt hardly represents the conditions in solution. Only the case where one of the ions of a binary salt, cation or anion, undergoes aggregation will be treated. The univalent counterion is assumed to remain unassociated. There may be several simultaneous processes occurring, and the aim of research in this field is to elucidate the predominant species and to determine the equilibrium constants for the various reactions. These are normally studied in the presence of a large excess of supporting neutral electrolyte such as $NaClO_4$ since this permits one to assume that the activity coefficients remain constant and accordingly the equilibria can be determined from measurements of the concentrations of the ionic species.

When there is only one such predominant ionic species, it may be possible to determine its aggregation number and its charge by light scattering. The supporting electrolyte makes it necessary to utilize multicomponent theory, but it also simplifies matters because, in a sufficient excess, the interaction coefficients become zero ($\partial \ln \gamma_i/\partial m_i = 0$). Starting with the above expression for the turbidity of a three-component system (9.3.18), Tobias and Tyree (1959) have shown that

$$\frac{Hc}{\tau_E} = \frac{1}{M_2} + \frac{10^3 Z_2{}^2 c}{2m_3 M_2{}^2} \tag{9.3.31}$$

The excess turbidity τ_E, as in the previous section, is with reference to a solution containing only the supporting electrolyte. The components are based upon Scatchard's (1946) definitions. Instead of the complex electrolyte PX_Z as component 2 in the presence of the supporting electrolyte BX as component 3, a new component 2 is defined by

$$PX_Z - (Z/2)BX \qquad \text{or} \qquad PX_{Z/2}B_{-Z/2} \tag{9.3.32}$$

as well as a new component 3 comprised of the remainder of BX. Now the addition of one mole of component 2 results in the addition of one mole of particles to the solution. Its molecular weight is $M_{PX_Z} - (Z/2)M_{BX}$. The molality of component 2, m_2, is the same as for the usual definition, i.e., PX_Z, but its activity is different. On the other hand, component 3 now differs in molality but not in activity from its usual definition. In addition to the zero value for the interaction coefficients, the above expression also assumes that

$$Z_2 m_2/2m_3 \ll 1 \tag{9.3.33}$$

where Z_2 is the charge of the polymeric cation. This condition is fulfilled as long as the charge is relatively small and there is a sufficiently high concentration of supporting electrolyte.

In many cases, the ionic polymerization occurs at such a low pH that there are two supporting electrolytes present, HX and BX. These are designated as components 3 and 5, respectively, and the appropriate expression for this case is

$$\frac{Hc}{\tau_E} = \frac{1}{M_2} + \frac{10^3 Z_2{}^2 c}{2M_2{}^2(m_3 + m_5)} \tag{9.3.34}$$

where component 2 is now defined by

$$PX_Z - \frac{Zm_3}{2(m_3 + m_5)}HX - \frac{Zm_5}{2(m_3 + m_5)}BX \tag{9.3.35}$$

Actually, m_3 and m_5 can be represented sufficiently closely by the molality of HX and BX on the usual basis.

Many systems are somewhat more complicated in that there may be complexing of the polymeric cation with the ions of opposite sign furnished by the large excess of supporting electrolyte. This has the effect of reducing the charge on the polymeric ion, thereby making the above conditions more applicable. When ions of opposite charge to the polyion react to form a complex, the effective charge becomes

$$Z_2{}^* = (Z_2 - J) \tag{9.3.36}$$

Component 2 is now redefined as

$$PX_Z - (Z_2{}^*/2)BX \tag{9.3.37}$$

for the three-component case, and component 3 is given by

$$m_3{}^* = m_3 - (Jm_2/2) \tag{9.3.38}$$

This leads to

$$\frac{Hc}{\tau_{\mathrm{E}}} = \frac{1}{M_2} + \frac{10^3(Z_2{}^*)^2 c}{2m_3{}^* M_2{}^2} \tag{9.3.39}$$

We will now consider the interpretation of the experimental data. The salt may be designated by

$$\{[PH_nX_J]X_{z+n-J}\}_N \tag{9.3.40}$$

where N is the aggregation number of the polyion. There are n hydrogen ions and J ions of X bound to each monomeric ion. The number of bound hydrogen ions (or hydroxyl ions) can be determined by electrometric titration. Therefore, the charge per monomer unit is limited to values

$$Z + n < Z_2{}^*/N < 0 \tag{9.3.41}$$

By working at high n (low pH), this range can be made quite small since Z and n have opposite signs. Furthermore, if perchlorate ion is chosen for B, the complexing is generally quite weak so that $J < 0.5$. It is convenient to plot a deviation function as a function of concentration

$$\frac{1}{N_{Z'}} = \frac{10^3 H''(\psi')^2 m'}{\tau_{\mathrm{E}}} - \frac{(Z')^2 m'}{2(m_3 + m_5)} \tag{9.3.42}$$

where m' is the molality of component 2 computed as monomer

$$m' = mN \tag{9.3.43}$$

and

$$\psi' = (\partial n/\partial m')_{T,p} \tag{9.3.44}$$

What is desired is to determine both the degree of polymerization or aggregation number N and the effective charge Z_2^* simultaneously from the data.

This is done by calculating $1/N_{Z'}$, using trial values of Z' selected from within the range given by (9.3.41) for various values of the concentration. A result for alkaline solutions of the complex anion prepared from $K_{18}Nb_{12}O_{37} \cdot 27H_2O$ (Nelson and Tobias, 1964) is shown in Fig. 9.19 where $1/N_{Z'}$ is plotted as a function of Z' with concentration as the parameter. The point of intersection of the curve leads to an aggregation number of about 6 and a charge per monomer unit of -0.3. This, along with other evidence, led to the suggestion that the niobate species has the approximate formula $[HNb_6O_{19}]^{-7}$ and that the charge is reduced to about -2 by the binding of potassium ions. An analysis based upon ultracentrifugation data gave similar results; an aggregation number of 5.8 and a charge per monomer unit of -0.2. The experimental limitations of the methods have been discussed in some detail by Tobias and Tyree (1960) and other studies have been reported by Angstadt and Tyree (1962), Craig and Tyree (1965), Nelson and Tobias (1963), Copley *et al.* (1965), and Hentz and Tyree (1964).

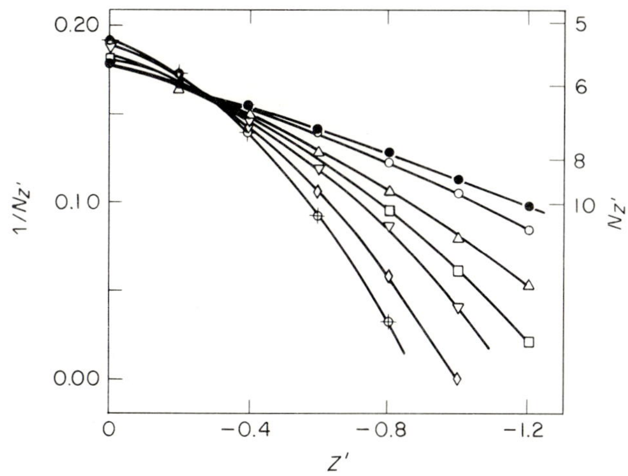

FIG. 9.19. Light scattering determination of $1/N$ as a function of Z' with concentration as parameter; ⊕ 0.354 *M*, ◇ 0.266 *M*, ▽ 0.177 *M*, □ 0.133 *M*, △ 0.0888 *M*, ○ 0.0444 *M*, ● 0.0266 *M* Nb (V) (Nelson and Tobias, 1964).

c. Surfactants. Surfactants or surface-active molecules are characterized by the presence of both a polar group, which exhibits an affinity for polar solvents such as water, and a nonpolar group which has a low affinity for such solvents. If the polar group ionizes, the surface-active molecule is said to be cationic or anionic depending upon the charge of the resulting complex ion. Sodium dodecyl sulfate $[CH_3(CH_2)_{11}OSO_3]^- Na^+$ is an example of an anionic surface-active molecule. The hydrocarbon chain is the nonpolar group: the sulfate is the polar group.

The great interest in this class of compounds is due to the detergency of their solutions, and this in turn depends upon the formation of micelles above a certain concentration termed the critical micelle concentration (CMC). A micelle is an aggregate of surface-active molecules in which the molecules are clustered so as to accomplish a minimum "contact" between the solvent and the nonpolar groups when the solvent is polar or between the solvent and the polar groups when the solvent is nonpolar. It is characterized by an aggregation number or micellar weight and, for an ionic surfactant, by a charge. Obviously, the micellar region should exhibit a marked increase in turbidity. Furthermore, the turbidity of ionic detergents should be sensitive to the ionic environment.

Some observed turbidity data for sodium dodecyl sulfate in water are shown in Fig. 9.20 for various concentrations of surfactant and in the presence of various concentrations of NaCl (Huisman, 1964). The sharp upturn of the turbidity denotes the point at which micelles first appear; this determines the critical micelle concentration. Obviously, the CMC varies with the concentration of added electrolyte. The micellar weight is usually assumed to remain constant beyond the CMC although it may depend upon the salt concentration. The solution is thought to consist of monodisperse micelles in equilibrium with a concentration of surfactant molecules equal to the CMC. Debye (1949) first proposed that this system could be treated as a simple two-component system and that the Debye equation could be applied as follows:

$$H(c - c^*)/\tau_{\mathrm{E}} = (1/m_a M_s) + 2Bc \qquad (9.3.45)$$

where c^* is the CMC, τ_{E} is the excess turbidity relative to the surfactant solution at the CMC, m_a is the aggregation number, and M_s is the molecular weight of the surfactant monomer. The micellar weight is obviously $m_a M_s$. This simple approach has been used successfully in determining the micellar weight of both ionic and nonionic detergents in polar solvents and of nonionic detergents in nonpolar solvents. In each of these cases, ionic effects are either absent or minimal. Heilweil (1964), Kuriyama (1962), and Elworthy and MacFarlane (1962) may be consulted for some examples.

Prins and Hermans (1956a) have considered the surfactant-solvent-salt system from the point of view of multicomponent theory under the assumption that the system is ideal, so that the Gibbs free energy of the system can be expressed in terms of concentrations. This assumption is equivalent to the approximation made by Tobias and Tyree (1959) that the interaction coefficients could be neglected in treating the aggregation of simple electrolytes in a high concentration of supporting electrolyte. The monomers and

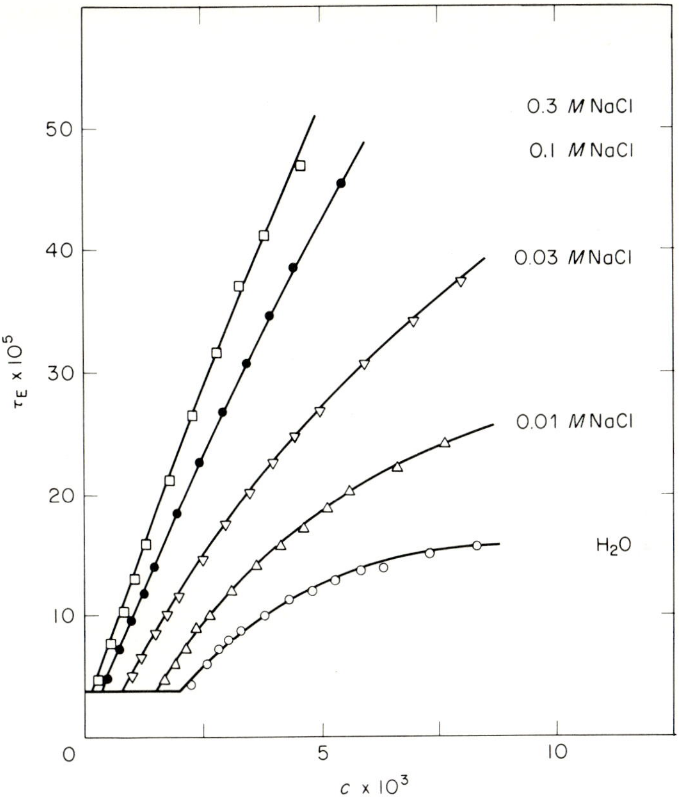

FIG. 9.20. Turbidity for solutions of sodium dodecyl sulfate at various concentrations of NaCl; 21°C; $\lambda = 436$ mμ (Huisman, 1964).

micelles are assumed to be separate components so the system contains four components rather than three. The micelles are also assumed to be monodisperse. The monomers are completely dissociated, but the micelles, on the other hand, are only partially dissociated. Accordingly, the micelle which consists of m_a monomer molecules will be assigned a charge of p. If all the ionic species are assumed to be univalent, this means that $m_a - p$ counterions are bound to the micelle; also that the effective degree of dissociation of the micelles is

$$\alpha = p/m_a \tag{9.3.46}$$

Consider a volume V containing N_0 water molecules, N_4 monomer ions, N_3 ions of added salt of the same sign of charge as the monomers, and N_m

micelles, each with charge p and containing m_a monomers. Then there will be

$$N_2 = N_4 + N_3 + pN_m \tag{9.3.47}$$

counterions, while the total surfactant content expressed in monomers is

$$N = N_4 + m_a N_m \tag{9.3.48}$$

The assumption of ideal behavior permits the Gibbs free energy to be expressed as

$$G/kT = \sum_j N_j(\mu_j{}^0/kT + \ln x_j) \tag{9.3.49}$$

where x_j represents the mole fraction

$$x_j = N_j/S; \qquad S = \sum_j N_j = N_0 + 2N_4 + 2N_3 + (p+1)N_m \tag{9.3.50}$$

In Prins and Herman's notation, the derivatives in the multicomponent expression for the turbidity are taken with respect to N_i. The derivatives of the chemical potential are actually the second derivatives of the total free energy, viz.

$$G_{ij} = (\partial^2 G/\partial N_i\, \partial N_j)_{T,p} = (\partial\mu_i/\partial N_j)_{T,p} \tag{9.3.51}$$

This results in the following expressions which must be substituted into the multicomponent equation:

$$G_{44}/kT = 1/N_2 + 1/N_4 - 4/S \tag{9.3.52}$$

$$G_{43}/kT = 1/N_2 - 4/S \tag{9.3.53}$$

$$G_{4m}/kT = p/N_2 - 2(p+1)/S \tag{9.3.54}$$

$$G_{33}/kT = 1/N_2 + 1/N_3 - 4/S \tag{9.3.55}$$

$$G_{3m}/kT = p/N_2 - 2(p+1)/S \tag{9.3.56}$$

$$G_{mm}/kT = p^2/N_2 + 1/N_m - (p+1)^2/S \tag{9.3.57}$$

There are a number of additional approximations to be made, viz.

$$\psi_m = m_a \psi_4 \tag{9.3.58}$$

$$N_3 \quad \text{and} \quad N_4 \ll m_a^2 N_m \tag{9.3.59}$$

$$\psi_m = (M_m/VN_A)(\partial n/\partial c)_{p,T} \tag{9.3.60}$$

With these

$$\frac{H(c-c^*)}{\tau_E} = \frac{g}{M_m}\left[1 + \frac{c-c^*}{2M_m}\,\frac{p^2+p-gp}{n_4+n_3} + 0(c-c^*)\right] \tag{9.3.61}$$

where

$$g = \frac{N_4 + N_3}{b_4 N_4 + b_3 N_3} \tag{9.3.62}$$

$$b_4 = 1 - \alpha + (\alpha^2/2) + (\alpha/2m_a) \tag{9.3.63}$$

$$b_3 = 1 - f\alpha + (f^2\alpha^2/2) + (f^2\alpha/2m_a) \tag{9.3.64}$$

$$f = (\psi_3/\psi_4) \tag{9.3.65}$$

Subscripts 3 and 4 denote salt component and the surfactant monomer; subscript m denotes the micelles; n_3 and n_4 are concentrations in moles per milliliter.

This shows that the Debye extrapolation procedure does not give the exact micellar weight, but an apparent one

$$M_{\mathrm{ap}} = M_m/g \tag{9.3.66}$$

Prins and Hermans (1956a) estimate that for a typical degree of dissociation these will differ by about 10%. It is clear that it should be possible to determine both m and p (remembering that $m_a = M_m/M$) from the intercept and slope of a Debye plot.

When the salt content is high, the CMC is frequently reduced to quite a small value so that $N_4 \ll N_3$ and then

$$\frac{H(c - c^*)}{\tau_{\mathrm{E}}} = \frac{1}{b_3 M_m}\left[1 + \frac{c - c^*}{2n_3 M_m}\left(p^2 + p - \frac{p}{b_3}\right)\right] \tag{9.3.67}$$

On the other hand, in the absence of salt ($N_3 = 0$)

$$\frac{H(c - c^*)}{\tau_{\mathrm{E}}} = \frac{1}{b_4 M_m}\left[1 + \frac{c - c^*}{2c^* m_a}\left(p^2 + p - \frac{p}{b_4}\right)\right] \tag{9.3.68}$$

It may be precarious to assume that the system behaves ideally in this latter case, particularly if there is appreciable ionization.

Prins and Hermans (1956a) also attempted to introduce an activity correction for the micelle. They assumed that the electrical double layer surrounding the micelle increases its effective volume. The activity coefficient in turn is calculated from the effect of the excluded volume upon the entropy of mixing. From this the interaction coefficient is obtained, and upon in-insertion in the light scattering equation (9.3.11) an additional correction is obtained as follows:

$$\frac{H(c - c^*)}{\tau_{\mathrm{E}}} = \frac{g}{M_m}\left[1 + \frac{c - c^*}{2M_m}\left(\frac{p^2 + p + gp}{n_4 + n_3} + 16v_m N_{\mathrm{A}}\right)\right] \tag{9.3.69}$$

where the excluded volume is

$$v_m = \tfrac{1}{6}\pi D_h^{\ 3} \tag{9.3.70}$$

and the effective diameter is given quite closely by

$$D_h = 2(a + [1/\kappa]) \tag{9.3.71}$$

Here a is the radius of the micelle and $1/\kappa$ is the effective thickness of the double layer as defined by (9.2.52).

Mysels (1954, 1955) developed expressions only very slightly different from those of Prins and Hermans (1956a), using a molecular model rather than a derivation based directly upon the multicomponent theory. In Mysels' theory, the quantity g in (9.3.61) is replaced by q where

$$q = (N_4 + N_3)^2/(N_4^{\ 2}\, d_1 + 2N_4 N_3\, d_2 + N_3^{\ 2}\, d_3) \tag{9.3.72}$$

$$d_1 = 1 - \alpha + (\alpha^2/4) + (\alpha/4m_a) \tag{9.3.73}$$

$$d_2 = 1 - (\alpha/2) + f(\alpha/2) + f(\alpha^2/4) + f(\alpha/4m_a) \tag{9.3.74}$$

$$d_3 = 1 - f\alpha + f^2(\alpha^2/2) + f^2(\alpha/4m_a) \tag{9.3.75}$$

$$f = (\partial n/\partial N_3)_{T,p}/(\partial n/\partial N_4)_{T,p} \tag{9.3.76}$$

The limiting expressions at high salt concentrations and at zero salt concentrations are identical. Princen and Mysels (1957) have discussed the difference between the two treatments and give explicit analytical solutions for p and m_a in terms of the intercept and slope of the Debye plot, but these will not be reproduced here. They have also considered the effect of dimerization of the monomers upon the turbidity. Significant errors may be introduced into the estimation of p and m_a if there is appreciable dimerization and if this is not taken into account.

On the basis of extensive experiments with the sodium alkyl sulfates which are anionic detergents (Prins and Hermans, 1956b) and with the quaternary alkyl ammonium bromides which are cationic detergents (Trap and Hermans, 1956), it may be sufficiently accurate to simplify the above expressions to

$$\frac{H(c - c^*)}{\tau_{\mathrm{E}}} = \frac{1}{M_{\mathrm{ap}}} + \left[\frac{p^2}{2(n_3 + n_4)M_m M_{\mathrm{ap}}} + \frac{8v_m N_{\mathrm{A}}}{M_{\mathrm{ap}}^2}\right](c - c^*) \tag{9.3.77}$$

In the above cases, the effect of this approximation on the calculated value of m_a was negligible and only affected p by a few percent.

Also, for these surfactants the difference between the true micellar weight and the apparent one (9.3.66) was comparatively small, and the marked influence of the salt upon the intercept of the Debye plot indicates that the micellar weight actually does increase significantly with added salt. The reality of this increase in micellar weight has been confirmed many times with both cationic and anionic surfactants so that this crucial problem in the theory of surfactants appears to have been decisively resolved by light scattering (Mysels and Princen, 1959; Tartar and Lelong, 1955; Kushner and Hubbard, 1955; Hutchinson and Melrose, 1954; Kushner *et al.*, 1957; Anacker, 1958).

Other light scattering studies have explored the effect of the specific ions of the supporting electrolyte upon the micellar weight. Both the charge and the chemical nature of the ion with charge opposite to the surfactant ion significantly influence the micellar weight. Thus the aggregation number of dodecyltrimethyl ammonium bromide increases in 0.5 molal supporting electrolyte from 38 to 84 as the counterion varies from IO_3^-, ClO_3^-, BrO_3^-, F^-, Cl^-, NO_3^-, to Br^- (Anacker and Ghose, 1963). On the other hand, the nature of the ion having the same charge as the surfactant ion appears to have little effect (Wasik and Hubbard, 1964).

The results obtained with the equations of Prins and Hermans and of Mysels have been compared with those from equilibrium ultracentrifugation measurements for sodium dodecyl sulfate in 0.4 *M* NaCl by Anacker *et al.* (1964). Both the aggregation numbers and charges obtained by these methods agreed quite well. Anacker and Westwell (1964) have shown, moreover, that if, in the Prins and Hermans theory, the monomeric surfactant is treated as if it were part of the supporting electrolyte rather than as an independent component, comparable results are still obtained. More recently, Emerson and Holtzer (1967) have obtained results for sodium dodecyl sulfate, as well as for dodecyltrimethylammonium chloride, bromide, and nitrate.

Kratohvil and Delli Colli (1968) have studied the light scattering of two bile salts which exhibit micellar properties similar to those of ionic detergents. The micellar weight increases with added salt, the aggregation numbers varying from 8 in water to about 50 in 0.5 *M* electrolyte solutions. However, there was little effect by the specific ions comprising the electrolyte. The values of the CMC agreed with those obtained by surface tension, and the micellar weights agreed with those determined by ultracentrifugation.

d. Polyelectrolytes. Polyelectrolyte solutions have been very actively studied by light scattering. The aim of such studies is to determine the molecular weight and the interactions of the polyelectrolyte in aqueous salt solutions.

A polyelectrolyte is a charged polymer containing ionizable groups and it derives its charge by ionization of these groups or alternatively by the binding of ions from the solution. The ionization may be dependent upon the pH and the ionic environment. The charge of polyelectrolytes containing both acidic and basic groups, termed polyampholytes, will vary from positive values at low pH through the neutral isoelectric point to negative values at high pH.

The solution properties of polyelectrolytes differ in the same way from uncharged polymers that electrolytes composed of ions differ from uncharged molecules. They dissolve in polar solvents, conduct electricity, and are profoundly affected by the coulombic forces between the charges that they possess. As an example of the latter effect, the presence of charges on the polyion sets up repulsive forces among the chain segments, which causes the ion to assume an extended configuration in solution rather than the relatively tightly coiled form exhibited by uncharged polymers. Increase of the ionic strength by addition of supporting neutral electrolyte gives rise to an electrostatic screening effect which permits the molecule to coil up more compactly. This is shown in Fig. 9.21 where the ratio of the polyelectrolyte concentration to the Rayleigh ratio is plotted against $\sin^2(\theta/2)$ for sodium poly(styrene-*p*-sulfonate) in various concentrations of aqueous NaCl (Takahashi *et al.*, 1967). The limiting slope and hence the radius of gyration decreases with increasing salt concentration in accordance with this point of view. Such configurational changes will also result in effects upon the second virial coefficient of the solution. Alternatively, the second virial coefficient may be influenced by changes in temperature as shown in Fig. 9.22 (Eisenberg, 1966) where a quantity proportional to Hc/τ is plotted against concentration for various temperatures between 19.1 to 45.6°C.

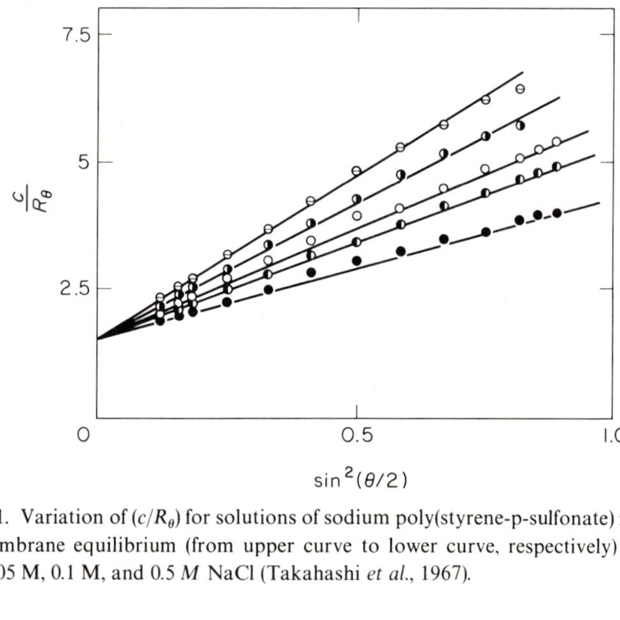

FIG. 9.21. Variation of (c/R_θ) for solutions of sodium poly(styrene-p-sulfonate) in NaCl which are in membrane equilibrium (from upper curve to lower curve, respectively) with 0.005 M, 0.01 M, 0.05 M, 0.1 M, and 0.5 M NaCl (Takahashi *et al.*, 1967).

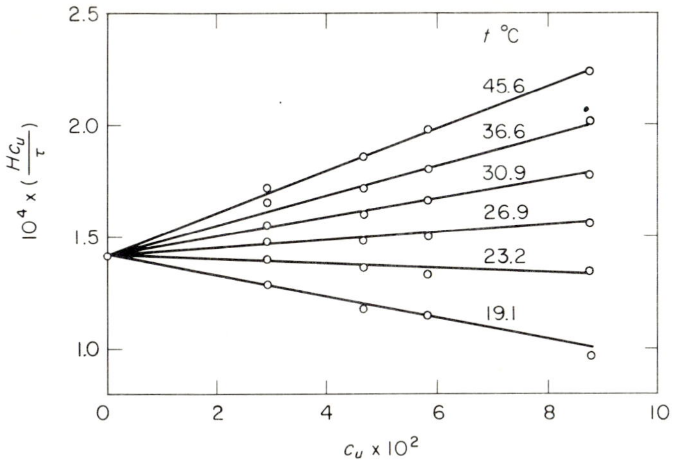

FIG. 9.22. Light scattering data of polyvinylsulfonate in 0.65 M KCl at various temperatures. c_u is the concentration in equivalents of monomer per liter of solution (Eisenberg, 1966).

This is for the three-component system potassium polyvinyl sulfonate–potassium chloride–water. The second virial coefficient is obtained from the slope of these curves and varies from negative values at the lower temperatures through the Flory temperature where the second virial coefficient is zero to positive values at still higher temperatures.

Polyelectrolytes exhibit the Donnan effect. If a polyelectrolyte solution is equilibrated with a salt solution through a membrane which is impermeable to the macromolecules but freely passes the solvent and the salt, the concentration of salt in the polyelectrolyte solution will always be less than in the simple salt solution with which it is in equilibrium. This is a consequence of the requirement that the chemical potential of the diffusible species be equal on both sides of the membrane. The Donnan equilibrium will be of help in interpreting the light-scattering data of such systems.

e. Polyelectrolytes; Proteins at Low Charge Density. It is sometimes possible to obtain the molecular weight of a protein directly from a Debye plot, provided there is a sufficient excess of supporting electrolyte. Edsall *et al.* (1950) were among the first to emphasize the importance of utilizing multicomponent theory in analyzing light-scattering data from protein systems. Their point of departure is Stockmayer's expression given by (9.3.16) or (9.3.18). They pointed out, for dilute solutions of the seven albumins with which they dealt, that the following equation is an adequate approximation:

$$10^3 H''\psi_2{}^2/\tau_{\mathrm{E}} = a_{22} - (a_{23}^2/a_{33}) \tag{9.3.78}$$

This equation can be expressed in terms of quantities more commonly used in the experimental treatment of such solutions by

$$\frac{HV^0g}{10^3\tau_E} = \frac{1}{M_2} + \frac{10^3}{M_2{}^2}\left[\frac{Z_2{}^2}{2m_3\varepsilon} + \beta_{22} - \frac{\beta_{23}^2 - (10^3Z_2{}^2g/2M_2m_3{}^2\varepsilon)}{(2/m_3\varepsilon) + \beta_{33}}\right]g \quad (9.3.79)$$

where

$$\varepsilon = 1 - (Z_2m_2/2m_3)^2 \quad (9.3.80)$$

Here Z_2 is the mean net proton charge per protein ion or the moles of "bound acid." It can be determined experimentally by a potentiometric titration, and its value can be adjusted by the addition of strong acid or base. The relations between the interaction parameter β_{ij} and the chemical potential, activity, and activity coefficient have been given earlier [(9.3.23) to (9.3.27)]. V^0 is the volume of solution containing 1000 grams of solvent, g is the weight of component 2 per gram of solvent, and m_3 is the molality of component 3.

The choice of components follows Scatchard (1946), just as for the polyions treated earlier. The solvent is denoted as component 1; the nondiffusible macromolecule, in this case the protein, is component 2; and the supporting uni-univalent diffusible electrolyte having an ion in common with the protein is component 3. The protein component is distinguished from the protein ion. It is defined so as to be electrically neutral and also so that addition of one mole of protein component involves only the net addition of one mole of ions to the system. This can be effected by adding $Z_2/2$ moles of diffusible ions with sign of charge opposite to that of the protein and removing $Z_2/2$ moles of the diffusible ion with the same charge when one mole of protein is added to the system. Thus, the net addition of moles of diffusible ions is zero and the solution remains electrically neutral after the protein component is added.

This will be illustrated by a numerical example for a solution of $10^{-4}\,m$ PCl_{20} in $10^{-2}\,m$ NaCl where P designates a protein ion. The total ionic concentrations present are $[P^{+20}] = 0.0001\,m$; $[Cl^-] = 0.012\,m$; $[Na^+] = 0.010\,m$. The value of m_2 is 0.0001 consisting of $0.0001\,mP^{+20}$ plus $0.001\,m\,Cl^-$ minus $0.001\,m\,Na^+$. The value of m_3 is $0.011\,m$ NaCl. Half of the chloride ion associated with the protein is counted as part of the protein component and half of it is counted as part of component 3. There is also a negative concentration of Na^+ associated with the protein component which must be compensated for by an addition to component 3.

Whenever Eq. (9.3.79) holds, the molecular weight can be obtained from the usual Debye plot of Hc/τ_E against c. However, unless the concentration of supporting electrolyte is sufficiently high, these curves will exhibit considerable curvature, and the data must be obtained at very low concentrations of proteins in order to carry out the extrapolation with sufficient accuracy. This is illustrated in Fig. 9.23 for bovine serum albumin in NaCl solutions of varying ionic strength. This effect of added salt upon the turbidity is similar to that observed for the heteropoly acids (Fig. 9.7).

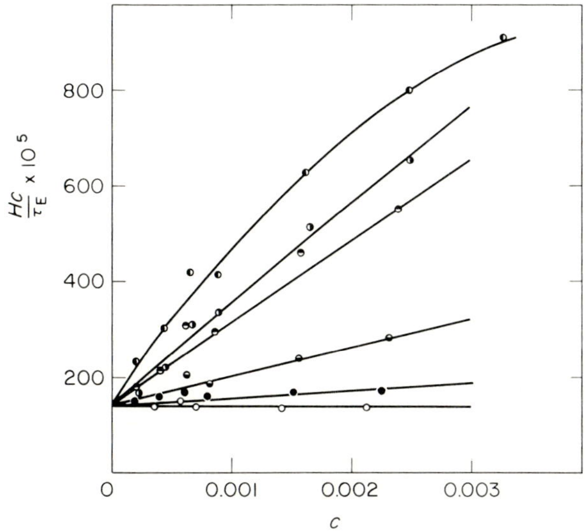

FIG. 9.23. Hc/τ_E for BSA in NaCl of ionic strengths (from upper curve to lower curve, respectively) of $<0.0001\ m$, $0.0001\ m$, $0.0003\ m$, $0.0015\ m$, $0.015\ m$, and $0.15\ m$ (Edsall *et al.*, 1950).

When the concentration of supporting electrolyte is high enough so that $|Z_2 m_2| \ll 2m_3$, the above equation reduces to

$$\frac{Hc}{\tau_E} = \frac{1}{M_2} + \frac{10^3}{M_2{}^2}\left[\frac{Z_2{}^2}{2m_3} + \beta_{22} - \frac{\beta_{23}{}^2 m_3}{2 + \beta_{33} m_3}\right]c \tag{9.3.81}$$

Here it has also been assumed that the solution is sufficiently dilute so that $V^0 = 10^3$ and $c = g$. Under these conditions the slope is just twice that of the corresponding osmotic pressure curve of π/c against c. The physical significance of each of these terms in the square bracket is well known in connection with osmotic pressure studies. The first term is the contribution to the osmotic pressure of the Donnan equilibrium in an ideal solution, i.e., when the interaction constants are zero. When this term predominates, this expression reduces appropriately to (9.3.31), used by Tobias and Tyree (1959). The third term is a measure of the protein salt interaction which can be interpreted in terms of the binding of salt ions to the protein (Scatchard and Bregman, 1959). Figure 9.24 compares the quantity in the square bracket of (9.3.81) obtained by light scattering (Edsall *et al.*, 1950) plotted as points with the corresponding quantity obtained by osmotic pressure measurements (Scatchard *et al.*, 1946). The agreement is well within the experimental error, particularly when one considers that there is often considerable variation between different preparations of bovine serum albumin. Additional experimental studies of this nature on proteins have been cited by Stacey (1956) and Kratohvil (1966).

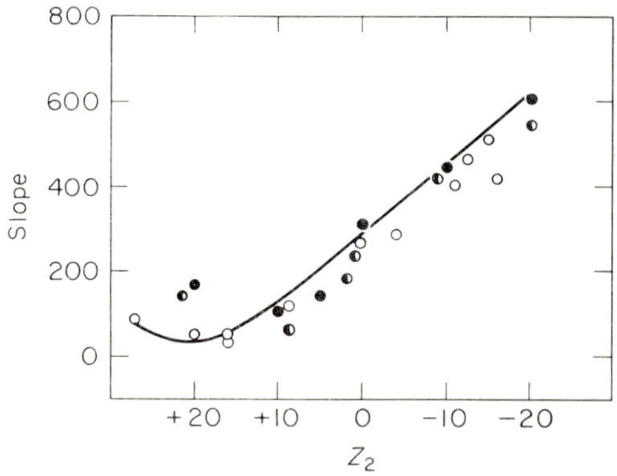

FIG. 9.24. Slope of the plot of Hc/τ against c for BSA in 0.015 m NaCl compared with the corresponding quantity obtained from osmotic pressure measurements. The light-scattering points were obtained with two different instruments. Full circles are for 0.018 M NaCl. The abscissa Z_2 is the mean net proton charge per protein ion (Edsall *et al.*, 1950).

f. Synthetic Polyelectrolytes; Use of Membrane Distribution Data. Strauss and his co-workers, in determining the molecular weights of synthetic polyelectrolytes and polyphosphates, have carried Stockmayer's equation to a higher approximation than Edsall *et al.* (Strauss and Wineman, 1958; Strauss and Williams, 1961; Strauss and Ander, 1962). Their result can be written (Ander, 1962)

$$\frac{H''V(\partial n/\partial g)_{m_3g}^2 g}{10^3\tau_E} = (1 - D)^{-2}\left\{\frac{1}{M_2} + \frac{10^3 g}{M_2{}^2}\left[\frac{\alpha^2 Z_2{}^2}{m_3} + \beta_{22} - \frac{a_{23}^2}{a_{33}}\right]\right\} \quad (9.3.82)$$

where g is the grams of polyelectrolyte per gram of water, α is the fraction of the groups on the chain which is ionized, and

$$D = a_{23}\psi_3/a_{33}\psi_2 \quad (9.3.83)$$

It is apparent that extrapolation of the Debye plot to infinite dilution will lead to a correct molecular weight only when $D = 0$. This was precisely the condition that Edsall *et al.* utilized to obtain their approximate relation. It corresponds to a low charge on the polyelectrolyte in a solution of high ionic strength. When D is finite, which is frequently the case for synthetic polyelectrolytes which have a very high charge density and a correspondingly strong interaction between polymer and salt, an apparent molecular weight will be obtained as given by

$$M_2^* = M_2(1 - D)^2 \tag{9.3.84}$$

In order to calculate D, the relation given by (9.3.19) was used, resulting in

$$D = (a_{23}/a_{33})(\psi_3/\psi_2) = -(\partial m_3/\partial g)_{\mu_3}[\psi_3/(\partial n/\partial g)]_{T,p} \tag{9.3.85}$$

The first factor on the right-hand side of this equation may be obtained from the distribution of component 3 on both sides of the membrane in a Donnan equilibrium experiment as given by

$$(m_3' - m_3)/g \tag{9.3.86}$$

where m_3' and m_3 are the salt concentrations in the dialyzing solution and the polyelectrolyte solution, respectively. The magnitude of D can be quite significant in some cases. For example, for one sample of sodium polyphosphate, Strauss and Wineman found that $(1 - D)^{-2}$ ranged from 1.28 to 1.42 as the supporting electrolyte concentration varied from 0.1 M to 0.4 M NaBr.

Timasheff and Kronman (1959) have shown that D is quite small for most protein systems which have been investigated by light scattering, thus justifying the calculation of molecular weights directly from the Debye plot as Edsall *et al.* (1950) had done. However, in some instances this cannot be done. In the system bovine serum albumin-urea or other protein solutions where there is a strong interaction between the protein and the low molecular-weight component, the molecular weight obtained from the simple Debye treatment may be in error by a factor greater than two if the value of D is neglected. In proteins, this interaction is usually conceived of as the result of the binding of a low molecular weight constituent by the protein. Alternatively, if the deviations of the molecular weight obtained from the extrapolation of Hc/τ_E are known, this can be coupled with an equation such as

$$D = -2\frac{\psi_3}{\psi_2}\frac{J}{\nu_3} \tag{9.3.87}$$

to determine the extent of binding (Timasheff, 1963). Here, J is the average number of particles of component 3 bound to a molecule of component 2 and ν_3 is the number of particles into which component 3 dissociates. Alternatively, if the solvent is preferentially bound (hydration), correspondingly similar effects are obtained, except that in this case D takes on positive rather than negative values.

It is possible to evaluate a_{23} from electromotive force measurements. If the activity of component 3 is measured in an appropriate cell, then the change in the electromotive force resulting from the addition of polyelectrolyte is

$$\frac{\partial E}{\partial m_2} = -\frac{RT}{F}\left(\frac{\partial \ln a_3}{\partial m_2}\right) = -\frac{a_{32}}{F} \tag{9.3.88}$$

where F is Faraday's constant (Ander, 1962). Of course, a_{33} may also be obtained from an appropriate cell by variation of m_3, so that a complete evaluation of each of the thermodynamic quantities in (9.3.16) can be obtained from electromotive force measurements, provided suitable electrodes can be obtained. Such measurements are considerably more accurate than membrane equilibrium studies, even when a suitable membrane can be found. Imai and Eisenberg (1966) have utilized potassium-sensitive glass electrodes for such a study.

g. Turbidity and Refractive Index Increment in Donnan Equilibrium. Still another formulation of the scattering equation by Vrij and Overbeek (1962) lends itself to the determination of the molecular weight of polyelectrolyte systems without the necessity of carrying out an actual chemical analysis of the supporting electrolyte on each side of a Donnan membrane experiment. This equation, which is equivalent to one also derived by Ooi (1958), is

$$\tau_E{}^* = \frac{32\pi^3 n^2 RT}{3\lambda_0{}^4 N_A} \frac{\left[\sum_{i=2} \psi_i(\partial c_i/\partial c_2)_{\mu_s,T}\right]^2 c_2}{(\partial \pi^*/\partial c_2)_{\mu_1,\mu_3,\ldots,T}} \tag{9.3.89}$$

where the subscript μ_s indicates that the chemical potentials of all low molecular weight components are kept constant. The osmotic pressure difference between the polyelectrolyte solution and a solution in equilibrium with it containing the diffusible components (Donnan equilibrium) is designated by π^*. The excess turbidity $\tau_E{}^*$ is the difference between the total turbidity and that of this latter solution, which can be measured directly. Eisenberg (1962) has obtained a similar result using a somewhat different notation.

The quantity in the square bracket of (9.3.89) can be measured in two ways. The refractive index increments ψ_i can be measured in the usual way, and then the quantities $(\partial c_i/\partial c_2)_{\mu_3,T}$ can be obtained from an analysis of the composition of the diffusible components in the Donnan equilibrium. These coefficients represent the amount of electrolyte component i to be added to maintain Donnan equilibrium when one mole of colloidal electrolyte is added. For salts having a common counterion with the colloidal electrolyte, the Donnan equilibrium will always require a lower concentration within the membrane so that this is sometimes referred to as negative adsorption. This procedure is equivalent to that used by Strauss *et al.*

A more direct and less laborious technique makes use of the following identity for the quantity in the square bracket of (9.3.89).

$$\sum_{i=2} \psi_i \left(\frac{\partial c_i}{\partial c_2}\right)_{\mu_s, T} = \left(\frac{\partial n}{\partial c_2}\right)_{\mu_s, T} \tag{9.3.90}$$

This is a general thermodynamic relation except for neglecting the effect of pressure upon the refractive index which can be assumed to be negligible in all practical cases. In any event, this can be treated exactly, should that be necessary (Huisman, 1964). This quantity may be determined directly by allowing a Donnan equilibrium to be established between two solutions with a difference in concentration Δc_2 of the colloid separating the solutions and measuring the difference in refractive index Δn. It does not require a chemical analysis of the solutions. Indeed, it actually constitutes an analysis of the negative adsorption $(\partial c_i/\partial c_2)_{\mu_s}$. Casassa and Eisenberg (1961) have utilized such refractometric data to calculate this quantity for protein solutions and have successfully compared the results with direct membrane distribution data and other measurements. Vink and Dahlström (1967) measured the differential refractive index for some polysaccharides in various aqueous solutions for both dialyzed and undialyzed solutions. They found that the most pronounced differences were obtained in those cases for which there was preferential binding of the permeable solute to the polymer.

Now, if the osmotic pressure is expanded in powers of concentration, (9.3.89) leads to an expression having the same form as the Debye equation,

$$H^*c/\tau_E^* = (1/M_2) + 2Bc + \cdots \tag{9.3.91}$$

This differs from the Debye equation in that the refractive index increment in the lumped constant H^* is obtained at constant chemical potential rather than at constant concentration of supporting electrolyte. Also, the turbidity of the salt solution in Donnan equilibrium rather than at the same salt concentration of the polyelectrolyte solution is subtracted from the total measured turbidity to obtain τ_E^*.

Casassa and Eisenberg (1960) have pointed out that this procedure based upon utilizing refractive index increments obtained at constant chemical potential of component 3 is equivalent to the reduction of the three-component system to a pseudo two-component system by an appropriate redefinition of the components. The polyelectrolyte component is defined as $X_{(1-\Gamma)p}PY_{-\Gamma p}$ where P is the polyanion of charge $-p$, X is the univalent counterion, and XY is the diffusible uni-univalent electrolyte. Γ is the negative adsorption coefficient $(\partial c_3/\partial c_2)_\mu$. Physically this means that when one adds one mole of the polyelectrolyte component to one side of a membrane system containing concentration c_3 on each side, this involves the addition of one mole of X_pP and the removal of Γ_p moles of XY. Hence, there is no net flow of simple salt XY across the membrane. This, in turn, means that there is no interaction between component 2 (so defined) and the salt or that $a_{23} = 0$. Then, the factor $(1 - D)^{-2}$ in (9.3.82) reduces to unity and the last term disappears so that an equation of the same form as a two-component system is obtained.

$$\frac{H''V(\partial n/\partial g)^2_{m_3 g}}{10^3 \tau_E} = \frac{1}{M_2} + \frac{10^3 g}{M_2{}^2}\left[\frac{\alpha^2 Z_2{}^2}{m_3} + \beta_{22}\right] \tag{9.3.92}$$

Operationally, one proceeds precisely as above, measuring the turbidity and the refractive index increment with respect to the dialysate. If these strictures are not followed, the Debye plot will lead to an apparent molecular weight and the apparent second virial coefficient which can be related to the true values by the following equations:

$$M_2{}^* = M_2\left[1 + \frac{(\partial n/\partial c_3)_{c_2}}{(\partial n/\partial c_2)_{c_3}}\left(\frac{\partial c_3}{\partial c_2}\right)_{\mu_s}\right]^2 \tag{9.3.93}$$

$$B^* = B\left[1 + \frac{(\partial n/\partial c_3)_{c_2}}{(\partial n/\partial c_2)_{c_3}}\left(\frac{\partial c_3}{\partial c_2}\right)_{\mu_s}\right]^{-2} \tag{9.3.94}$$

Also

$$\left(\frac{\partial n}{\partial c_2}\right)_{\mu_3} = \left(\frac{\partial n}{\partial c_2}\right)_{c_3}\left[1 + \frac{(\partial n/\partial c_3)_{c_2}}{(\partial n/\partial c_2)_{c_3}}\left(\frac{\partial c_3}{\partial c_2}\right)_{\mu_s}\right] \tag{9.3.95}$$

neglecting the effect of compressibility (Huisman, 1964).

Vrij and Overbeek (1962) have tested these relations with solutions of poly(methylmethacrylate) (PMA) half neutralized with sodium hydroxide in solutions of sodium halides. The results are shown in Table 9.12. The apparent molecular weights $M_2{}^*$ and second virial coefficients B^* obtained when the refractive index increment was taken at constant salt concentration vary with the particular sodium halide at the same concentration. On the other hand, constant values were obtained when the above correction factors involving the negative adsorption term were applied.

TABLE 9.12

APPARENT AND TRUE MOLECULAR WEIGHT AND SECOND VIRIAL COEFFICIENT OF PMA

Solvent	$M_2{}^* \times 10^{-3}$	$M_2 \times 10^{-3}$	$B^* \times 10^3$ (ml./g)	$B \times 10^3$ (ml./g)
0.1 *M* NaF	46.7	51.5	4.12	3.74
0.1 *M* NaCl	44.7	52.1	4.35	3.73
0.1 *M* NaBr	41.2	50.0	4.00	3.3
0.1 *M* NaI	37.2	50.7	5.00	3.66

A very powerful stratagem for obtaining the molecular weight, the second virial coefficient, and the negative adsorption, can be devised if it can be assumed that the negative adsorption on an equivalent weight basis is independent of the particular sodium halide used for the supporting electrolyte. Accordingly, (9.3.95) may be modified to

$$\left(\frac{\partial n}{\partial c_2}\right)_{\mu_3} = \left(\frac{\partial n}{\partial c_2}\right)_{c_3} + \left(\frac{\partial n}{\partial c_3}\right)_{c_2} \frac{M_3}{E_2}\left[\frac{\partial(c_3/M_3)}{\partial(c_2/E_2)}\right]_{\mu_s} \tag{9.3.96}$$

where E_2 is the equivalent weight of the polyelectrolyte. In Fig. 9.25 $(\partial n/\partial c_2)_{\mu_3}$ is plotted against $M_3(\partial n/\partial c_3)c_2$ for the above sample of PMA in the four sodium halides. The straight line which is obtained indicates that the negative adsorption is constant. Its value can then be found from the slope. The next step is to determine the values of M^* and B^* from the usual Debye plot.

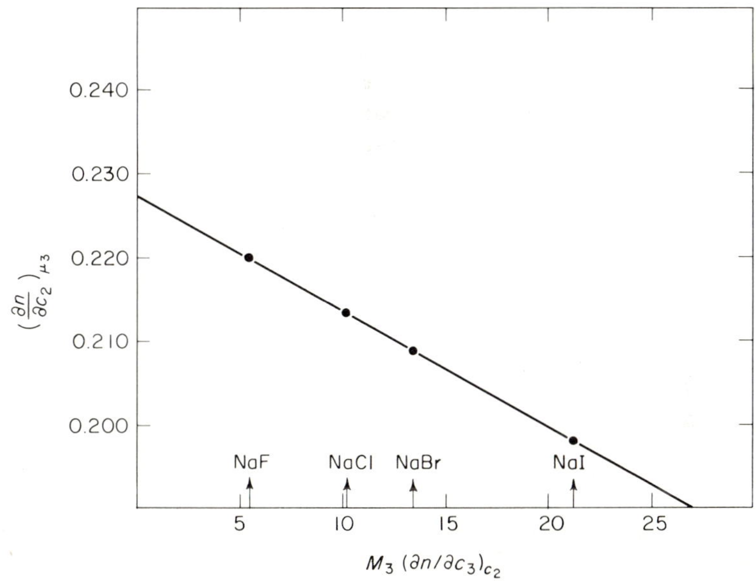

FIG. 9.25. Refractive index increment at constant chemical potential for PMA in 0.1 *M* sodium halide plotted against refractive index increment of the respective sodium halide (Overbeek *et al.*, 1963).

Then if $(M_2{}^*)^{1/2}$ and $(B^*)^{-1/2}$ are plotted against $M_3(\partial n/\partial c_3)_{c_2}$ in accordance with (9.3.94) and (9.3.95), the intercept at zero abscissa leads to the values of M_2 and B. The negative adsorption can also be determined from the slope of these curves.

This procedure does not entail any membrane experiments, requiring only conventional light scattering measurements. Vrij and Overbeek obtained a value of 50,000 for the molecular weight of PMA by this method, in excellent agreement with the values in Table 9.12 which were obtained with the aid of membrane measurements.

This method is particularly useful for the interpretation of surfactant solutions above the CMC (Huisman, 1964). Because there is an equilibrium between micelles and monomeric molecules, it is impossible to obtain true Donnan equilibrium. However, if the co-ions have negligible influence on the micellar weight and the interactions, this method can be applied to determine the correct micellar weights. In Fig. 9.26, Debye plots are shown for solutions of sodium dodecyl sulfate in 0.1 M NaF, NaCl, NaBr, and NaI and the values of $(M_2{}^*)^{1/2}$ obtained from each of the intercepts are plotted against the limiting values of $M_3(\partial n/\partial c_3)_{c_2=0}$ in Fig. 9.27. The straight line drawn through these points gives the correct molecular weight at the ordinate intercept.

Stigter (1960b, 1963) has connected the purely thermodynamic treatment of light scattering by a colloidal multicomponent system with a statistical thermodynamic theory based upon the virial expansion of the osmotic pressure and the distribution of diffusible components in the osmotic system. The osmotic pressure is represented by

$$\frac{\pi}{kT} = \rho_2 + \sum_{n\geq 2} B_n \rho_2{}^n \tag{9.3.97}$$

and

$$\frac{\rho_i}{\rho_i{}^*} = 1 + \sum_{n\geq 1} A_{in} \rho_2{}^n \tag{9.3.98}$$

where ρ_i is the number of molecules of supporting electrolyte of the ith kind per unit volume of solution inside the membrane system and $\rho_i{}^*$ is the number with which it is in osmotic equilibrium. The B_n's are cluster integrals involving the force potentials between solute molecules; the A_{in}'s involve the force potential between the solvent and the diffusible solute molecules. The final statistical equation is

$$\frac{H'' N_A \rho_2}{\tau_E}\left(\frac{\partial n}{\partial \rho_2}\right)^2_{T,p,\mu} = \frac{\rho_2}{\sigma_2}\left(\frac{\partial \sigma_2}{\partial \rho_2}\right)_{T,p,\mu} \times \left[\frac{\left(1 + \sum_{n\geq 1} A_{1n}\rho_2{}^n\right)\left(1 + \sum_{n\geq 2} nB_n\rho_2^{n-1}\right)}{1 - \sum_{n\geq 2} (n-1)A_{1n}\rho_2{}^n}\right] \tag{9.3.99}$$

where

$$\sigma_2 = \rho_2/\rho_1 \tag{9.3.100}$$

This equation has been applied to a system of colloidal spheres in an aqueous solution containing a uni-univalent simple electrolyte having an ion in common with the counterion of the colloidal electrolyte. The relevant potentials were obtained from the Verwey–Overbeek theory (1947).

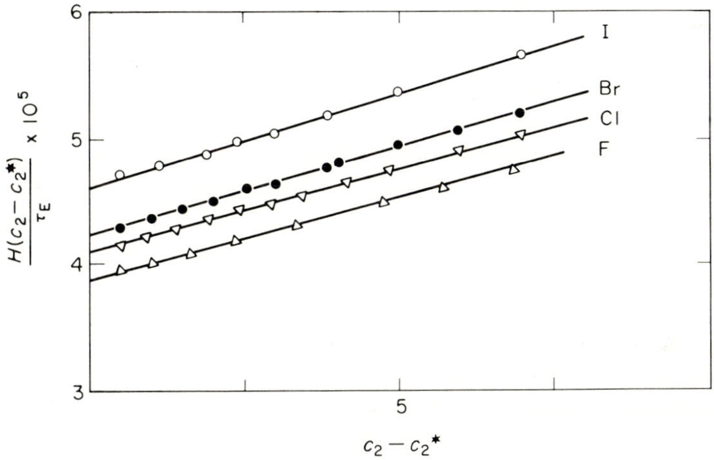

FIG. 9.26. Debye plots (Eq. 9.3.45) of sodium dodecyl sulfate in 0.1 M sodium halide (Huisman, 1964).

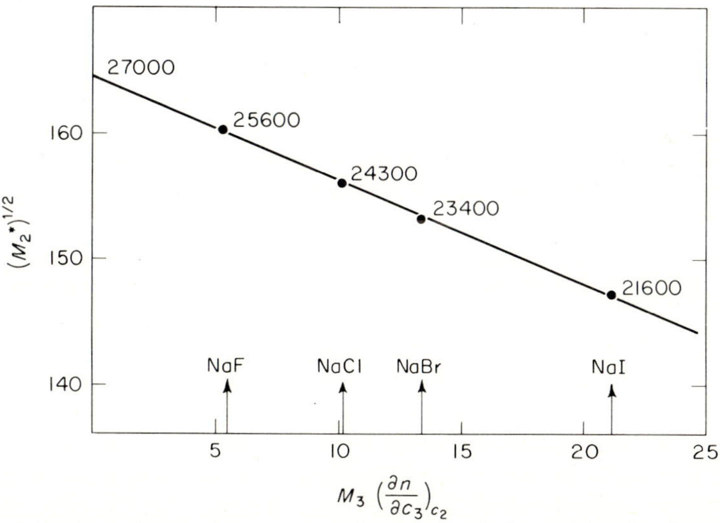

FIG. 9.27. Extrapolation of square root of apparent micellar weight of sodium dodecyl sulfate in 0.3 M supporting sodium halide to zero refractive index increment of electrolyte (Huisman, 1964).

9.4 Critical Opalescence

Much of the impetus for the early study of light scattering by liquids came from an interest in the appearance of a strong opalescence in the neighborhood of the critical point. The critical phenomenon had been discovered by Andrews (1869) in connection with his classic studies of the carbon dioxide isotherms. He found that, while isothermal compression of a vapor may involve a discontinuity between two phases, there is a critical temperature above which condensation does not occur and the substance passes from "vapor" to "liquid" by a sequence of states in which it remains perfectly homogeneous. Only below the critical temperature are the two phases capable of coexistence. Furthermore, there is a critical density such that at this density the substance will comprise one phase above the critical temperature but will separate into two phases, a vapor and a liquid, just below it. This critical temperature and critical density characterize the critical point. There is a completely analogous situation for certain partially miscible liquids for which the critical state is determined by a critical temperature and a critical composition. Here the critical temperature may be called either an upper or a lower consolute temperature, depending upon whether the heterogeneous system is stable below or above this temperature. Much of the quantitative work has been done with such solutions because the experiments can often be performed more easily and with greater precision than with pure substances.

There has been considerable discussion as to whether the critical phenomenon occurs at a discrete thermodynamic state or whether there is a range of composition over which the continuous transition at the critical temperature from a single to a two-phase system can be effected. The latter condition results in a "flat top" on the coexistence curve in the critical region. Atack and Rice (1954, 1955) have pointed out that the flat top found in the cyclohexane-aniline system may be attributed to the vanishing of the interfacial tension at a point at which the two phases still have different compositions, due to cancellation of energy and entropy effects. Attempts to find other systems in which this phenomenon occurs have either been inconclusive or unsuccessful. In this connection, perfluoromethylenecyclohexane-carbon tetrachloride has been very carefully studied to within 0.0001° of the critical temperature and there is no evidence of a flat region to within the above precision (Thompson and Rice, 1964).

Among the most spectacular of the critical phenomena is the very intense scattering of light, known as critical opalescence, which appears within a temperature range of 1 or 2° of the critical point. This was first reported for one-component systems by Altschul (1894) and von Wesendonck (1894) shortly thereafter and for binary liquid mixtures by Rothmund (1898). Brumberger *et al.* (1967) have reported critical X-ray opalescence in the binary liquid–metal system, Na–Li.

Smoluchowski (1907, 1912) and Einstein (1910) developed their theory of fluctuation scattering in order to explain this phenomenon. Einstein recognized that his equation was generally valid when the system was far removed from the critical point and, as we have already seen, he showed how, for an ideal gas, it reduced to Rayleigh's equation. Near the critical point of pure liquids, there is a dramatic increase of the isothermal compressibility and for partially miscible solutions the chemical potential becomes invariant to changes in composition. According to (9.1.14) and (9.2.6), this can account qualitatively for the increase in scattering near the critical point. At precisely the critical point, the Einstein–Smoluchowski equation predicts infinite scattering, a result which Smoluchowski proposed to obviate by retaining higher terms in the expansion of the Taylor series for the fluctuation in the free energy (9.1.10). Einstein, on the other hand, excepted the region in the close vicinity of the critical point from the domain of validity of the theory, remarking that in any case this region was not experimentally accessible.

The earliest experimenters found a second objection to the simple fluctuation theory in the neighborhood of the critical point. Kamerlingh-Onnes and Keesom (1908), Keesom (1911), and Fürth (1915) observed that, as the critical temperature was approached, the wavelength dependence of the scattered intensity decreased from the inverse fourth-power law predicted by the theory. Even with retention of the higher-order terms, as proposed by Smoluchowski, the theory continued to predict that the scattering would remain proportional to the inverse fourth power of the wavelength.

Still another deviation from the Smoluchowski–Einstein equations was observed by Rousset (1936). He found pronounced dissymmetry of the light scattered around 90° whenever there was deviation from the fourth-power dependence on the wavelength. Later work by Zimm (1950), Quantie (1954), and others has confirmed this effect.

9.4.1 The Ornstein–Zernike Theory

Ornstein and Zernike (1914, 1918) recognized that the breakdown of the Smoluchowski–Einstein equation in the neighborhood of the critical point could be attributed to the fact that the fluctuations in density and composition in the volume elements were no longer independent, but were correlated due to the intermolecular forces. In the critical region, the free energy required to effect a particular fluctuation is so small that the local perturbation due to fluctuations occurring elsewhere extends from a much greater distance. Accordingly, the scattering will no longer be incoherent since the spatial distribution of the fluctuations is no longer perfectly random. Instead of adding the scattering intensities associated with the time average fluctuation in each volume element, the time average of the scattered field strengths must be added vectorially, and the modulus of this sum must be squared.

The problem is completely analogous to that considered in the previous chapter where scattering by a heterogeneous system was treated with the aid of a correlation function. The Rayleigh ratio can be written

$$R_\theta = \frac{8\pi^4}{\lambda^4} \frac{(1 + \cos^2\theta)}{(\delta V)} \cdot \frac{\displaystyle\int_0^\infty \overline{\delta\alpha'_A \delta\alpha'_B} 4\pi s^2 \frac{\sin hs}{hs}\, ds}{\displaystyle\int_0^\infty 4\pi s^2 \frac{\sin hs}{hs}\, ds} \tag{9.4.1}$$

where $h = (4\pi/\lambda) \sin\theta/2$, $\delta\alpha'_A$ is the fluctuation in polarizability measured at point A, and $\delta\alpha'_B$ is the fluctuation measured at a second point B separated from A by the distance s. The bar across the product of these fluctuations indicates the average value. This expression is the same as for Rayleigh–Debye scattering for a radially inhomogeneous, spherically symmetric distribution of polarizability.

Ornstein and Zernike utilized the same correlation function (8.4.2) which proved so useful in the previous chapter in describing the spatial variation of the dielectric constant. In terms of polarizability, this can be written

$$\gamma(s) = \overline{\delta\alpha'_A \delta\alpha'_B}/\overline{(\delta\alpha')^2} \tag{9.4.2}$$

which leads to

$$R_\theta = (8\pi^4/\lambda^4)(1 + \cos^2\theta)[\overline{(\delta\alpha')^2}/\delta V]\Omega \tag{9.4.3}$$

where

$$\Omega = \frac{\displaystyle\int_0^\infty \gamma(s) 4\pi s^2[(\sin hs)/hs]\, ds}{\displaystyle\int_0^\infty 4\pi s^2[(\sin hs)/hs]\, ds} \tag{9.4.4}$$

This is a general expression valid both in the critical region and far removed from it. In the latter case, the correlation function is a step function whose value is unity over the volume of the fluctuation and zero outside of it. Then $\Omega = 1$ and the Einstein equation (9.1.1) is obtained directly, provided the transformation from Rayleigh ratio to turbidity is effected. Ω is a correction factor which accounts for the reduction of the scattering as a consequence of the correlations among the fluctuations. The value of the correlation function falls off from unity at $s = 0$ to zero at large distances. This vanishing of $\overline{\delta\alpha'_A \delta\alpha'_B}$ and of $\gamma(s)$ at wide separations follows from the decreasing effectiveness of the molecular forces at large distances with the concomitant increasing randomness among the fluctuations. For a one-component system, the Rayleigh ratio can be expressed as

$$R_\theta = (\pi^2/2\lambda_0^{\,4})(1 + \cos^2\theta)[\rho(\partial\varepsilon/\partial\rho)_T]^2 kT\beta_T\Omega \tag{9.4.5}$$

The correlation function can be evaluated from the angular distribution of the scattered intensity. The factor Ω can be evaluated experimentally at each angle with the aid of the above equation. Rewriting (9.4.4) as

$$\sin(\theta/2)\Omega = \frac{\lambda \int_0^\infty s\gamma(s)\sin(hs)\,ds}{\int_0^\infty 4\pi s^2[\sin(hs)/hs]\,ds} \tag{9.4.6}$$

and upon applying Fowler's transform we obtain

$$s\gamma(s) = \frac{\lambda \int_0^\infty \sin(\theta/2)\Omega\sin(hs)\,d\sin(\theta/2)}{\int_0^\infty 4\pi s^2[\sin(hs)/hs]\,ds} \tag{9.4.7}$$

Unfortunately, the practical evaluation of the integral in the numerator is limited by the range of $\theta/2$ which only varies between 0 and $\pi/2$. Furthermore, if the intensity falls off sharply with scattering angle, only a small angular interval in the forward direction may be available. Therefore, unless the integrand of (9.4.7) has dropped to a small value over this interval, special procedures are required to extrapolate the "tail" of the function to infinity (Fürth and Williams, 1954). The normalizing integral in the denominator may be obtained by noting that $\gamma(s)$ must reduce to unity as the parameter s goes to zero.

There are a number of general comments which can be made independently of the particular form of the correlation function. It is immediately apparent that there will be deviations from the usual inverse fourth-power wavelength dependence and the $(1 + \cos^2\theta)$ angular dependence. As long as hs is small over all values for which $\gamma(s)$ is significantly different from zero both the wavelength and angular dependence follow the usual pattern for a Rayleigh scatterer. Accordingly, the peculiar effects associated with critical opalescence become less pronounced in the neighborhood of zero scattering angle and at long wavelengths.

a. The Ornstein–Zernike Function. Ornstein and Zernike investigated the consequences of the following integrable correlation function:

$$\gamma(s) = [\exp(-s/l)]/s \tag{9.4.8.}$$

for which they obtained

$$R_\theta = \frac{(\pi^2/2\lambda_0{}^4)(1 + \cos^2\theta)[\rho(\partial\varepsilon/\partial\rho)_T]^2\, kT}{1/\beta_T + 8\pi^2 kT(l^2/\lambda^2)\sin^2(\theta/2)} \tag{9.4.9}$$

The length l is a measure of the rate at which the correlation function falls off with radial distance. The expression for the turbidity can be obtained by appropriate integration over all scattering angles (Debye *et al.*, 1962b).

A number of features immediately become obvious. Far removed from the critical point, where the compressibility is small, the second term in the denominator is insignificant and the above equation reduces to the Einstein–Smoluchowski formula. However, for a particular value of β_T, the relative significance of the second term depends upon l, λ, and θ in the manner discussed qualitatively above. As the critical point is approached, and the compressibility becomes large, the second term in the denominator becomes predominant. The scattering then follows an inverse second-power dependence on wavelength and the angular dependence is given by $(1 + \cos^2\theta)/\sin^2(\theta/2)$. The distance l, sometimes called the correlation length or the range of molecular forces, can be determined from the straight line plot of R_θ^{-1} vs. $\sin^2(\theta/2)$.

b. The Debye–Brillouin Equation. Debye (1960) has used a somewhat different approach to derive an equivalent expression. His point of departure is a line of reasoning proposed by Brillouin (1922) in which the thermal motion is conceived as a superposition of sonic waves much in the same fashion as has been found practical for the calculation of the specific heat of solids. Periodic fluctuations of the dielectric constant associated with these waves are responsible for the scattering. In this way, Brillouin was able to calculate the Einstein–Smoluchowski equation for a pure liquid.

Debye extended the treatment to the neighborhood of the critical point by considering not only the effect of the density fluctuations, but also their gradients. Such gradients will have a certain amount of surface energy associated with them. Under ordinary circumstances, the surface free energy is relatively small, but in the neighborhood of the critical point where the compressibility is large and the free energy to create a given fluctuation becomes correspondingly small, the influence of the local gradient becomes important. His result is

$$R_\theta = \frac{(\pi^2/2\lambda_0{}^4)(1 + \cos^2\theta)[\rho(\partial\varepsilon/\partial\rho)_T]^2 2kT\beta_T}{1 + (8\pi^2/3)(l^2/\lambda^2)[\sin^2(\theta/2)/(\tau - 1)]} \tag{9.4.10}$$

where τ is the reduced temperature

$$\tau = T/T_c \tag{9.4.11}$$

Debye (1963) has demonstrated the equivalence of the two expressions if an equation of state for a van der Waals gas is used.

c. The Persistence Length and the Range of Molecular Forces. If, in addition to the range of molecular forces l, a persistence length as defined by

$$L^2 = \frac{\int \gamma(s) s^2 \, dV}{\int \gamma(s) \, dV} \tag{9.4.12}$$

is introduced, it can be shown that

$$L^2 = l^2/(\tau - 1) \tag{9.4.13}$$

The persistence length measures the distance over which a certain fluctuation is maintained. Obviously, it becomes larger the nearer the critical temperature is approached. Since $(\tau - 1)$ can be made arbitrarily small, it becomes possible to measure by an interference effect a quantity l, which may be much smaller than the wavelength, λ. Munster (1960) has shown that in the general case for any arbitrary equation of state

$$L^2 = l^2 f(\tau) \tag{9.4.14}$$

Now we consider the ratio of the scattering at a particular angle relative to that scattered in the forward direction. For this

$$R_\theta / R_0 = \left(\frac{1 + \cos^2 \theta}{2} \right) \int_0^\infty \gamma(s) \frac{\sin(hs)}{hs} s^2 \, ds \Bigg/ \int_0^\infty \gamma(s) s^2 \, ds \tag{9.4.15}$$

It follows after expansion in powers of h that

$$\begin{aligned} \left(\frac{2}{1 + \cos^2 \theta} \right) \frac{R_\theta}{R_0} &= 1 - (h^2/6)L^2 + \cdots \\ &= 1 - (16\pi^2 L^2 / 6\lambda^2) \sin^2(\theta/2) \end{aligned} \tag{9.4.16}$$

Thus, the persistence length may be determined from the slope of a plot of R_θ vs. $\sin^2(\theta/2)$ in the neighborhood of $\theta = 0$.

Debye and Jacobsen (1968) have proposed that under favorable conditions, it may be possible to measure the persistence length by direct observations in an optical phase contrast microscope. This is because the relaxation time (Debye, 1965) as well as the amplitude of the concentration fluctuations increases as the critical temperature is approached. Desirable requirements for direct visual observation of the concentration fluctuations are a large difference in refractive index between the two components in order to produce refractive index fluctuations of large amplitude, a small diffusion coefficient in order to achieve long relaxation times, and a long range of molecular interactions in order to exhibit large persistence lengths. This was achieved with the system polystyrene ($M_w = 522{,}000$) in cyclohexane. At some point close to the critical temperature, the image in the phase contrast microscope would start to undulate and take on a rapidly changing mottled gray on gray appearance. This fluctuation field could easily be distinguished from the appearance of phase separation by the lack of fast movement in the latter case and by the fact that coalescence slowly occurred forming droplets large enough to be sharply resolved.

Finally, the range of molecular forces itself has been shown (Debye, 1963) to be the second moment of the interaction energy

$$l^2 = \int r^2 \varepsilon(s)\, dv \Big/ \int \varepsilon(s)\, dv \tag{9.4.17}$$

where $\varepsilon(s)$ is the potential energy of repulsion at distance s. The volume integral extends from s at the point of "contact" to infinity.

The formalism developed above in terms of the correlation function, the persistence length, and the range of molecular forces is equally valid for one-component systems or solutions. Debye (1959) has developed the theory for two-component systems in terms of the intermolecular potentials. Thus,

$$l^2 = \frac{l_{11}^2(W_{11}/w_1^{\,2}) + l_{22}^2(W_{22}/w_2^{\,2}) - 2l_{12}^2(W_{12}/w_1 w_2)}{W_{11}/w_1^{\,2} + W_{22}/w_2^{\,2} - 2W_{12}/w_1 w_2} \tag{9.4.18}$$

where

$$l_{ij} W_{ij} = \int \varepsilon_{ij}(s) s^2\, dv \tag{9.4.19}$$

$$W_{ij} = \int \varepsilon_{ij}(s)\, dv \tag{9.4.20}$$

Here $\varepsilon_{ij}(s)$ is the intermolecular potential and w_1 and w_2 the molecular volumes.

9.4.2 Experimental Verification

The resurgence of interest in critical opalescence during the past decade has been directed toward both the theoretical and the experimental aspects of the problem. The theorists have usually assumed a van der Waals model in which the pair interaction potential between particles consists of a finite size hard repulsive core and a weak long range attraction. The end product of the theory is to determine first the form of the correlation function and ultimately the angular dependence of the scattered light at small angles. In turn, inversion of the experimental angular dependence can, hopefully, lead to confirmation of the model.

The original exponential form proposed by Ornstein and Zernike for the correlation function (9.4.8) predicts a linear plot of the reciprocal of the Rayleigh ratio with $\sin^2(\theta/2)$ (or θ^2 at small angles) (9.4.9). This will be referred to as an O–Z plot. Such a plot permits evaluation of the range of molecular forces directly or through a plot of the relative value of the Rayleigh ratio as proposed by Debye (9.4.16). Fixman (1960) and Fisher (1964) have shown the equivalency of the Ornstein–Zernike theory and later treatments given by both Rocard (1933) and Debye (1959, 1960), all leading to a linear O–Z plot. Hemmer *et al.* (1964) and Hemmer (1964) have discussed the van der Waals model comprehensively and have shown that the O–Z theory is valid only when sufficiently removed from the critical region. The situation seems to be that the correlation function must be described by an expansion in which only the leading term is given by (9.4.8). As one approaches the critical point, more and more exponential type terms become excited and deviations from the O–Z linear plot will turn up.

An asymptotic s dependence for the correlation function different from the Ornstein–Zernike theory has been proposed by Green (1960). This predicts low angle scattering behavior which is strongly divergent from the O–Z plot. Frisch and Brady (1962) have also explored the effect of a conjectured correlation function upon the angular scattering with a view to exploring how deviations from the O–Z plot might follow from a quite different pair correlation function. They found that deviations from the O–Z plot were small except very close to the critical point and at very small angles, and they concluded, accordingly, that even an apparently linear O–Z plot may not necessarily imply the validity of the O–Z theory at the critical point.

The burden of proof has been upon the experimentalists but, unfortunately, the issue is most difficult to resolve because of the experimental difficulties as the critical point is approached. With the increasingly intense scattering in the critical region, the effect of multiple scattering may confuse the interpretation of the results. Also, the very difficult experimental problems associated with impurities and with maintenance of temperature control often make it difficult for different workers to agree on their results.

The experimental literature dealing with both light scattering and X-ray scattering from one- and two-component systems has been reviewed by Brumberger (1966) and by Chu (1964a, b, 1966). There is no clear evidence for the existence of deviations from the O–Z straight line plots in any of the one-component systems studied. For the two-component systems, reasonably good fits to the O–Z curves are found except at the largest and smallest angles or at temperatures very close to the critical temperature. However it is precisely under these conditions that the theory would be expected to fail. Brady *et al.* (1967) have discussed the deviations in the light of possible forms for the radial dependence of the correlation function other than that proposed by Ornstein and Zernike.

A second criterion which the O–Z theory predicts and which has been the subject of considerable discussion is that the extrapolated zero angle intensity should vary with $(T - T_c)^{-1}$. A model proposed by Widom (1962) predicts dependence upon $(T - T_c)^{-5/4}$.

a. Polymer and Detergent Solutions. Debye (1959) has also extended the above treatment to polymer solutions, showing for polymer coils, whose segment density follows a Gaussian distribution, that

$$l^2 = \overline{R_g^{\,2}}/6 \tag{9.4.21}$$

where R_g is the radius of gyration. This, in turn, is proportional to the root mean square end-to-end distance. This suggests for polymer solutions that the phenomena associated with critical opalescence should turn up at temperatures further removed from the critical point than for ordinary molecules. Of course, it should be kept in mind that for sufficiently large polymer molecules there may also be internal interferences that will produce dissymmetry even in highly diluted solutions far from the critical point. Debye *et al.* (1962a) have taken account of the latter effect by introducing the appropriate value for $P(\theta)$, the form factor, into the expression for the Rayleigh ratio.

The experimental values of l which have been obtained are surprisingly low compared to the size of the polymer coil (Debye *et al.*, 1962a; Debye *et al.*, 1960; McIntyre *et al.*, 1962). Indeed, the radius of gyration consistently turns out to be 2.5 to 3.0 times higher than the correlation length which presumably measures the range of molecular forces. In other regards, the polymer solutions behave in the critical region very much like binary solutions of low molecular weight liquids (Debye *et al.*, 1966).

De Gennes (1968) has proposed recently that l^2 is proportional to $M^{1/2}$ rather than to M, as indicated by (9.4.21). This is based both upon a theoretical argument and his finding that the $M^{1/2}$ dependence agrees better with the data of Debye *et al.* (1960).

An interesting study of the light scattering behavior of detergent solutions which separate into immiscible isotropic solutions has been made by Herrmann *et al.* (1966). The particular systems studied were solutions of various alkylethylene oxides. These solutions have been known for some time to exhibit a "cloud point" upon heating, and this had been attributed to a sharply increasing micellar weight with increasing temperature. However, in fact, the "cloud point" is critical opalescence associated with the approach to a lower consolute temperature and subsequent phase separation into two immiscible solutions. Indeed, this system is quite well behaved and exhibits a linear O–Z plot, as well as showing zero angle scattering intensity proportional to $(T - T_c)^{-1}$. Furthermore, the value of l was found to be extraordinarily large, $l = 140$ Å, compared to 30 to 40 Å for the polystyrene solutions studied by Debye *et al.*

b. Electrical Field Effects. The application of dc pulses to a binary solution near its critical solution temperature has been studied by Debye and Kleboth (1965). The electrical field alters the dielectric constant and causes a change in the free energy of mixing of the solution. This in turn affects the concentration fluctuations. Near the critical point, the free energy change associated with concentration fluctuations becomes very small, so that by going close enough to the critical point a condition may be reached where the thermal and electrical changes are of comparable magnitude. Rather large field strengths are required to obtain appreciable effects so that it was necessary to apply the field in pulses of short duration (100 μsec) in order to minimize the Joule heating.

There is a shift in the critical temperature due to the field given by

$$t = (Ww_1/k)[\phi_1{}^2(\partial^2\varepsilon/\partial\phi_1{}^2)]_{\text{crit}} \tag{9.4.22}$$

where t is the difference between the critical temperature in the absence of the field and in the presence of a field strength of E volts per centimeter. The electric energy density in vacuum is $W(= E^2/8\pi)$, w_1 is the molecular volume of component 1, ϕ_1 is the volume fraction, ε is the dielectric constant of the mixture, and k is the Boltzmann constant. Excellent agreement was obtained between the value of $t = 1.51 \times 10^{-2}$ calculated from this for nitrobenzene-2,2,4-trimethylpentane mixtures at 29.50°C and the measured value of 1.5×10^{-2}.

The effect of the electrical field on the turbidity can be shown to be

$$\begin{aligned} \tau &= \frac{1}{l} \ln \frac{I_0}{I} \\ &= \frac{32\pi^3}{3} \frac{w_1}{\lambda^4} \left(\frac{\phi_1}{n} \frac{\partial n}{\partial \phi_1} \right)^2 \frac{T_c}{T - T_c + t} \end{aligned} \tag{9.4.23}$$

where l is the path length, I_0 and I are the incident and transmitted intensities, respectively, T is the temperature at which the experiment is carried out, and T_c is the critical temperature. For small shifts of the critical temperature so that

$$t \ll \Delta T = T - T_c \tag{9.4.24}$$

it follows that

$$\frac{1}{l}\frac{\Delta I}{I} = \frac{w_1}{\lambda^4}\left(\frac{\phi_1}{n}\frac{\partial n}{\partial \phi_1}\right)^2 \frac{t}{(\Delta T)^2} \tag{9.4.25}$$

where ΔI is the change in the transmitted intensity due to the application of the field. This equation suggests that a plot of

$$\left(\frac{1}{l}\frac{\Delta I}{I}\right)^{-1/2} \quad \text{against} \quad \Delta T \tag{9.4.26}$$

should give a straight line. Some data for the system nitrobenzene-2,2,4-trimethylpentane are shown in Fig. 9.28 (Debye *et al.*, 1967). The circles represent experimental points obtained with sufficiently large pulse widths so that the equilibrium theory was applicable. The predicted linear relation

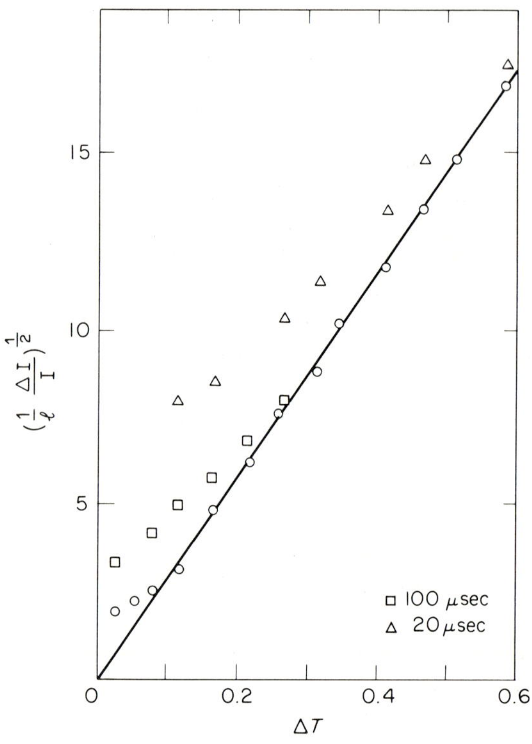

FIG. 9.28. Effect of electrical field on the turbidity of nitrobenzene-2,2,4-trimethylpentane near the critical point (Debye *et al.*, 1967).

is obtained to within 0.1° of the critical temperature. The deviations at lower increments are understandable both because of experimental error due to possible interception of forward scattered light and also to several assumptions in the theory which are not valid very close to the critical temperature. The deviations for the squares and triangles, which correspond to shorter pulse widths, are due to transient effects or to the relaxation times of the concentration fluctuations. The relaxation times of both the transmitted and scattered light can, in turn, be related to the diffusion coefficient in the binary mixture (Cooper and Mountain, 1968).

CHAPTER 10

Anisotropy

10.1 Scattering by a Small Ellipsoid

In 1897, Lord Rayleigh extended the dipolar scattering theory to an ellipsoid. As long as this is small compared with the wavelength, the external field is uniform over its extent. The problem is to calculate the perturbation of the field due to the presence of the ellipsoid. This had been solved for a static field by Maxwell (1873) who obtained the well-known result that the field within the ellipsoid will also be uniform, whatever the orientation of the ellipsoid, but will be rotated from the external field. This internal field may be described by a symmetric polarizability tensor for which the directions of the applied and induced fields will coincide whenever the applied field is parallel to one of the three mutually perpendicular axes of the ellipsoid. The external field may, accordingly, be resolved into the three components along the directions of these axes and the particle can be assigned three component polarizabilities. These completely define the internal field. Rayleigh assumed that the dipole associated with each of these oscillates synchronously with the exciting radiation and radiates in the usual manner.

For that component of the applied field along the semimajor axis, A, the polarizability α'_A is

$$\alpha'_A = \frac{V(m^2 - 1)}{4\pi + (m^2 - 1)P_A} \tag{10.1.1}$$

where m is the relative refractive index, and

$$P_A = \int_0^\infty \frac{2\pi ABC\, ds}{(s + A^2)^{3/2}(s + B^2)^{1/2}(s + C^2)^{1/2}} \tag{10.1.2}$$

B and C are the other semiaxes of the ellipsoid. The polarizabilities corresponding to the other components of the applied field are obtained from the analogous equations with alternation of the quantities A, B, and C. Osborn (1945) has computed the above elliptic integral for various shapes and has presented the results in both tabular and graphical form, arranged for convenient interpolation. When the ellipsoid degenerates to a sphere, $P = 4\pi/3$, and (10.1.1) reduces to the polarizability of a sphere, (3.2.3).

The scattering corresponding to each component of the incident wave will be the same as for an equivalent sphere whose polarizability is equal to the polarizability along the corresponding ellipsoidal axis. This equivalent sphere has a radius a' such that

$$(a')^3 = \frac{4\pi ABC}{3} \frac{(m^2 + 2)}{4\pi + (m^2 - 1)P_A} \tag{10.1.3}$$

The scattering can be obtained by inserting a' in Eqs. (3.2.6) and those that follow, or else by substituting the polarizability for the quantity $a^3(m^2 - 1)/(m^2 + 2)$ in the Rayleigh equation for a small sphere. Now, since the component of the incident field coincides with the axis of the ellipsoid, the angle, ψ, occurring in Eq. (3.2.6) becomes the angle between the axial direction and the scattering direction.

The three distinct polarizabilities along each of the ellipsoidal axes which have been described above and which will be considered in detail in the next two sections arise from the geometrical anisotropy of the ellipsoid which is assumed to be composed of an optically isotropic medium. However, this medium itself may be anisotropic, in which case the dielectric constant at optical frequencies is a symmetric tensor. Then the particle must be described by a polarization ellipsoid whose axes may or may not coincide with the axes of the geometrical ellipsoid. In any case, the polarizabilities along the axes of the polarization ellipsoid, which arise from a combination of optical and geometrical anisotropy, will describe the scattering as if they originated either from a corresponding isotropic ellipsoid or a sphere having a corresponding optical anisotropy. Formally, it is only necessary to specify the three principal polarizabilities.

10.1.1 Expressions for a Spheroid

The depolarization factor P_e corresponding to each of the axes A, B, or C can be integrated in closed form for the special case of a spheroid. Let A be the longer axis so that the eccentricity is defined by

$$e_s = \left(\frac{A^2 - B^2}{A^2}\right)^{1/2} \tag{10.1.4}$$

For oblate or flattened spheroids ($A = C > B$), the depolarization factor along the figure axis (B) is given by

$$P'_e = \frac{4\pi}{e_s^2}\left[1 - \left(\frac{1 - e_s^2}{e_s^2}\right)^{1/2} \sin^{-1} e_s\right] \tag{10.1.5}$$

For prolate or elongated spheroids ($A > B = C$) the depolarization factor along the figure axis (A) is

$$P'_e = 4\pi\frac{1 - e_s^2}{e_s^2}\left[\frac{1}{2e_s}\ln\frac{(1 + e_s)}{(1 - e_s)} - 1\right] \tag{10.1.6}$$

The depolarization factor corresponding to the two equal axes is

$$P''_e = (4\pi - P'_e)/2 \tag{10.1.7}$$

The case of a circular disk may be approximated by permitting $e_s \to 1$ for the oblate spheroid in which case

$$P'_e = 4\pi; \qquad P''_e = 0 \tag{10.1.8}$$

Similarly, for a very elongated spheroid ($e_s \to 1$)

$$P'_e = 0; \qquad P''_e = 2\pi \tag{10.1.9}$$

The polarizabilities may now be represented by

$$\alpha' = \frac{3V}{4\pi}L' \exp(-i\psi'_1) = \frac{V(m^2 - 1)}{4\pi + (m^2 - 1)P'_e} \tag{10.1.10}$$

$$\alpha'' = \frac{3V}{4\pi}L'' \exp(-i\psi''_1) = \frac{V(m^2 - 1)}{4\pi + (m^2 - 1)P''_e} \tag{10.1.11}$$

The phase functions ψ'_1 and ψ''_1 appear whenever the refractive index is complex.

Following Gans (1912) and Atlas *et al.* (1953), we can consider the effect of the orientation of the ellipsoid somewhat more explicitly. The linearly polarized incident radiation of unit intensity traveling along the z-axis has its electric vector vibrating at an angle β with the x-axis. The spheroid defines another coordinate system, ξ, η, ζ, where ξ is chosen in the direction of the figure axis, and η and ζ are along any pair of perpendicular diameters. The electric field at the spheroid has components f_ξ, f_η, f_ζ set up in the spheroid which are related to E by

$$f_\xi = \alpha' E_\xi, \qquad f_\eta = \alpha'' E_\eta, \qquad \text{and} \qquad f_\zeta = \alpha'' E_\zeta \tag{10.1.12}$$

The components of the electric field in the x, y, and z-directions are found to be

$$f_x = c\{(a - b)\alpha_1(\alpha_2 \sin\beta + \alpha_1 \cos\beta) + b\cos\beta\} \tag{10.1.13}$$

$$f_y = c\{(a - b)\alpha_2(\alpha_2 \sin \beta + \alpha_1 \cos \beta) + b \sin \beta\} \qquad (10.1.14)$$

$$f_z = c\{(a - b)\alpha_3(\alpha_2 \sin \beta + \alpha_1 \cos \beta)\} \qquad (10.1.15)$$

where

$$c = 3V/4\pi; \qquad a = L' \cos(2\pi\nu t - \psi_1'); \qquad b = L'' \cos(2\pi\nu t - \psi_1'') \quad (10.1.16)$$

Here α_1, α_2, and α_3 are the direction cosines of the ξ-axis with respect to the x, y, and z directions respectively, ν is the frequency of the radiation, and t is the time.

The geometry of the scattering is shown in Fig. 10.1. The direction cosines are given by

$$\alpha_1 = \cos \theta \qquad (10.1.17)$$

$$\alpha_2 = \sin \theta \sin \phi \qquad (10.1.18)$$

$$\alpha_3 = \sin \theta \cos \phi \qquad (10.1.19)$$

The Rayleigh ratio of the radiation scattered in a direction making an angle θ with the incident direction is

$$R_\theta = N(2\pi/\lambda)^4(f_y^2 + f_x^2 \cos^2 \theta + f_z^2 \sin^2 \theta) \qquad (10.1.20)$$

where λ is the wavelength in the external medium and N is the number of particles per unit volume, assuming incoherent scattering. The first term in the parentheses represents the intensity polarized parallel to the y-axis; the second and third terms together represent the intensity vibrating in the scattering plane. The component of the Rayleigh ratio parallel to the y-axis will be termed the vertical component and will be designated by

$$V = N(2\pi/\lambda)^4 f_y^2 \qquad (10.1.21)$$

The other component whose electric vector vibrates in the xy-plane is termed the horizontal component and will be designated by

$$H = N(2\pi/\lambda)^4[f_x^2 \cos^2 \theta + f_z^2 \sin^2 \theta] \qquad (10.1.22)$$

Furthermore, it is usually convenient to choose the direction of vibration of the incident radiation to be along either the x or y axis; i.e., $\beta = 0$ or $90°$. The incident radiation is then said to be horizontally or vertically polarized.[1]

After averaging f_x, f_y, and f_z over one cycle of the frequency ν and over the values of the direction cosines, we get for $\beta = 90°$

$$\overline{f_x^2} = (c^2/2)\overline{(\alpha_1\alpha_2)^2}[(L')^2 - 2L'L'' \cos(\psi_1' - \psi_1'') + (L'')^2] \qquad (10.1.23)$$

[1] The terms vertical and horizontal imply that the scattering plane is parallel to the surface of the earth. Although the terms perpendicular and parallel, which have been used earlier, are more apt, we will use the former terminology in this chapter because of the universal prevalence of V and H for the corresponding Rayleigh ratios.

$$\overline{f_y^2} = (c^2/2)\{\overline{\alpha_2^4}(L')^2 + 2[\overline{\alpha_2^2} - \overline{\alpha_2^4}]L'L'' \cos(\psi_1' - \psi_1'') + [1 - 2\overline{\alpha_2^2} + \overline{\alpha_2^4}](L'')^2\} \quad (10.1.24)$$

$$\overline{f_z^2} = (c^2/2)\overline{(\alpha_3\alpha_2)^2}[(L')^2 - 2L'L'' \cos(\psi_1' - \psi_1'') + (L'')^2] \quad (10.1.25)$$

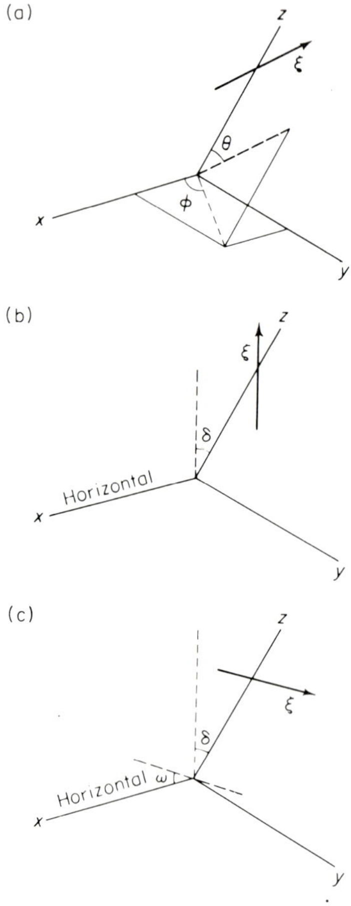

FIG. 10.1. Geometry of scattering for small spheroid. Incident beam propagates along the z-axis; ξ-axis is figure axis of scatterer. (a) General case. In (b) and (c) the figure axes are respectively vertical and horizontal. δ is angle of declination of beam, ω the angle made by horizontal x-axis and projection of ξ-axis (Atlas *et al.*, 1953).

If the xz-plane is taken as the scattering plane, these equations apply to vertically polarized incident radiation ($\beta = 90°$). The vertical component of the scattered radiation is given by $\overline{f_y^2}$ and the cross-polarized component by

$\overline{f_x^2} \cos^2 \theta + \overline{f_z^2} \sin^2 \theta$. When the incident radiation is horizontally polarized ($\beta = 0$)

$$\overline{f_x^2} = (c^2/2)\{\overline{\alpha_1{}^4}(L')^2 + 2[\overline{\alpha_1{}^2} - \overline{\alpha_1{}^4}]L'L'' \cos(\psi_1' - \psi_1'') + [1 - \overline{2\alpha_1{}^2} + \overline{\alpha_1{}^4}](L'')^2\} \quad (10.1.26)$$

$$\overline{f_y^2} = (c^2/2)\overline{(\alpha_1\alpha_2)^2}[(L')^2 - 2L'L'' \cos(\psi_1' - \psi_1'') + (L'')^2] \quad (10.1.27)$$

$$\overline{f_z^2} = (c^2/2)\overline{(\alpha_3\alpha_1)^2}[(L')^2 - 2L'L'' \cos(\psi_1' - \psi_1'') + (L'')^2] \quad (10.1.28)$$

Now $(\overline{f_x^2 \cos^2 \theta + f_z^2 \sin^2 \theta})$ represents the horizontal component of the scattered light and $\overline{f_y^2}$ is the cross component.

We will assume that the spheroids are randomly oriented with every possible orientation of the figure axis being equally probable. The mean values of the α's may be found by averaging over all possible values of θ and ϕ. The results are listed in Table 10.1.

TABLE 10.1
AVERAGES OF DIRECTION COSINES OF INTEREST[a]

	$\overline{(\alpha_i\alpha_j)^2}$	$\overline{\alpha_i^2}$	$\overline{\alpha_i^4}$
Random orientation	$\frac{1}{15}$	$\frac{1}{3}$	$\frac{1}{5}$

[a] $i \neq j$ for values 1, 2, 3.

The cross sections for scattering and absorption are given by (Gans, 1912)

$$\begin{aligned} C_{\text{sca}} &= (2\pi/\lambda)^4(8\pi/9)[|\alpha'|^2 + 2|\alpha''|^2] \\ &= (8\pi^3/\lambda^4)V^2[(L')^2 + 2(L'')^2] \end{aligned} \quad (10.1.29)$$

and

$$\begin{aligned} C_{\text{abs}} &= (8\pi^2/3\lambda)Im(-\alpha' - 2\alpha'') \\ &= (2\pi V/\lambda)(L' \sin \psi_1' + 2L'' \sin \psi_1'') \end{aligned} \quad (10.1.30)$$

The following ratios of the intensities of scattered radiation, designated as the polarization ratios, are convenient experimental quantities with which to work:

$$\rho_{\text{v}} = \frac{H_{\text{v}}}{V_{\text{v}}} = \frac{\overline{f_x^2} \cos^2 \theta + \overline{f_z^2} \sin^2 \theta}{\overline{f_y^2}} \quad (\beta = 90°) \quad (10.1.31)$$

$$(\rho_{\text{h}})^{-1} = \frac{H_{\text{h}}}{V_{\text{h}}} = \frac{\overline{f_x^2} \cos^2 \theta + \overline{f_z^2} \sin^2 \theta}{\overline{f_y^2}} \quad (\beta = 0°) \quad (10.1.32)$$

and

$$\rho_u = (H_v + H_h)/(V_v + V_h) \tag{10.1.33}$$

The quantities $\overline{f_x^2}$, $\overline{f_y^2}$ and $\overline{f_z^2}$ are defined differently for $\beta = 90°$ and for $\beta = 0°$, as described above. The quantities H_v and V_v are the horizontal and vertical components of the scattered light intensities (Rayleigh ratios) when the incident light is polarized vertically, and V_h and H_h are the corresponding quantities for horizontally polarized incident light.

For randomly oriented spheroids, Eqs. (10.1.23) to (10.1.28) reduce to

$$\underbrace{\overline{f_x^2} = \overline{f_z^2}}_{\beta = 90°} = \underbrace{\overline{f_y^2} = \overline{f_z^2}}_{\beta = 0°} = (c^2/30)[(L')^2 - 2L'L''\cos(\psi_1' - \psi_1'') + (L'')^2] \tag{10.1.34}$$

and

$$\begin{aligned}\overline{f_y^2}(\beta = 90°) &= \overline{f_x^2}(\beta = 0°) \\ &= \frac{c^2}{30}[3(L')^2 + 4L'L''\cos(\psi_1' - \psi_1'') + 8(L'')^2]\end{aligned} \tag{10.1.35}$$

so that

$$H_v \sim \overline{f_x^2} \tag{10.1.36}$$

$$V_v \sim \overline{f_y^2} \tag{10.1.37}$$

$$H_h \sim \overline{f_y^2}\cos^2\theta + \overline{f_x^2}\sin^2\theta \tag{10.1.38}$$

$$V_h \sim \overline{f_x^2} \tag{10.1.39}$$

where now $\overline{f_x^2}$ and $\overline{f_y^2}$ in these last four equations are evaluated for $\beta = 90°$. This means that the entire scattering pattern can be determined by two measurements at one angle. If that angle is chosen as $\theta = 90°$, then measurements of V_h and H_h yield $\overline{f_y^2}$ and $\overline{f_x^2}$ directly. At $\theta = 90°$

$$V_h = H_v = H_h \tag{10.1.40}$$

Only H_h varies with angle and this is given by

$$H_h = H_v\sin^2\theta + V_v\cos^2\theta \tag{10.1.41}$$

Two cases that are frequently encountered are lateral scattering ($\theta = 90°$) and back scattering ($\theta = 180°$). With the additional restriction that there is no absorption ($\psi_1' = \psi_1'' = 0$), one obtains

$$\rho_{\mathrm{v}} = \frac{(L' - L'')^2}{3(L')^2 + 4L'L'' + 8(L'')^2}$$

$$= \frac{(\alpha' - \alpha'')^2}{3(\alpha')^2 + 4\alpha'\alpha'' + 8(\alpha'')^2} \tag{10.1.42}$$

$$\rho_{\mathrm{h}} = \frac{(L' - L'')^2}{(L' - L'')^2} = 1 \tag{10.1.43}$$

$$\rho_{\mathrm{u}} = \frac{2(L' - L'')^2}{4(L')^2 + 2L'L'' + 9(L'')^2}$$

$$= \frac{2(\alpha' - \alpha'')^2}{4(\alpha')^2 + 2\alpha'\alpha'' + 9(\alpha'')^2} \tag{10.1.44}$$

for the 90° scattering, and

$$\rho_{\mathrm{v}} = \rho_{\mathrm{h}} = \frac{(L' - L'')^2}{3(L')^2 + 4L'L'' + 8(L'')^2} \tag{10.1.45}$$

$$\rho_{\mathrm{u}} = 1 \tag{10.1.46}$$

for the back scattering.

There are two limiting cases that are particularly worthy of note. First, when the polarizability along the figure axis of an oblate spheroid is relatively small ($L' < L''$) corresponding to a circular disk

$$\rho_{\mathrm{v}} = \tfrac{1}{8}; \qquad \rho_{\mathrm{u}} = \tfrac{2}{9} \tag{10.1.47}$$

for lateral scattering, and

$$\rho_{\mathrm{v}} = \rho_{\mathrm{h}} = \tfrac{1}{8} \tag{10.1.48}$$

for back scattering. On the other hand, for a relatively large figure axis ($L' > L''$, which for a prolate spheroid corresponds to a long circular cylinder)

$$\rho_{\mathrm{v}} = \tfrac{1}{3}; \qquad \rho_{\mathrm{u}} = \tfrac{1}{2} \tag{10.1.49}$$

for lateral scattering, and

$$\rho_{\mathrm{v}} = \rho_{\mathrm{h}} = \tfrac{1}{3} \tag{10.1.50}$$

for back scattering.

Lord Rayleigh (1918c) first deduced expressions such as these in an attempt to suggest that the depolarization of light scattered by gases might be due to molecular anisotropy. The lack of complete polarization of sky light at 90° can be attributed to factors other than anisotropy of the air molecules such as dust and haze, multiple scattering, and reflection from the surface of the earth. However, after Cabannes (1915) demonstrated in the laboratory that molecules of a dust free gas scatter a detectable amount of light, very careful observations by Lord Rayleigh's son (1918d) showed that for dust free air there was at 90° a cross component whose intensity was about 4% of the parallel component for vertically polarized incident radiation. This was verified by Cabannes (1921) and Gans (1921). Since then the depolarization of a large number of gases has been reported (Cabannes, 1929; Bhagavantam, 1942). Stein (1953) has calculated the principal polarizabilities of a number of paraffin hydrocarbons using the values of the bond polarizabilities. He found discrepancies between the depolarization calculated from these polarizabilities and the measured values and has suggested that these data be evaluated again. Such a study has been very carefully carried out by Rudder and Bach (1968), using a ruby laser. With helium and argon, they obtained $\rho_v < 3 \times 10^{-3}$ and 4×10^{-3}, respectively. Other spherical molecules, such as methane and argon, displayed very small but finite values ($\rho_v = 1.3$ and 1.6×10^{-4}, respectively). The values of ρ_v for H_2, D_2, N_2, and N_2O were found to be lower than the expected values. The Rayleigh ratios at 60° agreed with values calculated from the known refractive indices (3.2.33), and the angular variation also agreed with theory.

Atlas *et al.* (1953) have made detailed calculations for the backscatter and extinction coefficient from small spheroids. Their interest was in the radar detection of rain and hail at microwave frequencies. They compared the parallel-polarized backscatter and the extinction by randomly oriented water and ice spheroids relative to the values of the same quantities for equivolumic spheres. There is a very considerable increase in scattering with increasing eccentricity of the water spheroids. The effect is considerably less pronounced for ice, whose refractive index at microwave frequencies is much smaller than that of water ($m = 1.75$ for ice; $m = 8.9 - 0.69i$ for water at $\lambda = 10$ cm).

10.1.2 Expressions for an Ellipsoid

We now consider the randomly oriented ellipsoid with three unequal axes, A, B, and C, and three corresponding polarizabilities, α'_A, α'_B, and α'_C [(10.1.1), (10.1.2)] (Rayleigh, 1918c; Bhagavantam, 1942). For $\beta = 90°$

$$\overline{f_x^2} = \overline{f_z^2} = (1/30)(\alpha_A'^2 + \alpha_B'^2 + \alpha_C'^2 - \alpha'_A\alpha'_B - \alpha'_B\alpha'_C - \alpha'_A\alpha'_C) \qquad (10.1.51)$$

$$\overline{f_y^2} = (1/10)(\alpha_A'^2 + \alpha_B'^2 + \alpha_C'^2) + (1/15)(\alpha'_A\alpha'_B + \alpha'_B\alpha'_C + \alpha'_A\alpha'_C) \qquad (10.1.52)$$

and for $\beta = 0°$, $\overline{f_x^2}$ is the same as $\overline{f_y^2}$ above and $\overline{f_y^2}$ and $\overline{f_z^2}$ are the same as

$\overline{f_x^2}$ above. These formulations are for nonabsorbing dielectric media. The Rayleigh ratio is given by (10.1.20) and the polarization ratios at 90° are

$$\rho_v = \frac{\alpha_A'^2 + \alpha_B'^2 + \alpha_C'^2 - \alpha_A'\alpha_B' - \alpha_B'\alpha_C' - \alpha_A'\alpha_C'}{3(\alpha_A'^2 + \alpha_B'^2 + \alpha_C'^2) + 2(\alpha_A'\alpha_B' + \alpha_B'\alpha_C' + \alpha_A'\alpha_C')} \tag{10.1.53}$$

$$\rho_h = \frac{\alpha_A'^2 + \alpha_B'^2 + \alpha_C'^2 - \alpha_A'\alpha_B' - \alpha_B'\alpha_C' - \alpha_A'\alpha_C'}{\alpha_A'^2 + \alpha_B'^2 + \alpha_C'^2 - \alpha_A'\alpha_B' - \alpha_B'\alpha_C' - \alpha_A'\alpha_C'} = 1 \tag{10.1.54}$$

$$\rho_u = \frac{2(\alpha_A'^2 + \alpha_B'^2 + \alpha_C'^2 - \alpha_A'\alpha_B' - \alpha_B'\alpha_C' - \alpha_A'\alpha_C')}{4(\alpha_A'^2 + \alpha_B'^2 + \alpha_C'^2) + \alpha_A'\alpha_B' + \alpha_B'\alpha_C' + \alpha_A'\alpha_C'} \tag{10.1.55}$$

The corresponding values at other scattering angles can be readily obtained by retention of the factors $\cos^2\theta$ and $\sin^2\theta$ in (10.1.20). These reduce properly to (10.1.42) to (10.1.44) for the case of a spheroid where $\alpha_A' = \alpha_B' = \alpha''$ and $\alpha_C = \alpha'$. The polarization ratios at other angles of observation may be obtained by determining V_h, V_v, H_h, and H_v from the above expressions for $\overline{f_x^2}$, $\overline{f_y^2}$, and $\overline{f_z^2}$ with the aid of (10.1.21) and (10.1.22).

10.2 Effect of Anisotropy upon Intensity; Cabannes Factor

An ellipsoid will scatter more intensely than a sphere of the same volume; a sphere comprised of an anisotropic medium will scatter more intensely than an isotropic sphere of the same volume. The factor by which the Rayleigh ratio of the anisotropic particle is enhanced over that for the isotropic sphere is called the Cabannes factor. It will be designated $C(\theta)$ so that

$$R_\theta = R_{90}(1 + \cos^2\theta)C(\theta) \tag{10.2.1}$$

where R_{90} is the Rayleigh ratio at 90° of the small isotropic sphere for unpolarized incident light. The Rayleigh ratio may, in turn, be decomposed into four parts:

$$R_\theta = V_v + H_v + V_h + H_h \tag{10.2.2}$$

or

$$C(\theta) = (V_v + H_v + V_h + H_h)/(1 + \cos^2\theta)R_{90} \tag{10.2.3}$$

It follows for small isotropic particles where

$$V_v = R_{90}, \qquad H_v = V_h = 0, \qquad H_h = R_{90}\cos^2\theta \tag{10.2.4}$$

that $C(\theta)$ is unity. For small ellipsoidal particles [(10.1.21) and (10.1.22)]

$$V_v = N(2\pi/\lambda)^4\overline{f_y^2} \qquad (\beta = 90°) \tag{10.2.5}$$

$$V_h = N(2\pi/\lambda)^4\overline{f_y^2} \qquad (\beta = 0°) \tag{10.2.6}$$

$$H_v = N(2\pi/\lambda)^4[\overline{f_x^2}\cos^2\theta + \overline{f_z^2}\sin^2\theta] \qquad (\beta = 90°) \qquad (10.2.7)$$

$$H_h = N(2\pi/\lambda)^4[\overline{f_x^2}\cos^2\theta + \overline{f_z^2}\sin^2\theta] \qquad (\beta = 0°) \qquad (10.2.8)$$

The correction factor for anisotropy can now be expressed in terms of the polarization ratios with the aid of (10.1.51) and (10.1.52). For unpolarized incident light (King, 1923)

$$C_u(\theta) = \frac{1 + 3\rho_v(90) + [1 - \rho_v(90)]\cos^2\theta}{[1 - (4/3)\rho_v(90)](1 + \cos^2\theta)}$$

$$= \frac{1 + \rho_u(90) + [1 - \rho_u(90)]\cos^2\theta}{[1 - (7/6)\rho_u(90)](1 + \cos^2\theta)} \qquad (10.2.9)$$

where $\rho_u(90)$ is the polarization ratio at 90° for unpolarized incident light and $\rho_v(90)$ is the same quantity for vertically polarized incident light. The corresponding Cabannes factors for incident light linearly polarized in the vertical and horizontal directions are

$$C_v(\theta) = \{[1 - \tfrac{7}{6}\rho_u(90)](1 + \cos^2\theta)\}^{-1} \qquad (10.2.10)$$

and

$$C_h(\theta) = \frac{\rho_u(90) + [1 - \rho_u(90)]\cos^2\theta}{[1 - (7/6)\rho_u(90)](1 + \cos^2\theta)} \qquad (10.2.11)$$

These equations relate quantities directly accessible to experiment. Even though they ultimately depend upon the three principal polarizabilities α'_A, α'_B, and α'_C, these need not be known explicitly in order to account for the effect of the anisotropy upon the Rayleigh ratio. It is sufficient simply to know the polarization ratio.

The angular dependences of the polarization ratios are given by Melrose (1957).

$$\rho_u(\theta) = \rho_u(90) + [1 - \rho_u(90)]\cos^2\theta \qquad (10.2.12)$$

$$\rho_h(\theta) = \rho_u(90)[\rho_u(90) + 2\cos^2\theta\,(1 - \rho_u(90))]^{-1} \qquad (10.2.13)$$

At 90°, (10.2.9) reduces to

$$C_u(90) = \frac{3 + 9\rho_v(90)}{3 - 4\rho_v(90)} = \frac{6 + 6\rho_u(90)}{6 - 7\rho_u(90)} \qquad (10.2.14)$$

This result was first given by Cabannes (1920, 1929, p. 44). The general treatment, applicable to any angle, was developed by King (1923), who also carried out the integration over all angles in order to obtain the correction factor for the turbidity. This is

$$C_\tau = \frac{3 + 6\rho_v(90)}{3 - 4\rho_v(90)} = \frac{6 + 3\rho_u(90)}{6 - 7\rho_u(90)} \qquad (10.2.15)$$

The scattered light may be subdivided into an isotropic and an anisotropic contribution. Gans (1912, 1921) has derived the following expressions for the isotropic and anisotropic parts of the Rayleigh ratio at 90°

$$R_{\text{is}} = \frac{8\pi^4 N_{\text{A}}}{\lambda_0{}^4}(\overline{\alpha'})^2 \tag{10.2.16}$$

and

$$R_{\text{an}} = \frac{8\pi^4 N_{\text{A}}}{\lambda_0{}^4}\left(\frac{13}{45}\gamma^2\right) \tag{10.2.17}$$

Here the two parameters $\overline{\alpha'}$ and γ are the mean polarizability and the molecular anisotropy factor, defined by

$$\overline{\alpha'} = \tfrac{1}{3}(\alpha'_A + \alpha'_B + \alpha'_C) \tag{10.2.18}$$

and[3]

$$\gamma^2 = \alpha'^2_A + \alpha'^2_B + \alpha'^2_C - \alpha'_A\alpha'_B - \alpha'_B\alpha'_C - \alpha'_A\alpha'_C \tag{10.2.19}$$

In addition, a ratio of these quantities called the generalized anisotropy is defined by

$$\delta = \gamma^2/9(\overline{\alpha'})^2 \tag{10.2.20}$$

Considering only the polarization ratios at 90°, we obtain, with the aid of (10.1.53) and (10.1.55)

$$\rho_{\text{v}}(90) = \frac{(3/45)\gamma^2}{(\overline{\alpha'})^2 + (4/45)\gamma^2} = \frac{3\delta}{5 + 4\delta} \tag{10.2.21}$$

$$\rho_{\text{u}}(90) = \frac{(6/45)\gamma^2}{(\overline{\alpha'})^2 + (7/45)\gamma^2} = \frac{6\delta}{5 + 7\delta} \tag{10.2.22}$$

From these equations it can be readily deduced, for scattering at $\theta = 90°$, that the anisotropy results in an enhancement of the intensity of the scattered beam over the isotropic case by either $7/45\gamma^2$ or $13/45\gamma^2$ for each unit value of $(\overline{\alpha'})^2$. The first of these values is for vertically polarized incident light and the second is for unpolarized incident light. The above equations can be written alternatively as

$$7/45\gamma^2 = [7\rho_{\text{v}}(90)/(3 - 4\rho_{\text{v}}(90))](\overline{\alpha'})^2 \tag{10.2.23}$$

and

$$13/45\gamma^2 = [13\rho_{\text{u}}(90)/(6 - 7\rho_{\text{u}}(90))](\overline{\alpha'})^2 \tag{10.2.24}$$

[3] Coumou *et al.* (1964a) have $(2\gamma^2)$ for γ^2.

The fractional enhancement over the isotropic case due to the anisotropy is given by $(1 + (7/45)\gamma^2)$ and $(1 + (13/45)\gamma^2)$ respectively, leading directly to the Cabannes factor (10.2.15).

There is a complication whenever measurements of the depolarization are carried out on optically active systems (Mijnlieff and Zeldenrust, 1965). With unpolarized incident light, each of the components of the scattered light observed at 90° will have been rotated by an amount dependent upon the distance traveled through the cell and upon the specific rotation and concentration of the optically active solute. The observed depolarization will be given by

$$\rho_{u,m} = \frac{\rho_u(90) + \tan^2 \alpha_r}{\rho_u(90) \tan^2 \alpha_r + 1} \tag{10.2.25}$$

where α_r is the optical rotation. There may be quite significant differences between $\rho_{u,m}$ and $\rho_u(90)$ for optically active solutions. Since $\rho_u(90)$ is used to obtain the Cabannes factor for molecular weight determination or for determining the isotropic part of the scattering, this effect must be given appropriate consideration.

10.2.1 Krishnan's Relations

Krishnan (1938a, b) has established a relation among the three polarization ratios by utilizing a general law of reciprocity due to Lord Rayleigh (1900) which leads to

$$H_v = V_h \tag{10.2.26}$$

and from this to

$$\rho_u(\theta) = \frac{1 + 1/\rho_h(\theta)}{1 + 1/\rho_v(\theta)} \tag{10.2.27}$$

ᵗhermore, if the electric vector of the linearly polarized incident beam the angle ψ with the normal to the plane of scattering, Krishnan (1939)

$$\rho_\psi = \frac{H_\psi}{V_\psi} = \frac{1 + (\tan^2 \psi/\rho_h(\theta))}{\tan^2 \psi + (1/\rho_v(\theta))} \tag{10.2.28}$$

ᵗttered components polarized with their electric vectors ˡar, respectively, to this plane.

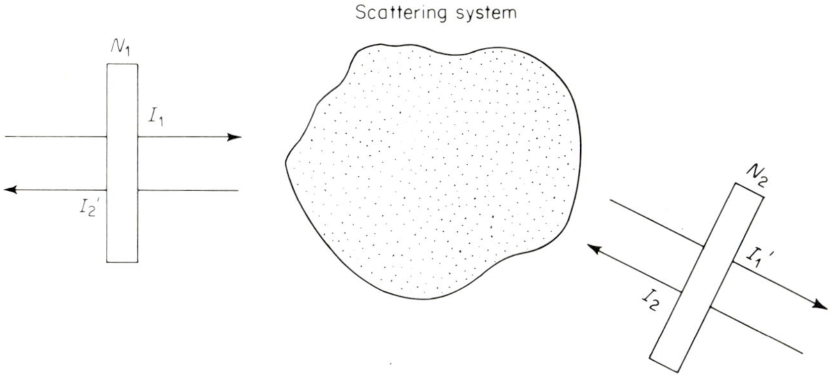

FIG. 10.2. Law of reciprocity. N_1 and N_2 are polarizers. If $I_2 = I'_1$, then $I'_2 = I_1$.

The law of reciprocity has been stated more explicitly by Perrin (1942) and can be described with the aid of Fig. 10.2. A monochromatic beam from a polarizer N_1, having an intensity I_1, is incident in a particular direction upon a system in which light can be scattered and absorbed. The emergent beam after passing through the polarizer N_2 has an intensity I'_1. There are two inverse beams which coincide with these but travel in the reverse direction. The first of these emerges from N_2 with intensity I_2; the second emerges from N_1 with intensity I'_2. Then, if $I_2 = I'_1$, it follows from the law of reciprocity that $I'_2 = I_1$.

This law is true only if the considered optical system is not affected by a reversal of time, so that the sense of propagation of the light is immaterial. There must be no movements, no electrical currents, no magnetic fields. Also there must be no frequency shifts or, at least, if these occur it should be possible to reverse them. For example, if there are frequency shifts as a result of movements in the system, these movements must be reversed appropriately for the inverse beam. When there is no mechanism for reversing the frequency shifts, such as in the case of fluorescence or Raman scattering, the law of reciprocity is invalid.

Perrin has shown in a general way that Krishnan's second relation follows directly from the law of reciprocity provided the scattering medium is symmetrical; i.e., for any large spherical volume, the center is a center of symmetry and any plane passing through the center is a plane of symmetry. The presence of optical rotatory power would vitiate this latter condition. However, the relation would be valid for a statistically isotropic distribution of small anisotropic elements.

10.3 Depolarization by Liquids

Light scattered by dense media, such as liquids, is generally depolarized but in this case the relation between the depolarization and the anisotropy of the particles or molecules is considerably more complicated. The field which acts upon each particular molecule is no longer approximated by the external field but consists, in addition, of the polarization field and the molecular field. The polarization field is created by the presence of the continuous, homogeneous medium around the particle and is itself subject to the action of the external field. The effect is to alter the field strength by changing the polarizability in accord with the Lorenz–Lorentz formula, but the direction of the field is not altered. The molecular field arises from the polarizability of those molecules immediately surrounding the point under observation. It is not isotropic and for that reason its contribution to the internal field will no longer be uniform so that even isotropic molecules, since they are now subject to a nonuniform field, will exhibit some depolarization. Thus, a liquid such as carbon tetrachloride exhibits some depolarization even though the molecules are isotropic, as evidenced by the lack of depolarization in the light scattered by its vapor. Yvon (1937), Fixman (1955), Zwanzig (1964), and many others have discussed the internal field for optically isotropic molecules. Theimer and Paul (1965) have claimed that significant depolarization would be obtained even for dense gases consisting of isotropic molecules, provided that $N\alpha' \rightarrow 1$ where N is the particle density and α' is the polarizability; i.e., $N > 10^{20}$ atoms per cm^3 or pressures of the order of 10 atmospheres.

If the molecules are themselves anisotropic, there are two contributions to the molecular field. As before, the polarizability of the neighboring molecules must be considered. This has been done by Weill (1961) and by Sicotte (1966, 1967). In this case, the anisotropic internal field depends upon the molecular anisotropy. Interestingly, the theory predicts that the generalized anisotropy δ (10.2.20) will decrease with dielectric constant and will therefore be lower for a liquid than for the corresponding vapor.

The second contribution to the molecular field arises because, when there are intermolecular attractions, the molecules are no longer randomly oriented and there will exist a correlation among the orientations. This will affect the internal field and in addition there will be a contribution to the scattering due to the "orientational anisotropy." Benoit and Stockmayer (1956) have calculated the latter effect with the aid of a correlation function which gives the probability that two molecules separated by a given distance will have their axes aligned. Shakhparonov (1961) has expressed this correlation function more explicitly in terms of the polarizability ellipsoid of the molecule and of the average value of the angle between the principal axes of polarizability ellipsoids of adjacent molecules. The final result leads to a correction factor to the isotropic scattering which is identical with the Cabannes factor so that it can continue to be used. Furthermore, Prins and Prins (1956) have shown that the Cabannes correction factor $C(\theta)$ holds equally well for the liquid model as for the gas. Prins (1961) has calculated the contribution of the orientational correlations to the scattering by aqueous sucrose.

Buckingham and Stephen (1957) have derived a general formula for the polarization of liquids in terms of the polarizability and the hyperpolarizabilities of the molecules. The hyperpolarizability tensors describe the departures from a linear law of the relation between the dipole moment and the electric field strength. For spherically symmetric molecules (such as the inert gases), the following very simple relation is obtained:

$$\rho_v = \left(3\frac{R_M{}^l - R_M{}^v}{R_M{}^v}\right) \Bigg/ \left(4\frac{R_M{}^l - R_M{}^v}{R_M{}^v} + \frac{10RT\beta_T}{V}\right) \tag{10.3.1}$$

where $R_M{}^l$ and $R_M{}^v$ are the molar refractions (Glasstone, 1946) of the liquid and gas phases, respectively, and β_T and V are the compressibility and molar volume of the liquid. For liquid argon it predicts $\rho_v = 0.12$ compared to 0.0004 measured for the gas (Bridge and Buckingham, 1966). This relation has not been checked experimentally.

One may subdivide the light scattered by liquids into an isotropic and an anisotropic contribution in much the same fashion as for media with independently scattering particles [(10.2.16) and (10.2.17)]. However, despite the efforts cited above for liquids, there exists an adequate theory only for the isotropic part of the scattering as given by the Smoluchowski–Einstein equation (9.1.14). The Cabannes factor enables one to separate the total scattering into these two parts. At 90°

$$R_{is} = R_{tot}(6 - 7\rho_u)/(6 + 6\rho_u), \tag{10.3.2}$$

$$R_{an} = R_{tot} 13\rho_u/(6 + 6\rho_u) \tag{10.3.3}$$

The value of R_{is} can be obtained either from the measured values of R_{tot} and ρ_u or from the Einstein–Smoluchowski equation; we have already seen (9.1.2) how the experimental results of Coumou *et al.* (1964b) have validated this relation.

Coumou *et al.* (1964a) have measured the anisotropic part of the scattering at $\lambda_0 = 0.546\ \mu$ for a number of liquids and have compared this with the corresponding quantity in the vapor phase. The molar isotropic and anisotropic Rayleigh ratios for the liquid and vapor ($R^l_{is,m}$, $R^v_{is,m}$, $R^l_{an,m}$, $R^v_{an,m}$) were obtained by multiplying the specific Rayleigh ratios by the corresponding values of the molar volume. The results for several liquids are given in Table 10.2. These show that for equal numbers of molecules the isotropic scattering intensity in the liquid is always relatively smaller than in the vapor. In most cases, the scattering in the liquids is only 4 to 10% of that in the vapors, clearly demonstrating the occurrence of destructive interference between the light scattered from different molecules in the liquid.

TABLE 10.2

MOLAR RAYLEIGH RATIOS IN LIQUID AND VAPOR PHASE

	$R^{l}_{\mathrm{is},m} \times 10^4$	$R^{\mathrm{v}}_{\mathrm{is},m} \times 10^4$	$R^{l}_{\mathrm{an},m} \times 10^4$	$R^{\mathrm{v}}_{\mathrm{an},m} \times 10^4$
Benzene	5.28	56.24	8.85	5.42
Toluene	5.58	79.37	14.0	8.32
Nitrobenzene	5.22	88.14	62.1	11.5
Carbon disulphide	7.23	40.35	43.3	14.1
Carbon tetrachloride	4.93	38.25	0.28	—
Cyclohexane	4.61	62.42	0.34	0.90
Isooctane	7.62	125.9	0.96	2.46
n-Hexane	5.85	73.30	—	—
n-Octane	6.46	125.9	1.45	4.09
n-Decane	7.41	192.7	2.27	5.51
Methyl ethyl ketone	2.81	34.91	0.95	1.72

The molar anisotropic Rayleigh ratios for the same compounds are also given in Table 10.2. Here the scattering in the liquid phase is the same order of magnitude as in the vapor. For rigid molecules, the anisotropic scattering is greater in the liquid phase; for the flexible molecules it is greater in the vapor phase. Also, the rigid molecules generally have a higher anisotropic scattering than the flexible molecules. Another recent set of Rayleigh ratio and polarization ratio data for liquids is reported by Weill (1958). For the same liquids, his data agree quite well with those reported in Table 10.2. Liquids not included above which have been studied by Weill are ether, dioxane, ethylbenzene, chlorbenzene, brombenzene, styrene, tetralin, and bromoform. Very accurate depolarization measurements using a laser have been obtained by Bridge and Buckingham (1964, 1966) for 27 gases and vapors including benzene. Dintzis and Stein (1964) have reported values for 11 gases and vapors.

The temperature dependence of the anisotropic part of the scattering can provide a useful tool to study structural effects in liquids. The molecular anisotropy factor in (10.2.17) is explicitly given by (10.2.19) only when due to random orientations of gas-like molecules. Another contribution may possibly arise from correlations in orientation among adjacent molecules. The so-called structures observed in liquids may influence this latter quantity, and if the liquid structure is temperature dependent, this should be reflected by a corresponding temperature dependence of the molecular anisotropy factor.

The Rayleigh ratio and depolarization of twelve liquids have been studied by Schmidt (1968) over a range of temperatures and at wavelengths of 0.436 and 0.546 μ. His results are presented in Table 10.3 as the temperature coefficient of the molecular anisotropy factor defined by

$$\gamma^2 = \beta_0 + \beta_1 t \tag{10.3.4}$$

where t is the centigrade temperature. For the liquids bromoform, dichloromethane, dibromomethane, and diiodomethane, the temperature dependence of γ^2 is zero; thus, the nearest neighbor orientations are probably random. In other cases, there is either a positive or negative temperature coefficient. An augmentation of γ^2 over that obtained from random orientation of the molecules can be interpreted as due to either a parallel or an antiparallel mutual orientation. On the other hand, perpendicular orientation of pairs of neighboring molecules would result in a decrease of γ^2 from that calculated from (10.2.19).

Coumou *et al.* (1964a) have also studied the scattering by binary mixtures of a liquid consisting of rigid anisotropic molecules with a liquid having nearly spherical molecules. Calculated values of both the isotropic and anisotropic scattering were obtained from the measurements of the total Rayleigh ratio and the depolarization. Figure 10.3 shows results for mixtures of nitrobenzene in cyclohexane and nitrobenzene in carbon tetrachloride.

Whereas the isotropic scattering is quite different for each system, the anisotropic scattering vs. mole fraction composition is identical for both systems. Indeed, these results clearly illustrate the difference in character between the two types of scattering. The anisotropic scattering appears to be independent of the composition fluctuations occurring in the liquid. Similar results have been obtained by Weill (1958) for a mixture of tetralin and decalin. A comparable study had also been carried out earlier by Powers and Stein (1953) with carbon disulphide and carbon tetrachloride.

TABLE 10.3
TEMPERATURE COEFFICIENT OF THE MOLECULAR ANISOTROPY FACTOR

Compound	β_1 (436 mμ)	β_1 (546 mμ)
Benzene	0.059 ± 0.002	0.061 ± 0.007
Aniline	0.045 ± 0.009	0.052 ± 0.003
Hexafluorobenzene	0.022 ± 0.010	0.032 ± 0.018
n-Hexane	-0.019 ± 0.002	-0.031 ± 0.002
n-Dodecane	-0.022 ± 0.009	-0.053 ± 0.002
2-Propanol	0.0077 ± 0.0006	0.0059 ± 0.0012
Dichloromethane	0.0	0.0
Dibromomethane	0.0	0.0
Diiodomethane	0.0	0.0
Bromoform	0.0	0.0

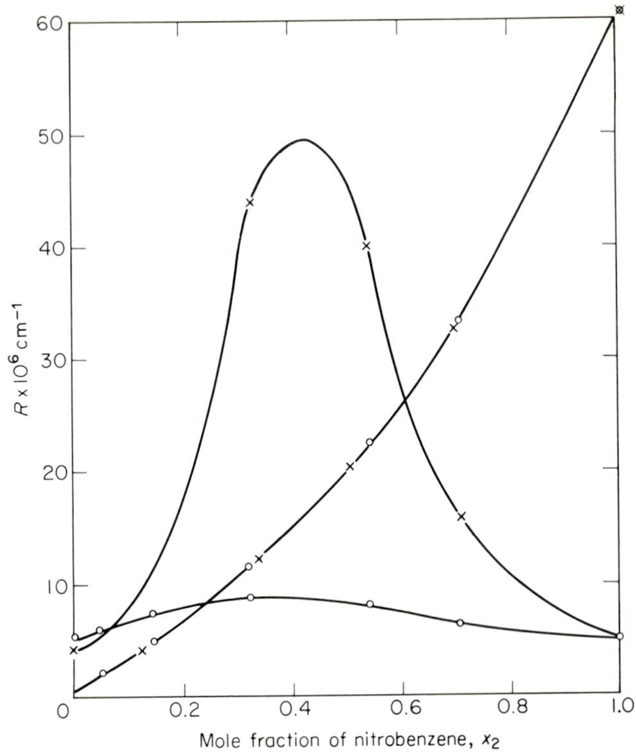

FIG. 10.3. Isotropic Rayleigh ratio of nitrobenzene in carbon tetrachloride (circles) and in cyclohexane (crosses) compared with the anisotropic Rayleigh ratio in both systems (circles and crosses) (Coumou *et al.*, 1964b).

Coumou *et al.* (1964a) have defined a molar anisotropic Rayleigh ratio of the anisotropic component at infinite dilution by

$$R^0_{\mathrm{an},m} = (dR_{\mathrm{an},m}/dx_2)_{x_2=0} \tag{10.3.5}$$

where x_2 is the mole fraction composition. The results are given in Table 10.4. The ratio of $R^0_{\mathrm{an},m}$ to $R^{\mathrm{v}}_{\mathrm{an},m}$ equals the factor $[(n_s^2 + 2)/3]^2$ where n_s is the refractive index of the solvent. This suggests that at infinite dilution, the local anisotropies in a liquid solution scatter like free gas molecules subject to an electric field vector $\mathbf{E}'$ given by the Lorentz relation

$$\mathbf{E}' = [(n_s^2 + 2)/3]\mathbf{E} \tag{10.3.6}$$

where $\mathbf{E}$ is the electric field vector of the incident light. This has the practical advantage of permitting the determination of the anisotropic scattering of a vapor from the molar anisotropic scattering in dilute solutions in a solvent consisting of nearly isotropic molecules.

TABLE 10.4
MOLAR ANISOTROPIC RAYLEIGH RATIOS AT INFINITE DILUTION

Anisotropic compound	$R^{l}_{an,m} \times 10^4$	$R^{v}_{an,m} \times 10^4$	Solvent	$R^{0}_{an,m} \times 10^4$	$\dfrac{R^{0}_{an,m}}{R^{v}_{an,m}}$	$\dfrac{n_s^2 + 2}{3}$
Benzene	8.85	5.42	Carbon tetrachloride	9.3	1.72	1.89
			Cyclohexane	9.8	1.82	1.81
			Isooctane	8.9	1.64	1.72
			Neopentane	8.9	1.65	1.60
			Methanol	9.0	1.66	1.58
Nitrobenzene	62.1	11.5	Carbon tetrachloride	22.9	1.99	1.89
			Cyclohexane	21.5	1.87	1.81
Carbon disulphide	43.3	14.1	Carbon tetrachloride	28.8	2.04	1.89
			Cyclohexane	28.9	2.05	1.81
3 Methyl-1 butyne	3.11	—	Carbon tetrachloride	2.60	—	1.89
1,5-Hexadiene	9.7	—	Cyclohexane	9.4	—	1.81

Bothorel (1968) and others have determined the molecular anisotropy factor, γ^2, of a huge number of organic liquids from the anisotropic part of the Rayleigh ratio at 90° (10.2.17). The latter quantity, in turn, is obtained from measurements of the total Rayleigh ratio and of the depolarization (10.3.3). Some of the experiments have been obtained with great accuracy by use of a laser beam (Lalanne and Bothorel, 1964).

The aim of this work has been to relate the molecular anisotropy factor to the geometric and electronic structure of the molecule—indeed, to assign partial values of this factor to each bond in the molecule. Then, knowing the molecular structure, the anisotropy of the molecule could be obtained by appropriate combination of the values for each of the bonds. In turn, the experimental value of the molecular anisotropy factor could be utilized to elucidate aspects of the molecular structure which might not be completely understood.

Clément and Bothorel (1964) have studied the effect of the solvent upon the measured values. Figure 10.4 is a plot of γ^2 against the number of atoms of carbon in normal alkanes for four different solvents as well as for the pure compound. Obviously, the solvent effects are significant. However, the relative effects in a particular solvent are similar. This is shown by the curve drawn through the solid circles. The molecular anisotropy factor for each species relative to that for normal dodecane is shown on this curve for each of the solvents as well as for the pure substance. The diameter of the full circles gives the range of uncertainty. Thus for any homologous series, the results may be compared upon some such relative basis.

The method can be illustrated by reference to the results for substituted aromatic compounds. When the hydrogen on the benzene ring is replaced, the excess anisotropy is given by

$$\Delta\gamma^2 = ({\gamma_s}^2 - {\gamma_0}^2)/{\gamma_0}^2 \qquad (10.3.7)$$

where ${\gamma_0}^2$ is the molecular anisotropy factor for benzene and ${\gamma_s}^2$ is the value for the substituted compound. The new bond will be assumed to be symmetrical about an axis of revolution as in benzene so that only two of the

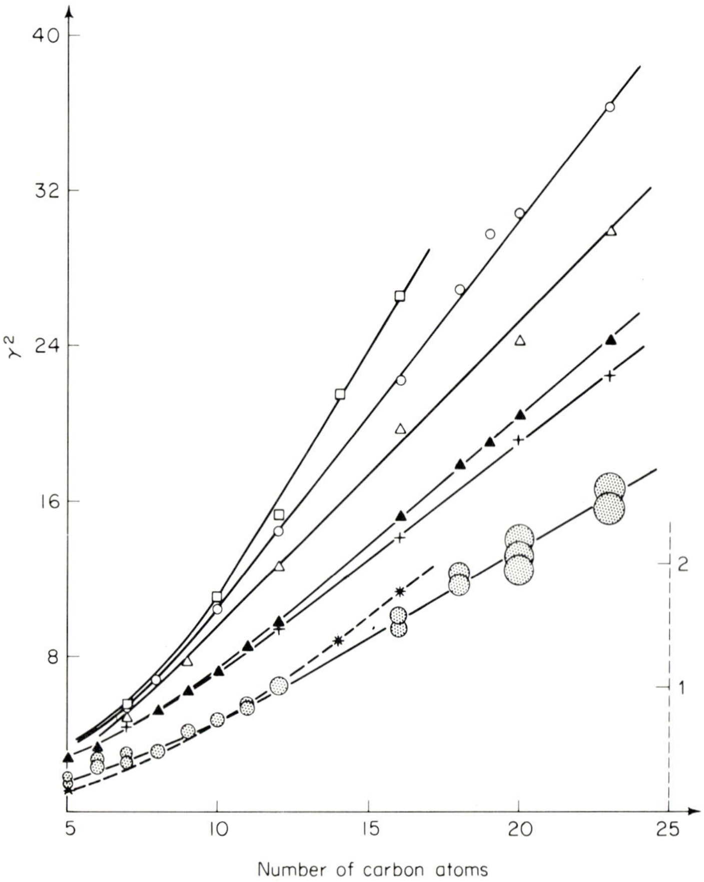

FIG. 10.4. Effect of solvent upon molecular anisotropy factor in normal alkanes. From the upper curve downwards; pure compound, heptane, pentane, cyclohexane, and carbon tetrachloride. Curve drawn through dark circles gives the anisotropy of the compound relative to that of a reference compound (dodecane) in the same solvent (Clement and Bothorel, 1964).

principal polarizabilities are equal. For such configurations, (10.2.19) reduces to

$$\gamma^2 = (\alpha_A' - \alpha_B')^2 \tag{10.3.8}$$

If the new bond is now assigned the anisotropy S_R, then for the substituted benzene

$$\Delta\gamma^2 = (\gamma_0 S_R + S_R{}^2)/\gamma_0{}^2 \tag{10.3.9}$$

and a numerical value of S_R can be obtained for each functional group from the experimental values of $\Delta\gamma^2$. The value of $\gamma_0{}^2$ for benzene is 36 Å^6.

The additivity effect of these substituents upon the molecular anisotropy of polysubstituted benzene has been verified for a large number of di-, tri-, tetra-, and hexa-substituted benzenes (Unanué and Bothorel, 1964) and for various polyphenylic compounds (Unanué and Bothorel, 1965). For example, the relative excess anisotropy of di-substituted ortho compounds is

$$\Delta\gamma^2 = (8S_R\gamma_0 + 7S_R{}^2)/4\gamma_0{}^2 \tag{10.3.10}$$

Aroney *et al.* (1965) have compared the molecular anisotropies obtained above by light scattering with those which can be extracted from the molar Kerr constant obtained from electric birefringence data. The values of $\gamma_0{}^2$ obtained by the latter technique were less than half of those from light scattering; e.g., $\gamma_0{}^2$ for benzene was 14.5 Å^6. However if the apparent anisotropies are all referred to benzene as a standard, then the anisotropies derived from light scattering and electric birefringence are in agreement with each other. Accordingly, these methods lead to a consistent technique for analysis of molecular structure even though the precise significance of the numerical value of $\gamma_0{}^2$ is doubtful.

10.4 Partial Orientation

Nonspherical or anisometric particles and molecules will be partially oriented if they are in a flowing system or if they are subjected to electric or magnetic fields. Flow orientation is due exclusively to the anisometry of the particle and the torque produced by the velocity gradient. Figure 10.5 illustrates the torque exerted on a rod, showing the tendency for it to line up with its axis parallel to the direction of the stream lines. In the case of a deformable material, the final shape of the particle will depend upon the flow. Magnetic and electric orientation may sometimes be somewhat simpler to analyze because the imposed fields are usually uniform and well defined. However complications may arise if the substance comprising the particle has an intrinsic electric or magnetic anisotropy. In any case, the orientation is governed by the torque acting upon the particles and the superimposed randomizing effect of the rotatory Brownian movement. This results in a distribution of orientations.

One optical effect of this phenomenon is the production of double refraction which is referred to as the Kerr effect, Cotton–Mouton effect, or the Maxwell effect, depending upon whether the applied force is electric, magnetic, or hydrodynamic. The latter effect is also commonly called streaming birefringence and is widely used to study colloidal and macromolecular systems (Cerf and Scheraga, 1952).

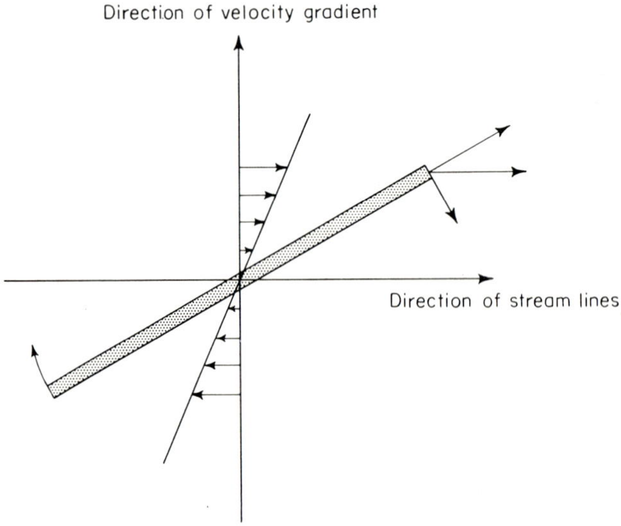

FIG. 10.5. Effect of a velocity gradient on a rigid rod (Cerf and Scheraga, 1952).

There will also be an influence upon the scattered light since this will depend upon the orientation of the anisotropic particles. It may, in turn, be possible to learn something about the shape of such particles by observing the result of well-defined orientational effects upon the light scattering. Figure 10.6 shows qualitatively how the scattering by rods or platelets oriented in a flowing system would vary with the direction of the primary beam, the direction of the electric vector, and the direction of observation. The blackened and open circles indicate whether the scattering is weak or strong respectively.

Figure 10.7 illustrates the effect of flow orientation on the scattering of an aqueous suspension of the rod-like particles comprising tobacco mosaic virus (Heller *et al.* 1961). The shearing force is exerted by rotation of the inner cylinder in a concentric cylindrical apparatus. The primary beam is parallel to the axis of the cylinder and the scattered beam is observed along the direction defined by the angle ω with respect to the reverse flow of the stream lines and the angle θ with respect to the incident beam. In the figure, the ordinate is the relative change in scattered intensity as a result of the flow and the abscissa is the shear rate. In both cases, $\theta = 90°$ but curve I is for $\omega = 90°$ while curve II is for $\omega = 72°$. Apparently the orientational effect of the flow does have an appreciable effect upon the light scattering for this system.

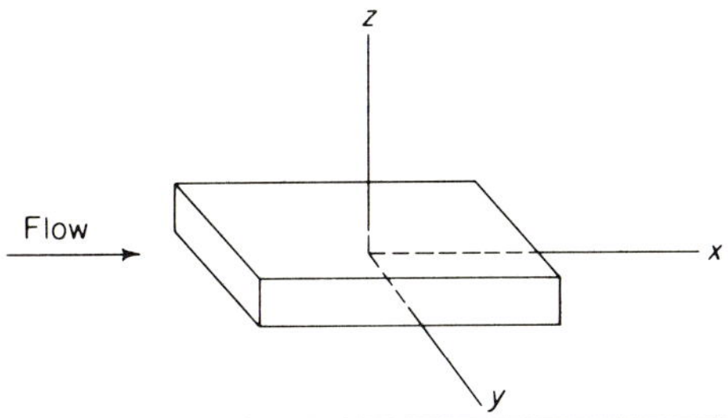

Direction of			Relative intensity	
Primary beam	Electric vector	Observation	Rods	Plates
X	Y	Z	●	○
X	Z	Y	●	●
Z	X	Y	○	○
Z	Y	X	●	○
Y	Z	X	●	●
Y	X	Z	○	○

FIG. 10.6. Qualitative effect of the directions of the primary beam, of the electric vector, and of the observation upon the intensity of the light scattered by rods and platelets oriented in a flowing system (Heller, 1959b).

If thin rods with axial ratios greater than 20 are assumed to behave as Rayleigh–Debye scatterers, then the partially oriented system will have a form factor obtained by averaging the value for a single particle over the distribution function for rotatory Brownian motion in the particular flow field. This is given by Okano and Wada (1961) as

$$P(\theta) = \int_0^{2\pi} \int_0^{\pi} [\sin^2(x \cos \eta)/(x \cos \eta)^2] f(\sigma, \phi) \sin \sigma \, d\sigma \, d\phi \qquad (10.4.1)$$

with

$$x = (2\pi/\lambda)L\sin(\theta/2) \tag{10.4.2}$$

where λ is the wavelength in the medium, L is the length of the rod, η is the angle between the axis of the rod and the half-scattering angle $\theta/2$, and σ and ϕ are the polar and azimuthal angles between the axis of the rod and the direction of flow. The probability distribution function of the rod $f(\sigma, \phi)$ was determined by Peterlin (1938). The unpolarized incident beam is perpendicular to the flow.

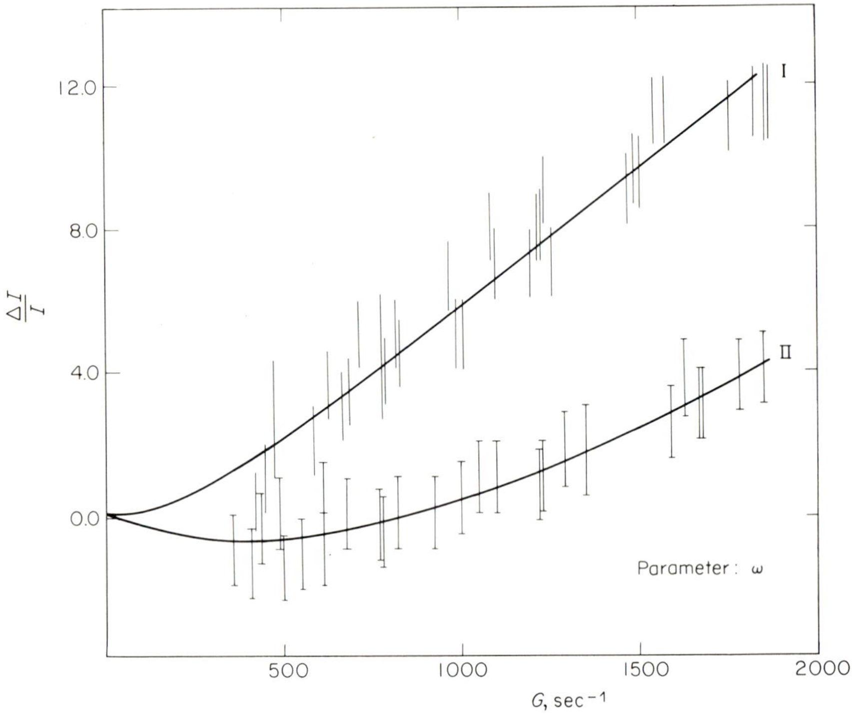

FIG. 10.7. Relative change in scattered intensity as a function of shear rate G for 0.028% tobacco mosaic virus suspension in 0.02 M phosphate buffer; I, $\omega = 90°$; II, $\omega = 72°$; $\lambda_0 = 0.5461\ \mu$; $\theta = 90°$ (Heller *et al.*, 1961).

Heller (1959b) has distinguished three distinct light scattering effects attributable to orientation which he designates dityndallism, conservative dichroism, and bidissymmetry. In defining dityndallism, the direction of propagation of the incident linearly polarized beam is assumed to be perpendicular to both the direction of flow and the direction of the velocity gradient, viz. along the y-axis in Fig. 10.6. The two polarized components of the incident beam are taken along the direction of flow and the direction of the velocity gradient and the intensity of the scattered light corresponding to these is designated I_e and I_0, respectively. Then the numerical value of the dityndallism is defined by

$$D = \Delta I_e - \Delta I_0 \tag{10.4.3}$$

where ΔI represents, for the appropriate incident component, the difference between the Rayleigh ratio of the system at flow and that at rest.

Conservative dichroism refers to the dependence of the turbidity or of the transmitted beam upon the direction of the electric vector of the incident linearly polarized beam. It is best observed in the absence of optical absorption, since it could be masked by true dichroism which is due to a difference in the extinction coefficients of the material to the two beams. Figure 10.8

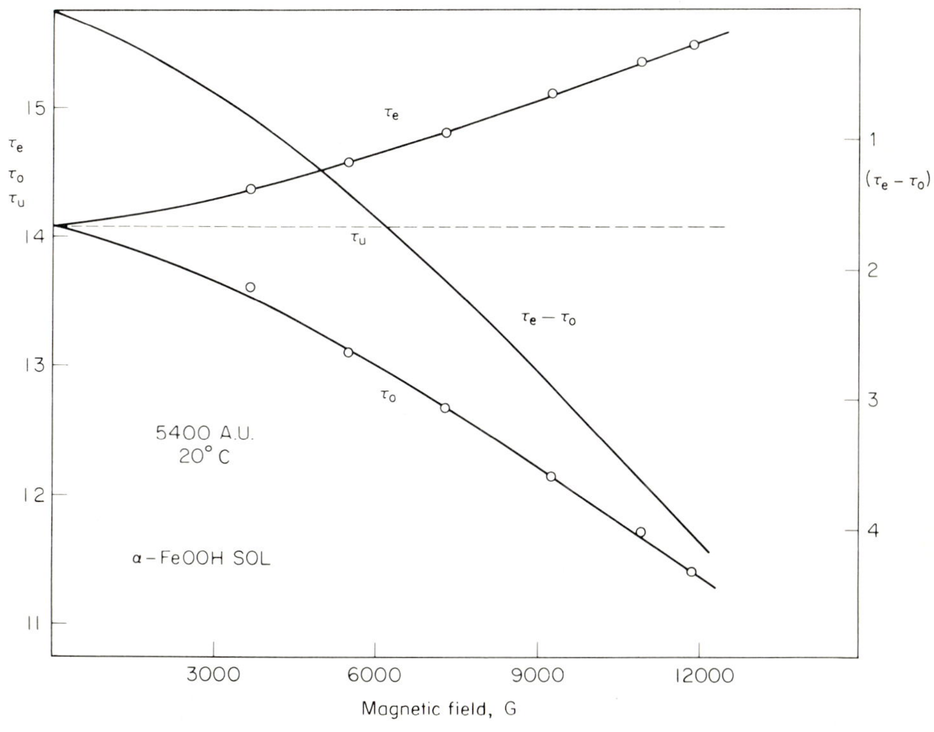

FIG. 10.8. Conservative magnetic dichroism of an α-FeOOH sol. τ_u is the turbidity of the unperturbed system and τ_o and τ_e are the turbidities of the oriented system when the electric vector of the incident beam vibrates along the direction of flow and along the direction of the velocity gradient, respectively (Heller, 1959b).

illustrates this effect for a colloidal solution of FeOOH in which the plate-like particles are orientated by a magnetic field. This effect produces a rotation of the plane of polarization of the incident beam. It can be easily distinguished from true optical activity because the amount of rotation depends upon the actual orientation of the incident beam.

Bidissymmetry is illustrated by Fig. 10.9 where the radiation pattern for spherical particles or randomly oriented anisotropic particles is schematically depicted on the left and that for oriented nonspherical particles is shown on the right-hand side of the figure. In both cases, longitudinal dissymmetry may be encountered. This consists usually of preferential scattering in the forward

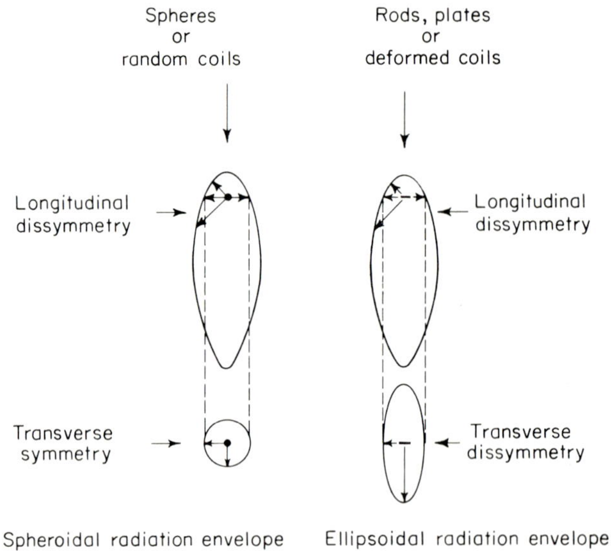

FIG. 10.9. Conventional dissymmetry (left) and bidissymmetry on orientation or deformation (right) (Heller, 1959b).

direction over that in the backward direction when the plane of observation includes the incident direction. We have seen that this is characteristic either of particles which are no longer small compared to the wavelength or of totally reflecting small particles. The cross sections of the diagrams in the lower pair of drawings apply for the plane which is perpendicular to the incident direction. Oriented anisometric particles will have a radiation pattern in this plane which is no longer circular and this is termed transverse dissymmetry. The three-dimensional radiation pattern, in this case, has ellipsoidal rather than spheroidal symmetry.

10.4.1 Flow Orientation of Flexible Macromolecules

The freely flexible macromolecule in a flowing system has been studied in some detail. The basic molecular model is a necklace in which each of the isotropic spherical beads (the monomers) is assembled by a random flight process. Except for being constrained to a particular link length, the beads are completely free and mutually independent in their arrangement. In flow, the beads are subjected to the frictional force, and the ultimate effect is that the macromolecule becomes both deformed and oriented.

Peterlin (1957) and Peterlin *et al.* (1958) have utilized the Rayleigh–Debye theory to calculate the form factor for such a configuration. The geometry is shown in Fig. 10.10 where the incident beam enters from the top of the page, y' is the direction of flow, and x' is the direction of the velocity gradient. The angle ϕ is between the direction of observation and the most probable direction x of the end-to-end distance within the macromolecule.

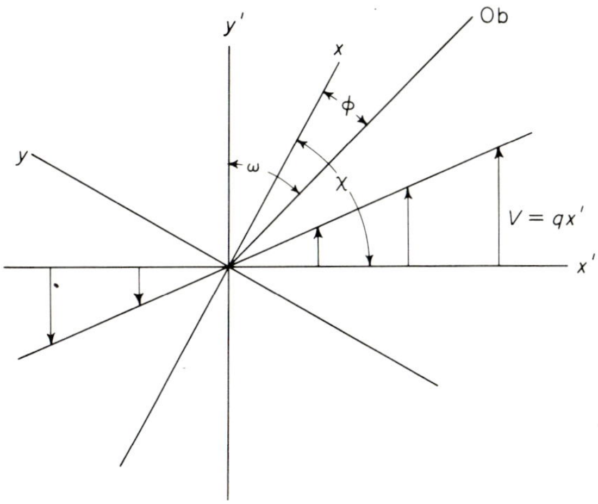

Fig. 10.10. Geometry for calculating scattering by a flexible molecule in a flowing system. The plane of the paper is perpendicular to the direction of the incident beam ($\theta = 90°$). x', direction of velocity gradient, q; y', direction of flow; Ob, direction of observation; x, most probable direction of end-to-end distance; V, velocity of flow (Peterlin *et al.*, 1958).

The final result can be represented by

$$P(\theta) = 2\int_0^1 (1 - u)\exp[-X(u + \zeta u^2)]\, du \tag{10.4.4}$$

where

$$\zeta = 2\beta^2/3 \tag{10.4.5}$$

$$X = X_0 \xi \tag{10.4.6}$$

$$X_0 = h^2 \overline{s^2}/6 \tag{10.4.7}$$

where as in the previous chapter $h = (4\pi/\lambda)\sin(\theta/2)$ and $\overline{s^2}$ is the mean square end-to-end distance of the macromolecule at rest given by

$$\overline{s^2} = (Z - 1)l^2 \tag{10.4.8}$$

with Z being the number of segments and l being the length of a link. In addition

$$\xi = \frac{\{1 + [\beta^2 + \beta(1 + \beta^2)^{1/2} \cos 2\phi] \cos^2(\theta/2)\}}{(1 + 2\beta^2/3)} \tag{10.4.9}$$

The quantity β is proportional to the velocity gradient q

$$\beta = (M/N_A kT)[\eta]\eta q \tag{10.4.10}$$

where M is the macromolecular weight, $[\eta]$ is the intrinsic viscosity, and η is the viscosity. The form factor $P(\theta)$ is related to the scattered intensity in the usual way by

$$I_u = \frac{16\pi^4(\alpha')^2}{\lambda^4 r^2} Z^2 P(\theta) \frac{(1 + \cos^2\theta)}{2} \tag{10.4.11}$$

where α' is the polarizability of each bead.

These authors have explored numerically the effect of flow upon the scattered light. In Figs. 10.11 and 10.12 the variation of the reciprocal of the form factor $P(\theta)$ is plotted against the flow parameter β for observation in the x and the y directions, respectively. This gives information of the deformation of the molecule by shear which is not complicated by the changes, with q, of the preferential direction of the end-to-end distance. In the one case, one is looking along the preferred axis; in the other, one is looking perpendicular to it. The effect upon the light scattering of the change in shape due to deformation by the flow is very pronounced. It decreases with increasing flow along x and increases when the scattered light is viewed along y. Figure 10.13 shows the variation of $P(\theta)^{-1}$ with flow when $\omega = 90°$ at various values of X_0. In view of (10.4.7), the parameter represents various values of the angle of observation θ at constant $\overline{s^2}$ (molecular weight) or various $\overline{s^2}$ values at constant θ. This figure permits X_0 to be determined if the intensity at a known flow is measured. Another useful graph is given in Fig. 10.14 where the longitudinal dissymmetry is plotted as a function of flow. In this case $\omega = 90°$ and D_θ is the ratio of the intensity at 60° to that at 120° for various values of $h^2\overline{s^2}/6$. It is possible to measure transverse dissymmetry in which θ is fixed at some particular angle, e.g., $\theta = 90°$, and

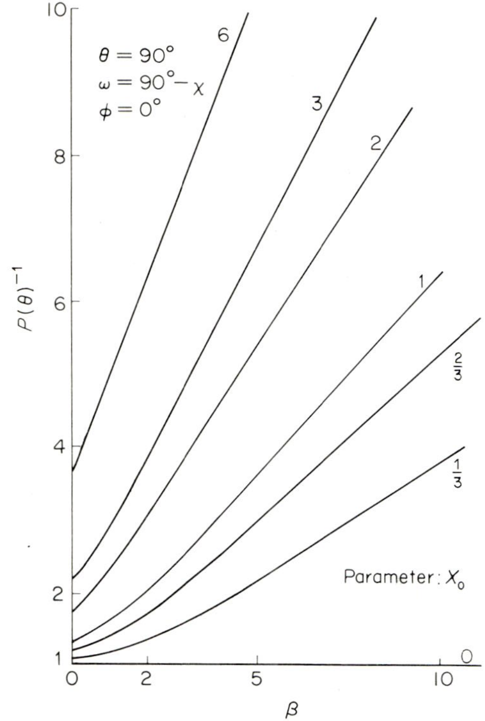

FIG. 10.11. Effect of flow parameter upon form factor of a flexible macromolecule for observation in the x-direction (Peterlin *et al.*, 1958).

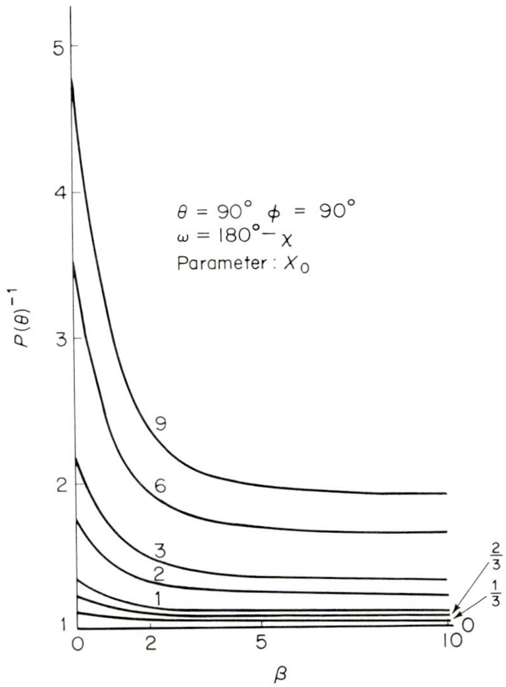

FIG. 10.12. Same as Fig. 10.11 for observation in the y-direction.

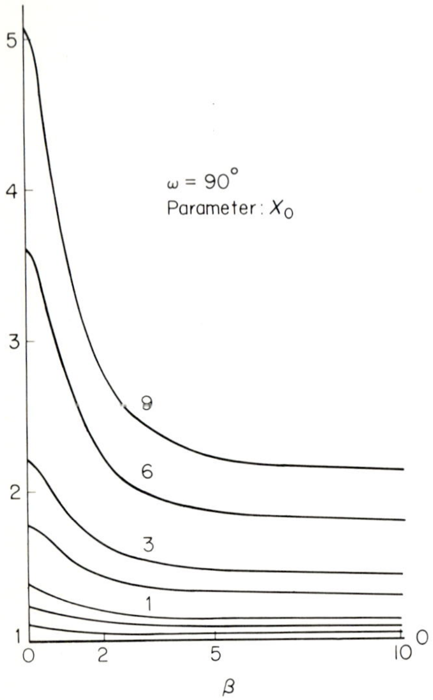

FIG. 10.13. Same as Fig. 10.11 for $\omega = 90°$.

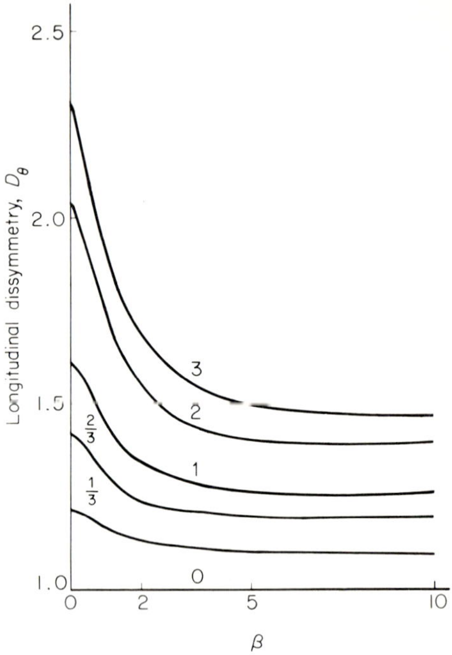

FIG. 10.14. Effect of flow parameter upon 60 to 120° longitudinal dissymmetry for $\omega = 90°$. Parameter indicated on curves is $(h^2 s^2)/6$ (Peterlin *et al.*, 1958).

the ratio of the intensity at $\omega = 120°$ is measured relative to that at 60°, again fixing the parameter $h^2\overline{s^2}/6$ at various values. This is illustrated in Fig. 10.15.

The above theory has been extended (Stevenson and Bhatnagar, 1958; Bhatnagar, 1963) to allow for anisotropic polarizability of the segments. Each segment is assumed to have two different polarizabilities, one in the direction of the length α_1' and one in the perpendicular direction α_2'. The final result is a correction term in (10.4.11) as follows:

$$I_u = \frac{16\pi^4}{\lambda^4 r^2}\left(\frac{\alpha_1' + 2\alpha_2'}{3}\right)^2 Z^2 \frac{P(\theta)}{2}(1 + \cos^2\theta - 2\varepsilon \sin^2\theta \sin 2\omega) \qquad (10.4.12)$$

where

$$\varepsilon = [\beta(\alpha_1' - \alpha_2')]/(\alpha_1' + 2\alpha_2') \qquad (10.4.13)$$

The effect of utilizing a more sophisticated and possibly more realistic molecular model to obtain the detailed structure of the macromolecule when it undergoes deformation in a hydrodynamic field has been explored by Bullough (1963b) and by Peterlin and Reinhold (1964) and Reinhold and

Peterlin (1965). This consists of assuming that the links connecting the individual segments in the chain have a Gaussian distribution of lengths. The effect upon the scattered intensity is rather slight, although it might just be possible to distinguish between the two models by light scattering measurements. However, the effect on the depolarization in the case of anisotropic polarizable segments is considerable, particularly at high shear rates. Accordingly, there should be little difficulty in distinguishing between the models if it is possible to observe the effects of shear at all. Furthermore, depolarization measurements could lead to values of the anisotropy in polarizability of the monomer units.

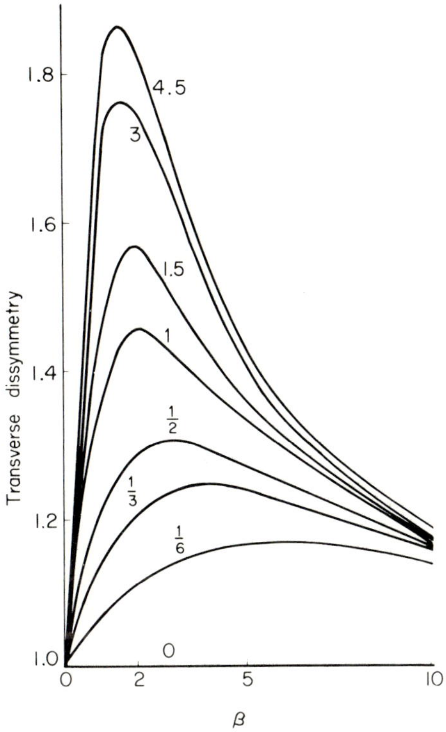

FIG. 10.15. Effect of flow parameter upon 60 to 120° transverse dissymmetry for $\theta = 90°$. Parameter indicated on curves is $(h^2\overline{s^2})/6$ (Peterlin *et al.*, 1958).

Nakagaki and Heller (1959) have considered the scattering by a macromolecule in a hydrodynamic field using a somewhat simpler model. They approximated the random coil by an equivalent spheroid. The major axis of the spheroid is placed in the direction of the most probable end-to-end distance of the macromolecule. The two minor axes are assumed to be equal to the smallest semiaxis of the ellipsoidal configuration which represents the macromolecule. The refractive index of the spheroid is taken as the volume average of the refractive index of the polymer and of the solvent enclosed by it, and its axial ratio in the hydrodynamic field is given by

$$p = (a/b)[(1 + \beta^2)^{1/2} + \beta] \tag{10.4.14}$$

where a and b are the largest and the smallest semiaxes of the macromolecule at rest.

These authors give the rather complicated expressions for the Rayleigh ratios for the scattered light for the two cases that the electric vector of the incident beam vibrates parallel and perpendicular to the plane of observation. The expressions are given as a function of the axial ratio, the flow parameter β, and the angles ω and θ. The latter as usual is the angle between the direction of observation and the primary beam, while ω is the angle between the plane of observation and the direction of flow.

Instead of working with the value of the dityndallism as defined by (10.4.3), the intensity ratio

$$\Delta' = I_0/I_e \tag{10.4.15}$$

was used. I_0 and I_e are the intensities for the flowing system when the electric vector of the incident beam vibrates along the direction of flow and along the direction of the velocity gradient respectively. The direction of the primary beam is perpendicular to the direction of flow and to the velocity gradient, i.e., along the y-axis in Fig. 10.6.

We consider the case $\theta = 90°$ and $\omega = 90°$ as an illustrative example. This corresponds to the incident beam along the y-axis in Fig. 10.6 and the scattered beam along the z-axis. Figure 10.16 shows the variation of Δ' with the flow parameter β for several values of the refractive index m. Other computations of this type indicate that the magnitudes are such that measurements of dityndallism might possibly lead to the determination of the molecular parameters of deformable macromolecules.

10.4.2 Rigid Rods in an Electric Field

Wippler (1954, 1955) has studied the light scattering of solutions of macromolecules or of suspensions of particles which have been statistically oriented by the action of an electric field. He first considered a suspension of rigid rods for which the distribution function is given by Benoit (1951)

$$p(\phi) = 4\pi f(\phi) = 1 + \sum_{1}^{\infty} a_n P_n(\cos\phi) \tag{10.4.16}$$

with

$$a_1 = (\mu/kT)E \tag{10.4.17}$$

$$a_2 = \frac{\mu^2}{k^2T^2} + \frac{\alpha_1' - \alpha_2'}{kT}\frac{E^2}{3} \tag{10.4.18}$$

where ϕ is the angle between the field direction and the axis of the rod, $f(\phi) \sin \phi \, d\phi$ is the probability that a rod has a value of ϕ between ϕ and $\phi + d\phi$, $P_n(\cos \phi)$ is the Legendre polynomial, μ is the permanent dipole moment carried by the rod, α_1' and α_2' are the principal polarizabilities which are parallel and perpendicular to the axis of the rod, k is Boltzmann's constant, and T is the absolute temperature.

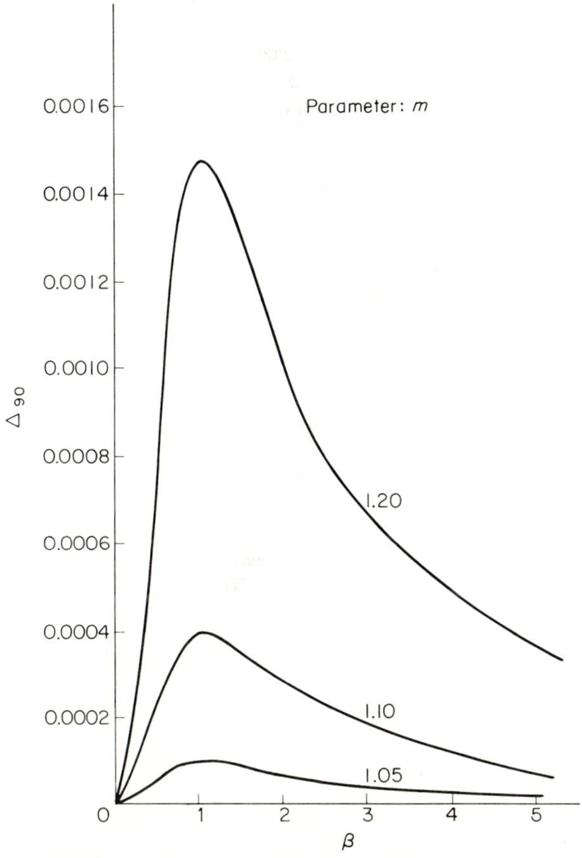

FIG. 10.16. Variation of intensity ratio Δ' of a flexible macromolecule with flow. The scattering is that of an equivalent spheroid with refractive index m; $\theta = \omega = 90°$ (Nakagaki and Heller, 1959).

The form factor is altered from that for a randomly oriented dispersion of rods (8.1.40) as a consequence of the applied field such that

$$\Delta P(\theta) = \frac{4\pi^2}{45}(1 - 3\cos^2 v)\frac{\mu^2}{3k^2T^2} + \frac{\alpha_1' - \alpha_2'}{3kT}\frac{L^2}{\lambda^2}E^2 \sin^2 \frac{\theta}{2} \qquad (10.4.19)$$

where v is the angle between the direction of the field and the direction s in Fig. 8.2 and L is the length of the rod. The direction s is the bisectrix to the angle external to the scattering angle. The formula has been obtained by truncating the distribution function at the square term and also by assuming that

$$(2\pi L/\lambda)\sin(\theta/2) \gg 1 \qquad (10.4.20)$$

It is interesting to note that the values of $\Delta P(\theta)$ corresponding to the field parallel ($\Delta' P(\theta)$; $v = 0$) and perpendicular ($\Delta'' P(\theta)$; $v = 90°$) to the direction $\theta/2$ obey the relation

$$\Delta' P(\theta) + 2\Delta'' P(\theta) = 0 \qquad (10.4.21)$$

This is no longer valid if the rod is deformed in the presence of the field so that this criterion can be used to distinguish between flexible and rigid rods (Wallach and Benoit, 1962).

This effect depends upon both the permanent dipole moment μ and the induced moments α_1' and α_2'. Therefore, light scattering in a static field can lead to an evaluation of one or another of these moments in the two extremes that the permanent dipole either predominates or is absent. If there is a significant contribution from both, the two mechanisms of orientation can only be distinguished by studying the temperature dependence of the phenomenon.

Using Peterlin's (1938) value for the distribution function in an alternating field, the change in the form factor is

$$\Delta P(\theta) = \Delta P_0(\theta)\left[\frac{1}{1 + (9/4)\omega^2\tau^2} + \frac{\cos(2\omega t - \delta_2)}{(1 + \omega^2 t^2)^{1/2}(1 + (9/4)\omega^2\tau^2)^{1/2}}\right] \qquad (10.4.22)$$

where $\tau = (3D)^{-1}$, D being the rotatory diffusion constant, ω is the frequency of the sinusoidal field, $\Delta P_0(\theta)$ is the value for a static field of the same effective field intensity (root mean square), and

$$\tan \delta_2 = 5\omega\tau/(2 - 3\omega^2\tau^2) \qquad (10.4.23)$$

If the orientation is due only to the effect of a permanent dipole,

$$\Delta P(\theta) = \Delta P_0(\theta)(1 + [\cos(2\omega t - \delta_1)/(1 + \omega^2\tau^2)^{1/2}]) \qquad (10.4.24)$$

where

$$\tan \delta_1 = \omega\tau \qquad (10.4.25)$$

Finally, the decay of the form factor upon suppression of the applied field is given by

$$\Delta P(\theta) = \Delta P_0(\theta)\, e^{-6Dt} \tag{10.4.26}$$

where now $\Delta P_0(\theta)$ is the value at the time the field is cut off.

Wippler (1955) has studied the effect of orientation in an electrical field upon the light scattering of a solution of tobacco mosaic virus. These particles are known to be rod-like and their size distribution can be estimated from electron microscopy. The system obeys Eq. (10.4.21) which demonstrates that the rods are rigid. Also, on the assumption that they do not carry a very large permanent dipole, the electrical anisotropy $(\alpha_1' - \alpha_2')$ was calculated from measured values of $\Delta P(\theta)$ (10.4.19).

Jennings and Jerrard (1965a, b, 1966) have noted, according to (10.4.22), that upon the application of an alternating electric field, the intensity change will consist of both a steady component and an alternating component varying at twice the frequency of the applied field. The additional information obtained from measurements of these two intensity terms permits evaluation of both the permanent dipole moment and the electrical anisotropy. They have carried out measurements on solutions of bentonite (1965a) and of tobacco mosaic virus (1966), obtaining in the latter case a value for the electrical anisotropy similar to that of Wippler (1955) as well as a value for the permanent moment.

The effect of an electric field upon light scattering of tobacco mosaic virus has also been studied by Stoylov (1966a, b), who proposed determining the particle size by using sufficiently high strength fields in order to completely orient the particles. Stoylov and Sokerov (1967, 1968) were able to determine the rotatory diffusion constant with the aid of (10.4.26) from the transient effects after removal of the electric field.

Wallach and Benoit (1962) and Jennings and Jerrard (1965b) have carried out an investigation upon the synthetic polypeptide poly-L-benzyl glutamate which has a helical structure which can be approximated by a rod. This structure also appeared to behave as a rigid rod if (10.4.21) is used as the criterion for rigidity. Because the value of $\Delta P(\theta)$ varied significantly with the frequency of the alternating field, the macromolecule appeared to have a significant permanent dipole. Indeed, at the highest frequencies, one can suppose that the contribution of the induced dipoles is not significant so that the term $(\alpha_1' - \alpha_2')/3kT$ in (10.4.19) can be dropped and the permanent dipole moment μ can be measured. The values obtained in this way were in excellent accord with values obtained from electric birefringence and from dielectric measurements. Finally, it is possible to relate the dispersion of $\Delta P(\theta)$ with frequency to the rotatory diffusion constant with the aid of expressions such as (10.4.22) and (10.4.24). The length of the rod can, in turn, be extracted from the rotatory diffusion constant if an assumption regarding the diameter of the rod is made. The particle lengths obtained in this way from the dispersion of $\Delta P(\theta)$ were in excellent agreement with those calculated from the radius of gyration obtained by low-angle light scattering and from those based upon a direct calculation from the molecular model.

10.4.3 Macromolecular Chains in an Electric Field

In considering the effect of an electric field upon the scattering by a randomly coiled macromolecule, each monomer comprising the macromolecular chain is represented by a polarizability spheroid with its permanent dipole directed along the axis of the chain. The orientation of a chainlet is assumed to be independent of its neighbors so that the effect of the field is simply to orient partially each of the chainlets and the net effect is the alteration of the statistical configuration of the macromolecule.

Wippler (1955, 1957) and Isihara *et al.* (1955) have derived the following form factor for this model:

$$P(\theta) = [2/(u^2 + v^2)^2]\{e^{-u}[(u^2 - v^2)\cos v - 2uv \sin v] - (u^2 - v^2) + u(u^2 + v^2)\} \tag{10.4.26a}$$

where

$$u = \tfrac{1}{2}(h^2 \lambda_N{}^2) = \tfrac{1}{2}h^2 Na^2\{\tfrac{1}{2}\sin^2 \nu + [\cos^2 \nu - \tfrac{1}{2}\sin^2 \nu]\,\overline{\cos^2 \phi}\} \tag{10.4.27}$$

$$v = hz_0 = hNa\,\overline{\cos \phi}\,\cos \nu \tag{10.4.28}$$

where $h = 4\pi \sin(\theta/2)/\lambda$, N is the number of segments in the chain, a is the segment length, ϕ is the angle between the axis of the segment and the electric field, and ν is the angle between the direction of the electric field and the bisectrix of the supplement of the scattering angle θ. The average values of $\cos \phi$ and $\cos^2 \phi$ will depend upon the field and the electrical properties of the segments.

If there is no appreciable permanent dipole, $v \to 0$ so that

$$P(\theta) = (2/u^2)(e^{-u} - 1 + u) \tag{10.4.29}$$

with

$$\lambda_N{}^2 = Na^2\left[\frac{1}{3} + \left(\frac{3\cos^2 \nu - 1}{45}\right)\left(\frac{\alpha_1' - \alpha_2'}{kT}\right)E^2\right] \tag{10.4.30}$$

which for small values of h reduces to

$$\Delta P(\theta) = -\frac{h^2}{6} Na^2 \left(\frac{3\cos^2 v - 1}{45}\right)\left(\frac{\alpha'_1 - \alpha'_2}{kT}\right)E^2 \tag{10.4.31}$$

In this case $P(\theta)$ has the same form in the presence of the field as in its absence. However, the numerical value of the form factor is different because of the different value for the parameter u (10.4.27). It is as if the length of the segments in the macromolecule were changed.

If the segments carry a permanent dipole, μ_0, which is significantly greater than the induced dipoles, the distribution of orientations is such that

$$\overline{\cos\phi} = \mu_0 E/3kT \tag{10.4.32}$$

$$\overline{\cos^2\phi} = \tfrac{1}{3} + (2\mu_0 E^2/45k^2T^2) \tag{10.4.33}$$

For small values of h, (10.4.26a) reduces in the case that $v = 0$ (the electric field is parallel to the bisectrix) so that

$$\Delta' P(\theta) = -\frac{h^2}{6}\frac{Na^2E^2}{2}\left(\frac{\mu_0}{3kT}\right) \tag{10.4.34}$$

and in the case that $v = 90°$ (the electric field is perpendicular to the bisectrix)

$$\Delta'' P(\theta) = \frac{h^2}{6}\frac{Na^2E^2}{45}\left(\frac{\mu_0}{kT}\right)^2. \tag{10.4.35}$$

As a consequence, the change in the form factor upon application of the field is quite insignificant when a field is perpendicular to the bisectrix compared to when it is parallel.

Wippler has also considered a model in which the individual segments carrying a permanent dipole are completely frozen and the effect of the field is to orient the rigid molecule which has the configuration of a broken line whose segments are oriented at random. Under these conditions

$$\Delta P(\theta) = -\frac{2\mu_1{}^2E^2Na^2}{9k^2T^2}\frac{3\cos^2 v - 1}{u^2} \cdot [e^{-u}(u^2 + 4u + 6) - 6 + 2u] \tag{10.4.36}$$

where

$$u = h^2Na^2/6 \tag{10.4.37}$$

For small h this reduces to

$$\Delta P(\theta) = -\frac{h^2}{6}\left(\frac{Na^2}{3}\right)^2\frac{\mu_1{}^2E^2}{k^2T^2}\frac{3\cos^2 v - 1}{6} \tag{10.4.38}$$

It may be possible to distinguish among these three cases. Only the latter model for a rigid macromolecule obeys the relation (10.4.21). Also, for the freely orienting model with only an induced dipole, $\Delta P(\theta)$ varies with the molecular weight while for both models (freely orienting and rigid) with a permanent dipole, $\Delta P(\theta)$ varies with the square of the molecular weight.

Wippler has carried out a number of experiments with $\theta = 90°$ and $v = 0$ and 90°. The electrical field was fixed at 45° to the incident beam and observations were carried out at $\pm 90°$ with respect to the incident beam. The system studied was the synthetic polypeptide poly-DL-phenylalanine. Significant variations were obtained in the scattered intensity for $v = 0$ but there was no change with field at $v = \pi/2$. This indicates that this macromolecule conforms to the model of a freely orienting chain in which the segments carry a permanent dipole. Figure 10.17 shows a plot of the relative increase of intensity with the square of the field strength. From the slope of this curve and Eq. (10.4.34), one can calculate the value of the dipole moment.

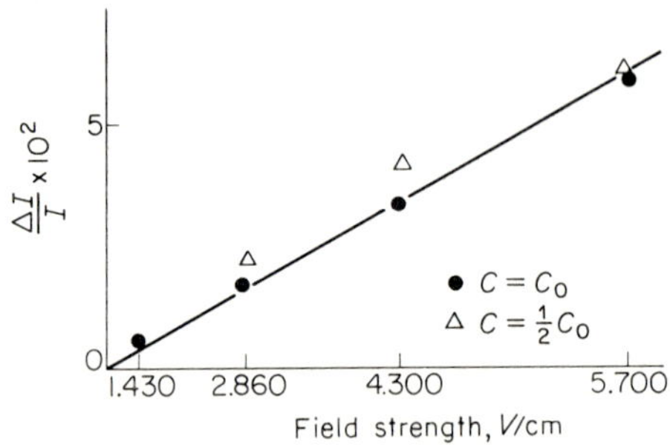

FIG. 10.17. Relative increase of scattered intensity with electric field strength for poly-DL-phenylalanine in benzene (Wippler, 1957).

Wallach and Benoit (1966) have extended the treatment to include the case of a freely orienting macromolecular chain for which successive segments have alternating dipole moments μ_1 and μ_2 which are oriented in different senses along the chain. This produces a resultant dipole off the chain backbone. The results for the two cases $v = 0$ and $v = 90°$ are given respectively for small h by

$$\Delta' P(\theta) = -\frac{h^2}{6} \frac{N^2 a^2 E^2}{2} \left(\frac{\mu_1 + \mu_2}{6kT} \right)^2 \tag{10.4.39}$$

and

$$\Delta'' P(\theta) = \frac{h^2}{6} \frac{N a^2 E^2}{45} \left(\frac{\mu_1{}^2 + \mu_2{}^2}{2k^2 T^2} \right) \tag{10.4.40}$$

The ratio of these two quantities is

$$R = -5N(\mu_1 + \mu_2)^2/4(\mu_1{}^2 + \mu_2{}^2) \tag{10.4.41}$$

When the molecule is rigid, this ratio is -2. When the dipoles are equal and directed in the same sense, it is $-5N/2$ as can be seen either from these equations or from (10.4.34) and (10.4.36). Stockmayer and Bauer (1964) have extended this treatment to an alternating field including relaxation effects. Wallach and Benoit have generalized their treatment for an alternating sequence of dipoles to include the case that the moments are no longer constrained to follow the chain contour.

10.4.4 Rigid Particles in a Magnetic Field

Premilat and Horn (1963, 1964, 1965) have shown that a magnetic field will alter the light scattering properties of suspensions of graphite in organic liquids, presumably by partial orientation of the particles. However, a quantitative interpretation of these results does not seem feasible in the absence of an adequate light scattering theory for highly refractive and absorbing particles of irregular shape.

Muray (1965) has observed the low angle scattering by suspensions of magnetic particles such as magnetite in the presence of multipole magnetic fields. The experimental set-up was very similar to that used by Stein and his collaborators to study the light scattering by thin polymer films (see Fig. 8.17). The intensity of the light was quite sensitive to the field strength and very beautiful symmetrical patterns were photographically recorded for the various multipolar fields.

10.5 Optical Anisotropy

The monomer units in the polymer chain have heretofore been assumed to be optically isotropic. However, if they are anisotropic, the scattering may be quite different and the anisotropy must then be taken into account.

Horn *et al.* (1951) have considered the case of a thin rod for which the form factor is given by (8.1.40) when the medium of which the rod is composed is isotropic. If it is anisotropic with two principal polarizabilities—one parallel to the axes of the rod and one perpendicular to this axis—each scattering element in the rod may be characterized by a polarization spheroid. The anisotropy of the scattering element is defined in the usual way by

$$\delta_0 = (\alpha'_A - \alpha'_B)/(\alpha'_A + 2\alpha'_B) \tag{10.5.1}$$

where α'_A and α'_B are the polarizabilities parallel and perpendicular, respectively, to the axis. The effect of the field due to neighboring elements was considered in addition to the incident field. Using methods developed by Sadron (1937) and Peterlin and Stuart (1939), rather complicated formulas were derived for H_h, V_v, and H_v. In Fig. 10.18, the effect of the anisotropy upon the particle scattering factor is shown as a function of $h^2 R_g{}^2$ where R_g is the radius of gyration and h as usual denotes $(4\pi/\lambda)\sin\theta/2$. Figure 10.19 shows the influence of the anisotropy upon the 45 to 135° dissymmetry. The effect of the anisotropy is sufficiently great so that erroneous results might be obtained if the dissymmetry method were used to obtain the length of the rod without taking the anisotropy into account.

Horn and Benoit (1953) have used the anisotropic rod model in order to determine the length of the tobacco mosaic virus. The anisotropy δ_0 of the scattering elements was first determined by extrapolating the polarization ratio ρ_v to zero scattering angle. Since for forward scattering ($\theta = 0$) there is no interference between the wavelets emitted from the various scattering elements, the polarization ratio depends only on the polarizability of each of the scattering elements. Accordingly, the rod scatters as a collection of small particles, each with an anisotropy factor δ_0 so that

$$\rho_v = 3\delta_0{}^2/(5 + 4\delta_0{}^2) \tag{10.5.2}$$

A value of $\delta_0 = 0.3$ was found for the tobacco mosaic virus. Then by comparing the angular distribution of the intensity with theoretical curves for $\delta_0 = 0.3$, agreement between the two was found for a rod length of 2300 Å. This, in turn, agrees quite well with the value determined by direct electron microscopic observation. A similar study by Mauss *et al.* (1967) elucidated the increase both in the pitch of the helix and in the optical anisotropy of DNA upon complexing with proflavine. Donkersloot *et al.* (1967) have extended the theory to the case of rods having a finite thickness in order to elucidate the structure of polymeric gels which were conceived of as randomly oriented collections of anisotropic rods.

Benoit (1956) has also explored the influence of the optical anisotropy upon the limiting slope of V_v at small scattering angles which may be used to determine $\overline{R_g{}^2}$ from the relation

$$V_v = V_v(0)[1 - (h^2/3)\overline{R_g{}^2} + \cdots] \tag{10.5.3}$$

However, when the rods are anisotropic, the limiting slope of V_v against $\sin^2(\theta/2)$ leads not to $\overline{R_g{}^2}$ but to

$$\overline{R_g{}^2}(1 - (4\delta_0/5) + (20/35)\,\delta_0{}^2 - \cdots) \tag{10.5.4}$$

In the case of randomly coiled macromolecules, the effect of the anisotropy of the chain elements upon the scattered light is not nearly as pronounced

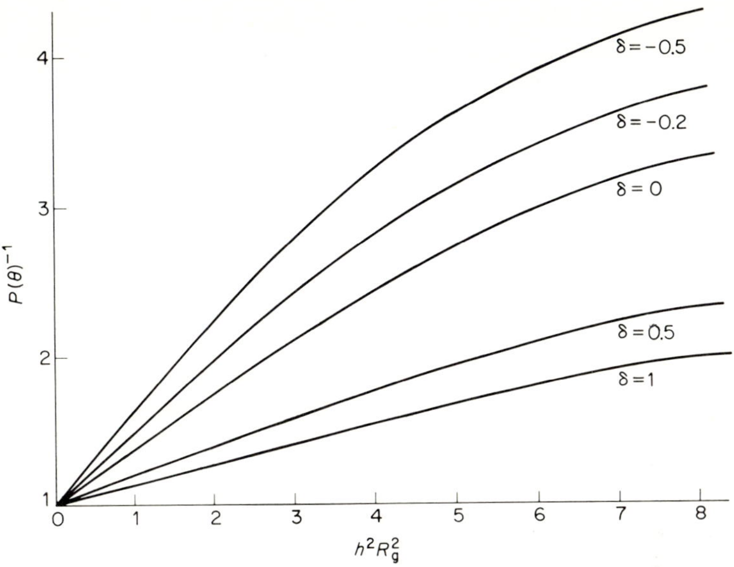

FIG. 10.18. Variation of form factor thin rod with $h^2R_g^{\ 2}$ for various values of the anisotropy factor (Horn, 1955).

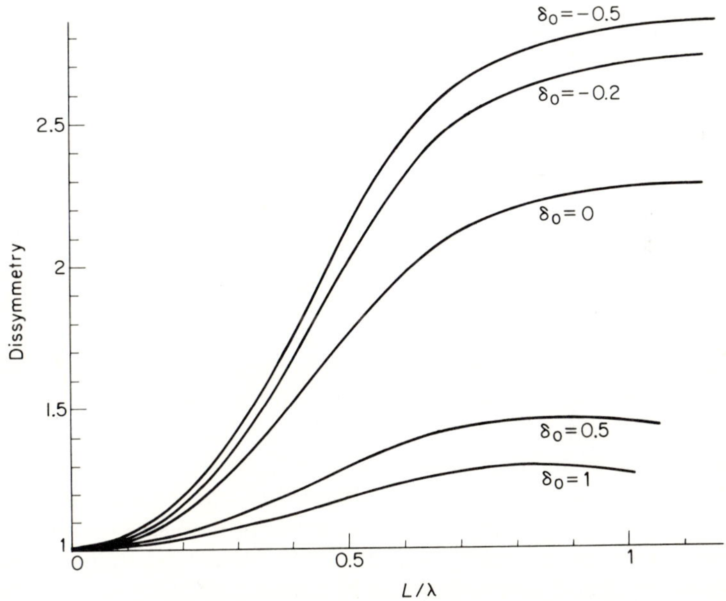

FIG. 10.19. Dissymmetry of a thin rod as a function of rod length to wavelength ratio for various values of the anisotropy factor, δ_0 (Horn *et al.*, 1951).

as for rods, particularly for molecules of high molecular weight. Benoit (1953b) has shown that the chain may be characterized as a rigid spheroidal particle with an equivalent anisotropy δ. If the orientations of the chain elements are completely random, $\delta^2 = \delta_0^2/z$; whereas if the chain is perfectly extended, $\delta = \delta_0$. If there is free rotation around a bond angle, $(180 - \theta)$, then

$$\delta^2 = \frac{\delta_0^2}{z^2}\left[z\frac{1+p}{1-p} - 2p\frac{1-p^z}{(1+p)^2}\right] \tag{10.5.5}$$

where z is the number of units in the chain and $p = (3\cos^2\theta - 1)/2$. This assumes that the axis of the polarization spheroid is directed along the chain axis. A somewhat more complicated expression results for restricted rotation.

Horn (1955) has calculated the Rayleigh ratios corresponding to H_h, V_v, and H_v for a randomly coiled chain. V_v is identical with the expression for the isotropic case, being directly proportional to the molecular weight and independent of the anisotropy. H_h differs from the isotropic case by a small term. Of greater interest is the expression for H_v:

$$H_v = \frac{4\pi^2}{\lambda_0^4 N_A} cn^2 \left(\frac{dn}{dc}\right)^2 \left(\frac{3}{5}\frac{M}{z}\delta_0^2\right) \tag{10.5.6}$$

Since M/z is the molecular weight of the monomer and δ_0 its anisotropy, H_v is independent of the degree of polymerization and is a direct measure of δ_0. Horn has verified these effects experimentally for various solutions of nitrocellulose.

10.6 Ellipsoids and Spheroids Comparable to the Wavelength

The scattering of electromagnetic waves by an ellipsoid of arbitrary size and optical properties may, in principle, be solved exactly, using the method of separation of the variables. The classical procedure would be to formulate the problem in ellipsoidal coordinates and to express the solution of the wave equation in a series of ellipsoidal harmonics (Whittaker and Watson, 1947). Möglich (1927) has done this for the Hertz vector of the scattered and internal fields and has written down the boundary conditions, thereby solving the problem in principle. However, an explicit solution for the scattering coefficients has not been obtained except for the special case of the backscattering by a perfectly conducting prolate spheroid at nose-on incidence. This was solved by Schultz (1950). Numerical studies have been carried out by Siegel *et al.* (1956) and by Senior (1967).

Stevenson (1953a, b) has obtained a solution for the ellipsoid problem using a different approach. He has shown that the general solution of electromagnetic scattering problems can be expressed formally as a power series in the ratio of the dimension of the scatterer to the wavelength. Each term in the series requires the solution of standard problems in potential theory; indeed, the first-order solution is precisely the quasistatic solution of Rayleigh. Thus, if the geometry of the scatterer is such that Laplace's equation can be solved in the appropriate coordinate system, the series can be continued arbitrarily, although the terms become increasingly complicated. The method is completely general, subject to the range of validity over which the particular order of the expansion applies.

The first three terms in the series for the ellipsoid have been derived explicitly by Stevenson. It turns out that at large distances from the particle, the second term vanishes so that only the terms for which the field is proportional to k^2 and k^4 appear (k is the propagation constant). Even for this approximation, the final results are quite complicated so that the reader is referred to Stevenson's paper (1953b) for the explicit formulas.

Mathur and Mueller (1955) have noted that Stevenson's third-order solution converges only for $2\pi b/\lambda = \alpha < 0.95$, where b is the longest semiaxis in the particle. They have compared both the first-order quasistatic theory and Stevenson's third-order solution with the exact solution of Schultz (1950) for a conducting prolate spheroid at nose-on incidence. Their results indicate that the first-order theory is inadequate for $\alpha > 0.3$, whereas the Stevenson third-order theory agrees with the exact theory up to $\alpha = 1$, at least for conducting spheroids at nose-on incidence.

The same authors have also made detailed computations using the third-order theory for particles with refractive indices corresponding to water in the microwave range ($m = 8.18 - 1.16i$ for $\lambda = 3$ cm and $m = 8.90 - 0.69i$ for $\lambda = 10$ cm). These include the backscatter efficiency for both prolate and oblate spheroids at sizes up to $\alpha = 1$, for eccentricities down to 0.1, and for different orientations of the spheroid with respect to the incident beam.

Greenberg *et al.* (1967) have applied the point-matching technique, which was discussed earlier in connection with spheres (Section 3.10), to a prolate spheroid whose figure axis is along the incident direction. The expansions were carried out in spherical coordinates rather than spheroidal coordinates. The number of terms retained in the truncated series was determined by the longest dimension.

Some numerical results are shown in Figure 10.20 where the scattering efficiency is plotted against the phase parameter $\rho = 2ka(m - 1)$ for various values of the axial ratio. Here a is the shortest semiaxis. The pattern is very similar to that for spheres showing an initial maximum in the so-called resonance region. A noteworthy aspect of this result is the increasing height of the broad maximum in the efficiency curve with increasing elongation of the spheroid as well as a deeper dip in the first minimum. This effect was also observed by Greenberg (1960) in earlier experimental studies of scattering by dielectric spheroids. There have been no comparisons between the results of this theory and that of Stevenson.

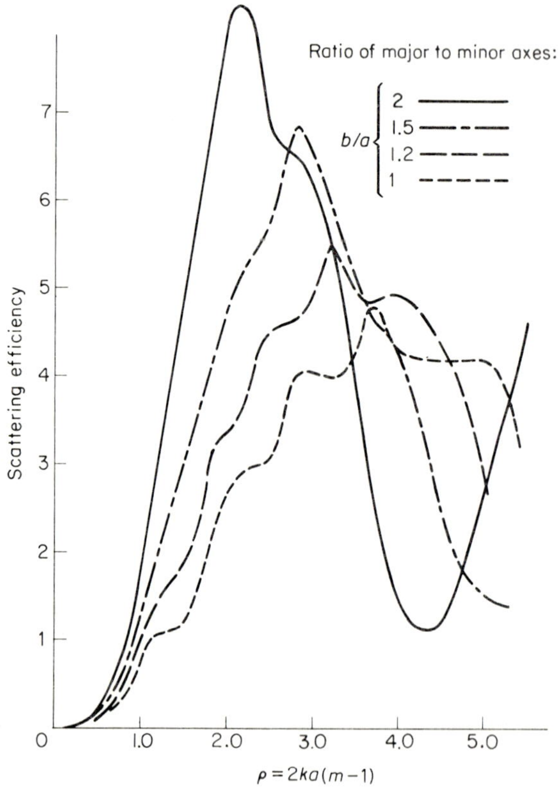

FIG. 10.20. Scattering efficiency of spheroid with $m = 1.5$ and various eccentricities using the point matching method (Greenberg *et al.*, 1967).

Atlas and Wexler (1963) have tackled the problem of elucidating the backscattering of spheroids by a direct experimental approach. Using a microwave backscattering range, they accumulated an extensive encyclopedia of experimental data on a set of ice and styrofoam oblate spheroids for 3.22 and 9.67 cm waves. The long semiaxes of the styrofoam prolate spheroids were $b = 1.27$, 2.54, and 3.81 cm, and the eccentricities (a/b) varied from 0.2 to 1.0 in steps of 0.1. A smaller number of ice particles in the same size range were studied. Results were obtained at all orientations of the spheroids.

For the smallest spheroids ($b = 1.27$ cm, $\lambda = 9.67$) the experimental results agreed reasonably well with the calculations of the quasistatic theory, but this was the maximum size for which such agreement was obtained. For larger particles, the patterns became more complex and the authors were unable to find any general patterns. This is certainly not surprising in view of the well-known complexity of the scattering pattern even for spheres.

References

Abbot, C. G., and Fowle, F. C. (1914). *Astrophys. J.* **40**, 435.
Aden, A. L. (1950). Electromagnetic scattering from metal and water spheres. Tech. Rept. No. 106. Cruft Lab., Harvard Univ., Cambridge, Massachusetts.
Aden, A. L. (1951). *J. Appl. Phys.* **22**, 601.
Aden, A. L., and Kerker, M. (1951). *J. Appl. Phys.* **22**, 1242.
Adey, A. W. (1955). *Can. J. Phys.* **33**, 407.
Adey, A. W. (1956a). *Can. J. Phys.* **34**, 510.
Adey, A. W. (1956b). *Wireless Engr.* **33**, 259.
Adey, A. W. (1958). *Electron. Radio Engr.* **35**, 149.
Adler, S. B., and Johnson, R. S. (1962). *Appl. Opt.* **1**, 655.
Agdur, B., Böling, G., Sellberg, F., and Ohman, Y. (1963). *Phys. Rev.* **130**, 996.
Aiken, H. H. (1947–1951). *Ann. Comput. Lab., Harvard Univ.*, **3–14**. The orders of the Bessel functions found in each volume are: **3**, 0, 1; **4**, 2, 3; **5**, 4–6; **6**, 7–9; **7**, 10–12; **8**, 13–15; **9**, 16–27; **10**, 28–39; **11**, 40–51; **12**, 52–63; **13**, 64–78; **14**, 79–135.
Aitken, J. (1892). *Proc. Roy. Soc.* **A51**, 408.
Albini, F. A. (1962). *J. Appl. Phys.* **33**, 3032.
Albini, F. A., and Nagelberg, E. R. (1962). *J. Appl. Phys.* **33**, 1706.
Alexandrowicz, Z. (1959). *J. Polymer Sci.* **40**, 91.
Altschul, M. (1894). *Z. Physik. Chem.* **11**, 578.
Anacker, E. W. (1958). *J. Phys. Chem.* **62**, 41.
Anacker, E. W., and Ghose, H. M. (1963). *J. Phys. Chem.* **67**, 1713.
Anacker, E. W., and Westwell, A. E. (1964). *J. Phys. Chem.* **68**, 3490.
Anacker, E. W., Rush, R. M., and Johnson, J. S. (1964). *J. Phys. Chem.* **68**, 81.
Ander, P. (1962). *Kolloid-Z. Z. Polymere* **185**, 102.
Andreasen, M. G. (1957a). *IRE Trans. Antennas Propagation* **AP-5**, 267.
Andreasen, M. G. (1957b). *IRE Trans. Antennas Propagation* **AP-5**, 337.
Andrews, T. (1869). *Phil. Trans. Roy. Soc. London* **159**, 575.
Angstadt, R. L., and Tyree, S. Y. (1962). *J. Inorg. Nucl. Chem.* **24**, 913.
Arago, D. F. J. (1811). *Oeuvres* **7**, 394, 432.
Arnush, D. (1964). *IEEE Trans. Antennas Propagation* **AP-12**, 86.
Aroney, M. J., LeFevre, R. J. W., and Saxby, J. D. (1965). *Australian J. Chem.* **18**, 1501.
Ashley, L. E., and Cobb, C. M. (1958). *J. Opt. Soc. Am.* **48**, 261.
Atack, D., and Rice, O. K. (1954). *J. Chem. Phys.* **22**, 382.
Atack, D., and Rice, O. K. (1955). *J. Chem. Phys.* **23**, 164.

ATLAS, D. (1964). *Advan. Geophys.* **11**, 317.

ATLAS, D., and GLOVER, K. (1963). *In* "Electromagnetic Scattering" (M. Kerker, ed.). Pergamon Press, Oxford.

ATLAS, S. M., and MARK, H. F. (1961). Report on molecular-weight measurements of standard polystyrene samples. *Comm. Macromolecules of I.U.P.A.C., Paris, July 1963*. Verlag Chem., Weinheim.

ATLAS, D., and WEXLER, R. (1963). *J. Atmospheric Sci.* **20**, 48.

ATLAS, D., KERKER, M., and HITSCHFELD, W. (1953). *J. Atmospheric Terrest. Phys.* **3**, 108.

ATLAS, D., HARPER, W. G., LUDLAM, F. H., and MACKLIN, W. C. (1960). *Quart. J. Roy. Meteorol. Soc.* **86**, 468.

ATLAS, D., BATTAN, L. J., HARPER, W. G., HERMAN, B. M., KERKER, M., and MATIJEVIC, E. (1963). *IEEE Trans. Antennas Propagation* **AP-11**, 68.

BAILEY, E. D. (1946). *Ind. Eng. Chem. Anal. Ed.* **18**, 365.

BAKER, M. C., LYONS, P. A., and SINGER, S. J. (1955). *J. Am. Chem. Soc.* **77**, 2011.

BAKHTIYAROV, V. G., FOITZIK, L., PERELMAN, A. Y., and SHIFRIN, K. S. (1966). *Pure Appl. Geophys.* (*Milan*) **64**, 204.

BARAKAT, R. (1963). *J. Acoust. Soc. Am.* **35**, 1990.

BARAKAT, R. (1965). *J. Opt. Soc. Am.* **55**, 992.

BARAKAT, R., and LEVIN, E. (1964). *J. Opt. Soc. Am.* **54**, 1089.

BARLOW, H. M., and BROWN, J. (1962). "Radio Surface Waves." Oxford Univ. Press, London and New York.

BARNES, M. D., and LA MER, V. K. (1946). *J. Colloid Sci.* **1**, 79.

BARUS, C. (1902). *Phil. Mag.* **4**, 24.

BARUS, C. (1908). *Am. J. Sci.* **25**, 224.

BATEMAN, H. (1914). "Electrical and Optical Wave Motion," Cambridge Univ. Press, London and New York (reprint by Dover, New York, 1955).

BATEMAN, J. B., WENECK, E. J., and ESHLER, D. C. (1959). *J. Colloid Sci.* **14**, 308.

BATTAN, L. J. (1959). "Radar Meteorology." Univ. of Chicago Press, Chicago, Illinois.

BATTAN, L. J., and HERMAN, B. M. (1961). *J. Geophys. Res.* **66**, 3255.

BEATTIE, W. H. (1965). *J. Polymer Sci.* **3A**, 527.

BEATTIE, W. H., and BOOTH, C. (1960a). *J. Polymer Sci.* **44**, 81.

BEATTIE, W. H., and BOOTH, C. (1960b). *J. Phys. Chem.* **64**, 696.

BEATTIE, W. H., and JUNG, H. C. (1968). *J. Colloid Interface Sci.* **27**, 581.

BEEBE, E., and MARCHESSAULT, R. H. (1964). *J. Appl. Phys.* **35**, 3182.

BEEBE, E., COALSON, R. L., and MARCHESSAULT, R. H. (1966). *J. Polymer Sci.* **13C**, 103.

BEIDL, G., BISCHOF, M., GLATZ, G., POROD, G., VON SACKEN, J. C., and WAWRA, H. (1957). *Z. Elecktrochem.* **61**, 1311.

BELLO, A., and GUZMAN, G. M. (1966). *European Polymer J.* **2**, 79.

BENOIT, H. (1951). *Ann. Phys.* (*Paris*) **6**, 561.

BENOIT, H. (1953a). *J. Polymer Sci.* **11**, 507.

BENOIT, H. (1953b). *Compt. Rend.* **236**, 687.

BENOIT, H. (1956). *Makromol. Chem.* **18–19**, 397.

BENOIT, H. (1963). *In* "Electromagnetic Scattering" (M. Kerker, ed.). Pergamon Press, Oxford.

BENOIT, H. (1966). *Ber. Bunsenges. Physik. Chem.* **70**, 286.

BENOIT, H., and DOTY, P. (1953). *J. Phys. Chem.* **57**, 958.

BENOIT, H., and GOLDSTEIN, M. (1953). *J. Chem. Phys.* **21**, 947.

BENOIT, H., and STOCKMAYER, W. H. (1956). *J. Phys. Radium* **17**, 21.

BENOIT, H., and WIPPLER, C. (1960). *J. Chim. Phys.* **57**, 524.

BENOIT, H., HOLTZER, A. M., and DOTY, P. (1954). *J. Phys. Chem.* **58**, 635.

BENOIT, H., ULLMAN, R., DE VRIES, A. J., and WIPPLER, C. (1962). *J. Chim. Phys.* **59**, 889.

Berner, A. (1965). *Acta Phys. Austriaca* **21**, 20.
Beyer, G. L. (1959). *In* "Technique of Organic Chemistry" (A. Weissberger, ed.), Vol. I, Pt. I, p. 191. Wiley (Interscience), New York.
Bhagavantam, S. (1942). "Scattering of Light and the Raman Effect." Chem. Publ., New York.
Bhatnagar, H. L. (1960). *J. Chem. Phys.* **32**, 674.
Bhatnagar, H. L. (1961). *J. Chem. Phys.* **35**, 999.
Bhatnagar, H. L. (1963). *J. Chem. Phys.* **39**, 1612.
Bhatnagar, H. L., and Heller, W. (1964). *J. Chem. Phys.* **40**, 480.
Bisbing, P. E. (1966). *IEEE Trans. Antennas Propagation* **AP-14**, 219.
Blank, A. (1955). *Trans. Chalmers Univ. Technol. Gothenburg* **168**.
Blum, J. J., and Morales, M. F. (1952). *J. Chem. Phys.* **20**, 1822.
Blumer, H. (1925). *Z. Physik* **32**, 119.
Blumer, H. (1926). *Z. Physik* **38**, 304, 920; **39**, 195.
Boël, M. (1966). Ph.D. Thesis: Polytechnic Inst. Brooklyn, New York.
Boll, R. H., Leacock, J. A., Clark, G. C., and Churchill, S. W. (1958). "Tables of Light Scattering Functions." Univ. of Michigan Press, Ann Arbor, Michigan.
Borch, J., and Marchessault, R. H. (1968). *J. Colloid. Interface Sci.* **27**, 355.
Born, M. (1933). "Optik." Springer, Berlin (reprint by Edwards, Ann Arbor, Michigan, 1943).
Born, M., and Wolf, E. (1959). "Principles of Optics." Pergamon Press, Oxford.
Bothorel, P. (1968). *J. Colloid Interface Sci.* **27**, 529.
Bouwkamp, C. J., and Casimir, H. B. (1954). *Physica* **20**, 539.
Bradford, E. B., and Vanderhoff, J. W. (1955). *J. Appl. Phys.* **26**, 864.
Bradford, E. B., and Vanderhoff, J. W. (1963). *J. Polymer Sci. Pt. C*, **3**, 41.
Brady, G. W., McIntyre, D., Myer, M., and Wims, A. (1967). *In* "Small Angle X-Ray Scattering: Syracuse Conference" (H. Brumberger, ed.). Gordon and Breach, New York.
Bremmer, H. (1949). "Terrestrial Radio Waves." Elsevier, New York.
Bridge, N. J., and Buckingham, A. D. (1964). *J. Chem. Phys.* **40**, 2733.
Bridge, N. J., and Buckingham, A. D. (1966). *Proc. Roy. Soc.* **A295**, 334.
Brill, O. L. (1967). X-Ray determination of particle size distribution. Ph.D. Thesis, Univ. of Missouri, Columbia, Missouri.
Brill, O. L., Weil, C. G., and Schmidt, P. W. (1968). *J. Colloid Interface Sci.* **27**, 479.
Brillouin, L. (1922). *Ann. Phys. (Paris)* **17**, 88.
Brillouin, L. (1949). *J. Appl. Phys.* **20**, 1108.
Brinkman, H. C., and Hermans, J. J. (1949). *J. Chem. Phys.* **17**, 574.
Brit. Assoc. for the Advan. of Sci. (1950). "Mathematical Tables, 6, Bessel Functions I: Functions of Order Zero and Unity." Cambridge Univ. Press, London and New York.
Brit. Assoc. for the Advan. of Sci. (1952). "Mathematical Tables, 10, Bessel Functions, II: Function of Positive Integer Order." Cambridge Univ. Press, London and New York.
Bromwich, T. J. (1919). *Phil. Mag.* **38**, 143.
Brown, W. F. (1950). *J. Chem. Phys.* **18**, 1193, 2000.
Brücke, E. W. (1853). *Poggendorf's Ann.* **88**, 363.
Brumberger, H. (1966). *In* "Critical Phenomena." *Natl. Bur. Std. Misc. Publ.* **273**.
Brumberger, H., Alexandropoulos, N., and Claffey, W. (1967). *Phys. Rev. Letters* **19**, 555.
Bryant, H. C., and Cox, A. J. (1966). *J. Opt. Soc. Am.* **56**, 1529.
Buckingham, A. D., and Stephen, M. J. (1957). *Trans. Faraday Soc.* **53**, 884.
Buley, E. R. (1967). *IEEE Trans. Antennas Propagation* **AP-15**, 677.
Bullough, R. K. (1960). *J. Polymer Sci.* **46**, 517.
Bullough, R. K. (1963a). *Proc. Roy. Soc.* **A275**, 271.
Bullough, R. K. (1963b). *Physica* **29**, 437, 453, 467.
Burberg, R. (1956). *Z. Naturforsch.* **11a**, 800.

BURKE, J. E., and TWERSKY, V. (1964). *J. Opt. Soc. Am.* **54**, 732.

BURKE, J. E., CHRISTENSON, E. J., and LYTTLE, S. B. (1964). *J. Opt. Soc. Am.* **54**, 1065.

BURMAN, R. (1965). *IEEE Trans. Antennas Propagation* **AP-13**, 646; also addendum to this, *ibid.* **AP-14**, 114 (1966).

BURMAN, R. (1966). *Electron. Letters* **2**, 66.

BURNETT, G. M., LEHRLE, R. S., OVENALL, D. W., and PEAKER, F. W. (1958). *J. Polymer Sci.* **29**, 417.

BUSHUK, W., and BENOIT, H. (1958a). *Can. J. Chem.* **36**, 1616.

BUSHUK, W., and BENOIT, H. (1958b). *Compt. Rend.* **246**, 3167.

CABANNES, J. (1915). *Compt. Rend.* **160**, 62.

CABANNES, J. (1920). *J. Phys. Radium* **49**, 129.

CABANNES, J. (1921). *Ann. Phys. (Paris)* (9) **15**, 5.

CABANNES, J. (1929). "La Diffusion Moleculaire de la Lumiere." Les Presses Universitaires de France, Paris.

CAMBI, E. (1948). "Eleven and Fifteen Place Tables of Bessel Functions of the First Kind, to All Significant Orders," Dover, New York.

CANTOW, H. (1956). *Makromol. Chem.* **18/19**, 367.

CARPENTER, D. K. (1966). *J. Polymer Sci. A-2* **4**, 923.

CARR, C. I., and ZIMM, B. (1950). *J. Chem. Phys.* **18**, 1616.

CASASSA, E. F. (1955). *J. Chem. Phys.* **23**, 596.

CASASSA, E. F., and BERRY, G. C. (1966). *J. Polymer Sci. A-2* **4**, 881.

CASASSA, E. F., and EISENBERG, H. (1960). *J. Phys. Chem.* **64**, 753.

CASASSA, E. F., and EISENBERG, H. (1961). *J. Phys. Chem.* **65**, 427.

CAULFIELD, D., and ULLMAN, R. (1962). *J. Appl. Phys.* **33**, 1737.

C.C.O.S.A. (1963). "The science of color." *Comm. Colorimetry Opt. Soc. Am., 1963*. Opt. Soc. of Am., Washington, D.C.

CERF, R., and SCHERAGA, H. A. (1952). *Chem. Rev.* **51**, 185.

CHEN, C. L. (1966). *IEEE Trans. Antennas Propagation* **AP-14**, 283.

CHEN, H. H. C., and CHENG, D. K. (1964a). *Appl. Sci. Res. Sect. B* **11**, 442.

CHEN, H. H. C., and CHENG, D. K. (1964b). *IEEE Trans. Antennas Propagation* **AP-12**, 348.

CHIN, J. H., SLIEPCEVICH, C. M., and TRIBUS, M. (1955a). *J. Phys. Chem.* **59**, 841.

CHIN, J. H., SLIEPCEVICH, C. M., and TRIBUS, M. (1955b). *J. Phys. Chem.* **59**, 845.

CHISTOVA, E. A. (1959). "Tables of Bessel Functions of the True Argument and of Integrals Derived from Them." Pergamon Press, Oxford.

CHOW, Y. (1960). *Appl. Sci. Res. Sect. B* **8**, 290.

CHROMEY, F. C. (1960). *J. Opt. Soc. Am.* **50**, 730.

CHU, B. (1964a). *J. Am. Chem. Soc.* **86**, 3557.

CHU, B. (1964b). *J. Chem. Phys.* **41**, 226.

CHU, B. (1966). *In* "Critical Phenomena." *Natl. Bur. Std. Misc. Publ.* **273**.

CHU, B., and TAN CRETI, D. M. (1967). *J. Phys. Chem.* **71**, 1943.

CHU, C. M. (1952). Scattering and absorption of water droplets at millimeter wavelengths, Ph.D. Thesis, Univ. of Michigan, Ann Arbor, Michigan.

CHU, C. M., and CHURCHILL, S. W. (1955). *J. Opt. Soc. Am.* **45**, 958.

CHU, C. M., CLARK, G. C., and CHURCHILL, S. W. (1957). "Tables of Angular Distribution Coefficients for Light Scattering by Spheres," Eng. Res. Inst., Univ. of Michigan, Ann Arbor, Michigan.

CHURCHILL, S. W., CLARK, G. C., and SLIEPCEVICH, C. M. (1960). *Discussions Faraday Soc.* **30**, 192.

CLAESSON, S., and OHMAN, J. (1964). *Arkiv Kemi* **23**, 69.

Clark, G. C., and Churchill, S. W. (1957). "Tables of Legendre Polynomials." Eng. Res. Inst., Univ. of Michigan, Ann Arbor, Michigan.
Clark, G. C., Chu, C. M., and Churchill, S. W. (1957). *J. Opt. Soc. Am.* **47**, 81.
Clausius, R. (1847a). *Crelle's J.* **34**, 122.
Clausius, R. (1847b). *Poggendorf's Ann.* **72**, 294.
Clausius, R. (1848). *Crelle's J.* **36**, 185.
Clebsch, A. (1863). *Crelle's J.* **61**, 195.
Clement, C., and Bothorel, P. (1964). *J. Chim. Phys.* **61**, 878.
Clough, S., van Aartsen, J. J., and Stein, R. S. (1965). *J. Appl. Phys.* **36**, 3072.
Coalson, R. L., Marchessault, R. H., and Peterlin, A. (1966). *J. Polymer Sci.* **13C**, 123.
Cohen, G., and Eisenberg, H. (1965). *J. Chem. Phys.* **43**, 3881.
Cooke, D., and Kerker, M. (1968). *Rev. Sci. Instr.* **39**, 320.
Cooke, D., and Kerker, M. (1969). *J. Opt. Soc. Am.* **59**, 43.
Cooper, M. J., and Mountain, R. D. (1968). *J. Chem. Phys.* **48**, 1064.
Copley, D. B., Banerjee, A. K., and Tyree, S. Y. (1965). *Inorg. Chem.* **4**, 1480.
Coumou, D. J. (1960). *J. Colloid Sci.* **15**, 408.
Coumou, D. J., and Mackor, E. L. (1964). *Trans. Faraday Soc.* **60**, 1726.
Coumou, D. J., Hijmans, J., and Mackor, E. L. (1964a). *Trans. Faraday Soc.* **60**, 2244.
Coumou, D. J., Mackor, E. L., and Hijmans, J. (1964b). *Trans. Faraday Soc.* **60**, 1539.
Coutarel, L., Matijević, E., and Kerker, M. (1967) *J. Colloid Interface Sci.* **24**, 338.
Craig, H. R., and Tyree, S. Y. (1965). *Inorg. Chem.* **4**, 997.
Dandliker, W. B. (1950). *J. Am. Chem. Soc.* **72**, 5110.
Dauchot, J., and Watillon, A. (1967). *J. Colloid Interface Sci.* **23**, 62.
Dautzenberg, H., and Ruscher, Ch. (1967). *J. Polymer Sci.* **16C**, 2913.
Davidson, N. (1962). "Statistical Mechanics." McGraw-Hill, New York.
Davies, C. N. (1945). *Proc. Phys. Soc. (London)* **A57**, 259.
Debye, P. (1909a). *Ann. Physik* (4) **30**, 57.
Debye, P. (1909b). *Math. Ann.* **67**, 535.
Debye, P. (1915). *Ann. Physik.* **46**, 809.
Debye, P. (1930). *Physik. Z.* **31**, 419.
Debye, P. (1944). *J. Appl. Phys.* **15**, 338.
Debye, P. (1947). *J. Phys. & Colloid Chem.* **51**, 18 (first presented in a lecture given at the Polytech. Inst. of Brooklyn, November 1944).
Debye, P. (1949). *Ann. N.Y. Acad. Sci.* **51**, 575; *J. Phys. & Colloid Chem.* **53**, 1.
Debye, P. (1959). *J. Chem. Phys.* **31**, 680.
Debye, P. (1960). *Conf. Non-Crystalline Solids, Alfred, New York, 1958*, pp. 1–25. Wiley, New York.
Debye, P. (1963). "Electromagnetic Scattering" (M. Kerker, ed.). Pergamon Press, Oxford.
Debye, P. (1965). *Phys. Rev. Letters* **14**, 783.
Debye, P., and Bueche, A. M. (1949). *J. Appl. Phys.* **20**, 518.
Debye, P., and Jacobsen, R. T. (1968). *J. Chem. Phys.* **48**, 203.
Debye, P., and Kleboth, K. (1965). *J. Chem. Phys.* **42**, 3155.
Debye, P., and Menke, H. (1930). *Physik. Z.* **31**, 797.
Debye, P., Anderson, H. R., and Brumberger, H. (1957). *J. Appl. Phys.* **28**, 679.
Debye, P., Coll, H., and Woermann, D. (1960). *J. Chem. Phys.* **32**, 939; **33**, 1746.
Debye, P., Chu, B., and Woermann, D. (1962a). *J. Chem. Phys.* **36**, 1803.
Debye, P., Woermann, D., and Chu, B. (1962b). *J. Chem. Phys.* **36**, 851.
Debye, P., Chu, B., and Kaufmann, H. (1963). *J. Polymer Sci. Pt. A* **1**, 2387.
Debye, P., Bashaw, J., Chu, B., and Tan Creti, D. M. (1966). *J. Chem. Phys.* **44**, 4302.
Debye, P., Gravatt, C. C., and Ieda, M. (1967). *J. Chem. Phys.* **46**, 2352.
De Gennes, P. (1968). *Phys. Letters* **26A**, 313.

DEIRMENDJIAN, D. (1957). *Ann. Geophys.* **13**, 286.

DEIRMENDJIAN, D. (1959). *Ann. Geophys.* **15**, 218.

DEIRMENDJIAN, D. (1960). *Quart. J. Roy. Meteorol. Soc.* **86**, 371.

DEIRMENDJIAN, D. (1963). Tables of Mie scattering cross sections and amplitude. Rept. R-407-PR. Rand Corp., Santa Monica, California.

DEIRMENDJIAN, D., and CLASEN, R. J. (1962). Light scattering on partially absorbing homogeneous spheres of finite size. Rept. R-393-PR. Rand Corp., Santa Monica, California.

DEIRMENDJIAN, D., CLASEN, R., and VIEZEE, W. (1961). *J. Opt. Soc. Am.* **51**, 620.

DENHAM, H., HELLER, W., and PANGONIS, W. J. (1963). "Angular Scattering Functions for Spherical Particles II." Wayne State Univ. Press, Detroit, Michigan.

DETTMAR, H.-K., LODE, W., and MARRE, E. (1963). *Kolloid-Z. Z. Polymere* **188**, 28.

DEVORE, J. R., and PFUND, A. H. (1947). *J. Opt. Soc. Am.* **37**, 826.

DEZELIĆ, G. (1966). *J. Chem. Phys.* **45**, 185.

DEZELIĆ, G., and KRATOHVIL, J. P. (1960a). *Kolloid-Z.* **171**, 42.

DEZELIĆ, G., and KRATOHVIL, J. P. (1960b). *Kolloid-Z.* **173**, 38.

DEZELIĆ, G., and KRATOHVIL, J. P. (1961). *J. Colloid Sci.* **16**, 561.

DEZELIĆ, G., and VAVRA, J. (1966). *Croat. Chem. Acta* **38**, 35.

DEZELIĆ, G., WRISCHER, M., DEVIDE, Z., and KRATOHVIL, J. P. (1960). *Kolloid-Z.* **171**, 42.

DEZELIĆ, G., DEZELIĆ, N., and TEZAK, B. (1963). *J. Colloid Sci.* **18**, 888.

DINTZIS, F. R., and STEIN, R. S. (1964). *J. Chem. Phys.* **40**, 1459.

DOBBINS, R. A., and JIZMAGIAN, G. S. (1966). *J. Opt. Soc. Am.* **56**, 1345, 1351.

DONKERSLOOT, M. C. A., GOUDA, J. H., VAN AARTSEN, J. J., and PRINS, W. (1967). *Rec. Trav. Chim.* **86**, 321.

DONN, B., and POWELL, R. S. (1963). *In* "Electromagnetic Scattering" (M. Kerker, ed.). Pergamon Press, Oxford.

DOREMUS, R. H. (1966). *J. Appl. Phys.* **37**, 2775.

DOREMUS, R. H. (1968). *J. Colloid Interface Sci.* **27**, 412.

DORSEY, E. N. (1940). "Properties of Ordinary Water Substance." Reinhold, New York.

DOYLE, W. T., and AGARWAL, A. (1965). *J. Opt. Soc. Am.* **55**, 305.

EDSALL, J. T., EDELHOCH, H., LONTIE, R., and MORRISON, P. R. (1950). *J. Am. Chem. Soc.* **72**, 4641.

EHL, J., LOUCHEUX, C., REISS, C., and BENOIT, H. (1964). *Makromol. Chem.* **75**, 35.

EHRENBERG, W., and SCHAFER, K. (1932). *Physik. Z.* **33**, 97.

EHRLICH, G., and DOTY, P. M. (1954). *J. Am. Chem. Soc.* **76**, 3764.

EIGNER, J., and DOTY, P. (1965). *J. Mol. Biol.* **12**, 549.

EINSTEIN, A. (1910). *Ann. Physik.* **33**, 1275 [English Transl. *in* "Colloid Chemistry" (J. Alexander, ed.), Vol. I, pp. 323–339. Reinhold, New York, 1926].

EISENBERG, H. (1962). *J. Chem. Phys.* **36**, 1837.

EISENBERG, H. (1965). *J. Chem. Phys.* **43**, 3887.

EISENBERG, H. (1966). *J. Chem. Phys.* **44**, 137.

EISENBERG, H., and FELSENFELD, G. (1967). *J. Mol. Biol.* **30**, 17.

ELIEZER, N., and SILBERBERG, A. (1967). *Biopolymers* **5**, 95.

ELWORTHY, P. H., and MACFARLANE, C. B. (1962). *J. Chem. Soc.* **102**, 537.

EMERSON, M. F., and HOLTZER, A. (1967). *J. Phys. Chem.* **71**, 1898.

EPSTEIN, P. S. (1956). *Rev. Mod. Phys.* **28**, 3.

ESPENSCHEID, W. F., KERKER, M., and MATIJEVIĆ, E. (1964a). *J. Phys. Chem.* **68**, 3093.

ESPENSCHEID, W. F., MATIJEVIĆ, E., and KERKER, M. (1964b). *J. Phys. Chem.* **68**, 2831.

ESPENSCHEID, W. F., WILLIS, E., MATIJEVIĆ, E., and KERKER, M. (1965). *J. Colloid Sci.* **20**, 501.

EVANS, L. B. (1963). Calculation of electromagnetic scattering functions for concentric cylinders with real index of refraction. Univ. Microfilms, Inc., Ann Arbor, Michigan.

EVANS, L. B., CHEN, J., and CHURCHILL, S. W. (1964). *J. Opt. Soc. Am.* **54**, 1004.
EVVA, F. (1952). *Z. Wiss. Phot.* **47**, 39.
EVVA, F. (1953a). *Kolloid-Z.* **133**, 79.
EVVA, F. (1953b). *Z. Physik. Chem.* (*Leipzig*) **202**, 208.
EVVA, F. (1954). *Z. Physik. Chem.* (*Leipzig*) **203**, 86.
EWART, R. H., ROE, C. P., DEBYE, P., and MCCARTNEY, J. R. (1946). *J. Chem. Phys.* **13**, 159.
EYKMAN, J. F. (1895). *Rec. Trav. Chim.* **14**, 177.
FAHLEN, T. S., and BRYANT, H. C. (1968). *J. Opt. Soc. Am.* **58**, 304.
FARADAY, M. (1857). *Phil. Trans. Roy. Soc. London Ser. A* **147**, 145.
FARONE, W. A. (1964). Ph.D. Thesis. Clarkson College of Technology, Potsdam, New York.
FARONE, W. A., and KERKER, M. (1966). *J. Opt. Soc. Am.* **56**, 481.
FARONE, W. A., and QUERFELD, C. W. (1965). Electromagnetic scattering from an infinite cylinder at oblique incidence. Rept. ERDA-281, U.S. Army Electron. Res. and Develop. Activity, White Sands, New Mexico.
FARONE, W. A., and QUERFELD, C. W. (1966). *J. Opt. Soc. Am.* **56**, 476.
FARONE, W. A., and ROBINSON, M. J. (1968). *Appl. Opt.* **7**, 643.
FARONE, W. A., KERKER, M., and MATIJEVIĆ, E. (1963). *In* "Electromagnetic Scattering" (M. Kerker, ed.). Pergamon Press, Oxford.
FAUGERAS, P. (1966). *Compt. Rend.* **262B**, 1195.
FEINSTEIN, J. (1951). *J. Geophys. Res.* **56**, 37.
FELSEN, L. (1967). *In* "Electromagnetic Wave Theory" (J. Brown, ed.), Pt. I. Pergamon Press, Oxford.
FENN, R., and OSER, H. (1962). Theoretical considerations on the effectiveness of carbon Seeding. USASRDL Tech. Rept. 2258. U.S. Army Signal Res. and Develop. Lab., Fort Monmouth, New Jersey.
FENN, R., and OSER, H. (1965). *Appl. Opt.* **4**, 1504.
FISHER, M. E. (1964). *J. Math. Phys.* **5**, 944.
FISHMAN, M. M. (1957). Light scattering by colloidal systems—an annotated bibliography. Tech. Service Lab., River Edge, New Jersey.
FISHMAN, M. M. (1958). Suppl. to above.
FIXMAN, M. (1955). *J. Chem. Phys.* **23**, 2074.
FIXMAN, M. (1960). *J. Chem. Phys.* **33**, 1357.
FLETCHER, A., MILLER, J. C. P., ROSENHEAD, L., and COMRIE, L. J. (1962). "An Index of Mathematical Tables," 2 Vols., 2nd Ed. Addison-Wesley, Reading, Massachusetts.
FOURNET, G. (1951). *Bull. Soc. Franc. Mineral. Crist.* **74**, 39.
FRANZ, W., and DEPPERMAN, K. (1952). *Ann. Physik* **10**, 361.
FRENKEL, S. YA., BARANOV, V. G., and VOLKOV, T. I. (1967). *J. Polymer Sci. Pt. C* No. 16, 1655.
FRIEDERICHS, K. O., and KELLER, J. B. (1955). *J. Appl. Phys.* **26**, 961.
FRIEDMAN, B. (1951). *Commun. Pure Appl. Math.* **4**, 317.
FRIEDMAN, B. (1962). "Electromagnetic Waves" (R. E. Langer, ed.). Univ. of Wisconsin, Madison, Wisconsin.
FRISCH, H. L., and BRADY, G. W. (1962). *J. Chem. Phys.* **37**, 1514.
FROESE, C., and WAIT, J. R. (1954). *Can. J. Phys.* **32**, 775.
FÜRTH, R. (1915). *Sitzber. Akad. Wiss. Wien Math. Naturw. Kl. Abt. IIa* **124**, 577.
FÜRTH, R., and WILLIAMS, C. L. (1954). *Proc. Roy. Soc.* (*London*) **A224**, 104.
GALLACHER, L., and BETTELHEIM, F. A. (1962). *J. Polymer Sci.* **58**, 697.
GANS, R. (1912). *Ann. Physik.* **37**, 881.
GANS, R. (1921). *Ann. Physik.* **65**, 97.
GANS, R. (1923). *Z. Physik.* **17**, 353.
GANS, R. (1925). *Ann. Physik.* **76**, 29.

GARBACZ, R. J. (1961). The bistatic scattering from a class of lossy dielectric spheres with surface impedance boundary conditions. Res. Foundation Rept. 925-5. Ohio State Univ., Columbus, Ohio, January 1961.

GARBACZ, R. J. (1962a). Electromagnetic scattering by radially inhomogeneous spheres. Res. Foundation Rept. 1223-3. Ohio State Univ., Columbus, Ohio, January 1962.

GARBACZ, R. J. (1962b). *Proc. IRE* **50**, 1837.

GARBACZ, R. J. (1964). *Phys. Rev.* **133**, A14.

GATES, D. W. (1966). *Science* **151**, 523.

GEIDUSCHEK, E. P., and HOLTZER, A. (1958). *In* "Advances in Biological and Medical Physics" (C. A. Tobias and J. H. Lawrence, eds.) Vol. VI, pp. 431–551. Academic Press, New York.

GERJUOY, E. (1953). *Commun. Pure Appl. Math.* **6**, 73.

GERMEY, K. (1964). *Ann. Physik.* **13**, 237.

GERMEY, K. (1966). *Ann. Physik.* **17**, 397.

GERMOGENOVA, O. A. (1963). *Bull. Acad. Sci. USSR Geophys. Ser. 4* (*English Transl.*) p. 403.

GIESE, R. H. (1961). *Z. Astrophys.* **51**, 119.

GIESE, R. H., DE BARY, E., BULLRICH, K., and VINNEMANN, C. D. (1962). Tabellen der Streufunktionem $i_1(\phi)$, $i_2(\phi)$ und des Streuquerschnittes $K(\alpha, \mathrm{m})$ homogener Kugelchen nach der Mie'schen Theorie. *Abhandl. Deut. Akad. Wiss. Berlin Kl. Math. Physik. Tech.*

GLADSTONE, J. H., and DALE, T. (1858). *Phil. Trans.* **148**, 887.

GLASSTONE, S. (1946). "Testbook of Physical Chemistry." Van Nostrand, Princeton, New Jersey.

GLEDHILL, R. J. (1962). *J. Phys. Chem.* **66**, 458.

GLOVER, K. M., and ATLAS, D. (1963). *J. Appl. Math. Phys.* (ZAMP) **14**, 563.

GOEHRING, J. B., KERKER, M., MATIJEVIĆ, M., and TYREE, S. Y. (1959). *J. Am. Chem. Soc.* **81**, 5280.

GOLDBERG, A. I., HOHENSTEIN, W. P., and MARK, H. (1947). *J. Polymer Sci.* **2**, 503.

GOLDSTEIN, M. (1953). *J. Chem. Phys.* **21**, 1255.

GOLDSTEIN, M. (1959). *J. Appl. Phys.* **30**, 493, 501.

GOLDSTEIN, M. (1962). *J. Appl. Phys.* **33**, 3377.

GOLDSTEIN, M. (1963a). *J. Appl. Phys.* **34**, 1928.

GOLDSTEIN, M. (1963b). *In* "Electromagnetic Scattering" (M. Kerker, ed.). Pergamon Press, Oxford.

GOLDSTEIN, M., and MICHALIK, E. R. (1955). *J. Appl. Phys.* **26**, 1450.

GOODRICH, R. F., HARRISON, B. A., KLEINMAN, R. E., and SENIOR, T. B. A. (1961). Studies in radar cross sections XLVII—diffraction and scattering by regular bodies—I., The sphere. Rept. No. 3648-1-T (AFCRL 62-40). Radiation Lab., Dept. of Elec. Eng., Univ. of Michigan, Ann Arbor, Michigan.

GOULD, R. N., and BURMAN, R. (1964). *J. Atmospheric Terrest. Phys.* **26**, 335.

GOVI, G. (1860). *Compt. Rend.* **51**, 360, 669.

GRAESSLEY, W. W., and ZUFALL, J. H. (1964). *J. Colloid Sci.* **19**, 516.

GREEN, A. E. S. (1959). *Proc. Intern. Conf. Nucl. Opt. Model, Tallahassee, Florida, 1959*, Florida State Univ. Studies No. 32.

GREEN, M. (1960). *J. Math. Phys.* **1**, 391.

GREENBERG, J. M. (1960). *J. Appl, Phys.* **31**, 82.

GREENBERG, J. M., and ROARK, T. P. (1967). *Astrophys. J.* **147**, 917.

GREENBERG, J. M., PEDERSEN, N. E., and PEDERSEN, J. C. (1961). *J. Appl. Phys.* **32**, 233.

GREENBERG, J. M., LIND, A. C., WANG, R. T., and LIBELO, L. F. (1967). *In* "Electromagnetic Scattering" (R. L. Rowell and R. S. Stein, eds.). Gordon and Breach, New York.

GUCKER, F. T., and COHN, S. H. (1953). *J. Colloid Sci.* **8**, 555.

GUCKER, F. T., and ROWELL, R. L. (1960). *Discussions Faraday Soc.* **30**, 185.

GUCKER, F. T., ROWELL, R. L., and CHIU, G. (1964). *Proc. Natl. Conf. Aerosols, 1st, Liblice, Czechoslovakia, October 1962*, K. Spurny, ed., p. 59. Czech. Acad. of Sci., Prague, Czechoslovakia.

GUCKER, F. T., CHIU, G., OSBORNE, E. C., and TUMA, J. (1968). *J. Colloid. Interface Sci.* **27**, 395.

GUINIER, A. (1939). *Ann. Phys.* **12**, 161.

GUINIER, A., and FOURNET, G. (1955). "Small Angle Scattering of X-Rays." Wiley, New York.

GUMPRECHT, R. O., and SLIEPCEVICH, C. M. (1951a). "Light-scattering Functions for Spherical Particles." Eng. Res. Inst., Univ. of Michigan, Ann Arbor, Michigan.

GUMPRECHT, R. O., and SLIEPCEVICH, C. M. (1951b). "Ricatti-Bessel Functions for Large Arguments and Orders." Eng. Res. Inst., Univ. of Michigan, Ann Arbor, Michigan.

GUMPRECHT, R. O., and SLIEPCEVICH, C. M. (1951c). "Functions of First and Second Derivatives of Legendre Polynomials." Eng. Res. Inst., Univ. of Michigan, Ann Arbor, Michigan.

GUMPRECHT, R. O., and SLIEPCEVICH, C. M. (1953a). *J. Phys. Chem.* **57**, 95.

GUMPRECHT, R. O., and SLIEPCEVICH, C. M. (1953b). *J. Phys. Chem.* **57**, 90.

GUREVICH, M. M. (1955). *Struct. Glass, Proc. Conf., Leningrad, 1953*, p. 202. Acad. of Sci. U.S.S.R. Press, Moscow-Leningrad.

GÜTTLER, A. (1952). *Ann. Physik* [6] **11**, 65.

HALD, A. (1962). "Statistical Theory with Engineering Applications." Wiley, New York.

HALWER, M. (1948). *J. Am. Chem. Soc.* **70**, 3985.

HAMMEL, J. J., and OHLBERG, S. M. (1965). *J. Appl. Phys.* **36**, 1442.

HAMMEL, J. J., MICKEY, J., and GOLOB, H. R. (1968). *J. Colloid Interface Sci.* **27**, 329.

HANSEN, W. W. (1935). *Phys. Rev.* **47**, 139.

HARNED, H. S., and OWEN, B. B. (1950). "The Physical Chemistry of Electrolytic Solutions," 2nd ed. Reinhold, New York.

HARRINGTON, R. F. (1965). *IEEE Trans. Antennas Propagation* **AP-13**, 812.

HART, R. W., and GRAY, E. P. (1964). *J. Appl. Phys.* **35**, 1408.

HARTEL, W. (1940). *Licht* **10**, 141, 165, 190, 214, 233.

HAVARD, J. B. (1960). "On the radiational characteristics of water clouds at infrared wavelengths." Ph.D. Thesis, Univ. of Washington, Seattle, Washington.

HEILWEIL, I. J. (1964). *J. Colloid Sci.* **19**, 105.

HELLER, W. (1959a). *Record Chem. Progr.* (*Kresge-Hooker Sci. Lib.*) **20**, 209.

HELLER, W. (1959b). *Rev. Mod. Phys.* **31**, 1072.

HELLER, W. (1963). *In* "Electromagnetic Scattering" (M. Kerker, ed.). Pergamon Press, Oxford.

HELLER, W. (1964). *J. Chem. Phys.* **40**, 2700.

HELLER, W. (1965). *J. Chem. Phys.* **42**, 1609.

HELLER, W., and MCCARTY, H. J. (1958). *J. Chem. Phys.* **29**, 78.

HELLER, W., and NAKAGAKI, M. (1959). *J. Chem. Phys.* **31**, 1188.

HELLER, W., and PANGONIS, W. J. (1957). *J. Chem. Phys.* **26**, 498.

HELLER, W., and PUGH, T. L. (1957). *J. Colloid Sci.* **12**, 294.

HELLER, W., and TABIBIAN, R. M. (1962). *J. Phys. Chem.* **66**, 2059.

HELLER, W., and VASSY, E. (1943). *Phys. Rev.* **63**, 65.

HELLER, W., and VASSY, E. (1946). *J. Chem. Phys.* **14**, 565.

HELLER, W., and WALLACH, M. L. (1963). *J. Phys. Chem.* **67**, 2577.

HELLER, W., and WALLACH, M. L. (1964). *J. Phys. Chem.* **68**, 931.

HELLER, W., NAKAGAKI, M., and WALLACH, M. L. (1959). *J. Chem. Phys.* **30**, 444.

HELLER, W., WADA, E., and PAPAZIAN, L. A. (1961). *J. Polymer Sci.* **47**, 481.

HELLER, W., BHATNAGAR, H. L., and NAKAGAKI, M. (1962). *J. Chem. Phys.* **36**, 1163.

HELSTROM, C. W. (1963). *In* "Electromagnetic Theory and Antennas" (E. C. Jordon, ed.). Pergamon Press, Oxford.

HEMMER, P. C. (1964). *J. Math. Phys.* **5**, 75.

HEMMER, P. C., KAC, M., and UHLENBECK, G. E. (1964). *J. Math. Phys.* **5**, 60.
HENTZ, F. C., and TYREE, S. Y. (1964). *Inorg. Chem.* **3**, 844, 873.
HERAK, M. J., KRATOHVIL, J., HERAK, M. M., and WRISCHER, M. (1958). *Croat. Chem. Acta* **30**, 221.
HERMAN, B. M. (1961). *J. Meteorol.* **18**, 558.
HERMAN, B. M. (1962). *Quart. J. Roy. Meteorol. Soc.* **88**, 143.
HERMAN, B. M., and BATTAN, L. J. (1961a). *Quart. J. Roy. Meteorol. Soc.* **87**, 223.
HERMAN, B. M., and BATTAN, L. J. (1961b). *J. Meteorol.* **18**, 468.
HERMAN, B. M., BROWNING, S. R., and BATTAN, L. J. (1961). Tech. Rep. #9. Inst. of Atmospheric Phys., Univ. of Arizona, Tucson, Arizona.
HERMANIE, P. H. J., and VAN DER WAARDEN, M. (1966). *J. Colloid Interface Sci.* **21**, 513.
HERMANS, J. J. (1949). *Rec. Trav. Chim.* **68**, 859.
HERRMANN, K. W., BRUSHMILLER, J. G., and COURCHENE, W. L. (1966). *J. Phys. Chem.* **70**, 2909.
HERTZ, H. (1889). *Ann. Physik.* **36**, 1.
HEYN, A. N. J. (1955). *J. Appl. Phys.* **26**, 519, 1113.
HILBIG, G. (1966). *Optik* **23**, 313.
HILL, T. L. (1959). *J. Chem. Phys.* **30**, 93.
HOBSON, E. W. (1931). "The Theory of Spherical and Ellipsoidal Harmonics." Cambridge Univ. Press, London and New York.
HODKINSON, J. R. (1963a). *Staub* **23**, 374.
HODKINSON, J. R. (1963b). *Brit. J. Appl. Phys.* **14**, 931.
HODKINSON, J. R. (1964). *J. Opt. Soc. Am.* **54**, 846.
HODKINSON, J. R. (1966). *Appl. Opt.* **5**, 839.
HODKINSON, J. R., and GREENLEAVES, I. (1963). *J. Opt. Soc. Am.* **53**, 577.
HOLDER, G. A., and WHALLEY, E. (1962). *Trans. Faraday Soc.* **58**, 2095.
HOLTZER, A. (1955). *J. Polymer Sci.* **17**, 432.
HORN, P. (1955). *Ann. Phys.* (*Paris*) [12] **10**, 386; Ph.D. Thesis, Strasbourg Univ., Strasbourg, 1954.
HORN, P., and BENOIT, H. (1953). *J. Polymer Sci.* **10**, 29.
HORN, P., BENOIT, H., and OSTER, G. (1951). *J. Chim. Phys.* **48**, 530.
HUISMAN, H. F. (1964). *Koninkl. Ned. Akad. Wetenschap. Proc. Ser. B* **67**, 367, 376, 388, 408.
HUTCHINSON, E., and MELROSE, J. C. (1954). *Z. Physik. Chemie* (*Frankfurt*) **2**, 363.
HUYNEN, J. R. (1958). Theory and design of a class of Luneberg lenses. Rept. LMSD-5108. Lockheed Missile Systems Div., Sunnyvale, California.
HYDE, A. J., and TANNER, A. G. (1968). *J. Colloid Interface Sci.* **27**, 179.
IKEDA, Y. (1963). *In* "Electromagnetic Scattering" (M. Kerker, ed.). Pergamon Press, Oxford.
IMAI, N., and EISENBERG, H. (1966). *J. Chem. Phys.* **44**, 130.
INN, E. C. Y. (1951). *J. Colloid Sci.* **6**, 368.
IRVINE, W. M. (1963). *Bull. Astron. Inst. Neth.* **17**, 176.
IRVINE, W. M. (1965). *J. Opt. Soc. Am.* **55**, 16.
ISIHARA, A., KOYAMA, R., YAMADA, N., and NISHIOKA, A. (1955). *J. Polymer Sci.* **17**, 341.
IVES, H. E., and BRIGGS, H. B. (1936). *J. Opt. Soc. Am.* **26**, 238.
IVES, H. E., and BRIGGS, H. B. (1937). *J. Opt. Soc. Am.* **27**, 181, 395.
JACOBSEN, R., and KERKER, M. (1967). *J. Opt. Soc. Am.* **57**, 751.
JACOBSEN, R., KERKER, M., and MATIJEVIĆ, E. (1967). *J. Phys. Chem.* **71**, 514.
JASIK, H. (1954). The electromagnetic theory of the Luneberg lens. Rept. No. AFCRC-TR-54-121.
JAYCOCK, M. J., and PARFITT, G. D. (1962). *Nature* **194**, 77.
JEFFREYS, H. (1959). "The Earth." Cambridge Univ. Press, London and New York.
JENNINGS, B. R., and JERRARD, H. G. (1965a). *J. Colloid Sci.* **20**, 448.

JENNINGS, B. R., and JERRARD, H. G. (1965b). *J. Chem. Phys.* **42**, 511.
JENNINGS, B. R., and JERRARD, H. G. (1965c). *J. Phys. Chem.* **69**, 2817.
JENNINGS, B. R., and JERRARD, H. G. (1966). *J. Chem. Phys.* **44**, 1291.
JOHNSON, I., and LA MER, V. K. (1947). *J. Am. Chem. Soc.* **69**, 1184.
JOHNSON, J. S. and RUSH, R. M. (1968). *J. Am. Chem. Soc.* **72**, 360.
JOHNSON, J. S., KRAUS, K. A., and SCATCHARD, G. (1960). *J. Phys. Chem.* **64**, 1867.
JONES, A. R., and FARONE, W. A. (1967). *J. Appl. Phys.* **38**, 4544.
JONES, A. R., and WOODING, E. R. (1965). *Electron. Letters* **1**, 171.
JONES, A. R., and WOODING, E. R. (1966). *J. Appl. Phys.* **37**, 4670.
JONES, D. S. (1955). *Phil. Mag.* **46**, 957.
KAMERLINGH-ONNES, H., and KEESOM, W. H. (1908). *Comm. Labor. Leiden* **104.**
KAMKE, E. (1948). "Differential Gleichungen," Vol. I, p. 440. Chelsea, New York.
KATTAWAR, G. W., and PLASS, G. N. (1967). *Appl. Optics* **6**, 1377.
KAWANO, T., and PETERS, L. (1963). Calculation of the echo area on a *n*-layer plasma cylinder and sphere. Res. Foundation Rept. 1116–33. Ohio State Univ., Columbus, Ohio.
KAY, A. F. (1956). *IRE Trans. Antennas Propagation* **AP-4**, 87.
KAY, A. F. (1959). *IRE Trans. Antennas Propagation* **AP-7**, 32.
KAYE, B. H., and ALLEN, T. (1965). *Analyst* **90**, 147.
KEANE, J. J., NORRIS, F. H., and STEIN, R. S. (1956). *J. Polymer Sci.* **20**, 209.
KEEN, B. A., and PORTER, A. W. (1913). *Proc. Roy. Soc.* **A89**, 370.
KEESOM, W. H. (1911). *Ann. Physik.* [4] **35**, 591.
KEIJZERS, A. E. M., VAN AARTSEN, J. J., and PRINS, W. (1965). *J. Appl. Phys.* **36**, 2874.
KEIJZERS, A. E. M., VAN AARTSEN, J. J., and PRINS, W. (1968). *J. Am. Chem. Soc.* **90**, 3107.
KEITEL, G. H. (1955). *Proc. IRE* **43**, 1481.
KELLER, J. B., and KAY, L. (1954). *J. Appl. Phys.* **25**, 876.
KELLER, J. R., MATIJEVIĆ, E., and KERKER, M. (1961). *J. Phys. Chem.* **65**, 56.
KELLY, R. E., and RUSSEK, A. (1960). *Nuovo Cimento* [10], **16**, 593.
KENNAUGH, E., and SLOAN, R. (1952). Effects of types of polarization on echo characteristics. Res. Foundation Rept. 389–15, AF 28(099)–90. Ohio State Univ., Columbus, Ohio.
KENYON, A. S., and LA MER, V. K. (1949). *J. Colloid Sci.* **4**, 163.
KERKER, M. (1950). *J. Colloid Sci.* **5**, 165.
KERKER, M. (1951). *J. Chem. Phys.* **19,** 1324.
KERKER, M. (1952). *J. Chem. Phys.* **20**, 1653.
KERKER, M. (1955). *J. Opt. Soc. Am.* **45**, 1081.
KERKER, M. (1958). *J. Polymer Sci.* **28**, 429.
KERKER, M., and LA MER, V. K. (1950). *J. Am. Chem. Soc.* **72**, 3516.
KERKER, M., and MATIJEVIĆ, E. (1961a). *J. Opt. Soc. Am.* **51**, 506.
KERKER, M., and MATIJEVIĆ, E. (1961b). *J. Opt. Soc. Am.* **51**, 87.
KERKER, M., LANGLEBEN, P., and GUNN, K. L. S. (1951). *J. Meteorol.* **8**, 424.
KERKER, M., COX, A. L., and SCHOENBERG, M. D. (1955), *J. Colloid. Sci.* **10**, 413.
KERKER, M., LEE, D., and CHOU, A. (1958). *J. Am. Chem. Soc.* **80**, 1539.
KERKER, M., KELLER, J. R., SIAU, J., and MATIJEVIĆ, E. (1961a). *Trans. Faraday Soc.* **57**, 780.
KERKER, M., KRATOHVIL, J. P., and MATIJEVIĆ, E. (1961b). *J. Phys. Chem.* **65**, 1713.
KERKER, M., KRATOHVIL, J. P., and MATIJEVIĆ, E. (1962). *J. Opt. Soc. Am.* **52**, 551.
KERKER, M., DABY, E., COHEN, G. L., KRATOHVIL, J. P., and MATIJEVIĆ, E. (1963a). *J. Phys. Chem.* **67**, 2105.
KERKER, M., FARONE, W. A., and MATIJEVIĆ, E. (1963b). *J. Opt. Soc. Am.* **53**, 758.
KERKER, M., KRATOHVIL, J. P., OTTEWILL, R. H., and MATIJEVIĆ, E. (1963c). *J. Phys. Chem.* **67**, 1097.
KERKER, M., FARONE, W. A., SMITH, L. B., and MATIJEVIĆ, E. (1964a). *J. Colloid Sci.* **19**, 193.

KERKER, M., KRATOHVIL, J. P., and MATIJEVIĆ, E. (1964b). *J. Polymer Sci.* **2A**, 303.

KERKER, M., MATIJEVIĆ, E., ESPENSCHEID, W., FARONE, W., and KITANI, S. (1964c). *J. Colloid Sci.* **19**, 213.

KERKER, M., COOKE, D., FARONE, W. A., and JACOBSEN, R. T. (1966a). *J. Opt. Soc. Am.* **56**, 487.

KERKER, M., FARONE, W., and ESPENSCHEID, W. (1966b). *J. Colloid Interface Sci.* **21**, 459.

KERKER, M., FARONE, W., and JACOBSEN, R. T. (1966c). *J. Opt. Soc. Am.* **56**, 1248.

KERKER, M., KAUFFMAN, L. H., and FARONE, W. A. (1966d). *J. Opt. Soc. Am.* **56**, 1053.

KERR, D. E. (1951). "Propagation of Short Radio Waves," Vol. 13 (M.I.T. Rad. Lab. Ser.). McGraw-Hill, New York.

KIESSLING, J. (1884a). *Meteorol. Z.* **1**, 117.

KIESSLING, J. (1884b). *Gotting. Nachr.* p. 226.

KING, L. V. (1913). *Phil. Trans. Roy. Soc. London Ser. A* **212**, 375.

KING, L. V. (1923). *Proc. Roy. Soc.* **A104**, 333.

KING, R. W. P., and WU, T. T. (1959). "The Scatter and Diffraction of Waves." Harvard Univ. Press, Cambridge.

KIRCHNER, F., and ZSIGMONDY, R. (1904). *Ann. Physik* **15**, 573.

KIRKWOOD, J. G., and GOLDBERG, R. J. (1950). *J. Chem. Phys.* **18**, 54.

KIRKWOOD, J. G., and SHUMAKER, J. B. (1952). *Proc. Natl. Acad. Sci. U.S.* **38**, 863.

KIRSTE, R., and POROD, G. (1962). *Kolloid-Z. Z. Polymere* **184**, 1.

KITANI, S. (1956). *Nippon Kagaku Zasshi* **77**, 1621.

KITANI, S. (1960). *J. Colloid Sci.* **15**, 287.

KLEINMAN, R. E. (1965). *Proc. IEEE* **53**, 848.

KLEINMAN, R. E. (1967). In "Electromagnetic Wave Theory" (J. Brown, ed.), Pt. 2. Pergamon Press, Oxford.

KLINE, M., and KAY, I. W. (1965). "Electromagnetic Theory and Geometrical Optics." Wiley (Interscience), New York.

KODIS, R. D. (1952). *J. Appl. Phys.* **23**, 249.

KODIS, R. D. (1959). *IRE Trans. Antennas Propagation* **AP-7**, S468.

KODIS, R. D. (1961). *J. Res. Natl. Bur. Std.* **D65**, 19.

KODIS, R. D. (1963a). In "Electromagnetic Theory and Antennas" (E. C. Jordan, ed.), Pt. 1. Pergamon Press, Oxford.

KODIS, R. D. (1963b). *IEEE Trans. Antennas Propagation* **AP-11**, 86, 703.

KOTTLER, F. (1952). *J. Phys. Chem.* **56**, 442.

KOUYOUMJIAN, R. G., PETERS, L., and THOMAS, D. T. (1963). *IEEE Trans. Antennas Propagation* **AP-11**, 690.

KRATKY, O. (1948). *J. Polymer Sci.* **3**, 195.

KRATKY, O., and POROD, G. (1949). *J. Colloid Sci.* **4**, 35.

KRATOHVIL, J. P. (1964). *Anal. Chem.* **36**, 458R.

KRATOHVIL, J. P. (1966). *Anal. Chem.* **38**, 517R.

KRATOHVIL, J. P. (1968). Macromolecular Structure, p. 59, Publication 1573, National Academy of Sciences, Washington.

KRATOHVIL, J. P. and DEZELIĆ, G. (1962). *Kolloid-Z. Z. Polymere* **180**, 67.

KRATOHVIL, J. P. and DELLI COLLI, H. T. (1968). *Can. J. Biochem.* **46**, 945.

KRATOHVIL, J. P., and SMART, C. (1965). *J. Colloid Sci.* **20**, 875.

KRATOHVIL, J., DEZELIĆ, G., KERKER, M., and MATIJEVIĆ, E. (1962). *J. Polymer Sci.* **57**, 59.

KRATOHVIL, J., KERKER, M., and OPPENHEIMER, L. (1965). *J. Chem. Phys.* **43**, 914.

KRATOHVIL, J. P., OPPENHEIMER, L., and KERKER, M. (1966). *J. Phys. Chem.* **70**, 2834.

KRAUSE, S. (1961). *J. Phys. Chem.* **65**, 1618.

KRISHNAN, R. S. (1938a). *Proc. Indian Acad. Sci. Sect. A* **7**, 21, 91, 98.

KRISHNAN, R. S. (1938b). *Kolloid-Z.* **84**, 2, 18.

KRISHNAN, R. S. (1939). *Proc. Indian Acad. Sci. Sect. A* **10**, 395.

KRONMAN, M. J., and TIMASHEFF, S. N. (1959). *J. Phys. Chem.* **63**, 629.

KUMAGAI, N., and ANGELAKOS, D. J. (1961). Electron. Res. Lab. Rept. No. 333. Inst. of Eng. Res., Univ. of California, Berkeley, California.

KURIYAMA, K. (1962). *Kolloid-Z. Z. Polymere* **180**, 55.

KUSHNER, L. M., and HUBBARD, W. D. (1955). *J. Colloid. Sci.* **10**, 428.

KUSHNER, L. M., HUBBARD, W. D., and PARKER, R. A. (1957). *J. Res. Natl. Bur. Std.* **59**, 113.

LALANNE, J. R., and BOTHOREL, P. (1964). *J. Chim. Phys.* **61**, 1262.

LAMB, H. (1881). *Proc. London Math. Soc.* **13**, 189.

LAMB, H. (1906). "Hydrodynamics," 3rd ed., Sect. 352, p. 574. Cambridge Univ. Press, Cambridge.

LA MER, V. K. (1948). *J. Phys. Colloid Chem.* **52**, 65.

LA MER, V. K., and BARNES, M. (1946). *J. Colloid Sci.* **1**, 71, 79.

LA MER, V. K., and GORDIEYEFF, V. A. (1950). *Science* **112**, 454.

LA MER, V. K., and PLESNER, W. (1957). *J. Polymer Sci.* **24**, 147.

LA MER, V. K., and SINCLAIR, D. (1943). Verification of the Mie theory. OSRD Rep. No. 1857 and Rept. No. 944. Office of Publ. Board, U.S. Dept. of Com., Washington, D.C.

LA MER, V. K., INN, E., and WILSON, I. B. (1950). *J. Colloid Sci.* **5**, 471.

LANDAU, L. D., and LIFSCHITZ, E. M., (1959). "Statistical Physics." Pergamon Press, Oxford.

LANGER, G., and PIERRARD, J. M. (1963). *J. Colloid Sci.* **18**, 95.

LARKIN, B. K., and CHURCHILL, S. W. (1959). *J. Opt. Soc. Am.* **49**, 188.

LATIMER, P., and BRYANT, F. D. (1965). *J. Opt. Soc. Am.* **55**, 1554.

LAX, M. (1950). *Phys. Rev.* **78**, 306.

LEE, W. C. Y., PETERS, L., and WALTER, C. H. (1965). *J. Res. Natl. Bur. Std.* **D69**, 227.

LETCHER, J. H., and SCHMIDT, P. W. (1966). *J. Appl. Phys.* **37**, 649.

LEVINE, S., and KERKER, M. (1963). In "Electromagnetic Scattering" (M. Kerker, ed.), Pergamon Press, Oxford.

LEVINE, S., and OLAOFE, G. O. (1968). *J. Colloid. Interface Sci.* **27**, 442.

LIANG, C., and LO, Y.T. (1967). *Radio Sci.* **2**, 1481.

LIBELO, L. F. (1962a). Extinction of light at normal incidence by dielectric cylinders with refractive index in the range $m = 1.15$ to $m = 1.30$ scattering coefficients. NOLTR 62–157. U.S. Naval Ordnance Lab., White Oak, Maryland.

LIBELO, L. F. (1962b). Scattering of light at normal incidence by dielectric cylinders with refractive index in the range $m = 1.4$ to $m = 3.0$. NOLTR 62–161. U.S. Naval Ordnance Lab., White Oak, Maryland.

LIBELO, L. F. (1962c). Light scattering by partially absorbing cylinders—Coefficients. NOLTR 62–142. U.S. Naval Ordnance Lab., White Oak, Maryland.

LIND, A. C., and GREENBERG, J. M. (1966). *J. Appl. Phys.* **37**, 3195.

LITAN, A. (1968). *J. Chem. Phys.* **48**, 1052; 1059.

LIU, B. Y. H., WHITBY, K. T., and YU, H. H. S. (1966). A condensation aerosol generator for producing monodisperse aerosols in the size range 0.036 μ to 1.3 μ. Particle Technol. Lab. Rept. Dept. of Mech. Eng. Univ. of Minnesota, Minneapolis, Minnesota.

LOCHET, R. (1953). *Ann. Phys. (Paris)* **8**, 14.

LODE, W., DETTMAR, H.-K., and MARRE, E. (1962). *Z. Chemie-Ingenieur-Technik.* **34**, 782.

LOGAN, N. A. (1962). *J. Opt. Soc. Am.* **52**, 342.

LOGAN, N. A. (1965). *Proc. IEEE* **53**, 773.

LORENTZ, H. A. (1879). *Wied. Ann.* **9**, 641.

LORENZ, L. (1880). *Wied. Ann.* **11**, 70.

LORENZ, L. (1890). *Videnskab. Selskab. Skrifter.* **6**.

LORENZ, L. (1880). *Wied. Ann.* **11**, 70.
LORENZ, L. (1898a). "Oeuvres Scientifiques," Vol. I, p. 301, Copenhagen, Denmark.
LORENZ, L. (1898b). "Oeuvres Scientifiques," Vol. I, p. 405, Copenhagen, Denmark.
LOUCHEUX, C., WEILL, G., and BENOIT, H. (1958). *J. Chim. Phys.* **55**, 540.
LOVE, A. E. H. (1899). *Proc. London Math. Soc.* **30**, 308.
LOWAN, A. N. (1945). "Tables of Associated Legendre Functions." Columbia Univ. Press, New York.
LOWAN, A. N. (1947a). "Tables of Spherical Bessel Functions," Vol. I and II. Columbia Univ. Press, New York.
LOWAN, A. N. (1947b). "Tables of the Bessel Functions $J_0(x)$ and $J_1(x)$ for Complex Arguments." Columbia Univ. Press, New York.
LOWAN, A. N. (1947c). "Tables of the Bessel Functions $Y_0(x)$ and $Y_1(x)$ for Complex Arguments." Columbia Univ. Press, New York.
LOWAN, A. N. (1948). "Tables of Scáttering Functions for Spherical Particles" (Natl. Bur. of Std., Appl. Math. Ser. 4). U.S. Govt. Printing Office, Washington, D.C.
LUNDBERG, J. L. (1969). *J. Colloid Interface Sci.* **29** (in press).
LUNDBERG, J. L., MOONEY, E. J., and GARDNER, K. R. (1964). *Science* **145**, 1308.
LUNEBERG, R. K. (1944). "Mathematical theory of optics." Brown Univ. Lecture Notes, Providence, Rhode Island (reproduced by Univ. Microfilms, Inc., Ann Arbor, Michigan, 1962).
LUNEBERG, R. K. (1948). Maxwell's equations in spherically symmetric media. Res. Rept. No. 172–8. Math. Res. Group, Washington Square College, New York Univ., New York.
LUZZATI, V. (1957). *Acta Cryst.* **10**, 33.
LUZZATI, V., CESARI, M., SPACK, G., MASSON, F., and VINCENT, J. (1961). *J. Mol. Biol.* **109,** 249.
LYNCH, P. J. (1963). *Phys. Rev.* **130**, 1235.
MACDONALD, H. M. (1902). "Electric Waves." Cambridge Univ. Press, London and New York.
MACDONALD, H. M. (1904). *Proc. Roy. Soc.* **A72**, 59.
MACDONALD, H. M. (1910). *Phil. Trans. Roy. Soc. London Ser. A* **210**, 113.
MACROBERT, T. M. (1945). "Spherical Harmonics." Dover, New York.
MCDONALD, J. E. (1960). *J. Meteorol.* **17**, 232.
MCDONALD, J. E. (1962a). *Quart. J. Roy. Meteorol. Soc.* **88**, 183.
MCDONALD, J. E. (1962b). *J. Appl. Meteorol.* **1**, 391.
MCINTYRE, D., WIMS, A., and GREEN, M. S. (1962). *J. Chem. Phys.* **37**, 3019.
MCLAY, A. B., and SUBBARAO, M. K. (1956). *IEEE Trans. Antennas Propagation* **AP-4**, 579.
MAECKER, H. (1949). *Ann. Physik.* **4**, 409.
MAILLIET, A.-M., and POURADIER, J. (1961). *J. Chim. Phys.* **58**, 710.
MALMON, A. G. (1957). *Acta Cryst.* **10**, 639.
MAL'TSEV, YU, V. (1959). *Opt. i Spektroskopiya* [*Opt. Spectry.* (*USSR*) (*English Transl.*) 7, 76.]
MAL'TSEV, YU. V. (1960). *Opt. i Spektroskopiya* [*Opt. Spectry.* (*USSR*) (*English Transl.*) **8**, 362].
MARCUVITZ, N. (1951). *Comm. Pure Appl. Math* **4**, 263.
MARGULIES, R. S., and SCARF, F. L. (1964). *IEEE Trans. Antennas Propagation* **AP-12**, 91.
MARON, S. H. (1959). *J. Polymer Sci.* **38**, 329.
MARON, S. H., and ELDER, M. E. (1963a). *J. Colloid Sci.* **18**, 107.
MARON, S. H., and ELDER, M. E. (1963b). *J. Colloid Sci.* **18**, 199.
MARON, S. H., and LOU, R. L. H. (1954). *J. Polymer Sci.* **14**, 29.
MARON, S. H., and LOU, R. L. H. (1955). *J. Phys. Chem.* **59**, 231.
MARON, S. H., and NAKAJIMA, N. (1960a). *J. Polymer Sci.* **47**, 157.
MARON, S. H., and NAKAJIMA, N. (1960b). *J. Polymer Sci.* **42**, 327.
MARON, S. H., ELDER, M. E., and PIERCE, P. E. (1963a). *J. Colloid Sci.* **18**, 733.
MARON, S. H., PIERCE, P. E., and ELDER, M. E. (1963b). *J. Colloid Sci.* **18**, 391.

MARON, S. H., PIERCE, P. E., and ULEVITCH, I. N. (1963c). *J. Colloid Sci.* **18**, 470.
MARON, S. H., PIERCE, P. E., and ELDER, M. E. (1964). *J. Colloid Sci.* **19**, 591.
MARSHALL, J. S., HITSCHFELD, W., and GUNN, K. L. S. (1955). *Advan. Geophys.* **2**, 1.
MARTIN, W. H. (1913). *Proc. Roy. Soc. Can.* **1**, III, 219.
MARTIN, W. H. (1926), *in* "Colloid Chemistry" (J. Alevander, ed.), Vol. I, p. 340. Reinhold, New York.
MASSOULIER, A. (1963). *J. Phys.* (*Paris*) **24**, 342.
MATHUR, P. N., and MUELLER, E. A. (1955). Radar back-scattering from nonspherical scatterers. Rept. No. 28. Illinois State Water Survey, Univ. Library, Univ. of Illinois, Urbana, Illinois.
MATIJEVIĆ, E., and KERKER, M. (1958). *J. Phys. Chem.* **62**, 1271.
MATIJEVIĆ, E., and KERKER, M. (1959a). *J. Am. Chem. Soc.* **81**, 5560.
MATIJEVIĆ, E., and KERKER, M. (1959b). *J. Am. Chem. Soc.* **81**, 1307.
MATIJEVIĆ, E., KERKER, M., and SCHULZ, K. (1960). *Discussions Faraday Soc.* **30**, 178.
MATIJEVIĆ, E. OTTEWILL, R. H., and KERKER, M. (1961). *J. Opt. Soc. Am.* **51**, 115.
MATIJEVIĆ, E., SCHULZ, K., and KERKER, M. (1962). *J. Colloid Sci.* **17**, 26.
MATIJEVIĆ, E., MATHAI, K. G., and KERKER, M. (1963). *J. Phys. Chem.* **67**, 1995.
MATIJEVIĆ, E., KITANI, S., and KERKER, M. (1964). *J. Colloid Sci.* **19**, 223.
MAURER, R. D. (1960). *Phys. Chem. Solids* **17**, 44.
MAURER, R. D. (1962). *J. Appl. Phys.* **33**, 2132.
MAUSS, Y., CHAMBRON, J., DAUNE, M., and BENOIT, H. (1967). *J. Mol. Biol.* **27**, 579.
MAXIM, L. D., KLEIN, A., MEYER, M. E., and KUIST, C. H. (1966). *Abstracts, Am. Chem. Soc. Meeting, September 1966*. New York, N.Y.
MAXWELL, J. C., (1865). *Phil. Trans.* **155**, 459.
MAXWELL, J. C. (1873). "A Treatise on Electricity and Magnetism," Vol. II, pp. 66–70 (3rd ed., 1892). Oxford University Press, Oxford.
MAXWELL-GARNETT, J. C. (1904). *Phil. Trans.* **203**, 385.
MAXWELL-GARNETT, J. C. (1906). *Phil. Trans.* **205**, 237.
MAYHAN, R. J., and SCHULTZ, F. V. (1967). *Radio Sci.* **2**, 853.
MECKE, R. (1920). *Ann. Physik.* **61**, 471; **62**, 623.
MEEHAN, E. J. (1968). *J. Colloid Interface Sci.* **27**, 388.
MEEHAN, E. J., and BEATTIE, W. H. (1960). *J. Phys. Chem.* **64**, 1006.
MEEHAN, E. J., and MILLER, J. K. (1968). *J. Phys. Chem.* **72**, 1523.
MEETEN, G. H. (1968). *Nature* **218**, 761.
MELROSE, J. C. (1957). The thermodynamic basis of light scattering theory. Ph.D. Thesis, Stanford Univ., Stanford, California.
MENTZNER, J. R. (1955). "Scattering and Diffraction of Radio Waves." Pergamon Press, Oxford.
METZ, H. J. (1963). *Phot. Korr.* **99**, 153.
METZ, H. J., and DETTMAR, H.-K. (1963). *Kolloid-Z. Z. Polymere* **188**, 28.
METZ, H. J., and DETTMAR, H.-K. (1963). *Kolloid-Z. Z. Polymere* **192**, 107.
MEVEL, J. (1958). *J. Phys. Radium* **19**, 630.
MEVEL, J. (1960). *Ann. Phys.* (*Paris*) [13] **5**, 265.
MICHAELS, A. I. (1958). Symposium on particle size measurement. Special Tech. Publ. No. 234, pp. 207–243. Am. Soc. for Testing Materials, Boston.
MIDZUNO, Y. (1961a). *J. Phys. Soc. Japan* **16**, 971.
MIDZUNO, Y. (1961b). *J. Phys. Soc. Japan* **16**, 1403.
MIE, G. (1908). *Ann. Physik.* **25**, 377.
MIJNLIEFF, P. F., and COUMOU, D. J. (1968). *J. Colloid Interface Sci.* **27**, 553.
MIJNLIEFF, P. F., and ZELDENRUST, H. (1965). *J. Phys. Chem.* **69**, 689.
MIKULSKI, J. J., and MURPHY, E. L. (1963). *IEEE Trans. Antennas Propagation* **AP-11**, 169.
MIRELÉS, R. (1966). *J. Math and Phys.* **45**, 127.

MITTELBACH, P. (1964). *Acta Phys. Austriaca* **19**, 53.
MITTELBACH, P. (1965). *Kolloid-Z. Z. Polymer* **206**, 152.
MITTELBACH, P., and POROD, G. (1961a). *Acta Phys. Austriaca* **14**, 405.
MITTELBACH, P., and POROD, G. (1961b). *Acta Phys. Austriaca* **14**, 185.
MITTELBACH, P., and POROD, G. (1962). *Acta Phys. Austriaca* **15**, 122.
MITTELBACH, P., and POROD, G. (1965). *Kolloid-Z. Z.Polymere* **202**, 40.
MIYAKE, A. (1960). *J. Phys. Soc. Japan* **15**, 875.
MÖGLICH, F. (1927). *Ann. Physik*. **83**, 609.
MONTROLL, E. W., and GREENBERG, J. M. (1952). *Phys. Rev.* **86**, 889.
MONTROLL, E. W., and GREENBERG, J. M. (1954). *Proc. Symp. Appl. Math.* **5**, p. 103. Am. Math. Soc. Providence, Rhode Island.
MOORE, D. M., BRYANT, F. D., and LATIMER, P. (1968). *J. Opt. Soc. Am.* **58**, 281.
MOREL, A. (1966). *J. Chim. Phys.* **63**, 1359.
MORGAN, S. P. (1958). *J. Appl. Phys.* **29**, 1358.
MORI, N., and KIKUCHI, H. (1957). *Denki Shikensho Ihô* **21**, 561.
MORI, N., and KIKUCHI, H. (1958). *Denki Shikensho Ihô* **22**, 209.
MORRISS, R. H., and COLLINS, L. F. (1964). *J. Chem. Phys.* **41**, 3357.
MORSE, P. M., and FESHBACH, H. (1953). "Methods of Theoretical Physics," Pts. 1 and 2. McGraw-Hill, New York.
MORSE, P. M., and RUBENSTEIN, P. L. (1938). *Phys. Rev.* **54**, 895.
MULLIN, C. R., SANDBURG, R., and VELLINE, C. O. (1965). *IEEE Trans. Antennas Propagation* **AP-13**, 141.
MUNSTER, A. (1960). *J. Chim. Phys.* **57**, 492.
MURAY, J. J. (1965). *Appl. Opt.* **4**, 1011.
MURPHY, E. L. (1965). *J. Appl. Phys.* **36**, 1918.
MYSELS, K. (1954). *J. Phys. Chem.* **58**, 303.
MYSELS, K. (1955). *J. Colloid Sci.* **10**, 507.
MYSELS, K. (1964). *J. Am. Chem. Soc.* **86**, 3503.
MYSELS, K., and PRINCEN, L. H. (1959). *J. Phys. Chem.* **63**, 1696.
NAKAGAKI, M. (1966). *Bull. Chem. Soc. Japan* **39**, 1689.
NAKAGAKI, M., and HELLER, W. (1959). *J. Polymer Sci.* **38**, 117.
NAKAGAKI, M., and SHIMOYAMA, T. (1964). *Bull. Chem. Soc. Japan* **37**, 1634.
NAPPER, D. H. (1967). *Kolloid-Z. Z. Polymer* **218**, 41.
NAPPER, D. H. (1968). *Kolloid-Z. Z. Polymer* **223**, 141.
NAPPER, D. H., and OTTEWILL, R. H. (1963a). *J. Colloid Sci.* **18**, 262.
NAPPER, D. H., and OTTEWILL, R. H. (1963b). *Kolloid-Z. Z. Polymer* **192**, 114.
NAPPER, D. H., and OTTEWILL, R. H. (1963c). *In* "Electromagnetic Scattering" (M. Kerker, ed.). Pergamon Press, Oxford.
NAPPER, D. H., and OTTEWILL, R. H. (1963d). *J. Phot. Sci.* **11**, 84.
NAPPER, D. H., and OTTEWILL, R. H. (1964). *Trans. Faraday Soc.* **60**, 1466.
NEGI, J. G. (1962a). *Geophysics* **27**, 386.
NEGI, J. G. (1962b). *Geophysics* **27**, 480.
NELSON, W. H., and TOBIAS, R. S. (1963). *Inorg. Chem.* **2**, 985.
NELSON, W. H., and TOBIAS, R. S. (1964). *Can. J. Chem.* **42**, 731.
NEUGEBAUER, T. (1943). *Ann. Phys.* **42**, 509.
NEWTON, I. (1706). "Optiks," (reprint). Dover, New York, 1952.
NEWTON, R. G. (1966). "Scattering Theory of Waves and Particles." McGraw-Hill, New York.
NICHOLSON, J. W. (1910). *Proc. London Math. Soc.* [2] **9**, 67.
NICHOLSON, J. W. (1912). *Proc. London Math. Soc.* [2] **11**, 277.
NOMURA, Y., and TAKAKU, K. (1955). *Tohoku Res. Inst. Res. Inst. Elec. Commun.* **7B**, 107.

Norris, F. H., and Stein, R. S. (1958). *J. Polymer Sci.* **27**, 87.
Ohba, Y. (1963). *Can. J. Phys.* **41**, 881.
Okano, K., and Wada, E. (1961). *J. Chem. Phys.* **34**, 405.
O'Konski, C. T. (1955). *Anal. Chem.* **27**, 694.
Olaf, J., and Robock, K. (1961). *Staub* **21**, 495.
Olaofe, G. O., and Levine, S. (1967). *In* "Electromagnetic Scattering" (R. L. Rowell and R. S. Stein, eds.). p. 237. Gordon and Breach, New York.
Onsager, L. (1936). *J. Am. Chem. Soc.* **48**, 1486.
Ooi, T. (1958). *J. Polymer Sci.* **28**, 459.
Oppenheimer, L. E. (1967). Ph.D. Thesis, Clarkson College of Technol., Potsdam, New York.
Orchard, S. E. (1965). *J. Opt. Soc. Am.* **55**, 1190.
Ornstein, L., and Zernike, F. (1914). *Proc. Acad. Sci. Amsterdam* **17**, 793.
Ornstein, L., and Zernike, F. (1918). *Physik. Z.* **19**, 134; **27**, 761.
Osborn, J. A. (1945). *Phys. Rev.* **67**, 351.
Oster, G. (1948). *Chem. Rev.* **43**, 319.
Oster, G., and Riley, D. P. (1952a). *Acta Cryst.* **5**, 1.
Oster, G., and Riley, D. P. (1952b). *Acta Cryst.* **5**, 272.
Ott, H. (1942). *Ann. Physik* **41**, 443.
Ott, H. (1949). *Ann. Physik* **4**, 432.
Ottewill, R. H., and Woodbridge, R. F. (1961). *J. Colloid Sci.* **16**, 581.
Overbeek, J. Th. G., Vrij, A., and Huisman, H. F. (1963). *In* "Electromagnetic Scattering," (M. Kerker, ed.), p. 321. Pergamon Press, Oxford.
Pangonis, W. J., and Heller, W. (1960). "Angular Scattering Functions for Spherical Particles." Wayne State Univ. Press, Detroit, Michigan.
Pangonis, W. J., Heller, W., and Jacobson, A. W. (1957). "Tables of Light Scattering Functions for Spherical Particles." Wayne State Univ. Press, Detroit, Michigan.
Pangonis, W. J., Heller, W., and Economou, N. A. (1961). *J. Chem. Phys.* **34**, 960.
Parfitt, G. D., and Wood, J. A. (1968). *Trans. Faraday Soc.* **64**, 805.
Partington, J. R. (1953). "An Advanced Treatise on Physical Chemistry," Vol. IV. Longmans, Green, New York.
Pedersen, N., and Malmstrom, L. (1964). Modification of electromagnetic scattering cross sections in the resonant region, J. K. Schindler and R. B. Mack, eds. Special Rept. No. 6. AFCRL, Hanscom Field, Massachusetts.
Penndorf, R. (1953). On the phenomenon of the colored sun. Geophys. Res. Paper No. 20. Air Force Cambridge Res. Lab., Bedford, Massachusetts.
Penndorf, R. (1958). *J. Phys. Chem.* **62**, 1537.
Penndorf, R. (1960). Scattering coefficients for absorbing and nonabsorbing aerosols. Tech. Rept. RAD-TR-60-27. Air Force Cambridge Res. Lab., Bedford, Massachusetts.
Penndorf, R. (1961). Atlas of scattering diagrams for $n = 1.33$. Tech. Rept. RAD-TR-61-32. Air Force Cambridge Res. Lab., Bedford, Massachusetts.
Penndorf, R. (1962a). *J. Opt. Soc. Am.* **52**, 896.
Penndorf, R. (1962b). *J. Atmospheric Sci.* **19**, 193.
Penndorf, R. (1962c). *J. Opt. Soc. Am.* **52**, 797.
Penndorf, R. (1963). *In* "Electromagnetic Scattering" (M. Kerker, ed.), Pergamon Press, Oxford.
Pernter, J. M., and Exner, F. M. (1910). "Meteorologische Optik." Braumüller, Vienna.
Perrin, F. (1942). *J. Chem. Phys.* **10**, 415.
Peterlin, A. (1938). *Z. Physik.* **111**, 232.
Peterlin, A. (1951). *Kolloid-Z.* **120**, 75.
Peterlin, A. (1957). *J. Polymer Sci.* **23**, 189.

PETERLIN, A. (1963). *In* "Electromagnetic Scattering" (M. Kerker, ed.). Pergamon Press, Oxford.

PETERLIN, A. (1965). *Makromol. Chem.* **87**, 152.

PETERLIN, A., and REINHOLD, C. (1964). *J. Chem. Phys.* **40**, 1029.

PETERLIN, A., and STUART, H. A. (1939). *Z. Physik.* **112**, 1, 129.

PETERLIN, A., HELLER, W., and NAKAGAKI, M. (1958). *J. Chem. Phys.* **28**, 470.

PETERS, L. (1965). *J. Res. Natl. Bur. Std.* **D69**, 231.

PETERS, L., and GREEN, R. B. (1961). *In* "Electromagnetic Effects of Re-entry" (W. Rotman and G. Meltz, eds.), pp. 133–141. Pergamon Press, Oxford.

PETERS, L., and SWARNER, W. G. (1961). Preprints. *Symp. Interaction of Space Vehicles with an Ionized Atmosphere, 1961.* Am. Astronaut. Soc., Washington, D.C.

PETERS, L., and THOMAS, D. T. (1962). *J. Geophys. Res.* **67**, 2073.

PETERS, L., SWARNER, W. G., and THOMAS, D. T. (1962). Further studies of the radar cross section of plasma-clad bodies *2nd Symp. Plasma Sheath—Its Effects upon Re-entry Commun. Detection*, Boston, Massachusetts, April 1962.

PETHICA, B. A., and SMART, C. (1966). *Trans. Faraday Soc.* **62**, 1890.

PETRO, A. J. (1960). *J. Phys. Chem.* **64**, 1508.

PFENNINGER, H. (1927). *Ann. Physik.* **83**, 753.

PFLEIDERER, J. (1959). *Optik.* **16**, 409.

PICOT, C., WEILL, G., and BENOIT, H. (1968). *J. Colloid Interface Sci.* **27**, 360.

PIERCE, P. E., and MARON, S. H. (1964). *J. Colloid Sci.* **19**, 658.

PIHL, M. (1963). *In* "Electromagnetic Theory and Antennas" (E. C. Jordan, ed.), Pt. 1. Pergamon Press, Oxford.

PILAT, M. J. (1967). *Appl. Opt.* **6**, 1555.

PLASS, G. N. (1964). *Appl. Opt.* **3**, 867.

PLASS, G. N. (1966). *Appl. Opt.* **5**, 279.

PLATZMAN, P. M., and OZAKI, H. T. (1960). *J. Appl. Phys.* **31**, 1597.

PLONUS, M. A. (1960). *Can. J. Phys.* **38**, 1665.

PLONUS, M. A. (1961). *IRE Trans. Antennas Propagation* **AP-9**, 573.

POROD, G. (1948). *Acta Phys. Austriaca* **2**, 133.

POROD, G. (1951). *Kolloid-Z.* **124**, 83.

POROD, G. (1952). *Kolloid-Z.* **125**, 51, 108.

POWERS, J., and STEIN, R. S. (1953). *J. Chem. Phys.* **21**, 1611.

PREINING, O. (1962). *Staub* **22**, 456.

PREMILAT, S., and HORN, P. (1963). *Compt. Rend.* **257**, 3348.

PREMILAT, S., and HORN, P. (1964). *Compt. Rend.* **258**, 6366.

PREMILAT, S., and HORN, P. (1965). *J. Chim. Phys.* **62**, 395.

PREMILAT, S., and HORN, P. (1966). *J. Chim. Phys.* **63**, 463.

PRINCEN, L. H., and MYSELS, K. J. (1957). *J. Colloid Sci.* **12**, 594.

PRINS, J. A., and PRINS, W. (1956). *Physica* **22**, 576.

PRINS, W. (1961). *J. Phys. Chem.* **65**, 369.

PRINS, W., and HERMANS, J. J. (1956a). *Koninkl. Ned. Akad. Wetenschap. Proc. Ser. B* **59**, 162.

PRINS, W., and HERMANS, J. J. (1956b). *Koninkl. Ned. Akad. Wetenschap. Proc. Ser. B* **59**, 298.

PRISHIVALKO, A. P. (1963). *Opt. Spectry. (USSR) (English Transl.)* **14**, 139.

PROBERT-JONES, J. R. (1963). *In* "Electromagnetic Scattering" (M. Kerker, ed.), p. 237. Pergamon Press, Oxford.

PROUDMAN, J., DOODSON, A. T., and KENNEDY, G. (1918). *Phil. Trans. Roy. Soc. London Ser. A* **217**, 279.

PRUD'HOMME, J., and SICOTTE, Y. (1968). *J. Colloid Interface Sci.* **27**, 547.

Pugh, T. L., and Heller, W. (1957). *J. Colloid Sci.* **12**, 173.

Quantie, C. (1954). *Proc. Roy. Soc.* **A224**, 90.

Querfeld, C. W. (1963). Private communication, USAERDA, White Sands, New Mexico.

Raman, C. V., and Ramanathan, K. R. (1923). *Phil. Mag.* **45**, 113, 213.

Raman, C. V., and Rao, K. S. (1923). *Phil. Mag.* **45**, 625.

Ramanathan, K. R. (1927). *Indian J. Phys.* **1**, 413.

Ramo, S., and Whinnery, J. R. (1960). "Fields and Waves in Modern Radio," p. 286. Wiley, New York.

Rapaport, E., and Weinstock, S. E. (1955). *Experientia* **11**, 363.

Ray, B. (1921). *Proc. Indian Assoc. Cultivation Sci.* **7**, 1.

Rayleigh, Lord (1871a). *Nature* **3**, 234, 264, 265.

Rayleigh, Lord (1871b). *Phil. Mag.* **41**, 107, 274, 447.

Rayleigh, Lord (1872). *Proc. London Math. Soc.* **4**, 253.

Rayleigh, Lord (1881). *Phil. Mag.* **12**, 81.

Rayleigh, Lord (1885). *Proc. London Math. Soc.* **17**, 4.

Rayleigh, Lord (1897). *Phil. Mag.* **44**, 28.

Rayleigh, Lord (1899). *Phil. Mag.* **47**, 375.

Rayleigh, Lord (1900). *Phil. Mag.* **49**, 324.

Rayleigh, Lord (1910). *Proc. Roy. Soc.* **A84**, 25.

Rayleigh, Lord (1914). *Proc. Roy. Soc.* **A90**, 219.

Rayleigh, Lord (1918a). *Phil. Mag.* **36**, 365.

Rayleigh, Lord (1918b). *Proc. Roy. Soc.* **A94**, 296.

Rayleigh, Lord (1918c). *Phil. Mag.* **35**, 373.

Rayleigh, Lord (1918d). *Proc. Roy. Soc.* **A94**, 453; **A95**, 155.

Read, B. E. (1960). *Trans. Faraday Soc.* **56**, 382.

Reichmann, M. E. (1959). *Can. J. Chem.* **37**, 489.

Reinhold, C., and Peterlin, A. (1965). *Physica* **31**, 522.

Reisler, E., and Eisenberg, H. (1965). *J. Chem. Phys.* **43**, 3875.

Remy-Battiau, L. (1962). "Mie scattering functions for spherical particles of refractive index $m = 1.25$." *Mem. Soc. Roy. Sci. Liege* (4) I. Tome **2**, Fasc. 7.

Rheinstein, J. (1962). Scattering of electromagnetic waves by an Eaton Lens. Tech. Rept. No. 273, June 1962. Lincoln Lab. Massachusetts Inst. of Technol., Lexington, Massachusetts.

Rheinstein, J. (1963). Tables of the amplitude and the phase of the backscatter from a conducting sphere. Rept. No. 22 G-16. Lincoln Lab., Massachusetts Inst. of Technol., Lexington, Massachusetts.

Rheinstein, J. (1964). *IEEE Trans. Antennas Propagation* **AP-12**, 334.

Rheinstein, J. (1965). *IEEE Trans. Antennas Propagation* **AP-13**, 983.

Richmond, J. H. (1965). *IEEE Trans. Antennas Propagation* **AP-13**, 334.

Richmond, J. H. (1966). *IEEE Trans. Antennas Propagation* **AP-14**, 460.

Riley, D. P., and Oster, G. (1951). *Discussions Faraday Soc.* **11**, 107.

Riseman, J. (1952). *Acta Cryst.* **10**, 193.

Robinson, R., and Stokes, R. (1955). "Electrolyte Solutions," Butterworths, London.

Rocard, Y. (1933). *J. Phys. Radium* **4**, 165.

Roess, L. C. (1946). *J. Chem. Phys.* **14**, 695.

Rose, H. E. (1953). "The Measurement of Particle Size in Very Fine Powders." Constable Press, London.

Rosen, J. S. (1949). *J. Chem. Phys.* **17**, 1192.

Rothmund, V. (1898). *Z. Phys. Chem.* **26**, 433.

Rousset, A. (1936). *Ann. Phys. (Paris)* **5**, 5.

Rowell, R. L. (1962). Extreme contour diagrams for intensity functions computed from the Mie theory. Rept. No. 1. Goessmann Lab., Univ. of Massachusetts, Amherst, Massachusetts.

ROWELL, R. L., WALLACE, T. P., and KRATOHVIL, J. P. (1968a). *J. Colloid Interface Sci.* **26**, 494.
ROWELL, R. L., KRATOHVIL, J. P., and KERKER, M. (1968b). *J. Colloid Interface Sci.* **27**, 501.
RUDDER, R. R., and BACH, D. R. (1968). *J. Opt. Soc. Am.* **58**, 1260.
RUDDUCK, R. C., and WALTER, C. H. (1962). *IRE Trans. Space Electron. Telemetry* **8**, 31.
RUEDY, R. (1944). *Can. J. Res.* **A22**, 53.
RULF, B. (1965). "Diffraction and scattering of electromagnetic waves in anisotropic media." Ph.D. Thesis, Polytech. Inst. of Brooklyn, Brooklyn, New York.
RULF, B. (1966). *J. Opt. Soc. Am.* **56**, 595.
RUSCH, W. V. T. (1964). *Can. J. Phys.* **42**, 26.
RUSCH, W. V. T., and YEH, C. (1967). *IEEE Trans. Antennas Propagation* **AP-15**, 452.
RYDE, J. W. (1931). *Proc. Roy. Soc.* **A131**, 451.
RYDE, J. W. (1946). *In* "Meteorological Factors in Radio-Wave Propagation," p. 169, The Physical Society, London.
SADRON, C. (1937). *J. Phys. Radium* **8**, 481.
SAKURADA, I., HOSONO, M., and TAMAMURA, S. (1964). *Bull. Inst. Chem. Res. Kyoto Univ.* **42**, 145.
SAMADDAR, S. N. (1963). *Appl. Sci. Res. Sect. B* **10**, 385.
SAMADDAR, S. N. (1970). *Nuovo Cimento* **66B**, 33.
SAXON, D. S. (1955). Lectures on the scattering of light. Sci. Rept. No. 9, Contract AF 19(122)-239. Dept. of Meteorology, Univ. of California, Los Angeles, California.
SAXTON, J. A. (1946). *In* "Meteorological factors in radiowave propagation," pp. 292, 306, The Physical Society, London.
SAXTON, J. A., and LANE, J. A. (1946). "Meteorological factors in radiowave propagation," p. 278, The Physical Society, London.
SAYASOV, YU. S. (1961). *Zh. Tekhn. Fiz.* **31**, 261 [*Soviet Phys. Tech. Phys.* (*English Transl.*) **31**, 189].
SAZONOV, D. M., and FROLOV, N. YA. (1965). *Zh. Tekhn. Fiz.* **35**, 990 [*Soviet Phys. Tech. Phys.* (*English Transl.*) **10**, 763].
SCATCHARD, G. (1946). *J. Am. Chem. Soc.* **68**, 2315.
SCATCHARD, G., and BREGMAN, J. (1959). *J. Am. Chem. Soc.* **81**, 6095.
SCATCHARD, G., and TICKNOR, L. B. (1952). *J. Am. Chem. Soc.* **74**, 3724.
SCATCHARD, G., BATCHELDER, A. C., and BROWN, A. (1946). *J. Am. Chem. Soc.* **68**, 2320.
SCHARFMAN, H. (1954). *J. Appl. Phys.* **25**, 1352.
SCHIFF, L. I. (1955). "Quantum Mechanics." McGraw-Hill, New York.
SCHMIDT, P. W. (1958). *Acta Cryst.* **11**, 674.
SCHMIDT, P. W., and BRILL, O. L. (1967). *In* "Electromagnetic Scattering," (R. L. Rowell and R. S. Stein, eds.), p. 169, Gordon and Breach, New York.
SCHMIDT, P. W., and HIGHT, R. (1959). *J. Appl. Phys.* **30**, 866.
SCHMIDT, P. W., WEIL, C. G., and BRILL, O. L. (1968). *In* "X-Ray and Electron Methods of Analysis," pp. 86–100, Plenum Press, New York.
SCHMIDT, R. L. (1968). *J. Colloid Interface Sci.* **27**, 516.
SCHMIDT, R. L., and CLEVER, H. L. (1968). *J. Phys. Chem.* **72**, 1529.
SCHULTZ, F. V. (1950). "Scattering by a prolate spheroid." Engineering Research Institute, University of Michigan, Ann Arbor.
SCHWARZSCHILD, K. (1901). *Sitzber. Bayer. Akad. Wiss. Math.-naturw. Kl.* (Muenchen) **31**, 293.
SEITZ, W. (1905). *Ann. Physik.* **16**, 746; **19**, 554.
SEITZ, W. (1906). *Ann. Physik.* **21**, 1013.
SENFTLEBEN, H., and BENEDICT, E. (1918). *Ann. Physik.* **54**, 65.
SENFTLEBEN, H., and BENEDICT, E. (1919). *Ann. Physik.* **60**, 297.
SENIOR, T. B. A. (1967). *IEEE Trans. Antennas Propagation* **AP-15**, 587.
SENIOR, T. B. A., and GOODRICH, R. F. (1964). *Proc. IEE* **3**, 907.

SESHADRI, S. R. (1964). *Can. J. Phys.* **42**, 860.

SESHADRI, S. R., MORRIS, I. L., and MAILLOUX, R. J. (1964). *Can. J. Phys.* **42**, 465.

SHAKHPARONOV, M. I. (1961). *Dokl. Akad. Nauk SSSR* **136**, 1162; *Proc. Phys. Chem. Sect.* (*English Transl.*) **136**, 189.

SHAKHPARONOV, M. I. (1962). *Zh. Fiz. Khim.* **36**, 2030 [*Russ. J. Phys. Chem.* (*English Transl.*) **36**, 1089].

SHAKHPARONOV, M. I. (1963). "Methods of Investigation of Thermal Motions of Molecules and the Structure of Liquids." Moscow Univ. Press, Moscow.

SHIFRIN, K. S. (1951a). "Light Scattering in Turbid Media." State Publ. House for Tech.-Theoret. Lit., Moscow.

SHIFRIN, K. S. (1951b). *Compt. Rend. Acad. Sci. USSR* **44**, No. 4.

SHIFRIN, K. S. (1954). *Compt. Rend. Acad. Sci. USSR* **77**, No. 4.

SHIFRIN, K. S. (1955a). *Proc. Vsesojusni Saschnuy Lesotechnichesici Inst.*, Leningrad, 1955, No. 1, p. 33.

SHIFRIN, K. S. (1955b). *Proc. Main Geophys. Obs., Leningrad* No. 46, p. 5.

SHIFRIN, K. S. (1955c). *Proc. Main Geophys. Obs., Leningrad* No. 46, p. 34.

SHIFRIN, K. S. (1957). Optical Studies of Cloud Particles. "Studies of Clouds, Precipitation, and Storm Electricity." Inst. of Appl. Geophys. of U.S.S.R. Acad. of Sci., Moscow.

SHIFRIN, K. S. (1961). *Proc. Main Geophys. Obs., Leningrad* No. 109, p. 179.

SHIFRIN, K. S. (1965). *Akad. Nauk. USSR* **18**, 690.

SHIFRIN, K. S., and CHAYANOVA, E. A. (1966). *Phys. Atmosphere and Ocean* **2**, 149.

SHIFRIN, K. S., and GOLIKOV, V. I. (1961). "Studies on Clouds, Precipitation, and Lightning Electricity," pp. 266–277. Inst. of Appl. Geophys. of the U.S.S.R. Acad. of Sci., Moscow.

SHIFRIN, K. S., and PERELMAN, A. Y. (1963). *Opt. Spectry.* (*USSR*) (*English Transl.*) **15**, 285, 362, 434.

SHIFRIN, K. S., and PERELMAN, A. Y. (1964a). *Geofis. Pura Appl.* **58**, 208.

SHIFRIN, K. S., and PERELMAN, A. Y. (1964b). *Opt. Spectry.* (*USSR*) (*English Transl.*) **16**, 61.

SHIFRIN, K. S., and PERELMAN, A. Y. (1965a). *Proc. A.I. Vocikov Main Geophys. Obs.* **170**, 37.

SHIFRIN, K. S., and PERELMAN, A. Y. (1965b). *Proc. A.I. Vocikov Main Geophys. Obs.* **170**, 3.

SHIFRIN, K. S., and PERELMAN, A. Y. (1965c). *Phys. Atmosphere and Ocean* **1**, 964.

SHIFRIN, K. S., and PERELMAN, A. Y. (1966a). *Tellus* **18**, 566.

SHIFRIN, K. S., and PERELMAN, A. Y. (1966b). *Opt. Spectry.* (*USSR*) (*English Transl.*) **20**, 75, 386.

SHIFRIN, K. S., and PERELMAN (1966c). *Phys. Atmosphere and Ocean* **2**, 149.

SHIFRIN, K. S., and PERELMAN, A. Y. (1967). *In* "Electromagnetic Scattering" (R. L. Rowell and R. S. Stein, eds.). Gordon and Breach, New York.

SHIFRIN, K. S., and ZELMANOVICH, I. L. (1964). *Opt. Spectry.* (*USSR*) (*English Transl.*) **17**, 57.

SHIFRIN, K. S., PERELMAN, A. Y., and PUNINA, V. A. (1966a). *Proc. A.I. Vocikov Main Geophys. Obs.* **183**, 3.

SHIFRIN, K. S., PERELMAN, A. Y., and PUNINA, V. A. (1966b). *Proc. A.I. Vocikov Main Geophys. Obs.* **183**, 19.

SHIOBARA, S. (1965). *J. Phys. Soc. Japan* **20**, 2301.

SHIOBARA, S. (1966). *J. Phys. Soc. Japan* **21**, 1380.

SHULL, C. G., and ROESS, L. C. (1947). *J. Appl. Phys.* **18**, 308.

SICOTTE, Y. (1964). *J. Chim. Phys.* **61**, 1086.

SICOTTE, Y. (1966). *J. Chim. Phys.* **63**, 403.

SICOTTE, Y. (1967). *J. Chim. Phys.* **64**, 583.

SICOTTE, Y., and RINFRET, M. (1962). *Trans. Faraday Soc.* **58**, 1090.

SIEGEL, K. M., SCHULTZ, F. V., GERE, B. H., and SLEATOR, F. B. (1956). *IRE Trans. Antennas Propagation* **AP-4**, 266.

SIEGER, B. (1908). *Ann. Physik.* **27**, 626.

SINCLAIR, D. (1947). *J. Opt. Soc. Am.* **37**, 475.

SINCLAIR, D., and LA MER, V. K. (1949). *Chem. Rev.* **44**, 245.

SMART, C. (1965). *J. Polymer Sci. Pt. A* **3**, 3015.

SMART, C., and VAND, V. (1964). *J. Opt. Soc. Am.* **54**, 1232.

SMART, C., and WILLIS, E. (1967). *J. Colloid Interface Sci.* **25**, 577.

SMART, C., JACOBSEN, R., KERKER, M., KRATOHVIL, J. P., and MATIJEVIĆ, E. (1965). *J. Opt. Soc. Am.* **55**, 947.

SMELLIE, R., and LA MER, V. K. (1954). *J. Phys. Chem.* **58**, 583.

SMOLUCHOWSKI, M. (1907). *Bull. Intern. Acad. Sci.* (*Cracovie*) 1057.

SMOLUCHOWSKI, M. (1908). *Ann. Physik.* **25**, 205.

SMOLUCHOWSKI, M. (1912). *Phil. Mag.* **23**, 165.

SOLEILLET, P. (1929). *Ann. Phys.* (*Paris*) **12**, 23.

SOMMERFELD, A., and RUNGE, J. (1911). *Ann. Physik.* **35**, 277.

STACEY, K. A. (1956). "Light Scattering in Physical Chemistry." Butterworths, London and Washington, D.C.

STEGUN, I., and ABRAMOWITZ, M. (1957). *Math. Tables and Other Aids to Computations* **11**, 255.

STEIN, R. S. (1953). *J. Chem. Phys.* **21**, 1193.

STEIN, R. S. (1963). *In* "Electromagnetic Scattering" (M. Kerker, ed.). Pergamon Press, Oxford.

STEIN, R. S., and RHODES, M. B. (1960). *J. Appl. Phys.* **31**, 1873.

STEIN, R. S. and WILSON, P. R. (1962). *J. Appl. Phys.* **33**, 1914.

STEIN, R. S., ERHARDT, P. F., CLOUGH, S. B., and ADAMS, G. (1966). *J. Appl. Phys.* **37**, 3980.

STEIN, R. S., RHODES, M. B., and PORTER, R. S. (1968). *J. Colloid Interface Sci.* **27**, 336.

STEINBERG, I., and KATCHALSKI, E. (1963). *Bull. Res. Counc. Israel* **11A4**, 379.

STEPHENS, J. J. (1961a). Total attenuation, scattering, and absorption cross sections of water drops for infrared radiation. Rept. 4-05. AFCRL 1011. Elec. Eng. Res. Lab., Univ. of Texas, Austin.

STEPHENS, J. J. (1961b). *J. Meteorol.* **18**, 348.

STEPHENS, J. J., and GERHARDT, J. R. (1961a). *J. Meteorol.* **18**, 818.

STEPHENS, J. J., and GERHARDT, J. R. (1961b). *J. Meteorol.* **18**, 819.

STERN, S. C., BAUMSTARK, J. S., SCHEKMAN, A. I., and OLSON, R. K. (1959). *J. Appl. Phys.* **30**, 952.

STERN, R. A. (1963). *J. Appl. Phys.* **34**, 2562.

STEUBING, W. (1908). *Ann. Physik.* (4) **26**, 329.

STEVENSON, A. F. (1953a). *J. Appl. Phys.* **24**, 1134.

STEVENSON, A. F. (1953b). *J. Appl. Phys.* **24**, 1143.

STEVENSON, A. F., and BHATNAGAR, H. L. (1958). *J. Chem. Phys.* **29**, 1336.

STEVENSON, A. F., and HELLER, W. (1961). "Tables of Scattering Functions for Heterodisperse Systems." Wayne State Univ. Press, Detroit, Michigan.

STEVENSON, A. F., HELLER, W., and WALLACH, M. L. (1961). *J. Chem. Phys.* **34**, 1789.

STIGTER, D. (1960a). *J. Phys. Chem.* **64**, 114.

STIGTER, D. (1960b). *J. Phys. Chem.* **64**, 842.

STIGTER, D. (1963). *In* "Electromagnetic Scattering" (M. Kerker, ed.). p. 302. Pergamon Press, Oxford.

STOCKMAYER, W. (1950). *J. Chem. Phys.* **18**, 58.

STOCKMAYER, W. (1964). *J. Am. Chem. Soc.* **86**, 3485.

STOCKMAYER, W. H., and BAUER, M. E. (1964). *J. Am. Chem. Soc.* **86**, 3485.

STOCKMAYER, W. H., MOORE, L. D., FIXMAN, M., and EPSTEIN, B. N. (1955). *J. Polymer Sci.* **16**, 517.

STOKES, G. G. (1852). *Trans. Cambridge Phil. Soc.* **9**, 399.

STOYLOV, S. P. (1966a). *J. Colloid Interface Sci.* **22**, 203.

STOYLOV, S. P. (1966b). *Collection Czech. Chem. Commun.* **31**, 2866, 3052.
STOYLOV, S. P., and SOKEROV, S. (1967). *J. Colloid Interface Sci.* **24**, 235.
STOYLOV, S. P., and SOKEROV, S. (1968). *J. Colloid Interface Sci.* **27**, 542.
STRATTON, J. (1941). "Electromagnetic Theory." McGraw-Hill, New York.
STRAUSS, U. P., and ANDER, P. (1962). *J. Phys. Chem.* **66**, 2235.
STRAUSS, U. P., and WILLIAMS, B. L. (1961). *J. Phys. Chem.* **65**, 1390.
STRAUSS, U. P., and WINEMAN, P. L. (1958). *J. Am. Chem. Soc.* **80**, 2366.
STRAZIELLE, C., and BENOIT, H. (1961). *J. Chim. Phys.* **58**, 675, 678.
STUART, H. A., and BUCKHEIM, W. (1938). *Z. Physik* **111**, 36.
SUBBARAO, M. K., and MCLAY, A. B. (1956). *Can. J. Phys.* **34**, 546.
SUBRAMANIAN, S. (1962). *Z. Physik. Chem.* (Frankfurt) **33**, 15 (1962).
SWARNER, W. G., and PETERS, L. (1963). *IEEE Trans. Antennas Propagation* **AP-11**, 558.
SWEITZER, C. W. (1927). *J. Phys. Chem.* **31**, 1150.
TABIBIAN, R. M., and HELLER, W. (1958). *J. Colloid Sci.* **13**, 6.
TABIBIAN, R. M., HELLER, W., and EPEL, J. N. (1956). *J. Colloid Sci.* **11**, 195.
TAI, C. T. (1956a). Theory of the cylindrical Luneberg lens excited by a line magnetic current. Tech. Rept. 678-3. Antenna Lab., Ohio State Res. Foundation, Columbus, Ohio; ASTIA Doc. No. 98815.
TAI, C. T. (1956b). The electromagnetic theory of the spherical Luneberg lens. Tech. Rept. 667-17. Antenna Lab., Ohio State Univ. Res. Foundation, Columbus, Ohio.
TAI, C. T. (1958a). *Appl. Sci. Res. Sect. B* **7**, 113.
TAI, C. T. (1958b). *Nature* **182**, 1600.
TAI, C. T. (1963). *J. Res. Natl. Bur. Std.* **D67**, 199.
TAI, C. T., and CHOW, Y. (1959). *ITA Engenharia* **2**, 71.
TAKAHASHI, A., KATO, T., and NAGASAWA, M. (1967). *J. Phys. Chem.* **71**, 2001.
TANFORD, C. (1961). "Physical Chemistry of Macromolecules." Wiley, New York.
TANG, C. C. H. (1957). *J. Appl. Phys.* **28**, 628.
TARTAR, H. V., and LELONG, A. L. M. (1955). *J. Phys. Chem.* **59**, 1185.
TEORELL, T. (1931). *Kolloid-Z.* **54**, 58, 150.
THEIMER, O., and PAUL, R. (1965). *J. Chem. Phys.* **42**, 2508.
THILO, G. (1920). *Ann. Physik.* **62**, 531.
THOMAS, A. S. (1964). Modification of electromagnetic scattering cross sections in the resonant region, Vol. I (J. K. Schindler and R. B. Mack, eds.), AFCRL Special Rept. No. 6, p. 125. Air Force Cambridge Res. Lab., Hanscom Field, Bedford, Massachusetts.
THOMAS, D. T. (1962). Ph.D. Thesis, Dept. Elec. Eng., Ohio State Univ., Columbus, Ohio; Also, Sci. Rept. No. 13, August. Antenna Lab., Dept. Elec. Eng., Ohio State Univ., Columbus, Ohio.
THOMPSON, D. R., and RICE, O. K. (1964). *J. Am. Chem. Soc.* **86**, 3547.
THOMSON, J. J. (1893). "Recent Researches in Electricity and Magnetism." Oxford Univ. Press, London and New York.
TIMASHEFF, S. N. (1963). *In* "Electromagnetic Scattering" (M. Kerker, ed.). Pergamon Press, Oxford.
TIMASHEFF, S. N., and KRONMAN, M. J. (1959). *Arch. Biochem. Biophys.* **83**, 60.
TIMASHEFF, S. N., DINTZIS, H. M., KIRKWOOD, J. G., and COLEMAN, B. D. (1955). *Proc. Natl. Acad. Sci. U.S.* **41**, 710.
TIMASHEFF, S. N., DINTZIS, H. M., KIRKWOOD, J. G., and COLEMAN, B. D. (1957). *J. Phys. Chem.* **79**, 782.
TOBIAS, R. S., and TYREE, S. Y. (1959). *J. Am. Chem. Soc.* **81**, 6385.
TOBIAS, R. S., and TYREE, S. Y. (1960). *J. Am. Chem. Soc.* **82**, 3244.
TODD, J. (1962). "Survey of Numerical Analysis." McGraw-Hill, New York.

TONKS, L., and LANGMUIR, I. (1929). *Phys. Rev.* **33**, 195, 990.
TRAP, H. J. L., and HERMANS, J. J. (1956). *Koninkl. Ned. Akad. Wetenschap. Proc. Ser. B* **59**, 190.
TREMBLAY, R., RINFRET, M., and RIVEST, R. (1952). *J. Chim. Phys.* **20**, 523.
TRINKS, W. (1935). *Ann. Physik.* **22**, 561.
TWERSKY, V. (1954). *J. Appl. Phys.* **25**, 859.
TWERSKY, V. (1964). *Appl. Opt.* **3**, 1150.
TWERSKY, V. (1967). *J. Math. Phys.* **8**, 589.
TYNDALL, J. (1869). *Phil. Mag.* **37**, 384; **38**, 156.
TYREE, Jr., S. Y., ANGSTADT, R. L., HENTZ, Jr., F. C., YOEST, R. L., and SCATCHARD, G. (1966). *J. Phys. Chem.* **70**, 3917.
ULLMAN, R., and BENOIT, H. (1962). *J. Chim. Phys.* **59**, 96.
UNANUÉ, A., and BOTHOREL, P. (1964). *Bull. Soc. Chim. France*, p. 573.
UNANUÉ, A., and BOTHOREL, P. (1965). *Bull. Soc. Chim. France*, p. 2827.
U.R.S.I. (1964). *Weather Radar Conf.*, C.R.P.L., *11th, 1964, Boulder, Colorado*, Nat. Bur. of Std., Boulder, Colorado.
VAN DE HULST, H. C. (1946). Thesis Utrecht. *Rech. Astron. Obs. d'Utrecht* **11**, *Pt. 1.*
VAN DE HULST, H. C. (1949). *Physica* **15**, 740.
VAN DE HULST, H. C. (1957). "Light Scattering by Small Particles." Wiley, New York.
VANDERHOFF, J. W., VITKUSKE, J. F., BRADFORD, E. B., and ALFREY, T. (1956). *J. Polymer Sci.* **20**, 225.
VASICEK. A. (1960). "Optics of Thin Films." Wiley (Interscience), New York.
VERWEY, E. J. W., and OVERBEEK, J. Th. G. (1947). "Theory of the Stability of Lyophobic Colloids." Elsevier, New York.
VINK, H., and DAHLSTROM, G. (1967). *Makromol. Chemie* **109**, 249.
VOISHVILLO, N. A. (1961). *Opt. Spectry. (USSR) (English Transl.)* **12**, 225.
VON IGNATOWSKY, W. (1905). *Ann. Physik.* **18**, 495.
VON SCHAEFFER, C. (1907). *Ann. Physik.* **23**, 163.
VON SCHAEFFER, C. (1909). *Sitzber. Koninglich Preuss. Akad. Wiss. (Berlin)* **11**, 326.
VON SCHAEFFER, C., and GROSSMAN, F. (1910). *Ann. Physik* **31**, 1955.
VON WESENDONCK, K. (1894). *Naturw. Rundschau* **9**, 210.
VRIJ, A., and OVERBEEK, J. Th. G. (1962). *J. Colloid Sci.* **17**, 570.
WACHTEL, R. E., and LA MER, V. K. (1962). *J. Colloid Sci.* **17**, 531.
WAIT, J. R. (1955). *Can. J. Phys.* **33**, 189.
WAIT, J. R. (1959). "Electromagnetic Radiation from Cylindrical Structures." Pergamon Press, Oxford.
WAIT, J. R. (1961). *J. Res. Natl. Bur. Std.* **B65**, 137.
WAIT, J. (1962). "Electromagnetic Waves in Stratified Media." Pergamon Press, Oxford.
WAIT, J. R. (1963). *Appl. Sci. Res. Sect. B* **10**, 441.
WAIT, J. R. (1965a). *J. Res. Natl. Bur. Std. D* **69**, 1307.
WAIT, J. R. (1965b). *Can. J. Phys.* **43**, 2212.
WAIT, J. R. (1965c). *J. Res. Natl. Bur. Std. D* **69**, 247.
WAIT, J. R., and JACKSON, C. M. (1965). *J. Res. Natl. Bur. Std. D* **69**, 299.
WAKASHIMA, H., and TAKATA, K. (1963). Rev. Kobe Univ. Mercantile Marine, Pt. II. Naval Marine Eng. Sci. Sect., No. 10, p. 81.
WALES, M. (1962). *J. Phys. Chem.* **66**, 1768.
WALKER, C. B., and GUINIER, A. (1953). *Acta Met.* **1**, 568.
WALKER, G. W. (1900a). *Quart. J. Pure Appl. Math.* **31**, 36.
WALKER, G. W. (1900b). *Quart. J. Pure Appl. Math.* **31**, 252.
WALLACE, T. P., and KRATOHVIL, J. P. (1967). *J. Polymer Sci.* **B5**, 1139.

WALLACE, T. P., and KRATOHVIL, J. P. (1968). *J. Polymer Sci.* **C25**, 89.
WALLACH, M. L., and BENOIT, H. (1962). *J. Polymer Sci.* **57**, 41.
WALLACH, M. L., and BENOIT, H. (1966). *J. Polymer Sci. Pt. A-2* **4**, 491.
WALLACH, M. L., and HELLER, W. (1964). *J. Phys. Chem.* **68**, 924.
WALLACH, M. L., HELLER, W., and STEVENSON, A. F. (1961). *J. Chem. Phys.* **34**, 1796.
WALSTRA, P. (1964a). *Koninkl. Ned. Akad. Wetenschap. Proc. Ser. B* **67**, 491.
WALSTRA, P. (1964b). *Brit. J. Appl. Phys.* **15**, 1545.
WALSTRA, P. (1965). *Brit. J. Appl. Phys.* **16**, 1187.
WALSTRA, P. (1968). *J. Colloid Interface Sci.* **27**, 493.
WALTER, H. (1957). *Optik* **14**, 130.
WALTER, H. (1959). *Optik* **16**, 401.
WALTON, A. G., and HLABSE, T. (1963). *Talanta* **10**, 601.
WASIK, S. P., and HUBBARD, W. D. (1964). *J. Res. Natl. Bur. Std. A* **68**, 359.
WATILLON, A., and DAUCHOT, J. (1968). *J. Colloid Interface Sci.* **27**, 507.
WATILLON, A., and VAN GRUNDERBEECK, F. (1956). *Bull. Soc. Chim. Belg.* **65**, 657.
WATSON, G. N. (1944). "Treatise on the Theory of Bessel Functions." Cambridge Univ. Press, London and New York.
WATSON, W. H. (1964). *IEEE Trans. Antennas Propagation* **AP-12**, 374.
WAXLER, R. M., and WEIR, C. E. (1963). *J. Res. Natl. Bur. Std. A* **67**, 163.
WEBER, H. H. (1963). *Kolloid-Z. Z. Polymere* **188**, 40.
WEIGEL, D., RENOUPREZ, A., and IMELIK, B. (1965). *J. Chim. Phys.* **62**, 125.
WEILL, G. (1958). Ph.D. Thesis, Faculty of Sci., Strasbourg.
WEILL, G. (1961). *Ann. Phys. (Paris)* **6**, 1063.
WELLMAN, P. (1937). *Z. Astrophys.* **14**, 195.
WESSLAU, H. (1963). *Makromol. Chem.* **69**, 213.
WESTON, V. H., and HEMENGER, R. (1962). *J. Res. Natl. Bur. Std.* **D66**, 613.
WHITBY, K. T., LUNDGREN, D. A., and PETERSEN, C. M. (1965). *Intern. J. Air Water Pollution* **9**, 263.
WHITTAKER, E. (1951). "A History of the Theories of Aether and Electricity," Vol. I. Harper, New York.
WHITTAKER, E. T., and WATSON, G. N. (1947). "A Course of Modern Analysis." Cambridge Univ. Press, London and New York.
WIDOM, B. (1962). *J. Chem. Phys.* **37**, 2703.
WILCOX, C. H. (1957). *J. Math. Mech.* **6**, 167.
WILES, S. T., and MCLAY, A. B. (1954). *Can. J. Phys.* **32**, 372.
WILHELMSSON, H. (1963). *Arkiv. Fysik.* **23**, 447.
WILHELMSSON, K. H. B. (1962). *J. Res. Natl. Bur. Std.* **D66**, 439.
WILKES, G. L., and MARCHESSAULT, R. H. (1966). *J. Appl. Phys.* **37**, 3974.
WILLIS, E., KERKER, M., and MATIJEVIĆ, E. (1967). *J. Colloid Interface Sci.* **23**, 182.
WILSON, C. T. R. (1897). *Phil. Trans. Roy. Soc. London Ser. A* **189**, 265.
WILSON, I., and LA MER, V. K. (1948). *J. Ind. Hyg. Toxicol.* **30**, 265.
WIPPLER, C. (1954). *J. Chim. Phys.* **51**, 122.
WIPPLER, C. (1955). Ph.D. Thesis, Strasbourg (published by Imprimirie Durand, Chartres, 1956).
WIPPLER, C. (1957). *J. Polymer Sci.* **23**, 199.
WIPPLER, C., DE VRIES, A. J., and BENOIT, H. (1959). *I.U.P.A.C. Symp. Macromol., Wiesbaden, Germany, 1959*, Commun. IIIC8. Verlag Chemie, Weinheim.
WOODWARD, D. H. (1964). *J. Opt. Soc. Am.* **54**, 1325.
WYATT, P. J. (1962). *Phys. Rev.* **127**, 1837.
WYATT, P. J. (1963a). *In* "Electromagnetic Scattering" (M. Kerker, ed.). Pergamon Press, Oxford.

Wyatt, P. J. (1963b). *J. Appl. Phys.* **34**, 2078.
Wyatt, P. J. (1964a). *Phys. Rev.* **134**, AB1.
Wyatt, P. J. (1964b). *J. Appl. Phys.* **35**, 1966.
Wyatt, P. J. (1964c). Modification of electromagnetic scattering cross sections in the resonant region, Vol. I (J. K. Schindler and R. B. Mack, eds.), Special Rept. No. 6. Air Force Cambridge Res. Lab., Hanscom Field, Bedford, Massachusetts.
Wyatt, P. J. (1965a). *Appl. Phys. Letters* **6**, 209.
Wyatt, P. J. (1965b). *J. Appl. Phys.* **36**, 3875.
Wyatt, P. J. (1966). *J. Appl. Phys.* **37**, 3641.
Yajnik, M., Witeczek, J., and Heller, W. (1968). *J. Polymer Sci.* **C25**, 99.
Yamakawa, H. (1967). *J. Chem. Phys.* **46**, 973.
Yang, J. T. (1957). *J. Polymer Sci.* **26**, 305.
Yee, H. Y. (1965a). *IEEE Trans. Antennas Propagation* **AP-13**, 818.
Yee, H. Y. (1965b). *IEEE Trans. Antennas Propagation* **AP-13**, 822.
Yeh, C. (1963). *J. Math. Phys.* **4**, 65.
Yeh, C. (1964a). *J. Opt. Soc. Am.* **54**, 1227.
Yeh, C. (1964b). *Can. J. Phys.* **42**, 1369.
Yeh, C. (1965). *J. Opt. Soc. Am.* **55**, 309.
Yeh, C., and Kaprielian, Z. A. (1963). *Can. J. Phys.* **41**, 143.
Yvon, J. (1937). *Actualités Sci. et Ind.* No. 543.
Zaiser, E. M., and La Mer, V. K. (1948). *J. Colloid Sci.* **3**, 571.
Zernike, F. (1915). L'opalescence critique. Ph.D. Thesis, Amsterdam.
Zernike, F. (1918). *Arch. Neerl. Sci. IIIA* **4**, 74.
Zernike, F., and Prins, T. H. (1927). *Z. Physik* **41**, 184.
Zimm, B. H. (1948a). *J. Chem. Phys.* **16**, 1093.
Zimm, B. H. (1948b). *J. Chem. Phys.* **16**, 1099.
Zimm, B. H. (1950). *J. Phys. & Colloid Chem.* **54**, 1306.
Zwanzig, R. (1964). *J. Am. Chem. Soc.* **86**, 3489.

Author Index

Numbers in italics refer to the pages where the complete reference is cited.

L

M

N

T

U

X-Y

Z

Subject Index

Q

R

S

Physical Chemistry

A Series of Monographs

Editor: **Ernest M. Loebl**

Department of Chemistry

Polytechnic Institute of New York

Brooklyn, New York

1 W. Jost: Diffusion in Solids, Liquids, Gases, 1952
2 S. Mizushima: Structure of Molecules and Internal Rotation, 1954
3 H. H. G. Jellinek: Degradation of Vinyl Polymers, 1955
4 M. E. L. McBain and E. Hutchinson: Solubilization and Related Phenomena, 1955
5 C. H. Bamford, A. Elliott, and W. E. Hanby: Synthetic Polypeptides, 1956
6 George J. Janz: Thermodynamic Properties of Organic Compounds — Estimation Methods, Principles and Practice, Revised Edition, 1967
7 G. K. T. Conn and D. G. Avery: Infrared Methods, 1960
8 C. B. Monk: Electrolytic Dissociation, 1961
9 P. Leighton: Photochemistry of Air Pollution, 1961
10 P. J. Holmes: Electrochemistry of Semiconductors, 1962
11 H. Fujita: The Mathematical Theory of Sedimentation Analysis, 1962
12 K. Shinoda, T. Nakagawa, B. Tamamushi, and T. Isemura: Colloidal Surfactants, 1963
13 J. E. Wollrab: Rotational Spectra and Molecular Structure, 1967
14 A. Nelson Wright and C. A. Winkler: Active Nitrogen, 1968
15 R. B. Anderson: Experimental Methods in Catalytic Research, 1968
16 Milton Kerker: The Scattering of Light and Other Electromagnetic Radiation, 1969
17 Oleg V. Krylov: Catalysis by Nonmetals — Rules for Catalyst Selection, 1970
18 Alfred Clark: The Theory of Adsorption and Catalysis, 1970
19 Arnold Reisman: Phase Equilibria: Basic Principles, Applications, Experimental Techniques, 1970
20 J. J. Bikerman: Physical Surfaces, 1970
21 R. T. Sanderson: Chemical Bonds and Bond Energy, 1970; Second Edition, in preparation
22 S. Petrucci, ed.: Ionic Interactions: From Dilute Solutions to Fused Salts (In Two Volumes), 1971

23 A. B. F. Duncan: Rydberg Series in Atoms and Molecules, 1971

24 J. R. Anderson: Chemisorption and Reactions on Metallic Films, 1971

25 E. A. Moelwyn-Hughes: Chemical Statics and Kinetics of Solution, 1971

26 Ivan Draganic and Zorica Draganic: The Radiation Chemistry of Water, 1971

27 M. B. Huglin: Light Scattering from Polymer Solutions, 1972

28 M. J. Blandamer: Introduction to Chemical Ultrasonics, 1973

29 A. I. Kitaigorodsky: Molecular Crystals and Molecules, 1973

30 Wendell Forst: Theory of Unimolecular Reactions, 1973

31 Jerry Goodisman: Diatomic Interaction Potential Theory. Volume 1, Fundamentals, 1973; Volume 2, Applications, 1973

32 Alfred Clark: The Chemisorptive Bond: Basic Concepts, 1974

33 Saul T. Epstein: The Variation Method in Quantum Chemistry, 1974

34 I. G. Kaplan: Symmetry of Many-Electron Systems, 1975

35 John R. Van Wazer and Ilyas Absar: Electron Densities in Molecules and Molecular Orbitals, 1975

In Preparation

Brian L. Silver: Irreducible Tensor Methods: An Introduction for Chemists

6
7
C 8
D 9
E 0
F 1
G 2
H 3
I 4
J 5